W9-APG-782

HUMAN KINETICS ONLINE STUDY GUIDE

How to access the supplemental online study guide

We are pleased to provide access to an online study guide that supplements *Introduction to Kinesiology, Third Edition.* This interactive study guide offers a variety of multimedia experiences that reinforce key material and allow you to develop and test your understanding of the diverse field of kinesiology.We are certain you will enjoy this unique online learning experience.

Accessing the online study guide is easy! Follow these steps if you purchased a new book:

1. Visit **www.HumanKinetics.com/IntroductionToKinesiology**.
2. Click the third edition link next to the corresponding third edition book cover.
3. Click the Sign In link on the left or top of the page. If you do not have an account with Human Kinetics, you will be prompted to create one.
4. If the online product you purchased does not appear in the Ancillary Items box on the left of the page, click the Enter Key Code option in that box. Enter the key code that is printed at the right, including all hyphens. Click the Submit button to unlock your online product.
5. After you successfully enter your key code, your online product will appear in the Ancillary Items box. On future visits to the site, all you need to do is sign in to the textbook's website and follow the link!

→ Click the Need Help? button on the textbook's website if you need assistance along the way.

How to access the online study guide if you purchased a used book:

You may purchase access to the online study guide by visiting the text's website, **www.HumanKinetics.com/IntroductionToKinesiology**, or by calling the following:

800-747-4457 . . . . U.S. customers
800-465-7301 . . . . Canadian customers
+44 (0) 113 255 5665 . . . . European customers
08 8372 0999 . . . . Australian customers
0800 222 062 . . . . New Zealand customers
217-351-5076 . . . . International customers

For technical support, send an e-mail to:
support@hkusa.com . . . . U.S. and international customers
info@hkcanada.com . . . . Canadian customers
academic@hkeurope.com . . . . European customers
keycodesupport@hkaustralia.com . . . . . Australian and New Zealand customers

HUMAN KINETICS
The Information Leader in Physical Activity & Health

5-2011

Product: Introduction to Kinesiology, Third Edition online study guide

Key code: HOFFMAN-NAX4EG-OSG

Your key code allows you access to the online study guide.

Access is provided if you have purchased a new book. Once submitted, the code may not be entered for any other user.

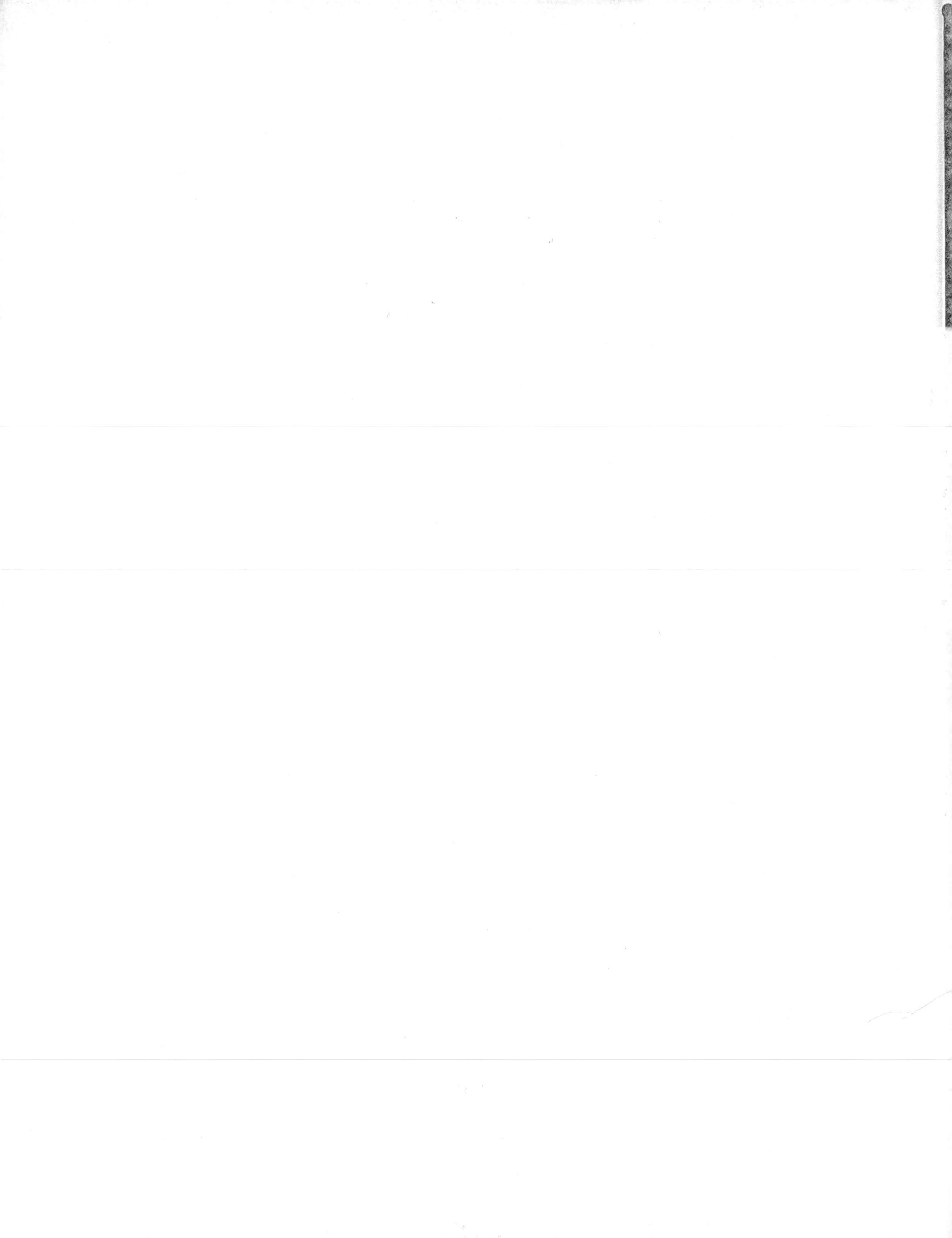

Introduction to Kinesiology

STUDYING PHYSICAL ACTIVITY

THIRD EDITION

Shirl J. Hoffman, EdD
Editor

Human Kinetics

Library of Congress Cataloging-in-Publication Data

Introduction to kinesiology: studying physical activity / Shirl J. Hoffman, editor. -- 3rd ed.
p. cm.
Includes bibliographical references and index.
ISBN-13: 978-0-7360-7613-5 (hard cover)
ISBN-10: 0-7360-7613-1 (hard cover)
1. Kinesiology. I. Hoffman, Shirl J., 1939-
QP303.I53 2008
612.7'6--dc22 2008022745

ISBN-10: 0-7360-7613-1
ISBN-13: 978-0-7360-7613-5

Permission notices for material reprinted in this book from other sources can be found on page xvii.

The Web addresses cited in this text were current as of May 2008, unless otherwise noted.

Acquisitions Editor: Myles Schrag; **Developmental Editor:** Holly Gilly; **Managing Editor:** Jillian Evans; **Copyeditor:** Joyce Sexton; **Proofreader:** Sarah Wiseman; **Indexer:** Betty Frizzell; **Permission Manager:** Dalene Reeder; **Graphic Designer:** Nancy Rasmus; **Graphic Artist:** Yvonne Griffith; **Photographer (cover):** Digital Vision; **Photographer (interior):** © Human Kinetics, unless otherwise indicated. Photos on page v (clockwise from left): courtesy of Middle Tennessee State University; Digital Vision; Photodisc; © Human Kinetics; courtesy of Joe Luxbacher; **Photo Asset Manager:** Laura Fitch; **Visual Production Assistant:** Joyce Brumfield; **Photo Office Assistant:** Jason Allen; **Art Manager:** Kelly Hendren; **Associate Art Manager:** Alan L. Wilborn; **Illustrator:** Jen Gibas; **Printer:** Courier

We thank Dick's Sporting Goods in Champaign, Illinois, for assistance in providing the location for the photo shoot for this book.

The paper in this book was manufactured using responsible forestry methods.

Printed in the United States of America 10 9 8 7 6

Human Kinetics
Web site: www.HumanKinetics.com

United States: Human Kinetics
P.O. Box 5076
Champaign, IL 61825-5076
800-747-4457
e-mail: humank@hkusa.com

Canada: Human Kinetics
475 Devonshire Road Unit 100
Windsor, ON N8Y 2L5
800-465-7301 (in Canada only)
e-mail: info@hkcanada.com

Europe: Human Kinetics
107 Bradford Road
Stanningley
Leeds LS28 6AT, United Kingdom
+44 (0) 113 255 5665
e-mail: hk@hkeurope.com

Australia: Human Kinetics
57A Price Avenue
Lower Mitcham, South Australia 5062
08 8372 0999
e-mail: info@hkaustralia.com

New Zealand: Human Kinetics
P.O. Box 80
Torrens Park, South Australia 5062
800 222 062
e-mail: info@hknewzealand.com

E4602

To Tom and Dale

"Without friends, no one would choose to live, though he had all other goods"

–Aristotle, *The Nicomachean Ethics*

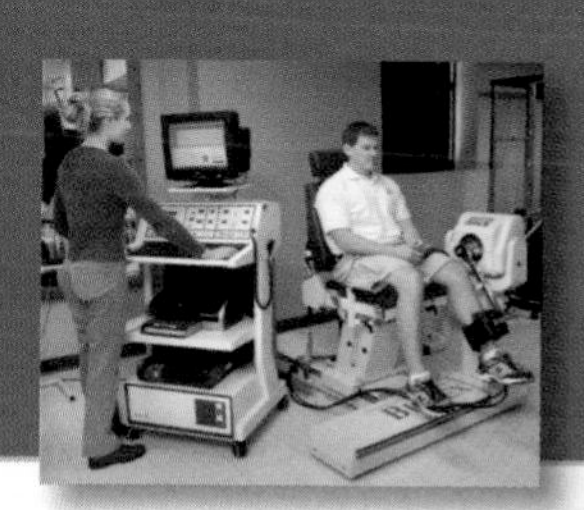

CONTENTS

PREFACE

For Instructors

Writing an introductory text for any discipline is a daunting task, but it is especially so in a young discipline undergoing rapid evolution. Substantial questions about the discipline's nature, future, organization, purpose, name, and relationship to the professions have yet to be answered to the satisfaction of all kinesiologists. In one way or another, all these issues converge in the writing of an introductory text. This text takes bold positions on many of these issues, such as the name for the discipline, the central phenomenon of study, and definitions of critical terms. Issues that seemed less critical for understanding the big picture of kinesiology were purposely kept in the background so as not to distract students from the primary theme of the text.

Even more problematic for authors of introductory texts is the changing demographic landscape of students enrolling in departments of kinesiology. Gone are the days when physical education professors could confidently assume that all students entering as majors in their departments were planning to be physical education teachers and coaches. Now it is just as likely that students have their sights set on careers in sport management, athletic training, cardiac rehabilitation, fitness counseling and leadership, aquatic leadership, rehabilitative exercise, sport information, adapted physical education, or related professions. Faculty are encountering many students who elect the kinesiology major as preparation for graduate study in physical therapy, chiropractic, podiatry, or medicine. Still others choose kinesiology as a course in liberal studies, without specific professional goals in mind. And of course, a substantial percentage of students come to the introductory course lacking any clear vision of what career they will pursue. The likelihood that students with a diverse range of needs and interests would use this book was very much in mind at every stage of the writing process.

I believe that a shared interest in and curiosity about physical activity are what unite students with these differing professional aspirations and also make kinesiology the logical discipline for them to study. This book is centered on the phenomenon of physical activity. Students who lack insight into its pervasiveness and indispensability, or who doubt its worth in their personal or social lives, will not easily be convinced that a field of kinesiology is needed; thus, much of the text is designed to convince them of the centrality of physical activity to our lives. Most students come to an introductory class in kinesiology with a wealth of knowledge and experience about physical activity, but that is no guarantee they will appreciate its variety, complexity, and elegance. Like art educators, who often must reorient students accustomed to viewing their world with category-hardened habits of seeing before they can view it from a fresh and novel perspective, so the teacher of this introductory course must assist students in viewing physical activity in new and exciting ways if they are to appreciate fully the importance of the discipline of kinesiology.

The overarching purpose of this text, then, is to help students appreciate the breadth and importance of physical activity, to orient them to the discipline of kinesiology and help them understand its relationship to physical activity, and to introduce them to the physical activity professions. The perspective on physical activity is both broad and narrow. In part I, especially chapter 2, the discussion ranges far beyond exercise, sport, or health-related physical activity. By embracing such an expansive view of physical activity and kinesiology, however, I am not suggesting that the emphasis currently given to sport and exercise in kinesiology be diminished. These topics have been and will continue to be the dominant forms of physical activity studied in the discipline. At the same time, the book will have missed its mark if it fails to expand students' conceptions of physical activity or fails to develop an appreciation for the potential contributions that kinesiology can make to fields far removed from sport and exercise.

Plan of the Text

The text is organized around a model of kinesiology as a discipline that draws knowledge from three sources: participating in and observing physical activity, studying and conducting research, and delivering professional services. By defining the discipline as an integration of knowledge associated with experience, formal study, and professional practice, the authors hope that students will recognize the critical importance of all three of these dimensions of the discipline.

Chapter 1 provides an overview of the book, offers key definitions, and presents the model around which the text is organized. The model is intended to explain to those unfamiliar with the traditions of our field how the field is organized, and to help those who already are members of the profession understand how their work fits into the larger whole. Part I (chapters 2 through 4) focuses on experiencing physical activity; it uses a wide lens to view the many ways in which our physical activity experiences contribute to our understanding of kinesiology. Chapter 2 introduces students to "the spheres of physical activity experience" as a way of helping them appreciate the pervasiveness and importance of physical activity, not only in sport and exercise but also in work, rehabilitation, daily living, and other spheres of existence. Chapter 3 directs attention to the importance of physical activity experiences, concentrating on the way kinesiologists ("physical activity experience specialists") design and manipulate physical activity experiences to bring about predetermined ends. Discussions of the benefits of physical activity experiences can, often quite unintentionally, lead students to appreciate only the physical aspects of physical activity. The holistic approach taken in chapter 4 is intended to pique students' interest in the subjective aspects of physical activity. It invites students to consider the deep emotional and spiritual meanings, as well as the many different kinds of knowledge, that they can glean from physical activity.

Part II (chapters 5 through 11) introduces students to the academic subdisciplines of kinesiology, portraying them as elements in three spheres of scholarly study of physical activity: the biophysical, sociocultural, and behavioral spheres. The authors of the chapters in part II, eminent scholars in their subdisciplines, provide a general overview of each area of study with an eye toward helping students understand the particular contribution it makes to the discipline of kinesiology. Each chapter gives a brief overview of major historical events in the development of the subdiscipline; the research methods used in the subdiscipline; what professionals such as biomechanists, exercise physiologists, and others do in the course of their professional work; and how students' present knowledge can form a foundation for more advanced study. Each chapter presents practical, real-world applications from the subdiscipline and is organized to help readers understand why the subdiscipline is important and how it may relate to a variety of professional endeavors.

The third edition is different in several ways. New authors make an appearance in both parts II and III. Scott Kretchmar has additional support from Cesar Torres for chapter 5 on philosophy of physical activity, and Katherine Jamieson joins Margaret Duncan in the newer edition of chapter 7 on sociology of physical activity. New also to this edition is Jennifer Caputo, who has written chapter 11 on physiology of physical activity. The text is better for their valuable contributions. Returning authors include Richard Swanson, Jerry and Kathi Thomas, Robin Vealey, and Kathy Simpson, all of whom have reworked their chapters in an effort to reach the critical audience of this text. Part II is ambitious in its effort to cover many subdisciplines in single chapters. We hope this edition makes these chapters more "digestible" and accessible to undergraduate students who may not have in-depth background in the various subdisciplines. In teaching the material in these chapters, I have found it helpful to incorporate lectures from colleagues in my department who are specialists in the relevant subdiscipline. Students enthusiastically welcome these visits.

The goal of part III (chapters 12 through 17) is to help students begin thinking about professions in which physical activity plays a central role. Throughout this section, authors have kept in mind that many students entering kinesiology departments may have tentative professional goals at best. To this end, chapter 12 takes a general approach to the physical activity professions, introducing students to what it means to be a professional and what will be required of them if they are to become professionals. This chapter also leads students through a series of steps intended to assist them in determining whether their interest, talent, and motivation suit them to a career in kinesiology.

Chapters 13 through 17 include a wealth of information about clusters of careers drawn from the spheres of professional practice. Each chapter focuses on a collection of professional opportunities within a sphere, including health and fitness, therapeutic exercise, teaching and coaching, and sport management. The authors of all these chapters explain the need for professions in the given area and offer a brief historical background of the sphere. In addition, each chapter includes an analysis of work settings in the sphere; up-close views of specific professions; an overview of educational qualifications needed to practice in the sphere; resources, case studies, and profiles of working professionals; and a section offering inside advice to those planning careers in that cluster of professions. All of these chapters have been updated and expanded for the third edition.

Returning authors include Jeremy Howell and Sandra Minor Bulmer, Chad Starkey, Kim Graber and Tom Templin, and Lori Miller and Clay Stoldt. A new face appears in the roster of authors in part III: chapter 16 has been coauthored by Joe Luxbacher. Brief biographies of all of the authors can be found in the back pages of the text. Again, in teaching the course, I have found that outside visitors representing various career areas of the physical activity industry are valuable adjuncts to the introductory course.

Features of the Text

Schematics guide the organization of the book and, while they may appear redundant, help students keep the big picture in mind as they read individual chapters. Key points and interesting sidebars are helpful supplements to the main text. As in the second edition, interactive exercises throughout the text involve students actively in their learning. These exercises engage students more deeply with the content, helping them put themselves into the picture described in a chapter. The study questions at the end of the chapter also have proven helpful in focusing students' attention on key points in the text. In the final analysis, however, the effectiveness of the text will depend on the creative uses to which it is put by the teacher. Rather than relying on the text as a single resource, teachers should integrate the various topics that it covers with challenging in-class and out-of-class projects that will encourage students to view themselves as professionals-in-training and take the initial steps toward identifying and attaining their career goals.

An addition to this edition of the book is the online study guide. This rich, interactive tool offers students a variety of learning features:

- **Interactive activities.** Using a variety of multimedia experiences, these activities engage students in the material. Scenario-based audio and video activities demonstrate an aspect of kinesiology and then require students to complete an activity based on the demonstration and explanation. Other activities involve drag-and-drop, selecting from a menu, or responding with short answers to photos and scenarios presented.
- **Study questions.** These questions reinforce key material throughout the text and help students comprehend the concepts.
- **Review of key points.** The online review allows students to test themselves on the key points with the use of interactive activities. Students have the option of printing out the key points as a study aid after completing the activity.

- **Activity feedback.** Students receive targeted information to correct or reinforce their answers to activities.
- **Web searches.** Students can seek out specific information on the Internet and then provide their findings on forms that can be printed out and turned in to you.

For Students

If you are like most students, you probably entered college having decided on a major before you knew a great deal about it. You may not have thought much about such questions as, Why is the major important? What types of careers do people in the major enter after graduating? What subjects do students study in this major? How is the course of study organized? Even though you have decided on or are considering a major in kinesiology, you may not know much about the field of study and may have only vague ideas about the career opportunities available. Not knowing the answers to these questions at the beginning of your undergraduate experience makes it difficult to put course work in perspective and could leave you questioning the value of many of the courses that you are required to take.

We wrote this textbook to address those questions and concerns. The book is intended to help you understand what the discipline of kinesiology is, why it is important, what kinesiology majors study, what types of knowledge you will acquire over the course of four years, and what types of careers are available to you. If you read and study it carefully, this book will help you reach an early decision not only about whether you want to major in kinesiology but also about what type of career most interests you. You may see no pressing need to decide on a career at this early stage of your college experience; but as you will see in this book, there are advantages to identifying early the career path that you want to follow.

Kinesiology may be a term that you have never heard before. In fact, you may be enrolled in a major called exercise and sport science, or physical education and exercise science, or something else. One of the confusing things about this major is that the discipline is known by many titles. This circumstance isn't all that uncommon for a young and evolving area of study such as kinesiology; the discipline simply hasn't been around long enough for everybody to agree on a single term! Over the past decade, however, more and more scholars and college and university departments have decided to use the term *kinesiology.* For this reason we have selected it as the title of this text.

Because kinesiology integrates knowledge from a variety of sources and draws on theories, concepts, and principles from many other disciplines, students often find it difficult to describe or explain the field to their friends who aren't familiar with it. This text presents the discipline in a new and exciting way. It will help you understand not only the enormous breadth of the discipline but also how the many parts of the discipline are tied together. In the pages that follow, you will learn that physical activity is the centerpiece of the kinesiology discipline. Indeed, this idea may be the most important point in the book. Whether you plan to be a sport administrator, athletic trainer, physical education teacher, or physical or occupational therapist, you should have an interest in and curiosity about physical activity.

Just as the discipline of psychology is the study of human behavior and the discipline of biology is the study of life forms, the discipline of kinesiology is the study of physical activity. But don't interpret *study* to mean merely reading books, writing papers, and taking exams. Kinesiology integrates knowledge from three different yet related sources: your experiences performing and observing physical activity, the formal study of physical activity, and professional practice centered in physical activity. This new approach will help you understand the interrelationships of the parts of the discipline.

How the Text Is Organized

Part I of the text will spark your thinking about physical activity, why it is important, and how it intersects our lives at a variety of different junctures that we call *spheres of physical activity experience.* If you think that sport and exercise are what kinesiologists study, you are correct; but you are not correct if you think that these are the *only* forms of physical activity that kinesiologists study. After reading this book you should come to appreciate the fact that kinesiology also contributes to our understanding and improvement of physical activity performances in venues far beyond the athletic field or exercise gym. It also will become clear that human experience in physical activity is a unique source of knowledge; that humans have a unique ability to plan and engage in physical activity experiences to improve the quality of their lives; that the interaction of our thoughts and emotions (subjective experiences) with physical activity may be as critical as the activity itself; and that the unique contribution of teachers, coaches, fitness leaders, sport managers, athletic trainers, cardiac rehabilitation therapists, and many other related professionals consists of their skill in manipulating physical activity experiences to achieve predetermined objectives.

Whereas part I invites you to reflect on knowledge you have gained through the experiences of participating in and watching physical activity, part II presents broad coverage of the knowledge that we can derive from the scholarly study of physical activity. The chapter titles in this part of the book outline the core body of academic (theoretical) knowledge that all kinesiology students should master before graduation. At your institution a single course may cover each of these areas, or one course may cover two or more areas. To help you better organize this vast body of knowledge in your mind, we have divided it into larger categories called *spheres of scholarly study.* Nationally recognized scholars who are particularly skilled at introducing you to their areas of specialized study have written these chapters. Keep in mind that each chapter is intended only to introduce you to the area of study, not to provide you with all the scholarly knowledge that you will need to master in that area before you graduate.

Part III introduces you to the world of professional practice. Students enrolled in kinesiology departments often have widely disparate career goals, ranging from careers as an athletic trainer or physical educator to becoming a corporate fitness leader or a cardiac rehabilitation specialist, among many other possibilities. Each of these is a form of professional practice. This part of the book groups the many possible careers into *spheres of professional practice* according to similarities in the types of work and general educational requirements. Outstanding scholars in various areas of professional practice wrote the chapters. These writers are eminently qualified to describe the different types of work that physical activity professionals do, as well as to offer details concerning what you can do to develop yourself into an outstanding professional.

Features of the Text

The book contains a great deal of information, perhaps more than you will be able to absorb in a single course. We have designed the text in a way that should help you absorb the material presented: Key points, interesting sidebars, interactive items, special elements, and some truly exciting photos should make this an enjoyable read. Don't skip over the objectives listed at the beginning of each chapter; they are your map for the pages that follow. Take time to read the many interactive items sprinkled throughout the text. These features will stimulate your thinking and reinforce what you have learned by reading the previous paragraphs. Don't miss the unique opportunity that they offer for learning.

Also, make sure to focus on the key points that you will find in every chapter. They are brief summaries of what your authors believe to be the most important points made in the previous few pages. Read each key point carefully and think back over what you have read. You will find also that many of the captions for photographs in this text are information packed. In many cases they constitute key points in their own right. The study questions at the end of each chapter will direct you to what the authors believe are some of the most important points they have covered.

Another don't-miss feature is the online study guide. Whenever you see an orange box in the margin of a page, that's a trigger to go to the study guide where you'll find a variety of multimedia experiences that'll help you learn, understand, and apply the information in the text. (If you're reading this as an e-book, you can simply click on the trigger and you'll go directly to the online activity). The study guide material is essential to your complete understanding of the content in your course. There are four different kinds of online activities:

These activities use a variety of multimedia to demonstrate an aspect of kinesiology. Audio, video, drag-and-drop activities, and other activity types make the content come alive.

Web searches point you to specific information on the Internet and then give you the opportunity to submit your findings to your instructor.

The study guide provides an online review of key points in each chapter and lets you test yourself on how well you understand the material. You can print out the key points as a study aid if you choose to.

These are repeats of the end-of-chapter questions in the text. Your instructor may ask you to complete them online and turn them in.

Accessing the Online Study Guide

- If you purchased a new copy of this book or an e-book, you'll find instructions for accessing the online study guide either on a key code page at the very front of the print book or in the e-mail you received as part of your e-book purchase.
- If you purchased a used copy of this book, you may purchase access to the online study guide by visiting **www.HumanKinetics.com/IntroductiontoKinesiology**, or by calling Human Kinetics:

United States 1-800-747-4457
Canada 1-800-465-7301
Europe .... +44 (0) 113 255 5665
Australia 08 8372 0999
New Zealand 0800 222 062

Good luck in your reading, and good luck in your career in kinesiology!

ACKNOWLEDGMENTS

The project of this sort is the work of many hands. I have been fortunate to play a central role in shaping its direction: a rare opportunity that has been thoroughly enjoyable. I would compare being editor of a work such as this to having been given the opportunity to drive a new automobile; the pleasure of steering it and guarding its fenders comes only if those at the factory have assembled a road-worthy product. My knowledgeable co-contributors and a team of highly competent people at Human Kinetics have indeed made this text—now in its third edition—a roadworthy product.

Although this book features the work of many authors, it isn't an anthology or a compilation: it is a textbook in every sense of the word. It is organized around a central theme; structured to accomplish specific purposes; and integrated in terminology, concepts, objectives, and graphics. Each author brings to the project a unique professional and academic background and a track record of achievement in his or her respective specializations. Their contributions have added a richness and breadth to the text that is largely responsible for the book's success. Achieving consistency and flow hasn't been easy, either for the authors, the editor, or the editorial staff at Human Kinetics. We will wait for your feedback to let us know how well we managed to pull it off. Ultimately the responsibility for pulling together the diverse contributions of authors, providing a structure for the text, ironing out inconsistencies in terminology, and interpreting material fell to the editor. Any errors that have been made, whether of commission or omission, are attributable to me, not to others who have given so generously of their time.

At the risk of inadvertently omitting a name I would like to give special acknowledgment to those who played key contributions in the production of *Introduction to Kinesiology.* My co-contributors caught "the spirit" of the project, sensed its importance, and produced highly readable and informative pages. The finished product is as much theirs as it is mine. Special credit goes to my colleague Janet Harris who shared editorial duties for the first edition and played a major role in giving structure to the text. The time I found to carry out the editor's role was extracted from the hours I might have spent with my wife, Claude Mourot. Her understanding and steadying influence was more imporant than she knows.

Only those who have been fortunate enough to author books published by Human Kinetics can appreciate the talents of its publication teams. Those specifically involved with this book are listed on the copyright page, and each did his or her work superbly. Pam Brown, a faculty member at University of North Carolina at Greensboro, and Lisa Morgan, an instructional designer at Human Kinetics, created the activities in the student study guide. They did an outstanding job of making the material in the book interactive and easier to learn.

CREDITS

Figure 2.3: Adapted from Human Resources and Social Development Canada (HRSDC), "Population Disabled by Age, Canada 2001-2026. Advancing the Inclusion of People with Disabilities," 2005.

Figure 2.6: Adapted from R.L. Zweigenhaft, 1997, "The empirical study of signature size," *Social Behavior and Personality* 5(1): 177-185.

Figure 2.10: Data from U.S. Department of Health and Human Services, 2000, *Healthy People 2010: Understanding and improving health,* 2nd ed. (Washington, D.C.: U.S. Government Printing Office), 22.

Figure 3.3: Adapted from U.S. Department of Commerce, Statistical abstract of the U.S., 2007, *Participation in selected sport activities: 2004,* p. 256.

Figure 3.9: Adapted, by permission, from R.B. Arnot and C.L. Gaines, 1984, *Sportselection* (New York: Viking Press).

Figure 6.2: SHULMAN, JAMES, L; THE GAME OF LIFE, © 2001 Princeton University Press, 2002 paperback edition. Reprinted by permission of Princeton University Press.

Figure 8.5: Reprinted, by permission, from R.A. Schmidt and C.A. Wrisberg, 2007, *Motor learning and performance: A situation-based learning approach,* 4th ed. (Champaign, IL: Human Kinetics), 153.

Figure 8.6: Adapted from R.M. Malina, 1984, Physical growth and maturation. In *Motor development during childhood and adolescence,* ed. J.R. Thomas (Minneapolis, MN: Burgess), 7. By permission of J.R. Thomas.

Figure 8.7: Adapted from K.M. Newell, 1984, Physical constraints to the development of motor skills. In *Motor development during childhood and adolescence,* edited by J.R. Thomas (Minneapolis, MN: Burgess), 108. By permission of Jerry R. Thomas.

Figure 8.8: Data from A. Espenschade, 1960, Motor development. In *Science and medicine of exercise and sport,* edited by W.R. Johnson (New York: Harper & Row), 330.

Figure 9.5: Reprinted, by permission, from Y. Hanin, 1997, "Emotions and athletic performance individual zones of optimal functioning model," *European Yearbook of Sport Psychology* 1: 29-72.

Figure 13.2: Reprinted, from R.J. Klein, 2008, *Healthy People 2010: Progress Review Focus Area 22-Physical Activity and Fitness Presentation.* [Online]. Available: http://www.cdc.gov/nchs/ppt/hp2010/focus_areas/fa22_2_ppt/fa22_paf2_ppt.htm [December 16, 2009].

Figure 13.5: From Centers for Medicare and Medicaid Services, Office of Actuary.

Figure 16.3: Based on the study of G.M. DeMarco, 1999, "Physical education teachers of the year: Who they are, what they think, say, and do," *Teaching Elementary Physical Education* 10(2): 11-13.

Figure 17.2: Based on the work of A. Meek, 1997, "An estimate of the size and supported economic activity of the sports industry in the United States," *Sport Marketing Quarterly* 6(4): 15-22.

Introduction to Kinesiology and Physical Activity

1

Shirl J. Hoffman and Janet C. Harris

Digital Vision

CHAPTER OBJECTIVES

In this chapter we will

- help you appreciate the pervasiveness and diversity of physical activity in human life;
- introduce you to ways of defining and thinking about physical activity;
- discuss the discipline of kinesiology and its relationship to physical activity;
- familiarize you with the types of knowledge concerning physical activity that are acquired through experience, scholarly study, and professional practice; and
- help you gain a preliminary understanding of what a profession is and of the career possibilities centering on physical activity.

If you haven't been to the gym or athletic field or engaged in hard labor today, you probably think you haven't been physically active. But you did manage to get out of bed, walk to the bathroom, get dressed, eat breakfast, and perhaps make your way to class, and each of these requires physical activity. These forms of physical activity are remarkable not only for their complexity but also for their variety. In fact, physical activity is so interwoven into your life that it's probably far easier to count the number of ways in which you weren't physically active today than the ways in which you were.

If you take a moment to reflect on how physically active you have been, you will see that your life is an endless universe of physical activity. You walk, reach, run, lift, leap, throw, grasp, wave, push, pull, and perform thousands of other movements as part of living a normal human existence. Physical activity is essential in your work, whether it is hard physical labor or consists of the low-energy tasks more common to professional practice. Physical activity is used to express ourselves in gesture, art, and dance. Our health depends on performing regular forms of vigorous physical activity, and we rely on various forms of recreational physical activity for fun and enjoyment.

> **DO it Activity 1.1**
>
> **Who's More Active?**
> Use this activity in your online study guide to consider the daily routines of two people and decide which one is more physically active.

Even more than this, physical activity is part of our human nature. It is an important means by which we explore and discover our world. It helps us define ourselves as human beings. A significant part of our lifetime is spent learning to master a broad range of physical activities, from the earliest skills of reaching, grasping, and walking to enormously complex skills such as hitting a baseball, performing a somersault, or playing the piano. Most of us master a broad range of physical activities at a moderate level of competence. Others concentrate on a limited number of skills, a focus that can lead to extraordinary performances ranging from the incredible accuracy with which a National Football League quarterback can throw a football to the amazing finger dexterity of a concert pianist.

In this chapter we will talk about physical activity in very general terms. Taking time to read the chapter carefully will help you appreciate the complexity, diversity, and importance of physical activity to human life. It will also provide a frame of reference for this text and for the field of study you have chosen. As a prospective physical activity professional, it is critical that you understand how the discipline of **kinesiology** is organized and how it relates to the phenomenon of physical activity. If you've been physically active throughout your life, you already have some knowledge of physical activity. This background will be of enormous benefit to you as you roll up your sleeves and begin to probe the depths of knowledge of kinesiology. But prior experiences can also hinder your understanding, especially when you are required to think about those experiences in new ways. At times you will have to set your assumptions aside so that you can examine physical activity from a fresh and exciting point of view. This may be more difficult than you might imagine.

Ways of Developing an Understanding of Kinesiology

Because people are now more aware than ever of the importance of physical activity to their cognitive, emotional, physical, and spiritual well-being, enrollment in college and university curriculums devoted to the study of physical activity is on the rise. There has been an explosion of career opportunities for college-trained professionals with in-depth knowledge of the scientific and humanistic bases of physical activity coupled with training in professional practice. Career possibilities now extend well beyond the traditional professions of teaching physical education and coaching. Physical therapy, cardiac rehabilitation, sport management, athletic training, and fitness leadership and management are just a few of the careers that are likely to require formal academic preparation in kinesiology.

With this growing recognition of the importance of physical activity in our daily lives has come the realization that it deserves to be studied just as seriously and systematically as other disciplines in higher education such as biology, psychology, and sociology. No doubt you've heard the word *discipline,* but you may not really understand what it means. A **discipline** is a body of knowledge organized around a theme or focus (see figure 1.1). Disciplines embody knowledge that learned people consider worthy of study. The focus of a discipline identifies what those who work in the discipline study. The central focus of biology, for example, is life forms; the focus of psychology is the mind and mental and emotional processes; and the focus of anthropology is cultures. Although debates are still being waged about the focus of kinesiology, and some would argue that it has not yet reached the level of a discipline, there is reason to believe that it has: It is now generally regarded as the discipline that focuses on physical activity.

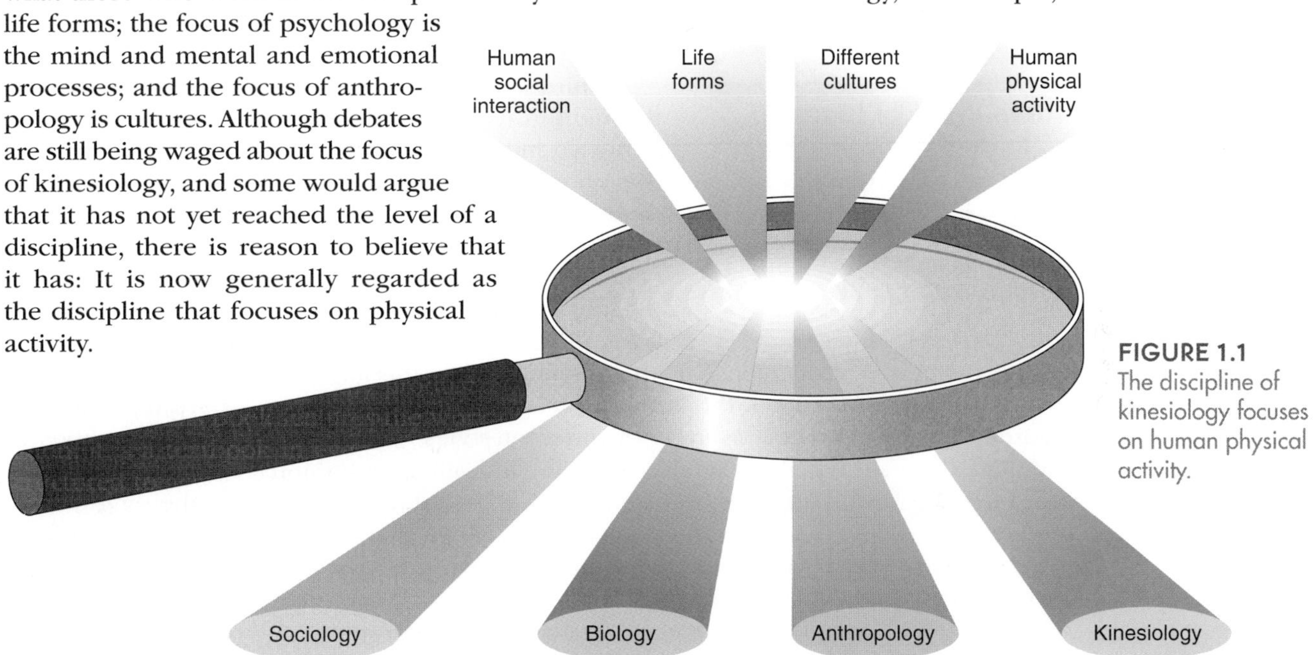

FIGURE 1.1 The discipline of kinesiology focuses on human physical activity.

Physical Activity: The Focus of Kinesiology

Physical activity is the main thing that distinguishes kinesiology from other fields of study in higher education. It's the very reason distinct administrative degrees are granted for this field of study in higher education (Newell 2007).

You will discover that people learn kinesiology in unique ways. You may have noticed in the college courses you've taken that all disciplines are not learned or studied in the same way. Art, for example, may be studied through reading, writing, and experimentation with artistic projects in the studio. People learn history, literature, and philosophy largely through reading, writing, memorization, and discussion.

Reading, writing, memorization, and discussion are also important in learning chemistry and biology, but these disciplines involve active participation in laboratory exercises as well. People learn kinesiology partly in the same ways they learn other disciplines and partly in different ways. There are three ways to learn about physical activity. One is by watching or performing, just as students in the disciplines of art or music learn to appreciate art or music by watching and listening or performing. Your direct experience with physical activity, then, is an important source of knowledge about physical activity and contributes to your understanding of kinesiology. Watching or performing physical activity may not seem as important as learning through scholarly study, but don't underestimate its value in developing your understanding of physical activity.

A second way of developing an understanding of kinesiology is through systematic scholarly study. This way of learning involves reading, thoughtful and careful analysis, and discussion with colleagues about theoretical and practical matters related to physical activity. It also involves experiences in laboratories. Mastering subjects in the kinesiology curriculum such as sport history and philosophy, motor development and learning, exercise physiology, biomechanics, and sport psychology requires these forms of study. Where does the knowledge contained in such subjects come from? Mostly from the work of research scientists and scholars in the field of kinesiology who have developed and added to the knowledge base through systematic research and scholarship. Our knowledge of sport history or philosophy of sport, for example, stems from the research of sport historians and sport philosophers. Scholars who conduct research in biomechanics laboratories at universities produce our knowledge of biomechanics of physical activity.

DO it **Activity 1.2**

Three Ways of Learning
In this activity in your online study guide, you'll distinguish among the ways of learning about physical activity.

A third way of learning about physical activity is through professional practice, although here the focus is not so much on learning about physical activity as it is on using physical activity to bring about predetermined ends. Professionals such as physical education teachers, personal trainers, and cardiac rehabilitation specialists, for example, systematically manipulate the physical activity experiences of students, clients, patients, and others whom they serve as a way of helping them achieve personal goals. Accomplishing this task requires a great deal of knowledge, but where does this knowledge come from? Some of it arises from applied research that tests the effectiveness of different styles of teaching, different methods of treatment, different types of exercise routines, and other professional routines. Some of it comes from hands-on experience—a knowledge base developed through helping clients, students, or patients to achieve their goals. And, as we shall see, knowledge gained from experiencing physical activity and the scholarly study of physical activity can be important in professional practice as well.

➤ Experiencing physical activity, scholarly study of physical activity, and professional practice centered in physical activity are the three sources of physical activity knowledge that constitute the discipline of kinesiology.

Thus exercise therapists, teachers and coaches, athletic trainers, and other physical activity professionals develop a rich knowledge base of physical activity through their personal experience engaging in and watching physical activity and through scholarly study about physical activity. This knowledge can serve as a foundation for professional practice knowledge. If this knowledge is grounded in careful, systematic observations of the effects of manipulations of physical activity on others such as students, patients, and clients, it often becomes incorporated into kinesiology curriculums offered by colleges and universities and is taught to students. As such it also is part of the discipline of kinesiology.

You will notice that we have carefully defined kinesiology as knowledge derived from experiencing physical activity, scholarly study of physical activity, and professional practice centered in physical activity. Such knowledge becomes part of the discipline of kinesiology *only when that knowledge is embedded in a college or university curriculum in kinesiology or used by kinesiologists in their physical activity research*. The reason for this is to clarify precisely what is part of the "official" discipline and what is not. People perform, study, and use physical activity in professional practice in many venues outside the college or university setting. These activities may be important and valuable in their own right, but they do not constitute the "doing" of kinesiology any more than the use of elementary psychological principles by a businessperson to motivate her sales force is "doing" psychology. The discipline of psychology remains tied to the college and university curriculum and to the research of psychologists. Similarly, people may use the principles of kinesiology outside the discipline, but kinesiology per se remains a function of curriculums and research in colleges and universities.

There is another reason for limiting our definition of kinesiology to knowledge contained in college or university curriculums or used in research. The knowledge that you will acquire in your major curriculum is more highly organized and more scientifically verifiable than the knowledge of physical activity that laypersons use. Universities employ rigorous methods to monitor the authenticity of the knowledge included in their curriculums and in the research of their faculty. Think about it. Would you have more confidence in the recommendations of a university kinesiologist who specializes in fencing than those of a lawyer who fences as a hobby? Would you have more confidence in the scientific accuracy of recommendations for exercise programs offered by an exercise physiologist than in that of statements offered by a television exercise guru who lacks formal training in kinesiology? Would you be more likely to trust the recommendations of a university specialist in pedagogy concerning how to organize a large group of young children for physical activity instruction than those of a volunteer coach who has no formal training in kinesiology? Given your decision to invest several years of hard work in preparing for a career in the specialized field of kinesiology, we might presume that your answer to all three questions is yes.

➤ Only knowledge about physical activity that is included in a college or university curriculum or used in research is part of the body of knowledge of kinesiology.

Many people experience physical activity (e.g., walk through a supermarket collecting groceries, play pickup basketball), study physical activity (e.g., read popular trade books on fitness or sports), or engage in some form of "professional" practice (e.g., volunteer as a youth league coach) outside the confines of the university curriculum, but these may not constitute "doing" kinesiology under the strict definition being used in this text.

Later in this chapter you will learn how all three components of physical activity fit together to make up the discipline of kinesiology. People gain the most complete knowledge of kinesiology from engaging in all three components—experience, scholarly study, and professional practice—although knowledge from each component is valuable in its own right. These three components are depicted in figure 1.2. This figure appears throughout the book to remind you of the component of physical activity currently being discussed in the text.

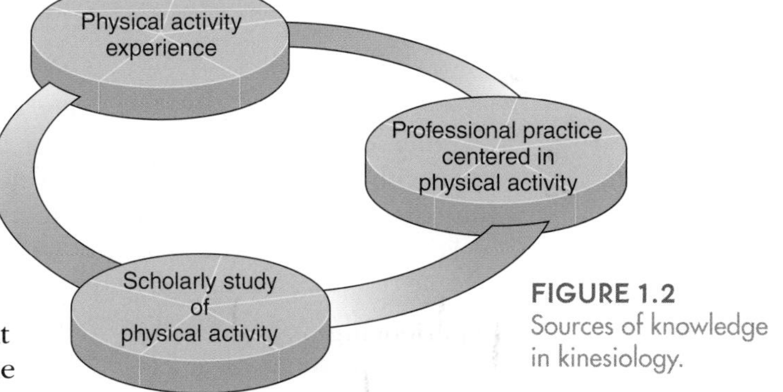

FIGURE 1.2 Sources of knowledge in kinesiology.

What Is Physical Activity?

We have already emphasized that the discipline of kinesiology focuses on physical activity. But what exactly *is* physical activity? At first you might think this a silly question. Everybody knows what it is, so why waste time defining it? But definitions are important, especially in scientific and professional fields where terms may be defined somewhat differently than they are in everyday language. These **technical definitions** ensure that people working within a science or profession have a common understanding. You may be surprised to learn that kinesiologists do not all agree on the technical definition of *physical activity.* Thus, before we go much further, let's be sure that *we* all have the same understanding of what physical activity is.

www. Web Search 1.1

Defining Physical Activity
In this activity in your online study guide, you'll document different definitions of physical activity.

In everyday life, almost any muscular action is considered physical activity. Throwing a javelin, driving a car, walking, performing a cartwheel, swimming, digging a ditch, hammering a nail, scratching your head—all are examples of physical activity, as are the kick you exhibit when the doctor taps your patellar tendon to test your reflexes, the blinking of your eye, the peristaltic action of your small intestine brought on by muscular contractions, the contraction of your diaphragm when you sneeze, and the action of your throat muscles when you swallow. But are all these muscular actions of equal

Defining Physical Activity

This is a good time to wrestle with some fundamental questions about physical activity. Although we are all familiar with physical activity, we rarely think deeply about it. So before you continue reading this chapter, think carefully about the following questions. Take time to compare your answers with those of your classmates.

1. How would you define physical activity?
2. What do you consider the most important characteristics of physical activity?
3. What methods can you think of that kinesiologists might use to study physical activity?
4. Name five professions that use physical activity.

After you have thought about or discussed these questions, rewrite your definition of physical activity. Keep your answers handy so you can see how they compare with the discussion that follows.

concern to kinesiologists? Not really. Although all are examples of human movement, they're too diverse in form and purpose for any single discipline to study. Indeed, if kinesiology focused on all forms of human movement, then kinesiologists would study everything that humans do, because living is moving!

For this reason, kinesiologists use a much narrower definition of physical activity than people typically use in everyday language. The discipline requires a definition that is neither too *inclusive* (e.g., all human movement) nor too *exclusive* (e.g., only human movement related to sports). The definition of physical activity used in this text takes its cue from Professor Karl Newell's definition of **physical activity** as *intentional, voluntary movement directed toward achieving an identifiable goal* (1990a). Notice three things about this definition: First, it does not stipulate anything about the energy requirements of the movements used to produce the activity. Large-muscle activities typically require the highest levels of energy, but the definition doesn't limit physical activity only to these. Swimming, lifting barbells, running marathons, and in-line skating are physical activities, but so are typing, handwriting, sewing, and surgery.

DO it **Activity 1.3**

Physical Activity—Yes or No? Use this activity in your online study guide to identify examples of physical activity.

Second, the setting in which physical activity takes place is irrelevant. Surely, shooting a basketball is a form of physical activity, but so is tossing a piece of paper into the wastebasket. Pole-vaulting is a physical activity, and so is jumping over a fence. Swinging a baseball bat is a physical activity, but so is swinging a sledgehammer. Just as physical activity takes place in many settings, it takes many forms. Wrestling and skiing are not similar to typing or performing sign language, but all are forms of physical activity. In chapter 2 you will discover the truly wide range of physical activity that is of interest to those in the field of kinesiology.

Third, according to this definition, simply moving your body doesn't constitute physical activity. The movements must be directed toward some purposeful end. This idea can be confusing, especially when we consider that the term *kinesiology* is derived from the Greek words *kinesis* (movement) and *kinein* (to move). **Movement** includes any change in the position of your body parts relative to each other. Obviously, it is impossible to perform physical activity without moving your body, but movement by itself does not constitute physical activity as we use the term in this text. One way to think about the relationship between movement and physical activity is this: *Movement is a necessary but not sufficient condition for physical activity.*

includes all physical activities

Only movement that is intentional and voluntary—purposefully directed toward an identifiable goal—meets the technical definition of physical activity that is used here. This excludes all involuntary reflexes and all physiological movements controlled by involuntary muscles such as peristalsis, swallowing, or blinking an eye. It also excludes voluntary movements that people perform without an intentional goal in mind. A thoughtless scratch of the head, an absentminded pulling of the earlobe, or the repetitive movements of a compulsive-obsessive psychiatric patient are examples of human movement that fall outside the technical definition of physical activity because these movements are not designed to achieve a goal.

➤ Physical activity is movement that is intentional, voluntary, and directed toward achieving an identifiable goal. This definition excludes human movements that are involuntary, such as reflexes, or those performed aimlessly and without a specific purpose.

As you read the literature in kinesiology, you will soon discover that kinesiologists define physical activity in various ways. This is why it is important for you to understand what is meant by physical activity in this text (see figure 1.3).

In the influential document titled *Physical Activity and Health: A Report of the Surgeon General,* physical activity is defined as "bodily movement that is produced by the contraction of skeletal muscle *and that substantially increases energy expenditure*" (italics added) (U.S. Department of Health and Human Services [USDHHS] 1996, p. 21). Let's take this definition apart. First, it limits physical activity to voluntary movement, which seems sensible because most of the time kinesiologists are interested in voluntary rather than involuntary movements of the body. At the same time the definition seems too broad, since it includes all movement (presumably whether purposeful or not) that "requires an expenditure of energy." Since all movements—intentional, random, or unplanned—involve some degree of energy expenditure, this definition includes virtually all skeletal or voluntary movement.

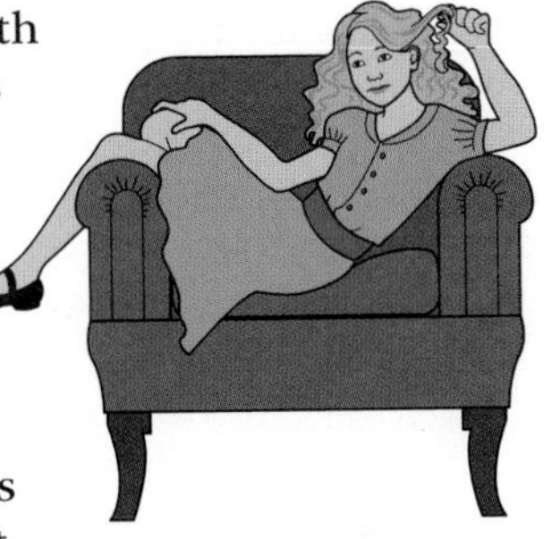

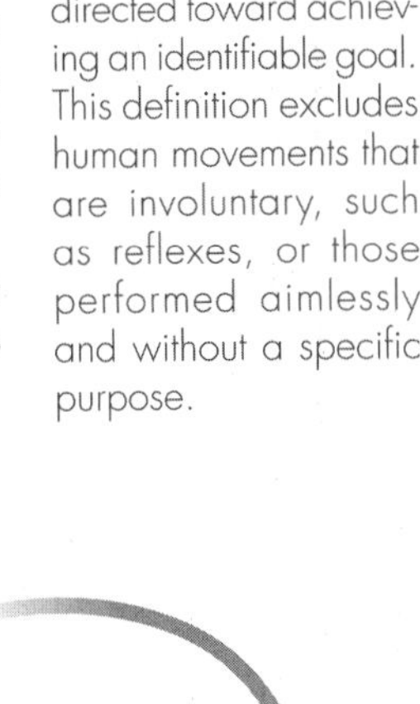

FIGURE 1.3 Definitions of physical activity. Only movement that is voluntary and goal oriented meets the technical definition of physical activity.

Obviously, kinesiologists are interested in vigorous forms of physical activity such as running, lifting weights, and exercising—forms of physical activity that substantially increase our energy expenditure. But kinesiologists are also interested in many types of physical activity that do not require substantial amounts of energy. For example, teachers and coaches often teach bowling or archery; occupational and physical therapists may spend large amounts of time teaching stroke patients or elderly residents of a nursing home how to use a knife and fork, to bathe, or to comb their hair. Motor control researchers may be interested in measuring

Physical Activity and "Doing"

The distinguished neurophysiologist Sir Charles Sherrington wrote the following in his classic book, *Man on His Nature:* ". . . all [humans] can do is to move things, and [their] muscle contraction is [their] sole means thereto" (Sherrington 1940, p. 107). What do you think Sherrington meant when he said all we can *do* is "move things"? Can you think of anything you can do without moving? What does your answer imply about the importance of physical activity in our lives?

Activity 1.4

Thinking More Deeply About Physical Activity

In this activity in your online study guide, you'll consider the goal of various types of physical activity.

fine motor skills or the speed at which individuals can move their arms under different experimental conditions. Obviously, none of these requires expenditure of "substantial" amounts of energy. Thus, the surgeon general's definition seems too restrictive because it eliminates many types of physical activity that interest kinesiology scholars and professionals.

Of course we could limit our technical definition of physical activity to include only those physical activities that relate directly to exercise and sport. Because kinesiology has its historical roots in exercise and sport and because most students entering the field have their sights set on a career in exercise or sport, this definition may be appealing.

But even though sport and exercise receive the most attention in the field of kinesiology—and they certainly take center stage in this text—the study of kinesiology includes a far wider range of physical activity than sport and exercise. For example, kinesiologists study basic postural mechanisms, the physiology and body mechanics of work, the development of reaching and grasping behaviors in infants, and the daily life support activities of the elderly. Physical education teachers teach children how to perform fundamental movement patterns such as hopping, running, and skipping, or expressive physical activities such as dance; and therapists working in rehabilitation programs teach patients to recover lost capacities to walk, sit, rise from a chair, or drive a car. As the range of occupations open to graduates of kinesiology departments continues to expand, clearly a technical definition of physical activity that reaches beyond exercise and sport is in order.

More About the Surgeon General's Definition of Physical Activity

Sometimes kinesiologists focus so tightly on the specific area of physical activity in which they work that they tend to define physical activity only in terms that make sense to them. Physiologists concerned about the health ramifications of not engaging in daily vigorous exercise wrote the surgeon general's report. Because it is well known that only forms of physical activity that "substantially increase energy expenditure" are likely to ward off heart and vascular disease, this limited type of physical activity served as the working definition for the final report. How might coaches using only their focused experiences of physical activity define physical activity? Physical therapists? Dance teachers?

What Is Kinesiology?

Kinesiology is a discipline or body of knowledge that focuses on physical activity. To reiterate: The discipline derives and incorporates knowledge from three different yet related sources:

- Experiencing (or doing) physical activity (experiential knowledge)
- Studying the theoretical and conceptual bases of physical activity (theoretical knowledge)
- Professional practice centered in physical activity (professional practice knowledge)

We have already mentioned that this book is organized around the unique ways in which we learn about physical activity. Not surprisingly, these also are the ways in which we approach the study of kinesiology. Typically, we tend to associate learning with reading, writing, discussion, and memorization. We have seen, however, that experience with physical activity, theories and concepts about physical activity, and professional practice all are important sources of knowledge for kinesiologists. They are also important sources of knowledge for you as a student who is studying kinesiology. For this reason, this text has been divided into three parts, each of which examines a distinct source of knowledge that is incorporated into the discipline of kinesiology.

Figure 1.4 depicts the three major dimensions of kinesiology. They will help you apply what you have learned. You will notice that each dimension corresponds to a different source of knowledge of physical activity as shown in figure 1.2. The part of the figure marked A1 represents disciplinary knowledge acquired from our hands-on experiences with physical

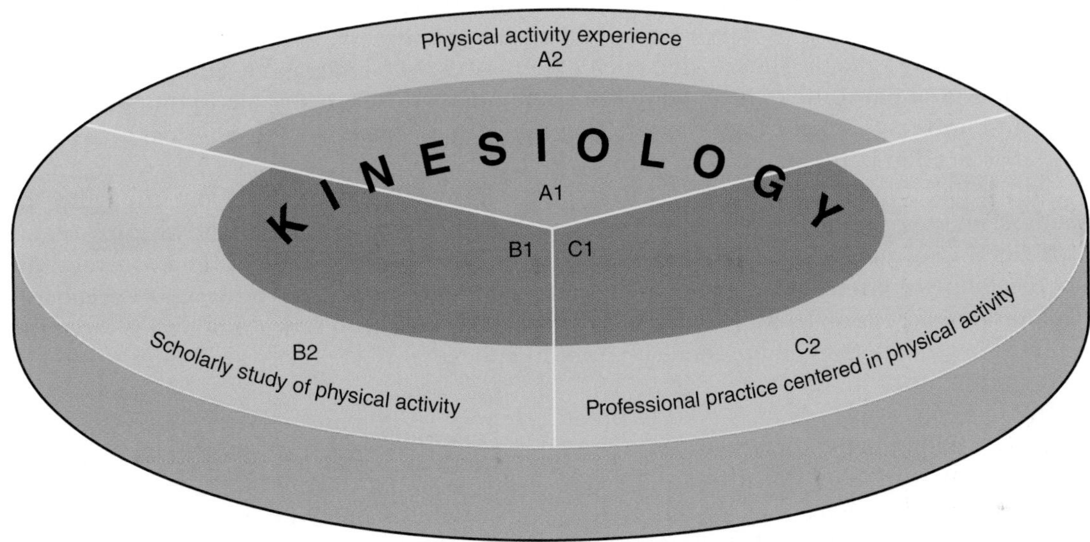

FIGURE 1.4 Discipline of kinesiology in colleges and universities.

A1 Knowledge gained through experiencing physical activity that is systematically incorporated into the discipline of kinesiology (e.g., college tennis class offered for academic credit, weight-training class offered for academic credit).

A2 Knowledge gained through experiencing physical activity that is not incorporated into the discipline of kinesiology (e.g., learning a dramatic stage movement, taking tennis lessons at a country club, playing Little League baseball, firefighting).

B1 Knowledge gained through scholarly study of physical activity that is systematically incorporated into the discipline of kinesiology (e.g., sport history, exercise physiology, motor development).

B2 Knowledge gained through scholarly study of physical activity that is not systematically incorporated into the discipline of kinesiology (e.g., research about playing a musical instrument, reading a popular book on fitness).

C1 Knowledge gained through professional practice centered in physical activity that is systematically incorporated into the discipline of kinesiology (e.g., knowledge gained in roles such as certified athletic trainer or elementary physical education teacher that is included in university kinesiology classes).

C2 Knowledge gained through professional practice centered in physical activity that is not systematically incorporated into the discipline of kinesiology (e.g., knowledge gained in roles such as certified athletic trainer or physical education teacher that is not included in university kinesiology classes).

activity. Usually, this involves performing physical activity, but we can also acquire knowledge by observing others perform. Students often acquire this **experiential knowledge,** described in part I of this book, through physical activity classes (e.g., classes in soccer, weight training, swimming) offered in kinesiology departments. The experiential knowledge included in a formal college or university kinesiology curriculum is part of the discipline of kinesiology. So far, experiential knowledge has not been divided into formal subdisciplines within the field of kinesiology, although each form of activity represents a specialized form of experience.

Of course, people may acquire experiential knowledge through physical activities outside a formal college or university curriculum; this kind of experiential knowledge is represented by part A2 of figure 1.4. This knowledge may be on the fringes of the discipline or not part of the discipline at all. For example, you might take swimming lessons at a private club, take tennis lessons on weekends at a recreation center, take martial arts lessons at a commercial center, work out on your own at a campus fitness center, play intramural football in a recreational league, or train for firefighting through the fire department. You can often learn extremely valuable things from such experiences; but, in and of itself, this knowledge is not a central part of the kinesiology discipline unless it is formally incorporated into college and university kinesiology classes. These aspects of experiencing physical activity are also discussed in part I of this text.

Part B1 of figure 1.4 represents disciplinary knowledge acquired from the scholarly study of theories and concepts about physical activity. This **theoretical knowledge** usually is divided into categories or **subdisciplines** in university curriculums. Taken together, the subdisciplines constitute the **"spheres of scholarly study"** examined in part II of this book. This disciplinary knowledge is derived from research and scholarly study and is taught in its most systematic and comprehensive form in college and university curriculums, usually in departments of kinesiology.

Because physical activity is such a broad category of human behavior, however, other university departments sometimes engage in the scholarly study of physical activity as well, as represented by part B2 in figure 1.4. For example, departments of drama, dance, and music sometimes teach about the scholarly aspects of physical activity, as do departments of engineering and medicine. This knowledge, however, is considered to be on the fringes of the discipline of kinesiology. Departments of kinesiology are the only academic units within colleges and universities that identify the unified study of physical activity *as their sole mission*. You have surely noticed many books and magazines on the shelves of your bookstore that deal with sports and fitness. People who lack in-depth education in kinesiology wrote many of these books, which therefore may have no basis in science or systematic analysis. Thus, we cannot consider them part of the discipline of kinesiology.

➤The discipline of kinesiology consists of experiential knowledge, theoretical knowledge, and professional practice knowledge. Experiential knowledge derives from experiencing physical activity; theoretical knowledge derives from systematic research about physical activity; and professional practice knowledge derives from and contributes to the process of delivering physical activity services.

Part C1 of the model in figure 1.4 represents disciplinary knowledge acquired from professional practice centered in physical activity—professional practice such as managing a fitness center, teaching physical education, engaging in personal training, or working in cardiac rehabilitation. This **professional practice knowledge** becomes part of the discipline when it is discovered or tested in preprofessional or professional settings and is incorporated by faculty into college and university kinesiology classes, usually classes focused on preparing students for specific physical activity professions. Professional practice knowledge usually deals with appropriate ways of manipulating physical activity experiences to bring about specific results. For example, personal trainers may manipulate the exercise experiences of their clients in one way to increase muscular strength and in another way to increase flexibility. Likewise, physical education teachers may manipulate physical activity experiences of their students in one way to achieve fitness and in a completely different way to develop skill. Part III of this book describes this component of physical activity knowledge in detail.

Of course, not all knowledge acquired through professional practice becomes part of the kinesiology curriculum. For example, a coach who is planning drills for her team may rely more on knowledge gained from people who coached her in the past than from the latest research on pedagogy, skill learning, and fitness. Such knowledge, of course, can be effective; after all, a coach is not likely to use it unless she has found it to be effective for preparing her players. But such knowledge is often flawed. The good results the coach observed may have less to do with her own actions than with other factors she failed to take into account. For this reason, some of the knowledge acquired by practitioners is not incorporated into the discipline. To indicate this, such knowledge is located on the perimeter of C2 in figure 1.4.

Professional practice knowledge is most valuable when combined with knowledge from the other dimensions of kinesiology (experience and scholarly study). Together, these can provide an important framework for conceiving of and using knowledge about professional practice. Sometimes people who possess only a fraction of the disciplinary knowledge (as represented in parts A1, B1, and C1) are hired to perform professional roles anyway. This can occur because the demand for physical activity professionals is so great that companies and institutions often hire those lacking adequate qualifications. School districts, for example, sometimes hire coaches who have little or no background in kinesiology. Likewise, fitness centers may hire personal trainers who lack adequate qualifications. Obviously, one can develop a modest level of competency in almost any profession without mastering the total body of knowledge. Through trial and much error, such a person may muddle through. This approach, however, can be dangerous. For example, you may have known or heard about laypersons who managed to learn enough about the law to represent themselves in court. But such a person runs a high risk of making a mistake. (This is the basis for the old saying "He who chooses to represent himself in court has a fool for a client.") Similarly, an unqualified individual who assumes the role of a physical activity professional lacks the informed judgment of one who has studied kinesiology.

Let's quickly review our description of kinesiology. It is a discipline or body of knowledge that concentrates on physical activity. The knowledge of kinesiology is acquired through experience (watching or performing physical activity), scholarly study of theories and concepts about physical activity, and professional practice centered in physical activity. Because kinesiology is a formal discipline, it typically is taught in a college or university by people highly trained in certain areas of specialization, whether this specialization is in performance, scholarly study and research, or professional practice. Thus, faculty who are able to instruct you about the broad knowledge base of the discipline teach in departments of kinesiology. The curriculums of these departments represent organized sequences of classes that are structured in ways to ensure maximal opportunities for integrating all three types of knowledge.

The Focus of Kinesiology: Exercise and Skilled Movement

Kinesiology focuses on two general categories or forms of physical activity: exercise and skilled movement (see figure 1.5). You know much about both forms of physical activity. In fact, your interest in one or both of these is probably responsible for your deciding to seek a degree in kinesiology.

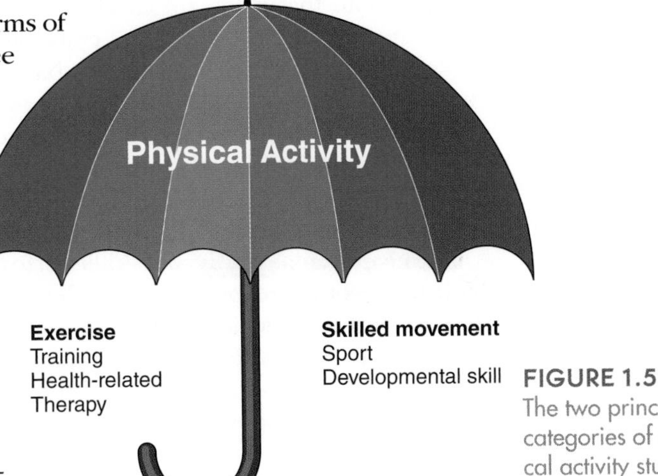

FIGURE 1.5 The two principal categories of physical activity studied by kinesiologists.

People engage in **exercise** to improve or regain performance, health, or bodily appearance. Running or lifting weights to increase your fitness (improve your health) or to lose body fat (change the appearance of your body) is exercise; so is weight training by bodybuilders hoping to increase the size and definition of their muscles to achieve an ideal "look." Working out to increase strength or cardiorespiratory endurance as adjuncts to healthful living also

➤ Kinesiology is a discipline that focuses on physical activity, especially two categories of physical activity: exercise and skilled movement.

is exercise. And so are the rehabilitation routines that patients undergo as they attempt to regain function following an injury or disease.

Because *exercise* involves many different types of physical activity, breaking the term down into three major categories is helpful:

- Exercise performed for the express purpose of conditioning your body to improve athletic or other types of performances is a specific type of exercise known as **training** (see chapter 3). Kinesiology graduates who embark on careers as conditioning specialists with university or professional sport teams focus on physical activity as training.
- Exercise undertaken specifically to develop or maintain a sound working body, free of disease and able to perform daily tasks and deal with emergencies, is known as **health-related exercise.** Kinesiology graduates who work as fitness leaders and personal trainers focus almost exclusively on physical activity as health-related exercise.
- Exercise also may be performed to restore capacities previously acquired or developed that have been lost because of injury, disease, or behavioral patterns. This type of exercise is **therapeutic exercise.** For example, postcardiac patients usually require physical activity regimens to help them regain cardiovascular health following a heart attack. Kinesiology graduates who embark on careers to work in these cardiac rehabilitation programs, or who work as athletic trainers or physical therapists, focus on physical activity as therapy.

➤ Exercise and sport are the principal forms, but not the only forms, of physical activity studied by kinesiologists. Exercise includes any physical activity performed to improve performance, health, or bodily appearance. Exercise is of three types: Training for improving performance in sport or other types of physical activity, health-related activity to maintain or improve health, or therapy to rehabilitate individuals from debilitating disease or injury. Skilled movement includes physical activities in which accuracy of direction, force, or rhythm are essential to attaining predetermined goals. Skilled movement is important in both sport and developmental skills.

Skilled movement is the second area of focus of kinesiology. Skilled movement involves performances in which accuracy of direction, force, and rhythm or timing are essential to accomplishing predetermined goals. Normally, people learn these qualities of physical activity through systematic practice. Factors normally associated with exercise such as strength, cardiorespiratory endurance, or flexibility, although important in executing many physical tasks, are not elements of skilled movement. (They are developed through training, health-related exercise, or therapeutic exercise. Chapter 3 addresses these differences.) Two categories of skilled movement are of primary interest to kinesiologists: sport and developmental skills.

Sport has long been of interest to scientists, teachers, and practitioners in kinesiology. We define **sport** in general terms as a form of physical activity in which a person performs skilled movements to achieve a goal in a manner specified by rules, usually in competitive contexts. People usually perform sport merely for the enjoyment it brings them. Note three things about this definition. First, the physical activity in sport is "skilled," which means that it is performed "efficiently" and "effectively." Not all forms of physical activity require a great deal of skill, but in every type of sport the advantage belongs to competitors who have learned to move their bodies in skillful ways. The soccer player who passes the ball deftly to her teammate, the golfer who strikes the ball squarely, and the gymnast who successfully completes a double rotation on dismount all are expressing skill in their performances. Second, note that rules are essential in sport. They exist for the sole purpose of creating the game. Without rules, players could do whatever they felt like doing at the time, and the game would soon break down. If a basketball player decides not to dribble the ball as she runs down the court, she is no longer playing the sport of basketball. Finally, note that the physical activities performed in sport tend to be framed in competition, either against other teams, against individuals, against established records, or against "personal bests." Rules create a level playing field for all competitors, ensuring that each has an equal chance to win the competition.

Developmental skills are skills performed in nonsport settings. For example, as part of their professional responsibilities, elementary physical education teachers teach 1st graders how to perform such fundamental movement patterns as skipping, throwing, or hopping. Acquiring these developmental skills at an early age may lead to high levels of proficiency in sport and in other activities in later years, but when they are taught they have no direct correlation with a specific sport. Likewise, graduates of kinesiology programs who go on to pursue careers in occupational or physical therapy or who work in nursing home facilities as physical activity specialists may spend a great deal of time teaching poststroke elderly patients the developmental skills required to eat, dress, and groom. The range of develop-

DO it Activity 1.5

Classifying Physical Activity

Categorize the forms of physical activity in your online study guide.

mental skills is enormous. Some kinesiologists study the characteristics of walking and running, some study the mechanics of grasping or reaching, and some work to improve the efficiency of movements used in industrial or military settings. In chapter 2 we explore this broad range of physical activity and discover how important it is in our daily lives.

Obviously, these categories of physical activity are not mutually exclusive. Some people engage in exercise and sport simultaneously. For example, you might compete in racquetball with the hope of getting good enough to win your city's championship but also intend to get enough exercise to improve your body's functioning or appearance. Individuals might participate in judo competition because they enjoy it but also participate because of the health benefits it brings. Use the categories as guides to understanding and appreciating the types of physical activity that concern kinesiologists, not as hard and fast distinctions.

(*a*) Exercise and (*b*) skilled movement are the primary focuses of kinesiology.

Bananastock (left) and Comstock (right)

Kinesiology and Your Career

One thing that makes kinesiology an exciting field is the diversity of professional opportunities available to graduates of kinesiology programs. Because undergraduate programs expose students to a diverse assortment of knowledge about physical activity, they are excellent preparation for a variety of careers. Preparing fitness leaders and consultants, teachers and coaches, cardiac and neuromuscular rehabilitation specialists, sport management specialists, athletic trainers, strength training specialists, and numerous other physical activity professionals may be the most popular mission of kinesiology; but other professions such as physical therapy, occupational therapy, podiatry, and medicine also are beginning to view kinesiology as appropriate preparation for study in their fields, especially at the graduate level. And increasingly, students are pursuing undergraduate kinesiology degrees as liberal studies subjects. These students may have no immediate plans to enter a physical activity profession but simply are interested in learning about physical activity.

Many undergraduates pursue master's degrees in kinesiology after graduation (see figure 1.6). Some, like those preparing to be physical education teachers, do so to become more knowledgeable about their profession and to meet special certification requirements. Others pursue graduate work to meet the educational requirements of a profession that is related to but not normally included in kinesiology. Increasingly, master's degrees are becoming the minimal requirement for entering some professions. Physical therapy and athletic training, for example, value the knowledge that students acquire in undergraduate kinesiology programs as critical **preprofessional** preparation for entering these fields. Sometimes students continue their studies beyond the master's level to obtain the doctoral degree in kinesiology so that they can become college or university faculty members or researchers.

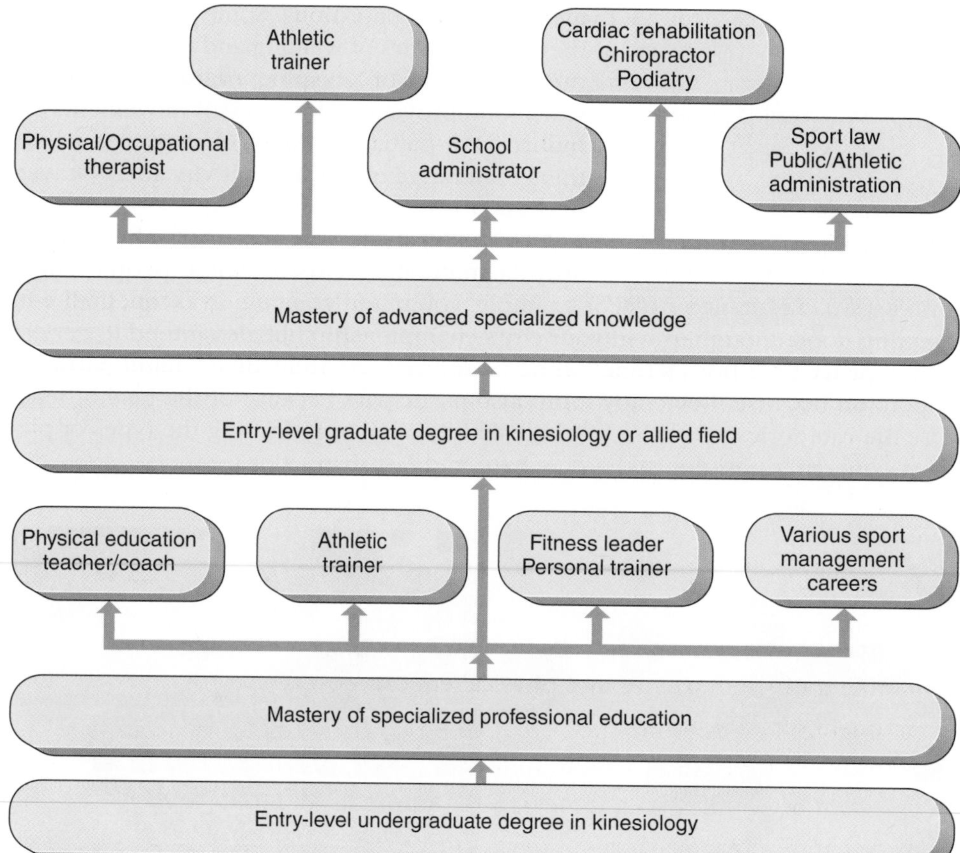

FIGURE 1.6 An undergraduate degree in kinesiology can be the starting point for many different physical activity careers.

Professionals trained in kinesiology share a common interest in and curiosity about physical activity in its broadest dimensions. At the same time, they also tend to develop specialized orientations depending on their professional roles. For example, practitioners working in the area of cardiorespiratory fitness may be most interested in vigorous, sustained forms of physical activity involving the large muscles of the body. They are especially interested in how such experiences can alter the physiological functioning of the body. Physical education teachers may entertain a more comprehensive perspective of physical activity. Because of the enormous breadth of their teaching assignments, physical education teachers tend to have a very broad view of physical activity, ranging from fitness and sport skills to dance and developmental skills. In addition, they may have a special interest in using physical activity to develop social responsibility and other desirable personal traits in children.

Professionals working in athletic training or rehabilitation exercise are particularly interested in physical activity as a medium of rehabilitation. Sport marketers have a different orientation to physical activity altogether. Their interests center on making physical activity appealing to paying audiences by promoting and staging performances that are attractive to the largest number of people. Thus, in addition to developing an understanding of and appreciation for physical activity, you should seek to develop a deep understanding of physical activity in the context of whatever specialized professional practice you choose to enter.

Why "Kinesiology"?

Scholars have debated at length the label that would best characterize an academic discipline focused broadly on physical activity. For many years, the label *physical education* was considered appropriate for the rather limited mission of kinesiology departments—preparing school physical education teachers and coaches. Today, physical education is generally understood to refer to instructional programs in sport and exercise performance offered by schools, colleges, and universities. Most colleges and universities continue to sponsor an activity program for the general student body. In most cases this program is referred to as "the physical education program" or the "activity instruction program." "Physical education" is also the name for the academic specialization within many kinesiology departments that prepares students to be physical education teachers. (Often this is known by the acronym PETE—physical education teacher education.)

Because departments originally called "physical education departments" now prepare students for a panorama of careers in physical activity, many believe that the term "physical education" fails to capture the broad essence of the field. Many names have been proposed for the discipline; but the emerging label, and the one used in this text, is kinesiology. Your department might be called exercise and sport science, kinesiology, human performance, health and human performance, human movement science, sport studies, exercise science, physical education, or any of the following departmental names:

- Department of physical education
- Department of health, physical education, recreation, and dance
- Department of physical education, health, and leisure sciences
- Department of physical education and fitness
- Department of exercise and health science
- Department of sport science and physical education
- Department of kinesiology
- Department of exercise and sport science
- Department of physical education and movement science
- Department of movement sciences and leisure studies
- Department of food, nutrition, and exercise science
- Department of human movement studies
- Department of sport studies

This diverse range of names can be confusing. Students and faculty in anthropology, sociology, psychology, and history don't have this problem. Why, then, does the problem exist

What's in a Name?

A growing number of departments at prestigious universities have adopted the term *kinesiology* because it best characterizes a discipline that is learned in many different ways and deals with many different forms of physical activity in diverse professional settings. Support for the label "kinesiology" has come from the American Academy of Physical Education, an honorary society of approximately 140 scholars. The 70-year-old academy undertook a two-year study of the matter in 1990 and decided to change the name of the organization to the American Academy of Kinesiology and Physical Education. Likewise, the National Association for Health and Physical Education in Higher Education recently changed its name to National Association for Kinesiology, Health and Physical Education in Higher Education to reflect the growing acceptance of "kinesiology" as a name for the discipline. Also of note is the recent establishment of the American Kinesiology Association. The organization is open only to academic departments in colleges and universities, not to individual members. Its stated purpose is to solidify the place of kinesiology in national education affairs, "to represent kinesiology when governmental agencies develop national policies that concern kinesiology, to draw on the knowledge of kinesiology to address public and professional policies," (and to) represent and promote kinesiology through participation in international associations and events.

in kinesiology? The answer is partly that kinesiology is a young, evolving discipline and that disciplines often need a long time to define themselves. Anthropology, biology, and history, for example, are very old disciplines. The problem also results from the fact that departments teaching kinesiology are often responsible for teaching other disciplines as well. Thus, on some campuses—usually at smaller colleges—the department may be named human kinetics and leisure studies or health, exercise science, and recreation to reflect the fact that the department offers degrees in leisure studies, recreation, or health as well as kinesiology. In one large southern university, kinesiology degrees are offered in the "Department of Nutrition, Food, and Movement Sciences," a name that reflects the fact that the department offers three different degree programs. Thus, you should be careful not to confuse the title of departments in particular universities with the title of the field as it is coming to be known across the country.

Although none of the various names used for our field is "wrong," most scholars emphasize the need for a single term broad enough to describe the discipline. Although not all scholars believe that *kinesiology* is the best name for the discipline (Locke 1990; Siedentop 1990), we believe that it is. So, regardless of the name of the department in which you are enrolled, we encourage you to refer to the discipline as kinesiology.

Allied Fields

Two of the largest professional organizations concerned with physical activity are the American Alliance for Health, Physical Education, Recreation and Dance (AAHPERD) and the Canadian Association for Health, Physical Education, Recreation and Dance. The membership of these organizations includes not only those with a central interest in exercise and sport but also those working in the fields of health, recreation, and dance.

Health problems arising from inactivity are of interest to physical activity professionals as well as health professionals; but health professionals are also interested in health problems that have little to do with physical activity, including issues related to sexually transmitted diseases, HIV (human immunodeficiency virus), smoking cessation, and drug abuse.

Recreation specialists are interested in physical activity as a leisure pursuit, but leisure pursuits stretch far beyond those in which physical activity is a primary concern. Travel, crafts, and nature study are examples.

Dance professionals also attend these meetings, and, although dancers share kinesiologists' interest in physical activity, they are interested in a much narrower slice of the physical activity pie. Expressive and artistic forms of movement form the centerpiece of the field of dance.

Combining this assortment of fields into a single professional organization may seem strange, but they have a long, historic relationship. For most of the 20th century these areas were represented in a single college or university department across the land; in many cases they are still housed in the same department. Today, each field has become more specialized and isolated from the others; each has its own professional organizations and journals. In many cases, each is taught in a separate department. But because of their shared histories and because they are often interested in the same problems, faculty from these areas frequently work together as colleagues.

Holistic Nature of Kinesiology

When you complete this introductory study of kinesiology, we hope that you will be convinced of the holistic nature of physical activity. **Holism** is a term underscoring the interdependence of mind, emotions, body, and spirit. Although some people think kinesiology deals exclusively with body movement, in reality it has a much broader reach. Physical activity involves our minds, emotions, and souls as much as it does our bodies. We find it convenient to speak of *physical* activity because the physical aspects are so easily observed, but physical activity also is *cognitive* activity, *emotional* activity, and even *soul* activity. Thus, studying kinesiology will take you far beyond the study of the biological aspects of physical activity. Kinesiology includes an analysis of the psychological antecedents and outcomes of physical activity; the sociological, philosophical, and historical foundations of sport and physical activity; the dynamics of skill development, performance, and learning; and the human processes involved in the teaching and learning of physical activity.

➤ Although the bodily aspects of kinesiology receive most of the attention, in a kinesiology program it is important to remember that human beings are multidimensional creatures with interrelated cognitions, emotions, body, and soul.

Think about the last time you went for a run or a brisk walk. It may have seemed so easy as you let your mind wander, enjoying the scenery of a lake or the beauty of a mountain trail, that you underestimated precisely how complex its underlying physiology and psychology were. Physical activity is so much a part of our everyday lives that we easily forget how wonderfully complex the human body is. Most students who enroll in kinesiology programs are curious about the functioning of the human body. In your kinesiology program you'll study anatomy and physiology along with taking advanced courses in exercise physiology, biomechanics, sport and exercise psychology, and motor behavior. These classes will help you understand and appreciate the many mechanisms and systems that enable our bodies to perform physical activity.

In addition, you may also be required to take a course in philosophy of physical activity. If so, you are fortunate, since this course will help you understand how truly holistic physical activity is. Thinking philosophically you will come to understand more fully the personal meaning you discover in physical activity. This thinking also will sensitize you to the myriad ethical issues associated with sport, exercise, and other types of physical activity. And, through your study of psychology and sociology of physical activity you will learn to appreciate how attitudes and social settings influence our interpretation of the significance of our physical activity experiences. Although it is easier to think and talk about our bodies or body parts such as the heart, muscles, or bones as though they were machines or instruments that our minds or souls "use" to achieve our purposes, our bodies are an indivisible part of our humanity. It may not be an exaggeration to say that no other discipline is so diverse in its aims, so interdisciplinary in its subject matter, or so complex in its organization as kinesiology.

The three-dimensional analysis of physical activity offered in this book—experience, study, professional practice—is designed to help you organize your thinking about the discipline of kinesiology and the broader field of physical activity. This approach will not only help you develop a framework for understanding physical activity and the physical activity professions but also help you understand the basis for your course work in kinesiology and assist you in planning and implementing career goals. Let's look briefly at each of the three dimensions.

Although body movement is the core focus of kinesiology, the discipline also recognizes the holistic nature of humans.

Brand X Pictures

Experiencing Physical Activity

Figure 1.7 shows the contribution of physical activity experience to your knowledge of kinesiology. By this point in your life, you have had millions of physical activity experiences. You've learned to master thousands of complex motor skills including tying your shoelaces, driving your car, playing Frisbee, maybe even performing cartwheels. Many of these forms of physical activity are essential in everyday life. Other experiences such as sport and leisure-time pursuits are discretionary—you choose to do them because they are fun. All of these experiences offer the potential to teach you something about physical activity, yourself, and the world around you.

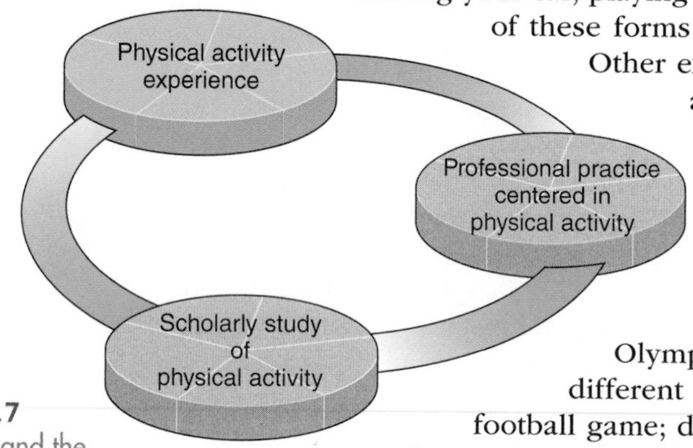

FIGURE 1.7 Experience and the other two sources of kinesiology knowledge.

Experiencing physical activity—performing or watching it—generates highly personal sensations, emotions, thoughts, and feelings. Watching an Olympic gymnastics championship is likely to produce different feelings than watching a dance performance or a football game; digging a ditch is likely to induce different feelings than swinging a golf club. Precisely how we experience physical activity depends on the nature of the activity itself, the context in which it is performed, and our own peculiar sets of attitudes and experiences. Watching an instructor demonstrate a golf swing when we are learning it is likely to be a completely different experience than watching Tiger Woods hit the ball.

Our experiences have a significant influence on the meanings we attach to particular physical activities. They also influence our future decisions about engaging in a particular activity again. Obviously, we have little choice about some types of physical activity; we must engage in physical activities associated with our work, in those required to eat, and in those required to maintain our personal hygiene, whether we enjoy them or not. Beyond

Inventory of Personal Experiences With Physical Activity

To think more about your personal experiences with physical activity, answer the following questions.

1. What physical activity can you perform that you are most proud of?
2. What sport or form of exercise do you enjoy the most? Why?
3. What is the hardest physical work you have ever done? What made it difficult?
4. What sport do you most enjoy watching? Which aspects do you find most enjoyable?
5. What is the most dangerous physical activity you have ever performed?
6. What is the longest distance you have run?
7. What physical activity would you most like to learn to perform well?
8. What physical activity have you performed that you never want to perform again?
9. Briefly describe how you felt when you won a close game. How did you feel when you lost?
10. What have you learned about yourself because of performing your favorite physical activity?

this, however, we have a great deal of discretion about which activities to perform, with whom, under what conditions, and for how long. Ultimately, our decisions about playing basketball or playing hockey, running 5 miles or swimming 50 laps, taking the elevator or climbing the stairs, mowing the yard or hiring someone else to do it, and seeking out physical activity or living the life of a "couch potato" will depend, to a large degree, on our personal experiences with those activities.

Thus, it is important to understand at the beginning that even though classrooms, laboratories, and libraries offer kinesiological knowledge, direct participation in physical activity is an important source of knowledge as well. By incorporating participation in physical activity as part of the formal course work, faculty in kinesiology departments add experiential knowledge to the overall body of knowledge that makes up the discipline of kinesiology.

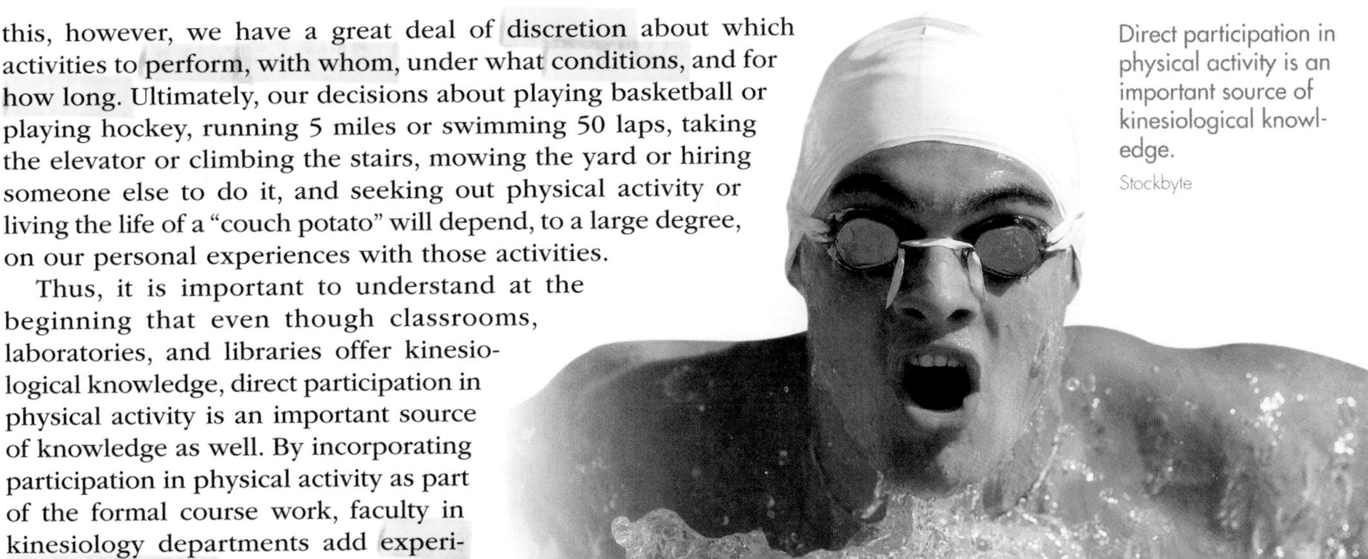

Direct participation in physical activity is an important source of kinesiological knowledge.
Stockbyte

When we participate in physical activities as part of the course work in kinesiology, we not only learn about the activity; we also learn about ourselves (and often others around us) as moving human beings. Climbing a rock face, skating on a frozen pond, or parachuting from an airplane can be the unique means by which we gain not only a greater understanding of the activity itself, but also knowledge of our own capacities and limitations and knowledge about those around us. Also, certain types of physical activities such as surfing, hiking, or long-distance running offer unique opportunities for reflective thought not normally available in other aspects of our lives.

Physical activity is such an ordinary aspect of our lives that often we fail to recognize how it intersects with our everyday experiences. We depend on physical activity when we work, play, cook our meals, drive our cars, type reports, or sign our names. As a student of kinesiology, you must appreciate the place of physical activity in each sphere or category of your life experiences. In this book, we refer to these various aspects of our everyday lives in which physical activity plays a distinct role as the **spheres of physical activity experience** (see figure 1.8). You will discover how physical activity penetrates all spheres of our personal lives.

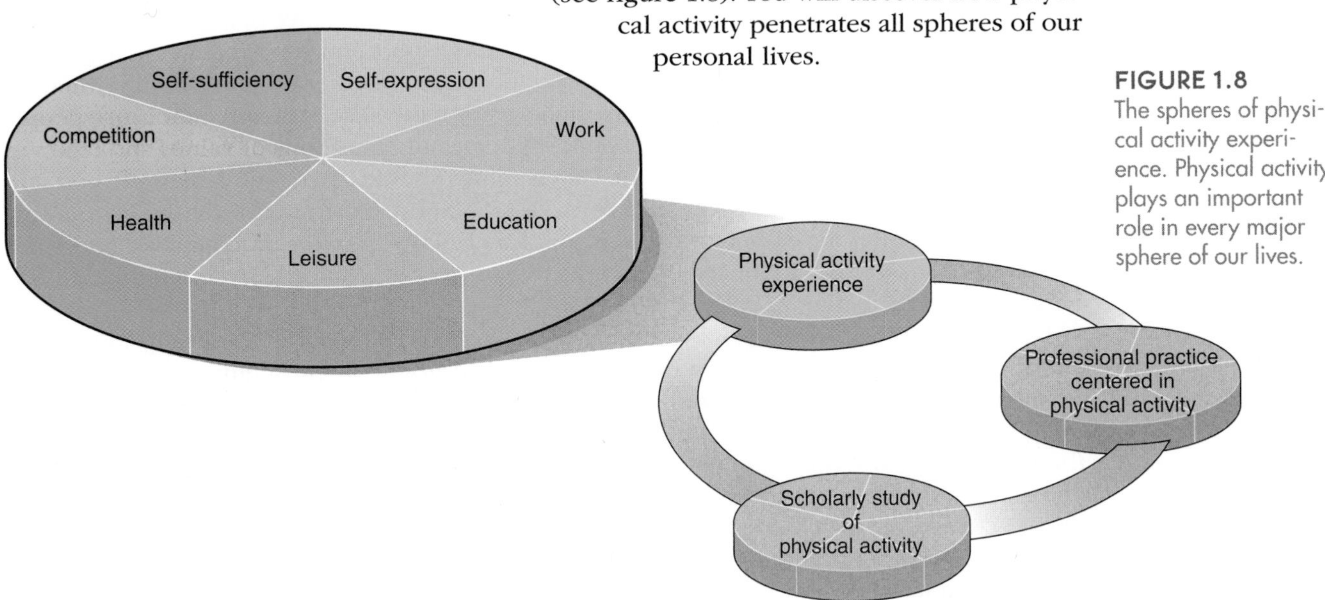

FIGURE 1.8 The spheres of physical activity experience. Physical activity plays an important role in every major sphere of our lives.

Scholarly Study of Physical Activity

Interest in studying and learning about physical activity appears to be at an all-time peak. At least 600 colleges and universities in the United States, and many more in other countries, have academic programs devoted to the study of physical activity (figure 1.9). In addition, unknown numbers of institutes and centers outside academe—from medical complexes, to military and space research programs, to industrial engineering centers—systematically investigate various dimensions of physical activity. And judging from the numbers of books on sport, exercise, and fitness flooding popular bookstores, laypeople have a keen interest in studying physical activity.

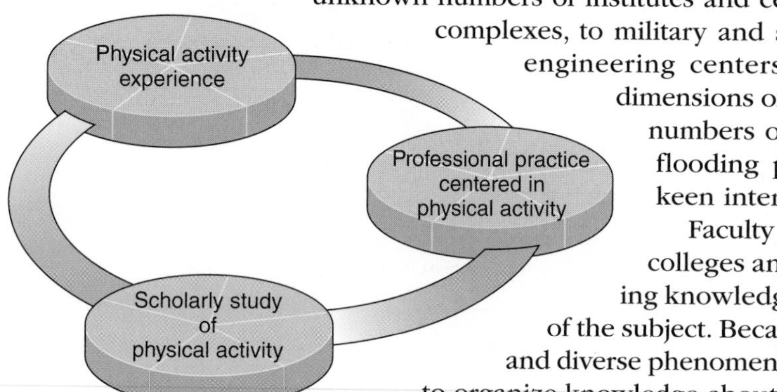

FIGURE 1.9 Scholarly study and the other two sources of kinesiology knowledge.

Faculty who design formal programs of study in colleges and universities are specialists in organizing knowledge to promote the fullest understanding of the subject. Because physical activity is such a pervasive and diverse phenomenon, kinesiologists have found it difficult to organize knowledge about it around a single framework that will help us study it systematically. In fact, debates continue about how to organize the discipline of kinesiology (Newell 1990b).

For the present, scholars have been content to divide the scholarly study of kinesiology into a number of subdisciplines. Each of these subdisciplines is related to some larger, older, and more established "parent" discipline such as psychology, physiology, sociology, biology, history, or philosophy. For instance, exercise physiology draws on basic concepts and theories from physiology; motor behavior draws on psychology; and philosophy of physical activity draws on philosophy. This means that kinesiology students must develop a working knowledge of the language and theories of a number of major disciplines and learn to apply them to physical activity. You will find this to be a formidable challenge!

You will also discover that each separate area of study addresses a different set of questions about sport and exercise. Exercise physiologists typically ask and answer questions related to short-term changes that occur because of exercise or long-term changes that occur because of training. Questions having to do with forces acting on and generated by the body, or ways of altering performance technique to improve physical functioning, are the province of biomechanics. Specialists in motor behavior study how human development affects motor skill performance, the neural mechanisms underlying the performance of motor tasks, and the effect of kinds of practice on learning of motor skills. Sport and exercise psychologists are most concerned about the mental aspects of performance, whether psychological techniques that help elite athletes attain optimal performances or motivational factors that cause people to drop out of exercise programs. Sport philosophers study questions of values and ethics as they relate to sport and exercise, such as "What values should guide our design of sport and exercise programs?" or "Can sport be a form of art?" Sport historians study the factors that have influenced the development of sport and kinesiology in different countries, and sport sociologists analyze the influence of such factors as gender, race, class, media, politics, and religion on participation in sport and exercise.

Over the past four decades, the subdisciplines have developed into specialized areas of study, each with its own place in undergraduate and graduate curriculums in kinesiology, its own academic and professional societies, and its own scholarly journals. For example, members of the Association for the Advancement of Sport Psychology meet annually to hear the latest research reports about the psychological aspects of sport and exercise. They keep abreast of developments in this subdiscipline by reading *The Sport Psychologist,* a journal devoted to applied psychological research in sport psychology, or *Journal of Sport and Exercise Psychology,* a journal about more theoretical aspects of the subdiscipline. Members of the North American Society for Sport History, the North American Society for the Sociology of Sport, the North American Society for the Psychology of Sport and Physical Activity, the

Activity 1.6

Specialized Areas of Scholarly Study

Distinguish among the areas of scholarly study in kinesiology in this activity in your online study guide.

Canadian Society for Psychomotor Learning and Sport Psychology, the American College of Sports Medicine, and the Canadian Society for Exercise Physiology, along with several other specialized societies, also meet annually to discuss the latest findings in their fields. Many kinesiology departments now offer master's degrees and doctoral degrees in sport and exercise psychology. We can see the same pattern of specialization for exercise physiology, biomechanics, pedagogy of physical activity, motor learning and control, sociology of physical activity, history of physical activity, and philosophy of physical activity.

Most kinesiology professors have received in-depth education in one or two subdisciplines rather than in all of them. Your professors, for example, may identify themselves as biomechanists, exercise and sport psychologists, exercise physiologists, or by some other title that denotes their special expertise. This process of specialization has advanced our knowledge of kinesiology by permitting researcher-scholars to narrow their scholarly focus on specific aspects of physical activity. The disadvantage is that communication among kinesiologists in the various subdisciplines is not always easy. Sometimes these divisions hinder efforts to integrate the different kinds of knowledge and apply them to professional practice.

Figure 1.10 shows the **spheres of scholarly study of physical activity.** The subdisciplines presented in this book are those that are dominant right now. Because kinesiology is a dynamic, rapidly expanding discipline, the subdisciplines that compose it will likely expand also. In fact, some have already moved toward dividing into smaller, more specialized subfields. Biomechanics, for example, could be further separated into anatomical biomechanics and sport biomechanics. Sport psychology and exercise psychology are developing into separate subdisciplines. Exercise physiology, the oldest of the subdisciplines, has divided into a number of specializations, including sport physiology, work physiology, exercise epidemiology, and exercise and nutrition. This process of expansion of the boundaries of scholarly knowledge and the increased specialization that follows likely will continue for many years. This development, of course, underscores the need for kinesiologists and physical activity professionals to continue to work at staying abreast of knowledge contained in the spheres of scholarly study.

University and college departments differ in how they cover each subdiscipline. Sometimes each is represented in the curriculum by a separate course (e.g., sport history, sport psychology, or motor learning). In some cases two or more subdisciplines may be combined in one course. A course in the sociocultural aspects of physical activity, for example, may include the study of history of physical activity, sociology of physical activity, and philosophy of physical activity. As kinesiology continues to flex its muscles and the knowledge base continues to

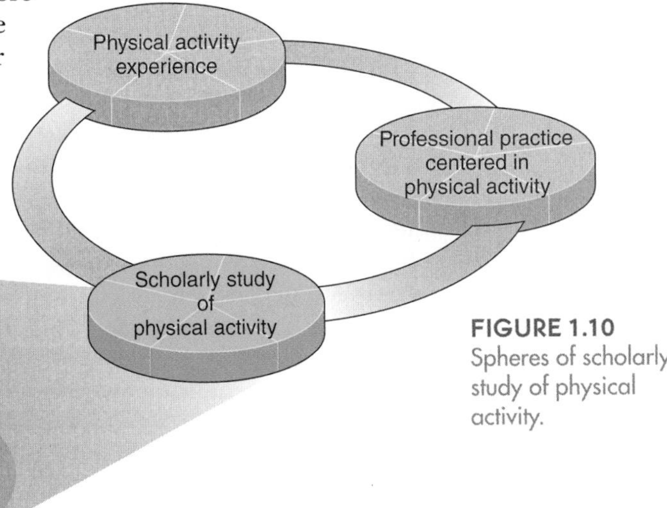

FIGURE 1.10 Spheres of scholarly study of physical activity.

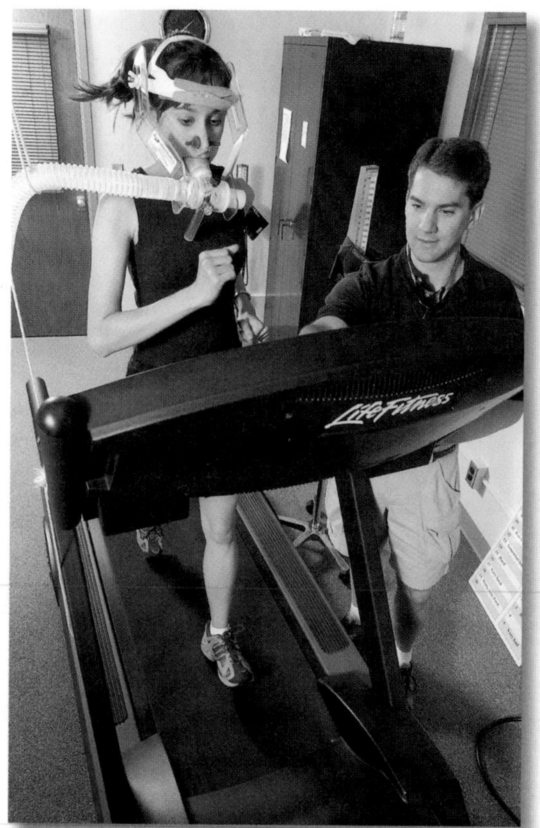

expand, the number of courses devoted to the scholarly study of physical activity will probably increase.

As you dig into the material in part II of this book, remember that the subdisciplines and spheres of scholarly knowledge are merely frameworks to help us study and organize our thinking about physical activity. Mastery of the knowledge in each subdiscipline will require you to think in slightly different ways, to master different theories, and to use different terminology. But as you move from a course in exercise physiology into a course on psychology of sport and exercise and then to a course in history and philosophy of sport, the temptation is to compartmentalize your thinking about the field. Unfortunately our most popular curriculum models encourage such thinking. But you can resist this kind of fragmented thinking by keeping in mind that kinesiology is a unified body of knowledge and that when you draw on this knowledge in professional practice you draw on the entire body of knowledge simultaneously, not in the tidy boxes in which you learned it in your undergraduate program. For the present this method of organizing the scholarly study of kinesiology seems to be the best way to approach it. As the discipline evolves, however, students and professors who are best able to make connections between facts, concepts, and principles drawn from different subdisciplines and different spheres of scholarly study will be in the best position to make significant contributions to the overall discipline.

Scholarly study is an important source of knowledge in kinesiology.

➤ Research and systematic analyses form the backbone of the scholarly study of physical activity. Our scholarly knowledge about physical activity has been organized into subdisciplines, each providing a unique perspective from which to view the dynamics and processes of physical activity.

Practicing a Physical Activity Profession

Nate Marshall lives to dance and perform gymnastics; eventually he wants to own a dance and gymnastics studio for youngsters. Jonathan Smith wasn't a football star, but he so much enjoyed the game that his professional goal is to become a coach at the college level. Yolinda Arbuckle was on the varsity basketball team in high school, but spent most of her junior year undergoing rehabilitation for a knee injury. Now she's interested in a career as an athletic trainer or physical therapist. Sandra Flynn was overweight and unfit until a friend got her interested in weight training and bodybuilding. Now she's a fit and confident young woman who would like to become a personal trainer to help others discover what she has learned. Jeff Armstong completed an undergraduate degree in business and has returned to school to complete a baccalaureate degree in kinesiology. He hopes to study sport management and become a college athletic director or manager of a major sport arena or stadium complex.

Like most college students, you probably didn't decide to enroll in college simply to learn about physical activity. You (and surely your parents!) hope that your studies and your degree will lead eventually to a job, preferably a job in the physical activity field. But you probably have many questions. What types of jobs are available to kinesiology graduates? What do these jobs entail? What are their requirements? Is special certification or a graduate degree required? What types of certifications are appropriate? What social environments surround the workplace? Can you be happy in this profession for the rest of your life? These and many more questions face students exploring careers in physical activity.

Part III of this book introduces you to the third dimension of kinesiology—professional practice (see figure 1.11). Professional practice may be envisioned as a process of putting knowledge to work.

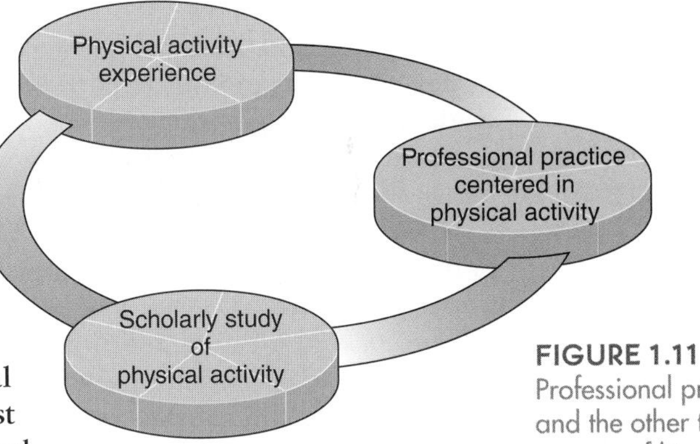

FIGURE 1.11 Professional practice and the other two sources of kinesiology knowledge.

When we enter professional practice, we learn new things about providing physical activity services to the public. For example, if you decide to become a physical education teacher, you will learn a great deal about how to structure lessons, how to provide practice experiences, how to organize a curriculum, and how to relate to students with special problems, all under the tutelage of a university specialist who has advanced training in the preparation of physical education teachers. As you pursue your career you will probably learn a lot that you were never taught in college classes about how to work productively and collaboratively with other teachers in your school. This knowledge, gained through hands-on experience under the supervision of a university specialist as well as on-the-job experience, is different from knowledge acquired through research, and it is different from knowledge acquired through your own direct experiences performing or watching physical activity. When college and university faculty incorporate this knowledge into classes, it is considered part of the discipline of kinesiology.

Becoming a Professional

Now let's turn to thinking about your career plans. One of the first decisions you will have to make is whether you really want to serve in a professional capacity. But what does it mean to be a professional? In chapter 12 you will learn that becoming a professional entails

Doing Some Preliminary Thinking About Your Career

It is not too early to begin preparing for your professional career. As a way of taking stock of how carefully you have explored a career in the physical activity professions, place a check in the appropriate box next to each question.

☐ 1. I have investigated the types of careers available to graduates of a kinesiology program.

☐ 2. I have investigated whether a master's degree in kinesiology or a related field is required to practice my chosen profession.

☐ 3. I know what courses will help me develop experiential knowledge of kinesiology.

☐ 4. I know what courses will help me develop theoretical knowledge of kinesiology.

☐ 5. I know what courses will help me develop professional practice knowledge.

☐ 6. I have a timetable for completing my undergraduate degree and a plan for preparing myself for a career in the field of physical activity.

☐ 7. I have consulted at least two physical activity professionals concerning how I can best achieve my career aspirations.

☐ 8. I have visited a professional setting similar to the one in which I hope to work when I graduate.

☐ 9. Given the choice between spending $25 on a hot CD or on dues to a professional organization, I would spend the money on dues to the association.

☐ 10. I have worked as a volunteer in the professional occupation that most interests me.

special responsibilities such as devoting yourself to serving others, giving priority to your clients' needs and interests, performing your duties in an ethical manner, and keeping abreast of developments in your field to ensure that your decisions are based on the best available evidence. Whereas those who specialize in scholarly study focus on discovery of knowledge and how to communicate it effectively to others, those who specialize in professional practice focus on enhancing the lives of those they serve.

Not all types of work qualify as professions. Usually professions require a formal college or university education in which one learns how to master a complex body of knowledge and to perform critical professional skills. In chapter 12 you will see how various types of work can be located on a continuum from strictly nonprofessional (e.g., day laborer) to strictly professional (e.g., surgeon). Most types of work are located between these two extremes. More important, you will learn about the expectations society holds for professionals, and you will have an opportunity to decide whether you really want to pursue a professional career.

Careers in the Physical Activity Field

> DO it Activity 1.7
>
> **Possible Careers**
>
> Identify careers that interest you and find out more about them in your online study guide.

Once you have made a commitment to becoming a physical activity professional, *you* must take primary responsibility for developing your career. A career is a lifelong pursuit that may move through various stages as you progress from one type of employment to another. For example, a career in the teaching profession may begin at the public school level and wind up at the community college or university level. On the other hand, you may spend an entire teaching career at one institution. A career in the fitness industry may take you from a position as a personal trainer at a local commercial center to a position in a corporate fitness program, and ultimately to the position of director of a hospital-linked wellness center. Career counselors now predict that young people graduating from college, on average, will change careers up to four times throughout their life spans. Of course, where you finish your career is not a pressing issue at this time, but getting a good start on your first career is.

One of the exciting aspects of kinesiology is the diverse range of careers that you may develop from a solid undergraduate education. Unlike your friends who may be enrolled in education, nursing, or accounting programs that offer training tailored to specific careers, you will find that a degree in kinesiology offers many options. The list in activity 1.7 includes a number of different professions that often have their roots in a kinesiology degree. Obviously, each of these may have specialized requirements *within* a kinesiology curriculum, but all can be viewed as professional applications of kinesiology. Each of these various occupations can be assigned to one of the **spheres of professional practice in physical activity** displayed in figure 1.12.

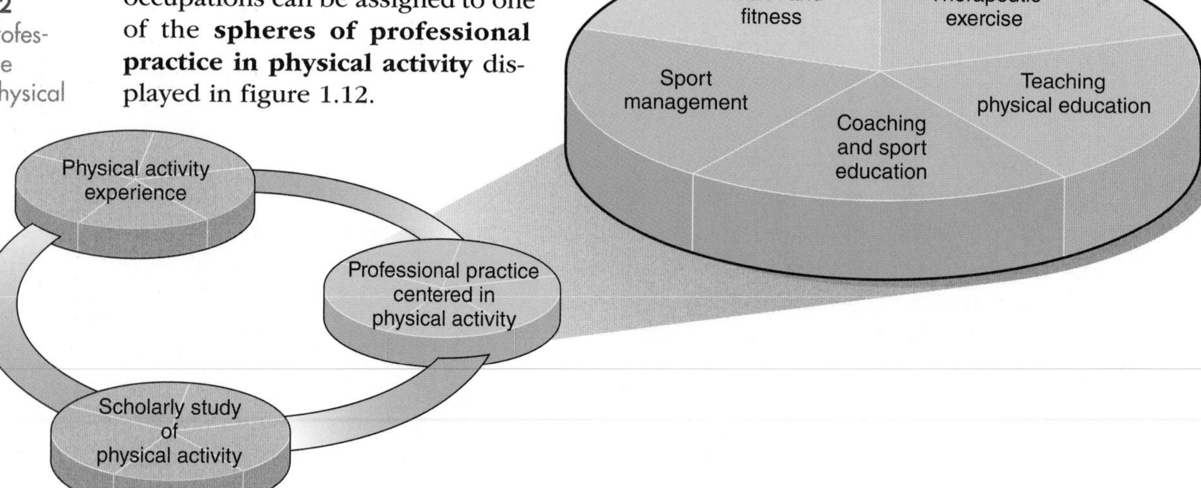

FIGURE 1.12 Spheres of professional practice centered in physical activity.

Part III of this book not only provides you with information concerning these professions but also helps you get a start toward making yourself marketable as a professional. From this point forward in your college experience, you need to realize that you are involved in a healthy competition, first with yourself as you work to achieve good grades on exams, papers, and projects. When you graduate, you will continue to compete (now against others), vying with other kinesiology graduates when you interview for professional positions. In many ways, this competition is the most important you will enter. How should you prepare? If you were preparing to compete for a world championship in the marathon, you probably would commit yourself totally to planning and implementing a strategic training program that offers you the best chances for achieving victory. If preparation is important for an athletic competition, how much more important is it for your lifelong career?

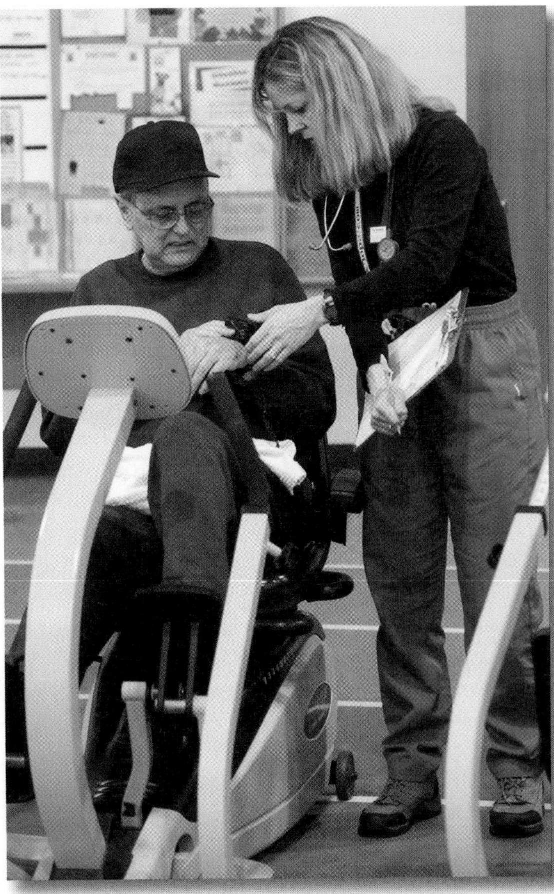

Graduates with kinesiology degrees are finding a wider range of career options open to them than ever before.

Throughout this book you will be reminded repeatedly of this important point: *You and only you are responsible for preparing yourself for your chosen career.* Your academic department is responsible for offering you a well-sequenced, relevant curriculum with learning experiences designed to achieve your career goals. Your professor is responsible for offering competent advisement and excellent teaching. But *you* are responsible for taking advantage of these resources to develop yourself into the best professional possible.

To become a marketable, sought-after professional, you must take advantage of experiences both on and off campus. Obviously, committing yourself to mastering the material of your course work is the first step, but it is only the first step. Hours spent outside the classroom in professionally relevant voluntary service work or summer employment are also critical. As a preprofessional you should show an interest in your profession that goes beyond excelling in course work. Becoming a student member of professional physical activity organizations is one way to demonstrate early commitment. The chapters that follow will encourage you to think on your own and act on behalf of your educational and professional development. It is never too early to take these vital steps.

➤ For most students, the goal of pursuing a degree in kinesiology is to obtain a job in a physical activity profession. Those who identify with a profession early, perform well in their course work, and seek out professionally related experiences outside the classroom will be the top competitors for jobs in physical activity professions.

Wrap-Up

So you've begun the journey to discover the field of kinesiology. The destination is clear: an understanding of what kinesiology is, how knowledge and competency in it are developed, and how it relates to your chosen career. The course of study will be long and at times difficult and will demand much of you, but the rewards of mastering the field that you are studying are well worth the time and effort. You've already learned a great deal. You know, for example, that kinesiology is the study of physical activity, and you know that physical activity has a technical definition that is quite different from our use of the term in everyday life. You also have learned that kinesiology is concerned with all forms of physical activity, but that the research of kinesiologists and the careers of those who graduate from programs in kinesiology tend to revolve around two general forms of physical activity: exercise and skilled movement.

You've also learned something about how the field of kinesiology is organized and that this textbook follows the same pattern of organization. One source of knowledge contributing to kinesiology is experience that comes from participating in and watching physical activity; this realm of experience can be broken down into seven spheres of physical activity experience. Another source of knowledge is scholarly study about physical activity, and this realm also can be broken down into seven spheres or subdisciplines. Finally, knowledge comes from professional practice centered on physical activity, and this realm, too, can be broken down into seven spheres of professional practice. Having read this introduction, you are now ready to explore physical activity as a form of human experience, an object of scholarly study, and a focal point for a professional career. Take some time now to review the key points of this chapter to prepare yourself for what lies ahead.

In the parts of the book that follow we will explore each dimension of the physical activity field in detail. Part I will help you appreciate the importance of experiencing physical activity as a source of enjoyment; as a means to health, education, and rehabilitation; and as a necessary ingredient of our lives at play and work. Part II contains an overview of the major facts, concepts, and theories in the subdisciplines that together make up knowledge gained from the scholarly study of physical activity. Each chapter offers a brief introduction to a separate subdiscipline. These chapters will give you a solid orientation for the in-depth study of those subdisciplines that you will undertake later in your program. Finally, part III focuses on professional practice centered in physical activity. The goals of that section are to introduce you to the most prominent physical activity professions and to point out ways in which undergraduate course work in kinesiology connects to them.

KEY▸ Activity 1.8

Review this chapter's key points in your online study guide.

QA Activity 1.9

These end-of-chapter questions and activities are also in your online study guide. Your instructor may ask you to complete them online and turn them in.

1. What is the difference between movement and physical activity? Give an example of an instance in which human movement does not meet the technical definition of physical activity.
2. What is meant when kinesiology is described as a holistic discipline?
3. What two general categories of physical activity receive the most attention in kinesiology? What are the subclassifications of these categories?
4. What are the three sources of knowledge of kinesiology?
5. List the spheres of physical activity experience.
6. List the spheres of scholarly study and the subdisciplines contained in each.
7. List the spheres of professional practice centered in physical activity, and give an example of a career in the physical activity professions for each sphere.

PART I

Experiencing Physical Activity

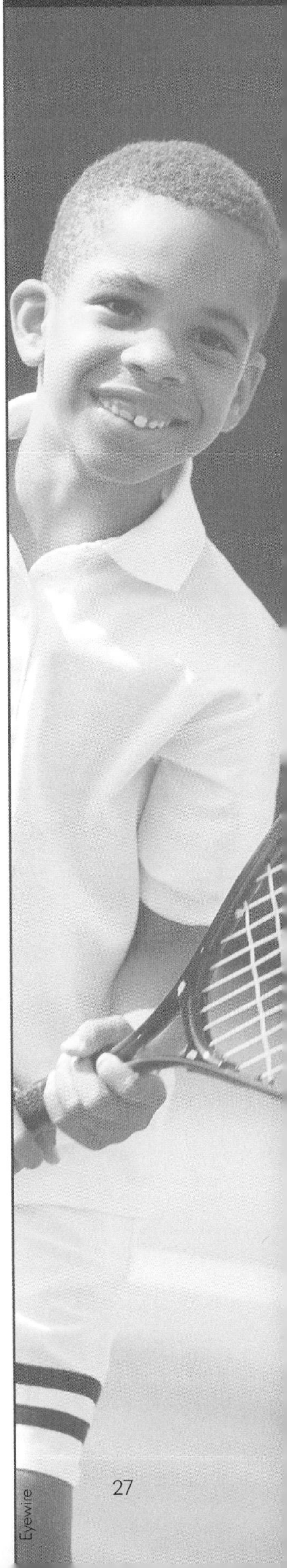

Adam Gopnik, an American sportswriter assigned to the 1998 World Cup matches, couldn't at first understand how Europeans could be so avid about a sport that often ends in a scoreless tie. But as he watched the games, experiencing them at first hand, he came to understand that soccer was more than simple entertainment; it mirrored the European's entire view of life. He wrote, "Soccer was not meant to be enjoyed. It was meant to be *experienced* [emphasis mine]. The World Cup is a festival of fate—man accepting his hard circumstances, the near certainty of failure. There is, after all, something familiar about a contest in which nobody wins and nobody posts a goal. Nil–nil is the score of life." As he experienced the game up close, Gopnik's eyes were opened to vistas he had never before seen. If he had been learning how to play rather than watch soccer, it is likely that he also would have experienced revelations, although different. "Doing" physical activity and observing it intently are unique experiences; both are completely different from studying *about* physical activity. Physical activity experience has many different dimensions. Consider how each of the following questions focuses on a distinct aspect of the term *experience.*

- Why are physical activity experiences important, and with what aspects of our lives do they intersect?
- What types of physical activity experiences have been most important in helping you master skill, training, and physical fitness?
- How have your thoughts, emotions, understandings of yourself and others, and the meaning you attach to physical activity been influenced by your physical activity experiences?

The chapters in this part of the book will help you answer these three related, but separate, questions. The first question deals with the part physical activity plays in the living out of your life. Obviously, if physical activity didn't contribute to our lives, there would be little justification for a field of study centered on it. To help you answer this question, chapter 2 describes the different ways in which physical activity penetrates our daily lives. The chapter will introduce you to the seven spheres of physical activity experience—distinct aspects of our lives in which physical activity plays an important role.

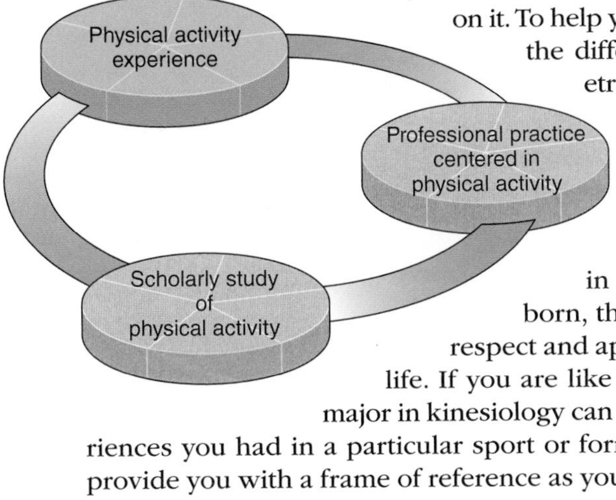

Even though you have been immersed in physical activity from the day you were born, this chapter should help you develop a new respect and appreciation for the role it plays in your daily life. If you are like most kinesiology majors, your decision to major in kinesiology can be traced to some particular, in-depth experiences you had in a particular sport or form of exercise. These experiences will help provide you with a frame of reference as you study this field; but you should remember

Eyewire

that as a student of kinesiology—a field that embraces all forms of physical activity—you should be interested in physical activity experience in all its various forms. Your program of study in kinesiology is designed to help you develop a refined curiosity about what physical activity is.

The second question refers to the particular effects experience has had on your physical capabilities. Much of your course of study in kinesiology will be aimed at showing you how physical experience can be manipulated to bring about specific changes in performance capacity or health, or increases in skill or development. Chapter 3 introduces you, in a general way, to the dynamics of this process. Skill in manipulating physical activity experiences intelligently and systematically is the exclusive domain of those who work in the field of kinesiology, just as skill in diagnosing various forms of disease is the exclusive domain of physicians. Kinesiologists are "physical activity experience experts." They understand the connections between types and amounts of physical activity experiences and specific outcomes, whether the goal is the development of skillful movement (skill); improved health (physical fitness); improved strength, endurance, and flexibility (conditioning); recovery from injury or disease (rehabilitation); or merely enjoyment. Knowing the most efficient, most effective, and safest way to achieve these goals is the kinesiologist's stock-in-trade.

The third question is directed toward your interior experiences in physical activity. Knowing how physical activity experiences affect our skill, endurance, strength, flexibility, health, and fitness is one thing; knowing how they affect our inner lives and why we enjoy them is another. Those who don't enjoy physical activity are not likely to do it voluntarily. Chapter 4 introduces you to this subjective dimension of experience. After reading and studying this chapter, you will understand that physical activity is not only a physical experience but an emotional, cognitive, and spiritual experience as well. How physical activity affects our thoughts and feelings, how these feelings may affect our future physical activity choices, and how we derive knowledge and meaning through physical activity are all part of what we call **subjective physical activity experience.** All of our involvement in physical activity ultimately has a bearing on this subjective domain.

By nature, our subjective experiences are difficult to describe, and sometimes they affect us in highly personal ways. Because of this we sometimes ignore this profound interior world of human experience, choosing instead to focus on observable dimensions of experience such as our time in a 5K race, how much weight we lost during an exercise program, how many points we scored, or how quickly we recovered from an injury. After reading chapter 4, you might wonder why we even refer to it as *physical* activity. Clearly, it is more than physical. (Perhaps *human activity* would be a better term.)

Physical activity experience is so prevalent in our lives that we tend to take it for granted. Yet it is the foundation of the study of kinesiology. In the following chapters we invite you to view physical activity experience from an entirely new perspective, beginning with the seven spheres of physical activity.

The Spheres of Physical Activity Experience

2

Shirl J. Hoffman

Photo by Steve Clarke. Dancers are Christal Brown and Gaetan Pettigrew.

CHAPTER OBJECTIVES

In this chapter we will

- broaden your understanding of the universe of physical activities,
- familiarize you with the ways physical activity is experienced in different social compartments called spheres, and
- introduce you to some of the potential benefits and limitations of physical activity.

Take a moment to examine the photograph of the two dancers on the previous page. What aspects of the performance do you find the most striking? Is it the height of the dancers' leaps or is it the coordination that you see between the two dancers, both shaping their bodies exactly the same way at the same time? Some will be attracted to the twisting motion, some to the unusual positions of arms and hands. You probably were struck by the expression on the female dancer's face. How do her movements relate to this expression? Do you see the photo of the movement as an example of grace and beauty or of sheer athletic skill?

Sport and exercise traditionally have been the primary focuses of kinesiology, but dance has a long history in kinesiology departments as well. On some campuses dance is taught and studied in the department of kinesiology or physical education, whereas on others it is housed in a separate department. Regardless of its academic home, dance—its beauty and grace as well as its technical and physiological aspects—remains an interest of kinesiologists. And so do hundreds of other forms of physical activity. The dancers in the photo help to underscore the main point of this chapter: Kinesiologists are interested in a broad range of physical activity, including physical activity that takes place in the arts. You may have arrived at this point in your college career because of a special interest in sport and exercise, and therefore you may have assumed that these are the only forms of physical activity that interest faculty and students in kinesiology. If so, prepare to broaden your horizons. For only as you come to understand and appreciate the beauty, elegance, drama, force, and precision of all forms of physical activity can you truly appreciate physical activity that occurs in sport and exercise settings.

How important is physical activity to your daily life? If you're like most people, you think seriously about the question only when your capacity for moving is limited due to disease or injury. For example, nothing makes us appreciate the importance of the ankle joint in walking quite like having a sprained ankle, or the importance of the thumb in grasping quite like having a broken thumb. Physical activity pervades our lives in thousands of ways and in countless forms and levels of intensity. Although kinesiologists traditionally have concerned themselves primarily with sport and exercise, they now recognize that the boundaries of kinesiology extend to all forms of exercise and skilled movement. Physical activity that occurs on sport fields and in fitness centers is important to physical activity professionals; but so is physical activity that occurs in the workplace, the rehabilitation center, the dance studio, the nursing home, and many other venues.

This chapter is intended to stretch your conception of physical activity. It will lead you on an expedition through a vast expanse of different types of physical activities done for different reasons and in different settings. The purpose, as we have said, is to expand your frame of reference for thinking about physical activity; but it also is to increase your appreciation of how critically important kinesiology—the subject matter of kinesiology—is to our daily lives.

One way to begin to appreciate the enormous variety of physical activities and the way they intersect your daily life is to ask yourself how often and in what ways you were physically active during the past week. In answering this question, you may first have thought about the various social situations you were in and then the role physical activity played in each situation. For example, you might think, "I worked at the restaurant 20 hours last week." With this as a framework to guide your recall, you might then proceed to identify the types and quantities of physical activity that were required. You might estimate, for example, that you walked back and forth from tables to the kitchen approximately 200 times, poured approximately 400 cups of coffee, prepared 200 salads, and collected credit cards or made change at the cash register approximately 50 times. Or you might think, "I played golf on Saturday," and with the framework of playing golf to guide your thinking you might recall that you walked about 4 miles and executed approximately 18 drives, 25 shots with midirons, 13 approach shots, and 30 putts.

But it isn't simply the amount of physical activity that is important; the variety of physical activity is important too. The types of physical activity and movements performed in working at a restaurant are much different from those involved in playing golf. So, as you relate this chapter to your everyday experiences, don't think only about how much physical activity you performed; think about the various kinds of physical activity as well.

As noted in chapter 1, physical activity is movement intended to accomplish specific purposes. As you act out your daily life, the purposes and goals you wish to achieve—whether the activity is something as simple as moving to the other side of the room or as complicated as juggling three balls—usually determine both the types and amount of physical activity that will be required. But when you stop to think about the enormous variety of goals and purposes that require some sort of physical activity, you can easily get overwhelmed. In order to get a handle on the importance and pervasiveness of physical activity in our lives, we need some system of organizing our personal experiences with physical activity—a conceptual framework.

In chapter 1 we used the term *spheres of physical activity experience* as a conceptual framework for classifying the different life experiences in which physical activity plays an important role (see figure 2.1).

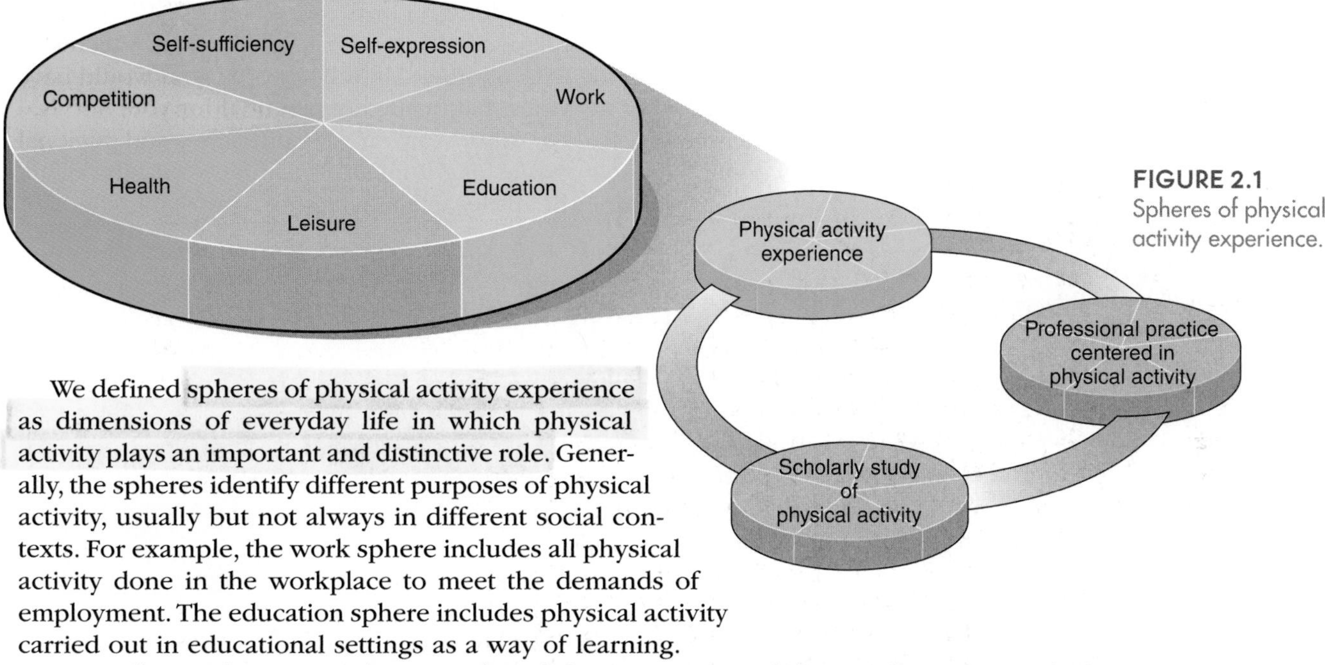

FIGURE 2.1 Spheres of physical activity experience.

We defined spheres of physical activity experience as dimensions of everyday life in which physical activity plays an important and distinctive role. Generally, the spheres identify different purposes of physical activity, usually but not always in different social contexts. For example, the work sphere includes all physical activity done in the workplace to meet the demands of employment. The education sphere includes physical activity carried out in educational settings as a way of learning.

You will note that some spheres are less definitive in terms of the social context in which the activity is usually performed. The self-sufficiency sphere, for example, includes physical activity carried out to survive and live an independent life. Although most of these activities are carried out in the home, they also can be performed at work, during leisure time, or in educational settings. Likewise, physical activity in the self-expressive sphere may occur in a variety of social contexts but always serves the purpose of allowing us to express our emotions. Keep these three things in mind as you study the spheres:

- The spheres are not intended to classify specific types of physical activities. They simply highlight the different compartments of our life experiences in which physical activity plays an important part.
- Some activities may be common to more than one sphere of experience. For example, if you run 3 miles each day, this activity may be assigned to the sphere of leisure but also be part of your health life (the health sphere) or your expressive life (the self-expressive sphere).

- Remember, the purpose of this chapter is to provide you with a general framework for thinking about the importance and pervasiveness of physical activity in your life, not to compartmentalize the activities themselves.

In the sections that follow, we will explore each of the seven spheres of physical activity experience—self-sufficiency, self-expression, work, education, leisure, health, and competition—by examining some activities that they encompass. We will discuss both positive and negative issues related to the spheres and learn about professionals who specialize in addressing some of those issues. Keep in mind that these spheres serve merely to help you look at the many ways and contexts in which you experience physical activity. We hope that this discussion will challenge you to think more deeply about the many levels of this field, levels that go far beyond the merely physical.

The Sphere of Self-Sufficiency

You devote a significant part of your life to taking care of yourself, and physical activity is an important means by which you accomplish this (see figure 2.2). Although tasks such as walking from your bed to the bathroom, eating breakfast, or brushing your teeth are perfunctory, they are critical to your ability to live independently or even survive. If you couldn't perform them you would have to rely on the help of others. In other words, you would have lost your self-sufficiency. There are other activities that, while not essential for your survival or independent living, also contribute to your self-sufficiency and personal comfort. These activities might include cleaning your apartment, washing the dishes, ironing clothes, or driving your car.

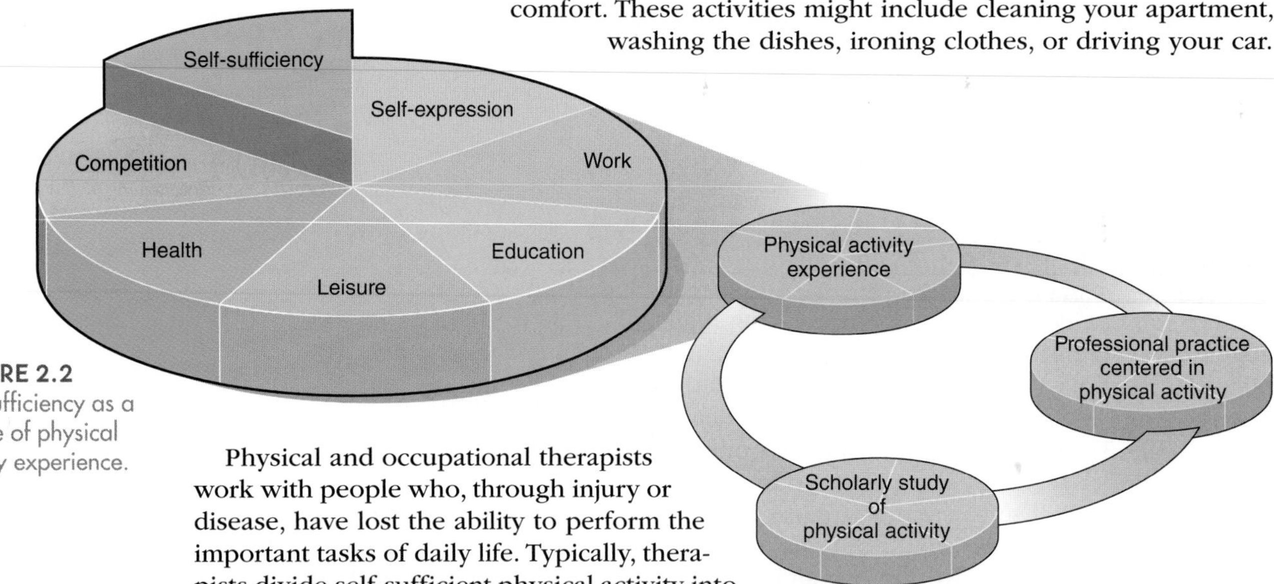

FIGURE 2.2 Self-sufficiency as a sphere of physical activity experience.

Physical and occupational therapists work with people who, through injury or disease, have lost the ability to perform the important tasks of daily life. Typically, therapists divide self-sufficient physical activity into two major categories: **activities of daily living (ADLs)** and **instrumental activities of daily living (IADLs).** The former include more personal behaviors related to grooming, using the toilet, dressing, eating, or walking. People who require daily assistance in these tasks are usually disabled due to injury or disease, or frail because of age. Activities classified as IADL include the less personal activities of telephoning, shopping, cooking, or doing laundry (Katz et al. 1963). Generally, IADLs are more physically demanding than ADLs. Health insurance companies often use these two classifications of physical activity in daily living to gauge the level of disability of patients and to determine the type of health support they require. For example, an older person may not qualify to collect long-term health care insurance unless a therapist documents that the individual is unable to perform two or more ADLs.

The physical activities we perform around our homes are often more complicated and demand more energy than typical ADLs and IADLs. Shoveling snow, fixing our automobiles, painting our apartments, repairing electrical fixtures, and similar tasks are examples of **home maintenance activities** that also reflect our level of self-sufficiency. All require relatively high levels of energy or skill or both. Although many people hire others to perform home maintenance activities for them (e.g., gardeners, painters, automobile mechanics, plumbers), the explosion of sales of do-it-yourself manuals and home improvement television shows suggests that more people are choosing to perform these household activities themselves.

➤ To live functional, independent lives, we must perform ADLs and IADLs, the latter of which tend to be more physically demanding. We also become self-sufficient by performing physical activities intended to maintain or improve the home. These activities are called home maintenance activities.

Limitations in Self-Sufficient Physical Activity

A variety of movements are required to carry out daily tasks such as housecleaning, doing laundry, bathing, cooking, opening jars, writing checks, and shopping for groceries. For example, using a vacuum cleaner includes the fundamental movements of walking, standing, grasping, reaching, pushing, and maintaining an upright posture. Remaining self-sufficient requires us to transport objects of different weights, ascend or descend stairways, or perform other fairly complex actions. When disease or injury compromises our movement capabilities, occupational and physical therapists can help us learn or relearn self-maintenance activities.

DO it **Activity 2.1**

Self-Sufficiency

Use this activity in your online study guide to practice distinguishing between ADLs and IADLs.

Before devising treatment plans, or "interventions," for rehabilitating people with physical disabilities, physical and occupational therapists must completely understand the physical activity requirements of each task. Obtaining this understanding usually involves a thorough analysis of the movements that a person must perform to carry out the task. Only then can the therapist decide which muscles the patient must strengthen or which movement patterns the patient must refine to regain self-sufficiency. These activity analyses can also help us understand the amazing complexity of what at first may seem to be simple tasks.

For example, an activity analysis of the seemingly simple self-care task of standing up from a seated position, a major challenge to the victim of a cerebral stroke, reveals at least four critical phases: (1) the feet must be well placed on the floor in a position to receive weight evenly divided on the two legs; (2) the trunk must be flexed forward at the waist while remaining extended; (3) the knees must move forward of the ankles; and finally, (4) the hips and knees must extend for final alignment (Carr & Shepherd 1987). By skillfully comparing the patient's movements to a model of correct performance, the therapist can design a training program to speed recovery in this fundamental task. Such activity analyses require an in-depth knowledge of body mechanics and anatomy, something the therapist learns from studying biomechanics, an important topic in kinesiology.

➤ Injury or disease can hinder a person's ability to perform daily physical activities. Physical therapists create therapeutic strategies based on activity analyses to help people recover their functioning within the limits of the disease or injury.

Self-Sufficiency and Aging

You may have given little thought to the importance of ADLs and IADLs in your life because you may have experienced little trouble performing them. But a surprisingly large portion of the population requires assistance in even these basic self-care tasks, a condition that deprives them of independence in their daily living. Limitations may result from accident, injury, congenital disorders, or aging. Aging is a major factor limiting performance of ADLs and IADLs. The population of those who are elderly is growing rapidly. Currently, 11% of the population are 65 years of age or older, and that group is expected to grow to 19% by 2030 (U.S. Bureau of the Census [USBOC] 2002a). One in eight Canadians was 65 years of age or older in 2001; but by 2026, one in five will be 65 or older.

The number of people with an impairment in one or more ADLs has been estimated to range from 11.8% of the population of 55- to 64-year-olds to nearly 50% of the 85-and-older population. Among the very elderly, the large-muscle activities of bathing and dressing are most often impaired (21.7% and 13%, respectively), followed by moving (8.9%), toileting

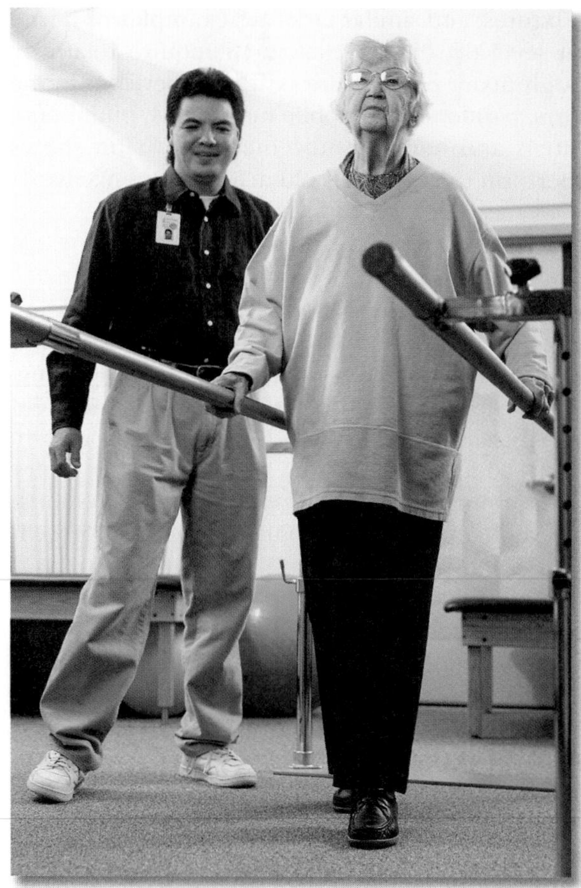

Physical activity professionals often conduct activity analyses to help improve functioning of diseased and injured patients.
Photodisc/Getty Images

(8.2%), and eating (2.7%) (Dunkle, Kart, & Lockery 1994). The most recent estimates suggest that over 6% of those 65 and older must have assistance to perform one or more of these ADLs, the rate varying by race and ethnicity. For example, although only 6.1% of white non-Hispanics over 65 require help with ADLs, 7.8% of Hispanics and 9% of non-Hispanic blacks do (U.S. Department of Health and Human Services [USDHHS] 2003). Figure 2.3 shows how disabilities are associated with increases in age. In 2026, approximately 54% of people with disabilities in Canada will be seniors versus 42% in 2001 (Human Resources and Social Development Canada [HRSDC] 2005).

A consistent finding is that physical limitations are not shared equally across the spectrum of elderly persons. Nonwhite, rural, elderly women earning less than $15,000 per year are most likely to report limitations. Thus, the effects of increasing age on physical activity can be traced to a number of critical social as well as biological factors.

Now the good news! Recent studies have documented a consistent decline in the rate of disabilities observed among those who are elderly. A general decrease in the percentage of disabled individuals over the age of 65 was observed from 1982 through 1999. The rate of decline in disabilities was greater in the 1990s than it was in the 1980s (Manton & Gu 2001). Also, reductions in nursing home use were observed for all age groups over 65, and they were largest for persons older than 85. Another encouraging sign was that the rate of increase in reported disabilities among black populations in 1982 through 1989 reversed itself in 1989 through 1994, and the

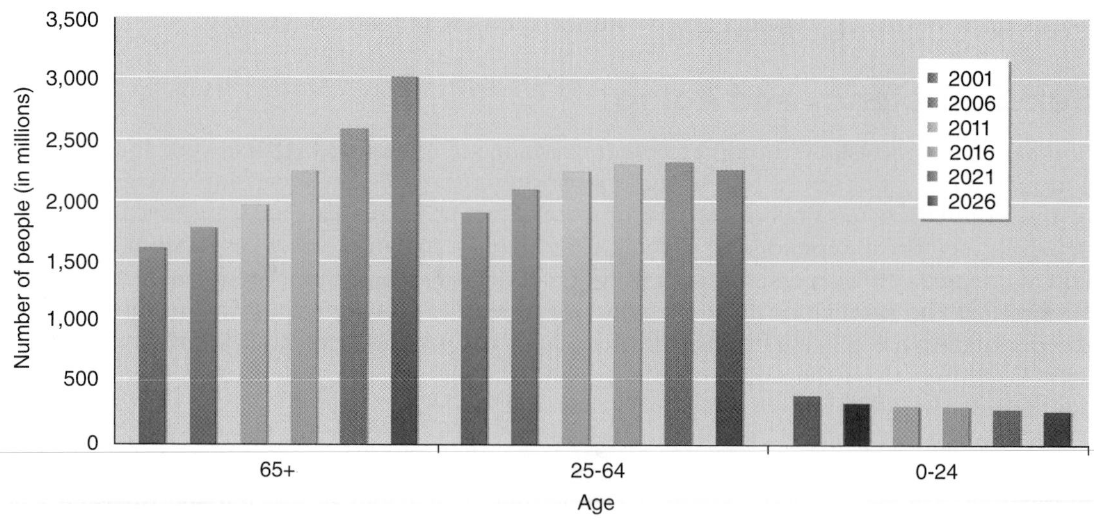

FIGURE 2.3 Disabled and projected disabled population in Canada, 2001-2021.
From Human Resources and Social Development Canada (HRSDC), 2005.

decline accelerated from 1994 to 1999. The trend of decline in reported disabilities across racial and economic groups continued through 2002, although the magnitude of decline was greater for those with lower incomes and those who were less culturally advantaged (Schoeni et al. 2005).

Older people are often injured in their attempts to perform ADLs and IADLs. Accidents experienced by elderly persons account for 43% of all home fatalities. Common accidents include falling on stairways or in bathtubs and suffering burns or scalds. Movement limitations cause most of these accidents, but environmental factors such as poorly lit stairways, frayed rugs, and poorly maintained homes play a part. Because older people realize that the risk of accidents around the home is high, many simply stop performing certain ADLs and IADLs. This situation results in a severely diminished quality of life (Czaja 1997). When persons who are elderly lose either the ability to perform self-care activities or confidence in their ability to perform them, they can suffer a devastating setback that deprives them of the satisfaction of being able to live up to their physical, mental, and emotional potential. The prospect of a growing population of individuals who are dependent on others to carry out their daily tasks is also undesirable from an economic standpoint. When major segments of the population lose their ability to perform ADLs and IADLs, the national health care system takes on a heavy burden. As a taxpayer, you have a real stake in the costs of Medicare and Social Security programs. An elderly population that requires assistance in the performance of ADLs has a significant effect on these programs. Estimates are that the decline in rate of disability among the population over 65 between 1982 and 1999 was enough to preserve Medicare solvency through 2070 and translated into a savings of $18.9 billion. Still, elderly persons account for 33% of total health care expenditures, largely because of their dependence on institutions (nursing homes) to provide self-care activities (Kane, Ouslander, & Abrass 1994).

➤ Limitations in the performance of ADLs and IADLs among elderly people require them to depend on others or institutions to perform the tasks of daily living. This problem is of great personal and economic importance.

These trends represent an awesome challenge to those in the physical activity profession who work with older populations. Although an increase in the number of elderly people over the next few decades is inevitable, an increase in the proportion of this population who are hampered by limitations in physical activity is not. Many of the people who suffer the most limitations in later years are those who failed to make physical activity a daily part of their normal lives—throughout their lives. Recreational activities and exercise programs designed and administered by physical activity professionals will play an increasingly larger role in preventing and rehabilitating these age-related disabilities.

Elderly people who remain physically active have fewer limitations.

Spotlight on Research: Physical Activity and Aging

A flood of research has underscored the positive effects of physical activity on aging, especially with respect to mobility. Here are some examples:

Study 1

Objective: To determine the extent to which leg strength and physical activity are associated with change in mobility in older people.

Subjects: 886 ambulatory older persons without dementia living in retirement communities.

Measured: Rate of change in mobility.

Findings: A higher level of physical activity was associated with slower rate of decline in mobility. Each additional hour of physical activity was associated with a 3% decrease in mobility decline. Each additional unit of leg strength was associated with an approximately 20% decrease in mobility decline.

Conclusions: Both physical activity and leg strength are important, but they appear to be independent predictors of mobility decline in older people (Buchman et al. 2007a).

Study 2

Objective: To determine whether physical activity modifies the course of age-related motor decline.

Subjects: 850 older persons with a range of educational and life experiences.

Measured: Baseline assessment of physical activity and administered motor tests (composite measure of nine strength tests and nine motor performance tests) for up to eight years.

Findings: Global motor function declined across groups, but each additional hour of physical activity at baseline was associated with a 5% decrease in rate of motor function decline.

Conclusions: Higher levels of physical activity are associated with a slower rate of motor decline in older persons (Buchman et al. 2007b).

Study 3

Objective: To determine the association between volitional walking and changes in walking ability and lower extremity function over one year.

Subjects: 800 community-resident females over 65 who were functionally limited but could walk unassisted.

Measured: Reported walking behavior and reported change in walking difficulty, usual and rapid walking speed, and lower extremity physical performance.

Findings: At the end of the study, one year later, walkers were 1.8 times more likely to maintain reported walking and showed less decline in customary speed or functional performance score than women who walked less than eight blocks per week.

Conclusions: The study provided evidence that even a small amount of regular walking can confer short-term protection from further mobility loss in functionally limited women (Simonsick 2005).

The Sphere of Self-Expression

The urge to express our inner feelings is one of the most basic human instincts. All of us would like to demonstrate, in one way or another, what is unique about us, what makes us special. Obviously, we are limited in the ways we can do this. People often hesitate to express themselves in speech, and few possess the talents of poets, songwriters, or artists. One way we can give outlet to our inner feelings is by moving our bodies (see figure 2.4).

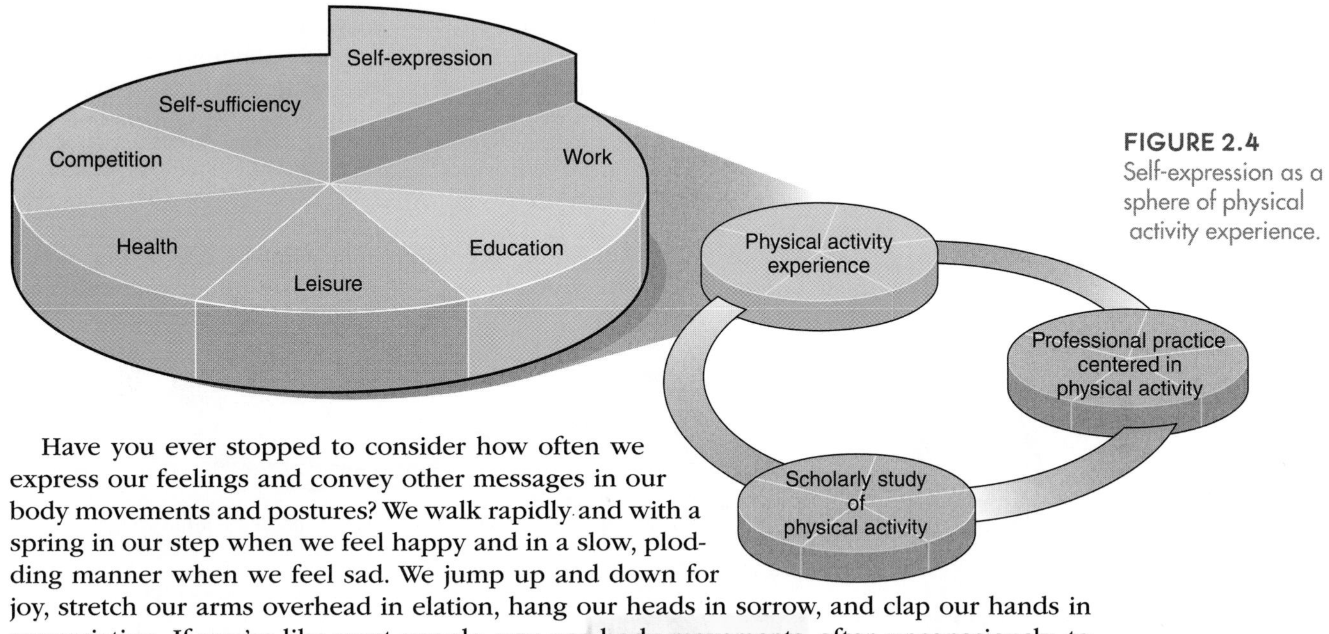

FIGURE 2.4 Self-expression as a sphere of physical activity experience.

Have you ever stopped to consider how often we express our feelings and convey other messages in our body movements and postures? We walk rapidly and with a spring in our step when we feel happy and in a slow, plodding manner when we feel sad. We jump up and down for joy, stretch our arms overhead in elation, hang our heads in sorrow, and clap our hands in appreciation. If you're like most people, you use body movements, often unconsciously, to express your inner emotional state. In all these instances, our body movements seem to be wired directly to our emotions so that changes in our movements occur without voluntary effort on our part. Although they point to interesting aspects of our human identities, these examples do not meet our technical definition of physical activity because they do not involve deliberate or voluntary movement directed toward a goal.

Many times, however, we deliberately use movements to express feelings and moods and to convey other messages. Have you ever thought about how much you depend on physical activity to help you communicate when you are talking? By using gestures, hand signals, and changes in body posture, we can deliberately and intentionally emphasize such verbal messages as describing shapes and showing directions. We also use physical activity intentionally to express our feelings and emotions in dance, religious liturgy, and various ceremonies. One could even argue that we express something about ourselves every time we move, whether in shooting a jump shot, running a race, or diving from a platform into a pool. Let's examine some ways in which we use physical activity to express ourselves.

Gestures

Gestures are movements of our hands, fingers, or other body parts that we use to communicate our intentions to others. We use them in place of or in conjunction with talking. Beckoning someone to approach by flexing and extending the index finger with the palm up, and holding up a hand with the palm away to signal someone to stop, are examples of gestures. Scientists who study nonverbal communication distinguish between three types of intentional gestures: emblems, illustrators, and regulators.

Emblems

Emblems are body movements, usually hand movements, that can be directly translated into words and are easily understood by those in the culture or subculture in which they are used (Morris 1994). People may use them with or without accompanying words, but even when used without words, emblems convey a great deal of information. For example, they can be used to communicate at a distance or in environments in which verbal communication is difficult or impossible, as in construction settings where crane operators receive messages from ground workers or in noisy athletic arenas where coaches must send signals to players on the court (see figure 2.5). Unable to talk to each other under water, scuba divers often use an elaborate set of hand signals to communicate. In noisy arenas or playing fields, referees and umpires use emblems to indicate to players, coaches, and audiences that a player has fouled, whether a base runner was out or safe, whether the tennis ball was in or out, and when time of the game has expired.

Illustrators

Illustrators are gestures that we use to illustrate or complement what we are saying. When you talk about yourself, you may point to yourself, and when you talk about someone else

FIGURE 2.5 Emblems used for sending messages in sport: *(a)* touchdown; *(b)* time-out; *(c)* safe; *(d)* foul; *(e)* fastball; *(f)* in (tennis).

in the room, you may point to him or her. People use illustrators to describe the motion of objects. If you were to tell someone about the path of a foul ball that narrowly missed your head when you were sitting along the third base line at the baseball park, you might use an illustrator gesture to describe it. Illustrators can also convey a particular tone in a verbal message. When a coach pounds the fist of his right hand into the palm of his left while talking, he is adding a sense of determination and seriousness to his verbal message.

Regulators

Regulators are body movements used to guide the flow of conversation. Hand and body movements used in greetings (shaking hands, waving, nodding) and in partings (waving, shaking hands, hugging, etc.) are examples of regulators. A person may also use regulators to signal to another party or parties that he or she is finished with the conversation. These signals include shifting weight from one foot to the other or turning toward the door. A person may use hand gestures, shifts in gaze, and head movements in conversations to signal to the other person that he or she has not yet finished talking or, conversely, that it is the other person's turn to talk.

Cultural Differences in Gestures

The meaning of a gesture often is specific to a particular culture. For example, to some Americans, a "cheek screw," in which a straightened forefinger is pressed against the center of the actor's cheek and rotated, may mean "cute." To others it may mean nothing. But in Italy it is regarded as a token of praise. Some of the most culturally specific gestures are regulators, especially those used for greetings. If you were to meet an acquaintance in America and he folded his arms across his chest when you extended your hand to shake his, you might think him rude, but this movement indicates a very respectful greeting in Malaysia. In Eskimo country, a hit on the shoulder doesn't mean that you are being challenged to a fight. It means "hello" (Argyle 1988). Even within cultures, the meanings attached to gestures can change over time. Thirty years ago, if you saw two Americans raising their right hands over their heads and slapping each other's palms, you might think it some form of secret greeting. Today, the "high five" has become a popular form of greeting or a gesture of congratulations, but it is gradually being replaced with bumping closed fists. Can you think of other relatively recent innovations in American gestures?

> ➤ We use physical activity as a form of communication and expression in combination with or in place of words. Gestures can supplement or substitute for spoken words.

DO it Activity 2.2

Self-Expression

In your online study guide, you'll analyze body postures to see what messages they contain.

Expressing Individuality While Working

Philosopher Jean Paul Sartre spent a great deal of time in the cafés of Paris. Inevitably his attention was drawn to the wait staff at his favorite café. He described a waiter's behavior this way:

> His movement is quick and forward, a little too precise, a little too rapid. He comes toward the patrons with a step too quick. He bends forward a little too eagerly; his voice; his eyes express an interest a little too solicitous for the order of the customer. . . . All his behavior seems to us a game. . . . He is playing; he is amusing himself. But what is he playing? We need not watch long before we can explain it; he is playing at being a waiter in a café. (Sartre 1956)

Over the next few days, take time to observe people at work in other occupations. Can you find other ways people express themselves through their work behavior? Do any of them "play at their work" as the waiter did?

Writing and Dance

Often, we express our feelings in the way we execute physical activity. We may combine expressive elements with movement used to accomplish specific tasks. For example, we may not run, swim, or lift weights specifically to express something about ourselves, but often we cannot help it. The manner in which a basketball player bounces the ball before shooting free throws and the movements she selects to propel the ball toward the basket involve instrumental and expressive movements. **Instrumental movements** are the critical movements required to attain the goal of the activity; **expressive movements** are idiosyncratic movements that are not required for goal attainment but that express something about the individual. No movements are more goal oriented or more reflective of our emotions, thoughts, and feelings than those we use for handwriting. Handwriting provides a permanent tracing of the movements one uses to produce it. Graphologists—analysts who study handwriting—

What Does Your Signature Say About You?

Psychologists have discovered that signature styles vary with such factors as self-esteem and status. For example, when self-esteem was manipulated experimentally by informing subjects that they had done well or poorly on a prior experimental task, those told they had done poorly used less space to sign their name than those told they had done well. Status also seems to affect the size of signatures. When the signatures of professors were compared to those of students, the professors' signatures were found to be significantly larger. Figure 2.6 shows the result of an informal record made by Karl Scheibe of the signatures he used to sign his name on the textbooks he purchased as a freshman in college and then again during each year of undergraduate school (Zweigenhaft 1977). He measured the size of his signature during four years of graduate school and later as a member of a college faculty. It is apparent that the amount of space used to sign his name systematically increased by a factor of three over an 11-year period. Scheibe and others believe the changes reflect the elevated status he enjoyed moving from undergraduate school to graduate school and eventually to a professorship. (Notice the large increase that occurred just as he was about to enter the professoriate.) Obviously, this wasn't a controlled study. Can you think of other factors that might have accounted for the change in Scheibe's signature? Could you design an experiment to test his hypothesis?

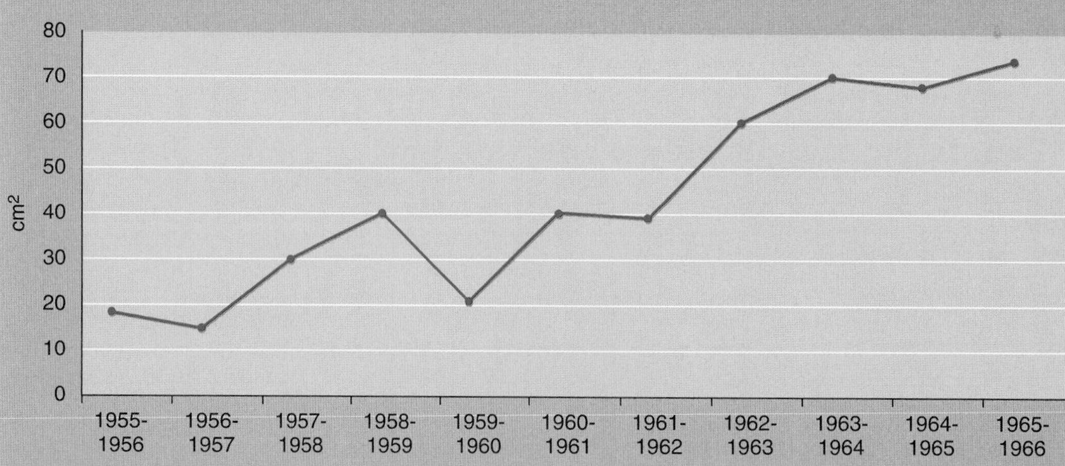

FIGURE 2.6 How life changes can affect the size of a signature.
Adapted from R.L. Zweigenhaft 1997.

believe that how we write is affected by our personalities and our emotional state at the time of the writing. Therefore they use handwriting samples to predict aspects of human behavior and emotions that may not be amenable to measurement using other means. Extreme right slant in writing, for example, is thought to reflect a more intense personality. Larger writing is associated with a more expansive, outgoing personality; small writing, with a more controlled personality. Although graphology has come under criticism by scientists on a number of counts, it continues to be used by marriage counselors to identify compatibilities, and by career guidance counselors and as a part of preemployment screenings to select executives. Separating instrumental movements from expressive movements may be impossible in such athletic performances as a twisting layup, a run around the bases, or a marathon run.

Sometimes, however, we employ physical activity deliberately to express sentiment and emotion with no instrumental goal in mind. In dance and in ritual, physical activity serves the express purpose of conveying feelings and symbolic meaning.

Dance has a long and interesting history. From the earliest periods of recorded time, people have used physical activity to express human emotion and meaning symbolically. One reason is that human movement enables us to communicate complex thoughts and feelings that are difficult or impossible to transmit verbally. Much the way painters use color and form, sculptors use texture and shape, and musicians use tones and rhythm, dancers use the mediums of force, time, and sequencing of movements to express aesthetic messages. The shape and dimensions of our bodies also affect the aesthetic characteristics of our dance. A leap by a tall, slender dancer will evoke much different emotions in observers than the same leap executed by a short, overweight dancer. In either case, the moving body can tell a story that is inexpressible in word or song.

The physical requirements of dance can be as exacting as—or even more exacting than—the most strenuous sport or exercise routines. The energy requirements of ballet, disco, jazz, Latin, modern, and tap dancing are approximately the same as the energy requirements for hunting, kayaking, gymnastics, climbing, and team sports (Ainsworth et al. 2000). Muscular and cardiorespiratory endurance, flexibility, strength, balance, agility, and coordination all are critical to dancing. Long and arduous practice and conditioning regimens are required to learn to achieve the standard body positions and movements required in ballet, for example. Likewise, the free-flowing, creative routines of modern dance, in which the dancer may have more freedom in interpreting the notations of the choreographer, require intense training and conditioning. Regardless of the form of dance performed, the result is the same: "the presentation of a significant emotional concept through formal movement materials . . ." (Phenix 1964).

Sometimes we dance not to communicate our feelings to others but simply as a way of giving outlet to our feelings, even when we are alone. Certainly, an audience, a partner, or a band is not required to enjoy dancing! Did you ever dance to music on the radio when no one else was in the room? Sometimes we dance spontaneously, merely to express joy and jubilation, as, for example, football players sometimes dance in the end zone after scoring a touchdown. You may like to dance because you enjoy social dances such as swing, tango, electric slide, or country line, or you may like to demonstrate your skill in moving your body in relation to a partner or to the beat of the music. Perhaps through dance you communicate something about yourself that you can't communicate in any other way.

➤ Dance is an art form that uses physical activity to express attitudes and feelings that may be difficult or impossible to express in normal verbal communication. Rituals often employ physical activity to express symbolically sacred values or beliefs.

The Sphere of Work

Work constitutes a significant portion of the total life experiences of most of us, and work usually involves physical activity (see figure 2.7). What different kinds of work have you done in your life? Have you worked as a construction laborer, a truck loader for a shipping company, a lawn maintenance technician, a pizza delivery person, a housepainter, a server at a restaurant, a kitchen worker for a fast food chain, an office assistant, or a checkout attendant at a grocery store? If so, what kinds of physical activity have these different jobs required?

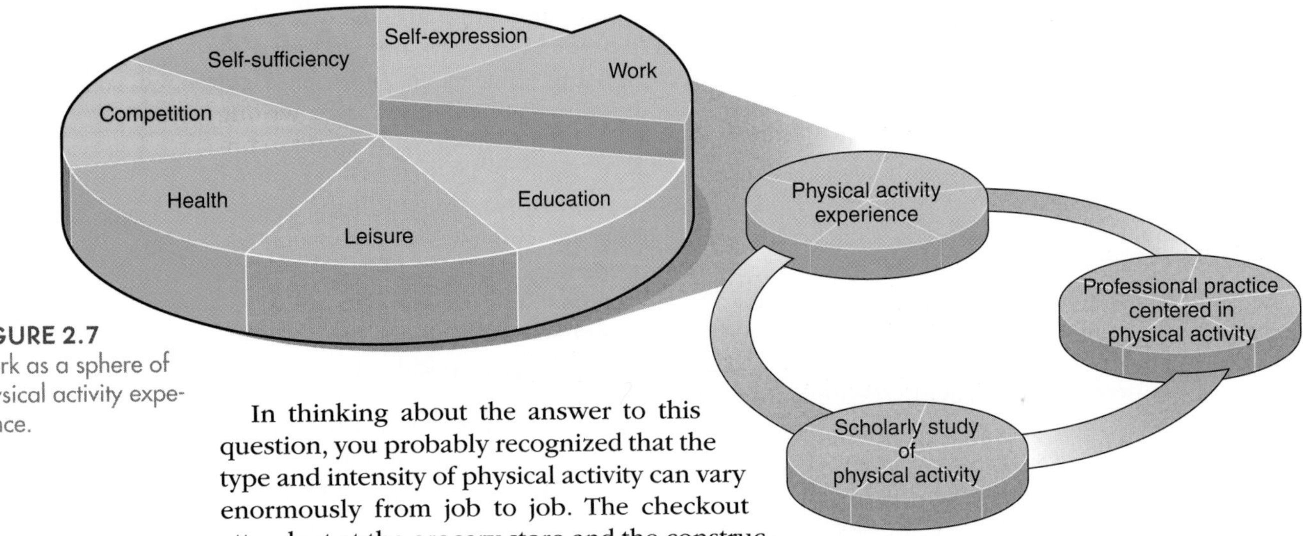

FIGURE 2.7 Work as a sphere of physical activity experience.

In thinking about the answer to this question, you probably recognized that the type and intensity of physical activity can vary enormously from job to job. The checkout attendant at the grocery store and the construction laborer both are engaged in physical activity, but they are not physically active in the same ways. Although you may have easily recognized the importance of physical activity to the work of the pizza delivery person or truck loader, you may not have appreciated the equal importance of physical activity (though of a different kind) in the work of secretaries, accountants, computer consultants, and graphic artists. The physical actions of typing, filing, and operating an adding machine, although not strenuous, require precise positioning of the limbs and fingers, movements that are essential to satisfactory execution of these jobs.

Throughout recorded history, societies have tended to associate manual labor, with its high demand for physical activity, with lower status, while according higher status to managerial and supervisory jobs that require little physical activity. Yet jobs requiring high levels of physical activity may offer a health bonus for workers. Increased levels of some forms of physical activity may ward off certain diseases associated with physical inactivity (sometimes called hypokinetic diseases, or low-physical-activity diseases). For example, researchers have found that moderate levels of physical activity, whether engaged in at work or during leisure time, promote good health and longevity. An early, classic study comparing incidence of heart attacks among drivers and conductors of English double-decker buses revealed that drivers, who were mostly sedentary (and had the easier, higher-status job) had higher incidence of heart disease than did conductors (lower-status workers), who were constantly in motion walking up and down stairs (Morris et al. 1953). This finding suggests that the types of work that may be attractive to us (those with a low amount and intensity of physical activity) may not be conducive to good health.

➤ As technology continues to shape the character of work, the amount of physical activity required on the job is likely to decrease, placing workers at higher risk for diseases brought on by physical inactivity.

As the workplace in the United States increasingly adopts technological innovation and the nature of work shifts from activity-intensive manufacturing, agriculture, and construction to service and professional occupations, the amount of physical activity required to perform work will continue to decrease. Current projections on the workforce show that this trend toward a sedentary workplace will continue at least through 2010. For example, the jobs of the future are much more likely to be found in the service sector, in which physical activity requirements tend to be more modest than they are in manufacturing, construction, or agriculture. Furthermore, as workers in the United States continue to log more working hours than in other countries (1821 hours in the United States in 2001 vs. 1467 hours in Germany, for example), and as the proportion of workers who commute 30 or more minutes to work increases (38% in 2000), they may not have the time required to exercise on a daily basis. This has led some health scientists to recommend that people put more emphasis on integrating physical activity into their working lives, for example by parking their cars farther from their offices or walking up stairs rather than taking the elevator (Dong, Block, & Mandel 2004).

The bottom line is that workers are expected to be at increased risk for diseases associated with reduced physical activity unless they increase their levels of activity during leisure (nonwork) time.

Physical Activity, Efficiency, and Injury in the Workplace

We can view work as a process in which workers trade physical activity for compensation. Employers, in turn, translate workers' physical activity into production of goods or services. The more efficient this production is, the greater the profit the employer will enjoy. Thus, workers' physical activity is the foundation on which the commercial enterprise rests.

Businesses can increase the efficiency of production in two ways. The first is to replace inefficient physical activity with lower-cost technology. For example, machines and robots now allow car manufacturers to produce many more cars per day at lower cost than they did when manual labor did all the assembly. But technology cannot replace all physical activity; in many jobs, the movement of workers remains critical to the efficient production of goods or services. In these cases, increases in production are sought via improvements in the efficiency and safety of workers' movements and working conditions. For example, the work of assembly line employees packaging television sets in boxes, of those picking fruit from trees or vegetables from fields, or of those operating machines in furniture and textile factories can be made more efficient and safer through changes in their movements, redesign of the machines with which they do work, or reconfiguration of the surroundings in which work is performed.

The physical activity professionals who specialize in improving the efficiency, safety, and well-being of workers are called **ergonomists** or **human factors engineers.** Although experts continue to discuss the differences between the roles of these specialists (some claim that human factors is more theoretical and relies more heavily on psychology and that ergonomics is more practical and relies more on anatomy and physiology), the trend is for these terms to be used synonymously (Kroemer, Kroemer, & Kroemer-Elbert 1994).

Ergonomists or human factors engineers apply their knowledge of exercise physiology, psychology, anatomy, and biomechanics in studying the demands of various jobs. (For this reason a background in kinesiology is excellent prepreparation for advanced study in human factors or ergonomics.) They then make recommendations regarding changes in movements used to perform the task, changes in the work environment, or changes in programs for training workers. Often, simply rearranging the location of objects to be assembled, adjusting the height of workers' chairs or tables, redesigning the shape of workstations, or modifying the order in which workers grasp components in the assembly process can result in marked improvement in worker efficiency. Usually, ergonomists are expert observers of the movements used in various kinds of work, and they make recommendations for change based on this critical analysis. Typical analyses and recommendations that an ergonomist might make in reference to two jobs are shown on page 45. Obviously, a thorough knowledge of the structure of the body as well as a complete understanding of how the body moves is essential to the work of ergonomists.

Productivity in the workplace suffers its greatest decline when workers are injured and miss work. This circumstance is especially troublesome when the injuries occur directly because of performing the physical activity required to carry out the job. Of course, workplace accidents remain a serious problem, but ergonomists mostly focus on injuries caused by the physical activity that a job requires. Many of these are musculoskeletal disorders that arise either from overexertion or from repetitive motion. In 1994, for example, 32% of all injury and illness cases that involved days away from work resulted from overexertion or repetitive motion. Overexertion injuries result from application of maximal forces by the body, usually when the worker is not physically capable of such exertions (e.g., weak back muscles) or is poorly prepared from a positional standpoint (e.g., lifting without bending the knees). Most of these injuries occur when a worker is lifting, pushing, or pulling objects or holding, carrying, or turning objects (USDHHS 1997, July).

Repetition injuries are those that result when workers repeat the same series of movements over long periods. These injuries are also known as **cumulative trauma disorders.** Usually they result from repeated physical stress to joints and muscles, which in turn damages tendons, nerves, and skeletal structures. Assembly line workers who package items perform the same arm, hand, finger, and trunk movements several hundred, or even several thousand, times each day. Even the small range of motion involved in typing can produce cumulative trauma disorder. Although our bodies are remarkably durable and adaptable, sometimes work asks too much of them, and they suffer deterioration or collapse.

One of the most common types of cumulative trauma disorders is **carpal tunnel syndrome,** a painful injury to the wrist caused by repetitive movements. The many tasks that can cause this injury include manual assembling, packing, typing, driving, lifting, hammering, or simply sitting at a desk for long periods performing clerical work. One way in which ergonomic specialists attempt to reduce the likelihood of this disorder is to redesign the tools or workspace to facilitate safe and efficient operational movements. For example, the photos in figure 2.8 show two types of pruning shears. In photo A, the handles are straight, which requires the wrist to be flexed to perform the task. This puts stress on the wrist and can induce a condition known as tenosynovitis. Designing the shears with bent handles (photo B) can allow maintenance of the wrists in the neutral position.

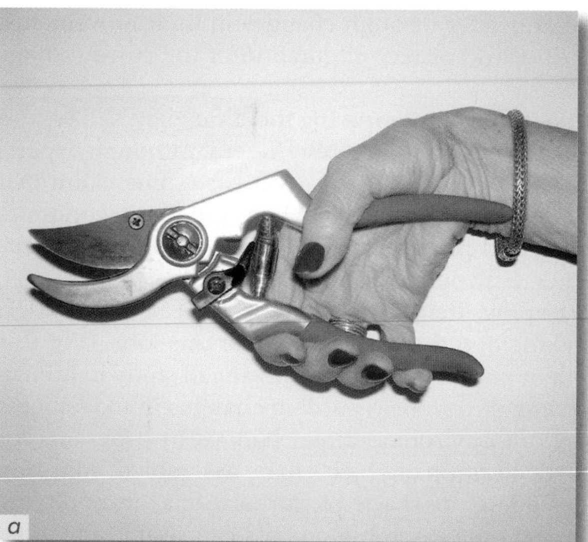

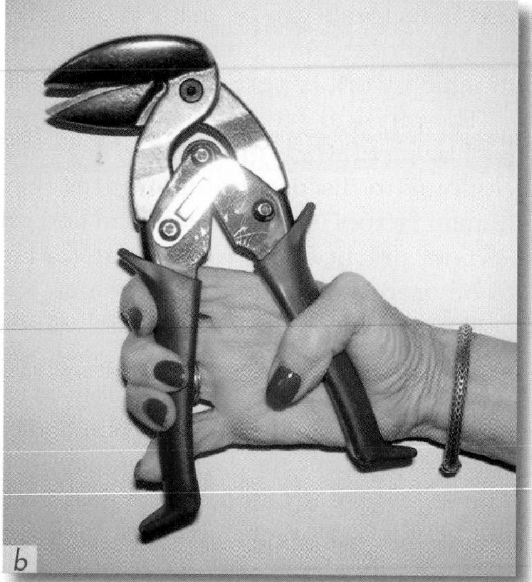

FIGURE 2.8 Human factors engineers have redesigned these shears by bending the handle to increase the efficiency of their use and to prevent a condition known as tenosynovitis. The motto of the human factors engineer is "Better to bend metal than to twist arms" (Helander 2006).

Some forms of work can lead to overexertion, which can produce severe musculoskeletal injury, most often in the back. Others require repetitive physical activity, which over long periods can damage muscles, tendons, and nerves. Such injuries usually require rehabilitation by physical and occupational therapists.

Physical and Psychological Demands of Work

What types of work do you think require the most physically demanding forms of physical activity? This question is important not only to employers but also to labor unions, insurance companies, and employees, as well as to ergonomists. To determine the physical demands of work, the Bureau of Labor Statistics applies a formula that takes into account the amount of weight that a person is normally required to lift on the job along with other factors such as the amount of stooping, climbing, or balancing.

Using these criteria and taking into account how many hours beyond 40 hours per week a job requires, *Jobs Rated Almanac* (Kranz 2002) presents a calculation for a total physical demand score for each of 250 occupations. A low score signals relatively low amounts of physical activity and exertion. Although the ratings suggest that high-status jobs such as managerial and professional jobs require relatively low levels of physical activity, this is not always the case. Some types of blue-collar work are fairly sedentary, whereas many white-collar jobs have high physical activity demands. For example, the jobs of musician,

Ergonomic Analysis of Two Jobs

Job 1

Task: Tasks investigated included nailing heels, trimming, hand application of dyes, and waxing used in the manufacture of shoes.

Work-related musculoskeletal problems involve: Upper extremities (hand, wrist, and elbow), shoulder, neck, and back.

Ergonomic risk factors: Awkward postures of the trunk, shoulder, and wrist; repetition, static exertions, and use of pinch grip.

Recommendations: Installation of height-adjustable swivel chairs and antifatigue mats, use of air-powered shears instead of scissors, raising and tilting machinery with added fixtures to relieve extreme work postures.

Job 2

Task: Tasks investigated included those of grocery store cashiers working at express checkout stands that involved frequent reaching, scanning, and keying tasks.

Work-related musculoskeletal problems involve: Neck, upper back, shoulder, lower back, buttocks, and legs.

Ergonomic risk factors: Repetition, awkward postures, excessive reach, and trunk flexion.

Recommendations: Adding a barrier at the far corner of the checkout stand to reduce excessive reach and trunk flexion and providing an adjustable keyboard to relieve postural stress.

Adapted from *Elements of Ergonomic Programs*, National Institutes for Occupational Safety and Health. Publication No. 97-117, U.S. Department of Health and Human Services, 1997.

photojournalist, physical therapist, undertaker, and fashion model have an average ranking of 175, whereas less prestigious jobs of office machine repairer, guard, electrical equipment repairer, automobile assembler, and railroad conductor have a mean ranking of 155.

From a total health perspective, we are also interested in the psychological stresses that various jobs place on workers. We know that the physical activity experienced in sport, games, and exercise during nonworking hours can help to reduce or eliminate stress. Our experience in the workplace suggests that high-prestige jobs that tend to require less physical activity may also be the most stressful. Are high-prestige jobs always the most psychologically stressful? *Jobs Rated Almanac* (Kranz 2002) also includes a ranking of jobs based on the stress levels they induce. These rankings are based on such considerations as whether the worker faces deadlines, is in competition with others, encounters physical hazards or uncomfortable environmental conditions, is expected to perform precise movements, or is responsible for the lives of others through his or her actions.

Of the jobs rated, the five least stressful (low score indicates low stress) were musical instrument repairer, florist, medical records technician, actuary, and forklift operator. The five jobs rated highest in stress were president of the United States, firefighter, senior corporate executive, racecar driver, and taxi driver. Although managers, supervisors, and professional workers do not always experience the highest levels of stress, the types of employment most likely to raise stress levels generally tend to be the jobs requiring low levels of physical activity. Thus, certain classes of workers—usually those responsible for making the major decisions for institutions and corporations—find themselves in a potentially health-threatening

Activity 2.3

Work

Use your online study guide to rank occupations according to how physically demanding and how stressful they are.

situation, subjected to this double whammy of high stress and low physical activity. These people may be especially vulnerable to heart attacks, hypertension, and strokes unless, of course, they pursue exercise outside of work.

The loss of key personnel because of death or disease is potentially devastating to businesses. To combat the health-eroding effects of high-prestige work, to ward off some forms of work-related injuries, and to ensure a vigorous and energetic workforce, many companies offer physical activity and fitness programs. In 1998 through 1999, 36% of companies with 50 or more employees offered worksite fitness programs, and another 22% made such programs available through health plans (USDHHS 2002).

With the advent of industrialization and technology, the physical activity requirements of work have diminished. The result has been the creation of jobs that threaten workers with higher levels of stress and lower levels of physical activity. Businesses have attempted to solve the problem by sponsoring exercise and sport programs for employees.

The Sphere of Education

Physical activity also plays an important role in the sphere of education (see figure 2.9). Education is essential both for the preservation of cultural traditions and for providing the knowledge and skills that enable society to progress. Physical activity is involved in all phases of education—from the eye movements required in reading and the wrist, finger, and arm movements required in writing to the more expansive forms of physical activity required in learning how to play a musical instrument or operate a power saw in vocational arts class. In almost every form of educational program, the aim is to change the behavior of those being instructed, in other words to change physical activity patterns of those being instructed.

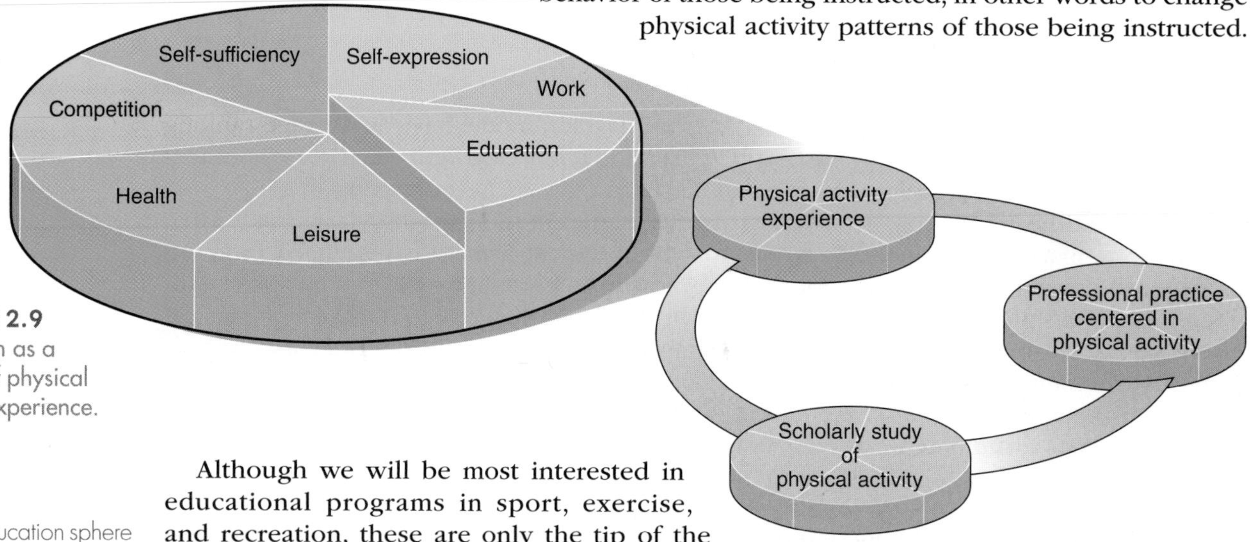

FIGURE 2.9 Education as a sphere of physical activity experience.

➤ The education sphere includes that aspect of our lives in which we set out to learn new skills or knowledge. Usually, physical activity plays an important role in this sphere, whether in connection with learning cognitive material or learning to perform physical skills.

Although we will be most interested in educational programs in sport, exercise, and recreation, these are only the tip of the instructional iceberg. Look around and you will see physical activity instruction occurring everywhere! Physical therapists teach patients how to walk; dental and medical school faculty teach students how to perform the intricate movements required in surgery; fathers and mothers teach their children how to dribble a soccer ball; industrial psychologists train employees in new assembly techniques; American Red Cross staff teach would-be lifeguards how to rescue swimmers; senior automobile mechanics train new mechanics how to use tools at the local garage; and business school instructors teach secretaries-in-training how to operate word processors. Anywhere physical activity is important, you will find some form of instruction, whether formal or informal, nearby. Instruction in physical activity may be as universal as physical activity itself.

Instruction in Sport and Exercise

Not long ago, instruction in sport and exercise was limited largely to public schools, college physical education programs, municipal recreation programs, and the military. Today, sport and fitness centers, resorts, corporations, hospitals, and clubs (tennis, golf, swimming) are also in the sport and fitness instruction business. Millions of children are taught how to play sports by their fathers or mothers or older siblings as well as youth sport team coaches. Velocity Sports Performance (with over 70 training centers across the United States devoted to specialized instruction of young athletes), sport camps offered each summer by famous coaches, and age-group swimming and private gymnastics schools all are involved in physical activity instruction. Adults, mindful of the adage "You can't teach an old dog new tricks," used to shy away from instruction in sport activities; but now they seek it out, often paying as much as $75 to $100 per hour for lessons from top skiing, scuba, fencing, squash, or fitness training specialists.

Private clubs and commercial spas, gyms, ice rinks, riding stables, dance studios, and martial arts centers now compete with school-based programs in the instruction business. Many large corporations offer recreation programs for their employees as a perk, and many of these include basic instruction in sports and exercise. Instruction through the media has also become a huge business. Television's Golf Academy Channel offers instruction in golf on a 24-hour basis. How-to books on every conceivable sport line the shelves of the megabookstores, and sport specialty magazines include regular instructional features.

Coupled with this interest in learning to perform sports and exercise is a growing demand for knowledge about how to live a healthful lifestyle. Corporate managers, keenly aware of the economic advantages of a healthy workforce, are making instructional programs available at the worksite, often in collaboration with their health maintenance organizations or health insurers. Typically, these classes use a variety of methodologies to integrate theoretical information about exercise, nutrition, and body mechanics with practical experiences in the exercise room. The explosion of interest in learning about sport and exercise has led to an expansion of instruction beyond the walls of schools and colleges.

Physical Education

Although instruction in sport, fitness, and exercise occurs in myriad formal and informal settings, it is most visible and accessible in physical education classes offered as part of the school curriculum (see chapter 15). Our educational system is based on the belief that it is in the best interest of a democracy to make available to its citizens a free and accessible program of education. Instruction in sport, exercise, and fundamental movements has long been considered an integral component of the educational enterprise.

In this section we will give cursory attention to the life experience you probably knew as "phys ed" or "gym class," and what physical activity professionals more appropriately call physical education. Because these programs have enormous potential for influencing the physical activity patterns of large segments of the population, they have attracted the attention of all physical activity professionals, as well as those in public health. For this reason, you—as a future physical activity professional—should appreciate this potential as well as some of the problems associated with school-based physical activity programs.

Chances are that you spent a significant part of your public school education experience in physical education class. But although most of us have attended physical education class, we have not all had the same experience. School districts do not all teach the same program of physical activities. Sometimes teachers within the same school district or even within the same school teach different subject matter and use different methodologies. Perhaps you were fortunate to have had a competent and enthusiastic teacher who offered carefully structured and sequenced classes. If so, you probably learned much about physical activity. On the other hand, you may have had a bored or unimaginative teacher who "threw out the ball" and offered little in the way of instruction. If that was your physical education experience, you probably learned less than your colleagues did who attended classes taught by more competent and energetic teachers.

Inevitably, your personal experiences in physical education will influence your conception of the subject. In any case, you should recognize that your experience—whether favorable or unfavorable—may not be an accurate reflection of what physical education is like in most schools.

➤ Physical education is the only near-universal program of sport and exercise instruction available to young people. For this reason it should be of the highest quality possible.

Effective educational programs are founded on clearly stated objectives. Objectives are vital maps for educators, pointing them toward the ends to be achieved with their instruction. They determine what and how teachers will teach. Students should also be aware of the objectives that are guiding the teachers' actions so that they can be partners in the learning experience. When you think back to your experiences in physical education classes, you may feel that the teachers' objectives were not always clear to you. Sometimes it may have seemed that the objective was developing fitness. At other times learning sport skills seemed to be the goal, and sometimes the teacher may have seemed most interested in developing the students' sense of fair play. What, then, *should be* the objectives of physical education?

School superintendents, principals, classroom teachers, parents, physical education teachers, and students all may have slightly different views concerning the objectives of physical education. Physical educators even disagree among themselves concerning the most important objectives of school physical education programs. As a result, experts have been unable to come to a universal agreement about what constitutes a physically educated person. In an effort to bring the factions closer together, the National Association for Sport and Physical Education (NASPE), working hand in hand with public school teachers and university professors, published a list of content standards for physical education programs. They are presented in the form of behaviors (see objectives of physical education on page 49) that should be demonstrated by a physically educated person (NASPE 2004).

Although developing physical fitness probably is at the top of most teachers' lists of objectives for physical education, only two of the seven NASPE standards focus on fitness and achieving a physically active lifestyle. Others emphasize the need to attain competency in a diverse number of activities, to experience social and psychological growth, to learn concepts about motor skills, and to have an opportunity for fun and self-expression. A growing trend is to view physical activity programs in schools as an opportunity to teach children social responsibility. You can begin to see the problem. Even this narrowed list of standards leaves a great deal of room for diversity in objectives, instructional approaches, and content. For the time being, school physical education programs will likely continue to aim at a cluster of different objectives, each requiring slightly different approaches and content. Individual teachers will also likely continue to emphasize those objectives that most closely align with their personal philosophies of physical education.

Physical education teachers work to meet varied objectives to develop physically educated people.

Objectives of Physical Education

A physically educated person

1. demonstrates competency in many movement forms and proficiency in a few movement forms;
2. applies movement concepts and principles to the learning and development of motor skills;
3. exhibits a physically active lifestyle;
4. achieves and maintains a health-enhancing level of physical fitness;
5. demonstrates responsible personal and social behavior in physical activity settings;
6. demonstrates an understanding and respect for differences among people in physical activity settings; and
7. understands that physical activity provides opportunities for enjoyment, challenge, self-expression, and social interaction.

Trends and forces in society influence what physical education teachers emphasize in their classes. This makes sense because public education must be responsive to the needs of society. For example, the growing popularity of sport during the early 20th century and the burgeoning interest among educators in the phenomenon of play as an educational experience were among the reasons that schools began to substitute sports in the physical education curriculum for the dull and often painful formal systems of exercises and gymnastics that were typical at the close of the 19th century (see chapter 5).

As the objectives of physical education show, physical educators teach for many objectives, but traditionally they have given priority in the curriculum to exercise and sports. Why is it important that public education pursue these objectives? Let's briefly examine each one.

Teaching Physical Education for Physical Fitness

The promotion of physical health through exercise has been an objective of physical education from its inception. The extent to which this objective is emphasized in public schools has varied according to pressure exerted by society. Today, fitness has been moved to the top of the agenda, primarily because public health officials have become alarmed at the overall poor fitness and sedentary lifestyles of children.

The population of "couch potatoes" (in the case of Internet addicts, they are now called cyber potatoes) is growing, and many of them are children who haven't yet reached middle school. You might be surprised to learn that elementary school children spend more time watching television than they do performing any other activity except sleeping (Dietz 1990). If you were like most children, by the time you graduated from high school you had spent 15,000 to 18,000 hours watching television compared with 12,000 hours attending school (Strasburger 1992)! (Approximately 32% of 2- to 7-year-olds and 65% of 8- to 18-year-olds have television sets in their bedrooms [Roberts et al. 1999].) Compounding the problem is the fact that watching television also is often accompanied by consuming high-calorie foods and beverages.

Education

Evaluate three case studies in this activity in your online study guide.

One effect of these sedentary lifestyles has been an increase in health problems associated with physical

Are You a Physically Educated Person?

On the basis of the NASPE standards and objectives for physical education listed on the previous page, do you consider yourself a **physically** educated person? Think back over your elementary and high school physical education experiences. Was your teacher enthusiastic? Did he or she model a physically active lifestyle? How often did you have physical education class? Was it a required subject in your school? Were you taught mostly lifetime sport activities (e.g., tennis, golf, archery) or team sports? What factors made the experience enjoyable or unenjoyable? What would you have done to increase its effectiveness?

inactivity. You will become painfully aware by reading this text that overweight and obesity have been declared by many health professionals as America's number-one health problem. Surveys show that only 42% of American adults over age 20 are at a healthy weight (USDHHS 2000). Latest statistics show that 32% of Canadian adults are overweight, including 15% who are obese (*Daily, The* 2002). What physical educators find especially alarming are the growing numbers of overweight and obese children and adolescents they see in their classes. Figure 2.10 shows how the proportion of overweight children has grown dramatically over the past 30 years, even among children as young as 6 years of age! Among Canadian children, an estimated 35% of girls and 38% of boys between the ages of 2 and 11 years are overweight; 17% of girls and 19% of boys were classified as obese in 1998 through 1999 (*Daily, The* 2002). Interestingly, these data revealed a greater incidence of overweight and obesity among children aged 2 to 5 than among children 6 to 11. Poor diet is a contributing factor, of course, but so is the lack of participation in regular vigorous physical activity. In Canada, for example, only 38% of obese children are physically active, whereas 47% of nonobese children are physically active (*Daily, The* 2002).

One might think that the public schools would have recognized the dangers posed by this state of affairs and increased the number of hours of physical education required each week. After all, as Professor Lawrence Locke, professor emeritus at the University of Massachusetts, has pointed out, "School physical education is an experience through which almost all of our population passes, at least in some form, and it is the single place where most of them encounter formal sport and exercise fitness activities before adulthood. Those encounters also take place at precisely the time when lifestyles are being examined and adopted, and when possibilities for the self are being accepted or rejected" (Locke 1996, p. 429). Although

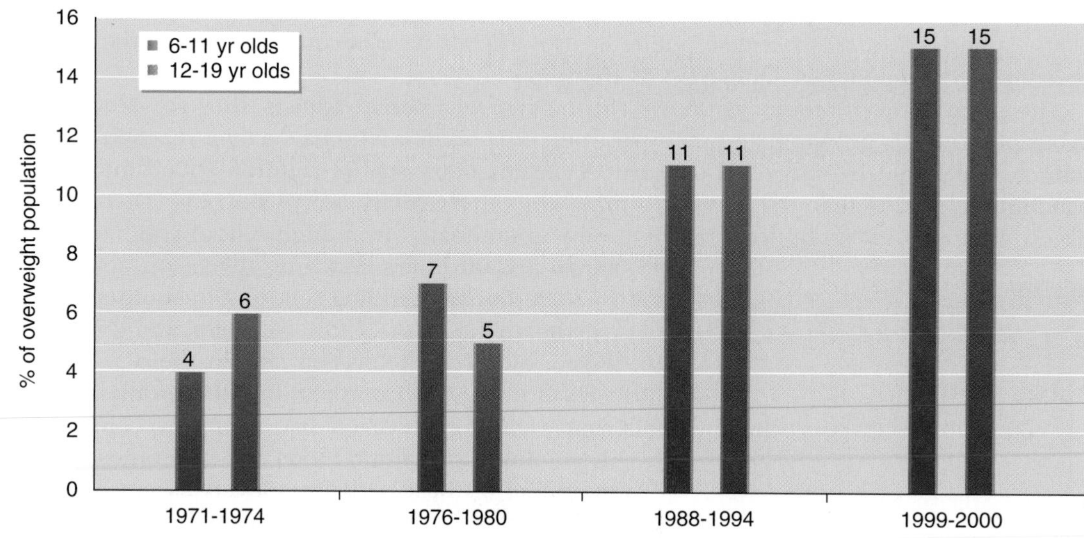

FIGURE 2.10 The proportion of overweight children has risen significantly over the past 30 years.

Data from U.S. Department of Health and Human Services 2000.

the logic of Professor Locke's argument is impeccable, the trend has been to cut rather than expand physical education programs because of pressures on schools to devote more time to math and science. For example, data reported from the School Health Policies and Programs Study showed that in 2000, the percentage of public schools requiring physical education in the elementary grades ranged from 50% to 52%. Dismal as this percentage might seem, it is significantly more than the 6% of schools that require physical education in the junior and senior years (National Center for Chronic Disease Prevention and Health Promotion [NCCDPHP] 2003). Naturally, with fewer schools requiring physical education, the number of students attending these classes has also dropped. In 1991, 42% of high school students attended physical education classes daily; in 1999, only 29% did so (NCCDPHP 2003).

A massive study involving 54,000 students from over 500 high schools showed that physical education requirements decline precipitously between 8th grade and 12th grade: 87% of 8th graders are required to take physical education while only 20% of 12th graders are required to do so. Moreover, access to physical education class is related to socioeconomic status (SES); only 49% of low-SES students go to schools where physical education is required compared to 59% of high-SES students. Contrary to popular opinion, varsity sports do not make up for the relatively low participation rates in physical education. Only 33% of girls and 37% of boys participate in varsity programs (Johnson, Delva, & O'Malley 2007, October).

Recognizing the future effect that childhood obesity could have on the national health budget, some state legislatures have now mandated daily physical education for all students, reversing the trend of the past few decades. Still, changes will have to be made in the way some physical education classes are conducted if they are to yield their intended health benefits. For example, best estimates are that only 38% of students in grades 9 through 12 were physically active during physical education class for more than 20 minutes three to five days per week (USDHHS 2002).

But there are some encouraging signs to indicate that the nation is beginning to take the obesity problem in children seriously. The Centers for Disease Control and Prevention (CDC), a national agency responsible for overseeing the health of the nation, called attention to the sedentary ways of the American population in 1996 with the publication of *Physical Activity and Health: A Report of the Surgeon General* (USDHHS 1996). This was preceded by the U.S. Department of Health and Human Services' publication in 1990 of *Healthy People 2000* (USDHHS 1990), a statement of health goals for the year 2000. After assessing the limited success at attaining these goals, the Department of Health and Human Services modified and restated them in *Healthy People 2010* (USDHHS 2000, November). All these reports give a great deal of attention to the problem of physical inactivity in young people and recommend an increase in the percentage of schools requiring physical education.

The Challenge to Physical Educators

1. Recent estimates are that 62% of children aged 9 to 13 years do not participate in any organized physical activity during nonschool hours and 23% do not engage in any free-time physical activity (USDHHS 2003).
2. In 2005, only 67% of high school students reported engaging in regular vigorous physical activity (USDHHS 2005).
3. A "midcourse report" showed that physical activity among students in grades 9 through 12 moved away from, rather than closer to, the Healthy People 2010 target. In 1999, the percentage of students participating in such activity was 27%; but in 2003 it had dropped to 25%, far short of the target of 35%. Vigorous physical activity and participation in daily physical education in schools among students in grades 9 through 12 also moved away from their targets ("Healthy People, Midcourse Report" 2005).

Teaching Physical Education for Sport Skill Development

Developing physically fit students is an important objective, but the objectives listed on page 49 go far beyond fitness. Sport education can help achieve some of these objectives, such as developing responsible personal and social behavior and providing opportunities for enjoyment, challenge, and self-expression. Sport also can be an important means for attaining and maintaining physical fitness. If students do not develop good habits of exercise by learning to incorporate moderate to intense forms of physical activity into their lifestyles for the remainder of their lives, making students fit while they are in school will achieve little. What good will we have done if we create a population of fit young people who eventually all slip into sedentary lifestyles after they graduate?

Coupled with this issue is the need to develop competency in these lifetime sport skills. People who have confidence in their ability to swim, play tennis, golf, in-line skate, or play squash are more likely to engage in these sports throughout their lives. Unless physical education classes are scheduled on a daily basis, teachers will have little chance to develop skill competencies in their students.

Because of this emphasis on lifelong patterns of physical activity, high school physical education programs should include a heavy concentration of lifetime activities such as tennis, golf, handball, or racquetball rather than team sports, which people are less likely to play following graduation. Yet, in spite of health and education leaders' emphasis on the importance of featuring lifetime sports in the curriculum, available data suggest that team sports, not lifetime sports, continue to dominate the American public school physical education curriculum. Basketball, baseball, volleyball, and soccer head the list of sports taught in high school curriculums. Swimming, jogging, tennis, racquetball, and hiking—physical activities that can be carried over into adulthood—are much less likely to be taught (USDHHS 1996). Clearly, much work needs to be done if physical education classes are ever to realize their potential for increasing the levels of physical activity of the general population.

➤ Surveys suggest that America and Canada are experiencing an epidemic of adult and childhood obesity. Increasing time allotted to physical education programs in the schools would seem to be one way to counteract this trend, but simply requiring physical education may not lead to a reduction of overweight and an increase in physical activity in youths.

Equally discouraging is a recent finding by researchers at Cornell University showing that increasing the physical education requirement—without changing the way physical education programs are organized or taught or increasing the general physical activity patterns of children when out of school—is unlikely to remedy the childhood obesity epidemic. Students who were required to attend physical education class were active only 31 more minutes per week than students whose schools had no requirement. While requiring physical activity increased the number of days per week that girls reported exercising vigorously or engaging in strength-building exercises, it was not associated with a lower incidence of overweight in students (Cawley, Meyerhoefer, & Newhouse 2007).

The Sphere of Leisure

Another area of our lives in which physical activity takes place is leisure (see figure 2.11). What do you do in your leisure time—play sports, exercise, read, attend concerts, watch television, hike? If you're like most people, you spend much of your free time in physical activity, and you tend to associate leisure with play. The term *leisure pursuits* is often used synonymously with *play* or *recreation*. But leisure isn't simply free time. Sometimes we choose to work in our free time. Have you ever given up an opportunity to spend a summer weekend (free time) camping or at the beach, instead volunteering to work an extra shift at your job? If so, was your work time "free time"?

As you can see, the terms *leisure* and *free time* are a bit more ambiguous than they first appear. Can you really work in your leisure time? You might have noticed that some people play softball or soccer with an intensity that seems to be anything but leisurely. Or, conversely, can people play (be at leisure) while they work? Perhaps. Artists, novelists, and professors sometimes seem to approach their work almost as if it were a leisure experience. Thus, a

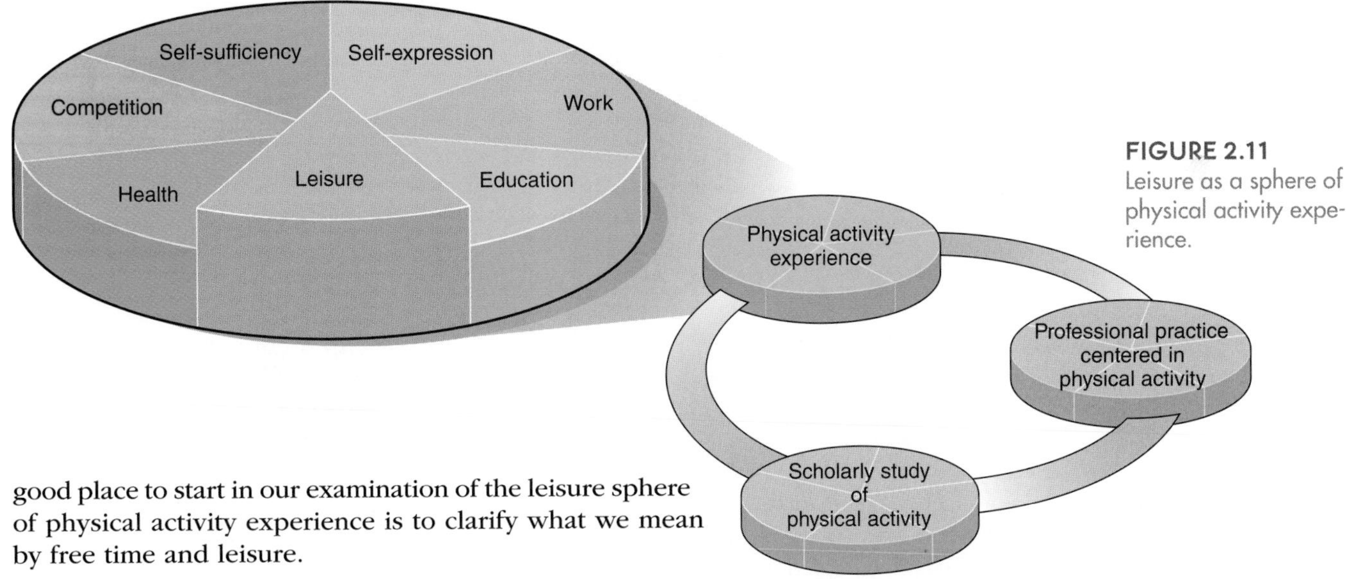

FIGURE 2.11 Leisure as a sphere of physical activity experience.

good place to start in our examination of the leisure sphere of physical activity experience is to clarify what we mean by free time and leisure.

Distinguishing Leisure From Free Time

When we are able to disengage from our everyday lives and participate in activities that really interest us, we often refer to this period as **free time.** We are free from our everyday routines and free to do what we want to do, when, where, and with whom we choose. Although we often use the terms *free time* and *leisure* synonymously, experts distinguish between the two.

Leisure theorists describe **leisure** not as free time, but as "a state of being" (De Grazia 1962). Philosophers, beginning with the ancient Greeks, have struggled to describe this state, but none have done so with complete success. In its purest state, leisure is a feeling that all is well with the world, a feeling of supreme contentment. Leisure can provoke feelings of celebration and wonder, creativity and discovery, excitement and reflection. Pieper (1952) described leisure as the "basis of culture," the psychological and spiritual disposition that is the fundamental prerequisite for the great works of art, music, theater, and philosophy.

➤ Leisure is a state of being, and free-time activities can help us attain this state. Large-muscle physical activities such as sport and exercise have the potential for nourishing and maintaining a leisure disposition.

> **DO it Activity 2.5**
>
> **Leisure**
>
> In your online study guide, review several scenarios and determine which scenario describes a person engaged in a leisure pursuit.

How do we achieve this state of being that we call leisure? Sometimes we achieve it while engaged in sedentary activities such as sitting beside a gurgling stream, reading a novel, or listening to music. At other times, we can achieve the state of leisure through large-muscle activities such as running, hiking, surfing, or skiing. Precisely what this state of being is or how it comes about may never be explained fully. One thing seems clear. Leisure has the effect of clarifying to ourselves and others who we truly are when we are free from the sense of obligation, anxiety, and pressure that can confront us in our everyday lives. And when we experience it, we look forward to experiencing it again and again.

To use our free time to achieve a state of leisure, we must be able to divorce ourselves psychologically from other aspects of our lives, a challenging task for many people. In addition, we must be sure to choose an activity that is conducive to a state of leisure. Although many of us choose to watch television during our free time, a study of teenagers showed that they do not find it a satisfying experience (Csikszentmihalyi 1990a, January). Hollow free-time experiences are not leisure experiences. They don't challenge us, stimulate our

Leisure is a state of mind that can be acquired in vigorous as well as more contemplative physical activity.
Brand X Pictures (left) and Stockbyte (right)

imaginations, or help us reveal our identities the way that true leisure experiences do. Thus, if we are to convert free time into leisure time, we must use it for participating in activities that tend to nourish and maintain the state of being known as leisure.

Physical Activity as Leisure Activity

The entire range of leisure pursuits—from sedentary activities such as chess or reading to large-muscle activities such as water skiing or softball—is the focus of study in the discipline known as **leisure studies or recreation.** Kinesiology concerns itself primarily with the large-muscle forms of leisure pursuits that require substantial physical activity. If this leads you to wonder whether the interests of kinesiologists overlap with those of leisure professionals, the answer is yes. Faculty and students in departments of leisure studies and departments of kinesiology both view sport and exercise as a legitimate part of their fields of study. Although it might seem that this blurring of the boundaries of the two disciplines would be problematic, it has existed for several decades and has not deterred the continuing expansion and refinement of each of these exciting areas of study.

Activities such as golf, folk dancing, softball, hiking, long runs in the park, and boating have great potential for putting us at leisure. But we can approach these in ways that make such a leisurely disposition impossible. For example, becoming upset over an official's call, cheating, intentionally injuring an opponent, and exercising under the compulsive stimulus of anorexia nervosa are not compatible with a leisure state. The context in which the activity takes place is also important. Is it possible for athletes to attain a state of leisure when 80,000 spectators are sending up a deafening roar from the stadium? In the final analysis, whether or not a sport or exercise is truly a leisure pursuit depends on its nature, the context in which it is pursued, and the motivations and attitudes that participants bring to the activity. Professionals in the leisure industry try to help people make choices about their free time that increase the state of being that we know as leisure.

At best, free-time activities of sport and exercise offer us only the potential for achieving the state of leisure. Whether or not a physical activity is truly a leisure pursuit depends on the nature of the activity and the context in which we pursue it. The challenge of the physical activity professions is to teach people to participate in free-time physical activity pursuits in ways that nourish the disposition known as leisure.

Because your future as a physical activity professional will likely be affected by participation trends in large-muscle **leisure activities,** it is worth taking the time to examine trends of participation in this important sphere. Remember, kinesiologists are particularly interested in leisure-time pursuits that involve relatively large amounts of physical activity. The proportion of adults engaging in light to moderate leisure-time physical activity for at least 30 minutes five times per week is surprisingly low, at or below 50% in all age categories, for men as well as women (CDC, National Center for Health Statistics 2007). The rate of participation for men is higher than that for women across all age groups, and participation declines as a function of age. You may not be as active in your leisure time today as you should be, but the data suggest that your activity levels are likely to decrease as you take on responsibilities of job and family.

What do people do in their leisure time? (Although competitive sports clearly fit under the leisure activity umbrella, we will look at their popularity under the competition sphere and focus here on large-muscle, noncompetitive forms of leisure activities.) In 2006, the most popular form of leisure-time physical activity in the United States was exercise walking (more than 82 million people [87%] over 7 years of age reported engaging in it at least once). Fifty-two percent reported exercising with equipment. Swimming (56%), camping (48%), bowling (45%), and fishing (40%) were also popular. Among the activities showing the largest increases between 2001 and 2006 were tackle football, mountain biking, and hiking; during the same period, decreases were observed in tennis (–5.1%), softball (–5.8%), volleyball (–8%), and skiing (–16%) (National Sporting Goods Association [NSGA] 2007).

➤ Although participation in some recreational leisure-time activities involving moderate to intense physical activity remains high, the rate of growth appears to have slowed and in some cases declined, whereas participation in more sedentary activities appears to be on the rise.

Because surveys of physical activity tend to vary widely, in terms of both the questions asked and the populations polled, research on leisure-time physical activities is, at best, an inexact science. With this caveat in mind, those tempted to think that Americans are fairly active in their leisure time should note that the same National Sporting Goods Association survey showed a marked decline in participation in many high-active leisure activities over the preceding 10 years. For example, participation in camping declined 10% between 1996 and 2006, exercise walking by 16%, exercising with equipment by 9%, bowling by 9%, and hiking by 15%.

Watching Sports

One form of sedentary leisure activity clearly on the rise is watching sports. People spend more than $1 billion on tickets to professional and amateur sporting events each year, approximately the same amount that they spend on all movies and 60% of the total that they spend on all theater, opera, and nonathletic entertainment sponsored by nonprofit organizations. Over 180 million spectators attended professional and college baseball, basketball, football, and professional hockey games in 2000, up from approximately 150 million in 1985. Annual spectator attendance had more than tripled in women's college basketball between 1985 and the turn of the 21st century (USBOC 2002a).

Such data, however, don't give the complete picture. For example, they don't take into account the 11.5 million fans who watched high school sports or the millions of others who

One of the purposes of spectator sports is to generate enthusiasm among fans.

watched youth sport teams and other amateur sports. Nor do they include the even greater numbers who watch sports on television. Estimates are that television sets in American homes are tuned to sports for an average of 180 hours each year ("Game Plans" 1994).

The growth in sport spectatorship represents something of a dilemma for the physical activity profession. On one hand, the jobs of many physical activity professionals—coaches, athletes, and athletic trainers—depend on the continued viability of sport as mass entertainment. Sport management, an increasingly popular career option for physical activity professionals, consists of specialists who plan, market, stage, publicize, and supervise mass spectator sport events. In a nutshell, their job is to make watching sports a profitable enterprise by increasing the number of people watching sport events. At the same time, the rise in mass spectatorship, a largely sedentary activity, works at cross-purposes with the profession's efforts to increase physical activity in the general population. When one considers the markedly upward trend in spectators alongside troublesome data presented in this chapter regarding declining or stable participation rates in exercise and sport activities, it is difficult not to gain the impression that we are becoming a nation of spectators rather than active participants.

Of additional concern to physical activity professionals is whether watching sports constitutes a viable leisure pursuit as that term is understood by leisure theorists. Does it nourish the psychological or spiritual state called leisure? Some believe that it can. Michael Novak's popular book *The Joy of Sports* (1976) is an eloquent testimony to the fact that watching sports may add to the quality of our leisure lives. On the other hand, an obsessive preoccupation with the fortunes of one's favorite sport team can be socially unhealthy, especially when it leads to neglect of more important issues such as human relationships. As enthusiasm for watching sports continues to grow, the profession will have the challenge of differentiating between healthy, constructive spectatorship and addictive, destructive preoccupation.

Although participation in many sports appears to have stabilized, and in several cases declined, interest in watching sports continues to grow, abetted by a thriving commercial and television market. This trend toward mass watching or mass vicarious participation through fantasy sport leagues as opposed to actually playing sports may not be compatible with national health efforts to increase the physical activity participation levels of the population at large.

Aging and Leisure Activities

At the 2007 World Masters Athletic Championships in Riccione, Italy, older athletes from all over the world came together for the 17th annual track and field competition. Nora Wedemo, an 87-year-old, won the medal in her age category for the 60-meter run (12.5 seconds). Eighty-five-year-old Carol Peebles won the 400 meters for her age group (2.30 minutes); 72-year-old Flo Meiler won the pole vault in her age category with a jump of nearly 7 feet; and 90-year-old Margareta Sarvana won the shot-put event for her age category with an effort of over 4 meters.

Not long ago, most people believed that life essentially ended at 65. Persons who were elderly faced the problem of finding sedentary ways to fill up their free time until they died. Thanks to advances in health maintenance and the growth of the field of gerontology, we now know that people over 65 have enormous potential to learn new skills, explore new vistas, and engage in forms of physical activity that most elderly people wouldn't have dreamed of engaging in 20 years ago.

According to one study, for example, nearly 7 million adults over age 75 participated in an exercise program at least one time in 1997. Amazing as it might sound, one and a half million people in this age category reported playing a sport at least once during the year (USBOC 2002b). As older populations continue to recognize the benefits of leading a healthy lifestyle—throughout the span of their lives—we can expect the numbers of physically active older adults seeking physically active leisure activities to rise over the next decade. Working with this population represents an exciting professional opportunity for today's students who are majoring in kinesiology.

The Sphere of Health

Attending to our personal and community health needs consumes a large part of our normal daily experiences. We bathe, dispose of our garbage, ensure that we drink clean water, eat proper foods, visit physicians when we are ill, and so forth to maintain our health, which in turn enables us to be productive citizens and enjoy life. All these health-related activities involve physical activity in varying degrees. We can't, for example, brush our teeth, bathe ourselves, or carry out the garbage without performing physical activity. But physical activity also intersects with this sphere in a much more direct way (see figure 2.12). We now possess hard, scientific evidence that physical activity performed in the right amounts and with sufficient frequency contributes to our health in many important ways.

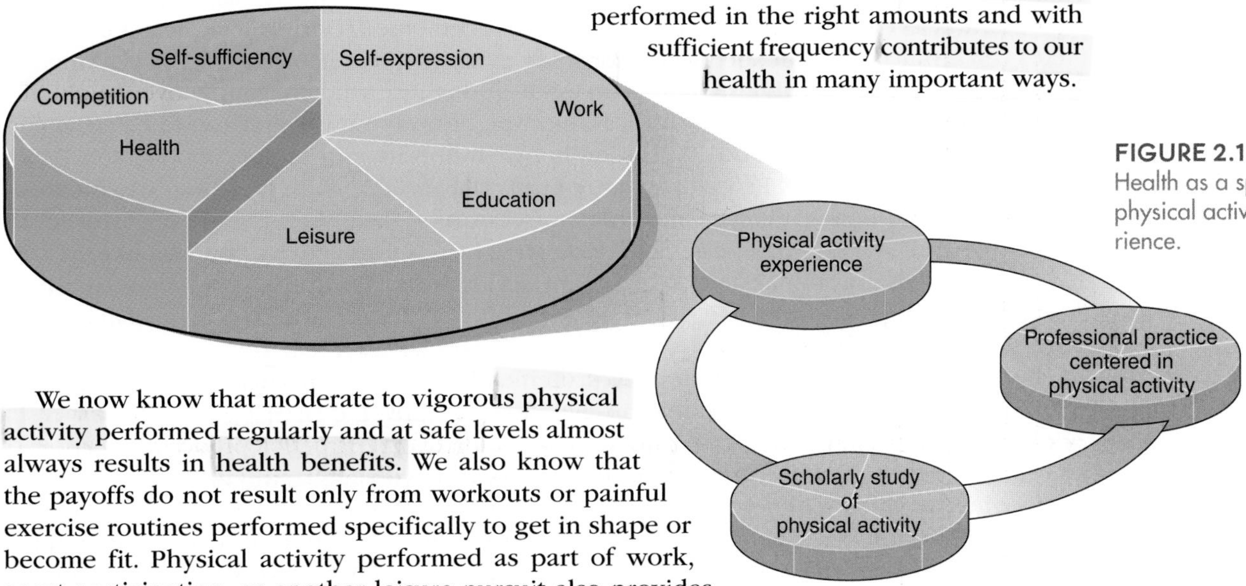

FIGURE 2.12 Health as a sphere of physical activity experience.

We now know that moderate to vigorous physical activity performed regularly and at safe levels almost always results in health benefits. We also know that the payoffs do not result only from workouts or painful exercise routines performed specifically to get in shape or become fit. Physical activity performed as part of work, sport participation, or another leisure pursuit also provides benefits. At the same time, it seems equally clear that the safest, most effective, and most efficient routes to attaining health benefits from physical activity are carefully designed programs supervised by exercise professionals who are well versed in the science of physical activity. As the public comes to understand this more, the demand for highly trained fitness leaders and consultants will continue to increase.

Physical Activity, Health, and the National Interest

Why is it important for Americans to maintain and improve their health through physical activity? For one thing, healthy people are vital to the national economy. A sick population not only incurs medical expenses but also drains the economy through lost productivity. Consider this: The United States spends a greater proportion of its gross domestic product on health care than any other industrialized nation. In 1990, total health care expenditures in the United States amounted to $696 billion. By 2000 the figure had risen to $1.3 trillion, and in all likelihood it will rise to $2.6 trillion by 2010 (USBOC 2002c).

> **www. Web Search 2.1**
>
> **The Great Disconnect**
> In this activity in your online study guide, you'll gather information on obesity trends on the Centers for Disease Control Web site.

The deaths of over 700,000 people each year because of heart disease cost the nation billions of dollars annually. Another 165,000 people die each year from strokes, and another 65,000 from diabetes (USDHHS 2003). Taken together, heart disease, stroke, and diabetes account for approximately 44% of deaths among those 65+ years of age and approximately 32% of all deaths in those 45 to 64 years of age. There is little doubt that this death rate could be reduced if the population was more physically active. As a taxpayer, you have a large stake in this issue. As health care costs rise in the private sector, the amount spent by state and federal governments on health care costs rises also. Obviously, these expenditures represent substantial drains on state and federal budgets, limiting the ability to fund education and other social programs. The massive spending also poses a potentially significant problem for future generations who will have to foot the bill. None of this, of course, measures the personal effect. Each incidence of illness brought on by lack of physical activity represents one person who has lost his or her potential for living an active, productive, enjoyable life.

Many of these diseases are related to obesity, which is a problem not only of childhood but also of adulthood. Sedentary lifestyles are a major contributing factor. For example, Americans spend nine times as many minutes watching TV or movies as they do on sport, exercise, and all other leisure-time physical activities combined (Dong, Block, & Mandel 2004). Across all ethnic and adult age groups, 31% of the American population can be classified as obese. (Keep in mind that these figures do not include those who are simply overweight; the percentage of the adult population that is overweight or obese is estimated to be a staggering 60% [USDHHS 2003].) The association between obesity and socioeconomic conditions has been well established. People who are poor and uneducated are more likely to be obese than those with higher incomes and levels of education. Likewise, people of lower levels of income and education are less likely to engage in physical activity and less likely to eat a well-balanced, nutritional diet. Regular physical activity performed at moderate to high levels of intensity is a front line of defense not only against obesity but also against heart disease, stroke, and diabetes.

The reason physical activity figures so prominently in the nation's health objectives is that it has been shown to prevent and help rehabilitate the costly diseases described earlier. In addition, physical activity has been associated with a reduced risk of colon cancer and appears to help lower the risk of developing non-insulin-dependent diabetes mellitus. Regular physical activity may enable those who are elderly to live independently for a longer time and avoid falls and other injuries. Regular exercise also lowers the risk of mortality for both younger and older adults. The response of U.S. health agencies has been to draft a list of national health objectives for the year 2010, which includes recommendations for increasing the level of physical activity among Americans. For example, two of the objectives are to reduce to 20% the proportion of adults who engage in no leisure-time physical activity and to increase to 30% the proportion of adults who engage regularly, preferably daily, in moderate physical activity for at least 30 minutes per day (USDHHS 2002). Clearly, the profession of kinesiology is well positioned to make a significant contribution to the attainment of national health goals.

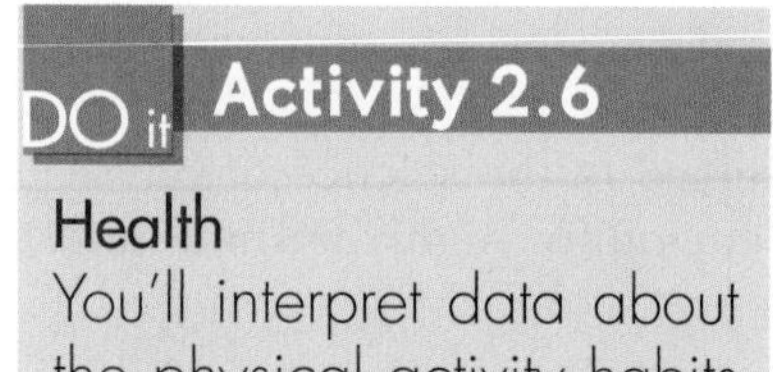

Health

You'll interpret data about the physical activity habits of adults in this activity in your online study guide.

The failure to make physical activity a part of our daily lives has led to a high incidence of deaths because of stroke, heart disease, and diabetes. Regular moderate to vigorous physical activity could prevent many of these deaths, yet large segments of the population remain sedentary.

Detriments of Physical Activity

At this point, a reality check might be helpful. Obviously, physical activity is a valuable adjunct to healthy living, but it is not an unqualified benefit. Just because some physical activity is good, it doesn't necessarily follow that more is better. What physical activities you select and how you go about performing them also can have detrimental effects. The modern physical activity professional must be acutely aware of the negative as well as positive aspects of physical activity.

Overexercise, whether in the form of weight training, running, aerobics, or repetitive motions in work settings, can result in stress fractures, strained muscles, inflamed tendons,

Injuries Plague Football Players

Few sports are more popular in North America or more dangerous than football. Football captures the imagination of North Americans in almost inexplicable ways from the time we are very young until we are old. Sometimes it is very difficult to objectively examine those aspects of our culture with which we are most familiar. It is like this with football, a sport associated with the highest number of direct catastrophic injuries for any sport as reported to the National Center for Catastrophic Injury Research (Boden et al. 2007). A person from another culture who wasn't familiar with the game might view with horror the violence and injuries that occur on the football field. But for those of us who grew up with the game, the injuries suffered in association with the game seem somehow "normal." We might wonder what a person unfamiliar with football would think of the following comment from former legendary National Football League defensive end Lawrence Taylor: "I don't like to just wrap the quarterback, I really try to make him see seven fingers when they hold up three [the reference is to the quick test usually given to see if a player has suffered a concussion]. So long as the guy is holding the ball I intend to hurt him. . . . If I hit the guy right, I'll hit a nerve and he'll feel electrocuted, he'll forget for a few seconds that he's on a football field" ("Nightmare in the Backfield" 2006, November 17).

Taylor's comments might sound heroic or even romantic to the football enthusiast, but when recent media accounts concerning the terrible, lifelong damage incurred by professional football players are taken into account, such talk seems nearly sociopathic. First-person accounts of men in their late 30s unable to walk unassisted due to repeated joint trauma, of elbows that don't work, or of fingers broken so many times they can no longer hold a coffee cup have flooded the sport pages. Perhaps of greatest concern is the scourge of brain concussions suffered by football players. The disorienting effects of a hard collision, often joked about by sportscasters as "having your bell rung," occur when the brain collides against the interior of the skull, injuring vital nerve tissue. The effects of repeated head trauma (often glorified by players as "head banging") are cumulative; and those who have spent many years playing the game sometimes suffer permanent disability including *dementia pugilistica,* a brain disorder more commonly associated with boxers. We hear heart-rending stories of professional football players in their 40s who are unable to keep a job or function as effective members of society as a result of mental disorders traced to repeated trauma they suffered playing a brutal game. Recently attention has shifted to high school football, where, because the athletes' brains are still developing, catastrophic head injuries can have even more serious and long-lasting effects. Although catastrophic head injuries remain relatively rare in the game (7.23 per year; the large majority at the high school level), the consequences (paraplegia, etc.) are tragic. More difficult to identify are the long-term effects of repeated "minor" head injuries, because the symptoms may not appear for years.

Engaging in physical activity produces unintended consequences.

1. Do you believe football is a body-threatening sport? Is it a violent sport?
2. If so, what factors do you believe contribute to the violence?
3. What recommendations would you make to lower the injury rate in football?
4. What are the implications for kinesiology majors who plan careers in athletic training and sports medicine? Coaching? Sport management? Sport psychology?

and psychological staleness. We have already noted that carpal tunnel syndrome, a painful and debilitating injury to the hand and wrist brought on by repeating the same hand motions day after day as in typing and assembly tasks, plagues thousands of workers. Sport participation and vigorous exercise bring, with all of their enjoyment and excitement, the dark side of injury. Over 4 million sport injuries were treated in hospital emergency rooms between July 2000 and June 2001. Sport- or recreation-related injuries accounted for nearly half of all nonfatal unintentional injury visits to emergency rooms by children aged 10 to 14 years during this period (CDC 2002). Approximately 715,000 sport and recreation injuries occur each year in school settings alone. Injuries are also a leading reason people stop participating in potentially beneficial physical activity (CDC 2006). Football, boxing, and hockey usually come to mind when we talk of sport injuries, but sometimes the damage brought on by long-term wear and tear, although more insidious, can be more traumatizing. Repeated injuries to the knees, fingers, wrists, hips, and backs take their toll on athletes not only in the weeks and months following the injury but in later years as well. Years of pitching a baseball can lead to rotator cuff injuries; stress fractures are common to gymnasts.

Some forms of contact sports such as American football have such alarming injury rates that we may well question whether the injuries incurred are indeed accidents. Indeed, we have come to expect them to occur, as evidenced by regulations in many schools requiring high school football teams to have a physician on the sideline and an ambulance waiting near the field. Habitual physical activity is also detrimental when people develop an unhealthy emotional dependence on exercise. Those with a physical activity addiction focus so single-mindedly on an activity that they neglect daily obligations important for a healthy and well-balanced life. When does devotion to an activity such as running or golf become addictive and harmful? No clear boundaries separate healthy from unhealthy exercise, but concern might arise when physical activity interferes with work or harms important personal relationships. Another signal is the onset of emotional disorders when the opportunity to engage in the activity is taken away. Many avid runners, for example, experience acute bouts of depression when they are injured and unable to run. High school wrestlers who must "make weight" to compete within a specific weight category often engage in harmful practices. Young women interested in gymnastics or figure skating can become preoccupied with weight loss and continue to exercise when body mass is already minimal. As with most other activities in our daily lives, the best prescription for a happy, healthful, and balanced life is to pursue physical activity with a sense of dedication and commitment, but in moderation.

➤ When pursued in moderation with an eye toward a balanced life, physical activity is desirable. When performed under circumstances that put the integrity of the body at risk or induce questionable behavior patterns and psychological states, it is undesirable.

In response to the rash of sport- and exercise-related injuries, kinesiology departments have cooperated in establishing excellent programs of athletic training and sports medicine. As you will see in chapter 14, preventing and treating athletic and exercise injuries is one of the fastest-growing specializations in the field of kinesiology.

The Sphere of Competition

Do you like to compete against others? Does the opportunity to compete with classmates for the highest grade, to compete with coworkers for the worker of the month award, or to compete against others for the league championship heighten your interest in the activity? Chances are that competition plays an important role in some part of your life. And there is a good chance that physical activity figures prominently in whatever form that competition takes (see figure 2.13).

Competition is not itself an activity but rather an organizing principle for activity. It can add to or detract from the enjoyment of activities, but it also usually leads to increases in level of performance. Skipping stones on a pond can be an exciting experience for children; when they decide to see who can skip a stone the farthest, the activity becomes even more exciting. As they become more intent in their efforts to outskip their playmates, their stone-skipping performances may also improve. Some people thoroughly enjoy shooting baskets alone on the playground or playing golf by themselves or in leisurely social contexts with

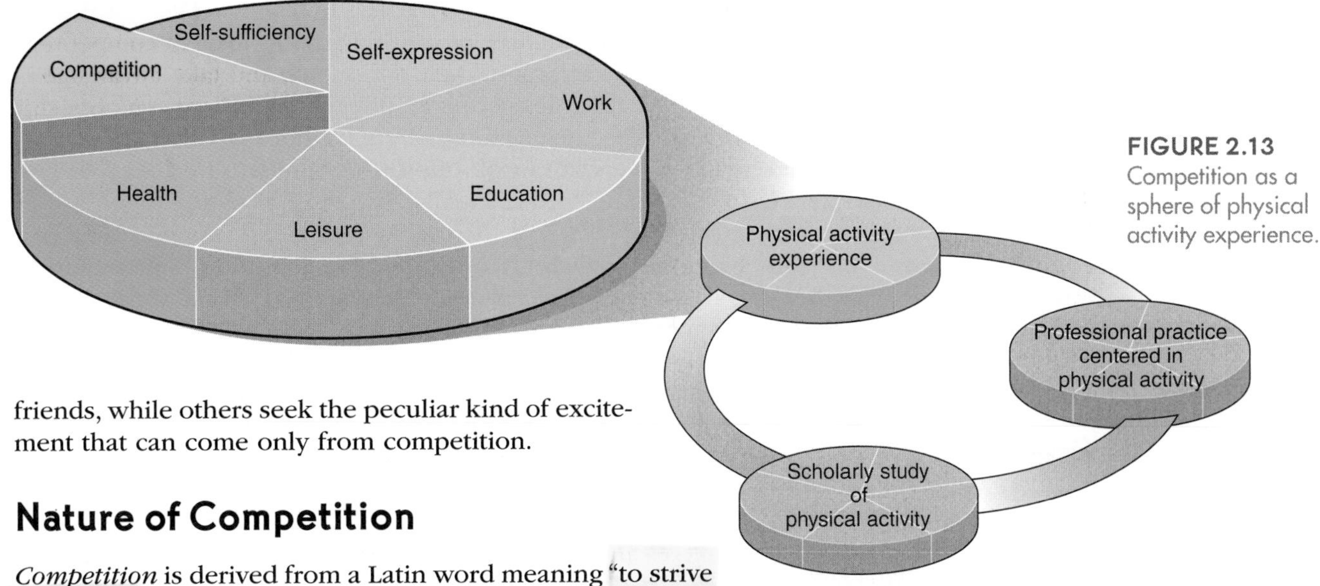

FIGURE 2.13 Competition as a sphere of physical activity experience.

friends, while others seek the peculiar kind of excitement that can come only from competition.

Nature of Competition

Competition is derived from a Latin word meaning "to strive together," but most of us think of it as "striving against." We compete *against* opponents. But if you think about it, competition between teams or individuals requires an element of cooperation or striving *together*. Cooperation is necessary to hold the game together. Sport philosopher Robert Simon defines competition at its best as "a mutual quest for excellence in the intelligent and directed use of athletic skills in the face of challenge" (Simon 2004, p. 24). Opponents who play their best help us to play our best. When players become spoilsports or give up, or when they cheat, the game falls apart.

When most people think of competitive physical activity, they think immediately of sports—physical activities that place a premium on skill, strength, endurance, or other physical qualities and in which one's performance is assessed in relation to a standard or the performance of others. Unlike physical activity in the other spheres of experience, most sports depend on the organizing principle of competition. Extract competition from most sports and you are left with goalless, aimless physical activity. For this reason we have designated a separate sphere for competition.

Competition can be added to physical activity in different ways (see figure 2.14). In **side-by-side competitive activities** such as golf, swimming, running, bowling, shot-putting, or cycling, no direct interaction takes place between competitors. Competitors do not interfere with each others' chances of succeeding; in fact, doing so is against the rules. In **face-to-face noncontact activities** such as

FIGURE 2.14 The four types of competition: *(a)* side by side; *(b)* face to face noncontact; *(c)* face to face contact; and *(d)* impersonal.

(a) Stockbyte, *(d)* Brand X Pictures

volleyball, baseball, tennis, or racquetball, competitors interact by trying to maximize their own chances and decrease their opponents' chances of winning. In this form of competition, players deliberately attempt to thwart the efforts of their opponents and take advantage of their competitive weaknesses, although the rules permit no direct contact between opposing players. In **face-to-face contact activities** such as football, rugby, hockey, or wrestling (and increasingly soccer and basketball), players thwart the efforts of opponents by direct physical manipulation (blocks, tackles, checks, and so on). Finally, **impersonal competition** occurs when participants compete against records set by themselves or others. Trying to set a new record for the time taken to swim the English Channel and trying to climb a mountain by a route never before taken are activities in which one's competitors (holders of the record) are not present. In fact, they may not even be alive!

Problems arise with competition when it becomes more than an organizing principle to increase our enjoyment or push us to better performances. Often it becomes an end in itself. When coaches and players elevate the goal of winning a contest above the sheer enjoyment of the activity or the basic values of friendship, caring, and cooperation, they convert competition into an end as opposed to a means to an end. Mihalyi Csikszentmihalyi, a psychologist who has taught us much about sport, says, "The challenges of competition can be stimulating and enjoyable. But when beating the opponents takes precedence in the mind over performing as well as possible, enjoyment tends to disappear. Competition is enjoyable only when it is a means to perfect one's skills: when it becomes an end in itself, it ceases to be fun" (Csikszentmihalyi 1990b, p. 46).

DO it **Activity 2.7**

Competition
Check your understanding of the four types of competitive activity in the online study guide.

Competition is a spice added to physical activity to make it fun and to push us to better performance. We are intrinsically fascinated by matching our talents with those of others or against impersonal standards. When competition becomes an end in itself rather than a means to an end, however, the activity may no longer be fun, and in some cases it can be harmful.

Identifying Good and Bad Competition

When competition goes well, sport is a beautiful thing. When competition goes astray, sport can be ugly. Briefly explain how you would determine if competition was contributing to or detracting from the beauty of sport in each of the following situations.

- A player's conduct after losing a game
- A player's conduct after winning a game
- How players conduct themselves before a game
- A coach's conduct before, during, and after a game
- The coach's behavior during the week before a game
- How a coach goes about recruiting a star athlete
- The crowd's reaction to an obviously incorrect call by a game official
- The crowd's reaction when their team wins the World Series or Super Bowl
- A parent's reaction to his or her child's failures in Little League baseball
- A university's policy regarding awarding scholarships to students

Popular Competitive Pursuits

Many people begin playing sports at an early age and continue to play throughout childhood. Recent estimates are that half of all children in grades 9 through 12 are members of at least one sport team run by a school or other organization (USDHHS 1996). In many cases this early contact results in a lifetime commitment to a sport. This circumstance accounts at least partly for the fact that approximately 90 million people aged 18 and older participate in at least one sport more than once a year (USBOC 2002d). Everywhere we look we see people playing sports—on playgrounds, on high school and college athletic fields, and especially on television. This pervasiveness of sport often leads people to conclude that participation in sport is on the rise. But is it? To the extent that we can accurately assess participation in sports, the best answer is yes and no.

A survey by the National Sporting Goods Association of sport participation rates of those 7 years old through 17 in the 10-year span from 1997 to 2006 presented a somewhat alarming picture (see table 2.1). Although participation increased across all sports nearly 10% a closer examination shows that this was due primarily to increases in tackle football (+44%) and ice hockey (+33%). During this period there were negligible increases in participation for baseball, soccer, and bowling and there were decreases—in some cases substantial—in participation in basketball, tennis, golf, softball, and volleyball (–38%) (NSGA 2007). When the view is narrowed to include only high school athletics, the picture is somewhat different (National Federation of State High School Associations 2007). School athletics offers excellent opportunities for young people to become engaged in competitive sports, and according to the National Federation of State High School Associations the number of participants has risen for the past 18 straight years. To a large extent these increases reflect a growing number of high school girls choosing to compete in sports (currently over 3 million). Although the annual increases have been modest (between 1.5% and 2% each year for the past three years), they are encouraging, given the fact that decreases were reported during 10 different year-to-year spans from 1978 to 1993.

Table 2.1 Changes in Sport Participation Rates for Children Ages 7 to 17 From 1997 to 2006

Sport	Percent change
Baseball	+3.5
Basketball	−12.8
Bowling	0
Tackle football	+44
Golf	−6.8
Ice Hockey	+33
Soccer	+2.7
Softball	−24
Tennis	−6.8
Volleyball	−38

Adapted from NSGA, 2007, National Sporting Goods Association: Sports Participation 2006. Available: www.nsga.org/public/pages/index.cfm?pageid=864. Accessed Sept. 2007.

➤ Participation in many kinds of sports has decreased dramatically over the past 10 years. This development is apparent in participation by both the 7-year-old to adult group and the 7-year-old to 11-year-old group. More encouraging participation rates are seen in high school athletics, particularly among high school girls.

Wrap-Up

We have covered a lot of ground in this chapter. We began by recognizing that although kinesiology focuses primarily on exercise and sport, kinesiologists often study a much broader spectrum of physical activity. By now you should realize that physical activity is important not only to our recreational lives but also to virtually everything we do. You should know about the various dimensions or spheres of daily living in which physical activity plays a vital role. We saw, for example, that physical activity is central to carrying out the daily chores through which we remain self-sufficient. It also is the means by which we express ourselves and do our work. Physical activity also plays a significant role in education, leisure, health, and our competitive pursuits.

In examining the spheres of physical activity experience, it becomes apparent that we rely on physical activity every day, not only to survive but also to live full and rich lives. Surely, physical activity has many practical benefits, although we do not seek these benefits through physical activity as much as we should. Having said this, it would be misleading to leave you with the impression that physical activity is important only because it is useful to achieving some external ends. Physical activity may be as important to us for the special types of personal experiences it offers as for any benefits we derive from participating. Put simply, life becomes richer as we immerse ourselves regularly in a variety of interesting physical activity experiences. We will explore this personal or subjective side of physical activity in chapter 4. But first let's take a detour to help you appreciate the vital importance of physical activity and learn how physical activity professionals use it in the service of their students, clients, and patients.

KEY▸ Activity 2.8

Use the key points review in your online study guide as a study aid for this chapter.

QA Activity 2.9

These end-of-chapter questions and activities are also in your online study guide. Your instructor may ask you to complete them online and turn them in.

1. Why are ADLs and IADLs important to kinesiologists who work with those who are elderly and disabled?
2. What type of physical activity professional is likely to be involved in treatment of an individual with carpal tunnel syndrome? What professional is likely to be involved in redesigning the workplace to reduce the risk of carpal tunnel syndrome?
3. What are gestures, and what purposes do they serve in our daily living?
4. Why is public school physical education important? What objectives do physical education teachers pursue?
5. List three health benefits of regular physical activity.
6. What does the element of competition add to physical activity? When is this helpful, and when might it be harmful?
7. Describe a situation in which physical activity may help nourish and maintain a state of leisure. Describe a situation in which physical activity may diminish the possibility of attaining a state of leisure.

The Importance of Physical Activity Experiences

3

Shirl J. Hoffman

AP Photo/Armando Franca

CHAPTER OBJECTIVES

In this chapter we will

- help you gain an appreciation for physical activity as it contributes to and expresses the signature of humanity;
- encourage your appreciation for the various ways that experience can expand this human potential for physical activity;
- help you develop insights into factors that should be taken into account in prescribing physical activity experiences;
- explain some basic principles governing the relationship among skill, practice, and learning and among physical capacity, training, and conditioning;
- discuss the relationship between physical activity experience and physical fitness; and
- discuss how heredity and other factors can modify the effects of experience.

Every once in a while an athlete comes along whose talents and skills and dedication to his or her sport are truly remarkable. Mia Hamm is one such athlete. A member of the University of North Carolina's four national championship teams, she went on to achieve fame as a star player for the U.S. Women's National Team when it won the Fédération Internationale de Football Association (FIFA) World Cup in 1991 and again in 1999 before 90,000 spectators in the Rose Bowl and tens of millions around the globe. That same year she broke the all-time international goal record by scoring her 108th goal in a game against Brazil. Five years later she extended that record by scoring her 150th goal. She played a major role in winning gold medals for the U.S. Olympic Team in 1996 and 2004. She has been identified by FIFA as one of the 125 greatest living soccer players. In 1999, the Nike Corporation named its largest building after her.

How did Hamm become such a great player? Giving the question some thought, you may have concluded that she possessed the innate abilities and inherited traits that are perfectly suited to the game. Certainly innate ability is a critical factor (a point considered later in this chapter), but a more likely explanation might be her extensive experience with the game. She was introduced to soccer when she was a toddler living with her family in Italy. Beginning at age 12 she became absorbed in soccer and two years later was a member of a regional team in Texas. At age 15 she became the youngest player ever on a U.S. national team. Hamm was engaged in the kinds and amounts of physical activity experiences that inevitably lead to improvement in performance of soccer. Why she chose to become engaged in soccer at such an early age is attributable to many factors. Certainly, personal factors are a major determinant, but so are social factors. For example, Hamm happened to have been born the same year the U.S. Congress passed Title IX, a law that opened the door for women to participate in athletics at all levels. Without this law she might never have had the opportunity to develop her full potential in this sport. In this chapter we will explore in a general way how our experience with physical activity can transform our bodies and our physical performance.

The word *experience* can have many different definitions, so we should first clarify precisely how we will use the term in this chapter. One definition is "individual reaction to events, feelings, etc." (Neufeldt 1988, p. 478). In this sense we may "experience" sadness, joy, beauty, and a host of other impressions. This version of experience, which we will call **subjective experience,** is a topic for the next chapter. This chapter centers on experience defined as any "activity that includes training, observation of practice, and personal participation" (Neufeldt 1988, p. 478). We will call this **activity experience,** which includes our actual physical performances or observations of physical activity. If you are *experienced* as opposed to *inexperienced* in skiing, you will have skied on many occasions in many contexts and probably with many different people. Mia Hamm, for example, has a tremendous depth of activity experience with the physical activities of soccer. Subjective experiences always accompany activity experiences, but here, for purposes of discussion, they are separated merely for convenience.

➤ Physical activity experience refers to our history of participation, training, practice, or observation of any particular physical activity. It differs from subjective physical activity experiences, which are our reactions, feelings, and thoughts about physical activity.

As you move through this chapter, you will be reminded repeatedly of the critical importance of experience in our efforts to expand our capacities to perform physical activity and to improve health and psychological well-being.

As central as physical activity is to human life, we do not come into the world readily equipped to do it. Birth may innately endow us with an ability to move, but developing control over our movements requires experience. Newborns spend a lot of time kicking their legs and waving their arms; these are aimless and spontaneous movements that occur without prompting or encouragement from their parents. Infants also move automatically in response to external stimuli. For example, holding a newborn in the prone position over water may elicit swimming-like movements of the arms and legs; applying pressure to the sole of the foot elicits a crawling pattern; and turning the head of a supine infant to one side causes the arm and leg on the same side to extend. Much experience and maturation are required to convert these spontaneous and automatic movements into what we know as physical activity.

In other words, physical activity experiences are the essential means by which we increase our capacity to perform physical activity. Think back to a time when you were first learning a skill. It is likely that your movements were jerky and inefficient. With practice, your movements smoothed out, allowing you to be not only more accurate but more forceful. Usually this involves an important change in the sequence of movements we use. Notice in figure 3.1*a,* for example, that the person skates in a stiff fashion. Although this increases her chances of keeping her balance, it severely limits the speed she is able to develop. The experienced skater shown in figure 3.1*b* moves the hip, trunk, legs, and arms in a way that enables him to generate maximal force.

a

b

FIGURE 3.1 Physical activity experience with skills leads to more powerful movements.

Almost everybody engaged in physical activities wants to be able to perform them faster, more forcefully, for a longer time, or more skillfully. The only way to do this is to expose ourselves systematically to more and more appropriate physical activity experiences. Such experiences enable people enrolled in fitness programs to become stronger or increase their cardiorespiratory endurance. Physical activity experiences also improve the skills of athletes, dancers, factory workers, military personnel, and students.

Physical activity experiences also are important because they are the means by which we achieve the expected health and psychological benefits of physical activity. These benefits usually go hand in hand with increased capacity for performing physical activity, but not always. As we will see, physical activity often leads to a reduction in body weight, a decline in percentage of body fat, and decreased stress levels, but it is possible to achieve these changes without significantly improving physical activity performance or, for that matter, without engaging in physical activity at all.

In part II of this text you will study the scientific bases for the effects of experience by learning how physiological systems respond to repeated performances of activities. You will appreciate at a deeper level why training experiences that stress the cardiorespiratory system will improve your capacity for running a mile in a shorter time or why your capacity for lifting a heavier weight will improve if you progressively increase the resistance against which your muscles work. You will also come to understand why we become better tango dancers, typists, or golfers by immersing ourselves in well-organized practice routines.

Obviously, not all types of physical activity experience induce positive changes. Poorly planned training or practice programs, for example, may not help you achieve your goals. In fact, they may waste your time and energy or, worse, retard progress toward your goals by causing injury or exposing you to unnecessary harm. Trained physical activity specialists are important safeguards against such dangers. As you progress through your undergraduate program you will come to view physical activity professionals such as physical therapists, coaches, physical education teachers, and personal trainers as experts in the systematic manipulation of physical activity experiences based on scientific principles, for the purposes of improving the skill, performance level, and health and well-being of those they serve. *(Manipulation* as the term is used here refers to skillful handling of physical activity experiences rather than exerting control over others.) Scholars and researchers in kinesiology also are in the experience business; through their work they identify the types of activity experiences that are most likely to produce the intended results. Many of the courses required for kinesiology majors center on ways to safely and effectively manipulate physical activity experiences to bring about positive changes in clients' capacities for physical activity.

➤ The central concern of all kinesiologists and physical activity professionals is the systematic manipulation of physical activity experiences based on sound scientific principles to bring about improvement in skill, performance, and health and well-being.

Physical Activity as a Signature of Humanity

Although it is tempting to think of physical activity experience as merely the mundane repetition of muscular activity, such a limited view wouldn't do justice to this far-reaching and important concept. Actually, physical activity draws on aspects of humanity that help define us as the highest order of living creatures. Were it not for this connection, physical activity experience would not work its marvelous effects. Let's step back from the topic of experience for a moment to explore how physical activity helps define our humanity and how our innately human capacities enable us to develop our physical activity capacity through experience.

DO it Activity 3.1

Human vs. Animal Physical Activity

Go to your online study guide to visually explore the uniquely human aspects of physical activity.

What characteristics set humans apart from animals or machines? Your first thought might be that we possess souls or a moral sense, or that we are able to engage in complex forms of social communication, or that we possess a capacity for high-level reasoning that animals and machines can never replicate. Good cases can be made for each of these arguments. What you probably didn't think about, however, is that our extraordinary capacity to translate complex mental concepts into precise and creative physical actions is also a defining human characteristic.

➤ A unique capacity for performing physical activity is one of the major features that contribute to the distinctive character of the human race.

But what about animals? Don't animals have a capacity to link movements with intentions and plans for moving; can't they perform physical activity? Animals do engage in goal-oriented, voluntary movements. Chimpanzees in West Africa, for example, use small stones as tools to crack palm nuts using the same motor patterns as the villagers from whom they (apparently) learned the technique (Kordtlandt 1989). Even relatively small-brained birds seem to be able to match movements with intentions. Some years ago, a species in the British Isles learned to rip open the cardboard tops of milk cartons and drink the cream, forcing milk companies in the region to redesign the containers delivered to people's porches. A small ground finch can uncover food using its feet to push aside stones weighing 14 times its body weight, which is equivalent to a 200-pound (75-kilogram) human's moving a 1.5-ton (1360-kilogram) rock, by bracing its head against a large rock for leverage (Gill 1989). Yet, though animals can perform goal-oriented movements, the human capacity for physical activity is of a different order than the capacity of these and other lower animals. This is a rather bold assertion; let's look at it more carefully.

Recall that physical activity is movement that is intentional, voluntary, and directed toward achieving an identifiable goal. Understanding the technical definition of physical activity is essential to understanding the discussion that follows. Having reviewed the definition, you are now ready to consider four ways in which physical activity may be considered unique to the human species.

Intelligence-Based Physical Activity

First, because humans are big-brained, highly intelligent creatures, their physical activity tends to be rooted in more intricate plans and directed toward more sophisticated goals than is the case with lower animals. Motor plans or mental images serve as dynamic maps to guide our movements in skills as simple as performing a standing long jump or as complex as piloting an airplane, performing brain surgery, or catching a Frisbee. Animals, by comparison, entertain relatively simple plans. A cheetah, for example, can easily outrun a human in a contest in which the goal is simply to run as fast as possible; but, being unable to formulate a clear understanding of goals and constraints imposed by rules, a cheetah will not be able to adapt its extraordinary running skill to the complexities of an Olympic relay race or to the rule-bound game of soccer.

Ethically and Aesthetically Based Physical Activity

Because humans are essentially spiritual creatures possessing unique moral and aesthetic senses, we can use our movements to express our imagination and moral reasoning. We can use our movements to express beauty, joy, wonder, and other deep and complex moods. This statement isn't to deny that other animals have emotional lives. Chimps, for example, can become so depressed when their mothers die that they sometimes die too (Heltne 1989). Although some chimps can use sign language to express emotions and can create basic paintings, animals cannot translate emotions through muscular actions into symbolic works of art or other elaborate expressions of sorrow, joy, or wonder. If thousands of chimpanzees were each given a hammer and chisel, by mere chance one of them might be able to create a recognizable work of art, but never something as profound as Michelangelo's *David* or Rodin's *The Thinker*. Animals may engage in elaborate mating dances, but the choreography is the product of instincts, not intentional expression of mood in movement as occurs, for example, in *Swan Lake*.

Flexibility and Adaptability of Physical Activity

Human physical activity is distinguished from that of other animals by virtue of the unique combinations of movements permitted by our anatomy. At first, this may seem an exaggeration. After all, elephants are equipped for performing far more forceful movements

Humans are able to connect physical activity to complex aesthetic sensibilities.
Plush Studios/Blend/Robertstock.com

than humans are. Greyhounds can surpass us in speed, dolphins in swimming, and monkeys in agility. In what ways, then, can we say that humans hold a movement advantage?

Two properties of our anatomies give us great advantage in moving. First, with our upright posture and bipedal gait we are the only animals whose forelimbs have been totally freed from assisting with walking, flying, brachiating, or swimming. The human body has a foot specially constructed to bear weight and give leverage to the leg, a pelvis specifically designed for attaching strong muscles needed to help maintain bipedal balance, and thighbones to permit long strides in walking. Our ability to walk on two feet has been described as "the most spectacular physical trait of human beings" (LaBarre 1963, p. 73).

Second, humans have available a dexterity of movement made possible by a unique complex of hands, arms, shoulders, and stereoscopic vision. The human hand possesses a true opposable thumb, a much more beneficial construction than that of the typical primate hand with its five fingers all operating more or less on the same plane. This arrangement allows us to perform delicate grasping, manipulating, and adjusting movements not possible for others in the animal kingdom. But this is not the only advantage. Our hands are positioned at the end of a series of long arm bones, joined to an amazingly movable shoulder girdle that a dog or cat can only dream of having. (Can you imagine a dog, cat, or horse being able to scratch its back with its forelimb?) This movable shoulder girdle enables us to position our hands through an enormously large path in space. This upper arm advantage is comple-

Humans and Animals at Play

One form of physical activity engaged in by both humans and animals is play. Cats, dogs, dolphins, monkeys, and otters play. Penguins play a sort of "king of the hill," rats wrestle, and ravens can throw sticks. Although the purpose that play serves is still being debated, it likely is essential for human survival. Which of the following types of play are likely to be a part of the lifestyles of both humans and animals?

Royalty-Free/Corbis (above) and iStockphoto/Rebecca Ellis (below)

Exercise play involving wide locomotor movements such as running, jumping and climbing

Object play involving use of fine movements to move objects in some way

Construction play in which materials are combined to create a new product, as in clay modeling

Social contingency play, motivated by the pleasure found in producing contingent responses in others (peek-a-boo, tickling, etc.)

Rough and tumble play, which involves vigorous physical contact accompanied by playful signals that distinguish it from hostile aggression (wrestling, pursuit games, etc.)

Fantasy play, in which real or imaginary objects are given properties different from those they actually possess

Games with rules, in which behavior is guided by explicit rules

Adapted from S.T. Parker, 1984, Playing for keeps: An evolutionary perspective on human games. In *Play in animals and humans,* edited by P.K. Smith (New York: Blackwell Science), 271-293.

mented by a facility for stereoscopic vision (lacking in many animals) that not only gives us advantages in depth perception but also allows us to perform most of our movements in our field of vision.

Ability to Improve Performance Through Planned Experience

The fourth—and perhaps the most significant—distinguishing characteristic of humans is our ability to improve our capacity for physical activity through planned, systematic practice and training. Only humans possess the intelligence that allows them to use physical activity in planned, systematic, and scientifically verifiable ways as a means of improving their health, performance, or skill or as a means of physical rehabilitation. Obviously, the cardiorespiratory efficiency of the young lion or eagle improves as its hunting range expands, but the driving force of its activity is hunger and survival, not a systematic conditioning plan to improve the physiological functioning of its body. Also, although some species of animals teach their young methods of hunting and fishing, the methods are relatively primitive, lacking the sophistication of human plans.

In summary, the capacity to link physical activity to complex plans and deep moral and aesthetic sentiments, an anatomy that permits these sophisticated thoughts and feelings to be played out in complex sequences of skeletal movements, and a highly developed ability to plan and implement activity experiences to improve physical performance have contributed immensely to the advance of civilization. In this sense, physical activity enables us to explore, test, and manifest our humanity.

Our rational capacity for systematically manipulating physical activity experience has given rise to the need for kinesiology, a discipline devoted to the intelligent application of experience to the solution of physical activity problems.

Human physical activity is distinguished from that of lower animals by four characteristics:

- Ability to match movements to sophisticated plans to achieve more complex physical activity goals
- Ability to apply physical activity to artistic creations
- A high degree of flexibility and adaptation in physical activity made possible by unique anatomical structures
- Ability to plan experiences that lead to improvement in performance

Factors Influencing the Kinds and Amounts of Performance Experiences

Because our potential for physical activity helps us to express what is unique about our humanity, it might seem that each of us naturally would maximize opportunities to engage in it and incorporate it into our lives. But that isn't the case. Although it is natural for us to engage in physical activity, as we saw in the last chapter, relatively few of us are inclined to do so, at least regularly and in its more vigorous forms. There is an enormous disconnect between what we know about the benefits of vigorous physical activity and our behavior patterns. Likewise, few of us are inclined to tap our full potential for developing skill in a variety of physical activities. We know that when people are competent in a sport, they enjoy engaging in it during their leisure time, yet not many people devote themselves to the considerable practice necessary to reach even a moderate level of competence. It's probably fair to say that average adults perform most types of motor skills at levels far below their capabilities.

➤ Although we posses a unique facility for performing, planning, and implementing physical activity experiences to improve our performance of physical activity, most of us are not inclined to explore this potential to the fullest.

Why this paradox? Why is it that we do not explore to its fullest potential something so fundamental to our human nature and so beneficial to us? This is one of the most pressing questions confronting the physical activity professions today; and as you will see when you study chapter 9, it is an area of study central to exercise psychology. Unfortunately, we have no simple answer. Many factors intervene in our lives to encourage us to explore physical activity and also to discourage us from doing so. Among the barriers that keep us from living physically active lives are lack of time, lack of access to an exercise or sport facility, and lack of a safe environment in which to be physically active. If you think back over your daily schedule for the past three days, you will probably see that one or more of these kept you from being as physically active as you could or should have been. But these aren't the only factors that influence our physical activity patterns. Let's look briefly at how two of many factors might affect the type and extent of our physical activity experiences: our social environments and individual circumstances (see figure 3.2).

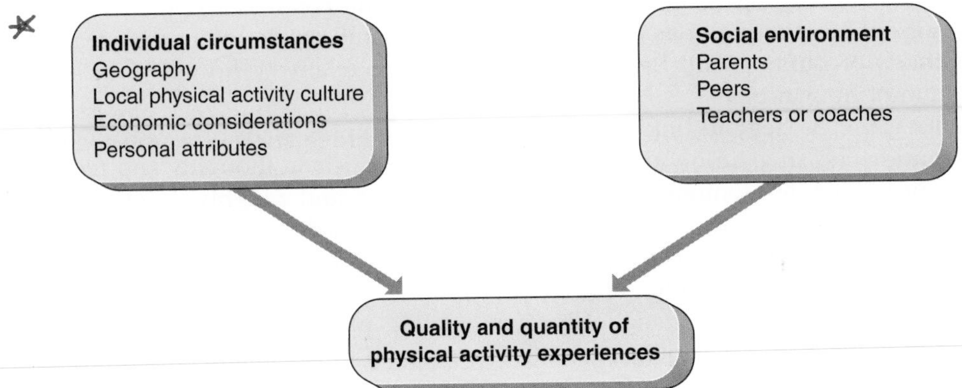

FIGURE 3.2 Many factors influence our decision to engage in physical activity.

Social Environment

The people with whom you interact on a regular basis can have a significant effect on the types and amounts of physical activity experiences you pursue, both as a child and as an adult. As a child you may have received substantial social support to engage in certain kinds of physical activity, causing you to be physically active in your younger years. Friendships and social alliances have been shown to have a strong influence on the amount of vigorous exercise undertaken by both young and middle-aged men and women (Sallis & Hovell 1990; King et al. 1992). If your girlfriend or boyfriend has a physically active lifestyle, you probably feel encouraged to do the same. If your boss plays tennis or racquetball, you might also be attracted to the game. Although each of us is ultimately responsible for our decisions regarding physical activity, those with whom we associate can affect our dispositions toward it.

Parents

When you were a child, did your parents lead a physically active life or were they largely sedentary? If they modeled an active lifestyle, took you with them to their workouts at an exercise facility, took a daily run in your neighborhood, or were avid tennis or squash players, they can take some credit for the fact that you are an active person too. If, on the other hand, they were "couch potatoes" who avoided sport, exercise, and other physical activity, you may have had to resist their adverse influence. The extent to which preschoolers, middle school children, and adolescents become involved in physical activity seems largely related to the extent of their parents' involvement in physical activity. Researchers have found, for example, that the activity levels of children who were participating in an obesity treatment program bore a remarkable correspondence to the activity levels of their parents (Kalakanis et al. 2001).

Parents who value physical activity and encourage their children—especially preschoolers and adolescents—to be physically active are more likely to have active children. One study has shown, for example, that children whose parents have a high perception of their (the children's) competence in a physical activity and view it as important are more likely to put forth effort in a running program at their school than those whose parents do not value it or do not have a high opinion of their child's competence (Xiang, McBride, & Bruene 2003). This finding is a good point for future parents to keep in mind. Merely showing support for their children's physical activity by transporting them to physical activity settings or assisting with the organization of the activity (e.g., serving as a Little League coach or sponsor of the cheerleading squad) may increase the probability that their children will be physically active (U.S. Department of Health and Human Services [USDHHS] 1996).

But how likely is it that our early childhood experiences will have an effect on our physical activity patterns later in life? Interestingly, researchers have not been able to discover a strong link between the amount of physical activity engaged in during youth and the amount engaged in during adulthood. We have reason to believe, however, that the *types* of activities engaged in during younger years may set a course for the selection of activities in later years. For example, one study showed that adolescents and adults who participated regularly in physical activity had become active participants by age 8 (Snyder & Spreitzer 1976). Another study showed that the best predictor of the sport experiences we will pursue as adults is those that we pursued as children (Greendorfer 1979).

Having said this, we must caution that a strong relationship between the extent of sport participation in youth and the extent of physical activity participation in adulthood has not been established. For example, participating in school or college sports does not appear to influence the amount of physical activity that one engages in as an adult (Brill et al. 1989; Cauley et al. 1991; Dishman & Sallis 1994). You are probably familiar with people who were star athletes in college and became physically inactive following graduation, just as you can probably think of individuals who did not lead physically active lives as children but became avid adherents of exercise and sport in later years. Obviously, we have much to learn about this connection. Until a more complete picture emerges, however, it seems best to ensure that constructive physical activity experiences are integrated into the early years of development.

Peers

The extent of your friends' involvement in physical activity also may have influenced your physical activity decisions. Do you hang around people who like to be physically active? If so, you are likely to be physically active too. For example, some researchers believe that the physical activity patterns of your peers may be as important as those of your family in determining whether you become involved in sport (Lewko & Greendorfer 1988). When we move beyond sport to consider the influence of peers on patterns of general (nonsport) physical activity, the precise influence of peers is less clear, although it appears to be an important determinant (Dishman, Sallis, & Orenstein 1985). (See "Spotlight on Research" on page 74.)

As you grew older, did you come into contact with more physically active peers, or did those around you become more inactive? Unfortunately, the statistical probability is that your

What Factors Have Influenced Your Physical Activity Patterns?

What leisure-time or competitive physical activities did you participate in as a child? Did you play Little League or soccer or football? Did you take private dance or gymnastics lessons? Did you ride bikes, climb trees, or roller skate with your friends in the neighborhood? Did physical education motivate you to engage in physical activity outside of class? Did you participate in the activities of the YWCA or the YMCA? Do you think that these early physical activities influenced your engagement in physical activity as an adult? Do you see similarities in the general types of physical activities that you pursued as a young child and those that you pursue now? What factors have contributed to your continuing (or discontinuing) engagement in these activities?

peer group became less rather than more active as you grew older. For example, a national survey revealed that 18% to 19% of 18-year-old high school students reported not being engaged in vigorous physical activity during the preceding seven days (USDHHS 1996).

If your peer group is immersed in online social networking, playing video games, or other sedentary pursuits, you will likely lead an inactive lifestyle too, rather than allow various types of vigorous physical activity to form the nucleus of your social life. As you grow older, peer groups may continue to affect your physical activity decisions. If you are married, your spouse and your spouse's friends are likely to influence your decisions to participate or not participate in physical activity (Loy, McPherson, & Kenyon 1978). Your relatives', friends', and social groups' predispositions toward physical activity, indirectly but predictably, will probably continue to influence your decisions throughout life.

Teachers and Coaches

If you were an athlete in school, your coaches—of youth sport teams, junior high school teams, high school varsity teams—probably had a significant effect on your physical activity decisions. Athletes, for example, tend to associate their participation in sport with the earlier influence of coaches, especially during adolescence (Ebihara, Ideda, & Myiashita 1983). Physical education teachers and coaches are influential because they are in a position to confirm or disconfirm a young person's competency in an activity. To have your father or mother remark that you performed well in a soccer game will probably encourage you to continue to develop your skills in that sport, but a similar remark from the coach is likely to be more influential.

➤ The people closest to you, including your parents, peers, coaches, and teachers, are major influences on the kind and amount of experience you have with a particular physical activity, especially when you are young.

Obviously, physical education teachers and coaches also can create social environments that discourage young people from seeking out physical activity experiences. In this sense, teachers and coaches may act as gatekeepers to the physically active life for thousands of children each year. Thus, it is not an exaggeration to say that the behavior of a teacher or coach may have profound and lifelong effects on a student's physical activity experience.

Spotlight on Research: Factors Influencing Physical Activity of Older Adults

A recent study (Wilson & Spink 2006) clarified the social influences affecting physical activity patterns in older adults. An important influence was the suggestions and affirmations regarding their activity from others, including friends. A second major influence was friends who were physically active. One study participant said, "I wouldn't have been doing it if it hadn't been for my friends doing it as well." Watching others who are active also appeared to be a motivating factor. One woman said, "I look over my balcony and see some of you people striking out for a walk and think I should be doing that." Finally, having someone—especially a health care professional—order participants to be more active physically was found to be a strong influence on their behavior.

Individual Circumstances

Individual circumstances, such as the availability and accessibility of facilities and play spaces, can determine the amount and kinds of people's physical activity experiences (Garcia et al. 1995; Zakarian et al. 1994). Geography is another important factor (see the sidebar on page 76). Those living in northern climates are more likely to develop competency in skiing, skating, and outdoor activities such as hunting, hiking, and fishing than they are in swimming or golf, simply because the weather dictates the availability of appropriate sport environments. Temperate climates with warm, sunny days tend to encourage jogging, in-line skating, and walking. Sometimes local cultural traditions emphasize particular sports. High school football is uniquely important in western Pennsylvania and Texas; boys' wrestling and girls' basketball are popular in Iowa. The game of the inner city is basketball, a cultural tradition owing at least partly to the fact that space isn't available for sports such as baseball and football.

Economic considerations can also dictate activity preferences. Generally, highly educated people in higher income brackets with high-status jobs tend to be more active than their poorer, less educated counterparts (see figure 3.3). Part of the reason for this disparity is that economic considerations often limit opportunities for participation in physical activities to those with higher income levels. The difference is especially noticeable in sports requiring expensive equipment, high admission fees, or, in cases such as swimming, transportation or admission to fee-based private clubs. The effect of income on participation in skiing and golf is particularly noticeable, with over 30 times more participants in skiing and 5 times more participants in golf from the upper income bracket as compared to the lowest. Although differences in participation rates are relatively small for exercise walking (an inexpensive activity), those at the highest income levels are three and a half times more likely to exercise with equipment, probably at health spas requiring fees that low-income populations cannot afford.

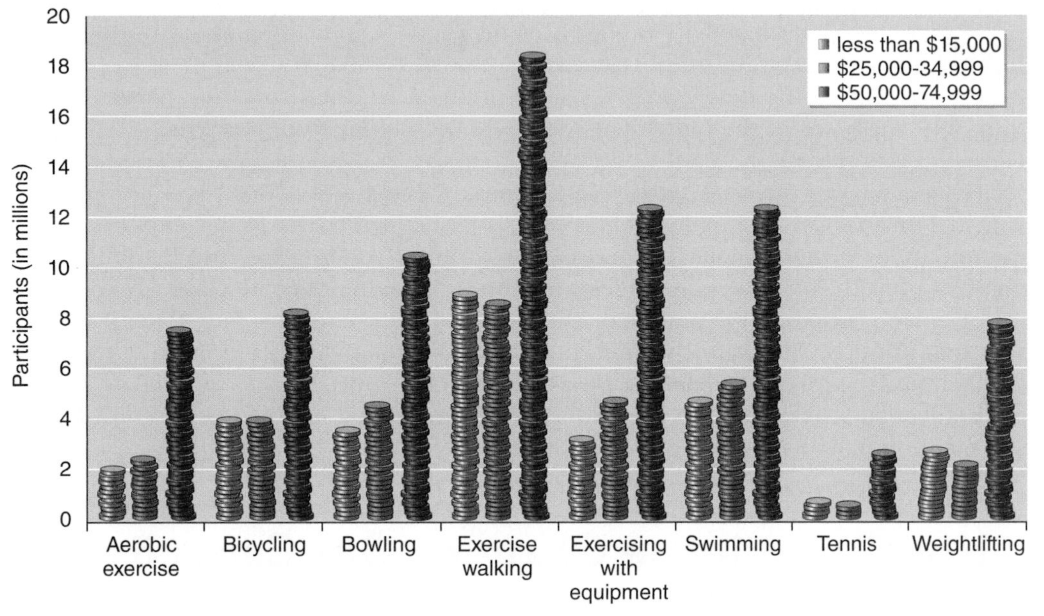

FIGURE 3.3 Influence of income on participation in selected sport and exercise activities.

Adapted from U.S. Department of Commerce, 2007.

➤ Often, our physical activity experiences—and consequently, those in which we develop proficiency—are determined by factors that lie outside of our control, such as climate, regional culture, and economic considerations.

Our own attitudes and personal attributes also affect our physical activity patterns. As we have noted, decisions to become involved in physical activities are not always outside your control. Your own perceptions, feelings, and decisions play a major part in the process. For example, a decision to prepare yourself for a marathon or train for bodybuilding competition may have to do with your personal attraction to certain unique characteristics of these activities. If you participate in hang gliding, it may have something to do with your willingness to take risks, the exhilaration you feel from heights, and your perceived ability to keep the glider in the air. As we will see in chapter 4, we tend to involve ourselves in activities that we enjoy, and only with difficulty can we pin down precisely those factors that cause us to enjoy some activities and shun others. Available evidence allows the suggestion that decisions to become involved in or avoid physical activities may have to do with our feelings of self-esteem and our perceptions of our competence in the activities. Nevertheless, an almost endless number of personal factors can affect these perceptions.

In the end, it may be as difficult to determine precisely why one person chooses to exercise regularly whereas another chooses to avoid exercise, or why one chooses to play tennis whereas another prefers water skiing, as it is to determine why some people like broccoli and

Where You Live Affects Your Physical Activity Patterns

- If you live in Colorado, you are nearly three times more likely to engage in snowboarding and nearly twice as likely to hike than the average person is.
- If you live in Montana, you are three times more likely to participate in alpine skiing and four times more likely to hunt than the average person.
- Illinois residents are more than one and a half times more likely to play baseball and softball than the average person.
- Delaware residents are nearly five times more likely to be volleyball players and nearly twice as likely to be bicycle riders compared to the average person.

National Sporting Goods Association, 2007.

➤After we have accounted for all the factors in our social and ecological environments that affect our decisions about physical activity, we are left to consider the indeterminable factors lying within us, such as our perceptions of ourselves, our competency in the activity, and the activity itself, that also may affect our decisions about physical activity.

some can't tolerate it or why some people like heavy metal music and others prefer classical. We have seen that physical activity reflects something of our humanity. It also seems to be the case that our physical activity preferences may reflect something about our uniqueness as individuals. Thus, as a future physical activity professional, you will need to pay careful attention not only to the developing research in this area but also to the particular needs, desires, and attributes of the individuals who will be seeking your services.

Let's stop and retrace the path we've taken to arrive at this juncture. We've seen that the capacity to elaborate physical activity into complex plans and actions is a unique characteristic of humans. Coincidental to this is our remarkable capacity to use experience to our advantage to learn motor skills; to increase our strength, endurance, and flexibility; and to improve our health. Yet evidence suggests that most of us do not come close to exploring the limits of our physical activity potential. This idea led us to consider in a general way some factors that affect our decisions to incorporate physical activity into our lives. These influences included the effects of parents, peers, and teachers and coaches, as well as individual circumstances such as geographical region, cultural traditions, and economic considerations, not to mention our own attitudes and dispositions.

How Experience Changes Our Capacity to Perform Physical Activity

As a student who is considering majoring in kinesiology, you should realize how vital it is that you become familiar with the processes and mechanisms by which performance experience improves our capacity to perform physical activity. You will learn about theories and research that consider such information in detail in courses such as biomechanics; exercise physiology; motor learning, development, and control; and sport psychology. Before you begin this advanced study, however, you will find it helpful to step back and look at the effects of physical activity experience in broader terms. This perspective will provide you with a conceptual framework upon which you can build a more specialized knowledge of the scientific bases of physical activity.

Think back to a particular form of physical activity in which you had a special interest. How did your experience with this activity affect your capacity to perform it on successive occasions? If it was a skill such as volleyball, the effects of your accumulated experiences probably were most evident in changes in your ability to perform the sequences of movements associated with serving, setting, or spiking. If the activity was cardiorespiratory training, the effects of your accumulated experiences may have been evident not so much in changes in the nature of your movements but in your improved capacity to perform the activity over longer periods at greater intensity. These represent the two fundamental effects of experience: (1) improvement in skill through practice, which is called learning, and (2) improvement in physical performance capacity through a type of exercise called training, which is termed conditioning (see figure 3.4).

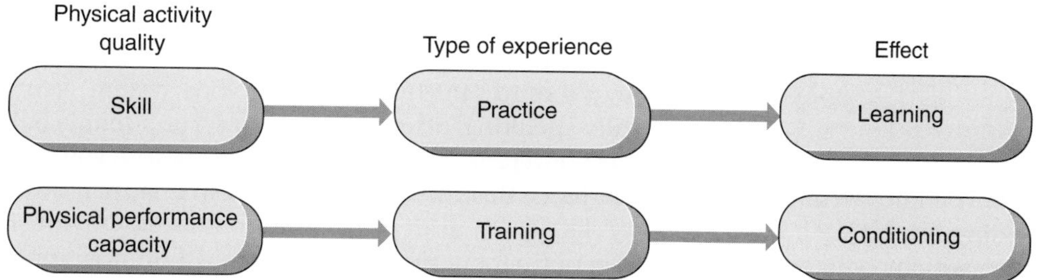

FIGURE 3.4 The nature of improvement in physical activity depends on the types of physical activity experience.

Skill, Practice, and Learning

Earlier we saw that infants face a lifetime challenge of recombining and refining their primitive motor responses into goal-directed voluntary actions to meet the demands of daily life. Throughout life we continue to confront the challenges of assembling sequences of movements and body positions to accomplish specific tasks. Activities in which performers attempt to attain specific goals by executing efficient, coordinated motor responses are called **motor skills.** Although each places distinctly different demands on our perceptual and motor processing systems, threading a needle, performing a cartwheel, and typing a letter are all motor skills. The quality of physical activity experience that underlies the performance of motor skills is **skill.** Skill is reflected in the efficiency and accuracy with which an individual is able to attain the goal of a motor skill.

Becoming a skilled mover is a never-ending process of gaining more control over our motor systems by gradually refining the nerve and muscle systems through performance experience. The type of performance experience engaged in for the express purpose of refining motor control function to improve skill is known as **practice.** Laypersons tend to think of practice as merely an endless repetition of the same movements by rote, but that isn't really how practice works. Practice of motor skills actually involves a lot of cognition. It involves higher brain function and integration of spinal neural activity as well as muscular action. (In fact, muscular action is merely the result of these higher cognitive processes.) Practice is a deliberate effort to "get it right" by modifying the erroneous movements we made on earlier trials and revising our cognitive plans for moving on the next trial. On each practice trial we vary the level of force or direction of our movements, adopt new strategies, and analyze feedback from our movements so that we may, for example, throw a ball farther or more accurately. This is why one scientist has referred to practice as "a particular type of repetition without repetition" (Bernstein 1967, p. 134).

➤ The type of physical activity experience that brings about changes in skill is called practice. The relatively permanent effects of practice are called learning.

Sometimes we refer to practicing as a process of acquiring experience, implying that we can collect and store our practice experiences. This metaphor is helpful. If we were unable to store memories of past physical activity experiences, each trial would be like our first. We would be in the same situation as people who have lost their short-term memory because of damage to an area of the brain known as the basal ganglia. Because such people can't store information about their previous response in memory, successive repetitions of a skill usually do not bring about learning. They may have played the piano 100 times, but each attempt is an entirely new experience. With a normally functioning brain, however, we can benefit from memories of our experiences and gradually reshape our motor responses in ways that allow us to attain skill goals more accurately and with greater efficiency.

The refinements in the nervous system that result from practice are referred to as **learning.** Although much has yet to be discovered about learning, scientists agree that learning is a central rather than a peripheral phenomenon. This is another way of saying that learning is a result of a rewiring of neural circuits in the brain and spinal cord rather than changes in peripheral organs such as bone or muscle. (Even though we often hear the term "muscle

Activity 3.2

You Be the Researcher
In this activity in your online study guide, you'll distribute a survey and chart the resulting data.

memory," it is the brain, not the muscles, that remembers.) We also know that these changes tend to be relatively permanent. If, after having played many years of tennis you do not return to the court for several years, the level of your play will deteriorate, but with a little practice you will regain it rapidly. Amazing, isn't it?

When you learn a motor skill, you actually remap the neural routes in your central nervous system. Neurologically speaking, after you learn to serve a tennis ball, you are a different person than you were before you mastered the skill. You know from your own experience that this remapping doesn't happen overnight. Depending on the skill, it may take weeks, years, or even decades to master. Obviously, the extent to which changes in skill become permanent depends a great deal on how much you practice the skill.

Physical Performance Capacity, Training, and Conditioning

Experience with an activity can bring about another type of change in our performance that is much different from learning. Improvements in physical activities sometimes have little to do with skill. In some situations, refining the accuracy or timing of your movements or coordinating them with features of the environment may be much less important than improving your capacity to exert greater force or increase the strength of muscles, the range of motion of your joints, or the length of time you can sustain an activity. These factors are not elements of skill as much as they are elements of physical **performance capacity.**

Consider the case of lacrosse players Katherine and Stacy. Both had devoted long hours of practice to the sport and had played in competitive leagues. Each had developed approximately the same level of skill in lacrosse, as evident by their equal proficiency in passing, catching, shooting, and guarding opponents. Then Katherine—a perceptive kinesiology major—began to supplement her practice with a rigorous training program designed to increase her physical capacity through development of muscular strength, muscular endurance, cardiorespiratory endurance, and flexibility, all of which she believed to be important in the sport of lacrosse. As a result, she became a more proficient lacrosse player than Stacy. Although both players had learned the *skills* of lacrosse to the same level, Katherine ended up the better player because she supplemented her skill development with improvement of other performance qualities that were important in the sport.

The supplemental types of experiences that Katherine engaged in were weightlifting, along with muscular and cardiorespiratory endurance and flexibility exercises. These are examples of a specific type of physical activity experience called training. The changes brought about by these experiences are known as **conditioning.** Training is physical activity experience designed to improve muscular strength and endurance, cardiorespiratory endurance, flexibility, and other aspects of performance that serve to condition the performer. (Do you recall our descriptions of the different types of exercise in chapter 1?) Whereas practice brings about improvement in skill through improvements in the accuracy and timing of movements, training brings about improvements in physical capacity through conditioning. Training increases our capacity to move our bodies more forcefully (strength), over longer durations (endurance), and in more flexible ways (flexibility). Generally, kinesiology majors study the experiences of practice that lead to learning in courses on motor behavior, and study the experiences of training that lead to conditioning in courses on physiology of physical activity.

➤ Physical activity experiences known as training are employed to develop such performance qualities as muscle strength and endurance, cardiorespiratory endurance, or flexibility. The state of having developed these qualities is known as conditioning.

Practice Versus Training

What are the differences between practice and training? Practice produces effects on memory, cognition, perception, and other central nervous system processes associated with problem solving. Training, on the other hand, produces effects that are largely peripheral to the central nervous system, usually on muscle, bone, soft tissue, and the cardiorespiratory system. Practice requires deliberate effort and intention by the performer to modify performance, but training per se generally requires little in the way of deliberate attention or problem solving. In fact, conditioning is often an unintended by-product of performing skills such as chopping wood or shoveling snow. Quite simply, some level of conditioning is inevitable—whether you intended for it to happen or not—if you repeat movements that appropriately stress muscles, tendons, and bones; elevate the heart rate to certain predetermined levels; or move your joints through large ranges of motion. On the other hand, mindlessly performing a skill, no matter how many times, is not likely to lead to learning.

Is practice or training the most appropriate physical activity experience for improving your performance in an activity? This depends on the relative importance of skill or physical performance capacity for success in the activity. If skill is the primary quality, practice is the most important experience; if physical performance capacity is the most important quality, training is the most important. If both skill and performance capacity are important, both practice and training are important. Even at this early stage in your education, you should be able to determine the extent to which an activity depends mostly on skill or physical performance capacity. Activities in which success depends primarily on learning precise and coordinated sequences of movements are located at the right end of the continuum in figure 3.5. Improvement in these activities is largely a product of practice. At the extreme left end of the continuum are activities that require little in the way of skill but depend heavily on factors such as strength, endurance, or flexibility. We improve our performance in these activities primarily by training experiences. Notice how activities might be distributed along the training-to-practice continuum.

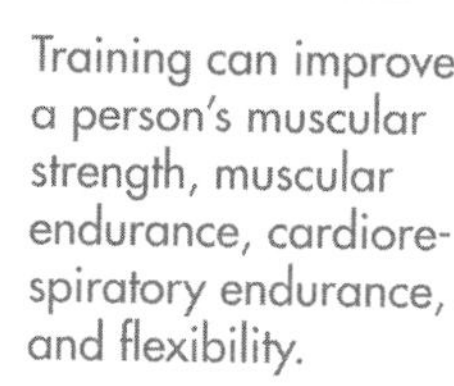

Training can improve a person's muscular strength, muscular endurance, cardiorespiratory endurance, and flexibility.

Digital Vision

FIGURE 3.5 Emphasis given to practice or training depends on the relative importance of skill or physical capacity factors in the activity.

DO it **Activity 3.3**

Practice or Training?
Distinguish between practice and training in this activity in your online study guide.

Exercise walking depends upon coordinated movements, but these usually have been learned by the second year of life. From that point on, improvement depends upon various aspects of performance capacity. An athlete's recovery from a postoperative rehabilitation program following knee surgery depends largely on training that develops strength rather than skill. The athlete has not forgotten ("unlearned") movements at the knee; rather, the reduced capacity of the muscles has weakened the movements. Gymnastics requires elements of skill but also depends largely on strength. Training and practice both are the appropriate experience in such activities and in basketball and golf, although strength plays less of a role in golf than in gymnastics. In rifle shooting or driving a car, success depends almost exclusively on learning to coordinate movements with visual information. (One would hardly try to improve one's driving by embarking on a weight training or long-distance running program!) Threading a needle is an extreme example of an activity in which improvement comes about through practice rather than training.

Because most physical activities incorporate both physical capacity and skill elements, improvement often requires both practice and training experiences. Rehabilitation programs designed to help individuals learn to walk following a stroke focus on practice; but because loss of muscular strength also may impinge on efforts to support the body, training may also be required.

➤ Generally, practice without training, or training without practice, is an incomplete formula for developing excellence in sport.

If your goal were to become a member of the varsity soccer team, much of your preparation would involve practicing the various skills of dribbling, tackling, and passing the ball. But because the game also requires you to run continuously, you would also need to train by running sprints and long distances to develop cardiorespiratory endurance. You also might train with weights, especially with your legs, to increase kicking force. Like many sports, soccer requires attention to both practice and training experiences.

Performance Experience and Physical Fitness

Training experiences are important not only for developing proficiency in a particular sport or rehabilitation activity but also for increasing overall capacity to perform activities of daily living. The term that we use to describe this capacity is **physical fitness.** As we learned in chapter 1, the type of exercise specifically intended to improve fitness, health, and well-being is called health-related exercise.

A physically fit person is one who has developed through exercise a capacity to perform the essential activities of daily living at a high level, has sufficient energy remaining to pursue an active leisure life, and is still able to meet unexpected physical demands that emergencies may impose. For example, our conception of a physically fit person is one who is physically able to carry out the demands of his or her job; play tennis, racquetball, or other sports on the weekend; and still be able to run downstairs in case of a fire, walk 5 miles on the interstate to get help when his or her automobile breaks down, or shovel snow from the driveway after a big storm.

Because this type of fitness is reflected in a capacity to *perform* physical activities, it is called **motor performance fitness,** which has been defined as "the ability to perform daily activities with vigor" (Pate 1988, p. 177). Motor performance fitness may be reflected in how many sit-ups or push-ups you can do; how long it takes you to run or walk a mile; or your ability to perform other activities that rely on strength, endurance, or flexibility. Because balance, agility, coordination, quickness of response, and similar factors are also important in performing daily activities, some kinesiologists include these as additional indicators of motor performance fitness. Unlike strength, endurance, or flexibility that are developed through training, these latter qualities are developed more by practice.

Definition of Physical Fitness

Although we all have a general idea as to what is meant by "physical fitness," the term is defined in many different ways. Here are a few examples:

- **Centers for Disease Control and Prevention:** Physical fitness is a set of attributes a person has with regard to his or her ability to perform physical activities that require aerobic fitness, endurance, strength, or flexibility and is determined by a combination of regular activity and genetically inherited ability (www.cdc.gov/nccdphp/dnpa/physical/everyone/glossary/index.htm).
- **Public Health Agency of Canada:** Physical fitness is a set of attributes that are either health related or performance (or skill) related. Health-related fitness comprises those components of fitness that exhibit a relationship with health status. Performance/skill-related fitness involves those components of fitness that enable optimal work or sport performance (www.phac-aspc.gc.ca/pau-uap/fitness/definitions.html).
- **Journal of the American Medical Association:** Five components of fitness include cardiorespiratory (heart and lungs) endurance, muscular strength, muscular endurance, body composition, and flexibility. Cardiorespiratory endurance is the ability to perform sustained physical activity, such as walking, swimming, or running. Muscular strength and endurance are linked and are improved through use of weight-bearing exercise, such as weightlifting or using resistance bands. The proportions of muscle, fat, and body water make up body composition. Flexibility is related to range of motion and is improved by gently and consistently stretching muscles and the connective tissues surrounding them ("Patient Page: Fitness" 2005, December 21).

What are the similarities and differences among these definitions of fitness? Which definition do you think comes closest to that discussed in the preceding paragraphs?

A second type of physical fitness, generally regarded by those in the health-related professions as more significant than motor performance fitness, is **health-related fitness.** This type of physical fitness refers to having developed, through physical activity experience, the traits and capacities normally associated with a healthy body, specifically in relation to diseases known to result from a physically inactive lifestyle (sometimes called **hypokinetic diseases**). Heart disease is one type of hypokinetic disease; high cholesterol, high blood pressure, and obesity also may be associated directly with an inactive lifestyle. Health-related fitness may be assessed directly using various technologies without actually assessing motor performance. Using a skinfold caliper to measure percentage of body fat is one example. In other cases, health-related fitness may be assessed while the subject engages in motor performances. For example, recording an electrocardiogram, blood pressure, and heart and respiration rates while a person exercises can give kinesiologists a fairly clear picture of a person's health in relation to cardiovascular-related hypokinetic diseases.

Measuring Physical Fitness

Most physical activity professionals, whether working in gymnasiums, clinics, fitness testing laboratories, fitness centers, or athletic conditioning programs, face the task of assessing the fitness of their clients. Because of this, knowing how to assess physical fitness is as important as knowing how to define it; in fact, how you decide to measure physical fitness is a clear reflection of your definition. How would you go about measuring fitness in a population of college students? If you approached the task by recording the number of sit-ups they could

Depth or Breadth of Experience: Generalists Versus Specialists

Do you have in-depth experience with a particular physical activity (a specialist) or a broad range of experiences with a number of different activities? If you are like most people, you are more of a generalist than a specialist. Limiting practice and training to one particular activity allows you to develop **depth of capacity** in that activity. On the other hand, exposing yourself to a broad range of different kinds of training and practice experiences increases your **breadth of capacity** for a number of different types of skills and activities.

Physical activity generalist: Physical activity generalists are individuals with experience in a broad range of skills and activities. A person who has developed low-average to above-average competency in, say, rock climbing, wrestling, ice skating, football, and baseball is an example of a generalist. The advantage that accrues to generalists is the enjoyment and satisfaction that come from being able to take advantage of opportunities to engage in a variety of activities. The disadvantage is that competence is not highly developed in any single activity, thus depriving the individual of the experiences that come from demonstrating excellence.

Physical activity specialist: Physical activity specialists are those who devote themselves to developing depth of capacity in a single activity or a narrow range of activities. The young girl who wants to become an Olympic gymnast must commit herself to training, practicing, and competitive schedules; she is unlikely to be found on the tennis court, golf course, or swimming pool on a regular basis. Tiger Woods is extraordinarily skilled in golf but apparently is not a great dancer. The advantages that accrue to specialists are the pride and satisfaction that come from being able to do one activity or a small number of activities at an above-average level. The disadvantage is that the individual misses out on opportunities to engage in a number of different activities. Concentrating efforts on a single physical activity can result in remarkable proficiency, as anyone who watches national- or international-class athletes can appreciate. These in-depth experiences can result in some amazing capacities.

- Through practice, Michael Kettman managed to spin 28 basketballs simultaneously.
- With intensive training, Anthony Thornton developed the capacity to walk a distance of 95 miles (153 kilometers) in 24 hours . . . *backward.*
- Chris Gibson developed the capacity to perform 3025 consecutive somersaults on a trampoline.
- Narve Laeret smashed 90 concrete blocks by hand in 1 minute.
- Practice and training allowed Ashrita Furman to walk more than 80 miles (over 23 hours) around a track in Queens, New York, balancing a milk bottle on his head.

These examples are striking testimony to the way in which in-depth training and practice experiences can improve the sophisticated nerve, muscle, and cardiorespiratory systems that allow us to perform.

perform in a minute, timing them on a 100-meter (109-yard) dash and in an agility test, or measuring range of motion of their spines or at their hip and shoulder joints, your conception of fitness is probably closely related to motor performance fitness.

If, on the other hand, you decided to measure their fitness by having them run on a treadmill for a specified period during which you monitored heart rate, blood pressure, oxygen consumption, and an electrocardiogram, and followed this with skinfold measurements to determine body fat and a test of blood cholesterol level, your definition is much more associated with health-related fitness.

Table 3.1 lists the various components assessed by motor and health-related fitness tests. As you can see, some of the motor fitness measures bear little relation to the health of the individual. Could an individual with a diseased cardiovascular system score high on a balance, strength, or agility test? Such a result is quite likely. As shown in the table, some components of fitness are assessed by both motor and health-related tests. Flexibility tests, for example, are a measure of motor performance capacity but also reflect something about the health of the joints. (Regular physical activity guards against the onset of diseases that limit motion, just as regular vigorous activity guards against diseases that destroy the circulatory system.)

Table 3.1 Components of Physical Fitness

Component	Motor performance	Health-related fitness
Anaerobic power	*	
Speed	*	
Muscular strength	*	*
Muscular endurance	*	*
Cardiorespiratory endurance	*	*
Flexibility	*	*
Body composition	*	
Agility	*	

* Indicates that the test measures this particular component of physical fitness.

Adapted, by permission, from R.R. Pate, 1988, "The evolving definition of physical fitness," *Quest* 40: 178.

Type of Physical Fitness

Which type of physical fitness should physical activity professionals be most concerned about? Generally, health-related fitness is of most interest to those in the medical community and, increasingly, to those in kinesiology. Given the growing interest among public health officials in the role of physical activity in disease prevention, the emphasis on health-related fitness is likely to continue. At the same time, it is reasonable to continue to conceive of physically fit persons as those who can effectively carry out their daily activities. This will become increasingly important as our population grows older. In light of this, assessment of both motor performance and health-related fitness would seem to be in order for most populations.

➤ Training experiences that improve our general capacity for performing daily activities and preventing disease processes associated with low levels of physical activity are known as fitness activities.

How Quality and Quantity Affect Physical Activity Experience

By now it should be clear that the route to improving your skill or physical capacity is selecting appropriate practice and training experiences. Merely engaging in random physical activity experiences is unlikely to bring about your goals for skill improvement or health enhancement. The type of physical activity experience must be matched appropriately with the desired performance improvement to bring about the intended effects. Thus, knowing how to match physical activity experiences with physical activity goals is a competency that practicing kinesiologists must master.

By studying exercise science, you will develop a clear understanding of the types of experiences and their underlying rationales. Two general principles can guide your thinking about the relationship between experience and physical activity: the principle of quality and the principle of quantity.

The **principle of quality** states that experiences that engage us in the most critical components of an activity are most likely to lead to increases in our capacity to perform that activity. It might be helpful to think of critical components as the elements of an activity that are most critical for performing the activity at a high level. As a first order of business, you must decide whether skill or performance capacity or both are critical to performance of the skill (see figure 3.4). For example, ask yourself how the principle of quality applies to lifting a heavy weight or serving a tennis ball. In the first case, your answer was probably "strong arm muscles" or, more appropriately, "strength in the large muscles of the entire body." In the second case, your answer may have been "good form," "the correct movements," or, more appropriately, "a coordinated pattern of muscle action that brings the hip, trunk, shoulder, and wrist joints into play at precisely the correct times to propel the ball forcefully into the opposite service court."

This principle suggests that if you want to be a weightlifter, your physical activity experiences must engage you in training exercises that focus on the critical component (strength). If you want to learn to serve a tennis ball, your physical activity experiences must engage you in practice that focuses on the movement pattern that you must learn. Running 4 miles a day will increase your cardiorespiratory endurance but will have minimal effect on arm strength. Thus running isn't an appropriate training experience for developing weightlifters because it fails to take into account the critical component of strength, just as pitching horseshoes is an inappropriate practice experience for learning the tennis serve because it involves a different movement pattern and skill goal.

According to the **principle of quantity,** when all other factors are equal, increasing the frequency of our engagement with the critical components of an activity usually leads to increases in our capacity to perform that activity. Generally, those whose activity experiences have engaged them *most often* in the critical components of an activity become most competent in that activity. Remember that the principle of quantity incorporates the principle of quality. Extensive experience—whether practice or training—that does not engage us in the most critical aspects of the activity is unlikely to result in substantial improvement in performance.

Whether your client is a postcardiac patient, an athlete recovering from injury, a teenager embarking on an exercise program to lose weight, a student in a physical education class, or an older adult hoping to improve her capacity for performing activities of daily living, as a physical activity professional you must first direct your attention to the critical components underlying the performance of the activity. The process of systematically identifying the critical components of an activity is called **task analysis.**

Let's take these concepts into the workplace. If Fred, a marathoner embarking on a training program to lower his time in the 26-mile, 385-yard (42.2-kilometer) event, asked you to plan training experiences that would help him achieve his goal, your first challenge would be to conduct a task analysis of the activity to identify its critical components. Quickly, you would conclude that the critical components in this activity relate more to physical performance capacity than to skill. This judgment, coupled with your knowledge of exercise physiology and clinical knowledge you have acquired as a practicing professional, would lead you to

Centering Practice on the Correct Critical Components

Based on your understanding of the principles of quality and quantity, which of the following coaching practices would you recommend? Think about whether the activities seem appropriate training experiences for the players. Specifically, what critical components of performance do they seem designed to develop? What critical components might a coach not have taken into account in preparing players for competition? What practice activities might have been more effective in developing skill?

- Basketball coach A devotes much of each practice session to having his players jump rope, lift weights, and shoot weighted balls at the basket.
- Volleyball Coach B has his players spend much of practice jumping sideways over benches, jumping rope, and tossing medicine balls.

conclude that Fred will need muscular endurance if he is to run for over 2 hours without fatiguing the muscles in his legs, and that he will need cardiorespiratory endurance if his circulatory system is to develop the capacity for delivering large amounts of oxygenated blood to the working muscles. Thus, training activities that emphasize these components would assume a priority in the training plan that you would prescribe.

When you begin to study kinesiology in depth, you will discover that the general principles of quality and quantity are subject to many qualifications. You also will discover that identifying the critical aspects of various physical activities and determining the appropriate types of practice or training experiences are not easy tasks. For example, what are the critical elements in batting a pitched ball? Is skill or physical capacity the primary consideration, or are both important? But beyond this basic question, is watching the ball the most important element? Developing force with the arms? Developing limb–target accuracy? Developing force with the hips and trunk? What are the critical elements that must be mastered by a patient learning to walk following a stroke? What components of a throwing motion must a pitcher rehabilitate following surgery? How much and what kinds of practice and training experience does each case require? Your competence as a physical activity specialist will depend a great deal on your skill at carrying out such analyses and using the results to plan appropriate practice or training experiences.

➤ Experience will improve physical activity capacity only if it engages us in the essential aspects or components of the activity (principle of quality). Generally, the more often our training or practice focuses on these components, the greater will be our performance of the activity (principle of quantity).

To understand this concept more deeply, consider the four possible choices a practitioner can make (see figure 3.6). The decision regarding which critical components (quality of activity) to emphasize can be correct or incorrect, as can the decision regarding the quantity of physical activity experiences to provide. When the practitioner selects the appropriate critical component or components and provides the maximal amount of engagement with these components, chances for success will be good (decision A). But if the practitioner provides maximal amounts of engagement in inappropriate critical components (decision B), the patient will be unlikely to make progress. In fact, large amounts of practice in inappropriate aspects of a skill or large bouts of training in inappropriate performance capacities may even lead to a drop in performance. In decision C, the practitioner has identified the correct critical components but has provided less than adequate amounts of practice, training, or therapy. Although the patient will make some progress, it won't be as substantial as it would have been if the practitioner had provided larger amounts of experience with these critical components. In decision D, the practitioner fails on both counts, providing infrequent engagement in inappropriate (irrelevant) critical components. Although we would not expect anything to come of these physical activity experiences, this decision may not be as harmful as decision B. Although practice or training was focused on inappropriate critical elements, at least only minimal amounts of experience were provided, making damage less likely than might be the case with decision B.

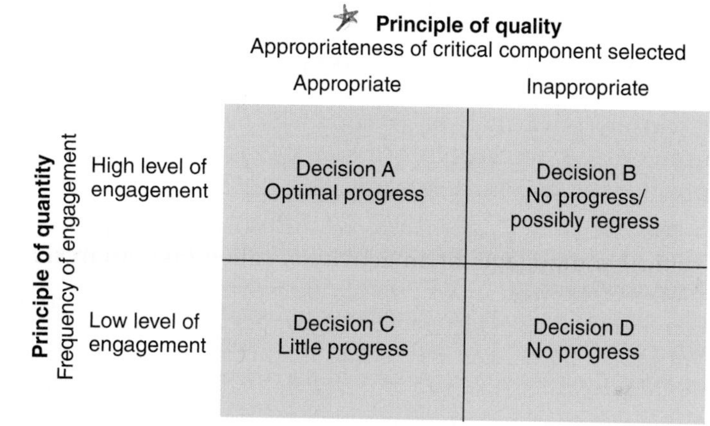

Principle of quantity: Frequency of engagement	Principle of quality: Appropriateness of critical component selected — Appropriate	Inappropriate
High level of engagement	Decision A Optimal progress	Decision B No progress/ possibly regress
Low level of engagement	Decision C Little progress	Decision D No progress

FIGURE 3.6 Making decisions about the quality and quantity of physical activity experiences.

DO it **Activity 3.4**

Best Route to Improvement

Determine the best approach to helping two athletes improve their performance in this activity in your online study guide.

➤ Kinesiologists, physical education teachers, and coaches constantly conduct task analyses to identify the critical components of physical activities; they use this information to design appropriate practice and training experiences.

Identifying Critical Components of a Physical Activity

How do physical activity specialists go about zeroing in on the critical components of an activity? In many instances, perhaps most, professionals use informal, intuitive task analyses developed over many years of practice. Like veteran physicians diagnosing patient ailments, experienced physical activity professionals can identify the critical elements of a task quickly without resorting to a checklist or other formal system. For beginners, however, a general framework is usually recommended. Some examples of basic frameworks for conducting task analyses of physical activities follow.

The first task is to identify the quality of experience that will bring about improvement. If the activity is located more toward the practice end of the continuum displayed in figure 3.5, the experiences should be those that are likely to improve skill. Thus the goal of your analysis will be to identify critical aspects of the performance that can be learned through carefully constructed practices. If the activity is located near the physical capacity-training-conditioning end of the continuum, your focus will be on training experiences that will promote conditioning appropriate for the activity.

Determining Skill Components Critical for Learning

If the physical activity in question falls near the skill practice-learning end of the continuum in figure 3.5, the analyst's first task is to identify components of the activity that must be practiced to maximize learning.

In making such important decisions, physical activity professionals often consult **motor skill taxonomies,** classification systems that categorize skills according to their common critical elements. Obviously, we can classify motor skills in many ways. We could, for example, classify them according to whether they require large movements or small movements, whether they are performed in the water or on land, whether they involve fast movements or very slow movements, or whether or not they require equipment. But it is doubtful that any of these factors could be considered critical because they do not appear to have direct implications for how to plan practice experiences. Remember, if a taxonomy is to be useful, it must at least supply us with hints about how to design practice experiences.

One of the most illuminating classification systems designed for analyzing motor skills in the past 40 years is a simple scheme that locates skills on a continuum from **closed skills** to **open skills,** as shown in figure 3.7 (Poulton 1957; Gentile 1972). The classification is based on the presumption that a critical component of all skills is the predictability of the environmental events to which performers must adapt their movements. In this view, performing a motor skill is essentially a matter of adapting one's movements to relevant objects, persons, or environments. For example, to execute the simple task of picking up a ball, performers must move toward the ball, reach their hands in the direction of the ball, and position their hands and fingers in such a way as to grasp the ball. Moving their arms or hands away from the ball or keeping their fingers in a stiff and extended position will not allow them to accomplish the goal. In this sense, then,

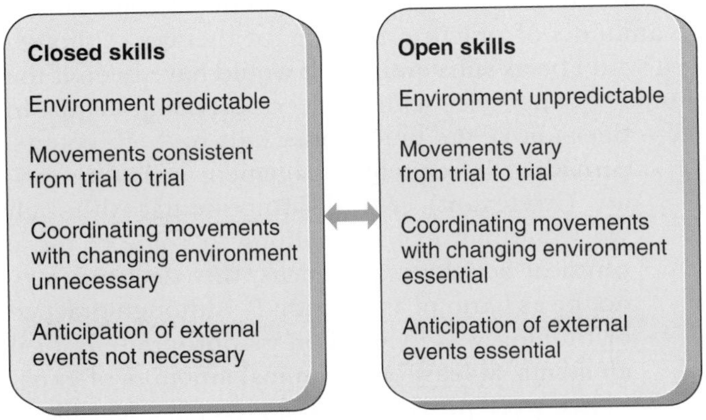

Closed skills	Open skills
Environment predictable	Environment unpredictable
Movements consistent from trial to trial	Movements vary from trial to trial
Coordinating movements with changing environment unnecessary	Coordinating movements with changing environment essential
Anticipation of external events not necessary	Anticipation of external events essential

FIGURE 3.7 The open skill–closed skill continuum.

the location of the ball in space at any given time, along with the shape of the ball, controls or regulates the movements of the performer.

If the ball to be picked up is resting on a table, its position is fully predictable before and during the movement. In this circumstance the movements that the performer must execute to achieve the goal are also predictable and may be planned before the actual trial. Motor skills performed in such highly predictable environments and requiring highly predictable movements are called closed skills. The appropriate practice experience therefore would be one that focuses on stereotypical and consistent arm, hand, and finger movements ("grooving in" the movement). Also, because the ball is stationary, performers need not be concerned about coordinating the timing of their movements to coincide with changes in the environment, as a baseball batter has to, for example. In this sense, performers are free to execute the movements in their own time.

If one were to change the goal of the task by requiring the performer to pick up the ball as it is rolled at various speeds across the table, a variable rather than a stereotypical motor response would be needed. In addition, performers could no longer plan and execute their responses in their own time. The start and duration of their movements would be controlled by the speed at which the ball is rolled. Now what was originally a closed skill has become an open skill. The skill is open in the sense that it requires performers to monitor a constantly changing environment so that they can adjust their movements accordingly.

Thus when the skill is "opened up," the ball controls not merely the direction and location of performers' movements, as was the case when it was stationary, but also determines when performers must initiate movements. (Thus, performers can make errors not only by being inaccurate in the direction of movements, but also by being early or late.) And if the ball's velocity and direction change on each trial, performers face an even more complex challenge. Because the ball might roll at high speed or low speed and in any direction, it could well topple over the edge of the table before the performer has time to decide on a response and execute it.

Thus, open skills present performers with unpredictable environments. This simple change in the structure of the task—creating a moving rather than a stationary environment to which performers must adapt their movements—makes for fundamental changes in the critical components of the task. And, whether the skill is to be performed in an open or closed environment must be taken into account in the planning of practice experiences.

The determining factor for locating any skill on the open skill–closed skill continuum is the predictability of the movements required to attain the goal. Figure 3.8 shows how some skills might be located on the continuum. Remember, skills near the closed end are performed in highly predictable environments. Thus, the movements required for successful execution are also highly predictable. Diving, sewing, and hitting a softball from a batting tee are examples of closed skills.

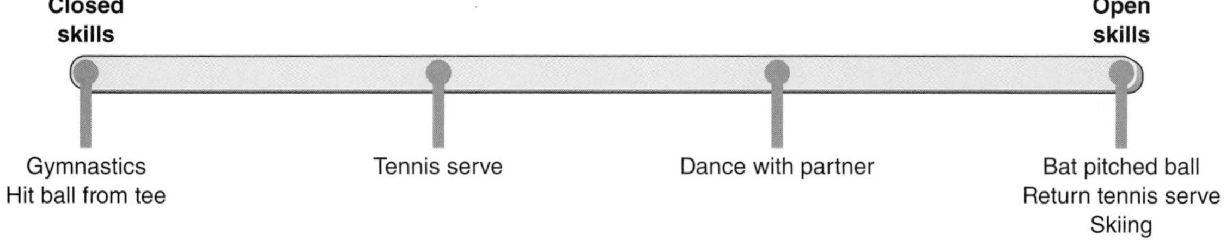

FIGURE 3.8 Possible locations of some skills on the open skill–closed skill continuum.

At the midpoint of the continuum are skills in which the movements to be executed are less predictable because the environmental events that performers must adapt to change location as the skill is being performed. Skills such as a tennis serve, hitting a softball pitched at very slow speeds, or dancing with a partner involve moderate amounts of unpredictability in the environment and consequently moderately unpredictable movements.

At the extreme open end of the continuum are activities in which the movements required are highly unpredictable because the locations of the environmental objects that performers must adapt to change in an irregular way. Skiing down a mountain would be placed near the open end of the continuum even though the environment (the terrain, trees, moguls, and so forth) remains stationary. In such skills, especially skills in which the body is moving swiftly, the relative motion between the performer and the environment creates the same element of unpredictability as occurs when the performer is stationary and the environment is moving.

➤Skills may be located on a continuum anchored at one end by skills in which performers must adapt their movements to fixed and therefore highly predictable environments (closed skills) and at the other end by skills in which performers must adapt their movements to changing and highly unpredictable environments (open skills).

Determining Practice Experiences Critical for Improving Skills

The types of practice experiences essential for bringing about improvements in skills vary depending on the location of the skill on the continuum. For example, skills near the closed end of the continuum are best practiced in situations in which the environment is structured the same way on each trial. Emphasis would be placed on grooving a theoretically correct technique that could be repeated trial after trial. Learning golf involves developing such a stereotypical pattern of movement. Practice regimens for skills near the open end would involve structuring the environment in various ways on each trial so that the direction and velocity of movement of the relevant objects, persons, or environment change on each trial. Thus, the goal of practice with open skills is to develop a flexible technique that allows one to adapt to a variety of environmental stimuli.

Referring back to figure 3.6, you can see that if a coach has her softball team spend a lot of time hitting balls from a tee (closed skill) rather than hitting pitched balls (open skill), her prescriptions fall within the upper right cell of the matrix (large amounts of engagement with inappropriate critical components). Such practice would not focus on the critical aspect of the task: coordinating the swing with the direction and speed of the ball. This would result in wasted time and effort or in learning inappropriate techniques. Also, practice for open skills typically concentrates on developing strategies for anticipating changes in the environment that are likely to occur. Baseball batters often do this when facing pitchers whom they have studied or watched on video. This approach provides performers with more time to plan and execute their responses. Obviously, skills located near the center of the continuum in figure 3.8 will require appropriate application of all these practice strategies.

Remember that the open skill–closed skill continuum focuses on only one critical component of motor skills. Other critical components may be present as well. Chapter 8 will introduce you to many other elements that you must take into account in planning practice experiences for skills.

Performers of open skills must match their movements to a changing environment.

Determining Experiences Appropriate for Training

Having considered how to determine the types of *practice* experiences essential for *learning* skills, how can we determine the types of *training* experiences likely to lead to the *conditioning* essential for improving physical capacity? Remember, unlike skill, physical performance capacity depends on training experiences involving such critical components as muscular endurance, cardiorespiratory endurance, strength, and flexibility. Knowing the critical components that underlie a particular activity permits you to plan appropriate training strategies for developing them.

How would you go about determining which types of physical activity experiences are essential for improving competitive performance in bicycling and performance in golf? The important question is, What are the critical components of golf and bicycling? You might approach the problem by locating the two activities on the continuum depicted in figure 3.5. You should recognize immediately that the key to successful performance in golf is primarily learning to perform a coordinated sequence of movements and that for the golfer, practice experiences are probably more important than training experiences. This is not to say that strength, endurance, and flexibility are not important in golf, only that they are less important than skill in this case. You will also realize that although success in bicycling probably depends to some extent on the development of a coordinated sequence of efficient movements, it is primarily determined by the contribution of such performance components as cardiorespiratory endurance, muscular endurance, strength, flexibility, and balance. These performance components, then, play a role in both skills, so both the golfer and the cyclist will require a certain amount of training. But what are the specific critical components in each skill around which training programs should be developed? The challenge at this point is to identify the critical performance components underlying each of the two tasks.

You can approach this question in two ways. One is to consult experts. Let's assume that you consulted a group of exercise physiologists or conditioning experts and asked each of them to rate the contribution of five performance components (on a scale of 0 to 3) to the performance of each of the two skills. Totaling the ratings of the judges for each component would result in a table such as table 3.2 (this table is based on the ratings of experts). Examining the table, we can see that four of the five physical capacity components listed were judged critical to bicycling, suggesting that any training regimen should focus on each of these. In the opinion of the experts, none of the components figured prominently in golf performance, something that won't surprise you. Although strength and endurance training may improve golf performance to some degree, golf is predominately a game of skill, hence more effectively improved through practice. In addition, an analysis of the predictability of movements required in golf would show that it is a closed skill, and this consideration would guide your planning of practice experiences.

➤ One way to identify the critical performance components underlying any physical activity is to ask experts to rate how important selected components are in the successful performance of the activity.

Table 3.2 Ratings by Experts of Physical Capacity Elements

Physical fitness	Jogging	Bicycling	Swimming	Handball or squash	Skiing (alpine)	Basketball	Tennis	Calisthenics	Golf	Softball
Cardiorespiratory endurance (stamina)	21	19	21	19	16	19	16	10	8	6
Muscular endurance	20	18	20	18	18	17	16	13	8	8
Muscular strength	17	16	14	15	15	15	14	16	9	7
Flexibility	9	9	15	15	14	13	14	19	9	9
Balance	17	18	17	17	21	16	16	15	8	7

Seven experts weighted the degree to which each sport activity developed cardiorespiratory endurance, muscular endurance, muscular strength, flexibility, and balance on a scale of 0 to 3. Ratings were combined for all seven judges (21 = highest score). A high rating for a sport for developing muscular strength is an indication that muscular strength is a critical component for that activity.

Adapted, by permission, from D.J. Anspaugh, M.H. Hamrick, and R.D. Rosato, 1991, *Wellness* (St. Louis: Mosby Year Book), 165. © McGraw-Hill Companies, Inc.

Activity 3.5

Binary Decision Tree

In this exercise in your online study guide, you'll experience a systematic approach to task analysis.

A second approach to determining critical performance components is to ask and answer a series of questions about the activity based on your knowledge of it and your understanding of kinesiology. Binary decision trees are often used by physicians when diagnosing diseases or by electricians when they troubleshoot a circuit defect. Binary decision trees focus your attention on the most important questions and can help you approach task analyses in a clearheaded, systematic way.

Obviously, the decision tree in the online study guide is not intended to be an exhaustive analysis of all the critical performance factors that might be important in an activity. Although it is a general taxonomic model, it will help you appreciate the many critical components underlying physical activities. The model will also lay the groundwork for learning in a more detailed fashion all that goes into determining how to match performance experiences with critical performance components.

➤ Binary decision trees that focus your attention on the most important questions to ask about a performance often can help identify the critical performance components underlying the task.

As important as physical experiences are in determining how well we are able to perform skills, they are not the only factor that we have to take into consideration. Think back to the case of Tiger Woods. Can we say that the only thing that separates him from most other professional golfers is that he has played so much golf? Probably not. He has a body build that lends itself to developing power in the swing, and he is able to move his hip, trunk, shoulder, and wrist joints at an amazingly high rate of speed. He also enjoyed the game from his youngest years and had a tremendous commitment to becoming a good golfer. Some of these qualities (certainly body build!) may have more to do with his genetic makeup than with what he has gained from practice and conditioning. All of this raises the following question: Does the depth of our physical activity—in and of itself—determine how proficient we will become in a sport, exercise, or other physical activity? Not at all. Although experience may account for most of the difference in proficiency between individuals, this judgment shouldn't cause us to ignore the fact that heredity contributes too; and this factor, of course, has nothing to do with physical activity experience. Let's take a brief side trip to explore the ways in which heredity can influence our physical activity experiences.

Heredity and Experience

You and a friend may have had about the same amount of competitive basketball practice, and both of you may have adhered to rigorous conditioning programs to improve your jumping ability and overall strength. Yet you may be a much better player than your friend. Why? Perhaps you were more dedicated and more highly motivated to succeed. You may have had the benefit of superior coaching. But part of the answer may also be that you were lucky enough to have inherited a greater proportion of the abilities required to play basketball. For example, you may be taller, faster, and more agile, or you may have what seems to be a natural facility for changing direction, jumping, or coordinating the movements of your arms and legs, all of which are important to success in basketball. Practice can modify most of these characteristics somewhat, but the genetic contribution of your parents may have played at least as important a role. Before we close the cover on this study of the role of experience, then, let's briefly consider ways in which the effects of heredity might modify the effects of physical activity experience.

Abilities as Building Blocks for Experience

Genetic predispositions that offer advantages or disadvantages for particular activities are called **abilities.** Although physical activity scientists still do not completely understand abilities, they believe that abilities are genetically endowed perceptual, cognitive, motor, metabolic, and personality traits that are susceptible to little or no modification by practice or training. For example, batting a baseball requires special visual abilities, surfing requires certain balance abilities, and orienteering requires spatial ability. People who possess great amounts of the unique abilities required for a particular activity have greater potential for success in that activity than those who did not inherit those abilities. Notice that we said that these individuals have greater potential. Usually, potential by itself is not sufficient for achieving high levels of skill or performance. The key to exploiting and eventually realizing potential is engagement in appropriate physical activity experiences.

➤Ultimately, how well we are able to perform an activity is determined by both our physical activity experiences and our abilities.

Think of abilities as the foundation on which we construct our experiences. Those with greater amounts of the abilities required in an activity have potential for higher achievement, but a person will not realize that potential unless he or she also capitalizes on opportunities for improving performance through practice and training. Thus, the sum of your abilities and your practice and training experiences will determine the highest level of competency that you are able to achieve in a particular activity. Sometimes, people inherit the abilities required by an activity but fail to exploit those abilities through practice and training (underachievers). Others appear to have little natural ability for an activity, but they compensate for this by unusually ambitious practicing and training schedules (overachievers).

Genetic factors also influence our body proportions, skeletal size, bone mass, limb length, limb circumference, and distribution of muscle and fat; and these features clearly have a bearing on our capacity to perform various physical activities. If you inherited a small frame, you probably won't excel in American football, rugby, or sumo wrestling, no matter how much you practice and train. On the other hand, if you inherited a very large frame, don't expect to make the Olympic team in gymnastics, ice skating, or diving—sports in which a compact body configuration is essential to rapid twisting and turning. Figure 3.9 shows some theoretically ideal body proportions for selected sport activities. Keep in mind that although inherited body characteristics may be a limiting factor for those aspiring to elite status as highly specialized performers, for most people this consideration is not important. Those hoping to achieve moderate levels of performance that will enable them to participate at a recreational or social level will find training and practice sufficient stimuli to achieve their goals.

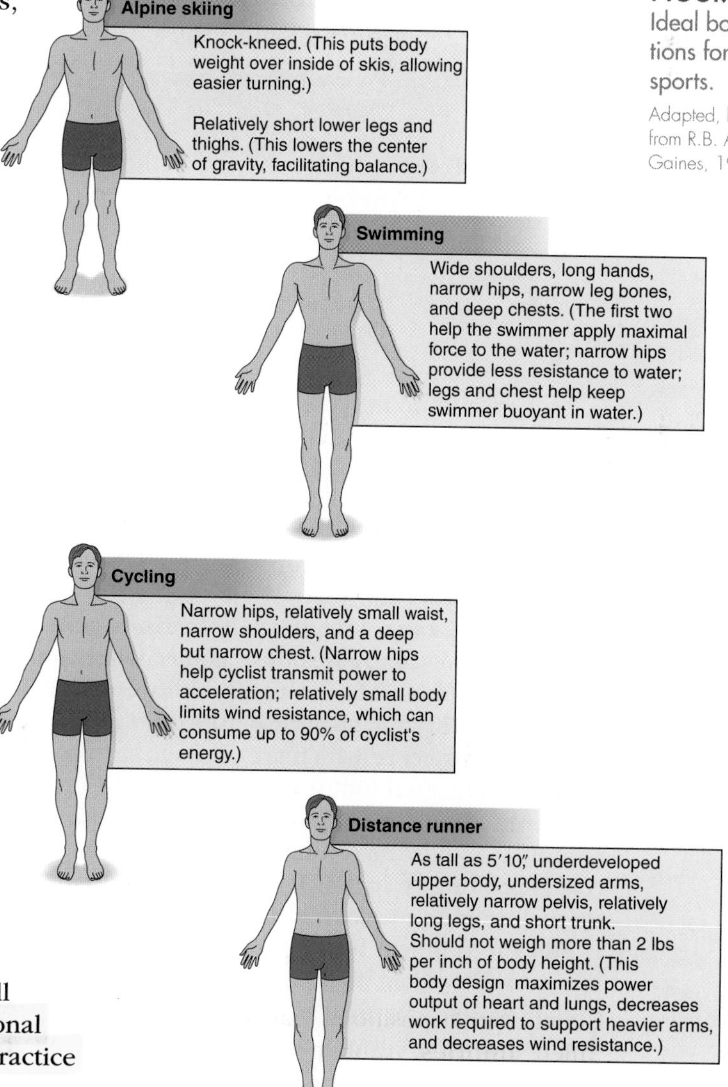

FIGURE 3.9 Ideal body proportions for selected sports.

Adapted, by permission, from R.B. Arnot and C.L. Gaines, 1984.

Interaction of Experience and Abilities

As we mentioned earlier, the contribution of inherited abilities to the performance of physical activity is much more complicated than most people tend to imagine. For example, genetics may not only supply us with the abilities required to perform certain activities (movement speed, hand–eye coordination, explosiveness, and so on) but also determine how our bodies respond to experience. One method scientists use to clarify the relationship between heredity and motor performance is to measure performance on a battery of tests by people with common genetic backgrounds. For example, correlating the performances of children with those of their parents on a reaction time test presumably would shed some light on the contribution made by hereditary factors. (A correlation of 1.0 is a perfect correlation [rarely found] and would suggest that almost all of the performance can be explained on the basis of heredity. On the other hand, a correlation of .50 suggests that approximately 25% of the performance [$.50^2$] is accounted for by common hereditary factors.) Table 3.3 shows some correlations found between various relatives and motor performance. Because we each inherit a different amount of the ability that allows us to profit from experience, we do not all respond in the same way to a given amount and kind of experience.

Table 3.3 Correlations for Selected Aspects of Motor Performance and Selected Relatives

Relatives	Reaction time	Movement time	Eye–hand coordination	Strength (push-up)	Strength (grip)	Endurance (sit-ups)
Parent-child	0.13	0.09	0.04	0.25	0.20	0.24
Siblings	0.37	0.33	0.32	0.38	0.29	0.34
Uncle-/aunt-nephew/niece	0.20	0.18	0.23	0.51	–0.12	0.54

Adapted from C. Bouchard, R.M. Malina, and L. Perusse, 1997, *Genetics of fitness and physical performance* (Champaign, IL: Human Kinetics), 324, 328.

➤ The upper limits of learning and conditioning are determined by both our physical activity experiences and inherited factors called abilities. Because experience is the only variable over which we have control, the best strategy for achieving excellence is to avail ourselves of the highest-quality physical activity experiences possible.

As a case in point, consider long-distance running, in which a runner's ability to deliver oxygen to the muscles (and use it) and to eliminate carbon dioxide, a phenomenon known as maximal oxygen uptake, largely determines performance. Endurance training generally increases runners' maximal oxygen uptake, but it doesn't affect everyone in the same way. A standard bout of conditioning may result in improvements in endurance performance of 16% for some individuals and 97% for others (Lortie et al. 1984). Why this wide variability of physiological response to training? Part of the reason may be that experience, in this case endurance training, interacts with various genetic traits. Thus, those lucky enough to have inherited this training-response ability will probably show greater improvement as a result of a standard bout of training than those who lack it.

Like the effects of training on conditioning, the effects of practice on learning may also be modified by inherited abilities, although the process appears to operate somewhat differently. Theoretically, each motor skill is supported by certain motor abilities, which are distributed unevenly among the population. If you are fortunate enough to have inherited the speed-of-response ability, for example, you will likely perform well in tasks that require this ability (e.g., sprinting or batting). If you inherited an ability called mechanical reasoning, you may do well in motor tasks that require this ability (e.g., the operation of complex equipment).

But the problem becomes more complicated when we look at the contribution of abilities to skills when they are practiced for long periods. Practicing skills usually leads to higher levels of performance, but scientists have discovered that the abilities required to be successful at low levels of performance may be different from those required to perform at very high levels. For example, cognitive abilities such as mechanical reasoning and spatial ability may be important in some skills during the initial stages of learning but not during later stages. On the other hand, motor abilities such as speed of movement or reaction time might not be important in early stages but are critical at higher stages of learning (Fleishman & Hempel 1955). Thus, the effects of practice, like the effects of training, interact with the influence of inherited traits. If you have inherited a large capacity for mechanical reasoning, you may develop skill rapidly during initial practice trials. If, however, you lack large capacity for speed of movement (and if speed of movement is a critical component of the task), you will proceed slowly during later practice trials.

The average person would be wise not to base his or her decision to engage in a particular activity solely on whether he or she seems to possess large amounts of requisite abilities for that activity. This might be a legitimate concern for the young person (and her parents) who envisions competing in the Olympics someday; but for most of us, our collection of inherited abilities—whatever they may be—coupled with practice and instruction will enable us to play most sports at an adequate level. Besides, none of us can ever know for sure what our abilities are. The best advice is to engage in the physical activities that you enjoy, realizing that regardless of your abilities, the only avenue to achievement will be through experience. In most cases, the greater the quantity of high-quality experience you have with an activity, the better your performance will be.

Wrap-Up

We began this chapter by making the obvious observation that the key to improving our physical activity performances is to accumulate physical activity experiences. Thus, we should engage ourselves and those we serve in experiences of the right quality and quantity. Knowing the specific types of experiences essential for learning skills or improving physical capacity in any activity or for developing physical fitness is a key element in the expertise of physical activity professionals. Although you will not be able to appreciate fully all the variables that come into play in prescribing physical activity experiences for clients until after you have completed your study of kinesiology, this chapter introduced you to a number of general factors that you must take into account. These include considering whether the activity consists of skills that must be learned through practice or whether the activity consists of physical capacities that must be conditioned through training. We also saw that any practice or training regimen must focus on the critical elements of the activity and provide in-depth experiences if performance is to improve. Finally, we looked briefly at the traits that we inherit as factors that interact with experience to bring about performance changes across individuals.

Regardless of ability, the only avenue to achievement is through experience. And in most cases, the greater your quantity of high-quality experience, the better your performance will be.

Photodisc/Getty Images

KEY Activity 3.6

Use the key points review in your online study guide as a study aid for this chapter.

QA Activity 3.7

These end-of-chapter questions and activities are also in your online study guide. Your instructor may ask you to complete them online and turn them in.

1. List four unique characteristics of human physical activity.
2. What factors influence our decisions regarding what physical activities we shall engage in and how physically active we shall be?
3. What type of activity can we improve by practicing? What do we call the improvement brought about through practice?
4. What is the principle of quality? What is the principle of quantity? Give an example in which a physical activity professional would use each principle to design an appropriate physical activity experience.
5. What is meant by an ability? In what way might abilities limit the level of proficiency that we attain in a physical activity?

The Importance of Subjective Experiences in Physical Activity

Shirl J. Hoffman

CHAPTER OBJECTIVES

In this chapter we will

- discuss four commonly overlooked truths about physical activity;
- describe the nature of subjective experience and its place in physical activity;
- explain factors that influence our personal experiences of physical activity—what makes some activities enjoyable, some boring, and some frustrating;
- discuss how our individual preferences, tastes, dispositions, activity history, and personal circumstances contribute to our subjective experiences in physical activity; and
- describe the subjective experience of watching sports and how various factors affect us as spectators.

Mary Hall McCarver had achieved what many would have considered to be a pinnacle of the athletic life: a scholarship to play basketball for one of the best university teams in the country, at the University of North Carolina–Chapel Hill. Then she admitted to friends and family that she wasn't enjoying playing high-level basketball.

"I had finally started admitting that basketball wasn't for me anymore," she said. She had invested heavily in the sport, having given up tennis and soccer in high school to play basketball, endured knee and ankle injuries, and devoted hours to perfecting her jump shot. Even as a small girl she had talked of playing basketball one day for the University of North Carolina. In her senior season in high school she averaged 25 points per game, made 60% of her shots, and set state records for 3-point shots. But very early in her freshman year she knew she couldn't continue. Two days before the first official practice, she quit. Her coach was puzzled: "I love Mary Hall but I couldn't understand it," she said. "This is one of the best places to play in the country." But Mary's friends knew she was unhappy, even though she wore a smile when she posed for the team photo. "Some people never lose their passion to play," she said, "but I lost it" (Cary 2005).

Was Mary crazy? How could she give up all those years of hard work and success? Can you imagine making the same decision?

Her story is important because it shows how easily we can be fooled by looking only at outward activity experiences. Examining physical activity in terms of performance characteristics or health benefits tells us little about what is happening inside the performer. If we are to become enthusiastic about a particular form of physical activity—whether basketball, aerobics, weight training, or taekwondo—something in the activity must contact our innermost selves in a way that causes us to return repeatedly to the activity. This relationship between physical activity and our inner lives—our thoughts and feelings—is the subject of this chapter.

Activity 4.1

Personal Inventory

Go to the online study guide to document your subjective experiences with physical activity.

You will notice that this chapter focuses primarily on physical activity experiences associated with sport and exercise (the competition and health spheres of experience). There are two reasons. First, our primary reason for playing sports or continuing to engage in exercise is that we enjoy these activities. Second, the focus is on activities in these spheres because traditionally they have formed the centerpiece of kinesiology. Kinesiologists are interested in physical activity of all kinds, but they are especially interested in sport and exercise.

Four Truths About Sport and Exercise

Mary isn't the first talented athlete to quit a sport because she simply didn't enjoy it. Her story is a dramatic illustration of the importance of subjective experiences in sport and other physical activities. Sport, of course, isn't the only form of physical activity in which subjective experiences play a key role. Some people like to exercise; others, no matter how substantial the health benefits, don't like to exercise and refuse to do it. All of this underscores four fascinating truths about sport and exercise:

- Physical activity is always accompanied by subjective experiences.
- Subjective experiences of physical activity are unique.
- We might not pay attention to the subjective experiences of physical activity.
- Physical activity will not be meaningful unless we enjoy it.

Physical Activity Is Always Accompanied by Subjective Experiences

The first truth is that physical activity is always accompanied by subjective experiences that are every bit as important as the objective or physical experiences discussed in chapter 3. When we talk about our experience with a particular activity, we are usually referring to our physical experience. For example, we think of an experienced runner as one who has spent much time running on the track, on the streets, and on country trails. In this sense, "experience" refers primarily to performing. The reason we naturally equate experience with performing is that the sheer physicality of human movement rivets our attention. When Serena Williams glides effortlessly across the court and returns her opponent's backhand, her movement represents a concrete image, a tangible event that we can observe.

But experiencing an activity means more than just performing it. Experiencing running is not just putting one foot ahead of the other in a mindless repetition of movements. It is not simply a matter of your muscles contracting and your heart pounding in a mechanical fashion. It also involves feeling your racing pulse and the warmth of your body, the sense of energy being released in your muscles, the pain of fatigue. It includes the memories that the experience impresses upon you and the changes in the way you think and feel about the act of running and about yourself.

Anyone who was fortunate enough to have watched Michael Jordan play basketball at his peak knows that his performance could generate dramatic emotions—amazement, awe, and even disbelief come to mind. But Jordan's performances also affected his own subjective state. His play did not involve robotic movements executed with little feeling or emotion. After making an outstanding shot or pass, he would raise his fist in triumph, leap in the air, or laugh. Those gestures are external signals that physical activity also involves subjective experiences and that performing activities can touch us at a deep emotional level.

It is easy to underestimate the importance of these subjective aspects. One reason is that physical activity has been linked historically to the scientific fields of biology and medicine. These areas are more objective than the humanities or social and behavioral sciences, which deal with intangible qualities such as values, feelings, and cognitions. Even those working in the biological or health-related fields now recognize that a complete understanding of physical activity must take into account the response of the entire person—the intellectual, emotional, and spiritual capacities as well as those of the physiological systems. In fact, the physical features of activity merely set the stage for our subjective experiences. George Sheehan, a cardiologist who wrote extensively about the spiritual side of running, described

For athletes such as Tony Hawk, performance is as much a subjective experience as it is a physical experience.
AP Photo/Denis Poroy

it this way: "The first half hour of my run is for my body. The last half hour is for my soul. In the beginning the road is a miracle of solitude and escape. In the end it is a miracle of discovery and joy" (1978, p. 225).

➤The interior and sometimes mysterious aspect of our lives, in which we collect, recall, and reflect on the feelings and meanings physical activity has for us, is the realm of subjective experience.

Subjective experience refers to the entire range of emotions and cognitions, dispositions, knowledges, and meanings that we derive from physical activity. Obviously, Mary McCarver didn't have the same subjective experiences playing basketball that Sheehan had running. Sheehan's experiences led him repeatedly to the tracks and trails where he ran throughout his adult life into old age. McCarver's experiences led her to stop playing basketball when she was still a teenager. This isn't to say that Sheehan's reaction was correct or that McCarver's reaction was wrong. The stories merely illustrate the fact that the feelings we associate with physical activity are very personal and often determine whether we will continue to engage in it.

Subjective Experiences of Physical Activity Are Unique

The second truth about physical activity is that the associated subjective experiences, particularly in the case of sport and exercise, are unique. Whether it is physical activity associated with work, education, or some other sphere of physical activity experience, the subjective experiences of moving our bodies are different from those associated with sitting or lying motionless. But not all physical activity stimulates the same subjective experiences. Executing a series of push-ups is a distinctly different subjective experience than putting a golf ball. The feelings associated with playing racquetball are different from those associated with running on a treadmill, and all these are different from the feelings associated with typing on a keyboard, even though all require physical activity. It is the unique subjective characteristics of sport and exercise that attract us. Would you work out in the gym, play sports, swim 5 miles each day, or run cross-country if those activities provided essentially the same human experiences as brushing your teeth, driving your car, or taking a final exam? Probably not.

➤One of the primary reasons we seek out exercise and sport is that they supply us with unique forms of human experience unavailable to us in our everyday lives.

We May Not Pay Attention to the Subjective Experiences of Physical Activity

A third truth about physical activity is that we can easily overlook the subjective experiences that accompany it. Sometimes we engage in physical activities without ever pausing to ask ourselves what thoughts and feelings the activity generates in us or how the activity fits within the larger scheme of our lives. Even though our subjective experiences may be the primary reason we engage in an activity, we sometimes become so involved in the competition or in achieving our goals that we lose sight of this important fact. Think back to the questions in activity 4.1 in the online study guide. Did you have to think much to come up with answers, or had you thought about these questions before? If not, why not?

The expressions on our faces often reveal the subjective experiences of physical activity.
AP Photo/Michael Sohn

Listening to Your Body

We have said that sometimes we don't pay attention to the subjective experiences of physical activity. Here's a different point of view. Philosopher Heather Reid says, "In contrast to such sedentary activities as taking tests or watching TV in which we try to ignore our bodies, sport forces us to *listen to our bodies*. All athletes know the importance of this skill. They take their pulses upon rising, gauge the tightness of their calf muscles, worry about every tingle in their throats. Even as they savor the exhaustion that follows a well-played game or a race well-run, even as they welcome the pain-numbing endorphins that miraculously quiet their bodies' screams midway through a hard effort, athletes neglect or ignore their bodies at their peril and they know it. What most athletes don't know but philosophical athletes do know is just how much can be learned from listening to their bodies. Listen closely enough to your body and you might discover a window to your soul" (Reid 2002).

Physical Activity Will Not Be Meaningful Unless We Enjoy It

Finally, a fourth truth about physical activity is that unless we are attracted to the subjective aspects of physical activity—unless we discover something enjoyable in it—the activity is unlikely to become personally meaningful to us. The physical, tangible side of physical activity provides the raw material for the experience, but it's the subjective experience that keeps us coming back for more. As we saw with Mary McCarver, accolades, recognition, prestige, and other factors external to the activity cannot fill the void. If the thoughts and feelings that accompany your engagement in physical activity don't cause you to invest in it with the deepest resources of your mind and will, the experience can be hollow and meaningless.

If you have ever ice skated in solitude on a frozen pond or cycled along a country road on a brisk fall morning, you have probably enjoyed the subjective pleasures that physical activity can bring. The same subjective pleasures may come to you when you're struggling through a difficult wrestling practice in a hot, humid room or when your lungs are bursting in the final stages of a tough aerobics workout. Engaging in physical activity can touch us emotionally, mentally, and spiritually. When we allow this to happen, our physical activity experiences are more meaningful.

Why Subjective Experiences Are Important

Learning about subjective experiences in physical activity is important for several reasons. For one, they can help clarify the bases of your career choices. You may be planning a career as a personal trainer or exercise consultant because you want others to know the kind of self-confidence that you have experienced through participating in a conditioning program. Or perhaps you have decided to become a physical education teacher or coach because of the good feelings that you experienced when participating in sport in high school.

Learning about subjective experiences can also help develop your skills as a physical activity professional. Physical educators can better serve their students if they have experienced the feelings associated with the clumsy beginning trials of a skill; personal trainers and fitness leaders who have suffered through the painful early stages of a training regimen are better able to give their clients assurance and encouragement. Such knowledge can be useful in designing programs and interventions that help people understand their subjective experiences and meanings in physical activity.

But the most important reason to study and learn about subjective experiences is that how we feel and what we think before, during, and after we engage in a physical activity largely determine whether or not we will make that activity part of our lives. Being attracted to mountain climbing rather than canoeing, aerobics rather than running, or baseball rather than track or field may be traced in some part to the subjective experiences that these activities have evoked in us. This chapter explores how physical activity can affect us in ways that can't be measured by stop clocks, strength gauges, or competitive points: The focus is on internal dynamics rather than external performance.

The Nature of Subjective Experiences

Imagine that you have just shot a 20-foot (6-meter) jump shot at the buzzer to win a close game against your team's arch-rival . . . that after an extremely tough practice you are sitting on a bench in the locker room, tired beyond belief but also surprised at how good it felt to endure the pain and pressure of that practice . . . that 10 years after graduating you return to your school for an alumni game, and as you sit on the bench in your old locker room, you reflect on what being a member of the team meant to you and how the experience has influenced your attitudes about the sport, about others, even about yourself.

These three short scenarios probably stimulate distinctly different feelings within you. In the first, your thoughts and feelings are likely to be intense and sharply focused on your game-winning shot. Those thoughts and feelings are immediate in that you live the experience during and shortly after performing the activity. The sensations of jumping and having the ball leave the hand, and the image of the ball arching toward the basket, are all indelibly impressed on your mind. You feel a deep sense of pride because you came through when the chips were down. You may also feel instant elation over contributing to a win for your team.

DO it **Activity 4.2**

Rate Your Feelings

In this activity in your online study guide, you'll rate your feelings about various physical activities.

In the second scenario, you may again feel pride about having gutted it out and finished the practice, but you might also think that you learned a valuable lesson about yourself. Although these thoughts and feelings are less immediate than those evoked by the first scenario, they are still in part a direct response to the activity just performed.

In the third scenario, your thoughts and feelings are nostalgic, mellow, and far removed from any specific activity. You could not have experienced this sense of meaning without having been deeply involved in the activity, but neither could you have discovered it without the time that has passed since.

These scenarios illustrate how multifaceted subjective experiences can be—they can be intense and immediate, or they can be past experiences that you relive repeatedly. We'll discuss each of these two kinds of experiences in the following sections.

Immediate Subjective Experiences

➤ All physical activity is accompanied by sensations that we can convert to perceptions, emotions, and knowledge. To get the full experience of a physical activity, we must be open to the emotional and cognitive impressions that the activity provides.

It is impossible to engage in physical activity without our movements creating immediate emotional and cognitive impressions as we do it. Our bodies are equipped with movement sensors called proprioceptors—sensory devices in tendons, ligaments, and muscles and in the inner ear that are stimulated by physical actions. They provide us with information about the body's movements and position in space that is just as important as the unending array of visual and auditory information that often impinges on us when we move, maybe more so. Another sensory apparatus in the circulatory system picks up biochemical changes in the blood that affect our perceptions of fatigue and effort. Thus, each time we engage in physical activity, complex sensory signals flood our nervous system. These signals not only are vital to our performance of the activity; they also provide raw data that will be transformed into perceptions, feelings, and knowledge. Of course, we can choose to ignore many of these sensations, just as we can ignore the vast array of subtle colors in a beautiful painting or the nuances of sound in a symphony recording. Remaining open to the sensory information that accompanies physical activity is the first step toward appreciating its subjective dimensions.

But staying attuned to this subjective experience isn't always easy, especially for novices who are concentrating on completing a demanding set of exercises or absorbed in the details of learning a new skill such as golf. When learning a new aerobics routine or attempting to use an exercise machine that we haven't used before, we must initially direct our attention to performing the movements of the activity. After we have become familiar with the activity, however, we can execute many of the movements without conscious attention, freeing our minds to think more deeply about the subjective aspects of performance. No doubt this is why experienced performers tend to give the richest and most elaborate accounts of subjective experiences in physical activity.

Replayed Subjective Experiences

Although an attention-demanding activity may seem to yield little in the way of subjective experiences while it is being performed, one may discover much later that it has produced a lasting effect. Subjective experiences often endure in memory for months, years, or even a lifetime. In one of the scenarios that opened this section, a basketball player returned years later to a memory-laden locker room and, perhaps for the first time, thought hard about what playing high school basketball had meant to him. A person might experience the same feelings when she thinks back to the day she earned a fitness award or to the early morning high school exercise program that taught her how to control her weight and eat nutritious meals. The process by which we reexperience the subjectiveness of physical activity is called **self-reflection.** Self-reflection refers to reliving a past experience. It is as though we replay a videotape that includes not only visual but also kinesthetic, auditory, and other impressions from an earlier time. Because of the passage of time, self-reflection often enables us to place specific subjective experiences within a framework that makes them more meaningful.

➤ The subjective experiences that accompany physical activity may have a profound effect on us during and immediately after our performances. Reflecting on these experiences months or years later often helps us put them into a more comprehensive and meaningful frame of reference.

We can reflect on our experiences in physical activity in various ways and for different reasons. Reflecting might be a means of summing up our total experience with the activity, accumulated over many years, to determine its overall value or worth to our lives. Sometimes when we are learning a skill, it helps to focus on how the previous attempt felt so that we can make adjustments to our movements. Sport psychologists often ask clients to replay experiences at times of anxiety as a way of calming them or giving them a boost in confidence. A tennis player, for instance, after losing the first set in a two-out-of-three match and then quickly falling behind in the second set, might replay previous successes that she has had under similar circumstances. If she can relive the feelings of the previous experience, she might be able to attain some of the same mental focus that led to her earlier victories.

Some forms of physical activity lend themselves naturally to concentrating on subjective experiences.
Photodisc

Components of Subjective Experience

When exploring a territory as broad and expansive as subjective experience, it's helpful to break it down into its various components. In the following sections, we'll discuss sensations and perceptions, emotions and emotional responses, and various ways in which we gain knowledge through subjective experience.

Sensations and Perceptions

Sensations are raw, uninterpreted data collected through sensory organs. Our experiences help us interpret these raw sensations into meaningful constructs or **perceptions.** Physical activity automatically stimulates a barrage of sensations. These include the feelings of contracting our muscles and moving our bodies, as well as feelings of an increased heart rate, breathlessness, or muscle fatigue. If we attend to them, these sensations provide us with an "online" report of our inner states. Organized into meaningful information by the process of perception, these sensations allow us to make ongoing corrections in our physical activity. For instance, perceptions of fatigue or discomfort usually form the basis of our decision about when to stop an exercise. Similarly, perceptions inform our judgment about how forcefully we are moving, how much resistance our movements are encountering, or, in the case of divers and gymnasts, how our bodies are oriented to the ground.

Sensations from outside our bodies also become part of the subjective experience. What we see or hear when we perform, even what we smell, all coalesce into a cohesive subjective experience. Football or soccer players who spend much time in close contact with the turf may automatically reflect on their playing experiences whenever they smell wet earth. For surfers, the sound of waves may trigger past visions of riding the waves, and the clang of a bell may cause boxers to think back on their experiences in the ring.

Emotions and Emotional Responses

➤ Physical activity is always accompanied by the internal sensations of moving our limbs and bodies; changes in our physiological functioning; and sensations from outside our bodies including visual, auditory, touch, and smell sensations. When organized into perceptions, these can give rise to emotions that we associate with particular physical activities.

Perceptions during physical activity can elicit many different internal reactions. Sometimes physical activity increases our level of excitement and motivation and at other times may dampen it. We might become angry, annoyed, surprised, pleased, disappointed, enthusiastic, and so on. Such reactions can stem from the feelings elicited by an activity, our impressions about the quality of our performance, or the outcome of a particular event. These subjective reactions are called **emotions;** they differ depending on the person and situation. For example, achieving a personal goal or performing an activity well may result in the emotion of elation or joy; failing to achieve a goal or performing poorly may prompt such emotions as disappointment or shame. Psychologists have identified over 200 different emotions, although they have yet to agree about precisely how to define them. What seems clear is that emotions involve a complex combination of psychological processes including perception, memory, reasoning, and action (Willis & Campbell 1992).

Knowledge and Subjective Experience

When someone mentions knowledge, what comes into your mind? Facts learned in history or science? Mathematical formulas? Story themes from great works of literature? These examples represent a specific type of knowledge derived from using logic, reason, and analysis. This type of knowledge is called **rational knowledge.** You'll rely on this type of knowledge as you study the theoretical aspects of physical activity such as exercise physiology, biomechanics, or sport history or when you read this book.

But a different type of knowledge derived from our subjective experiences in physical activity—**intuitive knowledge**—is equally important. **Intuition** is a process by which we come to know something without conscious reasoning. Describing this type of knowledge or knowing precisely how we mastered it is usually difficult. Intuitive knowledge gained from subjective experiences is usually personal in that it is knowledge about ourselves rather than about the exercise, the skill, or other people. Philosopher Drew Hyland (1990) identified three types of this intuitive, personal self-knowledge associated with participating in physical activity: psychoanalytic self-knowledge, Zen self-knowledge, and Socratic self-knowledge (see figure 4.1). Because knowledge gained through subjective experience is closely associated with our motivations for participating in physical activity, we'll discuss each of these types of self-knowledge in detail.

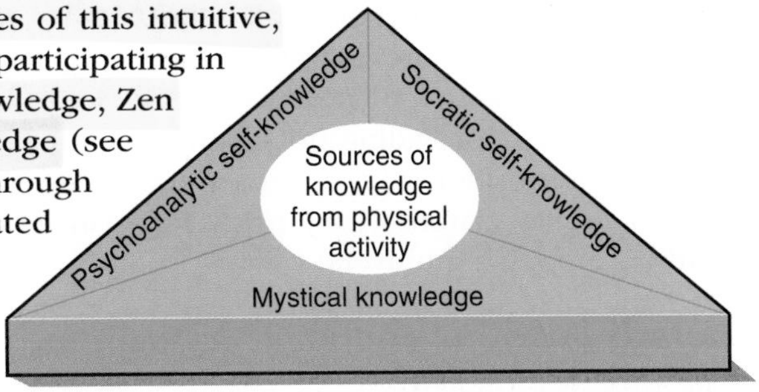

FIGURE 4.1 Three types of knowledge embedded in subjective experiences of physical activity.

Psychoanalytic Self-Knowledge

Psychoanalytic self-knowledge is knowledge about our deep-seated desires, motivations, and behavior. This form of self-knowledge relates primarily to the types of activities that we choose to participate in and the manner in which we pursue them. For example, when a young woman with an impoverished, disadvantaged background chooses to participate in a sport such as polo, golf, or squash, her choice may be a commendable effort to transcend social barriers. On the other hand, it may represent an attempt to deny her social roots. When a young man chooses contact sports over noncontact sports, it may reveal a love of competition and physical contact—or it could represent "a precariously controlled desire to physically dominate or even hurt others" (Hyland 1990, p. 74). That many active sportspersons probably don't access this type of knowledge doesn't mean that it isn't available to them. Some psychologists have suggested that participation in high-risk sports such as hang gliding, scuba, or motor car racing might be "propelled by personal needs to seek god-like experiences, which can activate an inflation of the ego" (Heyman 1994, p. 194).

The manner in which we choose to participate may also reveal psychoanalytic self-knowledge. Exercisers who are so concerned about the times of their runs or so intent on improving their performance that they can't enjoy their participation may have developed a narrow, single-focused pattern of behavior that may not serve them well in ordinary life, where they must meet many different responsibilities. A softball player who is a poor loser may have invested an inordinate amount of her self-worth in the game; the same may be true for a winner who rubs it in after the game is over. On the other side, someone who can't be serious about an exercise program or sport competition may be reflecting fear of failure. By not taking the game seriously, the player need not take failure seriously either. By carefully examining the activities that we choose to engage in, as well as our styles of involvement, we can learn a great deal about ourselves.

Mystical Knowledge

Mystical knowledge, termed Zen self-knowledge by Hyland, refers to subjective experiences available to experienced performers only in rare and special circumstances. Mystical knowledge touches on a dimension not experienced in ordinary life. Some would call the experiences that give rise to this type of knowledge *transcendent* to suggest that the experiences took them out of the real world. These subjective experiences can be so powerful that memories of them remain with performers for years, sometimes for a lifetime. The most frequently cited type of mystical experience is called **peak experience.** Research suggests that peak experiences tend to come involuntarily and unexpectedly (Ravizza 1984). Time usually seems to slow down, and specific features of the environment sometimes stand out in sharp contrast to the background. Usually, athletes report being totally absorbed in the activity.

Here is an example of a peak experience. Zinédine Zidane, one of the most prolific scorers in modern soccer, led his French team to the World Cup victory in 1998. He told reporters about this rare experience: "The game, the event, is not necessarily experienced or remembered in real time. I remember playing in another place, at another time, when something amazing happened. Someone passed the ball to me, and before even touching it, I knew what was going to happen. I knew I was going to score. It was the first and last time it ever happened" (Longman 2006).

Peak experiences provide no specific knowledge of their own. In some cases they may be windows through which performers claim to have a more extensive and inclusive world opened to them, providing not so much knowledge as a feeling of awe and reverence to which the performer may attach a religious or philosophical interpretation. In other cases performers may claim to have gained new insights into themselves as beings connected to an all-encompassing universe.

Socratic Self-Knowledge

Socratic self-knowledge is an understanding of the difference between what we know and what we don't know. We may translate this into the realm of exercise and sport as knowledge of what we can and cannot do. Knowing our performance limits can help us operate within the confines of our skills and abilities. When coaches refer to how important it is that athletes "play within themselves," they are giving testimony to the importance of Socratic self-knowledge.

➤ Performing physical activities can be a source of knowledge, including knowledge about our motivations for engaging in activity, knowledge about different dimensions of reality, and knowledge of our personal performance capabilities.

Ignoring personal limits can lead to disastrous performances and even injury. Weightlifters who injudiciously attempt to lift beyond amounts not recommended for their weight class, or ski jumpers who challenge heights for which they have no preparation, risk not only performance failure but physical harm. At the same time, as we saw in chapter 3, we must test these personal limits or we'll never improve. The purpose of practice and training is to expand the range of our performance limits systematically and gradually, thereby not only improving the quality of our performance but also reorienting our perceptions of our self-limits, with our newly developed ability to perform taken into account. One of the important roles of the personal trainer, coach, physical education teacher, physical therapist, or exercise leader is to help performers set realistic goals that will allow them to test the limits of their performances safely and productively.

Other Knowledge Gained From Physical Activity

Philosopher Scott Kretchmar (coauthor of chapter 5 in this text) provides a different description of the knowing that comes from performing physical activity. He identifies seven types of knowledge:

- Safety knowledge: knowledge about behaviors that influence physical well-being
- General performance knowledge: knowledge of factors that promote effective and efficient performance
- Specific performance knowledge: knowledge of factors that promote effective performance in specific sport, dance, work, or play activities
- Competitive performance knowledge: knowledge about one's own performance that is gained from competing with others
- Rules knowledge: understanding the rules (the "do's and don'ts") of a game
- Game spectator knowledge: knowledge about a game that enables one to be an intelligent spectator
- Sport history knowledge: knowledge about earlier events, persons, and social forces that influenced a particular activity
- Sport strategy knowledge: knowledge about techniques and tactics that can produce superior performance (Kretchmar 1994, pp. 141-142)

Talking About Subjective Experiences

Subjective experiences, by nature, are private experiences. And because many factors can affect them (including our success or failure in the activity; the conditions surrounding our participation; and the meaning given to the activity by our parents, personal trainers, coaches, or the media), two individuals participating in the same activity are unlikely to have the same subjective experiences. You cannot know precisely what the person riding the stationary bicycle next to you is experiencing; neither can that person know what you are experiencing. Because subjective experiences are interior by nature, the only way to compare our experiences with those of others is by talking about them. Some people have been able to describe their experiences quite eloquently. For example, former senator and outstanding New York Knicks basketball player Bill Bradley described his experiences on the court this way:

> In those moments on a basketball court I feel as a child and know as an adult. Experience rushes through my pores as if sucked by a strong vacuum. I feel the power of imagination and the sense of mystery and wonder I accepted in my childhood before life hardened. (1977, p. 57)

We have no reason to suspect that Bradley was exaggerating or fabricating his experiences, but it is impossible to validate his accounts. We can only know for certain what we experience; we cannot have the same certainty about the experiences of others.

If you've never before talked about your subjective experiences in physical activity, you may find it difficult at first. Most people are hesitant to share their deepest feelings—not only about physical activity but also about anything personal—with others, especially those whom they don't know well. You may worry that what you have to say about your subjective experiences will be misunderstood or seem silly to others. Such feelings are normal, because most of us have had little practice sharing such information, and we have not been encouraged to do so. Unfortunately coaches, personal trainers, aerobics instructors, and physical education teachers often are more concerned about the physical aspects of human performances than the subjective aspects. Thus they rarely encourage students to talk about them.

Even when we overcome our natural reluctance to talk about our subjective experiences, we often cannot find the right words to describe our feelings. Scott Kretchmar has suggested that words may not be able to adequately describe what we feel about physical activity. He pointed out that "the 'medium of exchange' in sport is 'feel'" and noted that "any meaningful distinctions in this realm typically outrun any verbal ability to refer to them" (1985, p. 101).

Philosopher Spencer Wertz tells of a former student, an elite gymnast, who told him how she once performed a floor exercise to the music of the "Lord's Prayer":

> She used the routine as her way of expressing praise both to God and to the abilities of the human body. The movements within the routine were mostly broad, sweeping movements. The arm movements were exaggerated and the leaps were extensive. Large movements are conventionally used to convey thanks, joy, and a general openness of the human spirit. A routine done to fast, lively music would contain shorter, jazzier movements than did the "Lord's Prayer" routine. It would be inappropriate and fail to convey what it had set out to through the choice of music. (Wertz 1991)

Nevertheless, words remain our primary medium for communicating our subjective experiences. Furthermore, struggling to put our feelings into words may help us come to grips with the deeper meanings that physical activity has for us. Sy Kleinman, a philosopher particularly sensitive to the meanings of physical movements, would agree: "Engagement in game, sport, or art, and a *description* of this kind of engagement enable us to know what game, sport, or art is on a level that adds another dimension to our knowing. . . ." (italics added; 1968, p. 31). Although you may feel inadequate to describe your physical activity experiences, such descriptions can be an important way to clarify to yourself why you engage in activity and what it means to you.

➤In attempting to communicate our subjective experiences in sport and exercise to others, finding the correct words to express our feelings is usually difficult. This activity is important, however, because it helps us better understand the meanings that we find in physical activity.

Intrinsic and Extrinsic Approaches to Physical Activity

By now it should be apparent to you that how you feel when and after you engage in a physical activity can determine whether or not you return to it in the future. But our subjective experiences are not the only factor that motivates us to participate.

Practical Approach to Physical Activity

One of the most often cited reasons for participating is the benefits that physical activity offers. Perhaps you recall listening to a lecture by your physical education or health teacher that outlined these benefits and encouraged you to "exercise for the health of it." Maybe you recall listening to your coach's speech at the athletic banquet suggesting that sport is important because it builds character. The promise of a paycheck probably motivates the physical activity that you perform at work; the promise of returning to normal functioning motivates patients who adhere to a grueling program of therapeutic physical activity.

These are examples of **extrinsic approaches to physical activity.** One who has an extrinsic orientation to physical activity professes to engage in it primarily because of something other than the subjective experiences of the activity itself. The physical education teacher who argues for her program because it develops fundamental skills in young children, or the recreational therapist who justifies shuffleboard as a means of psychological relaxation, is portraying sport in an extrinsic manner, as is the businessman who works out at the fitness center with his boss as a way of impressing his supervisor and boosting his chances for a promotion. When we approach an activity from an extrinsic standpoint, we do not value it primarily for its subjective experiences; we value it because we believe it contributes to some more important end.

Having an extrinsic approach to physical activity makes a great deal of sense in many instances. We perform self-sufficient activities for a practical (extrinsic) reason: to complete tasks around the house. We perform physical activity at work to earn a paycheck. We move in physical education class to learn a skill and meet the course requirements. But in the leisure, competitive, and self-expressive spheres, and often in the health sphere, an intrinsic rather than an extrinsic approach may make more sense. Intrinsic reasons for performing physical activities may be what keep us interested over the long term.

Running and playing golf and swimming and doing high-impact aerobics simply because we enjoy the experience of the activity are illustrations of having an **intrinsic approach to physical activity.** This approach also is called autotelic (*auto* = self, *telic* = end or goal) to underscore the fact that for us, the goal of the activity is the experience of performing the activity itself. When we approach weight training or mountain biking or any other activity as an end in itself, we are displaying an autotelic or intrinsic disposition toward the activity.

➤ Physical activity usually is accompanied by external benefits, whether in the form of health benefits from exercise, developing friendships through sport, or earning a salary through physical activity at work. Nevertheless, activities—especially those connected with sport and exercise—are more likely to become personally meaningful to us if we are attracted to the subjective experiences of the activity itself.

Physical activity results in many extrinsic benefits, many of them inevitable. The physiological benefits of physical activity, for example, are largely inescapable. When you pursue the subjective experiences of mountain climbing, for example, you automatically reap the benefits of increased muscular endurance and strength. But we should never presume that the mere fact of these benefits makes them the primary reason for engaging in the activity.

One way to think about the physiological, psychological, or sociological benefits of physical activity is to compare them to the nutritional benefits that we might gain after eating a meal at a gourmet restaurant. Although a nutritionist might be able to demonstrate that the meal served at the restaurant supplies us with a significant percentage of our daily dietary requirements, nutritional benefits are not the primary reason we go to such places. We are attracted not by the promise of improved health but by the unique subjective experience of eating at a first-rate restaurant. Whatever nutritional benefits accrue are merely icing on the cake. So too we may think of many of the extrinsic benefits of physical activity—for example, health or character or new friendships—as merely icing on the cake. Physical activity, espe-

Putting Thoughts and Feelings Into Words

Replay in your mind a notable physical activity experience—one that left an indelible impression on you, whether it occurred during exercise, sport, rehabilitation therapy, or another type of physical activity. Why did the experience make such an impression? Try to describe the experience in one word. Can you describe it in one sentence? In five sentences?

cially sport, exercise, and dance, are much more likely to become meaningful to us if we develop an attraction to their intrinsic qualities. Having said that, we should recognize that few people probably approach any activity entirely in an intrinsic or an extrinsic manner; with most of us it's usually a combination of the two.

Subjective Approach to Exercise for Health

People generally find it easier to appreciate the importance of subjective experiences when they play or watch sports than when they exercise. "We play sports for fun and enjoyment; we exercise for health," so the thinking goes. It is difficult to imagine being attracted to exercise by its subjective experiences when they often seem so painful. Wouldn't most of us prefer to have the improved strength, endurance, flexibility, longevity, and other benefits of exercise without the pain and discomfort required to attain them? Some of us might, but certainly not all. Exercisers do not always interpret the discomfort produced by physical exertion as a negative feeling. Often, the subjective experience of performing rigorous exercise is an algebraic summation of both pleasure and pain that results in an overall sense of satisfaction and well-being. Psychologist Mihalyi Csikszentmihalyi made this point in talking about the agony endured by a swimmer: "The swimmer's muscles might have ached during his most memorable race, his lungs might have felt like exploding, and he might have been dizzy with fatigue—yet these could have been the best moments of his life" (1990a, January, p. 4).

But how do we know that subjective experiences of exercise are important to us? Two lines of evidence suggest this. The first is that, although extrinsic factors may be the principal reason for initially undertaking an exercise program, intrinsic, subjective experiences often come to be the primary determinant over time. For instance, when people become committed to long-distance running, the prospective health benefits that may have prompted them to run in the first place often take a backseat to experiential outcomes such as feelings of achievement and self-development (Clough, Shepherd, & Maughan 1989). Psychologist Michael Sachs' description of his feelings about running reflected this idea: "Running has become much more than a means to the end of getting in shape; it has become the end itself" (1981, p. 118).

We can observe the second line of evidence in our everyday lives in the way that we make qualitative distinctions between the subjective experiences of strenuous work and strenuous exercise. "Muscles," someone has said, "don't know the difference between basketball and burglary." If this is true, when we exercise specifically for health and fitness we shouldn't care whether the exercise comes in the form of a stair stepper, a treadmill, a stationary bike, a brisk walk behind a wheelbarrow, digging a ditch, or moving heavy stones. (In fact, getting our exercise through work would make more sense because at its conclusion we would have accomplished something!) But most of us *do* care! We make important distinctions between these forms of activity, and our decisions reflect these distinctions. Have you known someone who hires out the jobs of painting the house, mowing the lawn, or raking leaves and at the same time *pays dues* to a local fitness center for the privilege of doing even more grueling forms of exercise there? Measured from a purely physiological standpoint, the physical activity of work may have the same healthful benefit as the physical activity of exercise, yet we tend to avoid the former and eagerly seek the latter.

➤ Although exercise provides participants with numerous health benefits, many people may engage in it because of the unique subjective experiences that it offers.

Activity 4.3

Subjective Exercise Experience Scale
Fill out the SEES in your online study guide.

Clearly, then, the subjective experiences gained from exercising appear to be unique, different from work, and influential in our decisions to engage in exercise. Scientists have yet to identify precisely how the subjective aspects of exercise differ from those of work. How does exercising make you feel? One way to assess your subjective responses to exercise is to complete the Subjective Exercise Experience Scale (SEES) (McAuley & Courneya 1994) in activity 4.3.

Internalization of Physical Activity

Have you ever decided to participate in an activity merely to please a friend or meet a class requirement? If so, the subjective experiences of the activity at first were probably incidental to your participation. But with time, your attitude toward the activity may have changed. Eventually, your decision to participate may have begun to center on something in the activity itself rather than extrinsic factors. Psychologists sometimes use the term **internalization** to refer to the gradual process by which something takes on intrinsic value, or passes from "a level of bare awareness to a position of some power to guide or control the behavior of a person" (Krawthwohl, Bloom, & Masia 1964, p. 27).

Most of us find the subjective experiences of work decidedly less enjoyable than the subjective experiences of exercise, even though both may place equal stress on the body.

AP Photo/Moscow-Pullman Daily News, Geoff Crimmins (left) and AP Photo/East Valley Tribune, Tim Hacker (right)

The process of internalization can be summarized in five stages (see figure 4.2). The first time you watched Major League Baseball, hiked up a mountain trail, embarked on a weight training program, or went fly fishing, you were probably merely *aware* of the new stimuli and physical activity that surrounded you. Later, after doing the activity many more times, you may have *responded* to the activity with positive feelings. Eventually, you might have come to value the activity to the point of *going out of your way* to seek it out. With continued involvement you might have begun to *conceptualize* and *organize* the importance of the activity by talking about the characteristics of the activity you found most attractive. If you were further attracted to the activity, you reached the final stage, *internalization*. This stage occurs when your behaviors demonstrate a commitment to the activity and you are able to integrate your beliefs and attitudes about the activity into a comprehensive philosophy or worldview.

As we progress from merely enjoying an activity to becoming engrossed in it, the activity can develop deep meaning for us, so deep that we incorporate it into our beliefs, attitudes, and personal identity. When this happens, we have *internalized* the activity.

You unenthusiastically attend a college badminton class because it is a requirement. A friend offers to tutor you on the fundamentals, and you accept her offer.

When your friend forgets to meet for your weekly lesson, you call her as a reminder. You decide to participate in a local tournament for novices. You begin to look forward to playing.

You join a local badminton club and attend matches, even though you must drive several miles to attend. You actively seek out others to compete with.

You discontinue your weekly game of racquetball in order to have more time for badminton. You discriminate between competitors, preferring to play more skilled players. You evaluate strategic approaches to the game, begin a training program, and buy instructional books.

Playing badminton is now an important part of your life; you participate in tournaments regularly, buy equipment, and other badminton players become part of your circle of friends. You can't imagine life without badminton.

FIGURE 4.2 Internalizing the game of badminton.

Factors Affecting Our Enjoyment of Physical Activity

It is not altogether clear how internalizing an activity is different from learning to enjoy it. Clearly we aren't likely to internalize an activity that we don't enjoy. Perhaps the first thing that comes to your mind when asked why you work out regularly at the fitness club, go in-line skating, or play lacrosse is that the activity is enjoyable. This idea sounds simplistic, but the concept of enjoyment is complex. Let's focus briefly on this concept so that we can better understand it and, in turn, understand how enjoyment attracts us to physical activity. What is it about physical activity that causes people to enjoy it?

Although a long list of factors is probably involved, here we'll focus on three general categories, including (1) factors related to the activity, (2) factors related to the performer, and (3) factors related to the social context in which the activity is performed.

Factors Related to the Activity

If you reflect on the physical activities that you enjoy and ask yourself why you enjoy them, the first things that come to mind are probably specific characteristics of the activities. For instance, you might enjoy tennis because you prefer individual sports to team sports and enjoy a game that makes you concentrate. You might prefer aerobics to running because you enjoy moving your body to music. In both examples, the enjoyment comes from a specific quality of the activity itself. Csikszentmihalyi (1990b) identified several characteristics of activities that make them enjoyable. The three that we'll discuss here involve (1) balance between the challenges of the activity and the abilities of the performer, (2) whether the activity provides clear goals and feedback, and (3) whether the activity is competitive.

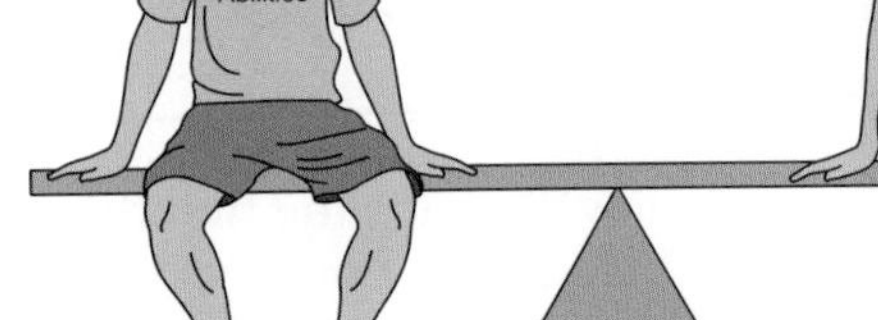

FIGURE 4.3 We are more likely to enjoy a physical activity when its challenges are balanced by our skills and abilities.

Evenly Matched Challenges

Imagine that two students have designed strength development programs using free weights. The first selects a weight that is very easy to lift (30 pounds [13.5 kilograms]), and each day he lifts this weight 10 times, never increasing the weight or the repetitions. Soon he becomes bored, realizes that he's not enjoying the activity, and quits. The second student decides to begin his weight training program by trying to lift 200 pounds (90.6 kilograms). Each day he comes to the gym and pulls with all of his might, but is unable to lift the weight off the ground. Soon he becomes frustrated and, like the first student, realizes that he's not enjoying the activity; he also quits.

Figure 4.3 shows that when our skills and abilities go far beyond the challenges of the activity, we usually experience boredom. On the other hand, when success in the activity requires skills and abilities that we lack, we often experience frustration or anxiety. Csikszentmihalyi claims that "enjoyment appears at the boundary between boredom and anxiety" (1990b), which is another way of saying that enjoyment requires a delicate balance between the challenges of the activity and the skills and abilities of the performer.

Clear Goals and Feedback

Part of the fun we get from playing sports results from testing our skills and abilities against the challenges of the game. Shooting the ball into the basket and putting the ball into the cup are goals that we try to accomplish by executing a series of coordinated movements. Enjoyment stems not so much from accomplishing the goal (e.g., the golf ball's being in the cup) as it does from our having attempted and succeeded at causing the ball to go into the cup. Unsurprisingly, the enjoyment of attaining goals is greatest when the goals are difficult to attain. If we attained the goal at every attempt, our enjoyment would soon vanish.

If attempting to attain the goal of an activity is central to our enjoyment of it, it makes sense that lack of a clear idea of what that goal is can detract from our enjoyment. This can happen when we find ourselves in a game—say cricket—in which we don't understand the rules or purpose of the game. Exercising on a piece of equipment about which we know little may be unenjoyable because we don't have a clear sense of what it is we are supposed to be doing.

Some degree of feedback is also central to our enjoyment of physical activity. Imagine yourself shooting free throws. You may have a clear idea of the goal, but each time the ball leaves your hands, the lights go out and you never know whether the ball went through the hoop. At first, the novelty of the situation might make it fun, but soon it would become dull. Knowing how we're doing relative to the activity goal is an important component of enjoyment. Activities such as table tennis or squash provide a lot of immediate feedback; others such as marathon running, in which the outcome occurs more than 2 hours after the gun has been fired, provide much less feedback.

Teachers, coaches, and trainers can play an important role in providing feedback to learners. After each trial or group of trials, the teacher usually provides a commentary on the major mistakes that performers made, along with recommendations for correcting them. This information, when coupled with encouragement, can affect performers' subjective experiences. For example, age-group swimmers are more likely to enjoy swimming when they perceive their coach as one who gives informative feedback along with encouragement following unsuccessful performances (Black & Weiss 1992).

➤ We are more likely to enjoy physical activities when the challenges of the activity match our abilities, when the activity has clear goals and is followed by feedback, and when the activity is arranged in a competitive framework.

Competition

According to Csikszentmihalyi (1990a), competition often heightens enjoyment of physical activity, something we have already noted in chapter 2. **Competition** is not an activity per se but an organizing principle that frames physical activity within a larger purpose. Usually, it is a way of relating to other people by comparing your performances with theirs. It also may involve comparing your performances to a standard such as par on a golf course or a time on a racecourse. Just as we can add poetry to language, color to art, or melody to sound to make them more enjoyable, we can add competition to physical activity to make it more enjoyable.

One problem with adding competition to activities is that it can become an end in itself. When coaches and players value winning above enjoyment of the activity or the basic individual and social values of friendship, caring, and cooperation, they are elevating competition to an end rather than a means. Csikszentmihalyi summarizes the issue:

> The challenges of competition can be stimulating and enjoyable. But when beating the opponent takes precedence in the mind over performing as well as possible, enjoyment tends to disappear. Competition is enjoyable only when it is a means to perfect one's skills; when it becomes an end in itself, it ceases to be fun. (1990a, p. 50)

Factors Related to the Performer

The second category of factors that influence our enjoyment of physical activities consists of those that lie within us. Whether we enjoy a particular type of physical activity depends a great deal on our dispositions and attitudes. **Dispositions** are short-term, highly variable psychological states that may be affected by a host of external factors. **Attitudes** are relatively stable mind-sets toward concrete objects that may be favorable or unfavorable.

Disposition

Csikszentmihalyi (1990b) identified three dispositions affecting enjoyment that are particularly relevant to physical activity: (1) how competent we feel in performing the activity, (2) the extent to which we are able to become absorbed in the activity, and (3) how much control we feel we have over the activity. Brief descriptions of each follow.

- *Perceived competency.* As a rule, we enjoy activities that we do well more than those that make us feel incompetent. Psychologists use the term self-efficacy to refer to how adequate we feel to perform a task (see "Spotlight on Research," p. 115). Research has shown that people who feel competent in exercise are more likely to adhere to a rehabilitative exercise program (Dishman & Sallis 1994), are more likely to engage in higher levels of intensity when participating in physical activities (Sallis et al. 1986), and will be more faithful in attending exercise programs (McAuley & Jacobson 1991). One study has shown that aerobic dancers with higher ratings of self-efficacy reported enjoying the activity more, and they exerted more effort than those with low ratings of self-efficacy (McAuley, Wraith, & Duncan 1991).

- *Absorption.* We all remember times when we were deeply engrossed in an exercise or sport and other times when we simply went through the motions. When we surrender to an activity and become absorbed in it, we lose consciousness of ourselves as distinct entities apart from the activity. Csikszentmihalyi (1990b) found that when people lose a sense of themselves while performing an activity, feelings of enjoyment increase, partly because when we aren't self-conscious, we aren't as likely to judge our own performances. Most of us have higher expectations for our performances than we are able to achieve; thus self-evaluation usually results in negative feedback, which is not enjoyable.

- *Perceived control.* Experiencing a sense of mastery or control over our environment is inherently enjoyable. Knowing that we can master a tough cross-country course, control the basketball (when dribbling, passing, or shooting), control our opponent in a wrestling match, or control our bodies in a difficult gymnastics skill adds to our sense of enjoyment of those activities.

At first glance, people who enjoy **sensation-seeking activities** such as hang gliding, parachute jumping, ski jumping, and rock climbing, in which so many factors are uncontrollable, would seem to contradict this principle. Yet, strange as it might seem, a sense of control is one of the factors that draws people to such high-risk sports (McIntyre 1992). In these cases enjoyment appears to stem not from the presence of threatening forces but from feelings of being able to control these forces through elaborate preparation, training, and painstaking adherence to safety procedures. After all, if we had complete control of our bodies and the surrounding environment, sports of all kinds would be boring because we could always accomplish our goals. A degree of risk and uncertainty is fundamental to our enjoyment.

Attitudes

➤ Everybody harbors certain temporary dispositions that affect his or her enjoyment of physical activity. Enjoyment tends to increase when we feel competent in the activity, when we become absorbed or lost in the activity, and when we feel we have control over our bodies and the environment.

Along with temporary dispositional factors, more stable and enduring attitudes also can affect our enjoyment of physical activity. Just as individual preferences, tastes, and attitudes toward food, movies, clothes, friends, cars, and the like vary widely, so do attitudes toward different activities. Why we vary so much in our tastes for physical activity is largely inexplicable, but there is no question that we do. In the late 1960s, Gerald Kenyon (1968) and his associates designed a scale for assessing attitudes toward physical activity. They used the scale in their surveys of individual preferences for a wide range of activities—in relation to which people could be either participants or observers—and were able to identify six categories of attitudes toward physical activity. Each represents a distinct preference for a particular form of physical activity based on "sources of satisfaction" that accrue to performers or observers:

- Physical activity as a social experience
- Physical activity for health and fitness
- Physical activity as the pursuit of vertigo
- Physical activity as an aesthetic experience
- Physical activity as a cathartic experience
- Physical activity as an ascetic experience

➤ People who have a natural affinity for social experiences are likely to seek out physical activities that provide opportunities for social interaction.

Physical Activity as a Social Experience Some forms of physical activity, such as team sports or group exercises, are intrinsically social events. That is, they normally take place in active social environments and involve interacting with other people. Team sports are the most social of sport activities, as are team aerobics or exercises done in health clubs where others are present. Other forms of physical activity such as rock climbing, surfing, or long-distance running incorporate less social interaction, although even those activities may provide opportunities for social engagement before, during, and after participation. Participants consistently identify social aspects of physical activities as positive experiences

(Neulinger & Raps 1972), and youngsters engaged in youth sports (Wankel & Krissel 1985), as well as elite ice skaters (Scanlan, Stein, & Ravizza 1988), have indicated that developing friendships and experiencing social relationships are important sources of enjoyment. Thus, individuals who enjoy high levels of social interaction are more likely to seek out physical activities that maximize such experiences.

Physical Activity and Special Moments

The social context in which physical activity takes place often provides a setting for some of life's most special moments. In his book *Final Rounds* (1996), James Dodson describes a trip to Scotland with his terminally ill father to play some of the historic courses there one last time. They had their hearts set on playing the most famous course, the Old Course at St. Andrews, but their names were never selected in the daily lottery that determines who is granted admission. After a few days of waiting, they settled instead for a twilight walk on the legendary course and played "air golf," an imaginary round using imaginary clubs and imaginary golf balls.

> Dad teed up his air Top-Flite, took his stance, and swung. "There," he said. "Right over the sheds. Just like 50 years ago."
>
> I teed up my air Titleist and asked, "How fast did that 50 years go by?"
>
> "Stick around. You won't believe it."
>
> I struck my shot and outdrove him, as usual, by at least a hundred yards.
>
> We walked down the fairway side by side. For a change I wasn't really thinking about all the greats who had walked this way to immortality. . . . I was thinking instead, how simply fine and proper it was that my old man and I were finally playing the Road Hole together. (Dodson 1996, p. 216)

What special moments centered around the social context of physical activity have you experienced?

Physical Activity for Health and Fitness Some people value physical activity for the contribution that it makes to their health and fitness. Feeling in shape, knowing that you can meet any physical demand that you might face in the course of a day, can give you a sense of confidence and well-being that adds immensely to your quality of life. But, as we have seen, many people simply enjoy engaging in the activities that lead to the development of fitness. The unique sensations that accompany the physiological response to vigorous and sustained exercise cause many people to return again and again to running trails and gymnasiums.

➤ Some people are attracted to activities requiring a great deal of strength and endurance, possibly because of the health benefits that such activities bring or because they enjoy pushing their bodies to their physiological limits.

Physical Activity as the Pursuit of Vertigo Some people are attracted to a certain category of physical activity that presents an element of risk or thrill, usually through the medium of "speed, acceleration, sudden change of direction, or exposure to dangerous situations, with the participant usually remaining in control" (Kenyon 1968, p. 100). The thrill that comes from disorientation of the body in such activities is called **vertigo.** People pursue vertigo through sensation-seeking activities that reorient the body with respect to gravity, as in amusement park rides, free-fall parachuting, bungee jumping, downhill skiing, ski jumping, or any other activity that creates a sense of danger, thrill, and intense excitement.

We do not know why some individuals seek out such activities whereas others, just as deliberately, avoid them. Petrie (1967) suggested that those who pursue vertigo may tend habitually to reduce perceptual input so that they experience a specific event at a lower level of intensity than would someone who habitually augments (exaggerates) perceptual input. Involvement in vertiginous activities may also represent an attempt to compensate

➤ People who find enjoyment in sensation-seeking activities also are attracted by the challenge of controlling the uncontrollable; people who find enjoyment in the aesthetic side of physical activity are likely to engage in gymnastics, diving, skating, and other movement forms that emphasize grace and beauty of motion.

for tedious experiences in the workplace (Martin & Berry 1974), may be a form of stimulus addiction (Ogilvie 1973, November), or may be the result of complex sociocultural factors (Donnelly 1977). Whatever the cause, our interest in such activities, as both participants and spectators, clearly shows no indication of declining.

Physical Activity as an Aesthetic Experience We normally associate **aesthetic experiences** in physical activities with dance, but some individuals, as either participants or spectators, perceive sport—and to a lesser extent, exercise—as providing certain artistic or aesthetic experiences (Thomas 1983).

The aesthetic element is especially prevalent in gymnastics, diving, ice skating, and other sports in which grace and beauty of movement are primary considerations in awarding scores to competitors. The aesthetic element is less apparent in sports in which the outcome is usually viewed as more important than the manner in which the athlete moves (e.g., football, basketball, or baseball). Nevertheless, it is common to hear sport broadcasters refer to a well-choreographed football play, speak about the dance-like characteristics of a basketball player leaping through the air, or even describe the movements of a tennis player as art in motion.

Exercise seems less likely to be a source of this type of subjective experience, although dance aerobics clearly contains a strong aesthetic component. Outdoor activities such as hiking, mountain climbing, or kayaking bring participants in contact with the beauty of the natural environment, and this element can add to the aesthetic experience of the activity. Women are more likely than men to engage in physical activities for the aesthetic experiences that they provide, whereas men are more likely to value activities for their vertiginous, ascetic, and cathartic elements (Smoll & Schutz 1980; Zaichowsky 1975).

Physical Activity as a Cathartic Experience Catharsis refers to a purging or venting of pent-up hostilities, either through attacking an enemy or some inanimate surrogate (object) in an aggressive fashion or through watching an aggressive event. Such a hypothesis owed to the work of behavioral physiologist Konrad Lorenz who wrote in his best selling book *On Aggression* that sport furnished a healthy safety valve for "militant enthusiasm" which he viewed as the most dangerous form of aggression (Lorenz 1966). Scientists have since questioned this "hydraulic hypothesis" of aggression generally (Berkowitz 1969) and its application to sport specifically. It is no longer viewed as a credible hypothesis for explaining our attraction to sports. For example, if watching sports lowered levels of aggression in spectators, we would expect to find lower levels of aggression in spectators following their attendance at a game than before the game begins, but research has shown just the opposite to be the case (Goldstein & Arms 1971).

But physical activity can be viewed as cathartic in another way. Scientists have discovered that a vigorous bout of physical activity can lower anxiety (Morgan 1982), thereby inducing a sense of relaxation and calm (but not necessarily reducing pent-up hostilities). Leisure theorists have long pointed to

Seeking sensation through dangerous physical activities is called pursuit of vertigo.
Photodisc/Getty Images

Spotlight on Research

The subjective experiences of physical activity can be influenced by a wide assortment of variables. One of the most obvious is how a person feels when he or she exercises or engages in sport. Another is the influence of various social variables such as parental and peer support. The following are two studies that explored these issues.

Study A

- *Purpose:* To examine changes in feeling states and self-efficacy (self-confidence) before, during, and after exercise.
- *Subjects:* 168 children aged 9 through 17.
- *Method:* Children walked at a fast pace on a treadmill for up to 1 hour. Their feeling states were tested at regular intervals during the exercise.
- *Results:* Higher self-efficacy during the exercise was associated with greater enjoyment of the task. Feeling states worsened as the exercise session progressed, especially for adolescents in early puberty, who felt significantly worse than those in middle or late puberty at 16 minutes and 20 minutes into the exercise. Here are some of the data from the study:

Differences in feeling states during exercise

	Early puberty	Middle puberty	Late puberty
Before exercise	4.42	4.53	4.23
At 4 minutes	2.81	3.15	3.12
At 12 minutes	0.95	1.24	1.71
At 16 minutes	0.11	0.69	1.38
At 20 minutes	−0.33	0.18	1.23

Scores range from a high of 5 ("very good") to −5 ("very bad").

Adapted from L.B. Robbins et al., 2004, "Exercise self-efficacy, enjoyment, and feeling states among adolescents," *Western Journal of Nursing Research* 26(7), 699-715.

Study B

- *Purpose:* To identify various social variables that predict children's involvement in physical activity.
- *Subjects:* 111 boys and girls in the 5th and 6th grades (Phase I) and 8th and 9th grades (Phase II) and their mothers.
- *Method:* 45-minute in-home interviews with children and mothers to assess the degree of vigorous physical activity. The study also used a standardized instrument to determine the presence of various social learning variables including, but not limited to, child's report of parent's involvement in physical activity, amount of exercise-related equipment at home, number of hours child spent watching television, parent's reported physical activity level, and parent's reported enjoyment of physical activity.
- *Results:* During the 5th and 6th grades, children's enjoyment of the activity was the only consistent predictor of physical activity levels, especially for girls. For boys, enjoyment, family and friend support, and modeling of physical activity by others were identified as predictors. Other variables had little or no bearing on physical activity level. In the 8th and 9th grades, peer and family social support were relevant predictors for girls; but for boys, personal interest in activity, including interest in sport media, were the most important predictors. Children's level of physical activity in 5th and 6th grades was not a predictor of level of physical activity in the 8th and 9th grade for girls, but it was for boys. Interestingly, father's enjoyment of physical activity was a predictor of physical activity for girls, and father's physical activity and self-efficacy for physical activity were predictors for boys.

➤ Participating in sport and exercise will not purge us of hostilities and aggression but may calm, relax, and refresh us.

the value of leisure activities (including sport) in promoting feelings of rest and relaxation. Sport and exercise are novel activities that provide us with a change of pace. This change may, by itself, recharge our batteries by shifting our attention from problems and worries. Thus, the noontime racquetball player may derive a sense of enjoyment, relaxation, and calm simply because he has directed his energies toward a new task and has become absorbed in its features, not because he has purged hostilities.

Physical Activity as an Ascetic Experience Physical fitness and training programs often require us to undergo pain, sacrifice, and self-denial and to delay gratification. These experiences are sometimes referred to as **ascetic experiences.** Elite athletes are accustomed to ascetic experiences—torturous training regimens intended to improve their capacity for performance. Ascetic experiences are also familiar to recreational athletes. Occasionally, you may see a runner, who by her attire and running style is obviously a novice, wheezing and groaning, and from all appearances suffering. Fitness clubs specialize in ascetic experiences; patrons with pained faces struggle to complete their exercise routines while a personal trainer shouts encouragement as they work through the pain.

DO it **Activity 4.4**

Attitudes Toward Physical Activity
Test your understanding of the six main reasons people engage in physical activity in your online study guide.

➤ Sometimes participating in exercise and sport can be painful and can lead to uncomfortable subjective experiences. Participants do not always interpret these sensations negatively, and, indeed, the discomfort may be a source of attraction.

You may be surprised to learn that not everybody interprets the ascetic experiences of breathlessness, fatigue, and muscle soreness as unpleasant. For example, in one study, 12- and 13-year-old soccer players told investigators that they enjoyed not only winning, learning new skills, and playing with teammates but also "working hard" and "feeling tired after practice" (Shi & Ewing 1993). In another study (Wankel 1985), exercisers who did *not* drop out of the exercise program reported experiencing physical discomfort more frequently than those who *did* drop out. In this study, middle-aged men who reported experiencing fatigue because of vigorous exercise also reported *decreased* feelings of distress and *increased* feelings of psychological well-being because of the exercise. Thus, for some people, the discomfort of physical activity is attractive rather than unappealing.

Factors Related to the Social Context

The third category of factors that can affect our enjoyment of physical activity concerns the nature of the social context in which the activity occurs. Have you ever competed in an important sport contest with a large crowd looking on? How did that compare with playing the same sport on the playground or in intramurals? Do you feel differently when you work out in a fitness center with many onlookers than you do when you work out alone? Most people find these experiences profoundly different. One isn't any better than the other; they are simply different.

How participants feel when they are engaged in physical activity may depend on the social conditions surrounding the activity as much as or more than the activity itself. The presence of others (parents, friends, strangers), the hype preceding a contest, players' feelings toward and relationship with an opponent, and the way media or coaches interpret a particular game to players can all affect how the players experience the sport. Running on a treadmill offers a different subjective experience than running through a park. Running when it is cold, dark, and raining offers a different subjective experience than running on a bright, warm summer day. Adding people to your exercise group changes the subjective experience of exercising just as adding competitors can change the social climate and the subjective experience of playing a game. Whether these changes add to or subtract from your enjoyment depends on your individual preferences.

DO it **Activity 4.5**

Your Attitudes

Complete a self-evaluation about your attitudes toward physical activity in your online study guide.

The social atmosphere created by the exercise leader can have a powerful influence on exercise class participants' enjoyment of the experience. Fox, Rejeski, and Gauvin (2000) reported on a study designed to test the effects of leadership style and group environment on enjoyment experienced by those attending a step aerobics class. The instructor, highly trained in aerobics, adopted either an "enriched style" that included high levels of interaction with class members (addressing them by name, providing positive reinforcement, orally rewarding effort, and so on) or a "bland style" (not addressing participants by name, avoiding conversation, not providing positive reinforcement, not rewarding effort, and so on). The investigators manipulated group environment by planting undergraduate students in the experimental sessions. The planted students established an "enriched" environment in one exercise group and a "bland" environment in another. In the rich environment, the students introduced themselves to all participants, initiated casual conversation, were compliant with the instructor's wishes, made positive remarks about the instructor, and so forth. In the bland environment, the students did not introduce themselves to other members, did not initiate conversation, were compliant but not enthusiastic or orally supportive of the instructor, and so forth. Thus, the study established two leadership styles and two group environment conditions. The investigators examined the effects of each condition as well as combinations of the conditions on participants' enjoyment. They also examined the effect of these variables on participants' intentions to take another exercise class.

As you can see by the data on this page, social environment and leadership affected both enjoyment and participants' intentions to enroll in another exercise class. Ratings of enjoyment were higher for the enriched leadership style than for the bland leadership style. Similarly, ratings were higher for the enriched group environment than for the bland group environment. But the most powerful effect was found for the condition in which participants were exposed to an enriched leadership style and exercised in an enriched group environment. Ratings of enjoyment were, on average, 22% higher for this condition than the other conditions. Intentions to enroll in another class were highest in this condition as well (approximately 16% higher than in the other conditions). The conclusion is easy to draw: Both the style adopted by the exercise leader and the social environment in which the exercise occurs can have a major effect on participants' enjoyment.

The Contribution of a Dynamic Leader and a Dynamic Social Environment to Enjoyment of Step Aerobics

		Leadership style	
Rating of enjoyment	Group environment	Enriched	Bland
Ratings from a high of 10 to a low of 1 with an 8 anchored by "enjoyed a lot"	Enriched	8.40	6.74
	Bland	6.24	6.26
Rating of interest in taking another class	Group environment		
0% = not interested	Enriched	70.87	60.00
100% = very interested	Bland	47.50	55.50

Adapted from L.D. Fox, W.J. Rejeski, and L. Gauvin, 2000, "Effects of leadership style and group dynamics on enjoyment of physical activity," *American Journal of Health Promotion* May/June, 14(5), 277–283.

➤ We never perform physical activities in a vacuum, and the social context that surrounds them can affect our sense of enjoyment. An example of social context occurs when we feel forced to engage in an activity rather than freely choosing to do it.

Another way in which the social context can affect our enjoyment is by influencing our sense of **perceived freedom.** As you learned in chapter 2, leisure theorists have long known that we enjoy activities more when we are free to choose them than when we feel obligated to do them. In fact, Bart Giamatti, former president of Yale University, and at the time of his death the commissioner of Major League Baseball, described leisure activity as "that form of nonwork activity *felt to be chosen, not imposed*" [italics added] (1989, p. 22). This sense of freedom to participate is not simply a matter of being free from actual coercion; it is feeling free from any subtle coercive forces that instill a sense of obligation. Getting up early on Saturday morning to lift weights with a friend because she has been begging you for weeks to join her robs you of your sense of freedom, and you may have a far different subjective experience in this situation than you do when you lift weights because you want to. Any social condition that creates within you a sense of obligation to engage in physical activity or to remain engaged in an activity can erode your enjoyment of it (although it doesn't have to).

Watching Sports as a Subjective Experience

One of the most influential people in the development of modern physical education theory was Jesse Feiring Williams, a medical doctor turned professor who devoted his life to preparing generations of physical education teachers, coaches, and professors. Like many scholars of his generation, Williams (1964) was skeptical of sports watching because he believed that it demanded only "simple sensory responses" and did not require "expressive, cooperative skill activity" that could be related to the purposes of education. Although one might question Williams' assessment, it is obvious that the rise of sports watching that so worried him in 1927 has continued into the 21st century and, for the most part, shows no sign of stalling as we discover more sports to watch and more ways to watch them. Figure 4.4 shows the dramatic rise in attendance at sport events between 1990 and 2006. The increases in attendance at National Collegiate Athletic Association women's basketball have been particularly dramatic. Whether all of this sedentary sports watching has been beneficial to the public or not, it is clear that it is a source of vivid subjective experiences for wide segments of the human community. Because of its connection to the physical activity professions, we should examine, if only briefly, the factors that affect our enjoyment of and attraction to sports watching.

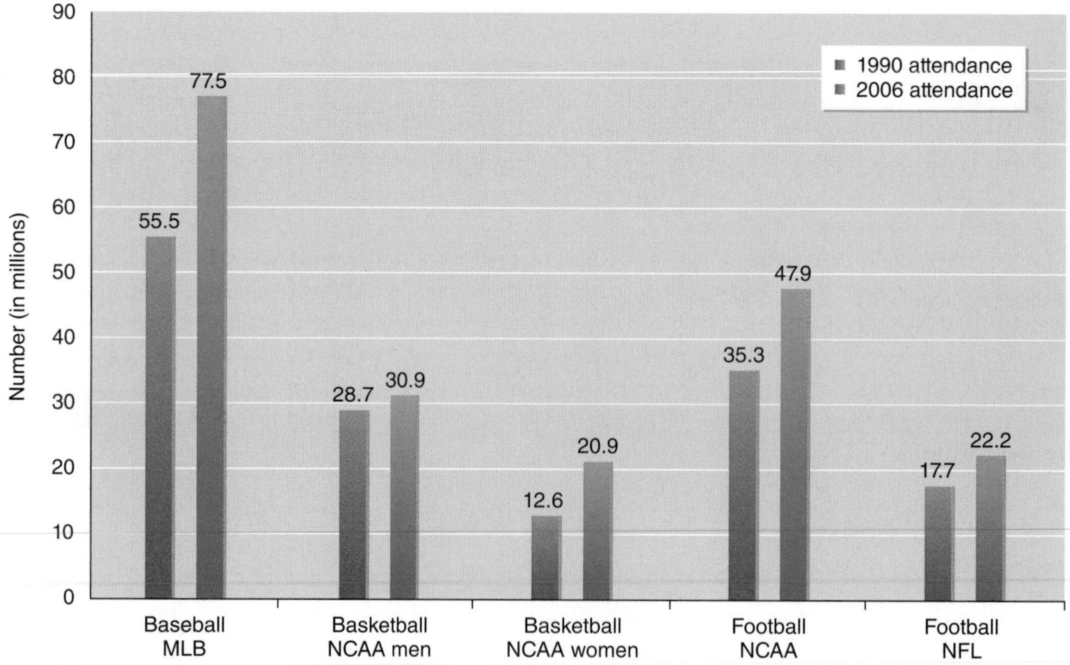

FIGURE 4.4 Attendance at selected spectator sports from 1990 to 2006.

Watching a sport contest on the local playground is unlikely to bring about the subjective experiences that are evoked in a sport spectacle such as the Olympic Games, and neither may evoke the same subjective experiences that come from watching sports on television. Although we often watch our peers play sports in informal settings, **sport spectacles** attract most of our attention. Sport spectacles are staged competitions designed and promoted for audiences and intended to evoke an entire range of human emotions by virtue of their grandeur, scale, and drama. Unlike pickup games on the playground, they are primarily entertainment.

Usually, we reserve the term *sport spectacle* for professional, collegiate, or international events that attract large numbers of spectators and have supporting casts of cheerleaders, bands, majorettes, officials, and the media. In spectacles, both players and spectators play defined roles; the players take on various roles associated with the competition, and the spectators take on various roles as consumers, supporters, or detractors. Thus, when spectators cheer, boo, do the wave, or sing the alma mater, they are participating in a spectacle, of which the sport contest is only one aspect.

Ways of Watching Sports

We can engage in spectating in many different ways, ranging from aligning ourselves with a team's performance and suffering extreme disappointment when the team loses or jubilation when it wins, to the other extreme of watching in a completely disinterested fashion.

Vicarious Participation

Vicarious participation is a form of watching in which observers participate in the contest through the powers of imagination. If, while watching a sport event, you notice yourself tensing your muscles or adjusting your body position in accordance with an athlete's movements, you probably are participating vicariously in the activity. Usually, vicarious participators are fans who identify with a particular player or team and have a stake in the outcome of the contest. The word *fan* is rooted in *fanum,* the Latin word for temple, which underscores a metaphoric link between religious zeal and the enthusiasm fans often have for their favorite teams and players. When their team wins, it is as though they have won; when their team loses, they experience profound disappointment.

➤ Vicarious participation in sport occurs when spectators imagine themselves performing the same activities as the athletes they are watching.

Disinterested Sport Spectating

Sometimes we watch sport events without great emotional investment, a phenomenon known as **disinterested spectating.** We are likely to watch games in this manner when we care little about the outcome of the contest. Watching a contest between two teams we are not familiar with or watching a game that we have attended only because a friend invited us may result in disinterested spectatorship. Those who watch sports as part of their employment, such as reporters, referees, hot dog vendors, and sometimes coaches, often do so in a disinterested fashion.

Nevertheless, it would be a mistake to assume that disinterested spectators derive no pleasure from watching contests. For example, coaches who watch competitions between teams other than their own tend to observe athletes' performances in a detached, objective manner, yet they may enjoy and admire the performances of those athletes. In these cases, spectators watch in much the same way audiences watch a symphony or a ballet: Enjoyment comes not from the drama of competition but from observing the skillful actions of the performers.

Factors Affecting Enjoyment of Watching Sports

If someone were to ask you if you like to watch sporting events, you might answer sometimes yes, sometimes no. Your answer depends on a number of factors. Few people enjoy watching their favorite team being trounced, and fewer yet are likely to enjoy watching a game that they know little about. Americans are unlikely to get excited about watching cricket or Australian rules football for the same reason that Australians and citizens of the United Kingdom are unlikely to enjoy American baseball. Our familiarity with sports and their importance in our culture often influence our attitudes toward them. Three factors play a major role in determining how much enjoyment we are likely to experience when watching a sport event: our knowledge of the game being played; our feelings toward the competing teams and players; and the extent to which the game entails a sense of drama, suspense, and uncertainty. Let's look a bit more closely at these factors.

Game Knowledge

Think about the sports you most like to watch. These sports are probably the ones that you know the most about. Knowledge about the game, what sport philosopher Scott Kretchmar (1994) calls **game spectator knowledge,** often determines our enjoyment as spectators. Game spectator knowledge is knowing about the game, including players, strategies, and competitive tactics. Only by having comprehensive knowledge of the activity can we fully appreciate the quality and significance of the athletes' performances. Such knowledge may come from reading about, watching, or participating in an activity, although it is not necessary to have played a sport to enjoy watching it. An individual who knows little about the game of lacrosse may enjoy watching a game without knowing the game's history, strategies, rules, or personalities. But it is unlikely that this person's enjoyment of the game will be as robust as that of the lacrosse fan who understands precisely what the players and teams are trying to do at every turn of the game.

Feelings Toward Competing Teams and Players

Think of the last time you watched a contest in which one of your favorite teams was competing against an opponent that, for one reason or another, you had come to dislike. What turn of events added to your enjoyment of the experience? Probably you enjoyed the experience more when good things happened to your favorite team (when they played well and scored points). Your enjoyment quotient probably also increased when bad things happened to the team that you disliked (when they played poorly, failed to score, or even suffered injuries). You probably experienced the highest level of enjoyment if your favorite team defeated the other team, and, conversely, you probably experienced the least enjoyment if the opposing team defeated your favorite team.

➤ Having a comprehensive knowledge of the players, rules, and competitive strategies adds to our enjoyment of watching sports, as do the feelings we harbor toward the participating teams.

These connections between enjoyment and feelings about the competing teams are the focal point of the dispositional theory of enjoyment that has been the subject of considerable research (Zillman, Bryant, & Sapolsky 1979) involving football, tennis, and basketball spectators. For example, researchers have discovered that spectators applauded the failed plays of the disliked team almost as much as they did the successful plays of the favorite team. Comparing reports by those who watched the Minnesota Vikings defeat the St. Louis Cardinals in a professional football game, they found that those who both liked the Vikings and disliked the Cardinals reported maximal enjoyment, whereas those who both disliked the Vikings and liked the Cardinals reported maximal disappointment (Zillman & Cantor 1976).

More recent research has shown that those who identify with a particular team are more likely to experience what researchers call the bask-in-reflected-glory feeling (BIRG), expressing greater personal joy when their team wins than do those who have identified with the team to a lesser degree. Some evidence suggests that the enjoyment brought about by BIRGing after a fan's favored team wins can lead to elevated self-esteem and an overall boost in self-confidence. By contrast, diminished self-esteem and self-confidence have been associated with losses by one's favored team. Our enjoyment of sport spectating also tends to increase when our favorite team defeats an opponent of high rather than low quality. In addition, we tend to BIRG and enjoy the game more when our favorite team plays much better than we had originally expected it to (Madrigal 1995).

Human Drama of Sport Competition

The sense of drama, suspense, and uncertainty that often accompanies sport contests also enhances our enjoyment of them. Usually, this sense of drama requires that the competing teams be equally matched in talent and ability, although some of the best drama in sport occurs when the underdog overcomes great odds to defeat a heavily favored opponent. Games in which one team far outplays the other (as in the typical Super Bowl game) tend to be less enjoyable, even when teams are evenly matched.

> **www. Web Search 4.1**
>
> **Sport Parents**
> In this activity in your online study guide, you'll search the Internet for information on parents' involvement in their children's sports activities.

Enjoyment also is related to the extent to which spectators perceive the contest to involve human conflict. Research by Zillman and his associates (Zillman, Bryant, & Sapolsky 1979) showed that when identical sport telecasts were viewed in the presence of broadcaster commentary that described the game as either involving bitter rivals who felt intense hatred for each other or involving friendly opponents, the former condition elevated spectators' enjoyment. Also, rough and aggressive play has been found to increase spectator enjoyment, presumably because it demonstrates that players are sincere in their efforts to win.

These findings demonstrate that in spite of the enormity of the sport spectator business, we know relatively little about the subjective effects of sports watching on the nation's collective psyche. Is watching sports a constructive use of our time? When the game is over, do we feel good about having watched it? Is it possible that sport spectating is a culturally hollow experience, little more than junk food for the human spirit? Is it sometimes a debasing experience? What changes might make sport spectating a more ennobling human experience? These important questions offer fertile ground for research by future kinesiologists.

Fans enjoy watching games more when a team they identify with is winning and a team they don't like is losing. They also find it fun to "bask in the reflected glory" of a winning team.

Reflecting on Your Subjective Experiences as a Spectator

One way to examine your experiences in watching sports is to interrupt your viewing of an event periodically to record your feelings at that moment. You can easily do this at home when watching sports on television, but you can also do it in your seat in the arena or stadium. Set a timer or wristwatch alarm for 10-minute intervals. Each time the alarm sounds, quickly jot down what you are feeling, why you think you are feeling as you are, and what is happening in the sport event that contributes to your feelings. You might also briefly note any effects that the social environment (people you are watching the game with at home, or the stadium crowd) might have on your feelings. After the contest is over, take time to review your notes and reflect on the total experience. How would you describe the overall emotional effect of the experience? What kind of knowledge, if any, did you gain from the experience? Would you describe the time that you spent watching the event as meaningful? Have you gained insight into the effects that sports watching has on you? How would you evaluate these effects—largely desirable or largely undesirable—of knowledge?

Wrap-Up

Let's take stock of what we have learned about physical activity in this chapter. The theme of the chapter is that physical activity involves more than moving our bones and muscles or increasing the flow of blood through our circulatory systems. When we move our bodies, unique sensations arising from internal and external sources give rise to emotions, thoughts, and other inner states. We have termed these sensations, along with our interpretations of their meanings, the subjective experience of physical activity. Subjective experiences of physical activity are important, not only because they play a significant role in determining our activity preferences but also because, for many people, they constitute the primary reason for engaging in physical activity in the first place. They also are keys to helping us learn about ourselves in physical activity environments.

The goal of most physical activity professionals is to help people develop an intrinsic orientation to sport and exercise, although in most cases participants' extrinsic orientation also plays a part. One way to think about an intrinsic orientation is as enjoyment. Anything that we value for its intrinsic qualities usually constitutes an enjoyable experience for us. Many factors can make sport and exercise more or less enjoyable, including those that lie within the structure of the activities themselves, those related to the dispositions and attitudes of the performers, and those associated with the social context in which physical activity is performed.

Finally, we examined briefly the subjective experiences associated with watching sports and saw that people have different ways of watching. Our subjective experience is affected by our knowledge of the game, our feelings about the competitors, and the drama of the competition, including its description by sport broadcasters. As yet, research concerning the benefit or harm of extensive sports watching has been limited. Because a significant part of our profession (sport management, athletic coaching, and administration) focuses on preparation of personnel for sports watching, we need to monitor its long-term effects carefully.

This brings to a conclusion our study of what it means to experience physical activity through performance and spectatorship. You are now ready to proceed with an in-depth study of the research knowledge about physical activity. As you become engrossed in the basic concepts of the scholarly subdisciplines of our field, do not forget that performing physical activity is an important source of our knowledge about it.

Activity 4.6

Use the key points review in your online study guide as a study aid for this chapter.

Activity 4.7

These end-of-chapter questions and activities are also in your online study guide. Your instructor may ask you to complete them online and turn them in.

1. Give an example of how an individual may *internalize* a daily run through the park.
2. List and describe three types of knowledge available to us from subjective experiences in physical activity.
3. What evidence exists to refute the notion that people don't like the sensations that accompany the hard physical effort required to exercise vigorously?
4. What types of physical activities might a person who values physical activity as an aesthetic experience engage in? A person who values physical activity as an ascetic experience? As a social experience?
5. What are the various ways in which one can watch sports? In your opinion, which ones add most to your knowledge of physical activity?

PART II

Scholarly Study of Physical Activity

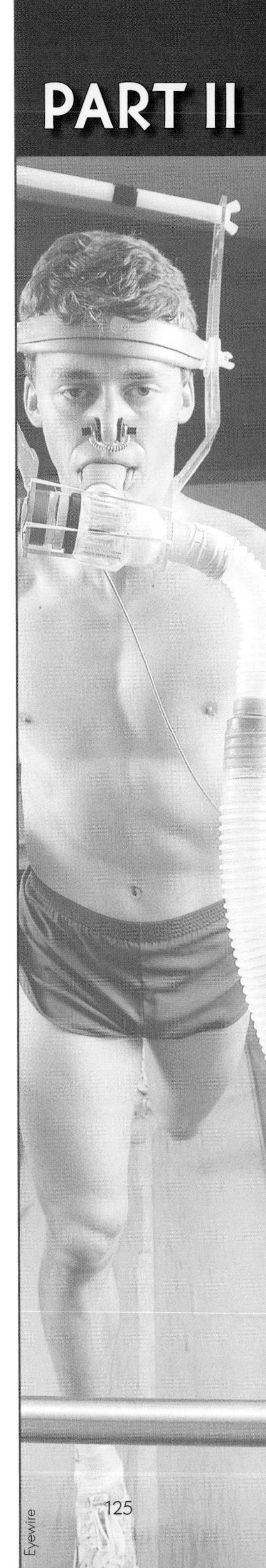

Although learning through physical activity experiences is an essential part of becoming a kinesiologist, one of the defining marks of a professional in this field is mastery of a complex body of knowledge about physical activity. Laypersons without any training in kinesiology who have been exposed to a variety of physical activity experiences may have learned much about physical activity and themselves. But it is unusual to encounter people who possess in-depth knowledge about the theoretical aspects of physical activity, including its sociocultural, behavioral, and biophysical dimensions, apart from those who have successfully negotiated a formal course of study in kinesiology. Some people who haven't formally studied kinesiology may *think* they know much about physical activity (and some, of course, may), but often their knowledge is incomplete, out-of-date, or simply untrue. As a kinesiology major, you—more than the typical person on the street, and more than your friends who might be studying biology, chemistry, history, or some other discipline—need to possess a solid knowledge base about physical activity.

Physical activity experience

Professional practice centered in physical activity

Scholarly study of physical activity

In this part of the book you will be introduced to seven areas of study, or subdisciplines, each of which deals with a different dimension of physical activity. Chapter 5 focuses on philosophy of physical activity, asking and answering questions about values, meaning, and ethics in sport and exercise. Chapter 6 is devoted to the history of sport and exercise. It also will provide you with a glimpse of how the field of kinesiology evolved. In chapter 7 you will examine relationships between sport and exercise and some of the pressing social issues of our day, such as racial and gender discrimination. Chapter 8 directs your attention to the performance of skilled movement and factors that affect it, including those related to human development. In chapter 9 you will study the psychological aspects of physical activity and will learn how this important part of our lives affects and is affected by sport and exercise.

Chapter 10 introduces you to the fascinating world of biomechanics, an approach to studying physical activity that will give you insights into the way in which forces affect and are affected by physical activity. Finally, chapter 11 reviews basic physiological aspects of physical activity, including the effects of exercise on muscular and cardiovascular systems and the physiological basis for training. Together, this knowledge, and the hundreds of concepts, theories, and principles embedded in it, will teach you much about physical activity. You may already have recognized that the organization of the discipline of kinesiology mimics the typical organization of liberal studies. Studying philosophy of physical activity, history of physical activity, and sociology of physical activity builds on and extends a way of thinking that you will develop by taking courses in the philosophy, history, and sociology departments. Similarly, motor behavior and psychology of sport and exercise build on and extend basic knowledge of psychology;

biomechanics builds on both biology and mathematics; and physiology of physical activity builds on general physiology.

Because each subdiscipline of kinesiology is, to some extent, an extension of another older discipline (e.g., psychology, sociology, biology, history, and so on), your past experiences with these disciplines will help you understand the theories and terminology used in the subdisciplines.

As you embark on your study, you may wonder at times whether what you are learning about is really an organized area of study. At times, kinesiology seems to be merely a collection of bits and pieces of theory and concepts borrowed from these older established parent disciplines. This notion can be troubling to those who prefer to think that kinesiology is an exclusive science that contains its own theories and scholarly methodologies. You need to recognize that the precise relationship between kinesiology and the parent disciplines and how the field of study should be organized are subjects of ongoing debate among kinesiologists. The most popular view is that kinesiology consists of unique knowledge about physical activity drawn from other disciplines, knowledge that those disciplines had largely ignored. If biologists, for example, had concentrated on studying and teaching about the effects of exercise, perhaps the subdiscipline of exercise physiology would not have evolved. If philosophers, historians, and sociologists had focused their energy on sport and exercise, the subdisciplines of philosophy of physical activity, history of physical activity, and sociology of physical activity may not have evolved. Because knowledge about physical activity was judged important, and because other disciplines ignored it, the 1960s witnessed the rise of the subdisciplines of kinesiology. Thus from its beginning, kinesiology has consisted of knowledge about physical activity gleaned from all the traditional disciplines and combined into a coherent discipline in its own right—the discipline of kinesiology. In addition, of course, the discipline of kinesiology includes unique knowledge gained from experiencing physical activity (see part I) and professional practice centered in physical activity (see part III).

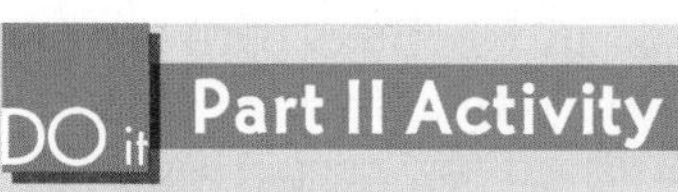

An Integrative Discipline

Before you begin part II, listen to this insightful conversation among experts in the field in your online study guide.

Although the chapters in this section are jam-packed with information, they actually represent a "once over lightly" survey of the scholarly subdisciplines. When you finish reading and studying them, you will not have mastered the contents. You will, however, have gotten a glimpse of what lies ahead of you in your program of studies, when you will have a chance to delve into each of these subdisciplines in greater depth.

5

Philosophy of Physical Activity

Scott Kretchmar and Cesar R. Torres

CHAPTER OBJECTIVES

In this chapter we will

- examine the nature of philosophic thinking;
- describe how philosophy fits into the field of kinesiology;
- review the history of the philosophy of physical activity;
- describe the research methods in the philosophy of physical activity; and
- provide an overview of knowledge in this domain by focusing on three issues: the nature of the person, the nature and value of play, and the ethics of sport.

AP Photo/Joe Imel

Late in the season, you are playing a basketball game against your rival high school. The contest will determine the league champion. Both sides have played well, and the outcome of the game has come down to the final seconds. Your squad is trailing by 1 point with only 3 seconds remaining. You have the ball out of bounds underneath your own basket, and your coach signals a time-out.

In the huddle by the side of the court, the coach outlines a most unusual plan while looking you directly in the eye. "When the referee hands us the ball," he says, "I want you to run to the sideline away from the ball, fall to the floor, and scream at the top of your lungs. When the players on the other team look to see what has happened, we will throw the ball to an open teammate right under the basket. I checked the rulebook," he continues, "and there is nothing at all illegal about this. Now go out there and win this game for Central High School, your fellow students, and fans!"

You are uncomfortable with the idea. Is this the right way to win a basketball game? Do you have an obligation to follow a coach's direct orders even when you think they may not be ethical? Is this kind of strategy simply "part of the game," and should the other team be prepared for such tactics? You wonder if anybody has answers to these kinds of questions. Maybe it all boils down to personal opinion—the coach's opinion against yours. What should you do?

Maybe sports aren't your thing, so here's another example: You are a junior high school physical education teacher. (If you are headed into a health or medically related field, you can substitute that setting.) You have several students (or clients) who are obese and clearly out of shape. You want to make a difference in their lives and get them involved in regular, healthful physical activity. Your goal is an ambitious one. It is nothing less than turning their lives around by promoting activity for a lifetime. You wonder how you will do it.

You decide to use a cognitive approach because they are old enough to understand both the importance of health and how activity is related to caloric expenditure and other health-related biological mechanisms. In short, you will motivate them to be healthy, and you will explain how activity combined with a good diet can get them there. You will have them set realistic goals and introduce them to exercises that are designed to meet their individual needs.

You begin your semester with energy and enthusiasm. The students seem fairly responsive. Some even report a loss of weight, an increase in energy, and general satisfaction with your class. After several weeks of exercise, your own testing procedures show them that their fitness measures are on the rise.

All too soon, however, the unit comes to an end. Your students are now on their own. You wonder if they will continue to be active without the support provided by your program. You wonder about the way you chose to motivate them. You presented movement as prudent, as something of a duty. What if you had been more playful? What if you had tried to present movement more as a delight than a duty? What is more important—the health of your students or the various joys and delights of living? You wonder!

The example that started this chapter actually occurred in a high school basketball game a few years ago, only the person who was the decoy went to the corner of the court, got down on all fours, and started barking like a dog. The other team was understandably distracted, the ball was passed in under the basket, the layup was made, and the home team pulled out a remarkable last-second victory. This tactic is now referred to as the famous "barking dog trick."

Because of that event, it is now illegal to distract opponents in this way. However, at the time of the incident, no rules were broken. In our version of this story, the coach underlined this fact as a partial justification for his plan. Some basketball people, in fact, argue that *all* legal methods for winning games should be at any team's disposal. If this view is defensible, such tactics are simply "strategies," and their deployment is nothing more than an example of one coach outsmarting another.

> DO it **Activity 5.1**
>
> **Test Your Critical Thinking Skills**
>
> Use your online study guide to develop arguments on both sides of this issue.

On the other hand, some individuals regard these strategies as clear examples of poor sportsmanship. They are desperate measures designed to get another "W," tainted victories gained by means of cheap tricks. For those who see matters this way, such victories are not a source of pride but more occasions for embarrassment. Which side has it right, and can philosophy provide an objective argument to show whether these kinds of tactics are ethical or not?

Kinesiology professionals understand the joy that comes from physical activity.
Photo courtesy of Joe Luxbacher

The dilemma the teacher faced in the second chapter-opening example is a core dilemma for all of us who work in the domain of physical activity. We love to move. We love to compete and dance and swim and bike. This is the play side of us. We see physical activity as a "jewel" that adorns our lives and makes them go better. From this perspective, movement is, and ought to be, presented as intrinsically valuable.

On the other hand, we know the importance of health and fitness. The obesity epidemic concerns us. The declining health and vitality of many of our citizens who lead sedentary lives bother us. This is the prudential, duty side of us. We see physical activity as a "tool" that allows us to live longer and get more out of life. From this perspective movement is, and ought to be, presented as extrinsically valuable.

Once again we can ask, which side has it right? Should we be playmakers more than purveyors of duty? Or should we be no-nonsense health promoters more than frivolous fun-time activity leaders? Can philosophic methods help us reach reasonable and forceful conclusions on these questions? Even if it should be a mixture of the two, what is the right balance? And finally, does philosophy give us more than one person's opinion over another's? Answers to these questions lay a foundation for central philosophic commitments that all movement professionals must make. They were identified in chapter 3 as among the most crucial issues we face—and indeed, they are. This is so because they color our curriculum, our interventions, our attitudes toward physical activity, the way we treat our students and clients—that is, virtually everything we do.

Introduction to Philosophic Thinking

Philosophic thinking involves reflecting more than it requires testing, measuring, or examining things with any of our senses. Physiologists test physiological responses, for example, by putting people on treadmills to find out how they react to different conditions. Biomechanists measure the forces generated by different movement techniques or various kinds of equipment. Psychologists examine the subjective reports of individuals who have been interviewed, and historians analyze various documents for information about our past.

Philosophers also learn a great deal by living in the world and experiencing it. Thus, in some ways they also test, measure, and examine. But they use this information from daily living for later reflections. In other words, the tools they use are those related to different ways of reflecting, using logic, speculating, imagining, or, simply, thinking! Later in this chapter, you will have a chance to learn about and practice four of these ways of reflecting.

But here it is important to notice two things: first, that philosophic methods typically do not include the gathering of data from controlled experiments, and second, that various types of reflection are used instead. A third issue is important as well.

It has to do with the validity and reliability of philosophic reflection. Contrary to what some believe, and perhaps at odds with some of your own doubts, philosophy can produce results that are as valid and reliable as anything discovered by the physical sciences. Of course not all philosophy fits this description, but that does not mean that everything in philosophy is unreliable and thus little more than guesswork or opinion.

If you doubt that this is the case, take the following two tests and see what you come up with. Our guess is that you will report the same exact two answers that we did, even though we are not measuring anything, and even though we have no data, no p values, no control groups, no double-blind research design, and nothing physical to work on. The first test is grounded in deductive logic. The second one requires discernment, intelligent recognition, or, more simply, good judgment.

Validity and Reliability Test 1

1. If it is true that basketball, baseball, soccer, and football are team sports, and
2. if it is true that John and Suzy are now playing one of these four sports,
3. then it follows that . . . (select one)
   a. John and Suzy are playing an individual sport.
   b. John and Suzy are playing a team sport.
   c. John and Suzy may or may not be playing a team sport.
   d. John and Suzy definitely are not playing a team sport.

The answer has to be *b* if the two premises are true. We "see" the right answer by following the deductive logic of this syllogism. Moreover, every sane person of reasonable intelligence could be expected to see the same exact thing.

Validity and Reliability Test 2

Here is the second test question, one that is very different from the first example. Philosophers, theologians, and secular ethicists have often said that it is "wrong to harm anyone without cause or justification." What is the status of this claim? (select one)

a. It is simply personal opinion; the opposite opinion that it is absolutely right to harm others in any and all circumstances is just as valid.
b. It should be followed in some parts of the world but should not be followed in others where different traditions exist. If some religion, for example, teaches that it is right to harm others without cause or justification, then we have to respect that tradition.
c. It is good moral advice and should apply to everyone regardless of background, nationality, and personal beliefs. Harming others without cause is wrong. Good people don't do that!
d. It is old-fashioned ethics that has no relevance in the modern world. We now see that it is acceptable to harm others, particularly if significant personal gains are produced.

If you think at all as we do, *c* has to be the preferred answer. Who would want to rally an argument to the effect that it is morally acceptable, without justification, to harm someone—for instance, to torture an innocent child? Who cares if there might be a cultural tradition somewhere in the world that condones such criminal behavior or if someone's opinion is that such torture is OK? If such a tradition exists, it is a bad tradition; and if someone holds such an opinion, we probably would be well advised to not listen to this "twisted" individual.

➤ Philosophers reflect more than they measure. They use the tools of logic, intelligent discernment, and speculation to make claims about everything from the nature of reality to proper human behavior. Like other areas of research, philosophy frequently uncovers plausible truths that need to be reviewed and revised.

We can say this confidently because a rational defense for torturing innocent children (or not so innocent children, for that matter) is almost unimaginable. Surely, the validity and reliability of this particular moral tenet on causing undue harm are firm, perhaps as firm as many findings of science at the .05 level of probability.

If you are thinking that these two examples are extreme, you are right. Much philosophy is not this cut-and-dried. It involves persuasion and plausibility more than proof-like validity and reliability. But most empirical science is like this too. How many of us have encountered a "scientific" recommendation about purportedly dangerous dietary or exercise practices only to read a few years later that more recent studies have shown them to be perfectly safe after all?

A few of us veterans in kinesiology remember when deep knee bends were part of most exercise regimens, when women were discouraged from running long-distance races on

Interviews With Practicing Professionals

Peter Hager
Assistant Professor, Department of Physical Education and Sport, The College at Brockport, State University of New York

Q: Why did you decide to study the philosophy of sport?

A: I was captured by philosophy's big questions during my senior year in high school. Questions like "Is there a God?" and "What is mind?" and "What is moral goodness?" occupied my mind throughout my college years and inspired me to pursue graduate work in philosophy. My love of sport and physical activity and my interest in the philosophic questions arising within these areas eventually led me to specialize in sport philosophy and become a philosophy of sport professor.

Q: What are some of the benefits of studying philosophy for students who are interested in kinesiology?

A: As a student and teacher of sport philosophy, I can attest to the value this area has for kinesiology students. While studying philosophy, students develop important critical thinking skills they will utilize in kinesiology-related professions and research, and as human beings living in today's complex world. Philosophy disciplines students to examine their own beliefs in greater depth and to develop well-reasoned arguments for them. It also helps them to become more open-minded and to entertain, examine, and possibly accept ideas, theories, and positions they may previously have ignored or discarded without good reason.

It is important for kinesiology students to contemplate the importance that sport and physical activity have for human beings in order to gain fresh perspectives about their potential benefits and meaningfulness. Such reflection also helps them become more aware of the significance their prospective professions have for people.

Q: Please describe an example of how students can apply the knowledge they gain from philosophy to their experiences with physical activity and sport.

A: In relation to sport, philosophy can, among other things, help students learn to keep winning in perspective. Winning's bloated importance in American society often influences athletes, coaches, and administrators to become absorbed in their own advantage seeking and lose sight of how their selfish actions negatively affect others. Understanding competitive sport as a unique type of activity, rather than as a form of battle or a means to personal gain, is essential if students are to learn how to attain excellence within sport and realize the important roles that fairness and good sporting behavior play in creating meaningful sporting endeavors.

---

account of their "delicate" constitutions, when joggers had to reach 60% of maximum heart rate in order to reap significant physiological benefits, when salt tablets were dispensed to athletes who had become dehydrated, and when weight training was almost universally discouraged in all hand–eye coordination sports, including basketball—all of these recommendations advanced on the supposedly bedrock foundation of scientific evidence.

The bottom line on validity and reliability is this. All of us who teach and conduct research in higher education, those who measure and those who reflect, those who produce data-driven conclusions, and those whose conclusions are based on logic and good judgment, more often circle in on the truth slowly and patiently than find it ready-made, whole, and durable. Sometimes this means taking two steps forward and one back.

A second important characteristic of philosophy is that it takes intangibles seriously. These intangibles include subjective *moods* or *emotions* like fear, hope, love, and anger. They include *ideals* or *values* like hope, happiness, and justice. They include *daily experiences* like hitting a golf ball on the sweet spot on the face of the golf club or swimming effortlessly across a lake.

One philosophy professor is said to have told his skeptical students that "an idea is just as real as a tree." The truth of this statement is evident to anyone who trusts the power of reflection. This is the case because one can reflect equally on physical objects like a baseball and subjective objects like "the experience of swimming." We can even turn our reflective gaze toward imaginary or fictional things like Purple People Eaters or the legendary Casey of "Casey at the Bat" despite the fact that neither the People Eaters nor Casey ever existed.

Thus, not only is our methodology different, but so too are the "things" to which these methods give us access. We reflect rather than measure, and we have a broader range of phenomena to study than do other scholars who are limited to physical objects that can be tested, measured, or examined.

Because philosophy operates differently, it can be considered a complementary discipline in the cross-disciplinary field of kinesiology. That is, we are players on a larger team of scholars who attempt to understand the body, movement, sport, games, and play. We cannot do it alone. We need chemists, biologists, physiologists, biomechanists, exercise psychologists, motor learning and control specialists, sociologists, historians, and perhaps even scholars of spirituality. All of us, it could be argued, should depend on one another because nobody has a corner on the truth. (See the interview with Peter Hager on page 131 for his views on why philosophy is needed in kinesiology.)

The evidence for this collaborative and ecumenical view of kinesiology comes from a theory of holism, a position that was recommended in part I of the text. Here are some of its tenets related to the place of philosophy among the other subdisciplines:

1. None of our domains is independent. Discoveries in one area are related to discoveries in another area.
2. None of our domains can be reduced to any other domain. Biology cannot be reduced to chemistry. Philosophy cannot be reduced to sociology or psychology.
3. None of our domains can claim initial causation. All causation is circular. A molecule or chemical reaction is not the start of a chain of events because it too was influenced by antecedent factors. A brain state is initiated no more by a pure idea than an idea is initiated by a pure brain state.

➤ Philosophy of physical activity involves reflection on both tangible and intangible objects in ways that provide insights about the nature and value of these phenomena. The validity and reliability of reflection range from the virtual certainty of some logical procedures to the legitimate doubts related to speculative assertions. Philosophy is a member of a cross-disciplinary team, one that provides unique information that complements and is complemented by data gained from other sources.

If these claims are true, and if we realize that we are all in interdependent areas of study, then the holistic representation of the spheres of physical activity is a bit different than the traditional representation we use in other parts of the book (see chapter 1, for example). While we can identify different "areas of study" listed in the big circle (see figure 5.1), none of them is entirely distinct. Thus, no lines divide them. They "fade off and into" one another—philosophy to philosophical psychology to psychology to psychological sociology and so on. The satellites that are attached to the big circle depict the ways in which scholarship and application are related. The sphere nearest the big circle represents the sum benefit of our cross-disciplinary field of study. A reciprocal relationship then exists between the two other parts. One leads to participation in physical activity—in work, exercise, games, play, and sport. The other leads to the workplace and the practical problems faced there. The interaction among all spheres is two-way, suggesting that theory informs practice and, conversely, that practice informs theory.

No one discipline rules the roost in the big circle. We are mutually dependent on one another, and we are the mutual beneficiaries of one another's work. Likewise out on the satellites to the right, no circle dominates the others. Theory, performance, and practical application are mutually dependent on one another and are likewise the beneficiaries of one another. This produces the holistic, democratic, and egalitarian notion of the discipline of kinesiology as described in chapter 1.

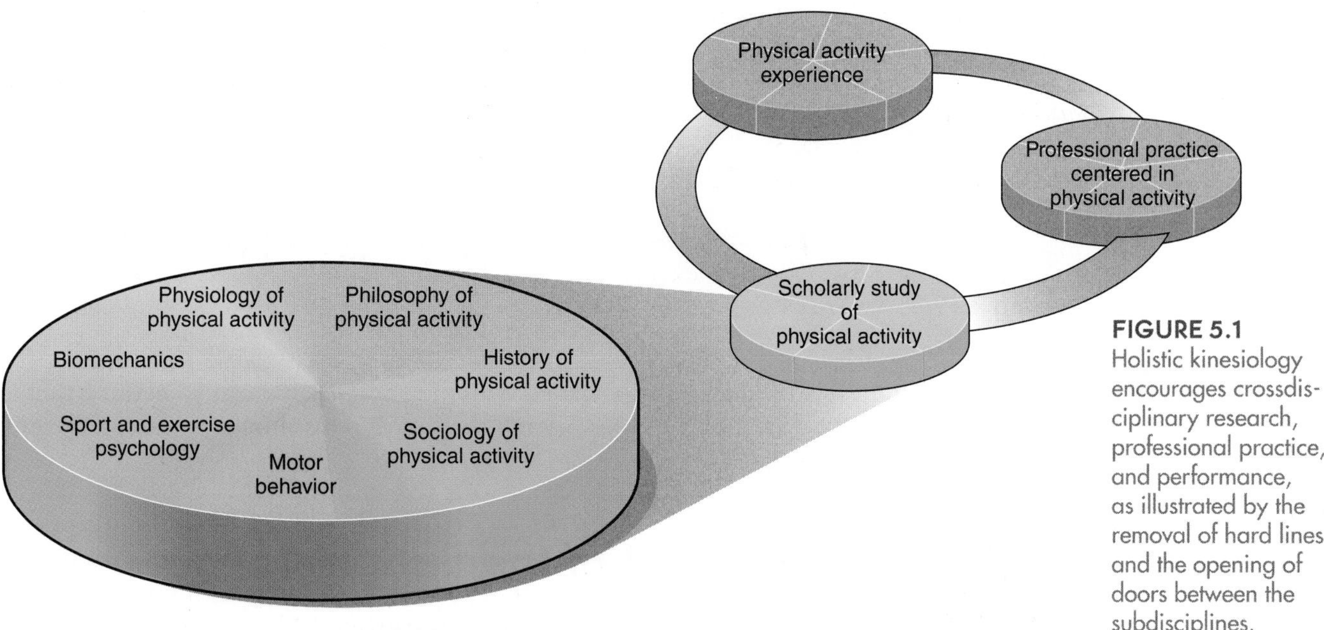

FIGURE 5.1 Holistic kinesiology encourages crossdisciplinary research, professional practice, and performance, as illustrated by the removal of hard lines and the opening of doors between the subdisciplines.

Philosophy of Physical Activity and Kinesiology

The goal of philosophy is no different than the goal of other members of the cross-disciplinary kinesiology team. In the broadest terms, it is simply to better understand the world and our lives in it. Specific to our field, the goal is to better understand human movement or, as it is identified in this text, physical activity. In order to be an effective, contributing member of the research team, philosophers reflect on three general kinds of issues:

- *The nature of physical activities such as sport, play, games, exercise, and dance, and the nature of human embodiment.* This domain includes "what is" questions. Here are some examples: What is the nature of sport? Is chess a sport? How about skate boarding? Are competition and play compatible in principle? Are they compatible in fact? What is the difference between dance and sport? What is the nature of skills? Is it possible to be creative in a rule-governed activity like sport, or is genuine creativity possible only in dance-like activities? Are you and other human beings a composite of mind and body? Is it possible to separate the two so that kinesiologists can work strictly on the body? What principles of capitalism are capable of influencing the nature and shape of sport in the Western world? Would the logic of socialism be likely to produce a variety of sport with a different flavor? Do intercollegiate athletics have an internal game logic and an identifiable educational value? If so, what are they and how are they related? In what sense is game logic potentially educative?
- *Knowledge and physical activity.* This domain includes "how do we know" questions—that is, questions about the theory of knowledge. Among the issues addressed here are the following: Is it possible to get an objective view of sport or dance, or are your reflections always biased or otherwise skewed? Do men and women typically perceive, value, and think differently? If so, what implications does this have in any quest for "the truth" in kinesiology? Is moving a form of knowing? If so, can it be an impressive way of knowing? What does it mean to be smart? How are smart, embodiment, and movement related?
- *Values connected with physical activities and embodiment.* This final area involves "should" questions. It includes questions about the worth of things, experiences, or possessions like health, love, and respect as well as concerns about proper behavior or what you ought to do. Examples include the following: What is the best way to rank professional values? Is health the top value in kinesiology, or is it something else? Is competition a good thing? How can a zero-sum activity, in which one side's victory requires the other side's defeat, be

Goals of Philosophy of Physical Activity in Kinesiology

1. To better understand the world and our lives in it
2. To understand the nature and value of physical activity, particularly in the form of exercise, sport, games, play, and dance
3. To understand what a person is and the role that physicality and movement play in how we come to know ourselves and our world

good? What is the value, if any, of the so-called extreme sports? How can something that is dangerous or painful be valuable? What is sportsmanship? Where should you draw the line between strategy and dirty tricks?

Development of the Subdiscipline

By the 1960s, American physical educators studying philosophy of physical activity were pursuing new directions by distancing themselves from educational philosophy and the single profession of physical education, and using philosophical techniques to produce new insights about physical activity itself—especially sport. This approach attracted a few scholars in other disciplines as well. The two most important physical educators in this vanguard were Eleanor Metheny (*Connotations of Movement in Sport and Dance,* 1965; *Movement and Meaning,* 1968) and Howard Slusher (*Man, Sport and Existence: A Critical Analysis,* 1967), both of them at the University of Southern California and both active in doctoral education. Philosopher Paul Weiss (*Sport: A Philosophic Inquiry,* 1969), then at Yale, was also at the forefront of this interest in sport per se.

Metheny conceptualized movement as a source of insight and meaning for individuals. Slusher used existential perspectives to examine ways in which sport enhances our authenticity as human beings as well as our freedom and responsibility. Weiss was concerned with excellence in sport.

In the middle of the country, physical educator Earle Zeigler mentored a group of graduate students at the University of Illinois, while another important center of doctoral education developed at Ohio State University under the leadership of Seymour Kleinman. Zeigler wrote about philosophic schools of thought and prominent philosophers while extrapolating implications of their work for the field of physical education. Kleinman focused on philosophic analyses of movement, dance, Eastern movement forms, and the bodily dimensions of being human.

The Philosophic Society for the Study of Sport (PSSS), now the International Association for the Philosophy of Sport (IAPS), was formed in 1972 following discussions at the Olympic Scientific Congress in Munich, Germany, among Weiss, Warren Fraleigh from the State University of New York at Brockport, and German philosopher Hans Lenk. Fraleigh was the key organizer of the association, and he continued to be centrally involved as it developed. The PSSS almost immediately set about developing a scholarly publication for the subdiscipline, and the first issue of *Journal of the Philosophy of Sport (JPS)* appeared in 1974 with Bob Osterhoudt as its editor.

Philosophy of sport grew slowly for two reasons. Philosophers and philosophy of sport classes were few and far between because many departments of physical education and kinesiology emphasized the physical sciences. Some schools that had multiple physiologists, biomechanists, and motor control specialists, for example, did not have even a single sport philosopher. Second, the cognate area of philosophy is uniquely conservative. Many parent discipline philosophers preferred not to be associated with a group that studied sport, and thus stayed away from philosophy of sport meetings and chose not to publish in *JPS*. Third, early philosophers of sport worked hard to separate themselves from education and other applied fields. As a result they produced very abstract literature that was often difficult to read and was perceived to have very little utility.

Much, however, has changed over the past 10 years, and this change is favoring the philosophy of sport. First, there is a greater appreciation for cross-disciplinary study and the role that philosophy, history, and the social sciences can play in this research. For example, new National Institutes of Health guidelines indicate that single-discipline research is often constrained and ineffectual and that collaboration across disciplines is to be encouraged. Second, parent discipline philosophers are beginning to show up more regularly at philosophy of sport meetings, and the literature is improving significantly in both quantity and quality. Philosophy of sport, for example, now has two refereed journals, each of which publishes semiannually. Third, we have seen philosophy of sport spread to universities around the world. Whereas early activity in the 1960s and 1970s was confined largely to North America, groups now exist in Great Britain, the Scandinavian countries, the Middle European countries, and South Korea and Japan.

Even with this positive activity afoot, the fate of sport philosophy in kinesiology departments in North America is uncertain. Because of the dominance of the biological sciences, the emphasis on large external grants, and the push for scientific specialization, among other factors, relatively few philosophy of sport doctoral programs remain in the United States. It is unclear where the philosophic leaders and researchers related to kinesiology will be produced in the years to come.

Research Methods in Philosophy of Physical Activity

Now that you know about the goals of philosophy of physical activity, as well as a bit of its history, let's look at how scholars produce knowledge in the subdiscipline. As you saw, philosophers use various techniques of reflection rather than measurement. Like all scholars, they ask questions. But, then, unlike many other researchers, they attempt to answer those questions by reflecting on experience, using logic, appealing to common sense, and deploying their conclusions in the everyday world to see how well they work. Nagel summarized the philosophic process by noting that philosophy "is done just by asking questions, arguing, trying out ideas and thinking of possible arguments against them, and wondering how our concepts really work" (Nagel 1987, p. 4).

You don't need to be able to conduct scientific experiments or work through mathematical proofs. You do, however, need to be able to think carefully and precisely, and you can improve this skill with practice, just as people improve writing skills or athletic skills by working on them. For some of us, reflection may be a relatively new experience. This is so because in our fast-paced, rapidly changing world, finding time for contemplation is sometimes difficult. And in our modern science-dominated world, measurement tends to be trusted more than reflection. (As one humorist put it, "In God we trust. All others data!") Because of these sentiments, you may not have had sufficient opportunity to practice your skills of reflection.

In truth, the gap between those who measure and those who reflect is not as great as it seems. Most philosophers think that good empirical experimentation is crucial and that the gains made by the physical sciences have been both impressive and useful. Philosophers use scientific findings to put parameters on their own work. For instance, in the philosophy of mind, philosophers are very much aware of advances made in neurophysiology. Accordingly, philosophic claims about the nature of mind need to be consistent with the findings of brain science. It could be the case that intelligence is the product of an incredibly complex computer-like brain. Or perhaps it is better to model thinking on chaos theory and dynamical systems. Perhaps intelligence is a single generic capability. Or maybe we have multiple intelligences that have been developed in specific environments to solve specific problems. It is possible, in other words, that we have multiple IQs—one for spatial problems and another for motor problems, music problems, mathematical issues, language dilemmas, and so on—rather than a homogeneous aptitude that allows us to call ourselves "smart." Philosophers need to know these things and use them to constrain and direct their philosophic reflections on the so-called theory of mind.

On the other hand, scientists necessarily lean on philosophic methods and such intangibles as meaning. They reflect on various matters throughout the inductive scientific process. First of all, they employ values to determine what is worth looking at and what is not. They settle on issues that are "important," "significant," "of potential value for humankind," or "crucial for the advancement of science" in general or in their particular branch of scientific inquiry. Then, they reflect in formulating their problem, coming up with potential hypotheses, and developing a research design. They imagine different possibilities and use logic as well as empirical data in deciding to test one way or another. Ultimately, they reflect on their data as the data begin to come in. They ask themselves, "What do these data mean? Are they ambiguous? Could they mean two or more different things? What are their implications for future study? The meanings that come to mind in answering these questions are recorded as part of the findings and then are discussed even more thoroughly in the "Discussion" of what the findings mean and what their implications are for future study.

➤ The gap between those who reflect and those who measure is not as great as some have thought. Both philosophers and scientists are affected by the realities of the physical world, on the one hand, and the force of ideas and meanings on the other.

If scientists are good at these kinds of value judgments and reflections on meanings, their work goes better. If they are not, their research never gets off the ground, and their findings are never published in respectable journals. In short, scientists who are unable to reflect skillfully and are not good at uncovering the right meanings associated with their data are not likely to be very good scientists.

Even though scientific and philosophic methods overlap in some ways, they are not the same. Reflective reasoning processes are central to philosophical research, and four of them are highlighted in this section of the chapter (see figure 5.2) (Kretchmar 1994, 1996):

- Inductive reasoning
- Deductive reasoning
- Descriptive reasoning
- Speculative reasoning

Inductive Reasoning

Inductive reasoning moves from the examination of a limited number of specific examples to broad, general principles or conclusions. For example, you could use inductive reasoning to examine a metaphysical question on the nature of exercise. You would begin by developing a list of examples of exercise activities (e.g., rhythmic aerobics classes, strength training to achieve better performance in football, weightlifting to develop muscles for a bodybuilding contest, mowing the lawn, walking to work because you don't own a car, taking a brisk walk at lunch each day to improve your health), deciding whether all of them are genuine examples of exercise, eliminating any that are questionable, and then identifying their general or common characteristics—in particular those features that distinguish exercise from activities like dance, sport, and play. You will need to be sure your examples are noncontroversial or "safe." Otherwise, your inductive conclusions are likely to be wrong. Also, you will need to be sure that your generalizations are neither too narrow (e.g., exercise requires the use of equipment) or too broad (e.g., exercise requires a person who exercises). When conclusions are too narrow, they are typically inaccurate. (Many types of exercise require

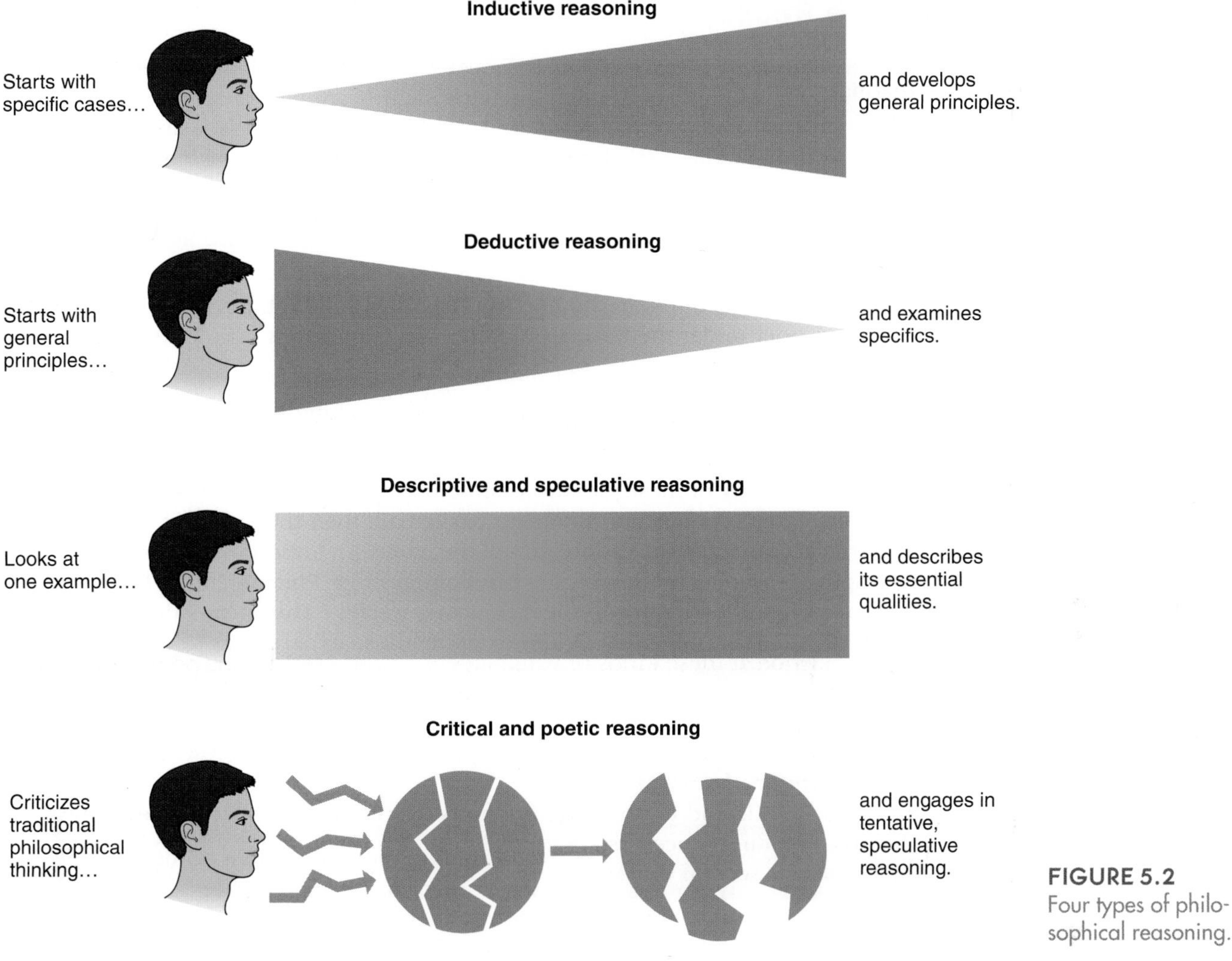

FIGURE 5.2 Four types of philosophical reasoning.

equipment, but some do not.) When they are too broad, conclusions are unhelpful. (Sport, dance, and play require performers too, so this characteristic does not tell us much about what is unique to exercise.)

Deductive Reasoning

Deductive reasoning starts with one or more broad premises to identify conclusions that follow from them. You might use deductive reasoning to determine whether a particular activity you happen to be observing is a sport. Let's say that you are attending your first chess tournament and are curious about whether or not it should be identified as a sport. You develop a major and minor premise as follows: (Major premise) If it is true that sport requires the employment of physical skills to solve its game problems, (Minor premise) and if it is true that chess does not require physical skills in order to achieve checkmate, (Conclusion) then it necessarily follows that chess is not a sport. Of course, the validity of this reasoning depends on the accuracy of the two premises. Someone could argue with the first premise and suggest that because chess is competitive and is a game, it deserves to be identified as "sport" even though physical skills are not required.

Descriptive Reasoning

Descriptive reasoning involves looking at one example of an event and describing its essential qualities. For example, you might imagine slightly altered forms of basketball and then, through a process of descriptive reasoning, determine whether these are still the same game. What do you think about the following variations on the sport of basketball? Would each one still be basketball?

- Changing the goal of the game to one of achieving a tied score at the end of the contest
- Eliminating all restrictions on the means of scoring a basket, thus making it legal to use a step ladder under the basket or stand on a teammate's shoulders
- Playing on an outdoor court rather than in an indoor facility
- Playing a computer version of the game, thus eliminating the need to move around on the court

Negative answers to any of these variations should help you zero in on what is central to this activity. If the variation does not eliminate basketball, then the variation in question is probably not crucial to the identity or integrity of the game. An important research technique in which descriptive reasoning is often used is **phenomenology.** Phenomenologists are guided by the requirement that they "return to the things themselves." This means that they should get back in touch with their experiences and examine those experiences to determine what exactly they mean. Variations help phenomenologists distinguish meanings that are central from those that are peripheral, just as we saw with the basketball example.

Speculative Reasoning

➤ Inductive reasoning begins with specific cases to develop broad, general principles. Deductive reasoning begins with broad factual or hypothetical premises in order to determine more specific conclusions that follow from them. Descriptive reasoning begins with one example of some phenomenon and then varies it to see how dramatically it changes. Change (or its absence) allows a more accurate description of the central characteristics of the item being examined. Speculative reasoning uses inductive, deductive, descriptive, or imaginative reasoning in making claims that may or may not be true, but that are extremely difficult to demonstrate or otherwise defend.

Speculative reasoning is more ambitious than the other three forms of reflection. It can employ inductive, deductive, and descriptive reasoning but it uses them to make claims that cannot be easily or clearly justified. Because of this, many philosophers choose to avoid speculation. They believe that it gives philosophy a bad name. Others counsel against speculation because it can be used for good or ill. Some have speculated, for instance, that certain races are inherently superior to others, or that males ought to have more rights than females. Self-serving propaganda, in other words, can be dressed up to look like serious philosophic speculation. And in an era of political correctness, insightful philosophic speculation can be dismissed as nothing more than propaganda.

But these worries aside, the long tradition of philosophy is full of speculation, much of it directed toward the improvement of the human condition. Idealists, in particular, like to imagine a better world with better people and better values. Descriptions of these ideal conditions may be highly speculative. Nevertheless, they too can carry a degree of plausibility because they resonate with human experience.

DO it **Activity 5.2**

Four Types of Philosophical Reasoning
Identify examples of each in this activity in your online study guide.

From experience, for example, we know that unconditional love has something to do with the good life. This claim is a relatively modest one that can be found in most secular and spiritual traditions around the world. It is also one that can be defended by inductive, deductive, and descriptive methods. But to claim that love, understood in a certain way, is *the* key to good living or the essence of God's will is speculative. It is unlikely, in other words, that normal powers of logic or intelligent discernment will be able to demonstrate that this is true—at least not in the foreseeable future. In a similar fashion, it would be difficult to imagine how someone could forcefully defend the claim that the highest value of intercollegiate sport resides in the pursuit of excellence. As such, this proposition (as attractive as it may be) is speculative in nature.

Applying Inductive, Deductive, and Descriptive Philosophic Reasoning to Philosophic Questions

Many philosophers use inductive, deductive, and descriptive approaches jointly in the process of researching a single problem. We will use the second issue introduced at the start of the chapter, the one that raises questions about using intrinsic, play-oriented methods to promote persistence in activity over extrinsic, rational, duty-oriented methods. See if you can use each method to make some headway. We will try to get you started on each one:

- For inductive reasoning, develop a list of examples of individuals at play—that is, people who are fully engaged in some activity in the spirit of play. Once you have your set of examples in place, ask yourself the following question: What is common to all of these?—that is, what general features make all these examples of play and not work, labor, or duty?
- For deductive reasoning, start with premises: If it is true that play promotes future participation because it has the qualities of (fill in this blank with the qualities you identified using inductive reasoning), and if it is true that calisthenic exercises typically lack these qualities, then it follows that . . . (it is up to you to draw the most accurate conclusion based on the content of premises 1 and 2).
- For descriptive reasoning, let us suppose that you picture 3rd graders in the following ways:

  a. They are involved in make-believe play in a sandpile behind the school. They are making "mountains," "rivers," and "towns." It is difficult but interesting.

  b. Now imagine the same children involved with the sand doing "anything they want." There seems to be no particular goal, no objective. They are just "playing" or, as we say, "fooling around."

  Now ask yourself the question: Was anything important lost when the project was varied from "making something" to "fooling around"? What was lost? Was this loss important to the spirit of play?

Applying Speculative Reasoning to Philosophic Questions

Speculative reasoning can also be used in dealing with intrinsic and extrinsic motivational methods. Some might claim that the intrinsic should be emphasized over the extrinsic because the good life is one that is full of experiences that are satisfying in their own right. That is, we should spend our time doing things that are self-justifying rather than investing too much time in things that lead to what we really want. Thus, games, sport, dance, and other movement activities should be presented, cultivated, and experienced more as play than as a healthful duty. Life, in other words, should be more about play than about work.

What kinds of arguments can you rally for these speculative claims? Can you say why such speculations may be useful even if they cannot be proven true? Do you, as a future professional in the field of kinesiology, believe that much of your future success may be tethered to your ability to promote play?

Nevertheless, such claims may still be accurate even if they cannot be demonstrated, as was the case for many years with the law of gravity. Its requirements ruled the world for millions of years even though they were not known. In addition, and importantly, speculative claims can inspire and otherwise affect people for the good. People in many spiritual traditions, for example, cannot prove that their beliefs are true but may still be moved and otherwise inspired by them.

Overview of Knowledge in the Philosophy of Physical Activity

It is now time to look at several important topics in the philosophy of physical activity. While scholars have produced a complex, broad literature related to human movement, we will be able to examine only a few examples of the insights produced by this fascinating subdiscipline. Specifically, we will examine

- the nature of the person—specifically the mind–body relationship,
- the nature of sport and its relationships to work and play,
- the values promoted by physical activity, and
- ethical values and sport.

The "Person Problem" (How to Understand Mind and Body)

Some sport philosophers believe that you should begin work on understanding human movement with the adjective (human) before you get to the noun (movement). We will follow that general pattern here. We will look at who it is that moves before we get to the nature and value of movement forms like play, games, sport, and competition.

Good reasons exist for dealing with philosophic knowledge in this sequence. It can be argued that we must understand the person before we can understand how and why people are attracted to games and play. For practitioners, the same logic holds. If you are going to use play or exercise as an intervention to promote human welfare, you had better first know what it is to be human. The subsequent intervention, in short, must be appropriate for the individual. So it is with the individual that we should begin our analysis.

Philosophers have taken many positions on what human beings are. Human beings have been called social animals, symbolic creatures, sexually driven individuals, the product of economics, composites of two elements (mind and body), composites of three things (mind, body, and spirit), or simply complex machines and thus little more than a bundle of atoms. While each of these positions may merit some attention, we will focus on only three of them in this chapter: materialism, dualism, and holism. If you understand these three, you will have a reasonably good grasp of fundamental positions that can be taken on the nature of the human being. And you should have a good grasp of the implications these positions have for kinesiology and the movement professions.

Materialism

Perhaps the simplest and most straightforward position on human nature is captured by materialism. Philosophers call it a monistic position because the person is said to be made of only one thing—namely, atoms. In fact, all that truly exists in the world is atoms and void. The human being, including the body, the blood, the wiring, the brain—all of the human being—is nothing more than a complex machine. The brain can be likened to an incredibly sophisticated computer. The body is a combination of very complex lever systems, motors, wiring, and plumbing.

Consciousness for materialists is often thought of as a kind of a sideshow. While these philosophers may acknowledge that subjective experiences are real, materialists do not believe that they have any power or efficacy. Ideas and values are a product of brain states, and it is the brain states, now and in the future, that make things happen.

This is a no-nonsense kind of philosophic position. What you see is what you get! The person appears to be an impressive material object . . . and is! Nothing more. Nothing less. You and I do not need to rely on spooky substances like mind or spirit to account for human behavior. Neither should we discount the considerable capabilities of this powerful machine. Once we gain the ability to measure and understand all parts of the organism, all human capabilities will be accounted for in terms of atoms and void.

Materialistic accounts of persons were popular as early as the time of Plato and before, but they became ever more prominent during the period of the scientific revolution in the 15th century. The careful skepticism inherent in scientific inquiry, the emphasis on measurement, the onset of the Copernican revolution and the realization that the earth is not the center of it all, the rise of Newtonian physics and the increased appreciation of causal relationships, the invention of the microscope—all of these things and others encouraged us to think of persons as simply one kind of physical object among others.

Many physical educators and exercise scientists adopted the assumptions and perspectives associated with materialism and began to study the person as a mobile machine. Physiologists, biomechanists, motor control and motor learning experts, sport psychologists, and even some sociologists and philosophers of sport began looking for the holy grail of materialism, often referred to as "underlying mechanisms"—those multiple physical causes that, when pasted together, would explain the whole.

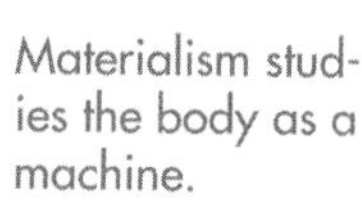

Materialism studies the body as a machine.

Dualism

Some philosophers, while appreciating the simplicity and honesty of materialism, doubt that consciousness can be cast aside so quickly and easily. In other words, they doubt that human behavior can be accounted for by looking only for underlying mechanisms or looking solely for material causes. While these philosophers acknowledge that human beings have a body that works very much like a machine, they argue that people are also endowed with a mind, with consciousness, with a steady stream of very concrete emotions and ideas. Moreover, these subjective states have power. Ideas of love and hope and justice, for example, can (and do) cause changes in our behavior.

When these philosophers are asked how many parts go into the production of a human being, they answer "two"—mind and body. Thus, instead of being monists like the materialists, they are identified as dualists. Descartes, perhaps the most articulate and forceful of the modern dualists, said that a person was composed of two substances, one that was extended in space (body) and one whose nature it was to think (mind).

Dualism was (and still is) an attractive philosophic position. It is attractive because our intuitions and daily experiences suggest that our thoughts must count for something. We have an idea, we draw up a plan reflectively, and then we act on it. We have an exciting idea, and everything from our heart rate to our outlook on life is affected. Dualism, in short, can be attractive because it takes ideas seriously.

Many physical educators and exercise scientists have been dualists because they believe that consciousness is real and that nonphysical ideas somehow invade the world of motor behavior—of swimming across pools and catching baseballs and performing pirouettes. Many expressed their respect for both of our human elements and their close interrelationships, as did Juvenal when he argued for "a sound mind in a sound body" or Jesse Feiring Williams when he proposed an idealistically flavored "education through the physical."

Holism

A third position has been receiving a considerable amount of attention more recently. In some ways it can be considered a middle position between monistic materialism and mind-body dualism. It takes both the mechanistic and thoughtful sides of our human nature seriously, but it refuses to separate them. In other words, it argues that our physical nature, all the way from our anatomy to our genes, is always shaped and influenced by emotions and ideas. Likewise, our subjective experiences are always shaped and influenced by our physicality. In short, we never get the one without the other. Thus, neither the physical nor the thoughtful side of our nature has any independence. We are thoroughly ambiguous—whether one is looking at an injured knee or at a child playing soccer. Both phenomena are a composite of "bodymind"!

➤ How we deal with people—how we cure them, educate them, or teach them sport skills, for example—is affected by what we think a person is. If we labor under false impressions of human nature, the risk of inappropriate, ineffectual, or even dangerous interventions increases. Or, to put it another way, we are unlikely to understand health, movement, the body, games, and play unless we understand what a person is. That question lies at the very heart of our field.

Philosophers like William James, John Dewey, and Maurice Merleau-Ponty are frequently identified as champions of holism. They all claim that it is unhelpful to think of a person in terms of parts or dichotomies like mind and body. They all believe that intelligence works at many levels from mere reflex behavior to creativity. Thus behavior is more or less mechanistic or more or less creative. But no behavior is simply mechanical, and no actions are the result of an independent mind giving instructions to a dependent body. Once again, all behavior is ambiguous.

DO it **Activity 5.3**

Practical Applications

Analyze the philosophic approaches of three fitness professionals in your online study guide.

Kinesiologists and physical educators have seen the relevance of such a position for their own work. Among other things it has expanded the significance of physical activity as something that counts as creative, intelligent behavior in its own right, without the addition of theory or other propositions. It has increased the significance of body in understanding human nature and behavior. Holists, unlike materialists and dualists, do not say that we *have* a body. Rather they argue that we *are* our bodies. Or more accurately, we are simply bodymind! Finally, holism has expanded the kinds of interventions that are designed to improve health or repair injuries. Rather than working on the plumbing (our cardiovascular system), or fixing the hinge (e.g., one of our knees), or manipulating the computer (our brain), we can intervene at multiple levels—from our embodied ideas and aspirations to our idea-shaped cells. Holistic medicine, for example, is now taken seriously in many quarters. A bad back, for example, can be addressed by any number of interrelated therapies from meditation and yoga to chiropractic manipulations and surgery.

Sport, Competition, and Their Relationships to Play and Duty

Sport is virtually everywhere in today's society. It is not surprising, then, that it occupies a prominent place in physical education programs across the nation as well as in the lives of millions of citizens. As a kind of game, sport has a unique structure, one that can be experienced in multiple ways. Here you will look at its structure, its connection to competition, and both its duty-like and its play-like potential. You will see that, at its best, sport allows us to test our capabilities while thoroughly enjoying the doing.

➤ Sport is a game in which motor skills are tested. Its rules specify a goal to be achieved and limit the means that participants can use to reach the goal. These rules exist for the sole purpose of creating the game.

Games and Sport

Games are like no other activities in the world. They are created by a set of rules that specify a goal to be achieved and limit the means that participants can use to reach the goal. These rules exist for the sole purpose of creating the game; they would be absurd in ordinary life (Suits 1978). In other words, games are artificial problems (Kretchmar 2004; Morgan 1994). The built-in inefficiencies structured by the rules of games are different from the inefficiencies encountered in many other realms of life—for example, education, politics, and business—in which limits on efficiency come mainly from ethical constraints. For instance, we cannot copy someone else's writing and call it our own. That is an act of plagiarism, and it is a serious breach of ethics. Game rules, however, handicap us in gratuitous or unnecessary ways that have nothing to do with ethics. Think about the ordinary activity of shopping for groceries. This activity could become a game, for example, if you restrict yourself to buying products whose brand names start with a vowel. You could make your game even more complex by giving yourself only 30 minutes to go over your shopping list. This method is not the most efficient way to shop for groceries, but it does turn shopping for groceries into a game. Nearly any activity can be converted into a game . . . even reading this textbook!

Any activity can be turned into a game.

In light of this definition of games, we can describe sport as a game in which motor skills are required to reach its goal. In National Football League football, for example, the objective is to get the ball over the goal line. In competitive situations, the objective is not just to do that but to do it more times than the other team does. Furthermore, the rules prohibit certain actions that might otherwise increase a team's ability to accomplish the goal. To illustrate,

- offensive linemen are limited in the ways in which they can block opponents,
- only certain players are eligible to receive forward passes,
- a player who is advancing the ball must stay within the boundaries of the field,
- each team may have only 11 players on the field at one time,
- teams may not use mounted horsemen to outrun their opponents, and
- players may not use brass knuckles or clubs to beat their opponents into submission.

The built-in inefficiencies structured by the rules of football not only make attaining its goal more difficult but also provide the sport with its distinctiveness. A sport is a game primarily meant to test our motor skills.

Significance of Rules Rules can be thought of as formal types of game cues. They tell us the structure of the test, that is, what should be accomplished and how we should accomplish it. In this sense, rules create a problem that is artificial yet intelligible. Only within the rules of the game of, say, basketball or baseball do the activities of jump shooting and fielding ground balls make sense and take on value. It is precisely the artificiality created by the rules, the distinctive problem to be solved, that gives sport its special meaning. That is why getting a basketball through a hoop while not using a ladder or pitching a baseball across home plate while standing a certain distance away becomes an important human project. It appears that respecting the rules not only preserves sport but also makes room for the creation of excellence and the emergence of meaning. Engaging in acts that would be considered inconsequential in ordinary life also liberates us a bit, making it possible to explore our capabilities in a protected environment. To keep that liberation and exploration appealing, we often make rule changes to recalibrate a test that has became either too easy or too difficult. We like our sports to more or less match our ever-changing potential.

Significance of Skills While sport can be understood in abstract terms, the actual attainment of the goal requires the deployment of a variety of motor skills. For instance, successful participation in soccer requires proficiency in kicking, passing, running, marking, and tackling, among other motor skills. Indeed, soccer expertise involves a blend of these motor skills. Rather than being generic in nature, soccer skills require specialization. The ways soccer players kick the ball are markedly different from the ways rugby or football players do, and all of these are different from how we kick pebbles down the street or sidewalk. This is so because sport skills develop out of the relationship between the goal, on the one hand, and the means allowed and prohibited to pursue it on the other.

These relationships are established in the rulebooks of different sports (Torres 2000). Soccer's kicking skills are the result of players' inventiveness in negotiating the test established by the rules, which prohibits the use of their hands, allows for passes in any direction, and does not impose a time limit on ball possession, among many other stipulations. It should be clear that the rulebook of each sport indicates the set of motor skills that this unique game is designed to test. Thus a sport's set of motor skills provides it with idiosyncratic characteristics that differentiate it from other sports. Moreover, motor skills are very important because they represent the standards of excellence by which players evaluate their performance.

Significance of Competition Any sport, or any game for that matter, can be organized both competitively and noncompetitively. This is easier to see in the so-called parallel test such as bowling, archery, or track events. You can bowl by yourself or test how fast you can run the 100-meter dash in the absence of competitors. It is more difficult to see in sports in which the artificial test is provided, in part, by others. In soccer, wheelchair basketball, and tennis, for example, you facilitate the test of another test taker who at the same time facilitates the test for you. This, however, does not mean that both parties are automatically engaged in a contest. Think about scrimmages or games between parents and their children. Although fully engaged in their respective tests, these people are not competing. What is missing in these examples of parallel and simultaneous testing is the central element of competition—that is, the attempt to determine athletic superiority.

Thus, competition requires that two parties share the same test and that they commit themselves to surpassing each other's performances (Kretchmar 1975). When individuals enter the world of competitive sport, they enter a community that defines itself in terms of the unique test they share. Thus, in competition, players learn not only if they are able to do well on the test provided by the sport, but also how they will fare in comparison to their opponents. Testing evaluations and competitive comparisons here go hand in hand.

At one level, competition determines winners and losers. A more encompassing view of competition highlights the value of performance and cherishes process as much as the outcomes that result from the play (Torres & McLaughlin 2003). Excellent play, in other words, transcends mere wins and losses and can even become the central value of competition (Simon 2004).

Competition has value beyond just winning.
Photodisc/Getty Images

Play and Duty in Sport

We have seen that sport is a goal-oriented activity in which we accept rules—imposed limitations on how we can go about attaining that goal. This is done primarily to test our proficiency in a given set of motor skills. We have also seen that this activity can be organized just to evaluate our proficiency in solving the test or, in addition, to compare our skills to those of an opponent. That is, sport is not necessarily competitive.

A second distinction is important as well. Sport may be encountered as a chore or as play. In what follows we will explore the differences between duty-like and play-like modes of experiencing sport and physical activity.

Duty-Like Sport One of the most common arguments for including sport and physical activity in educational settings and used to justify personal involvement in these activities focuses on their power to bring about beneficial effects. That is, sport and physical education are presented as valuable because they help us advance in concrete and desirable ways, from improving our health and teaching civil values to fostering national pride and combating sedentary living and obesity. This account is clearly based on utility, a rationale for activity that many professionals in the field have emphasized for years. When sport and physical activity are articulated in this way, it is likely that we will approach them as tools, as activities that we must do. In short, they become a form of duty.

Regarding sport and physical activities as chores that produce external rewards tends to diminish their intrinsic worth. The more we remind ourselves that our participation in sport and physical activity produces good results, the less we are prone to cavort in their intrinsic meaning and value. Have you ever wondered why time and again we use the word "workouts" when referring to our sport and physical activity experiences? This says something about what we expect them to be like—namely, work! Wouldn't it better to describe instances of physical activity and sport as "playouts"?

➤ Duty-like sport and physical activity emphasize instrumental effects and their capacity to bring about beneficial effects such as health, moral values, and nationalistic sentiments. When the focus is on external rewards, sport and physical activity are more likely to be encountered as a form of duty. Play-like sport and physical activity are self-contained. As a way to live these activities, play accentuates their intrinsic values and ends. At play, sport and physical activity are experienced as delightful, absorbing, and meaningful, and they allow us to show our human potential.

By stressing a "this is good *for* you" approach, duty-like sport and physical activity might not offer the most effective strategies for motivating people to get up and move. This is true because we tend to avoid whatever we *live* as a monotonous duty. Even the most committed and cautious among us would welcome the addition of some delight and sensuosity to his or her "workout." Duty-like sport and physical activity serve the professional need to defend against the charge that they are frivolous and unimportant. In the process, however, they become not only a "workload" that we often cannot wait to escape but also activities that are only indirectly valuable. Their value lies in what they produce, not in what they allow us to be and to experience.

Play-Like Sport If you asked a normal sampling of people how play ranks in comparison to education, religion, family, business, or government, most would probably place it close to the bottom. This is so because play is normally presented as a trivial, infertile, and senseless activity reserved mostly for children. In the face of the delightful and mesmeric power of play, it is paradoxical that many still want to confine it to childhood and consider it at best an ornamental aspect of life. Are we adults frightened of play? Do we really understand it?

Although found during engagement in an activity, play is not the activity itself. Rather, play is the manner in which we approach, embrace, and experience the activity (Meier 1980). Play requires an autotelic attitude in contrast to all forms of instrumental or utilitarian orientations toward the world. This means that while at play, we are not interested in the payoffs that might potentially come from our participation in the activity but rather in its intrinsic values and ends. To put this in a different way, at play the doing is what really counts. For example, a play-like encounter with soccer is experienced as utterly absorbing. Even if extrinsic rewards and goals were of concern when the activity started, they recede into the background as we take great pleasure in just embodying our soccer skills. In those moments, soccer is lived as "good for nothing" and as activity that we like for what it is and allows us to feel. At play we simply let go.

Play-like sport and physical activity, and for that matter any activity lived in a playful spirit, become something we *want* and *like* to do . . . sometimes desperately! We often forget about time, the necessities of life, and duties as we surrender to the magnetism of the playground. Play, then, is a way of embracing the world that includes a reciprocal relationship between a willing player and an inviting playground, where the value of the activity is experienced as intrinsic to itself (Kretchmar 2005; Torres 2002). This means that we can predispose ourselves to find play, but not guarantee it. And, as professionals, we can help others predispose themselves to find play in any variety of potential playgrounds—but we cannot force the experience on them. We can set an example by respecting play, by being patient in waiting for the call of play in likely playgrounds.

As you undoubtedly know from your own experiences, play-like sport is characterized by freedom, spontaneity, creativity, sheer joy, and personal meaning. You are also aware that it is fragile and temporary. Play-like sport cannot be dictated but, as already suggested, can be coaxed out of hiding more often through ingenuity and care. Don't you think that fostering play-like sport and physical activity can enhance our chances that people will fall in love with these activities and stick with them for the rest of their lives?

Play-like elements are potentially found in both competitive (from the elite level all the way to recreational leagues and low-key youth sports) and noncompetitive sports. Many casual and professional athletes are clearly drawn to the process of doing the activity, an internal focus, rather than just to its instrumental value. Similarly, play does not recognize differences of age, gender, race, religion, socioeconomic status, or motor skill development. It is not the exclusive property of any particular group. While alive we are always at risk of finding or being found by play, or both (Kretchmar 1994). The magnetism of play and the way we live an activity while at play make it an essential component in our lives. No wonder some have recommended that we live our lives fostering as much play as possible (Morgan 1982; Lasch 1979; Pieper 1952).

Two Potent Combinations

1. The combination of sport or physical activity and play offers powerful incentives to get us out of our chairs and into the realm of movement. When the doing becomes intrinsically meaningful and sensuously enchanting, we are more likely to continue with the activity. The artificiality of sport seems to be especially attractive. We love to solve problems created for the sole purpose of discovering whether or not they can be solved.

2. If the playful sporting test is appealing, many of us encounter yet another layer of meaning when competition is added to the mix. This combination offers not only the uncertainty and tension of discovering our chances in the sport but also the ambiguity and drama of learning how we will fare in comparison to our opponents. When competitive sport is experienced playfully, we can focus on the doing, develop friendships with fellow sportspeople, and aim for excellence, not just victories.

Creating Movement Playgrounds: How Is It Done?

While teaching an introductory unit on skateboarding, what sorts of things might you do to enable your students to find play? What allows these strategies to generate play? (Please be sure to address, at minimum, the issues of skill development, repetition, feedback, and the amount of time devoted to this unit.) Can your students play in ways that are compatible with, or even advance, your goals as a teacher? If this is possible, think of some examples.

Values Connected With Physical Activity

Given past and current scandals, what do you think about rewarding players or teams for being good sports by giving extra points at the end of games based on referees' judgments of players' behavior? Did you ever wonder what your responsibility would be if you were to discover that a teammate had been cheating? Should you tell your coach? League officials? Would it be acceptable just to keep quiet about it? Are there strong reasons to eliminate dodgeball from school physical education programs? Will physical activity become antiquated if science and technology solve most of our health-related problems? To answer these questions, you have to examine your values.

www. **Web Search 5.1**

A League of Their Own
Investigate the All-American Girls Professional Baseball League in this activity in your online study guide.

All of us use values—our conceptions about the importance of things—to make decisions, both in personal and in professional matters. Values can be moral or nonmoral. Moral values refer to our character and how we ought to behave; nonmoral values refer to objects of desire such as happiness, ice cream cones, and good health. In this section we will look at several central nonmoral values promoted by the field of physical activity. Then we will examine the connection between ethical values and sport—a fascinating topic that has attracted a considerable amount of philosophical study. Philosophers of physical activity have also studied how aesthetic and sociopolitical values relate to sport. Although there is not space to discuss these relationships here, you'll look up additional information about them in the activity in your online study guide.

Values Promoted by the Field of Physical Activity

Nonmoral values are things that people consider desirable. Your values help you define the good life and steer a course toward achieving it (Kretchmar 2005, p. 207). Are there values that are especially central in physical activity, and if so, what are they? In his analysis of major values promoted by physical activity programs, Kretchmar (2005, pp. 232-245) identifies four:

➤ Each of the four values promoted by physical activity programs—fitness, knowledge, skill, and pleasure—would lead to a different sort of program if used as the central guide or priority for planning and intervention.

- Health-related physical *fitness*
- *Knowledge* about the human body, physical activity, and health practices
- Motor *skill*
- Activity-related *pleasure or fun*

Each of these values supports a different approach to developing physical activity programs. Physical fitness provides direction for those that are most interested in health and appearance. From this perspective, physical activity professionals should be primarily concerned about such things as assessing fitness, measuring fitness changes resulting from physical activity, and promoting physically active lifestyles.

Knowledge is the primary value for those who believe that information is a precursor to improved behavior—for example, for staying healthy, looking good, and performing physical skills better. This view would encourage more research in scholarly subdisciplines of kinesiology. Physical activity professionals should become well grounded in these subdisciplines on the assumption that having more of this knowledge is good for its own sake and would also make them better practitioners.

Motor skill is valued by those who like to move well and who know the joy of proficiency. Some might rank this goal highest because the pleasures of skilled movement might encourage people to stay physically active and thus improve fitness. Intrinsic motivations, in other words, lead to extrinsic gains. For those who highly value motor skill, activity professionals should concentrate on teaching appropriate movement techniques, for example, in school and college physical activity classes, focusing on particular sports such as golf or softball.

Valuing pleasure resonates well with people's conceptions of the good life. Fun and enjoyment are important for just about everyone. Enjoyment grows when a situation is complex enough to present stimulating challenges but simple enough so as not to overwhelm people's capabilities (Csikszentmihalyi 1990b). We become bored if things are too easy and anxious if the situation is too difficult. Activity

The joy of proficiency is an outcome of valuing motor skill.
Comstock

Activity 5.4

Using Values to Develop Physical Activity Programs

Evaluate the approaches of four professionals in your online study guide.

professionals who see the world from this perspective should make sure that games are simpler for beginners and more complex for advanced players, and that they are suited to the motor skill level of the players. We have already seen that competition notches up complexity even further, and it probably increases fun as long as the challenges do not overwhelm the players. Another way of varying the complexity is to make sure that opponents are evenly matched.

The four values of fitness, knowledge, skill, and pleasure do not exist in isolation from one another. All are important in physical activity, and each one influences the others (Kretchmar 2005, pp. 245-246). Consequently, the main task we face is not to pick one or two to the exclusion of the others, but to weigh their relative merits and rank them. Once we do that, we can use this as a guide for setting priorities in physical activity—priorities for experiencing it as performers and spectators, for studying it, and for engaging in professional practice centered on it.

To illustrate the latter, professionals might use their value rankings to prioritize the topics they cover in discussions with people who control purse strings (e.g., city government officials, school boards, representatives of private foundations that donate to worthy causes) in an effort to encourage more spending for physical activity programs. If fitness is emphasized, then the profession becomes largely a means to an end or, as Kretchmar argued, an "auxiliary profession," albeit a very important one. If values like skill and knowledge are emphasized, the field takes on a liberal arts character. Movement, in other words, contributes directly to the good life by freeing us to move well and to understand. Kretchmar clusters a kind of knowledge (wisdom- or encounter-understanding) with skill as his own choice for the two highest-priority values. Fun, fitness, and theoretical knowledge are ranked lower. This profile presents physical education as a liberal arts subject matter, one that is aimed more at improving the quality of one's life than at meeting various day-to-day health and recreation needs. His thoughts about the intrinsic and extrinsic values associated with each value are important in his rationale for his rankings. He cites three criteria in his rationale:

- Intrinsic values are better than extrinsic values because intrinsic values have immediate, direct worth, whereas extrinsic values lead only indirectly to an intrinsic value.
- Among intrinsic values, those involving broader satisfaction and contentment are better than those involving more limited pleasure.
- Among intrinsic values involving broader satisfaction, those involving longer-term meaning and coherence in our lives are better than those involving more short-term or immediate pleasure.

Ethical Values and Sport

Given that in elite levels of sport, so much rides on winning and securing large amounts of money, athletes, coaches, and administrators may be tempted to give less attention to questions of morality. Cheating, shaving points, fixing games, using performance-enhancing drugs,

Ranking Activity-Related Values

Before reading further, rank, according to your opinion about their relative importance, Kretchmar's four values: health-related physical fitness; knowledge about the human body, physical activity, and health practices; motor skill; and pleasure.

Why did you rank them the way you did? How might these rankings influence your choice of favorite physical activities, your central interests in studying physical activity, and your physical activity career choices?

Ethics and the Use of Performance Enhancers

Although athletes have used performance-enhancing drugs for a long time, these drugs became more prevalent in the late 1950s when anabolic steroids came on the scene. They have been on the rise ever since (Todd 1987). Although precise figures are difficult to find, data from the U.S. National Football League show that drug suspensions were much higher in 1995 and 1996 than they were in the previous decade (Boeck & Staimer 1996). In this section we'll look at the use of performance enhancers—drugs as well as other things—from ethical perspectives (Osterhoudt 1991, pp. 122-126; Simon 2004, pp. 69-90).

The main ethical problem is that performance enhancers give athletes an unfair advantage over opponents. You might ask, then, which substances improve performance unfairly, and which substances improve it fairly? Your first answer might be that natural substances (e.g., food, testosterone) are fair, whereas unnatural substances (e.g., steroids, amphetamines) are not.

The natural–unnatural distinction, however, is problematic. Technological breakthroughs in facilities and equipment—long-body tennis rackets, artificial track surfaces, sophisticated basketball shoes, and better-engineered racing bicycles—are unnatural. Yet performance enhancers such as these would generally be considered fair, even though limits are sometimes placed on technology (e.g., golf balls and baseballs could be engineered to travel farther, but limits are placed on this attribute). On the other hand, testosterone is a natural substance. Nevertheless, its use is typically considered unfair when it is taken in doses beyond what a person might produce in his or her own body (Brown 1980).

One solution is to create a level playing field by giving all competitors access to whatever performance-enhancing aids they want. Another possibility is to distinguish between performance enhancers that increase a body's capabilities and those that just make it possible to use more effectively whatever capabilities the body already has (Perry 1983). But neither of these options, for a variety of reasons, has proved satisfactory.

Another view suggests that performance enhancers should be prohibited if they put the health of an athlete or others at risk or if they are shown to make athletes dangerous role models for young children. But this approach raises serious questions about the rights and freedoms of athletes to make informed choices about their own health. It raises equally complicated questions about where, when, and under what conditions adult freedoms should be curtailed for the well-being of less mature or less well-informed individuals.

Finally, there is the matter of privacy and individual rights. When is drug testing acceptable, and when is it an ethically unwarranted intrusion on an athlete's freedom as a human being? If we test athletes for recreational drugs that don't enhance performance or for illegal drugs that don't enhance performance, some would argue that we are invading their privacy in unacceptable ways (Thompson 1982).

Clearly, difficult ethical issues are connected with athletes' use of performance enhancers. Uncertainties remain concerning which aids are morally unacceptable, although performance enhancers that present health risks to athletes seem to be viewed as less acceptable than others are. Additional investigations are needed to examine the ethics connected with using different kinds of aids to enhance sport performance.

and employing different kinds of craftiness often make the headlines. Sometimes these headlines are about misbehavior in youth sport and recreational leagues. It is not unusual to hear about dishonest behavior, brawls, and verbal abuse in these settings. In virtually all sporting environments, it seems that our capacities for moral reasoning are severely tested.

Ethics is about determining what is right and wrong, and what ought and ought not to be done. To put it another way, ethics is concerned with questions about how we should live our lives. In sport, this means formulating defensible standards of behavior. While thinking ethically you must make an effort to be impartial, consistent, and critical. As you know, our emotions, biases, and tendencies to give ourselves a break can bias our thinking. The moral

Competing against others who play hard and require the same of you supports guidelines for ethical sport.

point of view is universal, one in which you do not count more than anyone else. Thus, bending the rules simply because "everyone does it" or intentionally injuring the best opposing player in a soccer match to send her to the sidelines for the rest of the game is difficult to justify. On the basis of universality, we would not want to be the recipient of such actions; and, because no one counts as more important than anyone else, it would not be right for us to prohibit rule bending or intentional harm but exempt ourselves from these restrictions. Consequently, societal understandings about right and wrong and the force of self-interest do not provide solid foundations for ethics.

In sport, decisions about right and wrong actions require that we keep in mind our conceptual analysis of competition. The more we understand the nature of competitive sport, the clearer and stronger our deliberations will be about what actions by athletes, coaches, and administrators would promote its health and well-being. These deliberations will provide principles about how these folks ought to behave. Sometimes ethical considerations in sport also include spectators and other consumers of sport. The goal is to produce the best possible world of sport for everyone.

When considering competitive sport as a reciprocal challenge in the pursuit of excellence and when using the moral guideline of creating the greatest good for the most people, we can generate the following basic behavioral guidelines:

- Follow the rules of the sport not simply because they are the rules but rather because they are the foundations of the artificial problem you find special. Avoid cheating, which alters, and even destroys, the sport and vitiates the legitimacy of results.
- Respect your opponent as someone who not only facilitates the contest but also makes possible the creation of athletic excellence. See your opponent as a partner who belongs to the same community of contestants and shares your interests and passion.
- Strive to bring out the best performance in one another. This honors each person's motor skills and the sport's standards of excellence.
- Recognize and celebrate athletic excellence, your own as well as your opponents'.
- Seek opponents who are close to you in ability and who compete with personal resolve to win without slacking off during the contest.
- Care about your opponent's well-being as much as your own. Your opponent is integral to the contest, and a victory is fully meaningful when opponents are at their best.
- Remember that how you play says as much about you as an athlete as the scoreboard does.

Winning Strategies

What kinds of strategies do you believe should be acceptable in sport? What about flopping in soccer to get a foul, or intentionally fouling opponents in basketball in order to stop the clock, or intimidating a batter in baseball by intentionally throwing a pitch toward his or her head? What principles can you develop to distinguish between acceptable and unacceptable sporting actions? Also, can you come up with ideas to discourage morally unacceptable strategies? Should we levy harsher penalties? Use incentives? Provide more education?

In the high school basketball scenario at the beginning of the chapter, you saw that a dubious strategy cast doubt on the value of the victory and on any excellence shown by the winning team. Was this a tainted victory? It did not appear to be so for the coach, who argued that the "barking dog trick," which did not violate any rules, was not an instance of cheating. For him, it was simply clever and effective strategy.

➤ Ethical reflection in sport typically requires that we critically examine societal understandings about right and wrong and tendencies that favor self-interest. When evaluating the best course of action in a given situation, we must also consider what is good for sport and the interests of everybody.

However, even if the barking dog trick was not cheating, it still raises moral questions. Although not against the rules at the time it happened, it did subvert the central purpose of competitive sport in several ways. First, the coach's strategy clearly undermined the set of motor skills that shape basketball. Is getting down on all fours and barking like a dog a basketball skill? Is this a skill at which we want our players to excel? Second, the barking dog trick disregarded the standards of excellence that define basketball. Was this an example of basketball at its best? Third, by implementing this strategy, the coach showed a greater concern for winning than for the way in which winning was achieved. Perhaps even more important, his players were treated merely as means to an end and were not given a chance to demonstrate legitimate basketball superiority. Similarly, opponents' interests were not taken into consideration. In short, the barking dog trick diminished victory by trivializing athletic competition, subverting basketball excellence, and not considering the good of everyone.

Wrap-Up

Philosophy offers not just a body of knowledge but more importantly an intriguing way of thinking about reality. Philosophy of physical activity helps us reach new insights about various forms of physical activity and what it means to move and move skillfully. It also includes explorations of the structure of games, sport, and play and their relationships, as well as knowledge that can be derived from experiences with physical activity. Values connected with physical activities are of much concern in the subdiscipline. Philosophy of physical activity encourages using reasoning skills and insights to positively influence people's lives as much as your own. For example, you might bring your knowledge of play to bear on important decisions that you have to make to motivate people to exercise. Similarly, your ethical analyses might help you decide the best policy concerning the use of performance-enhancing drugs in a league for which you serve as the executive director.

If you find the research methods and knowledge in this subdiscipline interesting, take time to leaf through some of the texts on the philosophy of physical activity, as well as *Journal of the Philosophy of Sport*. You might enjoy an introductory course on the topic. You'll learn to think more carefully and precisely about the nature of things, and this will help you be a better performer, scholar, and professional.

Activity 5.5

Use the key points review in your online study guide as a study aid for this chapter.

Activity 5.6

These end-of-chapter questions and activities are also in your online study guide. Your instructor may ask you to complete them online and turn them in.

1. Describe the main goal of philosophical study of physical activity and the three major kinds of issues it most commonly tackles.
2. Compare the strategies of early philosophers of physical activity to the ones philosophers have used in the last 30 years. Where and why do they differ?
3. Describe the four reasoning processes that are central research tools in philosophical studies of physical activity.
4. Discuss the concept of blended unity of mind and body, its implications for school physical activity programs, its implications for research on physical activity, and its implications for the well-being of competitive athletes. Why is this an improvement over dualistic views of mind apart from body?
5. Explain the relationship between rules and skills in sport. Why is it relevant to competition?
6. Discuss duty-like and play-like sport and physical activity and their implications for how people relate to and experience these activities.
7. Elaborate on the four values promoted by the field of physical activity and their implications for designing physical activity programs.
8. Provide five examples of morally strong sport contests. What makes them so?

6

History of Physical Activity

Richard A. Swanson

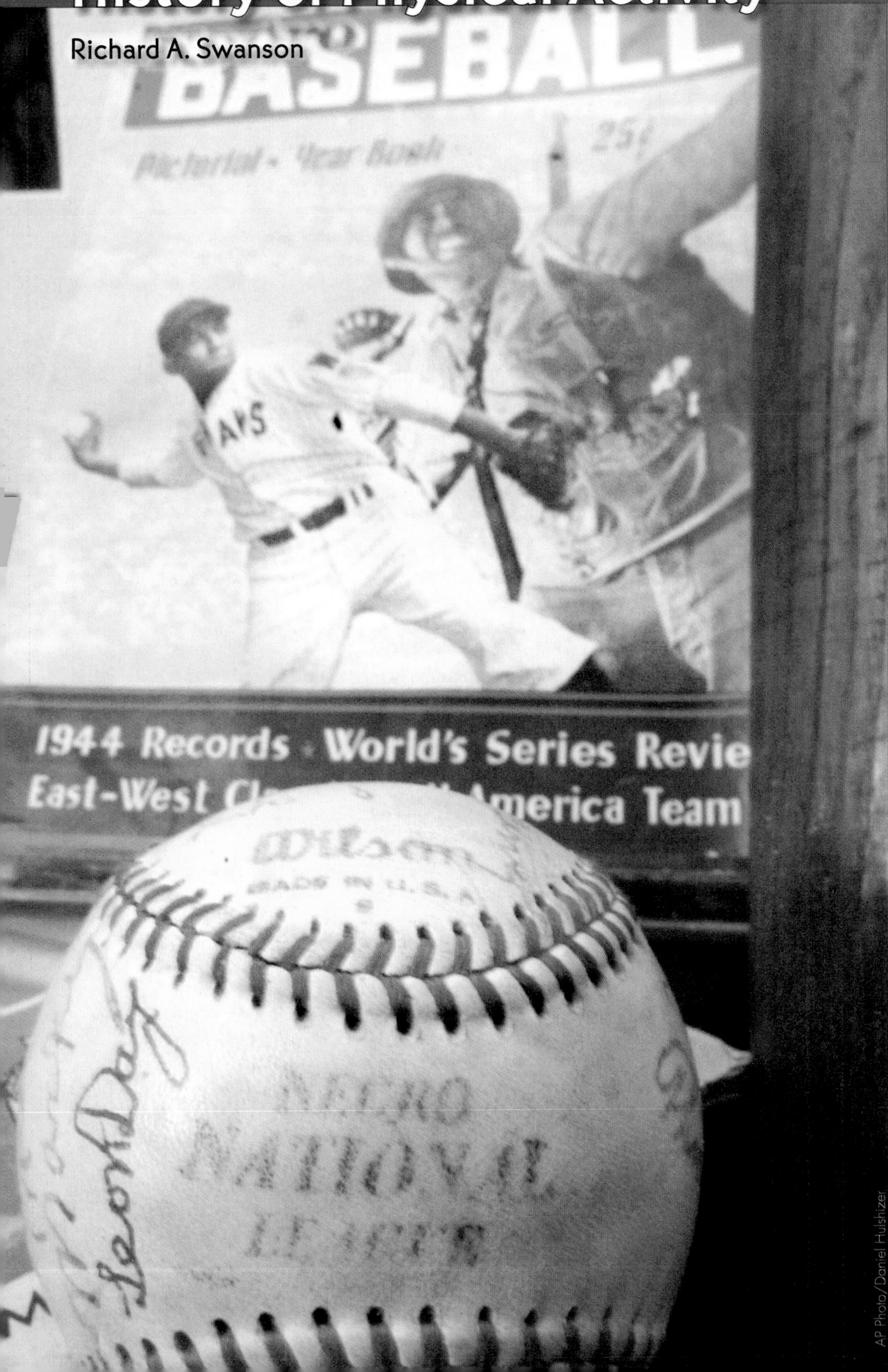

AP Photo/Daniel Hulshizer

CHAPTER OBJECTIVES

In this chapter we will

- explain what a physical activity historian does,
- describe the goals of history of physical activity,
- describe how the subdiscipline of history of physical activity developed,
- explain how research is conducted in history of physical activity, and
- explain what research tells us about physical activity in American society from the industrial revolution to the present.

You are making your first visit to the National Baseball Hall of Fame and Museum in Cooperstown, New York. Looking at photographs and displays depicting early 20th-century major league professional baseball, you notice that there are apparently no African American players. At another display, you learn of the existence of organized Negro Leagues from the 1920s through the 1950s. Your curiosity is aroused, and you move back and forth between the separate displays. In one photo you see white major league players in a large stadium, filled with spectators who, for the most part, appear to be white. In another, you see African American players engaged in a game in a small, wood-frame stadium, and the majority of the fans in attendance are of the same race. A third photo shows several white players in suits enjoying a meal in an obviously expensive restaurant. Another picture depicts a group of African American players in their game uniforms eating sandwiches while standing next to a bus in the parking lot of a small restaurant where a sign in the window reads "Whites Only!" Earlier in your museum visit you had read quotes of several famous 19th- and early 20th-century figures who referred to baseball as the national pastime and as the most democratic of our institutions, in which a person was judged only by his performance on the playing field. Yet here was evidence that questioned such lofty pronouncements.

You wonder why there were racially segregated professional baseball teams and leagues. Were there any opportunities for white and black players to compete together? Did such practices reflect the preferences of most players? Was it the fans who dictated such an arrangement? Did individual team owners or managers ever attempt to change the system? When and why did racial segregation in professional baseball end? What were the ramifications for both major league and Negro League baseball with the coming of integration? To what extent might the integration of baseball have affected the wider civil rights movement of the 1950s and 1960s? You wonder about the validity of many of the records and statistics of the major league game that were established when so many great African American players were excluded.

Why History of Physical Activity?

Consider what life would be like if you didn't have a memory. You wouldn't be able to recall your experiences such as the 10 years that you spent training with a swim team, the pickup basketball games you played at your local neighborhood courts, the time you broke your leg falling off your bike and went to an exercise rehab program for several months, or the details about the fortunes of your favorite professional football team. Memories give you insights about how things came to be the way they are. No one has perfect memory, and no one can predict the future with total accuracy; but our recollections help us act intelligently and develop reasonable plans for the future.

History offers broad and detailed insights that go far beyond our own memory. It gives us the opportunity to develop a more extensive "memory" than we could ever acquire independently. History consists of a vast collection of information—mostly events that occurred before we were born, often in geographic regions and societies different from our own. The ancient Greek Olympic Games, 18th-century peasant ball games in Europe, and 20th-century American basketball all influenced the physical activities that we take part in today. Studying history gives us windows to the past and magnifying lenses to look closely at things that we find especially interesting. It helps us understand how and why our current physical activities are structured the way they are, allows us to compare them with physical activities from earlier periods, and gives us the tools to look toward the future from new vantage points.

A kinesiologist who is knowledgeable about history of physical activity would probably interpret such things as the current fitness craze, the popularity of spectator sports in colleges and universities, or the availability of opportunity in sport across racial, ethnic, and gender lines quite differently from someone who lacks this information. A person who has studied history of physical activity would look at the snapshots in our opening scenario and have answers for the questions that we asked. For example, the historian would know that racial attitudes and practices have affected all social institutions in America. The segregation within professional baseball reflected segregation in the broader society, and it took courageous action on the part of several people to bring about racial integration in this visible public arena.

Knowledge of history also helps us make better decisions today. A good example of this would be decisions by sport organizations to ban players, coaches, and managers from contact with gamblers. Major League Baseball was the first to adopt such a policy following the 1919 World Series when players on the heavily favored Chicago White Sox were accused of accepting money from gamblers in exchange for purposely losing the Series to the Cincinnati Red Stockings. The public outcry about the loss of the integrity of the game threatened to reduce public confidence in the honesty of the game, which, in turn, could result in a decrease in ticket sales. With that lesson in mind, other sport organizations such as the National Football League, the National Basketball Association, the National Hockey League, the National Collegiate Athletic Association, and the U.S. Olympic Committee adopted similar policies. The memory of the 1919 World Series scandal has informed decisions by these organizations to fine, suspend, and dismiss player and management personnel for associating with gamblers in any manner.

www. Web Search 6.1

History of Basketball
Use this activity in your online study guide to document the development of the game.

American society's current interest in exercise, health, and fitness is a phenomenon that's hard to miss. Exercise studios and health clubs dot the landscape; and videos, books, and magazines on these topics are available in stores in almost every neighborhood shopping center. But if you asked teenagers or those in their 20s about the origins of this trend, most would probably draw a blank.

Kinesiologists would know that interest in exercise picked up in the 1950s because of a variety of influences, such as Cold War fears, President Eisenhower's heart attack, and evidence suggesting that American children were less physically fit than their European counterparts were. They would also know about an earlier period in American history when exercise was also prominent—the late 19th century (Park 1987a).

➤History of physical activity teaches us about changes as well as stability in the past, and this helps us understand the present and also make reasonable decisions for the future.

Kinesiologists would be able to point out that American fascination with exercise has fluctuated over the years. This information is important: Is the current, relatively high level of interest in exercise likely to decrease in the future, in line with the fluctuating pattern that we've seen in the past? If so, this decrease will have serious negative implications for the health of many Americans. Can we do anything to prevent this possible downward cycle? Or do we now have such strong scientific evidence confirming the health benefits of exercise that we will break the roller-coaster pattern of the past? We could not even raise these questions without having an understanding of history of physical activity.

Your previous physical activity experiences, academic studies, and even observations of professional practice have already given you some insight into history of physical activity. From *experience,* you may have participated in competitive team sports such as basketball, volleyball, and soccer and noticed that these sports are incredibly popular in Western societies. If you have ever wondered how this came to be, you have been on the verge of historical inquiry. Or perhaps you have noticed gender inequities in physical activity and wondered what historical events led to this situation. If you are a man, maybe you at some time have feared being teased for your lack of athletic ability. If you are a male athlete, you may have noticed uneasily how much more financial support you received to develop your athletic talents in relation to your female counterparts. If you are a woman, perhaps at times you have felt strangely out of place in a competitive environment, or maybe limited financial support has been available to help you improve your athletic skills. In either case, if you have ever wondered why such situations exist, you were on the verge of delving into history.

From previous *scholarly study,* you are probably familiar with a lot of historical information that will help you understand the history of physical activity. Most of the historical research on physical activity deals with ancient Greece and Rome, Europe and Great Britain, and North America. Knowing something about the overall history of these societies will make history of physical activity easier to learn and remember. For example, if you know that nationalism and national patriotism swept through Europe in the 19th century, then it's not surprising to learn that in countries such as Germany and Sweden, people thought that special gymnastics systems could help build a healthy, strong citizenry that could keep their nations strong. Making this connection helps you remember when, and at least one reason why, these gymnastics systems came into being.

Finally, from watching physical activity professionals in action, you may have observed some history in the making. Perhaps you know people who work as athletic trainers, sport marketers, personal trainers, and sport physical therapists. These professionals are part of an important historical trend. Not long ago, career opportunities for people in the physical activity field were limited to professors, physical education teachers, and coaches. But in recent years, professional opportunities have greatly expanded, and this is a historical phenomenon.

As your knowledge of history of physical activity broadens, you will see that it is related to both philosophy of physical activity and sociology of physical activity. As you can see in figure 6.1, all three are part of the sociocultural sphere of scholarly study—that is, the sphere of research that focuses on the social and cultural aspects of physical activity.

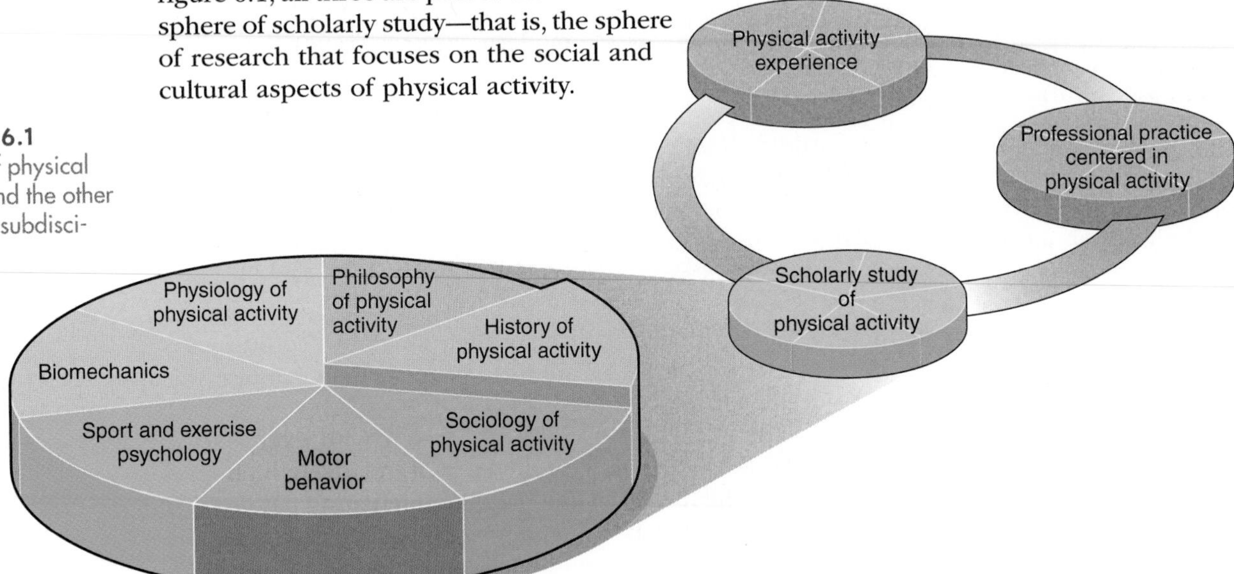

FIGURE 6.1 History of physical activity and the other scholarly subdisciplines.

What Does a Historian of Physical Activity Do?

Historians who study physical activity are usually college or university faculty members; but a handful are librarians, consultants for book publishing companies, archivists in charge of special collections of documents, or museum curators. We will focus on faculty members. They engage in typical activities consisting of teaching, research, and professional service. They often teach broad survey courses on history of physical activity, and some of their classes focus more narrowly on topics such as the ancient Olympic Games, physical activity in colonial America, or sport and ethnic relations in the United States. Examples of professional service include giving a community presentation about the history of a region's minor league baseball team, participating on an educational TV panel to discuss the history of gender relations in American sport, and helping a city's nonprofit historical society put together a museum display about the history of rehabilitative exercise.

Libraries are perhaps the most important research tools for historians. If you want to study the physical activities of African American slaves before the Civil War in the United States, for example, you could locate library sources such as records of interviews with former slaves who were asked to recall their experiences under slavery, published accounts from white slaveholders written during the time of slavery, and accounts from northern whites who visited plantations where slaves worked (Wiggins 1980). Most historians take great care to develop solid working relationships with librarians. Sometimes library staff act almost as gatekeepers—deciding how much access you should have to rare and valuable materials.

Historians of physical activity sometimes apply for research money from their own colleges or universities, as well as from outside organizations that specialize in supporting historical research. They often use the money to travel to libraries in distant cities or pay research assistants to help collect information. When asking for research money, they must write clear, persuasive proposals. Funding organizations usually require applicants to describe such things as the topic they want to investigate, why it is important, the sort of information they will collect, the sources where they'll find this information, what makes them competent to conduct the study, how much money they'll need, and what they'll buy with it.

Kinesiology and the History of Physical Activity

While historians of physical activity are academic specialists in their own right, with specific goals that help to guide their research, the fruit of their labors often finds application in a variety of other professional roles. For instance, many of the modern exercise machines found in college recreation centers, athletic training facilities, and commercial fitness centers are based upon models developed by an early American physical education leader, Dudley Allen Sargent, director of the Hemenway Gymnasium at Harvard University in the late 19th and early 20th centuries. Later, exercise specialists and engineers took the ideas and principles upon which Sargent had based his inventions and adapted them to contemporary knowledge and construction materials unavailable in an earlier time. Some of Sargent's ideas were, in turn, based upon principles first developed by earlier inventors that he undoubtedly learned of through the work of historians. Athletic coaches often create new playing strategies and training techniques by adapting those of an earlier generation of coaches whose work they have studied.

Knowledge of the history of sport, games, and exercise likewise gives the creative physical education teacher another tool with which to build interest in the lessons being presented by helping students discover the development and evolution of an activity. Scholars and scientists doing research in any of the other kinesiology subdisciplines described in this book also rely upon the work of earlier researchers in their field and often use the work of physical activity historians who have analyzed this past work and summarized it within a

Historical information paves the way for modern innovation.

HBS Archives Photograph Collection/Baker Library Historical Collections/Harvard Business School (left) and iStockphoto/David Lewis (right)

broader historical context. One example of the latter would be examining how the first generation of scholars in a given field resolved any ethical considerations related to the use of human or animal subjects and whether those issues have changed as the field has matured. An understanding of past practices can help one avoid earlier mistakes, make adaptations that work within a new scheme, and gain insights that sometimes lead to entirely new ways of thinking about a specific problem.

Kinesiology practitioners also benefit from the theoretical work of historians. An understanding of the goals of physical activity historians and an example or two of how those goals guide their work might lead us to consider ways of creating new opportunities for exercise, recreation, and sport in various settings.

Physical activity historians have two main goals, as shown on page 159. The first is to identify and describe patterns of change and stability in physical activity in particular societies or cultures during specific periods. Historians provide an enormous number of factual details to give readers a thick, robust description of such things as what it was like to be involved in the particular physical activities being studied, the values and attitudes that people had about those activities, people's broader lives, general societal values and attitudes prevalent at the time, how people's physical activities fit into their overall lives, and how they changed their physical activities. Because written historical descriptions require many pages, we can't include examples here.

Comparing Sport: Then and Now

A lot of movies show people playing sport a long time ago. Here are just a few examples:

Cinderella Man (boxing),
A League of Their Own (female professional baseball),
Field of Dreams (baseball),
Hoosiers (basketball),
Seabiscuit (horse racing),
Kansas City Bomber (roller derby),
Breaking Away (bicycling)

There are many others. Think about film coverage that you have seen showing people playing a sport a long time ago. Compare this with the way this sport is played today.

Sometimes changes occur slowly, giving us the impression that things have stayed the same. For example, the rules, strategies, styles of play, and equipment in many of our sports changed a lot during the 20th century. Because these shifts occurred relatively slowly, however, the games remained recognizable. At other times, changes take place much more rapidly and dramatically. For example, improvements in bicycle design in the United States more than a hundred years ago contributed to a rapid rise in bicycling—the bicycle craze in the 1890s (Hardy 1982).

The other main goal of physical activity historians is to analyze patterns of change and stability in physical activity in particular societies or cultures during specific periods. Most historians go beyond description to demonstrate relationships and influences in an effort to explain why things occurred. They look for individuals, groups, events, and ideas that helped bring about changes or contributed to maintaining stability. For example, Hardy's *How Boston Played: Sport, Recreation, and Community 1865–1915* (1982) focuses on the city-building process that took place in Boston over several decades after the Civil War. During that time the population of the city mushroomed as European immigrants and people from the American countryside arrived to take jobs created by the industrial revolution. The growing population led to major problems such as overcrowded tenements, poor health, neighborhood destruction, crime, ethnic conflict, and political and moral disorder—all serious threats to traditional ways of life. Hardy believed that Bostonians were looking for ways to revitalize their sense of connectedness to one another, and sport and recreation played an important part.

Based on what happened in Boston during this period, Hardy created three categories of ways in which Bostonians responded to the problems in their city: escape from the city and its problems, reform of the city and its problems, and accommodation to new forms of city life through developing a renewed sense of group identity (pp. 197-201). Hardy used these categories to examine the complex ways in which people sought community through building new playgrounds and parks, expanding sport and exercise programs in schools and universities, joining community-based sport clubs, riding bicycles, and conferring hero status on top athletes.

Understanding how this earlier generation of Bostonians utilized physical activity to benefit the entire city could well stimulate modern-day community leaders to work with various practitioners such as physical educators, recreation specialists, exercise consultants, landscape architects, land developers, and urban planners to create new and exciting opportunities for current and future generations of citizens.

➤ Most physical activity historians are college and university faculty members; through their research, teaching, and service, they pursue the two main goals of the subdiscipline.

History of the Subdiscipline

A handful of American scholars studied history of physical activity in the early 20th century, and even in the late 19th century at least one physical educator viewed this as important. Even so, it wasn't until the 1960s that the subdiscipline began to develop a recognizable identity in North America.

One of the earliest American reports to emphasize history of physical activity was Edward Hartwell's "On Physical Training" (1899), which included information about ancient Greece, Europe and Great Britain, and the United States (Gerber 1971). Hartwell believed that professionals who would be charting a course for the future should know about the past, and as early as 1893 he pointed to the need for a textbook on history of physical education. Several decades would pass, however, before such texts would begin to appear.

Historian Frederic Paxson wrote a classic article in 1917 titled "The Rise of Sport." On the basis of the well-known frontier thesis developed by his graduate school mentor Frederick Jackson Turner, Paxson argued that the disappearance of the American western frontier spurred the increased popularity of sport because it served some of the same escape or release functions as the frontier did (Pope 1997, pp. 1-2; Struna 1997, pp. 148-149). This safety valve function and Paxson's belief in the gradual, inevitable social evolution of sport were central ideas in the research of many sport history scholars for decades to come.

Goals of History of Physical Activity

1. To identify and describe patterns of change and stability in physical activity in particular societies or cultures during specific periods
2. To analyze patterns of change and stability in physical activity in particular societies or cultures during specific periods

Whether the focus is on the 1890s bicycle craze in the United States, the shrinking attention to physical activity in the education of young boys in the ancient Roman Empire, or the U.S. middle-class notion in the late 19th century that exercise could improve young men's morality, a central concern is identifying shifts and continuities. These trends are always examined over specific periods and in relation to the broader societies in which physical activities took place.

In the 1920s and 1930s, physical educators published several textbooks on the history of physical education, as well as articles in professional journals on the development of physical education programs in schools and colleges. These books and articles probably helped open the door in college and university physical education departments to recognition of the importance of history of physical activity.

Seward Staley (1937), a professor of physical education at the University of Illinois, opened the door wider when he suggested that courses on the history of physical education should include a focus on the history of sport. Following World War II, graduate students began studying sport history at the University of Illinois under his direction (Struna 1997).

In 1940, historian Foster Rhea Dulles wrote his widely read *America Learns to Play: A History of Popular Recreation*. Dulles went beyond Paxson's safety valve and evolutionary perspectives to point out more positively that popular 20th-century American recreational pursuits—including sports—were well liked because they were satisfying to the American public (Struna 1997). Furthermore, a broad range of people had access to them, indicating to Dulles that American democracy had come of age (Pope 1997).

After World War II, historian John Betts completed an important doctoral dissertation titled "Organized Sport in Industrial America" (1952). He stressed that sport played an active role in society by helping bind people together, and he highlighted the ways in which late 19th-century entrepreneurs harnessed a variety of new manufacturing processes, new forms of transportation, and new forms of communication to shape sports into profitable commercial enterprises (Pope 1997; Struna 1997). In sharp contrast to Betts' active model, a passive model of sport was offered a decade later by journalist Robert Boyle in *Sport—Mirror of American Life* (1963). Boyle viewed sport as a mirror that simply reflected society. Both Betts' and Boyle's works had considerable influence on the subsequent work of sport historians.

The ranks of sport historians began to grow in the 1960s, but no scholarly association was in place to help the subdiscipline advance. The first attempt to remedy this occurred in 1962 when Marvin Eyler, Seward Staley, and Earle Zeigler worked to develop a section for sport history in the College Physical Education Association (soon renamed the National College Physical Education Association for Men [NCPEAM]), a professional organization for college and university physical education faculty members and administrators. Because this was the only formal organization for sport historians at the time, the NCPEAM *Proceedings* became an important published record of their work (Berryman 1973; Struna 1997).

Scholars in the discipline of history became more interested in studying sport when Eugen Weber, a prominent historian, presented "Gymnastics and Sports in Fin-de-Siècle France: Opium of the Classes?" at the 1970 meeting of the American Historical Association (AHA) and published it the following year in *American Historical Review* (Berryman 1973). The program of the 1971 AHA meeting included an entire section on sport history. The North American Society for Sport History (NASSH) was established in 1972, and it was soon recognized as the main scholarly association for sport historians in the United States and Canada. The North American Society for Sport History was designed to attract a broader scholarly membership including women, Canadians, international scholars beyond North America, and sport historians from disciplines beyond physical education (Struna 1997). It began publishing *Journal of Sport History* in 1974, and a report issued in 1985 showed this publication to be the seventh most frequently cited history journal among hundreds in existence (Struna 1997).

Research Methods in History of Physical Activity

Research in sport history expanded in several directions during the 1970s and 1980s (Pope 1997; Struna 1997). Scholars began using two new analytical frameworks, or sets of general concepts or ideas, to make sense of the historical information they collected. The first framework, **modernization theory,** emphasized that the rise of modern sport occurred during the industrial revolution as American society shifted away from agricultural and local economies toward city-based industries rooted in science and technology. Sports changed from relatively unspecialized games to highly organized contests involving many rules and

Interviews With Practicing Professionals

Photo courtesy of Sherry Salyer

Sherry L. Salyer, EdD

Senior Lecturer and Director of Undergraduate Studies, Department of Exercise and Sport Science, University of North Carolina at Chapel Hill

Sherry L. Salyer has taught at the University of North Carolina at Chapel Hill since 1994. She also serves as director of undergraduate studies in that department as well as assistant dean for the academic advising program in the College of Arts and Sciences and the General College at the University. Previously, from 1974 to 1992, she taught physical education at the middle school level in the Winston-Salem/Forsyth County Public School System in North Carolina. She earned her bachelor of science and master of arts degrees in physical education at Appalachian State University, a master of arts degree in English, and the PhD degree in exercise and sport science at the University of North Carolina at Greensboro.

Q: How did your knowledge of history of physical activity help when you taught in middle school, and how did you bring historical topics into classroom activities?

A: At the middle school level my bulletin boards exhibited sport and exercise news, especially around such significant events as the Soccer World Cup, U.S. Tennis Open, Olympic Games, NCAA tournaments, and professional team sport events such as the baseball World Series, football's Super Bowl, and National Basketball Association playoffs. Often, articles and photos depicting the history of these events were included. Class discussions were held around these and other events or personalities important to the particular sport. One significant topic was Jackie Robinson and the racial integration of professional baseball and how that was an early part of the modern civil rights movement. I gave the students an assignment of preparing a notebook of newspaper and magazine articles and photographs related to the Games. My bulletin board material included information on the symbols, pledge, philosophy, and events of the Olympic Games as well as past and current personalities associated with this major international competition. Each of these assignments, discussions, and bulletin board displays was designed to both inform the students and stimulate their interest in sport and exercise. Of course, I hoped that this might also result in a lifelong interest in exploring the history and background of not only sport and exercise but other subjects as well.

Q: How have you used history of physical activity in your classes at the college level?

A: At the university I was able to share my public school teaching experiences with the undergraduate physical education majors in my teaching methods and curriculum courses. Now I make the history of physical activity a significant part of my course in the foundations of exercise and sport science. For example, a discussion of Dr. Franklin Henry and his 1964 article challenging the physical education profession to begin identifying its unique "body of knowledge" acquaints students with a seminal event that led to a radical reformation of the profession into the present discipline of kinesiology. Another example of my use of history occurs when the students and I examine personalities such as Dr. Kenneth Cooper, Jane Fonda, Jack LaLanne, Jim Fixx, and others who transformed and stimulated interest in physical fitness over the past half century. Likewise, the 50th anniversary of Althea Gibson's victory in what is today the U.S. Tennis Open and the rise of Cullen Jones as one of the few nationally or internationally elite African American swimmers has afforded wonderful openings into broad-ranging discussions of race and sport.

Dr. Salyer's career at two different levels of the educational spectrum provides an excellent example of how a knowledge of the history of physical activity can enrich the learning experience and stimulate in students a lifelong interest in the subject.

specialized playing positions. Allen Guttmann pioneered the use of modernization theory in his book *From Ritual to Record: The Nature of Modern Sports* (1978).

The second conceptual framework emphasized **human agency,** suggesting that people were actively involved in developing or "constructing" their own sports. Research focused on the details of how this occurred (Pope 1997; Struna 1997). One of the earliest works to use this approach was Stephen Hardy's *How Boston Played: Sport, Recreation, and Community 1865–1915* (1982). Hardy pointed to many local struggles among middle-class and working-class groups—often tinged with ethnic distinctions—that occurred as Bostonians went about structuring their sports and recreational activities.

Other scholars expanded the focus on human agency to look at gender differences. For example, in *Cheap Amusements: Working Women and Leisure in Turn-of-the-Century New York,* Kathy Peiss (1986) examined working women's culture between 1880 and 1920, emphasizing their newfound recreational activities outside the home in dance halls, nickelodeons, social clubs, and amusement parks.

An important research direction inaugurated in the 1980s dealt with exercise and health (Struna 1997). This trend was sparked by the American public's growing interest in exercise and physical fitness in the late 20th century and by the expansion of advanced course work on exercise in college and university kinesiology and physical education programs. One of the earliest studies was James Whorton's *Crusaders for Fitness: The History of American Health Reformers* (1982). Whorton showed the role of exercise among the many practices advocated in 19th- and early 20th-century reform movements aimed at improving the health of the American public.

In the 1990s, physical activity scholars continued to use the approaches to historical research that had been pioneered in the 1970s and 1980s. In addition, they increasingly linked their work to research in other social science disciplines, especially anthropology, economics, and sociology (Struna 1997).

Building on a small base of scholarly knowledge gathered earlier in the century, the history of physical activity subdiscipline grew in size, sophistication, and scope from the 1960s onward. Scholars initially focused on sport; but beginning in the 1980s, exercise and physical fitness received greater attention.

Remembering what you've learned about the goals of history of physical activity, as well as the chronology of the development of the subdiscipline, let's look at how scholars produce this knowledge. How do they find out what happened, who was involved, when and where things took place, and why? The process has similarities to what detectives and attorneys do when they hunt for evidence that helps reconstruct important details of a crime. It involves locating evidence; critiquing and examining evidence; and piecing it together in a coherent, insightful framework that explains how and why things occurred. We can divide this process into three stages: finding sources of evidence; critiquing the sources; and examining, analyzing, and synthesizing the evidence.

Thinking From a Historical Perspective

Scholars in the discipline of history had little interest in studying sport history until the 1970s. Why did these historians ignore the topic for so long? What might have been their concerns? Can you think of any challenges or roadblocks they might have faced in doing historical research in the field? What about sport and activity itself might have made historians slow to identify sport history as a viable research topic?

Finding Sources of Evidence

Think about the many different physical activities that currently exist in our society. If historians 200 years in the future wanted to study these, what sorts of items would be useful? All kinds of things come to mind: aerobic workout videos; popular books, articles, and magazines about exercise and physical fitness; business records of professional sport franchises; videos of televised sporting events; school physical education curriculum guides; videos of dance concerts; policy statements from the National Collegiate Athletic Association; and the 1996 U.S. government report *Physical Activity and Health: A Report of the Surgeon General*. The list could go on and on.

Sport historians of the future will study today's relics to draw conclusions about the role of sport in today's world.

AP Photo/Jens Meyer

These items are all primary sources of information, meaning that they were produced in the society and period being studied. But where would scholars find them? We can only guess where these things might be stored in the future, but we know where historians look today—libraries, archives (storage facilities for documents from important people and organizations), and private collections (Struna 1996a). Sometimes researchers interview people to tap their memories of life in earlier times.

Materials are scattered throughout the country. For example, the Avery Brundage Collection (Brundage was president of the International Olympic Committee from 1952 to 1972) is at the University of Illinois. The collection contains items such as letters to and from Brundage, books, pamphlets, scrapbooks, Brundage's notes on various topics, speeches given by Brundage, and policies from the International Olympic Committee and other sport governing bodies (Guttmann 1984). Copies of *Boston Medical and Surgical Journal* from the 19th century are located at the National Library of Medicine in Bethesda, Maryland, as well as at a library at the University of Chicago (Struna 1996a). Local historical societies are often rich sources of information, as are archives located in the library at your own college or university.

DO it **Activity 6.1**

Primary Source Material

Distinguish between primary and secondary source material in this activity in your online study guide.

Historical reports written by people in later periods—secondary sources—are also useful for finding primary sources. Examples of secondary sources include a journal article written in the 1980s about African Americans who participated in intercollegiate athletics in the 1930s and 1940s, or a book-length report produced in the 1990s on physical activity in the American colonies. Authors include detailed notes pointing to the nature and location of specific information (primary sources and other secondary sources) used in their investigation. Most libraries have electronic databases to help you locate sources, and some of these are accessible through the Internet.

Critiquing the Sources

After you locate primary sources, you must scrutinize them carefully for authenticity and credibility. Sources that are authentic and credible make your historical research believable. Let's concentrate on written documents such as newspaper articles, books, and letters. Determining authenticity involves such matters as who wrote the piece and when it was written. You might learn, for example, that a document was written many years after the events it describes took place or that the author listed on the document was not the person who wrote it. For example, if a letter dated in the 17th century uses the phrase "didn't get to first base," you can be sure that the piece was written much later because this term comes from American baseball, a game that didn't exist in the 17th century (Shafer 1980, p. 131). If ink used to publish a report wasn't manufactured until several years after the date listed, then you know that the piece was written later. (Such a determination would require a technical specialist.)

Determining the credibility of written documents helps you interpret them. First, the *rule of context* encourages you to make sense of a document's language in relation to what the words meant to people in the society in which it was produced. Second, the *rule of perspective* requires you to examine an author's relationship to the events that he or she describes and how the author obtained the information. Finally, the *rule of omission* or *free editing* reminds you that you always have partial records of events—not the complete events themselves; locating multiple sources to add more details to your evidence is important (Struna 1996a).

Examining, Analyzing, and Synthesizing the Evidence

➤ Historical research involves finding sources containing evidence about past events, critiquing the sources for authenticity and credibility, and analyzing the data contained in the sources to learn how and why things happened.

After you've located authentic and credible sources, you should examine the evidence to find information that addresses the tentative hypotheses or research questions that you established at the beginning of your investigation. The goal is to describe events in detail and then analyze them to learn how and why things took place. Historians accomplish this by placing events in an analytical framework that uses (1) trends and relationships in the events themselves or (2) theoretical models from the social sciences. You have already seen examples of analytical frameworks—the reestablishment-of-community framework, the modernization framework, and the human-agency framework. There are probably as many approaches to analysis and synthesis as there are researchers. The main goal is to piece together the evidence to gain a detailed understanding of what happened, as well as how and why (Shafer 1980; Struna 1996a).

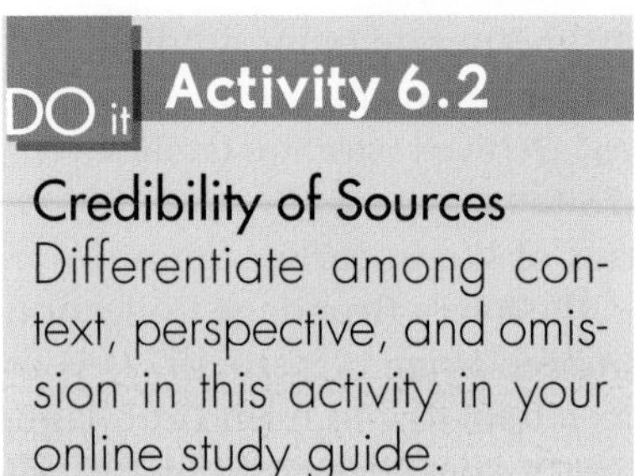

Credibility of Sources

Differentiate among context, perspective, and omission in this activity in your online study guide.

Overview of Knowledge in History of Physical Activity

Now that you know what historians of physical activity do, let's look at some of the fascinating information they've discovered. The body of knowledge is large and complex; it deals with many societies worldwide and covers many different periods. Here we will touch on a few of the highlights.

If you were a boy living in ancient Greece, for example, during the 6th century B.C.E. in the city-state of Sparta, you would have received more extensive and harsher physical activity training than you would have in the city-state of Athens, where you would have gotten a more balanced physical and intellectual education. In both places, however, your physical activity instruction would have had the goal of preparing you for warfare. And although the ancient Greeks considered the Olympic Games similar to warfare in competitiveness, this prestigious event for full-fledged citizens was distinct from the brutal gladiators' contests—mostly among slaves and convicts—staged several hundred years later in the Roman Empire. And the Roman contests were very different from the informal, rough team ball games played by European peasants in the Middle Ages and also distinct from high-profile women's gymnastics in the Olympics of the late 20th century. We haven't even mentioned other forms of physical activity such as dance and exercise. Understanding the ways in which physical activities have fit into societies around the world since the beginning of history is indeed a monumental task.

In North America we know a lot about the history of physical activity in Western civilizations—especially ancient Greece and Rome, Europe, Canada, and the United States. We must recognize, however, that people in Africa, Asia, Australia and New Zealand, the Indian subcontinent, Central and South America, the Caribbean region, and the Pacific Islands developed rich physical activity traditions. These include a variety of martial arts in Asian societies, sumo wrestling in Japan, ball games played in high-walled courts in native Central American civilizations, an African martial art brought to the Atlantic and Caribbean coasts of South America, and a sport in Afghanistan involving hundreds of mounted horsemen who simultaneously try to gain possession of a goat or calf carcass. Historical information exists about all of these. In this text, however, we will focus on the history of physical activity in North America.

British and European colonists began arriving in North America in greater numbers beginning in the early 1600s, swelling an earlier trickle. Initially, people worked hard to establish themselves in unfamiliar and often harsh environments. These hardships, coupled with strict religious prohibitions against idleness and amusements in some areas of the colonies, meant that participation in sport and games was limited. Nevertheless, people sometimes engaged in their own versions of traditional British and European sport and recreational activities. Participation grew as the colonists became more established and religious sanctions were lifted. Activities included horse racing, fishing, hunting, sailing, boat racing, golf, team ball games, sleigh races, skating, foot racing, boxing, wrestling, animal baiting (in which a wild animal such as a bear was tied by a chain and attacked, often by dogs or rats), cockfighting, and billiards. Some of these were commercialized and yielded profits for promoters; gambling was also common (Baker 1988; Lucas & Smith 1978; Struna 1996b).

We'll look more closely at physical activity in North America during three periods:

- 1840-1900: Industrialization, growth of science and technology, immigration, urbanization, democratization, and westward expansion characterized North American society; the American Civil War occurred.
- 1900-1950: The growth of consumerism, immigration, and democratization characterized North American society; a major economic depression and two world wars occurred.
- 1950-2000: The growth of electronic communication, the growth of global trade, immigration, and democratization characterized North American society; the Cold War, Korean War, and Vietnam War occurred.

For each of these periods, we'll examine

- participation in physical activity,
- physical activity professions, and
- scholarly knowledge about physical activity.

Rich historical physical activity traditions exist throughout the world, including sumo wrestling in Japan.
AP Photos

1840-1900: Industrialization and Westward Expansion

From the mid-1800s to 1900 the economy, population, and geography of the United States and Canada expanded. New developments in science and technology fueled the growth of business and industry. As city populations mushroomed, major sanitation problems, crowded living conditions, lack of space for physical exercise, and unsafe factories threatened everyone's health. The advent of more leisure time sparked an interest in building city parks and playgrounds, as well as in developing more commercial entertainment and amusements such as spectator sports. Many new schools and colleges opened, with both physical training and student-run intercollegiate sport programs. The citizenry started buying more products in response to newspapers, magazines, and eventually movies that offered glimpses of the good life. Westward expansion continued, highlighted by the completion of the transcontinental railroad in the United States. Massive numbers of European and British immigrants arrived, along with Asians on the West Coast. The Civil War (1861-1865) exacted a terrible toll of death, destruction, and disruption, but it led to legal affirmation of the ideal of racial equality, an important goal in American society ever since.

Participation in Physical Activity

Liberal religious and philosophical currents had begun to flow together in the early decades of the 19th century, emphasizing that a human being's body, mind, and soul were an integrated whole. This perspective considered it important to keep one's body healthy to achieve peak mental functioning as well as the highest possible level of moral reasoning. Some believed that good health practices indicated a morally righteous person in the eyes of God. This thinking, combined with squalid conditions in overcrowded cities and a nationalistic desire to make the United States self-sufficient, led to widespread efforts throughout most of the 19th century to improve people's health. Proper diet and exercise were among the recommended practices (Berryman 1989; Whorton 1982).

Physicians, writers, and teachers, addressing primarily the middle class and social elites, called for more vigorous physical activity—especially for boys and men. They coupled their notions of the importance of health reform with ideas from England about the value of sport for developing moral character and manliness. As the century progressed, physical activity grew increasingly attractive and popular.

Many professionals also recommended moderate exercises for improving the health of girls and women. Straining too hard in exercise or competitive sports, however, was usually considered unladylike and dangerous for women's physical well-being and future childbearing capabilities (Vertinsky 1990). By the end of the 19th century, some girls and women were involved in vigorous physical pursuits.

European systems of gymnastics, especially those from Germany and Sweden, gained an important American foothold during this time. German cultural societies called Turnvereins eventually offered well-equipped gymnastics facilities in many North American cities. In the United States alone, there were 148 of these by 1867 serving 10,200 members (Swanson & Spears 1995, p. 128). The centerpiece was exercise on heavy equipment such as parallel bars, vaulting horses, and hanging rings. By contrast, the Swedish system involved calisthenics and exercises with lighter equipment such as dumbbells, long wooden rods, and weights. Elements from both systems, as well as other exercise programs developed by Americans, grew increasingly popular in YMCAs, YWCAs, and city parks and playgrounds.

Sports were popular with many of the new immigrants who arrived in the United States and Canada from 1840 onward. They played American baseball, and when American and Canadian football evolved from rugby and soccer later in the century, they played that too. Youngsters often learned American sports and games at local playgrounds, and many civic leaders hoped that immigrants' participation would speed their assimilation into American society. But immigrants also sought to keep the cultures of their homelands alive through participation in physical activities. Many belonged to special clubs in which sport, exercise, and other forms of recreation were often prominent. Examples in the city of Boston included the Boston Turnverein, featuring gymnastics and catering to German immigrants; the Caledonian Club, devoted to Scottish culture and including an array of sports and games such as caber toss, races, and pole vault; the Irish Athletic Club, the Irish-American Athletic Club, the Ancient Order of Hibernians, the Boston Hurling Club, and the Shamrock Hurling Club—all focusing on Irish sports and games, including oarsmen's regattas, stone throwing, and hurling; and cricket clubs formed by West Indian immigrants (Hardy 1982, pp. 136-138).

Other immigrants took part in sports and amusements that, at the time, were considered less than respectable by members of mainstream society. These included prize fighting, billiards, bowling, cycling, and wrestling. The Irish dominated prize fighting for several decades, and later in the century boxing gyms could also be found in African American neighborhoods (Gorn 1986). After the Civil War, African Americans found their way into mainstream sports—professional baseball and horse racing, for example—but by 1890, racial prejudice forced most of them into segregated leagues and organizations.

In school and college physical activity classes, gymnastics exercise was preeminent. The Swedish and German systems were common, as were exercise programs developed by Americans that focused on calisthenics and lightweight equipment. The primary focus of such programs was health through fitness as promulgated by such medically trained early physical education leaders as Dr. Edward Hitchcock (Amherst College), Dr. Dudley Allen Sargent (Harvard), and Dr. Delphine Hannah (Oberlin College). Sports were not usually included in classes but instead existed as extracurricular activities. Originally low-key intramural activities, they were a rich supplement to often-stodgy exercise classes (Rudolph 1962; Smith 1988). Student-run intercollegiate competition appeared about midcentury and gathered steam in the 1870s. The most common intercollegiate sports were American football, crew, baseball, and track and field.

Boxing was a popular amusement for many immigrants during the second half of the 1800s.

AP Photo

College and university administrators believed that intercollegiate sports could be helpful for recruiting new students, attracting alumni contributions, and promoting a sense of community among an increasingly diverse student body. According to thinking that arrived on North American shores from England, sport could also develop leadership, vigorous manliness, and upstanding moral character, all qualities considered desirable for college men. For all these reasons, athletics were believed too important to be left in the hands of students. Campus administrators, faculty, and coaches gradually took control.

During this era fewer sport opportunities were available for college women. They engaged in occasional intercollegiate competition and somewhat more frequent intramural competition. Many female physical educators, however, began including sport activities in their classes in the 1870s and 1880s (Swanson & Spears 1995). Women who wanted to compete faced societal concerns about overly strenuous and highly competitive physical activities for women and girls. They avoided sports in which injuries were more likely (football, for example). Basketball, tennis, golf, and baseball were popular. Women's sports usually took place in secluded locations with only a few spectators present. A delicate balance had to be struck between making competition available to college women and keeping the activities acceptable to faculty and campus administrators (Park 1987b).

Beyond schools and colleges, by the 1860s many amateur baseball teams were on their way to becoming professionalized; common practices by that time included charging admission and paying players. Baseball had an avid following, and newspapers gave it the most coverage of any sport (Adelman 1986, p. 174). In the 1880s, the most popular sports were horse racing, baseball, and prize fighting.

➤ American interest in physical activity—for example, gymnastics exercises, baseball, football, crew, track and field, horse racing, boxing, bicycling, and less well-known sports tied closely to people's ethnic origins—grew throughout the 19th century.

American athletes also competed in a new amateur event developed by Europeans and first held in Athens, Greece, in 1896—the modern Olympic Games. The United States fielded a team of 10 male track and field competitors—all collegians or ex-collegians—who participated in the fledgling Games; they chalked up nine victories in 12 events, along with five second-place finishes. Three other American athletes, two pistol shooters and one swimmer, also competed in these first Olympic Games, resulting in two additional victories in the sharpshooting events.

Curriculum Changes in Physical Education

In the late 19th century, physical education programs emphasized health and the activities focused on calisthenics; sports were not usually included in men's physical education classes. Why do you believe there was such an emphasis? What types of activities have your physical education classes in high school and college included? What do you think caused the change between what was offered in the 1890s and what is offered today?

Physical Activity Professions

An identifiable physical activity profession didn't appear in the United States until late in the 19th century. Nevertheless, a variety of practitioners in earlier decades focused at least some of their work on physical activity. These practitioners included physicians, successful athletes, journalists, educators, ministers, health reform advocates, business entrepreneurs, and a handful of European gymnastics specialists who immigrated to the United States. Some of them wrote about the physical, intellectual, and moral benefits of exercise and sport. Some developed exercise programs in schools and colleges. Others worked to establish various European gymnastics systems on American soil. Some wrote popular self-help manuals on achieving healthy lifestyles that included how-to information on exercise routines. A few became professional athletes in sports such as horse racing, pedestrianism (long-distance walking), boxing, and baseball. Some tried to sort out the best training techniques for athletic success. Some became coaches, and some bankrolled professional sports.

By the 1880s a climate of intense interest in exercise and sport had developed in the United States. Numerous professions were beginning to organize in response to the growing need for people with specialized knowledge and skills in technologically based businesses and industries. Efforts were under way to develop more stringent educational standards for new practitioners.

In this atmosphere, the physical education teaching profession took root (Park 1987a), the first recognizable physical activity profession in the United States. In 1885 the American Association for the Advancement of Physical Education was formed by about 60 people who wanted to have an organization that would promote the new profession. After several name changes, it is today known as the American Alliance for Health, Physical Education, Recreation and Dance, an important professional association for practitioners in several fields.

➤ The earliest identifiable American physical activity profession—teaching physical education—was established in the late 19th century during a period of high interest in physical activity among the general public.

Several physical education teacher training programs were inaugurated, primarily in the Northeast, and they varied in length from a few weeks to about two years. Some of the most famous were the Sanatory Gymnasium (opened near Harvard in 1881 by Dudley Sargent and renamed the Normal School of Physical Training in 1894), the Harvard Summer School of Physical Education (opened in 1887 by Dudley Sargent), and the Boston Normal School of Gymnastics (opened in 1889 by Mary Hemenway and Amy Morris Homans). Around the turn of the century, several four-year bachelor's degree programs were initiated for the preparation of physical education teachers.

Scholarly Knowledge About Physical Activity

Scientific discoveries in the 19th century produced much new information about human anatomy and physiology. By the 1840s improved microscopes permitted more detailed studies of phenomena such as oxygen transport in blood, energy transformation in muscles, and the anatomy and functioning of the nervous system (Park 1987a, p. 144). Later in the century, scholars who wanted better understanding of the biological effects of physical activity on the human body occasionally used such information. Educators also sometimes incorporated this knowledge into physical education teacher training programs.

For example, Edward Hartwell, an associate in physical training and director of the gymnasium at Johns Hopkins University, published an article in 1887 titled "On the Physiology of Exercise." George Fitz, an instructor and researcher in physiology and hygiene at Harvard University, established a four-year undergraduate program in "Anatomy, Physiology, and Physical Training." Not far from Harvard, the Boston Normal School of Gymnastics—a two-year physical education teacher training institution that attracted mostly female students—offered a curriculum focused on Swedish gymnastics that also included a variety of basic science courses taught by well-known scholars from major universities around Boston. This excursion into the biological sciences by early leaders in physical education is important because it demonstrates the value that they placed on developing a scientific base for the emerging physical education teaching profession (Gerber 1971; Park 1987b).

Despite this new interest in the sciences, professional programs focused on learning physical activities and how to teach them outnumbered scientific curriculums. By the end of the 19th century the emphasis was clearly on the professional programs, as well as on the positive social values that students could learn through participating in play and sport. Investigations into the biological mechanisms underlying physical activity were not completely curtailed, but it was not until the 1960s that scientific information again moved to center stage in college and university physical education curriculums (Park 1987a).

➤ Scholarly knowledge about physical activity—mostly information from the biological and physical sciences—became important in a few professional teaching training curriculums in the late 19th century, but in most programs this subject took a backseat to learning physical activities and practical knowledge about how to teach them.

1900-1950: Consumerism, Immigration, Democratization

In the first half of the 20th century, American industry perfected mass production and churned out a wide variety of consumer goods from automobiles to canned food to refrigerators. With readily available products, entrepreneurs stepped up their advertising to encourage the public to become enthusiastic consumers. Big-time intercollegiate sports and professional baseball and football leagues became major parts of the growing entertainment industry and benefited from the rise of this consumer culture. Many new immigrants—especially from southern Europe—arrived on American shores, adding more ethnic diversity to an already multifaceted population. American society made progress toward the democratic ideal of equal opportunity and social justice: Women got the right to vote, and African Americans were hired in greater numbers for federally funded jobs. Full achievement of social equity, however, remained distant and elusive. The severe economic depression that began in 1929 left personal, lifelong scars on massive numbers of Americans who suffered through it. And many other horrible and unforgettable disruptions occurred during World War I (1914-1918) and World War II (1939-1945).

Participation in Physical Activity

Competitive sports were in the limelight during the first half of the 20th century. Professional baseball, professional boxing, horse racing, and American collegiate football—all men's sports—were especially popular with fans. Professional baseball's American League and National League joined forces in 1903 to keep player salaries in check and reduce competition for fan loyalty. American professional football and basketball gained a degree of prominence by the 1930s, although they were still fledgling enterprises. World War II temporarily put a damper on men's professional athletics because most healthy young men entered military service. This spurred Chicago Cubs owner and chewing gum magnate Philip K. Wrigley to organize the All-American Girls' Baseball League, which played in the Midwest from 1943 to 1954.

American football dominated intercollegiate athletics during this era. Major college games received widespread news coverage, and they were occasions for lively partying and revelry. By 1903 Harvard had a stadium that could seat 40,000 fans, and by 1930 seven of the "concrete giants" on college campuses could hold over 70,000 people (Smith 1988, p. 169; Rader 1990, p. 182).

Competitive sports for boys were also common in high schools and elementary schools, on playgrounds, and in youth service agencies such as YMCAs. By the 1930s elementary school competition began to decline when physical education teachers complained about the high stress of competition. Many people still wanted competition for youngsters, however, and new

The Golden Age of Sport: The 1920s

The 1920s, sometimes referred to as the "Roaring Twenties," was a decade of change and excess. Whether one measures the change in terms of clothing fashion, political opportunity, technology, or popular culture, there is no doubt that a seismic shift in American society took place. Change was apparent as women exercised their newly acquired right to vote as well as to raise the hemlines of their skirts. Excess might be measured by stock market and real estate speculation, as well as by personal spending as America became increasingly a consumer nation. Sport became one of those products consumed by a people hungry for entertainment, be it on the stage, on the silver screen, on the radio, or in the stadium. And, where once the titans of business and industry were the childhood heroes and role models, now the Thomas Edisons, Henry Fords, and John D. Rockefellers were forced to make room for sport figures such as Babe Ruth and Gertrude Ederle and movie stars like Charlie Chaplin and Mary Pickford. The middle class was growing; times were good for larger numbers of Americans, and increasing numbers had more leisure time to watch and follow their favorite teams and individual sport stars.

While earlier decades had their individual athletic heroes, such as the boxer John L. Sullivan in the 1880s and baseball pitcher Cy Young in the 1890s and early 1900s, the "twenties" brought them in greater abundance and across a wide array of sports. Babe Ruth and Lou Gehrig (baseball), Bill Tilden and Helen Wills (tennis), Bobby Jones (golf), Gertrude Ederle and Johnny Weismueller (swimming), Red Grange (football), and Jack Dempsey and Gene Tunney (boxing) were representative of this sudden explosion of athletic excellence and public interest in sport. To accommodate demand, larger and grander stadiums were constructed by both professional baseball clubs and college football programs. New York's Yankee Stadium was opened in 1923 with more than 50,000 seats to meet the demands of fans flocking to see the great Ruth hit his mammoth home runs. New college football stadiums such as those at Ohio State University (1922) and the University of Illinois (1923) exceeded 60,000 seats to handle the masses of people who arrived by train and private automobile from near and far.

Individual athletic heroes like Red Grange brought an explosion of public interest in sport and led to the construction of new college football stadiums to handle the enormous crowds of spectators.
AP Photos

Fueling this phenomenon was a media industry just beginning to realize the revenue potential available through the exploitation of commercial sport. Newspapers had begun to recognize this around the turn of the 20th century when the sport page made its debut in many large cities to meet the growing demand for Major League Baseball coverage. By the 1920s many of these big-city papers began hiring more talented writers such as Grantland Rice *(New York Herald Tribune)*, John Drebinger *(New York Times)*, Shirley Povich *(Washington Post)*, and Paul Gallico *(New York Daily News)*, who further popularized the stars of the sport world through their coverage of the contests and feature stories on the athletes. The introduction of commercial radio in 1924 brought a whole new dimension to sport coverage; and by the end of the decade, announcers such as Graham McNamee were recreating football and baseball games for listeners in millions of homes across the nation. Contrary to the initial fears of college administrators and baseball club owners, the radio brought in countless numbers of new fans whose interest led them to purchase tickets to see in person what they had heard on the airwaves.

In the mid-1930s, writer Paul Gallico, looking back on the tremendous growth of interest in sport during the twenties, dubbed it "The Golden Age of Sport." Today we might refer to it more accurately as "The *First* Golden Age of Sport," since we have certainly been in a similar "sport-crazed" era for the past four decades or more. With the arrival of cable television and its numerous dedicated sport networks beginning in the 1980s, sport-specific magazines and almanacs, newer and ever more grandiose stadiums and arenas, and the creation of new and increasingly extreme competitive activities, it is fair to say that we are today in an unprecedented age. Nevertheless, we can trace the roots of today's sport mania to that amazing decade some 80 years ago.

programs designed by city recreation departments and organizations such as Little League Baseball, Inc. (formed in 1939) and Biddy Basketball (formed in 1950) filled the void.

Girls' and women's sports were subdued in comparison to major boys' and men's events with their accompanying publicity and hoopla. A few women achieved national and international success in sports such as tennis, basketball, golf, long-distance swimming, and track and field. For the most part they trained and competed outside the educational system in settings such as private country clubs and industry-sponsored leagues. For example, in 1930 Mildred "Babe" Didrikson was offered a job as a stenographer at Employers Casualty Insurance Company in Dallas so that she could play for the company's Golden Cyclones basketball team. The company also sponsored her to compete at the Amateur Athletic Union's national track and field championships in 1932. Her phenomenal success led her to the 1932 Olympics, where she won two gold medals and a silver (Guttmann 1991).

In schools and colleges during the 1920s, female physical educators turned toward intramural sports and low-key "extramural" competition for girls and women. They wanted to avoid the stress of high-level competition and encourage all girls and women to participate. Although a few institutions continued to sponsor elite sports for females, "fun and games" became the goal at most schools and colleges, greatly curtailing the development of elite female athletes in educational settings. Things didn't begin to change until the late 1950s.

Public interest in competitive sport was so pronounced in the early decades of the 20th century that sports replaced gymnastics exercises as the centerpiece in most school and college physical education curriculums. A large number of physical education teachers believed that student participation in sport would help develop high moral character and other qualities needed by a good citizen. And students seemed more interested in physical activities that were sparked with competitive excitement, compared with the tedium of repetitious, traditional gymnastics exercises.

Men's military service in World War II gave rise to the All-American Girls' Baseball League, which played in the Midwest from 1943 to 1954.

AP Photo/Richmond Times-Dispatch, Don Long

But exercise did not completely disappear from physical education classes. Teachers sometimes replaced the old gymnastics systems with expressive movements designed to communicate ideas and feelings, or with sport-related exercises such as shooting baskets or batting baseballs. By the 1920s, "corrective" physical education classes also offered special exercises to students with posture, fitness, and health challenges (Van Dalen & Bennett 1971, pp. 460, 461, 468).

The U.S. military recognized the need for physical training during World War I, when approximately one-third of U.S. draftees were initially declared unfit to serve. Moreover, recruits with musculoskeletal incapacities such as flat feet and backaches deluged overseas hospitals. Recreational sports—organized by about 345 military athletic directors—were widespread and popular with the troops (Murphy 1995; Rice, Hutchinson, & Lee 1969).

Physical education teachers on the home front added military training such as marching drills and calisthenics to school and college physical education programs. People could also exercise by following an instructor on the radio. Recognition of the poor physical condition of the troops served as a wake-up call after the war, spurring the expansion of school and college physical activity programs.

Racial discrimination eliminated most African Americans from the highest levels of competitive sport by 1900. In spite of this, a few were active in boxing (Jack Johnson and Joe Louis were heavyweight champions); a small number participated in intercollegiate athletics at predominantly white schools; a few competed in the National Football League until 1933 when racial barriers were raised; and a handful took part in the Olympic Games (track star Jesse Owens won four gold medals in the 1936 Berlin Games). African American baseball players had no choice but to play for African American teams until Jackie Robinson signed with the Brooklyn Dodgers in 1945 and played on one of their minor league teams in 1946; he moved up to the Dodgers in 1947.

➤ Sports—in elite-level competition, community recreation programs, and school and college physical education classes—were Americans' favorite physical activities in the first half of the 20th century. Americans paid less attention to exercise, although interest picked up during the two world wars in order to improve physical fitness.

Cultural heritage also influenced the sport involvement of people in many other racial and ethnic minority groups. For example, baseball, basketball, and boxing served as a middle ground for second-generation Jews, providing at the same time a place to celebrate Jewish heritage and a place to assimilate into mainstream American society (Levine 1992, pp. 3-25, 270-274). First-generation Japanese immigrants who arrived early in the 20th century often used traditional perspectives from their homeland to make sense of American sports, sometimes recalling samurai (historic Japanese knight-warriors) principles of courage and honor (Regalado 1992).

World War II renewed people's interest in exercise. After the war, educators expanded many school and college physical activity curriculums in response to weaknesses observed in the fitness of wartime military recruits, and sports were again at center stage (Swanson & Spears 1995).

Physical Activity Professions

Bachelor's degree programs in physical education grew in the early 1900s, expanding to about 135 by 1927 (Park 1980). Master's degrees were available just after the turn of the cen-

Sport and Ethnicity

Think of as many examples as you can of groups of people with a particular ethnic heritage getting together to

- play or watch sports associated with their own cultural heritage and
- play or watch sports that are widely popular in the United States.

Is there any overlap between your two lists? If so, why do you think there are such overlaps?

tury, and doctoral programs began in the 1920s. Undergraduate programs focused on training physical education teachers, continuing the trend that had started in the late 19th century. Graduate programs offered advanced training for teachers, as well as the academic preparation to become a college or university faculty member. The quality of these programs improved as research on physical activity picked up in the late 1920s.

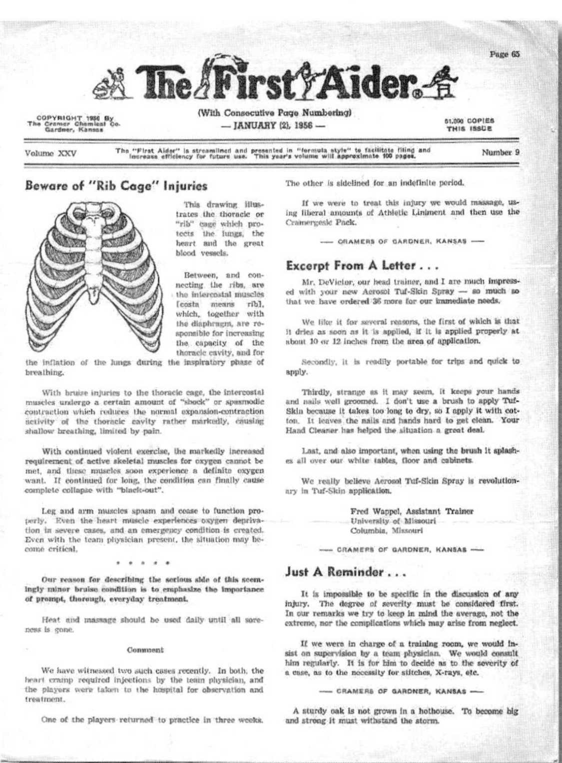

Page 65

The First Aider®

COPYRIGHT 1956 By The Cramer Chemical Co. Gardner, Kansas

(With Consecutive Page Numbering)

— JANUARY (2), 1956 —

51,000 COPIES THIS ISSUE

Volume XXV

The "First Aider" is streamlined and presented in "formula style" to facilitate filing and increase efficiency for future use. This year's volume will approximate 100 pages.

Number 9

Beware of "Rib Cage" Injuries

This drawing illustrates the thoracic or "rib" cage which protects the lungs, the heart and the great blood vessels.

Between, and connecting the ribs, are the intercostal muscles [costa means rib], which, together with the diaphragm, are responsible for increasing the capacity of the thoracic cavity, and for the inflation of the lungs during the inspiratory phase of breathing.

With bruise injuries to the thoracic cage, the intercostal muscles undergo a certain amount of "shock" or spasmodic contraction which reduces the normal expansion-contraction activity of the thoracic cavity rather markedly, causing shallow breathing, limited by pain.

With continued violent exercise, the markedly increased requirement of active skeletal muscles for oxygen cannot be met, and these muscles soon experience a definite oxygen want. If continued for long, the condition can finally cause complete collapse with "black-out".

Leg and arm muscles spasm and cease to function properly. Even the heart muscle experiences oxygen deprivation in severe cases, and an emergency condition is created. Even with the team physician present, the situation may become critical.

* * * * *

Our reason for describing the serious side of this seemingly minor bruise condition is to emphasize the importance of prompt, thorough, everyday treatment.

Heat and massage should be used daily until all soreness is gone.

Comment

We have witnessed two such cases recently. In both, the heart cramp required injections by the team physician, and the players were taken to the hospital for observation and treatment.

One of the players returned to practice in three weeks. The other is sidelined for an indefinite period.

If we were to treat this injury we would massage, using liberal amounts of Athletic Liniment and then use the Cramergesic Pack.

— CRAMERS OF GARDNER, KANSAS —

Excerpt From A Letter . . .

Mr. DeVictor, our head trainer, and I are much impressed with your new Aerosol Tuf-Skin Spray — so much so that we have ordered 36 more for our immediate needs.

We like it for several reasons, the first of which is that it dries as soon as it is applied, if it is applied properly at about 10 or 12 inches from the area of application.

Secondly, it is readily portable for trips and quick to apply.

Thirdly, strange as it may seem, it keeps your hands and nails well groomed. I don't use a brush to apply Tuf-Skin because it takes too long to dry, so I apply it with cotton. It leaves the nails and hands hard to get clean. Your Hand Cleaner has helped the situation a great deal.

Last, and also important, when using the brush it splashes all over our white tables, floor and cabinets.

We really believe Aerosol Tuf-Skin Spray is revolutionary in Tuf-Skin application.

Fred Wappel, Assistant Trainer
University of Missouri
Columbia, Missouri

— CRAMERS OF GARDNER, KANSAS —

Just A Reminder . . .

It is impossible to be specific in the discussion of any injury. The degree of severity must be considered first. In our remarks we try to keep in mind the average, not the extreme, nor the complications which may arise from neglect.

If we were in charge of a training room, we would insist on supervision by a team physician. We would consult him regularly. It is for him to decide as to the severity of a case, as to the necessity for stitches, X-rays, etc.

— CRAMERS OF GARDNER, KANSAS —

A sturdy oak is not grown in a hothouse. To become big and strong it must withstand the storm.

Before the National Athletic Trainers Association (NATA) was formed in 1950, a newsletter called the *First Aider* was the primary source of training information for coaches.

Despite the popularity of sport during the first half of the 20th century, college and university physical education curriculums didn't include much course work to prepare students for the occupations of coach or athletic trainer. Coaches most often came from the ranks of successful athletes. If they worked in a school or college, they were often required to teach physical education classes as well, so they earned the necessary college degrees and professional teacher certifications.

The few athletic trainers on the scene during this period had little formal education in health care. In colleges and universities, they often began as gymnasium jacks-of-all-trades with custodial responsibilities as well as other duties such as repairing equipment and facilities, laundering clothes, driving the team bus, and maintaining outdoor playing fields. They sometimes met their counterparts from other institutions at intercollegiate athletic events and shared training techniques (Smith 1979). Because elite sports were primarily for boys and men, most coaches and athletic trainers were men.

The Cramer Company, founded in 1922, identified a commercial niche and began selling liniment for sprains as well as other products for athletic training. In 1933 Cramer inaugurated the *First Aider,* which soon became a popular newsletter aimed at helping high school coaches understand training techniques. For many years Cramer also sponsored educational seminars for athletic trainers. The profession grew, but it wasn't until 1950 that a professional organization was founded—the National Athletic Trainers Association.

The outbreak of World War I sparked interest in physical therapy. By the end of the war, U.S. military reconstruction aides—an entirely female corps of therapists trained to help the war wounded—were receiving training in massage, anatomy, remedial exercise, hydrotherapy, electrotherapy, bandaging, kinesiology, ethics, and the psychological effects of injuries. In 1921 the reconstruction aides formed the American Women's Physical Therapeutic Association, the organization that eventually became the present-day American Physical Therapy Association, the main professional association for physical therapists (Murphy 1995, pp. 54, 71).

Teaching physical education continued to be the main profession for which students were prepared in college physical education programs during the first half of the 20th century; bachelor's degree programs increased in number, and master's and doctoral programs came on the scene.

Scholarly Knowledge About Physical Activity

At the beginning of the 20th century, a number of scholars were investigating topics such as neuromuscular fatigue, the vascular effects of exercise, kinesiology (today called functional anatomy or biomechanics), body measurements and proportions, the psychological aspects of play, and the history of physical education and sport (Park 1981). Not until the late 1920s, however, did research on physical activity gain much visibility. Faculty members in physical education as well as other disciplines conducted investigations. The prestigious Harvard Fatigue Laboratory, focused on research in exercise physiology, opened in 1927. *Research Quarterly,* a scholarly journal devoted to physical activity, began in 1930. One issue in 1934 contained the following topics: "physiology of respiration, reflex/reaction time, measurement of motor ability, effects of temperature on muscular activity, and test construction" (Park 1980, p. 5). Because of the importance of teacher training in college and university physical education departments, physical educators often did research that could be applied to teaching. For example, studies of techniques for assessing motor skills could be used to help teachers develop ways to evaluate student progress.

➤ Research on physical activity started to expand in the late 1920s. Most physical educators studied topics relevant to teaching physical education; a few examined other aspects of physical activity.

A handful of physical educators began doing research on aspects of physical activity not centrally applicable to teaching, including areas such as motor ability, motor capacity, physical fitness, and exercise physiology. Research on these topics continued to grow in the 1940s and 1950s, but it was not until the 1960s that such research mushroomed.

1950-2007: Electronic Communication and Globalization

In the second half of the 20th century, electronic communication expanded at a breathtaking pace. Television—in its infancy in 1950—became a common fixture in most homes in less than two decades. Digital computers developed from massive, room-sized arrays of vacuum tubes with limited capabilities to tiny but enormously powerful collections of memory chips and electronic displays. Communications satellites were placed in orbit, greatly simplifying intercontinental communication. Air travel became faster and easier. As in the past, each of these technological advances affected sport and exercise in numerous dimensions. Global trade expanded, accompanied by greater worldwide political and economic interdependence. New immigrants continued to arrive in the United States—especially from Asia and Latin America. Progress was made toward the American democratic ideal of equal opportunity and social justice: Federal civil rights legislation mandated greater equality in education, business, and housing; and professional and intercollegiate sport became more racially integrated during this period. Nevertheless, complete achievement of this goal remained elusive. The sometimes-frightening Cold War between communist and capitalist countries was a worldwide reality for over 40 years. The Korean War (1950-1953) and the Vietnam War (1960s-1975) were hot, disruptive episodes in the Cold War. But communism unraveled after the fall of the Berlin Wall in 1989, leading to vast global changes in political and economic relations.

Participation in Physical Activity: 1950-2007

A dramatic increase in health-related exercise began in the United States in the early 1950s and continued for several decades. A number of events contributed to this trend. The Korean War again underscored the importance of physical fitness for military preparedness. The USSR's 1957 launch of Sputnik—the first earth satellite—graphically displayed the continuing threat of communism. A highly publicized 1953 report showed that over 50% of American children failed a strength and flexibility test, compared with a failure rate of less than 10% by children from European countries. After President Eisenhower's heart attacks in the mid-1950s, people were well aware that his doctor prescribed exercise for recovery, a radical idea at the time. And in the early 1960s, President Kennedy continuously demonstrated an active lifestyle—touch football, tennis, swimming, sailing, horseback riding, badminton, and general exercise (Rader 1990, p. 242).

By one estimate, commercial health clubs increased from 350 in 1968 to over 7000 in 1986 (Rader 1990, p. 243). By another, the number of health clubs and physical fitness centers was about 24,000 in the late 1980s (Eitzen & Sage 1993, p. 390). Marathon races throughout the United States increased from less than 25 in 1970 to over 250 in 1996. Between 1980 and 1996, road races covering shorter distances increased from 4100 to over 20,000 (Eitzen & Sage 1997, p. 316). Adult membership in commercial and nonprofit health, racket, and sport clubs reportedly grew by a whopping 51% between 1987 and 1996, increasing from about 13.8 million to about 20.8 million (Carey & Mullins 1997).

The proportion of children and adults who said that they did something on a daily basis to help keep physically fit rose from 24% in 1961 to 59% in 1984. Conversely, adults who said that they had a sedentary lifestyle dropped from 41% in 1971 to 27% in 1985 (Blair, Mulder, & Kohl 1987; Ramlow, Kriska, & LaPorte 1987; Stephens 1987). But recent surveys conducted by the Centers for Disease Control and Prevention using more precise criteria for vigorous physical exercise indicate that in 2004, more than 50% of American adults 18 years of age and older were not engaging in the recommended amount of activity. Moreover, the level of engagement in vigorous physical activity in leisure time was significantly less for females, ethnic minorities, and rural populations. In 2005, almost one-third of adults 18 years of age and over engaged in regular leisure-time physical activity. Adults in families with incomes above twice the poverty level were more likely to engage in regular leisure-time physical activity than adults in lower-income families (34% compared to 20%-22%; age-adjusted) (National Center for Health Statistics [NCHS] 2007).

Americans' well-established fascination with sport grew more widespread—both for participants and for spectators—in the latter half of the 20th century. For example, between 1950 and 2006 the number of youngsters, worldwide, competing in Little League Baseball, Inc., increased by more than 140 times, from 18,300 to approximately 3,000,000. This increase was staggering in light of the much smaller growth in the population of 10- to 14-year-olds in the United States (from 11,119,000 to 20,528,072 between 1950 and 2000) (U.S. Bureau of the Census [USBOC] 1961, 2007a; Little League Baseball, Inc., www.littleleague.org). In the last two decades of the century, the number of youngsters playing soccer sponsored by the American Youth Soccer Organization, Soccer Association for Youth, and the United States Youth Soccer Association grew by 16 times, from 888,705 in 1980 to 14,240,000 in 2006 (National Sporting Goods Association 2007).

Elite athletic competition for girls and women reappeared in high schools and colleges beginning in the late 1950s. In 1972 the U.S. Congress approved Title IX of the 1972 Education Amendments to the 1964 Civil Rights Act, requiring equal opportunity for males and females in educational programs—including athletic programs. Between 1971 and 2007, participation in athletic competition among high school girls increased more than 10-fold. Similar dramatic increases occurred in the number of female players in colleges and universities (National Federation of State High School Associations 2007).

In the 1950s African Americans trickled back into elite professional athletics, and by the 1960s this trickle grew to a fast-moving stream. African Americans constituted about 10% of the players in professional baseball, basketball, and football in the late 1950s. These proportions increased in the 1960s, and by 1994, 77% of the basketball players were African American, along with 65% of the football players and 18% of those in baseball (Eitzen & Sage 1997, p. 264; Leonard 1998, p. 212). By 2006 the number of African American players in Major League Baseball had fallen to

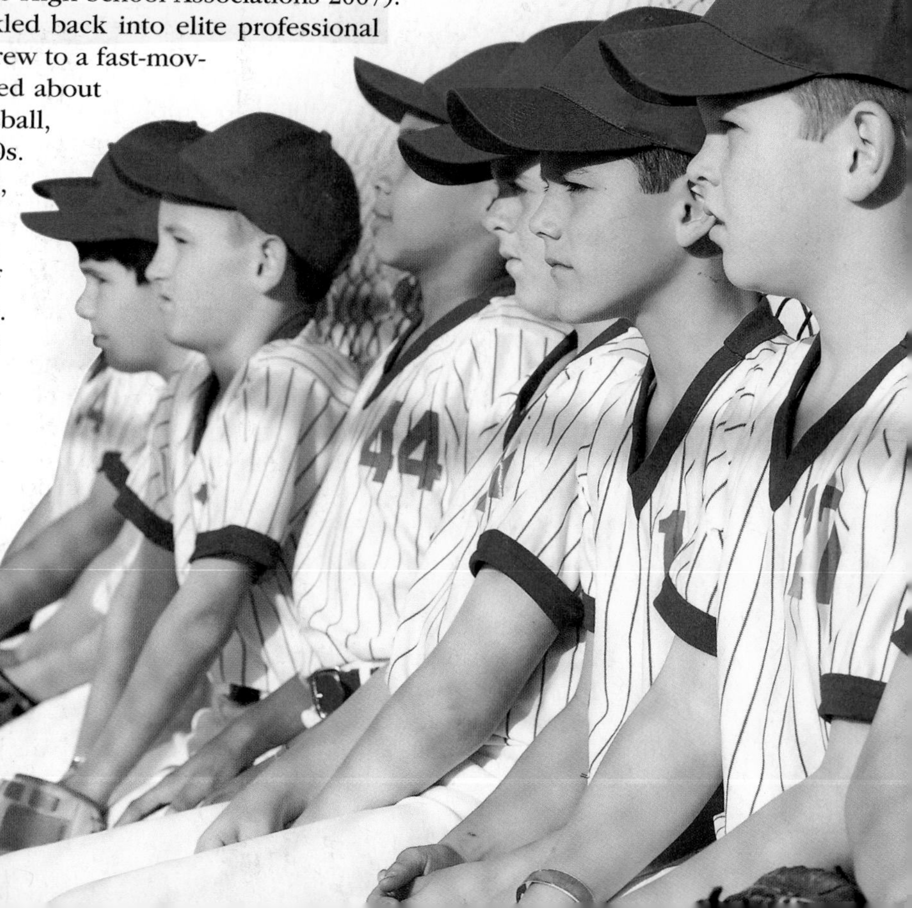

Organized sport for youngsters mushroomed in the second half of the 20th century.
Brand X Pictures

8.4%, although the number of front office personnel, coaches, and managers had significantly increased (Lapchick 2007). There is a general consensus among many that the attraction of football and basketball has been largely responsible for the decreasing numbers of African American youth playing baseball in recent years. To counteract this trend, Major League Baseball has begun programs designed to sponsor youth baseball programs in many of the inner cities of America. In 2007, Latinos constituted more than 25% of Major League Baseball players, and the number of Japanese and Korean players on major league teams steadily increased during the first decade of the 21st century.

The result of this growth in high-level sport competition was an enormous increase in levels of performance. Between 1896 and 1956, Olympic track and field performances (times, distances, heights) increased 130%. Between 1908 and 1948, swimming times improved an average of 2.2% every Olympiad (figure 6.2).

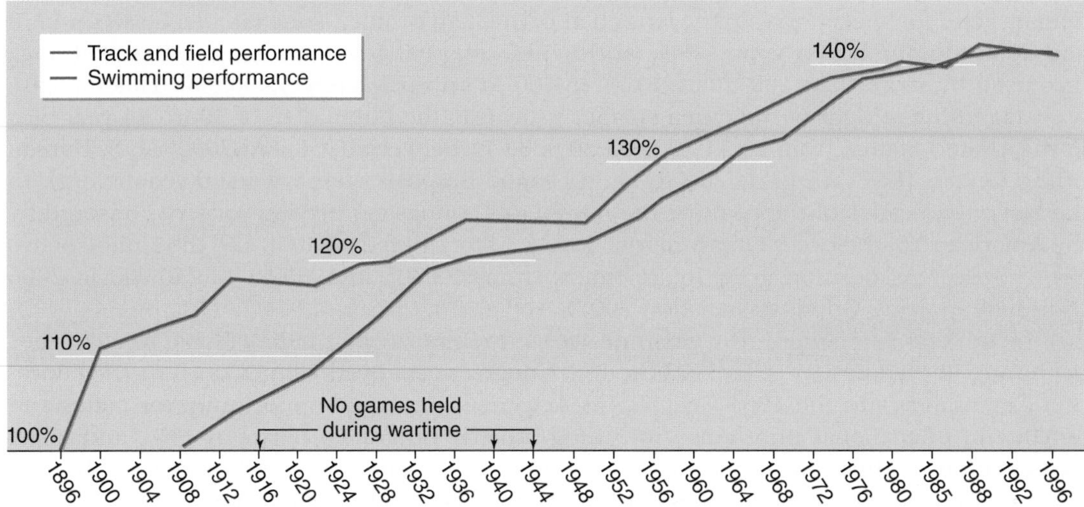

FIGURE 6.2 Dramatic improvements in athletic performance have occurred over the 26 modern Olympiads.

Reprinted by permission from J.L. Shulman 2001.

Enthusiasm for outdoor physical recreation skyrocketed during this period. Between the mid-1960s and the mid-1970s, the increase in interest was so rapid that it sometimes outstripped manufacturers' ability to supply people with proper clothes and equipment. Cross-country skiers increased from about 2000 in 1964 to 500,000 in 1974. Backpacking at Kings Canyon and Sequoia National Parks in California increased 100% between 1968 and 1971. Visitor days at New Hampshire's White Mountain National Forest zoomed from 234,000 in 1968 to 482,000 in 1972. Bike path mileage in the United States went from almost none in 1965 to over 15,000 miles in 1974 (Wilson 1977). By the first decade of the 21st century, participation continued at a high level, although there were signs of changing patterns of activity choices. One annual study found an increase in participation across 22 outdoor activities from 159 million Americans (age 16 and older) in 2004 to 161.6 million in 2005, but the number of separate outings decreased by 11%. "Bicycling and fishing accounted for the bulk of the decline in outings, with approximate outing decreases of 800,000 and 300,000, respectively. Despite that, both these sports ranked among the top five for most outings in 2005: bicycling (3.1 billion), trail running (1.3 billion), fishing (1 billion), hiking (800 million), and camping (347 million)." This study concluded that individuals are looking for less commitment-heavy activities, citing an "8 year decline of 22.5% in overnight backpacking and a significant increase in snowshoeing (83%) and trail running (22%). Finally, the study concluded that the activities with the broadest appeal were those with easy access, ease of learning, can be completed in a day, and require less specialized technical gear (American Recreation Foundation 2007).

Americans also increased their spectator involvement. Attendance at Major League Baseball games was four times greater in 1993 than it was in 1950. Professional football attendance mushroomed to 7.4 times its earlier size during the same period. And attendance at professional basketball games in 1993 was a whopping 9.6 times larger than it was in 1960. These increases far outstripped a more modest U.S. population growth over the same period (USBOC 1966, 1996).

Although soccer is a preeminent spectator sport throughout most of the world, it didn't gain much of a foothold in the United States as an entertainment event until the 1990s. The final rounds of the 1994 men's World Cup were held on American soil, and in 1996 entrepreneurs developed a relatively stable men's professional league—Major League Soccer—with teams that held special allure for Latino spectators. In 1991, the United States won the first Women's World Cup with hardly anyone in the nation noticing. After losing to China in 1995, the Americans won the gold medal at the 1996 Olympic Games before the largest crowd to ever witness a soccer game on U.S. soil. The following year, this same team again won the World Cup before 100,000 spectators in the Rose Bowl and a worldwide television audience numbering in the millions. These successes led to the creation of the professional Women's United Soccer Association (WUSA), which was moderately successful but fell into financial exigencies following the 2003 season. Nevertheless, the league did much to promote women's sport, especially soccer, in North America.

DO it **Activity 6.3**

Participation Timeline
Match events with the correct time periods in this activity in your online study guide.

By the early 1960s, Roone Arledge, head of sport programming at ABC, was incorporating the highly innovative and popular instant replay technique into sport broadcasts. He unveiled additional novel production technologies—for example, close-ups of players and fans, multiple cameras, and directional microphones—in 1970 with the inauguration of ABC's *Monday Night Football* (Gorn & Goldstein 1993, p. 239). These innovations made televised sports much more exciting to watch and drew new spectators. The commercial success of sports on TV—primarily men's team sports—provided vast amounts of money for professional and big-time intercollegiate athletics.

➤ The enthusiasm of Americans—both as direct participants and as spectators—for a widening array of sports and exercise mushroomed in the second half of the 20th century.

The Internet took off with phenomenal growth in the 1990s. By 2007, anyone with a computer attached to high-speed Internet service had instantaneous access to worldwide video broadcasts of sport events as well as news and discussions about sport through electronic chat rooms and Web sites.

Physical Activity Professions

At midcentury, most American college students who majored in physical education—the name used almost universally at the time—became high school teachers, and some also coached. Things began to change in the mid-1960s with the development of kinesiology—the body of knowledge about physical activity that included numerous scholarly subdisciplines. It soon became apparent that many careers other than high school teaching and coaching were centered in physical activity. Let's look at a few examples.

With the growing American interest in exercise and physical fitness, it is little wonder that new health clubs, cardiopulmonary rehabilitation programs, worksite health promotion

Attitudes of Adults Toward Exercise Have Changed

Fifty or 60 years ago, most Americans perceived regular participation in exercise and sport to be the exclusive province of young people engaged in high school, college, or professional sport. Today, large numbers of people pursue such activities throughout much of their lives. What historical events and social changes in our society have influenced this shift in attitude and practice over the past several decades? How have you seen this shift in attitude and practice within your own family or circle of acquaintances?

Career opportunities in physical activity professions were more plentiful during the second half of the 20th century.

Bananastock

programs, and health maintenance organizations began offering physical activity programs. People with specialized training in exercise began working in a variety of jobs such as cardiopulmonary rehabilitation, personal training, aerobics instruction, and strength training. Some of them rapidly advanced to management and business ownership roles.

The National Athletic Trainers Association was founded in 1950 to support the growing number of specialists who worked to prevent and treat athletic injuries. The membership was almost entirely men until 1972 when Title IX (requiring equal opportunity for males and females in educational settings) became law, encouraging women to enter the field in greater numbers. As of 2007 there were about 362 NATA-accredited degree programs—a mixture of master's and bachelor's programs—and about 32,236 NATA-certified athletic trainers (Shannon Leftwich, National Athletic Trainers Association Board of Certification, personal communication, September 17, 2007; www.nata.org).

At midcentury there was little by way of standard educational preparation for most careers in sport management and leadership. As you already know, the sport industry experienced enormous growth during the second half of the 20th century. As a result, there were more jobs in sport settings, and the variety of jobs was wider. Job opportunities were available, for example, with intercollegiate athletic departments, professional athletic teams, city stadiums, and participant-oriented nonprofit organizations such as Boys and Girls Clubs and YMCAs. Some jobs didn't even require a college degree (e.g., ticket seller in minor league baseball, YMCA program leader), whereas some required extensive high-level leadership experience or advanced degrees (e.g., big-time intercollegiate athletic director, city arena or stadium manager). This unevenness made it difficult to accredit degree programs or certify practitioners.

Nevertheless, in 1987 the National Association for Sport and Physical Education (affiliated with AAHPERD, the American Alliance for Health, Physical Education, Recreation and Dance) and the North American Society for Sport Management jointly published curriculum guidelines for bachelor's and master's degree programs in sport management. These were revised in 1993 and published as voluntary accreditation standards.

Beginning in the 1960s, physical therapy continued to grow as a profession. The number of physical therapy jobs in the United States in 1960 was estimated to be 8000; this grew over 19-fold by 2006 to 156,100 (U.S. Bureau of Labor Statistics [USBLS] 1961, 2008). In the 1950s students were trained in undergraduate physical therapy degree programs or in older, hospital-based certificate programs designed for people who had completed bachelor's degrees in other fields. By the late 20th century, hospital-based programs were phased out, undergraduate programs were in decline, and master's programs were on the rise. In 1981, accredited physical therapy curriculums included 100 undergraduate degree programs, seven certificate programs, and eight master's programs (USBLS 1982, p. 166). In 2007, there

DO it Activity 6.4

Professions Timeline

Match aspects of the development of physical activity professions with time periods in this activity in your online study guide.

were 210 accredited physical therapy programs in the United States, including 31 offering the master's degree and 179 offering the doctorate (American Physical Therapy Association 2007). Since 2002, all physical therapy programs seeking accreditation are required to offer degrees at the master's level or above (www.apta.org). Increasingly, the doctor of physical therapy (DPT) is becoming the accepted entry-level degree, the number of accredited DPT programs having grown almost fivefold between 2002 and 2005.

➤ Most college students majoring in physical education in the middle of the 20th century entered the same profession—physical education teaching; by the end of the century, students had a wide array of physical activity careers to choose from.

Scholarly Knowledge About Physical Activity

Scholars developed new knowledge about physical activity at a feverish pace during the final 35 years of the 20th century, a trend that is continuing today. You will learn a lot more about this in the chapters of this text on the other subdisciplines—each has a section on history. Here we'll focus on the big picture.

By the 1950s researchers had already produced a lot of information about physical activity. Problematically, knowledge was scattered throughout many different research journals, and no one had taken the time to make it easily accessible. Warren Johnson, professor of health education and physical education at the University of Maryland, attempted in 1960 to remedy this by publishing a book titled *Science and Medicine of Exercise and Sports*. The book contained 36 essays summarizing current knowledge. Another early attempt at organizing research knowledge appeared in the May 1960 supplement to *Research Quarterly*, edited by several scholars headed by Raymond Weiss of New York University; each paper summarized research concerning a different aspect of physical activity (Karpovich et al. 1960).

➤ Beginning in the 1960s, the discipline of kinesiology grew rapidly and branched into numerous scholarly subdisciplines.

Calls from politicians and educators in the early 1960s for better quality control in college and university teacher education programs resulted in scrutiny and criticism of physical education curriculums because they were mostly teacher education programs at the time. In response to this, physical education faculty began charting new directions for their discipline. Franklin Henry, a physical education professor at the University of California and top-notch scholar in exercise physiology and motor learning, published a paper outlining his views on the nature of the discipline titled "Physical Education: An Academic Discipline" (1964).

Henry envisioned a body of knowledge about physical activity that would draw from "such diverse fields as anatomy, physics and physiology, cultural anthropology, history and sociology, as well as psychology" (p. 32). He also believed that the discipline should be cross-disciplinary. By that he meant that it should bind together knowledge from various disciplines to form a distinct, new body of knowledge concerning human movement.

DO it Activity 6.5

Knowledge Timeline

Connect aspects of the development of scholarship related to physical activity in this activity in your online study guide.

Henry's vision of the discipline spurred rapid changes during the rest of the 1960s and 1970s. A nucleus of physical educators who were already heavily engaged in research began pushing to make scholarship more central in university physical education departments. They continued their own research and trained a new cadre of doctoral students who expected to be centrally involved in inquiry throughout their own careers. They communicated frequently with scholars in other disciplines—for example, physiology, medicine, psychology, history—who were also studying physical activity. This exchange led to rapid growth in the size and scope of the body of knowledge.

The knowledge also became more specialized, and this soon led to the formation of subdisciplines. By 1970 the earliest ones were clearly identified, and most of the others were on the drawing boards. You already know the names of the subdisciplines and how they fit together to form the scholarly source of knowledge in the discipline. Throughout the rest of the 20th century, scholars formed many new associations and worked with publishers to inaugurate a host of new research journals to support the growing subdisciplines. In the 1980s and 1990s, the amount and quality of research continued to increase, and the subdisciplines became stronger and more identifiable (Massengale & Swanson 1997; Park 1981, 1989).

Wrap-Up

History extends your "memory" with fascinating, evidence-based stories about how and why physical activities such as exercise and sport came to be shaped the way they are. Knowledge of the past gives you important, broad understanding about the present that you can use to make better-informed personal and professional decisions for the future. As demonstrated in this chapter, historical knowledge is utilized by kinesiology practitioners in a variety of ways across many specialties and professions. Since the 1970s, scholars have gone beyond earlier descriptive approaches to put more emphasis on analytical frameworks that help answer questions about why things happened as they did. So far, this subdiscipline has focused mostly on sport, but since the 1980s scholars have given greater attention to exercise.

If you are intrigued by what you have learned here, you might have fun venturing into sport history textbooks, taking an introductory course in sport history, looking through the book-length research studies mentioned in this chapter, and learning through the major journals such as *Journal of Sport History* and *International Journal of the History of Sport.* This interesting and provocative subdiscipline will give you a new outlook that will enrich your understanding of physical activities as you step into the future.

Use the key points review in your online study guide as a study aid for this chapter.

QA **Activity 6.7**

These end-of-chapter questions and activities are also in your online study guide. Your instructor may ask you to complete them online and turn them in.

1. List and discuss the goals of history of physical activity.
2. Describe the expanding research directions in history of physical activity from 1970 to the present.
3. List and discuss three ways in which a kinesiology practitioner might utilize a knowledge of physical activity history in her or his area of specialization.
4. Describe participation in physical activity in the United States during the following three periods:

   1840-1900

   1900-1950

   1950-2000
5. What factors led to the bicycle craze of the 1890s? Give some examples of evidence that there was, indeed, a greatly increased interest in bicycling during that time.

6. Describe professional practice centered in physical activity in the United States during the following three periods:

   1840-1900

   1900-1950

   1950-2000

7. Describe scholarly knowledge about physical activity during the following three periods:

   1840-1900

   1900-1950

   1950-2000

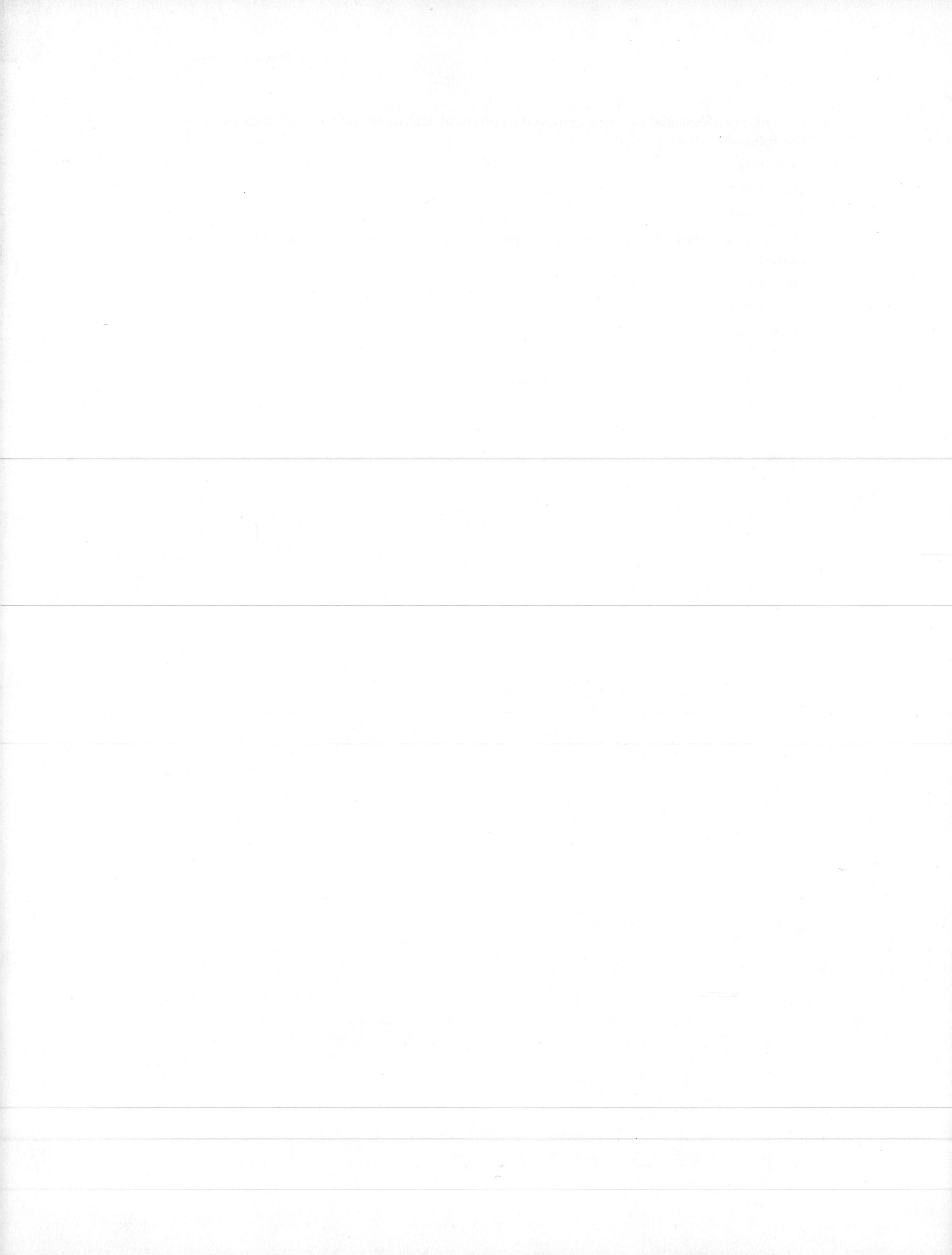

7 Sociology of Physical Activity

Margaret Carlisle Duncan and Katherine M. Jamieson

CHAPTER OBJECTIVES

In this chapter we will

- explain what a sociologist of physical activity does;
- identify the goals of sociology of physical activity;
- discuss how sociology of physical activity came into being;
- explain how research is conducted in sociology of physical activity; and
- examine what research tells us about inequitable power relations relevant to physical activity, especially gender relations, ethnic and racial relations, and socioeconomic relations.

You're having coffee with your roommate at the local bookstore, and a couple of people at the next table are having an intense conversation. You can't help overhearing, and you're interested, because they're talking about the beginner soccer program their kids are in. As you listen, you gather that many of the coaches in the program are parents who are volunteering their time and expertise. Your neighboring coffee drinkers are complaining to each other about how the parent-coaches seem to be very competitive and regimented in the ways they organize team sessions and interact with participants. You gather that the head coach in particular seems extremely competitive, having been overheard speaking ill of the less skilled participants and calling the participants "losers" unless they work hard enough to beat all the other teams. You hear the complaints that their kids are not having fun and are constantly afraid of getting in trouble. Hearing these people, who are clearly frustrated, talk about what's going on in their soccer program makes you start to wonder about what these programs ought to focus on. Should the emphasis be on learning new skills, having fun, and making friends? Or should it be on winning? You start to think about where the coaches' attitudes and beliefs come from. Is it appropriate for children to be taught to strive for excellence at such a young age? Is this how the win-at-all costs sporting mentality begins? You wonder what you would do if you were in the place of these people. Would you speak to the coaches or to the sport director at the community center? Would you withdraw your child from the program? Would another local soccer program be better, or would you encounter the same behaviors?

It would be difficult to think of physical activities that do not somehow involve interaction with other people. Try to imagine it—even if you exercise at home, how might you still be interacting with others? There are many kinds of social interaction, especially in this highly technologized world where it is nearly impossible to escape social interaction. Without these opportunities for social interaction, our physical activities would be very different. Most of us would probably engage in basic movements such as walking, running, and jumping, but if in the process we never interacted with others—never watched them, never communicated with them, never played with them—each of us would probably develop a unique style of moving that would be unfamiliar to anyone else.

In such an absurd, asocial world, sport would be impossible; without communication we would have no way to decide on game rules. People also wouldn't have a mutually shared concept of exercise (i.e., physical activity done to improve human functioning or appearance), and they probably wouldn't be consciously aware that people's health can improve with moderate physical activity on a regular basis. Some individuals would probably do strenuous physical activities such as running fast or lifting heavy loads, and if they did these things often enough their health would probably improve. But other people would never learn about this idea. Exercise, sport, dance, and many other specific forms of physical activity have been brought into being, and are continuously being modified, through social interaction. Without it, these activities wouldn't exist.

What Is Sociology of Physical Activity?

Our social life is made up of mutually influential social practices (everyday behaviors) and shared beliefs; these are so commonplace that we rarely think about them. Whether we're playing soccer, coaching a basketball team, rehabilitating a broken leg, cheering for our favorite football team, or working out at a local gym, our social practices as athletes, coaches, exercisers, and spectators are influenced by commonly held beliefs in our society. And in turn, it is through our social practices that we come to have beliefs or understandings that we share with others. Yet few of us spend time analyzing the social arrangements—social practices and shared beliefs—that underlie physical activity. When we do look more closely, though, we gain insights about ourselves and our culture. It is the job of sociologists of physical activity to give this a closer look.

Although the subdiscipline of sociology of physical activity focuses primarily on sport, there has been a growing interest in exercise and in the ways in which people in our society conceive of the human body. By exploring how we, as a culture, view our bodies, we may come to a renewed understanding of what it means to be human.

Let's look again at our opening scenario to see what advice a well-educated kinesiologist with knowledge of sociology of physical activity might give to the parent who questions the current competitive model as the best model for children. This advice would probably be different from that of a layperson without such training. For example, a kinesiologist would know that sport organizations hold a range of different beliefs about competition and winning. Although some might seem to reflect a win-at-all-costs mentality, others adhere strongly to a much different set of priorities, such as learning fundamentals, engaging in creative human movement, and having fun. Some programs value good sport conduct more highly than they do winning a game. In some programs, no one keeps score.

Kinesiologists with a background in sociology are aware of the dominant competitive model of sport as well as the kinds of daily human interactions that make particular ways of organizing and engaging in physical activities more socially valued than others. Well-informed kinesiologists also recognize that there are varied social and cultural contexts for physical activity and that, because of this, different forms of sport, exercise, and recreational activities may hold different values at different times among different groups of people. Recognizing this diversity of social and cultural values applied to physical activities, many organized physical activity programs have proposed alternative models, especially for young children. In the opening scenario, a sociologically informed kinesiologist would likely advocate a less outcome-centered, more process-centered sporting experience for kids, one that promotes skill building, self-awareness, and responsibility for one's actions.

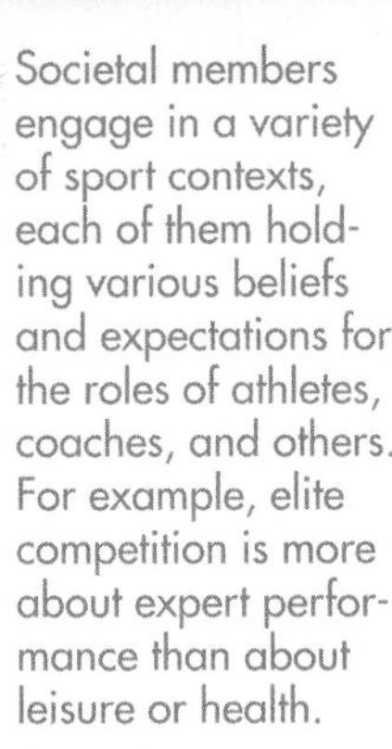

Societal members engage in a variety of sport contexts, each of them holding various beliefs and expectations for the roles of athletes, coaches, and others. For example, elite competition is more about expert performance than about leisure or health.

Comstock

Through your previous physical activity experiences, scholarly studies, and observations of professional practice, you are already familiar with some aspects of sociology of physical activity. Your *experiences* have likely exposed you to a number of issues that sociologists of physical activity study. Perhaps you played on the girls' basketball team in high school and wondered why the boys' scores showed up on the television news but yours didn't, or why the boys' games were packed to capacity but bleachers sat empty at your games. Or maybe as a young boy you felt pressured to compete, to show coordination, and to be stronger and play better than your female counterparts. Perhaps you watched sports from around the world on television—soccer, sumo wrestling, cricket—and wondered how and why those sports developed as they did. Maybe you noticed how racial, socioeconomic, and geopolitical (e.g., geographical, cultural, political) issues affect physical activity participation. You have likely noticed your ethnic or racial group's dominance or lack thereof in the sports you enjoy most. All of the issues that you have recognized through experience are topics that sociologists of physical activity investigate.

From your previous course work or scholarly studies, you may be familiar with basic principles from sociology, cultural anthropology, economics, political science, and communications. All of these will help you understand sociology of physical activity. You don't have to know all these disciplines in depth, but the more information you have, the more insightful you will be about the social side of physical activity. For example, if you know a little about the general nature of socialization—the process of learning about your own society and how to get along in it—you will find it easier to understand how societal members become *socialized* into dominant sports or into subcultures centered in activities such as surfing, triathlons, mountain climbing, skateboarding, rugby, and exercise related to social or political causes (e.g., Race for the Cure).

➤ Sociology of physical activity focuses on the shared beliefs and social practices that constitute specific forms of physical activity (e.g., sport, exercise). Accordingly, sociological information adds to the breadth of knowledge of a well-educated kinesiologist.

As you get to know more about socialization, however, you will likely notice details and nuances of physical activity involvement that might lead you to ask more complex questions. For example, you might become interested in the extent to which, and the ways in which, youngsters contribute to socializing their parents into participating in physical activity, rather than being content to think that socialization flows in only one direction—from parents to children.

Even your observations of professional practice have lent you some insight into sociology of physical activity. Perhaps you have noticed that there are few female athletic trainers for American football teams, that many people cannot afford to belong to a country club, that fitness professionals often promote achievement of a dominant body ideal, and that jobs in college and professional coaching remain largely held by white males. All these observations are ripe for sociological study, and sociologists of physical activity have already documented and analyzed many of them.

Sociology of physical activity joins philosophy and history of physical activity as subjects that deal with the sociocultural side of sport and exercise. Like philosophers of physical activity and historians of physical activity, sociologists of physical activity use their expertise to study the social life and patterns associated with physical activity. By examining physical activity from yet another standpoint, they bring new insights into our understanding of the human experience of being physically active. Figure 7.1 shows how sociology of activity fits into the broader context of physical activity studies.

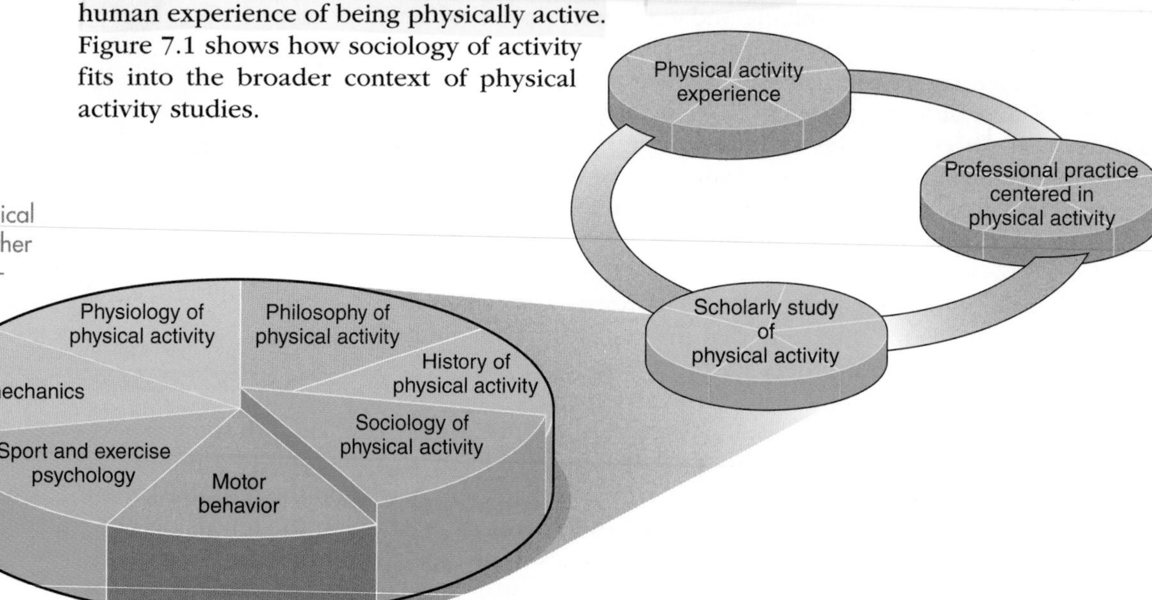

FIGURE 7.1 Sociology of physical activity and the other scholarly subdisciplines.

What Does a Sociologist of Physical Activity Do?

Sociologists of physical activity are usually faculty members in colleges or universities whose teaching, research, and service advance a scholarly and practical understanding of the complex social and cultural contexts in which societal members experience physical activity. They may teach an introductory course on sociology of physical activity and offer courses on specialized topics such as the Olympic Games; interscholastic and intercollegiate athletics; or the ties between physical activity and phenomena such as gender relations, violence and aggression, or the mass media (e.g., television, magazines, the Internet).

Sometimes sociologists of physical activity combine teaching with professional service by encouraging their students to assist in physical activity programs located in nearby neighborhoods. In so doing, they help local communities and foster a sense of civic responsibility in future kinesiology professionals. Faculty service activities might also include consulting with a community group about the role of physical activity in agendas for community-wide

Activity 7.1

Reflecting on Your Physical Activities
Document your experiences in this activity in your online study guide.

change (e.g., immigrant settlement issues and cultural forms of physical activity; city planners and physical activity space, access, and use priorities). Given these examples of scholarly engagement, it should be clear that sociologists of physical activity cannot carry out their research in a typical laboratory. Even if it were possible to bring a "slice of life" (e.g., fans watching a National Football League [NFL] football game, an aerobic exercise class) into a lab for study, in most cases such analyses wouldn't be useful. An unrealistic result would be obtained because when you move such an event into a lab, you alter the very thing that you're trying to study. You would be studying an aerobic exercise class in a lab rather than an aerobic exercise class in ordinary life. Sociologists usually prefer to study their subjects in the field—for example, professional golfers on the Ladies Professional Golf Association tour, a television production crew covering a downhill ski race, the political processes that underlie the preparation of elite Canadian athletes for international competition, or preferred physical activity practices among immigrant groups.

As part of this desire to understand everyday life and physical activity engagement, sociologists of physical activity spend a lot of time making contacts and gaining the confidence of the people whom they wish to study. A sociologist of physical activity wanting to study a National Collegiate Athletic Association (NCAA) Division I men's basketball team, for example, would probably need permission from the university, the athletic director, the coach, and the team members themselves. He or she would need to develop especially good rapport with the coach and the team members to give them confidence in the worth of the project. The quality of such relationships has an important bearing on the quality of the research. Sociologists of physical activity usually study ongoing social life as it ordinarily occurs; they don't often bring it into a laboratory.

Sociology of Physical Activity in Kinesiology

Sociology of physical activity in kinesiology typically advances three main goals, as shown on page 189. The first is *to look at physical activity with a penetrating gaze that goes beyond our common understanding of social life.* This is a habit that must be developed, and acquiring it involves having a dash of skepticism about common assumptions. If you remember that the complex workings of social life are usually not readily observable, it will be easier to form the habit of looking further into your own physical activity experiences, knowledge, and professional practice. For example, on the surface it appears that sport and exercise are personal choices available to all societal members, but if you look more closely you'll find that a host of external social conditions influence the physical activity choices that people make.

To illustrate, in the United States, becoming an Olympic competitor in sports such as figure skating, gymnastics, or speed skating often requires a number of resources, including societal value for that particular sport; development of experts and training facilities in that sport; financial commitment of the athlete and family throughout training; a formal set of rules, guidelines, and benchmarks for elite performance in that sport; and enough international interest that the sport remains an Olympic event. Kinesiologists who lack a sociological understanding of physical activity might misinterpret trends in Olympic participation as indications of natural interest or willingness to work hard among particular groups (e.g., men more than women; whites more than other racial-ethnic

www. **Web Search 7.1**

Apply Your Sociological Imagination
Search the Internet for material that will help you respond to four scenarios in your online study guide.

groups). But a kinesiologist who is able to *see* what some sociologists refer to as chains of interdependence will be much better able to respond to the physical activity needs of an ever-changing, diverse societal membership.

Building on this sociological lens is the second goal of the subdiscipline, *to identify and analyze patterns of change and stability in physical activity.* Once you start looking beneath the surface of social life, your goal is to identify changes as well as ongoing regularities. In keeping with our example of Olympic involvement, sociological analyses indicate that the Olympic Games continue to reflect the sporting traditions of the major nations. Despite Pierre de Coubertain's desire to revive a modern Olympic movement that would truly be about cross-cultural exchange, the Games continue to privilege large, economically advanced countries in sporting displays of nationalist propaganda. Yet change has occurred: More women and more countries currently participate in the Olympic Games than in earlier times (see table 7.1); new sports have been added to the Games; professional athletes are now allowed to compete in particular sports; and environmental issues are often part of planning discussions for Olympic venues. Whether you agree with the current state of affairs in the Olympic Games is not the issue for sociologists; rather, it is crucial to be able to identify shifting trends and stabilizing forces in the physical activity experiences of societal members.

Table 7.1 Number of Women's Events and Female Athletes in Olympic Games From 1980 to 2000

Year	Number of women's events/total	% women's events	Number of female athletes/total	% female athletes
Moscow 1980	50/203	24	1125/5217	21
Los Angeles 1984	62/221	28	1567/6797	23
Seoul 1988	72/237	30	2477/9627	25
Barcelona 1992	86/257	33	2845/9905	28
Atlanta 1996	97/271	35	3624/9905	28
Sydney 2000	120/300	40	4254/11,116	38

Data abstracted from reports by Australian Sports Commission and International Olympic Committee.

Given that you have learned to see physical activity in new ways, and that you are able to identify patterns of change and stability, you are now ready to engage the third goal of sociology of physical activity: *to critique physical activity programs in order to identify problems and recommend changes leading to the enhancement of equality and human well-being.* As you have probably already guessed, sociologists point to the constant stream of social interaction that goes on in any society as the most important source of societal problems, although they don't deny that psychological and biological factors have important influences in our lives. They study social interaction, and more importantly, they evaluate it—pass judgment on its contributions to overall human well-being and equitable social relationships (e.g., relationships among racial and ethnic groups, men and women, or people at different socioeconomic levels). Returning to our Olympic Games example, sociologists have critiqued the Games as an exclusionary physical activity space that privileges societal members who happen to be from the best-resourced countries, teams, and families.

> **DO it Activity 7.2**
>
> **Goals of the Subdiscipline**
>
> In this activity in your online study guide, you'll review three scenarios and identify which goal each represents.

Additionally, as already mentioned, the Olympic Games typically require host cities to utilize public

Goals of Sociology of Physical Activity in Kinesiology

1. To look at physical activity with a penetrating gaze that goes beyond our common understanding of social life
2. To identify and analyze patterns of change and stability in physical activity
3. To critique physical activity programs in order to identify problems and recommend changes leading to the enhancement of equality and human well-being

resources, land space, and human resources to build the facilities necessary to host a sporting event that will last only two weeks. Some suggest that hosting such an event will create a local interest in sport and exercise, thereby increasing the physical activity of societal members. Sadly, there has been no evidence of this, yet there has been evidence that watching the hypermediated (i.e., multiple forms of ongoing media attention), flashy, entertainment-oriented Olympic Games may lead to increased television viewing even after the Games have ended. Clearly, for kinesiologists with a sociological understanding, these consequences of the Olympic Games would be considered problematic and would move them to modify the Games toward more beneficial outcomes for all societal members, not only those who compete in or profit from the Games.

➤ Sociology of physical activity in kinesiology looks beneath the surface of social life to see things from different angles; kinesiologists with this knowledge are able to point out social problems in an effort to encourage people to make changes that may enhance human well-being and promote equitable social relationships.

In keeping with the third goal, sociologists of physical activity are not oriented toward supporting or reinforcing status quo physical activities such as present exercise programs and competitive sports. For example, a former Olympic athlete-turned-coach may start an Olympic training center where there previously was none; a biomechanist might design an exercise machine that meshes with the human body to improve elite performance. Both of these are examples of working within the status quo—working to improve performance in activities as they are currently structured. But a sociologist might advocate changing the structure itself by questioning the desirability of fierce competition between nations or treating the human body like a piece of machinery that needs to be meshed with other pieces of machinery.

History of Sociology of Physical Activity

It is important to note that the subdiscipline is relatively young. It began to take shape in North America in the mid-1960s with the notable influence of English physical educator Peter McIntosh's 1963 book *Sport in Society.* His analysis of the social significance of sport encouraged several North American scholars to channel their careers toward this emerging area (Sage 1997). Subsequently, in 1964, the International Committee of Sport Sociology was founded, and in 1966 it inaugurated a scholarly journal, *International Review of Sport Sociology,* which was joined in 1977 by *Journal of Sport and Social Issues.* In 1978, the North American Society for the Sociology of Sport was formed under the leadership of physical educators Susan Greendorfer and Andrew Yiannakis. The society held its first annual conference in 1980 and began publishing an academic journal, *Sociology of Sport Journal,* in 1984. Both the society and the journal continue to be crucial, active scholarly spaces.

➤ In the 1970s, 1980s, 1990s, and 2000s, sociology of physical activity expanded as scholars explored a wider variety of topics, especially those concerning sport and societal inequities related to socioeconomic, gender, racial, and ethnic inequalities; globalization and regional and national differences; exercise and societal conceptions of the human body; disability and ability; and obesity.

Since the mid-1970s the number of scholars in sociology of physical activity has increased, as has the range of theories and research methods employed by these scholars. The hottest topics have focused on social inequities—especially those connected with gender, race, ethnicity, wealth, sexual orientation, and different cultures around the world. Research interests also include globalization and regional and national differences; exercise and societal conceptions of the human body; disability and ability; and obesity.

Research Methods in Sociology of Physical Activity

Now that you know something about the subdiscipline and its historical roots, you may be wondering how sociologists of physical activity produce their knowledge. What methods do researchers use to answer important questions about the social side of physical activity? They gather both quantifiable data (things you can count or measure) and qualitative data (e.g., interviews, direct observations of social life, written documents, artifacts). We'll look at six methods: survey research, interviewing, thematic analysis, ethnography, societal analysis, and historical research. Of course, sometimes two or more of these are used in the same study. Some studies even use both qualitative and quantitative methods (and are called **mixed methods** research).

Survey Research

Doing survey research involves using questionnaires that are completed directly by respondents or filled out by a researcher during brief, highly structured interviews. Questionnaires are used to collect data from a large sample of people. The largest survey project in the country is the U.S. census, conducted every 10 years. Political polls about voting preferences are also surveys. Investigators have conducted surveys dealing with numerous topics related to physical activity, including youngsters' opinions about what led them to become involved in sport, former collegiate athletes' thoughts about leaving competition at the end of their four years of eligibility, and college athletic directors' opinions about the characteristics of successful coaches.

Interviewing

Researchers use interviews when they want broader and deeper information than they can get through a questionnaire or when they want information about activities that would be difficult or impossible to observe themselves. Because interviews are time-consuming, studies often focus on a small number of people. Interviews may be conducted one-on-one or in a group. One form of group interviewing that is increasingly being used in our field is **focus group** research. This method is particularly valuable when the researcher wants to gain insights into people's shared understandings and to see how individuals are influenced by others in a group situation (Gibbs 1997). Typically, interviews are tape recorded and later transcribed. Researchers often ask relatively open-ended questions and then probe to get details. One investigator asked top high school, college, and professional athletes to discuss their initial involvement in sport, their participation over the years, and their disengagement from sport. Another asked young people about the characteristics of their athletic heroes.

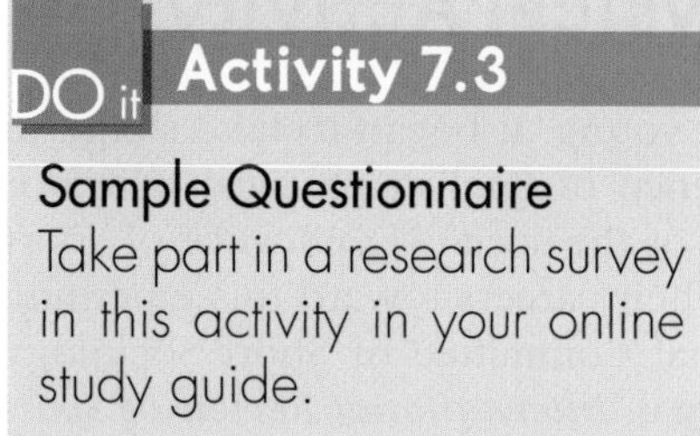

Sample Questionnaire

Take part in a research survey in this activity in your online study guide.

Thematic Analysis

Researchers use thematic analysis, sometimes called content analysis or textual analysis, to investigate cultural material such as magazine and newspaper articles, photos, the verbal and visual content of television programs, interview data, sporting events, and sport celebrities (Birrell & McDonald 2000). This procedure involves examining the material and then categorizing the content in various ways. For example, magazine photos of female and male Olympic athletes were analyzed using the categories of physical appearance, poses, position of the body, emotional displays, camera angles, and groupings of people. A researcher often ends up with several main categories or themes along with a set of subthemes within each. Sometimes investigators count the number of times each theme and subtheme occur. In other cases they are more interested in qualitative data, such as the richness and complexity of the themes. The themes help to organize a large mass of data into a manageable number of categories that the researcher can analyze further in relation to theories of interest.

➤ Research methods used in sociology of physical activity include survey research, interviewing, thematic analysis, ethnography, societal analysis, and historical analysis.

Ethnography

Researchers using ethnography spend many months or even years observing in a particular social setting. They "hang around" while ordinary day-to-day events take place, often taking part themselves, talking with people about what's happening, and keeping careful field notes so that they can remember details. Although observation is their primary source of information, they may also look at local documents or use interviews and questionnaires. Researchers analyze most of the data that they collect using thematic analyses. Ethnography has been used to study a variety of sport settings, including minor league ice hockey, baseball in the Dominican Republic, women's softball, the Olympic Games, women's professional golf, rodeo, and boys' Little League baseball.

Societal Analysis

In societal analysis, the researcher's goal is to examine the sweep of social life, usually from the perspective of a broad social theory. This method, of course, isn't the only one tied to theory—all good social research is theoretically based. The theories used in societal analysis, however, are extremely broad—they attempt to explain the most fundamental ways in which societies operate. Examples of these include Marxism (and its many derivatives), modernization, structural-functionalism, figuration theory, various strands of feminism, various cultural studies frameworks, and various forms of postmodernism. These theories are extremely complex; massive collections of scholarly papers and books support, refute, or expand on each of them. Illustrations of societal analysis in sociology of physical activity include a study of modernization that focuses on sport in preindustrial and postindustrial societies (Guttmann 1978) and an investigation, focused on social class, of societal constraints and human freedoms in sport (Gruneau 1983).

Activity 7.4

Which Method?
Identify examples of each of the six research methods in this activity in your online study guide.

Historical Analysis

Social scientists interested in large-scale social change frequently incorporate historical research into their work. Historical analyses often are part of a larger study involving theoretical explorations of the kind just noted. Examples of historical research by sociologists of physical activity include an examination of the ways in which development of sport in Canada was related to socioeconomic inequities (Gruneau 1983) and a study of unruliness among soccer spectators in England from the 1890s to the 1980s (Dunning, Murphy, & Williams 1988).

➤ Research in sociology of physical activity involves collecting quantitative and qualitative data using a variety of different methodologies.

Methods of Research: An Example of How the Research Process Works

Which research method an investigator uses depends on the topic that she or he wants to explore or the questions that she or he wants to answer. In qualitative research, sometimes the method that the researcher first chooses does not elicit the information needed. In such cases, the researcher may choose to alter or supplement the original research method. A research project that one of my undergraduate students and I conducted clearly illustrates this situation (Duncan & Robinson 2004).

We were interested in finding out whether there were differences between white female body ideals and African American body ideals. Our first strategy was to conduct a textual analysis of health and fitness magazines written for African American women. We planned to compare the themes in photos and articles that appeared in these magazines with the themes in publications written primarily for white readers. We were in for a bit of a surprise! In our geographic region, there were no health and fitness magazines specifically aimed at black female audiences. We had to change our plan.

Our second strategy was to choose general interest African American magazines that included articles on health and fitness. We found about 10 magazines that met our requirements. As we analyzed the pictures and themes, we soon discovered that we needed more information than the magazines themselves could provide. We discussed other ways of attacking the problem and finally agreed that we would interview a group of African American women. This way we could ask questions directly and follow up on interesting themes. The most efficient method of interviewing was to conduct a focus group, and we thought that we might get more representative results from a focus group interview than from interviewing each woman one at a time. Because this was an exploratory study, we chose a small sample of young black women (nine college-age individuals on campus) who were part of my undergrad student's network of friends and acquaintances.

With lunch and gift certificates as incentives, we invited the group to assemble on their own turf, a peer mentoring office where students of color hung out. I asked my student, who was a young African American woman, to pose all the questions from a list that we had previously created. As the white outsider, I wanted to minimize my presence so that the group would feel more comfortable about speaking frankly. We displayed the 10 magazines that we had analyzed earlier and asked our group to page through them. What did they think of the images? Were the models realistic? Would they like to look like the women on these pages? Did the models represent black body ideals? Were they different from the models found in white magazines? As soon as the group members started talking, we turned on the tape recorder.

The discussion ended two and a half hours later. Using the recorded tape, we transcribed (put in writing) the entire conversation—90 single-spaced typewritten pages altogether! During the discussion, the women commented on a variety of topics: the difference between body ideals for African American women and white women, the appearance of the models in the magazines, beauty ideals, physical shape, skin color, hair color, facial features, confidence, sexiness, and preferences of African American men. The two of us studied the transcript and generated a list of the most prominent themes—flexibility in African American body ideals, the idea that beauty comes from within, lack of a support network for working out, the holistic nature of health and fitness, economic barriers, and several others. Ultimately, we concluded that for this group of African American women, body and beauty ideals differed significantly from the typical white body and beauty ideals.

If you were the researcher studying this topic, what methods would you use? Would you consider any other qualitative methods? How about quantitative methods like surveys? Could you design a lab experiment that would give you the answers you need?

Overview of Knowledge in Sociology of Physical Activity

This section is meant to be a snapshot of the knowledge produced by scholars in the subdiscipline, and as such cannot do justice to the richness and variety of the knowledge that scholars have produced. In fact, the body of knowledge is far-reaching and especially complex because it concerns many different societies throughout the world, and social life is not the same everywhere.

Sociologists of physical activity often look at these areas from the standpoint of power relationships or the ways in which external, dominant cultural beliefs about social life, morals, valued behaviors, and so on influence everyday choices for physical activity. Power relations are difficult to see in our everyday lives, but once you learn to recognize them, you may begin to see them everywhere. You'll be able to look more carefully, with a more penetrating gaze, at various fascinating social aspects of physical activity.

For sociologists, **power** is the ability to do what you want without being stopped by others. Money, prestige, body size and strength, and information are major sources of power, hence major sources of inequity. If you look closely at kinesiology contexts—gyms, playgrounds, athletic fields, hiking or biking trails, and so on—you will see that it is not overt, bullying kinds of power that are typically exhibited, but rather subtle, built-in, almost hidden forms of power that most influence physical activity involvement. For example, there are no signs saying that the cardio room and group exercise classes are for women, but in most fitness facilities, it is women who populate these spaces. When a male enters these spaces, he may be teased by some or may be celebrated by others who are happy he broke the gender barrier. Both of these responses reflect a somewhat subtle cultural knowledge that we all share to some degree. This "knowingness," as it is referred to by some scholars, is a form of power in society because it reinforces the way societal members think things *ought to be* organized (e.g., males on the free weights and females in the cardio room).

With specific reference to physical activity, power inequities sometimes affect our participation as performers as well as our opportunities for involvement in influential leadership roles such as health club owner, coach, athletic director, or cardiac rehabilitation program director. People with more power can usually get involved more easily. Those with less power often face

Overall quality of life and opportunities for future advancement may depend on gender, ethnicity, and socioeconomic status.

© DigitalVision (right)

social barriers that make involvement more difficult. For example, African American players in the NFL have a more difficult time getting jobs in coaching and administrative positions in the league than do whites. Furthermore, movies, television, music, and physical activities frequently display—and thus reinforce—current power differences in American society.

Whether we participate directly in activities such as sport, dance, and exercise or watch others performing them, the activities express subtle information that tends to support and reinforce status quo power differences. For example, telecasts that highlight the physical appearance of female athletes—with special focus on their clothes and hairstyles—contribute to undermining, and thus trivializing, their athletic accomplishments.

An understanding of power in society is crucial for kinesiologists as they identify the best ways to promote meaningful physical activity for all societal members. Jennifer Hargreaves (2001) suggests that "stories of sport are almost exclusively stories of those in power" (p. 8). Thus, kinesiologists might ask, What are the hierarchies of bodies that emerge in physical education classes, rehabilitation settings, or school-based sport programs? What are the social structures that reinforce societal privilege of particular bodies (e.g., laws, policies, school curriculums, expert knowledge)? And, most importantly, how do physical educators, athletic trainers, fitness professionals, and coaches contribute to (or resist) these processes of bodily privilege? In the remainder of this overview, we will relate sociological analyses directly to the ability of kinesiologists to promote quality physical activity, including sport and exercise, for all societal members. Power relations underlie social inequalities; they affect people's quality of life, their chances for a better life in the future, and their opportunities for engagement in physical activity. Accordingly, sociologists produce research that examines the influence of systems of power (e.g., gender, race, social class) on the everyday experiences of societal members. We'll focus on three areas within these systems of power: participation, leadership, and cultural expression.

Gender Relations

Gender is different from sex. As a system of power, gender is not an identity you have (e.g., male or female), but rather a set of norms or expectations about how we should behave, and these are linked to societal understandings of sexuality and procreation. At birth, babies are assigned to a sex category based on the appearance of the genitalia (Lorber 1994). This assignment process, and the subsequent differential treatment of youngsters based on different sex categories, constitute the gendering process. Genders are not natural, biological categories—they are socially defined. You can't inherit a gender—you have to be assigned to it and learn it.

In American society, men typically have more power than do women, although this relationship changes when race, social class, and sexual orientation are considered (e.g., white, middle-class women will exercise more societal power than will a Mexican working-class male). The situation changed somewhat during the 20th century, but many inequalities remain. In the field of physical activity, inequalities prevail because of beliefs about the appropriateness of certain forms of physical activity for each gender.

Participation

In the United States, many girls play in organized youth sports, and many of their older sisters play on high school and college teams. This occurrence is relatively recent. The passage of Title IX in 1972, requiring that women be provided with equitable opportunities to participate in sport, is perhaps the single most significant factor in accounting for the huge increase in girls' and women's participation in sport over the last three decades. In 1970, only 1 out of every 27 high school girls played varsity sports. Since Title IX was enacted in 1972, female high school athletic participation has increased by 904%, and female college athletic participation has increased by 456% (Women's Sports Foundation 2008). In 2006 to 2007, girls' participation in high school athletics exceeded 3 million for the first time, with 3,021,807 females participating in high school sports according to the National Federation of State High School Associations survey (NFHS 2007).

In North American society, sports are deemed important avenues for exploring and confirming masculinity. They are thought to sharpen a number of qualities traditionally considered appropriate for men and boys, such as toughness, aggressiveness, roughness, working well under pressure, and competitiveness. You probably know girls and women with some of these qualities, a point that shows the indistinct boundaries between the two genders. Nevertheless, in our society people still conceptualize these qualities as masculine.

Although many sports attract large numbers of both males and females, some are more gender specific (table 7.2). For example, girls and women less often engage in boxing, football, wrestling, or weightlifting contests. American boys and men almost never play field hockey, although it's a sport for men in international competition. If a woman or girl you know started training to be a football player, what would you think? Many Americans would judge this negatively. But what makes it inappropriate? Although many females are now developing muscular strength and endurance through weight training, few take part in sports requiring great strength, physical domination of an opponent through bodily contact, or the manipulation of heavy objects. Furthermore, only in recent decades have women taken up competition in strenuous, long-distance runs, including marathons. An early, cutting-edge analysis showed that the sports we judge more appropriate for females, such as gymnastics or tennis, entail little body contact, the manipulation of light objects, and aesthetically pleasing movements (Metheny 1965).

Table 7.2 American Participation in Sports and Physical Activity by Gender

	% of total participants	
Sports and physical activities	Male	Female
Aerobic exercise	26.2	73.8
Baseball	77.9	22.1
Basketball	68.8	31.2
Bowling	51.3	48.7
Exercise walking	38.2	61.8
Exercising with equipment	46.3	53.7
Golf	76.6	23.4
Soccer	58.3	41.7
Softball	48.2	51.8
Swimming	46.4	53.6
Tennis	47.1	52.9
Volleyball	41.5	58.5
Ice hockey	75.2	24.8
Skiing (alpine)	55.6	44.4
Skiing (cross-country)	57.8	42.2
Tai chi, yoga	12.8	87.2
Working out at club	44.1	55.9

Adapted, by permission, from National Sporting Goods Association, 2004, *Sports participation in 2001: Series I and series II* (Washington, DC: NSGA).

➤ Although many more girls and women are participating in physical activities today compared with several decades ago, they tend to participate in sports that are considered socially appropriate, involving less body contact, prominent aesthetic dimensions, and less extreme strength development.

The continued emphasis on masculinity in sport often results in downplaying women's competition. Examples include fewer resources for female sports (e.g., less travel money, lower coaching salaries, fewer publicity funds, less media exposure, fewer corporate sponsorships) and sometimes team nicknames that reflect physical ineptitude, such as the Teddy Bears, Blue Chicks, Cotton Blossoms, or Wild Kittens (Coakley 2007, pp. 243-249; Eitzen 2006, pp. 40-43).

Leadership

Our knowledge of gender inequities in physical activity leadership pertains primarily to sport, but we will also briefly examine leadership in other physical activities. We know that opportunities for American women in coaching, sport telecasting, and officiating are growing but are still more limited than they are for men (Acosta & Carpenter 1994, 2008; Coakley 2007). Moreover, from the 1970s into the 1990s, sharp decreases occurred in the percentage of female coaches of female high school and intercollegiate teams and in the proportion of female administrators heading women's intercollegiate programs. For example, in 1977 to 1978, 79% of the coaches of female basketball teams were women; by 1991 to 1992 that

➤ Women occupy a relatively small proportion of coaching and leadership positions in sport.

number had dropped to 64% (Acosta & Carpenter 1994). The percentage of female administrators heading women's intercollegiate programs went from 90% in 1972 to only 19% in 2004. In addition, 18% of all women's programs have no female administrators at all (Coakley 2007, pp. 255-258).

Why are more men now coaching and administering women's teams and programs? Researchers have postulated a number of reasons. To begin with, the larger number of male players constitutes a larger pool of qualified candidates. In addition, the fact that women's coaching salaries still lag behind those of men may cause some of the top female players to seek careers outside athletics. One estimate of the gender ratio among applicants for coach-

Interviews With Practicing Professionals

Photo courtesy of Suzanne Speer

Keri Flynn
Athletic Trainer

Keri Flynn puts her knowledge of kinesiology to use as head certified athletic trainer at a high school in Greensboro, North Carolina. She earned her BS in athletic training from the State University of New York at Cortland in 2006 and is currently studying to earn a master's degree in sport and exercise psychology. Flynn has gained professional experience in Division III collegiate athletics, junior and senior high school athletics, summer sport camps, and regional Senior Games. Additionally, she is active in the National Athletic Training Association (NATA) as well as in regional professional associations (Eastern Athletic Training Association, New York State Athletic Training Association).

Q: What connections do you see between sociological issues and athletic performance?

A: I have devoted much of my career to studying and caring for the athletic body—helping elite performers to stay healthy and well. I'm an athletic trainer, but more broadly, I'm also a kinesiologist who understands and appreciates the social-cultural aspects of the field. I realize that what goes on outside the locker room, off the field, or outside of the gym may affect an athlete's physical well-being. In my experience, it is clear that many performance and wellness issues stem from facets of life that influence both the physical and social-cultural aspects of the body. It took countless hours on and off the field to recognize that each athlete has their own story to tell, and being willing to hear it has allowed me to establish a professional camaraderie built on trust that is the key to success in my career.

Q: How does your knowledge of sociology help you in your work as an athletic trainer?

A: I understand how race, social class, and gender influence my interactions with coaches, school officials, students, athletes, family members, and others. Thus I am careful not to reinforce inequalities in my own professional practice. On a daily basis, I negotiate my statuses as a young, white female in a social space that is largely male, racially diverse, and mixed in age. Moreover, as with most athletic trainers, my schooling was largely in the natural sciences, which of course is helpful, but I feel strongly that a purely medical approach to the physically active body is inadequate for an athletic trainer. The body is much more than physical; it is cultural and multidimensional, constantly in a process of change, and subjected to external judgments based on the social labels surrounding it.

ing jobs in women's collegiate basketball was five or six men to one woman (Marcy Weston, personal communication, September 1996). In addition, men occupy most of the influential athletic leadership positions (e.g., collegiate athletic directors, leaders in the Olympic movement). When they make decisions about hiring coaches or appointing people to important committees, they likely use their network of connections with other men in athletics, usually referred to as the old-boy network, to find candidates. Finally, men may favor applicants with stereotypical masculine qualities (e.g., those who are aggressive, dominating, physically tough), a preference that puts many well-qualified women at a disadvantage (Coakley 2007, pp. 256-258).

Cultural Expression

Knowing that gender inequalities exist in American society and that all physical activities have expressive dimensions, it's not surprising that our physical activities frequently demonstrate such inequities. For example, in some forms of dancing (e.g., the waltz, many country-western dances) men usually move forward and lead women, who usually move backward. This technique communicates a subtle message about men's superiority.

Our most popular sports, the ones we see constantly on TV, are celebrations of heterosexual manhood centered on aggressiveness, roughness, and the ability to dominate physically. Girls and women have most often been relegated to marginal positions as spectators of boys' and men's competition or as players in their own less popular versions of these and other sports. One of the messages communicated by this is that females don't matter much in "real sports." Coaches sometimes criticize weak or ineffective plays by boys and men by calling them "a bunch of girls" or "sissies" (Coakley 1998, p. 236). Intended to spur better performance, such comments send a message that females are inferior. Coaches communicate a similar message when they tell players that they "throw like a girl." These examples show ways in which sports tend to reinforce ideas about the acceptability of heterosexual male dominance.

Girls and women are not the only ones shortchanged by the gendered attitudes that prevail in sports. Our society often questions boys and young men who aren't interested in athletics or lack ability. They are called "sissies," and their sexual orientation may be challenged with taunts of "gay," "fag," or "queer" (Coakley 1998, p. 236). In high school, boys' popularity is closely connected with being good in athletics. A group of white college graduates who had not played varsity sports in either high school or college said that as boys they experienced social ostracism and name-calling because of their poor athletic ability. One said, "I identified sports as a major aspect of what I was supposed to be like as a male, which oppressed me because I could not do it, no matter how hard I tried" (Stein & Hoffman 1978, p. 148).

➤ Many sports serve as vehicles for exploring, celebrating, and giving privilege to masculinity; because of this they express ideas that are problematic for girls and women as well as for boys and men who are not athletically inclined.

Danica Patrick has become successful in the male-dominated sport of Indy car racing.
AP Photo/Rob Carr

Have you ever thought about the degree of coverage of men's and women's sports in the media (e.g., television, newspapers, magazines)? About 8.7% of sport on television focuses on women's athletics, whereas men's competition constitutes 88.2% of coverage (Duncan, Messner, & Willms 2005). In U.S. newspapers, the space devoted to women's athletics remains less than 15% of the total for all sports (Coakley 2007). This disparity trivializes women's athletic accomplishments and sends an underlying message that women's events aren't worth much.

Masculinity and Femininity Examined

Think about situations you have encountered in physical activities in which questions occurred about a male's masculinity or a female's femininity. What do these examples show us about our shared beliefs concerning appropriate activities for boys and men and for girls and women? What do they show us about gender inequalities in these activities? What do they show us about our society's definitions of masculinity and femininity?

Ethnic and Racial Relations

Race is a historically, culturally, and socially defined category of social difference typically marked by phenotypical variance among people (e.g., skin color, body shape, facial features, hair type, etc.). It is crucial to understand that race is socially defined based on the characteristics we select; it is not a natural, biological category. Like gender, race is not a fixed identity that you have, but rather a shifting condition of social life that you experience every day. We live in a society that is organized around arbitrary beliefs about gender, race, and social class; thus kinesiologists must understand how to work through these social conditions and inequalities to make meaningful physical activity available to all societal members.

Ethnicity refers to cultural heritage. People who share important and distinct cultural traditions, often developed over many generations, are classified as an ethnic group. Ethnic markers include such things as language, dialect, religion, music, art, dance, games and sports, and style of dress. Latinos and Latinas, Jews, Inuits, Amish, and Cajuns are examples of ethnic groups in the United States. Sometimes race and ethnicity overlap. For example, African Americans are defined as a racial group, yet they also have distinct cultural traditions. In addition, the African American population itself includes enormous ethnic diversity (e.g., distinct differences from African tribal cultures, heritages from the various African nation-states, influences from different Caribbean cultures, and a variety of European colonial influences).

In the United States, members of racial and ethnic minorities have typically held less power than the white majority. Although society made strides toward equality during the 20th century, many difficulties remain. Physical activity, of course, is not immune to such problems. Our focus will be mostly on sport, and we'll highlight comparisons between African American and white males because these groups have been most fully researched.

Participation

In elite team sports and in track and field, African American male athletes are overrepresented in relation to their proportion in the general population. For example, in 2004, about 13% of the U.S. population was African American, whereas their numbers were much higher in major league professional sports: 76% in basketball, 68% in football, and 17% in baseball. Baseball and hockey are major league sports in which African American males are underrepresented (9% and less than 1%, respectively) (Coakley 2007; Lapchick & Brenden 2006). The high participation rates probably relate most to American social structure (Eitzen & Sage 1997, pp. 254-269). Whether they are local playground heroes or major league players,

African American male athletes receive much public applause from African American men and women alike. Some of the most visible and popular African American role models are top male athletes, and their status may give young African American boys the incentive to spend long hours practicing their sport skills.

But this doesn't account for African Americans' high participation rates in a few sports and their low involvement in most others. Opportunities, and lack of opportunities, probably influence this situation (Eitzen & Sage 1997). African Americans tend to be more visible in sports with easy access to facilities and coaching, primarily in school and community recreation programs. Sports requiring private coaching, expensive equipment, empty land for playing fields, or facilities at private clubs—for example, tennis, golf, swimming, soccer, or gymnastics—are less popular. The extent to which these various factors influence African American athletic involvement remains unclear. The success of African American athletes in historically segregated sports (e.g., Tiger Woods in golf, Venus and Serena Williams in tennis, and Bill Lester in NASCAR) raises new questions about access, identities, and underlying reasons for sport involvement. Figure 7.2 presents a breakdown of sport-related activities by race.

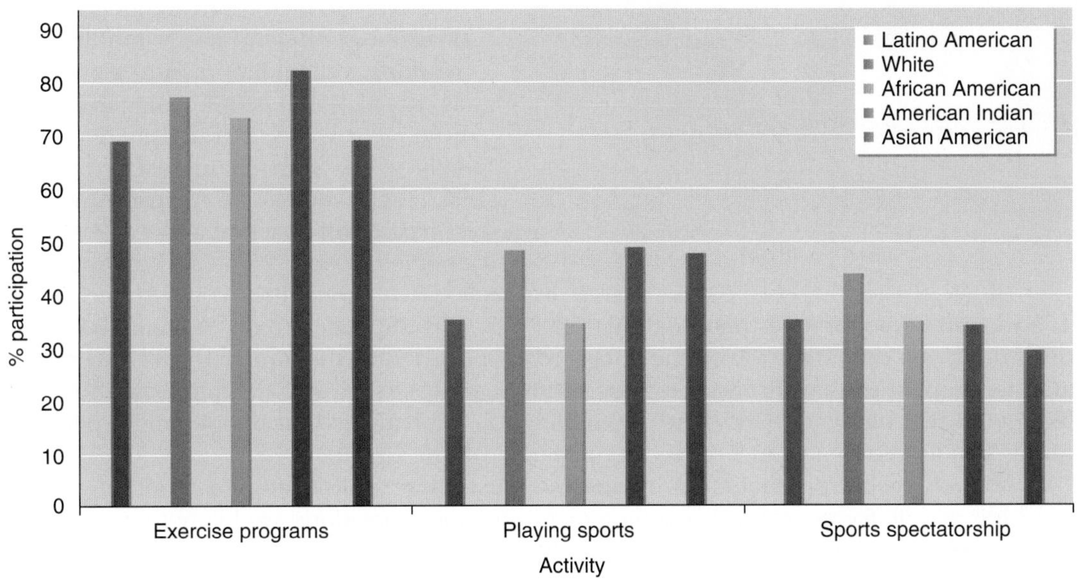

FIGURE 7.2 Percentage of U.S. adult participation in exercise programs, playing sports, and sport spectating by race.

Recreational physical activities in which competition is not essential, as well as exercise programs geared toward improving health and appearance, are probably subject to some of the same factors. For example, surfing, backpacking, exercising at a health club, swimming, and skiing require various combinations of expensive equipment, facilities, lessons, and travel, limiting involvement by those in lower income families. Because a disproportionate number of African Americans live at the poverty level, it is understandable, though not justifiable, that they tend to be underrepresented in such activities (USBOC 2007b).

➤ Large numbers of African American athletes play major team sports and participate in track and field, but representation remains disproportionately low in most other sports in U.S. society.

Leadership

If our society had complete racial equality, then in sports with high proportions of African American players—football, basketball, and baseball—we would expect to find rather high proportions of African American coaches. But this is not the case. In 2000, African Americans represented only 13% of Major League Baseball managers, 6% of NFL head coaches, and 31% of National Basketball Association (NBA) head coaches (Eitzen & Sage 2003). Among NCAA Division I-A intercollegiate programs in 2000, African Americans were head coaches of fewer than 5% of the football teams, 21.6% of the men's basketball teams, and 9% of the women's basketball teams. In Division I baseball, there was only one African American coach (Eitzen & Sage 2003, pp. 302-303). Indeed, the 2007 Super Bowl game was a historic moment, as

Super Bowl XLI made history because the head coaches of both teams were African American.

AP Photo/David Duprey

not only one but both teams were led by African American head coaches, Tony Dungy and Lovie Smith. Some credit has been given to a recent NFL rule requiring that at least one minority candidate be interviewed when any team is searching for a new head coach.

In addition, few minorities officiate games; they made up only 33% of NBA basketball officials in 2000. This is a poor showing when we remember that in 2000 a whopping 78% of the players were African Americans (Eitzen & Sage 2003, p. 302).

There are two major explanations for these leadership inequities. First, racist stereotypes featuring erroneous conceptions of inferior intellectual capabilities of African Americans are probably operating to keep them from attaining athletic leadership roles in which thinking and working well under pressure are viewed as crucial. When people doing the hiring have racist views, African Americans are likely to be shortchanged (Shropshire 1996). Ricky Stokes, an African American assistant basketball coach at Wake Forest University, put it this way: "As an African American assistant coach, sometimes you can be labeled as a recruiter. A lot of times people think that's all you're supposed to do. Sometimes people don't think you can do the job on the floor—coaching, teaching the fundamentals, and, taking another step, speaking to booster clubs." Although Stokes expressed optimism about the likelihood that he would land a head coaching job, he realized that people might perceive hiring him as taking a chance (Ross 1996, March 10, pp. C1-C2).

A second reason for leadership inequities is that white athletic administrators in charge of hiring people for entry-level coaching or management positions are likely to learn about candidates from their professional friends in other athletic programs, most of whom are also white and may think in similar ways. Because ethnic and racial minorities are not usually involved in these old-boy networks, athletic administrators can easily overlook them. Relatively few people from marginalized ethnic and racial groups hold leadership positions in programs featuring expensive sport and exercise regimens (Coakley 1998; Eitzen & Sage 1997; Shropshire 1996). A quick tour of facilities such as health clubs, soccer fields, skating rinks, ski areas, gymnastics schools, and tennis courts would probably reveal only a handful of coaches and administrators of color.

➤ Few African Americans reach important sport leadership positions, even in team sports that boast many African American players.

Cultural Expression

On one hand, African Americans take pride in the athletic prowess and creativity of African American men and women who have become sport heroes. Conversely, the very sports that make African Americans heroes can also contribute to their racial and ethnic stereotyping.

Expressive behavior by African American athletes may be one way of dealing with barriers to achieving manhood that African American males face in our society (Majors 1990). Because their access to education, jobs, and power is restricted, African American males often work on proving their masculinity in other ways. One of these is constructing an expressive "cool pose" consisting of "styles of demeanor, speech, gesture, clothing, hairstyle, walk, stance, and hand-shake" (Majors 1990, p. 111). Cool pose sends a message that the person has a potent, interesting lifestyle that deserves respect. Sports are a ready-made arena for cool pose. Examples include football end-zone dances and spikes; Tommy Smith's and John Carlos' raised, black-gloved fists on the victory stand at the 1968 Olympics; Muhammad Ali's

Assessing the Ethnic and Racial Mix in Sport

Think about the local physical activity programs with which you're familiar in schools and community recreation centers (e.g., physical education classes, high school and college varsity athletic teams, rhythmic aerobics classes at local YMCAs, pickup basketball games on inner-city courts, gymnastics instruction at private centers, youth sport leagues formed by groups such as the American Youth Soccer Organization or Little League Baseball, Inc.). From your observations, what ethnic and racial minorities are represented among the people in leadership roles in these programs (e.g., supervisors, directors, coaches, officials)? How do the proportions of different ethnic and racial minorities in these positions compare with the proportions of ethnic and racial minorities among the players? What are possible reasons for the similarities or differences?

boasting, poetry, and dancing in the ring; and Julius Erving's gravity-defying moves to the basket that often began at the foul line (Majors 1990). Many current African American athletic stars continue to create images rooted in the traditions of cool pose. In addition, cool pose is threaded through a variety of stylish gestures and "walks" that are embedded in the ordinary activities of African Americans and in distinctive dances such as hip-hop.

Young African American boys and men develop their own renditions of cool pose at playgrounds and school athletic facilities. Their style is a crucial part of their sports—especially basketball (Carlston 1983; Kochman 1981; Majors 1990; Wilson & Sparks 1996). They challenge each other with their creative, flashy moves, assertively daring other players to top them. Their goal is to heighten their reputation by combining effective game skills with a spectacular look—a look that we can even see in a range of media images designed to make profitable a particular form of black masculinity through black athleticism. (See Neal 2006 for further reading.)

Unfortunately, sport also communicates ideas that help maintain racial and ethnic inequities. For example, the fact that many top athletes are African American seems to send an upbeat message to African American boys and young men that they can make it to the pros if they work hard enough. But this message is misleadingly positive. Although 43% of African American high school varsity basketball and football players believe that they have a chance of playing in the pros, a minuscule number make it. And because their constant athletic practice usually gets in the way of acquiring other job skills, they have little to offer employers when they fail to make it in athletics (Coakley 2007, pp. 289-290). White youths in higher socioeconomic brackets who dream of playing professional sports also usually end up short of their goal, but because they don't face racial discrimination and are from higher-income families that have more contacts with potential employers, they are usually in better positions to pursue other job opportunities.

Racial Stereotyping

Racist stereotypes are sometimes used in television portrayals of white and African American athletes. In two articles published in 1989 and 1996, Derrick Jackson studied the comments of TV sportscasters about African American and white athletes. He noticed that sportscasters tended to recognize white basketball and football players for their brains: intelligence, thoughtfulness, and strategy. African Americans, on the other hand, were noted more for their brawn: physical skills, moves, and muscular strength. Such reactions reinforce a prominent racist stereotype. It is heartening to learn, however, that although brains were still mentioned more often in 1996 television portrayals of white athletes, this tendency was not as strong as it was in 1989. Unfortunately, over the same period the "brawn gap" between African Americans and whites doubled. African Americans were twice as likely to receive praise on television for their physicality in 1996. What is crucial to understand from Jackson's research is that our knowledge of racial categories changes over time and place and thus, as consumers of sport, we need to be aware of the racialized rhetoric used to convey sporting stories.

American Indian Mascot Controversy

AP Photo/Heather Coit

One of the ongoing controversies in the sporting world is the use of American Indian team names and mascots for sport teams. A well-known example is the Cleveland Indians and their mascot named Chief Wahoo. If you are a baseball fan, you've probably seen the Cleveland Indians baseball logo. If not, you can find it on the official Cleveland Indians' Web site (http://indians.mlb.com/index.jsp?c_id=cle). As you will see, the logo features the head of Chief Wahoo, whose toothy grin, hook nose, fire-engine red complexion, and single-feather headdress "identify" him as an American Indian.

Critics of the sport teams who adopt American Indian imagery argue that such representations create false and degrading images of American Indians. They believe that the "rituals" carried out by teams or fans (e.g., the tomahawk chop or the dance done by Chief Illiniwek) show disrespect for Indian culture. Some critics contend that the Indian names and mascots contribute to the miseducation of Americans by producing a kind of cultural illiteracy about indigenous (native) people (Staurowsky 1999, November). They argue that the symbols of Indian life featured in sporting imagery—the war paint, masks, costumes, drumming, chants, and ceremonies—are neither authentic nor accurate. They are what some critics call pseudo-Indian symbols (King 2004) and just feed long-standing stereotypes of American Indians.

A number of institutions and schools—such as Cornell, Dartmouth, Marquette, Stanford, and Syracuse—have seen merit in these arguments and have changed their names and mascots accordingly. Others, including professional sport teams (e.g., the Cleveland Indians, Washington Redskins, and Atlanta Braves) have not.

Those who defend American Indian symbols (and virtually all defenders are white people) assert that their appropriation of Indian imagery is a tribute. To be called a warrior or a brave is a compliment because these words suggest that American Indians are courageous and tough. In addition, defenders argue that it is their right to adopt and subsequently own, for marketing purposes, such American Indian imagery (Staurowsky 2004).

Most American Indians themselves are clear on their position. The violent and aggressive images associated with others' ideas of American Indians offend them. They believe that they own their culture and that non-Indians should not have the right to use it for their particular, often commercial, purposes. They contend that the Indian names, mascots, and logos have the effect of distorting and trivializing their culture. They see Chief Wahoo as a racist caricature of an American Indian. In short, they argue that such American Indian imagery contributes to the long history of oppression that characterizes white dealings with American Indians (King 2004).

Web Search 7.2

Native American Mascots
Research both sides of the issue in this activity in your online study guide.

The brain versus brawn manner of describing athletic performance is just one way to convey cultural messages through bodily performances. This takes on new formations as we consider how race intersects with gender, social class, sexuality, and ability (Coakley 2007; Douglas & Jamieson 2006; Eitzen 2006).

Michael Jordan, probably the most visible athlete in the world in the 1990s, provides much food for thought in connection with our discussion of expressiveness. A collection of articles devoted exclusively to this athletic superstar demonstrates the ways in which he was constructed as a media icon who transcends race (Andrews 1996). Part of this involved stripping away some of his "blackness"—his racial identity. For example, advertisements accentuated his physical, athletic black body (a common media portrayal of African American athletes). At the same time, however, the threat of heightened sexuality stereotypically associated with black bodies in American society was defused by frequent references to his wholesomeness and strong family ties. Interestingly, these mediated images of Jordan took hold so well that even following his divorce, he retained a "wholesome" image. Such mixed messages probably contribute to his popularity among whites and African Americans alike (McDonald 1996). Media portrayals of Jordan give us a collage of meanings, some lending positive support to a distinctive African American culture and others reinforcing old ideas that help to maintain racial inequalities. Sorting out which are which is not easy.

➤ African American athletes who engage in "cool pose" express creativity, strength, and pride associated with masculinity; however, mass media professionals (e.g., journalists, broadcasters, advertisers) sometimes use racial stereotypes to characterize African American athletes and may refer to them in ways that strip away some of their racial identity.

Socioeconomic Relations

Wealth, education, and occupational prestige are all ingredients of socioeconomic status. People with more money, more education, and greater occupational prestige have higher socioeconomic status and usually more power. To illustrate, it's not hard to recognize the many advantages of being a corporate chief executive compared with being a custodian at the corporate headquarters. The chief executive can afford a luxurious lifestyle, whereas the custodian might have to scrimp to make ends meet. The chief executive can send his children to the most expensive universities, whereas the children of the custodian might have to work to put themselves through college. The chief executive can make major decisions affecting future directions of the company, whereas the custodian might only be able to make recommendations to his boss about better ways to do cleanup work.

Abundant evidence indicates that the income gap between rich and poor is widening in the United States. For example, after inflation is considered, between 1968 and 1994 the average income among the top 20% of households grew by 44%, whereas among the bottom 20% it grew by only 7% ("Income Gap" 1996, June 20).

This vast and growing income disparity in the United States has important implications for many aspects of American life including education, recreation, health care, and physical activity. For example, among industrialized nations, citizens of countries with wider income gaps generally are less healthy (Siedentop 1996). Much evidence also indicates that poor people are more likely to

People with a high economic status can take advantage of opportunities not available to people with lower incomes.

Bananastock

suffer from health problems. In fact, one of the poorest demographic groups in the United States is African American women; not coincidentally, this group is most at risk for poor health (National Center for Health Statistics 2002).

Yet, in the United States we have seen an enormous growth in private health and exercise clubs that charge membership fees and cater to the upper middle class and the wealthy. Meanwhile, a decline has occurred in publicly funded programs in schools and recreation departments that are more accessible to middle- and working-class people (Wankel 1988). For example, it is increasingly common for students to pay user fees to participate in varsity athletics in the public schools and for families to pay user fees for youngsters to compete in community youth sport programs (Coakley 2007). Most U.S. citizens could benefit from physical activity, yet barriers are in place that make programs less accessible to low-income people. Table 7.3 shows how participation in certain kinds of physical activity varies by level of income.

Table 7.3 Participation in Selected Sport Activities by Household Income

Physical activities	% total participants in the activity by household income: 0-$25,000	% total participants in the activity by household income: $75,000 and over
Aerobic exercising	24.4	37.5
Baseball	26.6	37.0
Basketball	27.8	35.4
Bowling	30.2	30.9
Exercise walking	32.1	30.4
Exercising with equipment	21.2	40.4
Fishing (freshwater)	35.8	23.3
Football (tackle)	35.6	29.4
Golf	15.1	48.6
Hunting with firearms	33.2	23.0
Martial arts	28.5	36.8
Running or jogging	26.3	36.3
Tennis	22.0	49.6
Skiing (alpine)	12.5	47.6
Skiing (cross-country)	32.3	40.1
Swimming	27.4	36.2
Working out at club	17.0	43.7
Sailing	16.1	50.3

Adapted, by permission, from National Sporting Goods Association, 2004, *Sports participation in 2001: Series I and series II* (Washington, DC: NSGA).

Like the United States, Canada has experienced a widening of the gap between the haves and the have-nots. In 2002 a study showed that the net worth of the top 25% of Canadian families rose by 14% between 1994 and 1999 and that their financial wealth increased by 40%. During the same period, however, the bottom 75% of Canadian families reported no gain in either area (Abbate 2002). Not surprisingly, in Canada the health outcomes of poverty follow the pattern described in the United States. Aboriginal peoples in Canada and the poor, compared to the rich, are much more likely to be obese; to suffer from diabetes, tuberculosis, and heart disease; and to have a higher mortality rate (Kruzenga 2004). Impoverished Canadians face similar barriers to exercise and physical activity. Community user fees have risen, as have student user fees. Those who are already without resources have less access to sport participation ("User Fees" 2003, November).

Participation

Research indicates that people's level of education is clearly tied to participation in physical activity. One study showed that highly educated groups were 1.5 to 3.1 times more likely to be active than were the least educated groups (Stephens & Caspersen 1994). Salaried professionals are much more likely to participate in corporate fitness programs than hourly workers are (Eitzen & Sage 1997). To the extent that educational attainment is related to economic resources, these data may also reflect availability of time and resources for participation in sport (Coakley 2007).

Income is also linked to the types of sport and recreational activities in which people take part. Wealthy adults tend to participate in individual activities such as tennis, golf, and skiing, whereas those with lower incomes tend to play team sports such as football, basketball, and baseball (see table 7.3). At least two explanations can account for this difference. Equipment and facility costs are so high in some sports that these sports don't attract people with low incomes. In addition, the working hours of higher-income professionals often fluctuate, making it more difficult to schedule regular competition with a team. Blue-collar workers tend to have standard working hours, a circumstance that makes it easier to schedule regular team sport competition (Eitzen & Sage 1997, pp. 244-245). Expensive physical activities that lend themselves to flexible scheduling may have more appeal to higher-income people. Examples include skiing, scuba diving, mountain climbing, hang gliding, and white-water kayaking.

➤ Socioeconomic status influences the types of physical activities to which people have access; physical activities requiring expensive equipment, facilities, and coaching are mostly beyond the reach of people at lower income levels.

We can point to many athletes who grew up in poverty and eventually acquired enormous wealth and fame as professional players. Many people thus think that talented athletes have a sure path to upward social mobility. But only a tiny minority of mostly male athletes successfully travel this route. In 2004, for example, the 1035 major league professional players in football, men's and women's basketball, and baseball represented only 0.19% (less than 1%) of high school athletes playing these sports. About 16,000 players from colleges and universities were eligible to be drafted as rookies in the NFL in 2004, but only about 250 ended up on the final team rosters (Coakley 2007, p. 344).

Leadership

The people who control elite sports are generally quite wealthy. At the top are the owners of professional team sport franchises, sport media moguls, and corporate executives. In 2004 the individual net worth of U.S. pro team owners was estimated to range from a high of $21 billion (former Microsoft executive Paul G. Allen, who at the time was the fifth richest person in the world and now owns the Portland Trailblazers and the Seattle Seahawks) to a mere $1.2 billion (Robert L. Johnson, the founder of BET [Black Entertainment Television], who was approved by the NBA to operate a new sport franchise, the Charlotte Bobcats, in 2004-2005). Most were sole or partial owners of several other businesses, and many were (or had been) corporate chief executives. In 2002, three of the most powerful people in sport as identified by the editors of *Sporting News* were Michael Eisner (Chairman and CEO of Disney), Rupert Murdoch (Chairman and CEO, NewsCorp), and Ed Snider (Chairman, Comcast-Spectacor) (*SportingNews.com* 2002).

Increasingly, corporate heads are being hired as athletic directors in NCAA Division I college and university programs, and their salaries and other financial compensation have to be high enough to lure them away from private business. Top coaches are also well paid. In 2006 the average NCAA football coach's salary was $950,000 per year not including benefits, perks, and incentives. Bobby Bowden's total salary from all sources in 1985

Volunteer youth sport coaching positions are often filled by people with average incomes who have little opportunity to assume leadership roles at elite levels.

was $174,500. In 2006 it was $1,691,900. The average income of NCAA Division I basketball coaches whose teams made the 2006 NCAA tournament was $800,000, but those in the highly visible conferences (Atlantic Coast Conference, Big 12, Big East, Big 10, Pacific 10, and South Eastern Conference) earned an average of $1.2 million. In 2006 the University of Alabama signed a football coach to a contract that pays him $32 million over eight years. The University of North Carolina hired Butch Davis as football coach in 2006 under arrangements that will pay him a sum in one week equal to the average household annual income in the state. (By comparison, the average salary of presidents of public universities in 2006 was $397,000 [Dotingg 2008].) Some people believe that these incomes are justified because many top college coaches are in contention for jobs with professional teams that can offer huge sums of money.

Few opportunities are available for the less affluent to occupy important leadership roles. In the commercial enterprises of our most popular spectator sports, the goal is to make money—either for profit (e.g., professional sport franchises, media companies) or to improve the program (e.g., nonprofit, big-time intercollegiate sports). People who are particularly adept at doing this (e.g., winning coaches; astute collegiate athletic directors; efficient professional franchise managers, television executives, and production personnel; forward-looking team owners) are usually rewarded with higher incomes or more profits. Less affluent people have few opportunities to occupy important leadership roles in this high-stakes business.

People with average incomes have much better chances to assume leadership roles in lower-level collegiate programs, high school athletics, and community recreational programs such as organized youth sports and adult athletic leagues. The competitive structure of these programs often mimics elite athletics, of course. For example, youth sport teams often take the names of major professional squads. Procedures involving cutting players from teams, having playoffs, and naming all-star teams are derived from elite college and professional sports. People at the lowest income levels don't participate in sport as frequently and are therefore not likely to advance to positions of leadership even in lower-level programs and leagues.

➤ Wealthy people occupy influential leadership positions in our popular spectator sports and in some physical activities. Those at the lowest levels of socioeconomic status rarely find themselves in positions of leadership.

Leaders in other types of physical activity programs—for example, noncompetitive recreational activities, instructional programs, and health-related exercise programs—also range from people who are extremely wealthy to those with modest incomes. For example, owners of health club chains, ski resorts, and golf courses are probably among the most affluent. Because these are private businesses, specific financial information is difficult to obtain. As in any industry, their profitability undoubtedly varies widely. People who manage or teach in such facilities are probably from the middle and upper middle classes. Most individuals at the low end of the socioeconomic continuum probably don't participate enough to make it into the leadership ranks in these types of programs.

Cultural Expression

Sports such as golf and sailing, as well as physical activities that are not essentially competitive such as skiing and scuba diving, are sometimes used to mark economic affluence. Belonging to private country clubs or indulging in vacations at fancy sport and recreational resorts lets others know that you have enough time and money to enjoy luxurious physical activities.

In similar fashion, well-exercised, lean, taut bodies serve as status symbols in white North American society. They adorn magazine and television advertisements, model the latest designer clothes, and appear in popular movies and television shows. A thin, well-contoured "hard body" accentuated by spandex tells others that you have the time, money, and self-discipline to shape yourself according to today's difficult-to-attain standards. These ideals are readily observable in almost any magazine or movie. Having a personal trainer sets you off as even more affluent. Most people can't afford all the body work—exercising, dieting, beauty treatments, plastic surgery, and other procedures—needed to achieve this look. The look thus becomes a scarce commodity—something that marks you as wealthy and stylish.

Sport sends other messages that support socioeconomic inequities. For example, winning is the most prevalent organizing theme in newspaper stories and telecasts of sporting events

(Kinkema & Harris 1998). Winning is usually attributed to self-discipline, talent, and hard work. If an athlete or a team doesn't win, then we assume that the player or the team was lazy or lacked talent and so didn't deserve to win. Such beliefs underscore the American conception of merit—we often link hard work and talent to financial success. The flip side is that if someone fails financially, it must be because she or he isn't talented or didn't work hard. This reasoning allows us to hold the belief that the rich and poor both deserve whatever money they have. The point here is not that merit is a bad idea. The problem is that this logic often leads us to overlook the societal barriers (e.g., poor nutrition, neighborhood gang violence, poor access to libraries and computers, dysfunctional families, lack of child care) that prevent poor people from developing themselves to the fullest and becoming valuable members of society (Coakley 1998, p. 292).

➤ A well-sculpted, lean body and participation in expensive sports are often markers that a person is wealthy. In addition, sport expresses other ideas that reinforce the socioeconomic status quo, such as the notion that winners and losers deserve what they get and the value of obedience and teamwork.

Finally, our popular team sports also send messages about the importance of obedience and teamwork. These qualities are valuable for most workers. Although some employers may need employees who can break out of traditional molds and be creative, most simply need people who can work well with others and follow directions. Because sport tends to reinforce obedience and teamwork, it helps maintain our current economic system.

Attributions of Failure: Lack of Merit or Lack of Money?

Think about the things coaches typically tell players when a team loses a game. You can recall your direct experiences or talk to other people about theirs. How often did coaches tell players that they didn't work hard enough, and that if they had practiced harder or given more in the game, victory would have been theirs? How often did coaches tell players that they obviously tried hard but lost to a team that was just more talented? What other reasons did coaches offer for why the team lost? Did any (or all) of these reasons make the players feel that they really didn't deserve to win?

Wrap-Up

Sociology of physical activity takes us beyond common, everyday understandings of sport and exercise. It illuminates patterns of change and stability, identifies social problems, and urges modifications aimed at enhancing equality and human well-being. This subdiscipline has so far emphasized sport; but interest in exercise, fitness, and societal conceptions of the human body is increasing. Although researchers have analyzed relationships and trends in social life, for the most part they have not completely nailed down causal factors underlying these trends and relationships. This circumstance is not unusual in the social sciences because it is difficult to bring a slice of social life under the strict laboratory controls needed to demonstrate causation.

If you are interested in what you have learned here, you might enjoy looking through some of the sport sociology textbooks, taking an introductory course in this subdiscipline, and exploring major journals such as *Sociology of Sport Journal, Journal of Sport and Social Issues,* and *International Review for the Sociology of Sport.* You may also wish to review research reports released by major sport studies research centers, such as the Centre for Research into Sport and Society at Leicester University, the National Research Institute for College Recreational Sports and Wellness at Ohio State University, and the Center for the Study of Sport in Society at Northeastern University. Sociology of physical activity provides information that kinesiologists may use to increase our understanding of our experiences as participants, spectators, and professionals, as well as the experiences of others. It also helps us think more clearly about the changes we would like to make in these physical activity programs, as well as the things we would like to keep the same.

Finally, sociology of physical activity is an ongoing reflection on the changing social conditions in which kinesiologists do their work. Thus, a question that is always relevant for kinesiologists is, What do I need to know about this social, political, and cultural context for physical activity so that I do not mistakenly reinforce existing social inequalities?

Activity 7.5

Use the key points review in your online study guide as a study aid for this chapter.

Activity 7.6

These end-of-chapter questions and activities are also in your online study guide. Your instructor may ask you to complete them online and turn them in.

1. List and discuss the goals of sociological study of physical activity.
2. Give an analysis of the expanding research directions in sociology of physical activity from 1970 to the present.
3. List and discuss the six research methods commonly used for sociology of physical activity.
4. Describe the ties between participation in physical activity and power relationships based on gender, race and ethnicity, and socioeconomic status.
5. Describe the ties between leadership in physical activity programs and power relationships based on gender, race and ethnicity, and socioeconomic status.
6. Describe the ties between physical activity expressiveness and power relationships based on gender, race and ethnicity, and socioeconomic status.
7. Identify the contrasting beliefs that underlie the American Indian mascot issue. Which argument would a sociologist of physical activity be most likely to advance? Why?

8

Motor Behavior

Jerry R. Thomas and Katherine T. Thomas

Digital Vision

CHAPTER OBJECTIVES

In this chapter we will

- explain what a motor behavior researcher does;
- present the goals of motor behavior, including motor learning, motor control, and motor development;
- explain the research process used by scholars in motor behavior; and
- present some of the principles of motor learning, motor control, and motor development

Your roommate, an accounting major, is on the college track team. For the past six weeks the coach has had the team training in the swimming pool as part of the off-season program. Runners practice standing starts and sprints across the pool, and field athletes are practicing throwing and jumping in the pool without the shot, javelin, and hammer. All the athletes have been working at a water depth that places just their heads out of the water.

Your roommate is excited to be on the track team and believes in the coach's rationale for pool practice—that the water will provide resistance training for all the muscles used in each event. First, you wonder how swimming and track are related, but you know that weight training is helpful to most athletes. Because pool work is resistance training, it is like weight training. You wonder whether the resistance of water will equal the resistance of weights. Your roommate says that weight training does increase strength but that the movements are not the same as the movements in track and field. You know that the lifts used in weight training train a specific set of muscles in an optimal way. So you wonder why most strength and endurance training methods do not include the pool technique. Finally, you wonder if other reasons related to the sport techniques explain why none of your classes has presented this technique. What other factors could you consider as you decide whether or not this type of practice will help the team?

This chapter introduces the subdiscipline of motor behavior and establishes the relationship of motor behavior with other scholarly subdisciplines in the discipline of kinesiology. You'll begin to understand why motor behavior is of interest to performers who want to improve movement *expertise,* to scientists who contribute to *scholarly study,* and to practitioners in *professional practice* who seek to assist others in developing motor skills.

What Is Motor Behavior?

Have you ever wondered why you enjoy certain physical activities? Often, being skilled helps us enjoy an activity. If you pitch a great fastball, run a quick 5K, have a killer volleyball spike, or can hit a terrific backhand in tennis, you are more likely to enjoy these activities. Someone with less skill may find the activities frustrating and less enjoyable. Of course, there are notable exceptions, such as golf, a sport on which some of us spend many hours and a great deal of money demonstrating that we have very little skill!

Think about a skill you like and do well, for example dribbling a basketball. How did you learn to dribble? What types of practice experiences worked best for you? How did you learn to coordinate the movements of your feet, your hands, and the ball? Remember those early years when you dribbled slowly, couldn't turn quickly, could dribble only with one hand, and sometimes lost control of the ball? In what ways did your brain and nervous system develop and adjust so that you could improve your control and coordination? Across the life span, practice is an important factor in learning skills. Most people improve motor skills and never understand how the nervous system adapts, how it develops or controls movement, or how best to use practice to improve performance. The study of motor behavior focuses on how skills are learned and controlled and how movement changes from birth through the end of life.

People believe that practice improves performance; we see evidence of this in homes, schools, and communities. Parents urge children to practice playing the piano; athletes ask coaches for help practicing a sport skill; and students perform practice exercises to learn keyboard typing. Clearly, some forms of practice are superior to others. *What, how,* and *how much* are important questions about practice. Another important question is, What can be changed with practice? For example, is the expert better because of practice, or was she born with special talent? Research in motor behavior tells us that your parents and teachers were partly right when they said, "Practice makes perfect." More precisely, they might have said, "*Correct* practice makes perfect"; research also tells us that many factors influence how well the practice works.

So, is your roommate's coach using good practice technique? The answer is probably no—the practice routine that the coach developed will not improve performance in running or field events. The athletes may become stronger, but this routine may interfere with the timing of their movements, which is critical to most complex skills. (In other words, practice that improves timing would help your roommate and the other athletes on the track team.) Athletes use many special practice drills hoping to improve their playing performance, but only practice conditions that are similar to actual game performance will benefit performance in athletic contests. This principle, called **specificity of practice,** is one of the most important and easy-to-replicate principles in motor behavior. In addition, other factors such as talent and confidence will influence the skill that the team demonstrates at meets. We will say more about specificity of practice and other principles and theories within motor behavior later in this chapter.

Teachers and coaches should be experts on practice, because much of their training focuses on how to organize and conduct practice. But we all might encounter situations that call for knowledge about practice and skill acquisition. Consider these examples:

- Your golf partner asks you to watch her swing and advise her on how to avoid slicing her drive (of course, you know that the less skilled a golfer is, the more likely it is that she will share her ideas about the golf swing).
- You volunteer for a youth league football program. The topic of discussion is how best to group kids for fair competition.
- You drive to school every day and realize that many days you don't remember a single event during your drive (and you also realize that hundreds of others are driving in the same scary way).
- You work part-time at an assisted living center. You see that many of the residents have difficulty going from sitting to standing, balancing, and grasping small objects. You wonder why this is so, and if they can regain these skills.
- You and your dad are watching a National Basketball Association game, both of you marveling at the graceful moves and expert skills of the players. How would you explain some players' attaining such a high level of expertise?

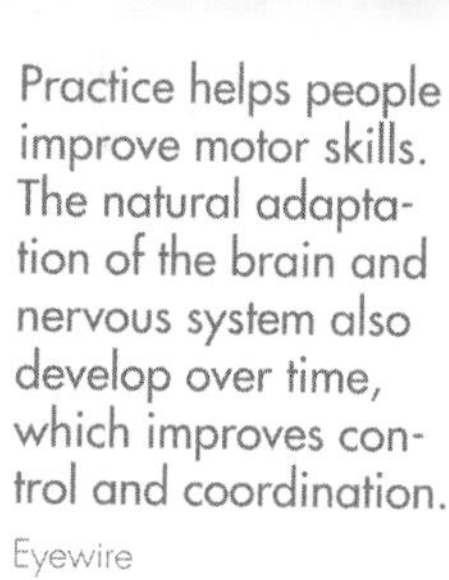

Practice helps people improve motor skills. The natural adaptation of the brain and nervous system also develop over time, which improves control and coordination.

Eyewire

You have probably faced questions similar to these. If you're like most people, you weren't entirely sure what to do or what to say. Although most of us can talk on a superficial level about motor skill performance—"Did you see how long he was off the ground? He floated through space to dunk that ball!"—few people really understand how a person controls, learns, and develops motor skill over time. Motor behavior provides answers to many of these questions and helps us understand the process of the development, control, and learning of motor skills (Thomas 2006; Ulrich & Reeve 2005). Understanding will allow you to answer questions and plan practice.

College students may see themselves merely as students in a class rather than as scholars engaging in *scholarly study.* But you have accumulated a large body of knowledge that you can use to learn about motor behavior. For example, you may recognize some of the theories and research methods used in motor behavior from your introductory, developmental, or experimental psychology course. The study of motor behavior began as a branch of psychology that used movement or physical activity to understand cognition. Information from biology or zoology is incorporated into motor behavior as heredity, aging, and growth influence physical activity (e.g., see Thomas & Thomas 2008). In addition, researchers apply principles and laws from physics to their study of humans in motion. So, what you know about physics, biology or zoology, and psychology will help you understand motor behavior.

You have some applied experience with motor behavior. Remember your physical education teachers or coaches asking you to practice passing a volleyball or swinging a golf club or dribbling a soccer ball repeatedly? They gave you feedback about your practice: "Bend a little more at the knees"; "Follow through with a smooth stroke"; "Keep the ball as close to your feet as possible." They were implementing the motor behavior principles that correct practice and appropriate feedback improve performance. Of course, some of us can recall adults providing less helpful information, such as "Practice harder" or "Hit the ball." Motor behavior guides us in providing better situations for learning and practice, as well as understanding why some cues and feedback are better than others.

The subdiscipline of motor behavior is part of the behavioral sphere within the study of physical activity, along with psychology of sport and exercise (see figure 8.1). As people practice physical activity, they *experience* the essence of motor behavior by trying to control and learn movements. Scholars of motor behavior *study* how motor skills are learned, controlled, and developed to assist people as they practice and experience physical activity. Professionals who try to improve individuals' physical skills use the knowledge from motor behavior research in professional practice. Coaching and teaching physical education are two of many professions that base their practice on motor behavior knowledge. Other examples include the gerontologist who wants to improve the motor skills of people who are elderly, the physical therapist who rehabilitates the movements of injured patients, and the athletic trainer who attempts to prevent injuries and rehabilitate injured players.

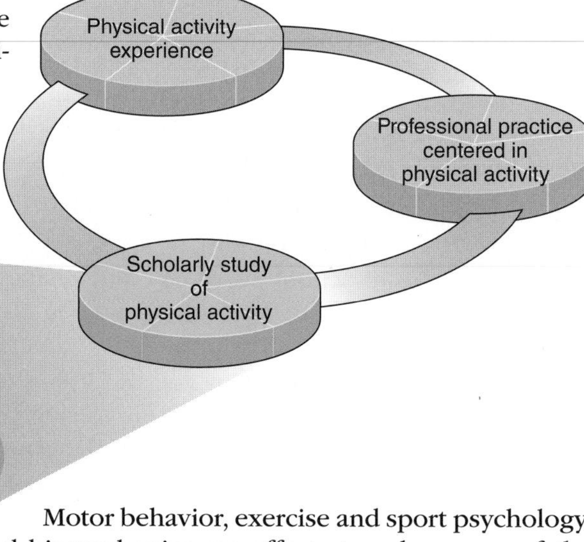

FIGURE 8.1 Motor behavior of physical activity and the other scholarly subdisciplines.

Motor behavior, exercise and sport psychology, and biomechanics are efforts to solve some of the same problems and use some of the same methods and equipment. The pedagogical areas of physical education and physical education for individuals with disabilities (adapted physical education) should be based on what we know about how people learn and control movements. Motor behavior also differs specifically from sport psychology in that sport psychology typically studies elite athletes in competitive settings whereas motor behavior studies people of all skill levels (including elite athletes).

Finally, the pedagogical areas of physical education and adapted physical education use motor behavior knowledge to design instructional programs and evaluate motor skill. For example, a physical education teacher may use motor behavior information to plan a series of instructional classes for 6th graders to improve soccer dribbling techniques. Motor behavior is part of the behavioral sphere of kinesiology and is linked to the other behavioral subdisciplines—psychology of sport and exercise and pedagogy of physical activity. Motor behavior is also closely associated with biomechanics from the biophysical sphere of study.

What Does a Motor Behaviorist Do?

Scholars who study motor behavior are often employed at universities, where they teach and do research and service, but some work in research facilities not associated with universities. Motor behavior research can be conducted in a university laboratory, in a clinical setting (e.g., a hospital), or in an industrial or military setting. For example, a researcher might ask the question "Does pushing an elevator button multiple times cause the elevator to arrive sooner?"

Motor learning and motor control research has industrial and military applications. Researchers might study, for example, how night vision glasses influence pilots' control of helicopters or what the optimal method is for training workers to assemble a product. Besides conducting research, industrial and military researchers also write grants and perform professional services, such as reviewing manuscripts for scholarly journals.

Medical and educational researchers often study motor learning, motor control, and motor development. They might study how the nervous system changes across age in the control of movement, what the best methods are for teaching rehabilitation protocols to patients in physical therapy, how the game performance of youth soccer players can be improved, or what the causes are of movement deterioration with Parkinson's disease.

The duties of a motor behaviorist at a university typically include research in the area of learning, control, or development; teaching; and service. Scholars may write grants to support research or programs related to practice. Teaching duties could include courses in motor behavior (learning, control, and development) or related courses such as biomechanics or sport psychology, research methods, measurement and evaluation, or pedagogy and youth sport. Service for a university faculty member might include evaluating motor disorders or managing a program for individuals with motor disorders and doing workshops or clinics about motor disorders.

The subdiscipline of motor behavior produces knowledge of how people achieve motor skills across the human life span. The subdiscipline involves three areas: motor learning, motor control, and motor development (a developmental view of motor learning and motor control). One goal of motor behavior is *to understand how motor skills are learned.* The second goal is *to understand how motor skills are controlled.* Motor learning research deals

Motor Learning Theory in Practice

Think back to a physical education class in which your teacher introduced a new skill. How did the teacher present the information—demonstration, instructions, an audiovisual aid—and how was practice organized? For example, did you practice a part of the skill and then receive more instruction? Who demonstrated? Compare this class to a class in which you and the other students were doing a well-learned skill. How did the practice differ in the two situations? For a new skill, teachers often explain how you will use the skill and then demonstrate the skill and present instructions so that you will understand what to do first. Next, you will practice the skill several times. After seeing some consistency in your performance, the teacher may present more instructions and demonstrations.

For a well-learned skill, the teacher may begin by having students use the skill in a game or game-like situation. Instruction, demonstration, and feedback are more likely to focus on strategy and less likely to include information and correction of the skill.

Contrasting Motor Control and Learning

Remember when you were learning to drive a car? Name one task that was difficult for you to master (e.g., using the clutch, slowing down to turn without stopping completely while signaling, parallel parking). Now think about the last time you drove somewhere familiar, for example, to school today. Sometimes we cannot remember anything about a familiar drive.

Notice how your skill has changed as a result of learning to drive and then continuing to practice. When you first learned to drive you had to think about every movement—moving your foot from the accelerator to the brake, turning the steering wheel, using the turn signal. Later you thought only about where you were going and the traffic. Now, on an often-traveled route, you only occasionally monitor the traffic. Motor learning helps us understand how we learn skills like driving a car so that the skill becomes this automatic. Can you identify other skills that you have learned so well that you don't think about the skill at all during or after the performance? In other words, driving your car on a familiar route is well learned and automatic. Can you think of another situation when you have to think about a well-learned skill? For example, when traffic chokes the road or a storm occurs, you must pay more attention to driving (not to mention driving to school while talking on your cell phone, fixing your hair, and eating breakfast). When task demands change, as in this example, we sometimes move from the automatic control of movements to conscious control of movements.

You can see that motor control and learning are related. Motor control is essential for every movement—from poorly skilled to well skilled. Motor learning is responsible for the shift from poorly skilled to highly skilled movements. One way to judge when a movement is well learned or highly skilled is to recognize that the movement is automatic, or executed without conscious control. We don't have as much practice driving in storms and heavy traffic, so that type of driving is not well learned and therefore not automatic.

with "the acquisition of skilled movements as a result of practice," whereas work in motor control seeks to "understand the neural, physical, and behavioral aspects of movement" (Schmidt 1988, p. 17); both are often evaluated across the life span, resulting in a developmental view. This brings us to the third goal of motor behavior, which is *to understand how the learning and control of motor skills change across the life span.*

Motor learning and motor control are not distinctly different areas, but they tend to ask slightly different questions. The goals of motor learning can be summarized as understanding the influence of feedback, practice, and individual differences, especially as they relate to the retention and transfer of motor skill. The goals of motor control are to understand how to coordinate the muscles and joints during movement, how to control a sequence of movements, and how to use environmental information to plan and adjust movements. In addition, scholars are often interested in exploring how motor learning and motor control vary with different age groups, from children to senior citizens.

When you think of motor behavior, you may think of sport skills. But consider the many other types of movements that people use in their daily activities:

- Babies learning to use a fork and spoon
- Dentists learning to control the drill while looking in a mirror
- Surgeons controlling a scalpel; microsurgeons using a laser while viewing a magnified TV picture of the brain
- Children learning to ride a bicycle or roller skate
- Students learning to use a computer
- Teenagers learning to drive
- Dancers as they perform carefully choreographed movements

- Pilots learning to control an airplane
- Young children learning to control a crayon when coloring

All these activities and many others involve motor behavior and are of interest to researchers and practitioners in this area. Thus, the goals of motor behavior are important not only in sport but also in the total field of physical activity. As we saw in part I, understanding the learning, control, and development of movements plays an essential role in our culture and society. For example, the first clue parents may have that their baby is developing normally may be a reflex or motor milestone, such as reaching. Individuals have probably been asking questions about motor behavior since the beginning of time; scientists have been seeking the answers to those questions for over a hundred years.

Goals of Motor Behavior

1. To understand how motor skills are learned
   a. To explain how processes such as feedback and practice improve learning and performance of motor skills
   b. To explain how response selection and response execution become more efficient and effective
2. To understand how motor skills are controlled
   a. To analyze how the mechanisms in response selection and response execution control the body's movement
   b. To explain how environmental and individual factors affect the mechanisms of response selection and response execution
3. To understand how the learning and control of motor skills change across the life span
   a. To explain how motor learning and control improve during childhood and adolescence
   b. To explain how motor learning and control deteriorate with aging

History of Motor Behavior

Research has a long history in all three areas of motor behavior (for reviews, see Thomas 2006 or Ulrich & Reeve 2005), but the reason for the research in each area has changed dramatically. For example, in the late 1800s and early 1900s, researchers were primarily interested in using motor skills as a means to understand the mind; motor skills were a tool to examine cognition (Abernethy & Sparrow 1992). Adams (1987) identified five themes from this early work that have persisted through the years: knowledge of results, distribution of practice, transfer of training, retention, and individual differences. These themes are the foundation of motor learning. Although the themes persist, the focus is now on the motor skills rather than simply using motor skills to understand cognition.

Similarly, motor control research can be traced to research in the late 1800s on the "spring-like" qualities of muscle (Blix 1892-1895), and Sherrington's (1906) seminal work on neural control is still useful in explaining how the nervous system controls muscles during movement. In both cases, the purpose was to understand not motor behavior but the biology.

The World War II era, from 1939 to 1945, was one of great interest in motor behavior research (Thomas 1997) because the military needed to select and train pilots (Adams 1987). Thus motor skills were viewed as a necessary component in military efforts, but not because understanding motor skills was an important area of study.

Beginning in the 1960s and increasingly in the early 1970s, motor behavior evolved as a scholarly subdiscipline of kinesiology. Scholars doing motor behavior research were no longer neurophysiologists or psychologists; they were specialists in physical activity. Franklin Henry's (Henry & Rogers 1960) paper on **memory drum theory** was the first major theoretical paper (landmark study; see Ulrich & Reeve 2005, pp. S63-S64) from the discipline of kinesiology (at the time called physical education). Henry's theory stated that reaction time was slower

Interviews With Practicing Professionals

Photo courtesy of Catherine Stevermer

Catherine Stevermer, PT
Physical therapist

Catherine Stevermer is a physical therapist who obtained her BS in physical education from Iowa State University and received her MS in physical therapy from the University of Iowa.

Q: How do physical therapists use their knowledge of the concepts of motor behavior to help patients?

A: Many aspects of rehabilitation involve matching an individual's abilities to movement characteristics and the environment where they will be performing movement. My undergraduate course work in motor behavior provided important background for concepts I use daily. As a physical therapist, I need to modify skill components to accommodate a patient's limitations for learning purposes, adjust treatment environments and techniques according to a patient's age, and design rehabilitation programs based on the neuromuscular system's control of movement.

Q: What are some examples of this?

A: In a hospital setting, physical therapists provide practice scenarios for patients to relearn movements (such as sit-to-stand or walking) so that patients can function in real-world settings. Initially, this requires teaching a skill such as crutch walking by breaking it into smaller parts. As patients use crutches proficiently, therapists add complex activities such as stair climbing or obstacle negotiation. Therapists evaluate whether patients are ready for more challenging tasks by determining where patients are in the learning process.

Another example is after a stroke, patients are relearning movement; attention deficits caused by neurological damage may be compounded by age-associated impairments (i.e., vision and hearing) that interfere with treatment sessions. To reduce this effect, therapists may use a distraction-free treatment area and focus on one skill at a time. Therapists often provide feedback after a skill is completed so that older adults are not dual-tasking in the initial stages of learning a skill.

Q: How do motor behavior concepts apply to physical therapy for athletes?

A: Physical therapists can use techniques based on neuromuscular concepts with athletes in a sports medicine setting. For example, it is possible to facilitate muscle activation through positioning, to use contract-relax techniques for more effective stretching, and to incorporate proprioception through weight-bearing activities. As patients gain neuromuscular control, therapists simulate the learning environment by incorporating agility activities to speed recovery.

My practice as a physical therapist has shown that the concepts from motor behavior (including motor learning, motor development, and motor control) are essential.

for complex movements because those movements took more planning time. For example, movements with several segments—involving moving from one position to a second position and then to a third position—required a longer reaction time than did single-segment movements because the brain required more time to specify the needed information for the more complex movement. The current work on motor programs (central representations in the brain of the plan for movements) evolved from Henry's memory drum theory.

➤ The subdiscipline continues to evolve because of research, technology, and differing theoretical models. Motor learning and motor control are distinct but are linked by a common goal—to understand human movement.

Current research in motor control and learning usually focuses on understanding how the neuromuscular system controls and repeats movements. One purpose of this research is to understand and potentially develop treatments for diseases and injuries such as Parkinson's disease and spinal cord injuries. Another is to improve performance in sport and physical activity.

A developmental approach to motor learning and motor control (also known as motor development) originated in developmental psychology and child development. Motor development grew from the "baby biographies," many done before 1900, that described the changes in reflexes and movements of infants. The early work used twins to establish the role of environment and heredity in shaping behavior (Galton 1876; Bayley 1935; Gesell 1928; McGraw 1935, 1939; Dennis 1938; Dennis & Dennis 1940). As with motor control and learning, the initial research in motor development did not take place because scholars were interested in motor skills. Instead, they were using motor skills to understand other areas of interest.

In the 1940s and 1950s, developmental psychologists lost interest in the developmental aspects of motor skill. If it had not been for three motor development scholars—Ruth Glassow, Larry Rarick, and Anna Espenschade—the area might have died out. The emphasis of their work was different from that of the developmental psychologists in that it focused on how children acquire skills—for example, how fundamental movement patterns are formed and how growth affects motor performance. These three scholars maintained the developmental nature of motor learning and motor control in their research through the 1950s and 1960s.

Just as research in motor learning and motor control increased around 1970, developmental research addressing the questions of motor learning and motor control also became popular during that period (Clark & Whitall 1989; Thomas & Thomas 1989). Motor development

Motor Expertise

Research on **motor expertise** often compares novices and experts to determine how they differ within various sports. Motor expertise is a recent and appropriate addition to motor development because age and experience are associated with increases in expertise. The interesting point occurs when a younger person is more expert than an older person is, when, for example, our grandchildren can fix the DVD player, hit the golf ball out of the sand trap, or ride a skateboard and we, in spite of our wisdom and age, cannot! By the way, your chapter authors play golf in the low 80s; if it's warmer than that, we won't play.

Research on motor expertise compares experts and novices. What differences do you see?

AP Photo/Lennox McLendon (left) and AP Photo/Steve Pope (right)

was considered part of the subdiscipline of motor behavior because the same topics were studied developmentally. But two research themes in motor development continued from the years before 1970—the influence of growth and maturation on motor performance and the developmental patterns of fundamental movements. Growth clearly influences the performance of motor skills, and some of the studies on growth relate growth to performance. Because growth cannot explain all the improvement in motor performance during childhood, developmental scientists have turned to motor control, motor learning, and biomechanics for more information. Thus, currently we see three lines of motor development research: the study of motor learning in children; the study of motor control in children; and the influence of growth on motor learning, control, and performance.

Research Methods in Motor Behavior

Research begins with a question that is stated as a hypothesis. Hypotheses are undoubtedly familiar to you. The question helps the scientist decide who will participate in the research project, what type of task they will do, and what the researcher will measure to answer the question. This is the same process used in most research in kinesiology. Understanding how movements are learned, how they are controlled, and how they change across the life span is complicated. In fact, to answer one question it is usually necessary to perform many experiments. Motor behavior researchers have concentrated on the use of techniques to measure movement speed and accuracy. Motor control and motor learning researchers use technology similar to that used by researchers in biomechanics. During the past 40 years, technology (e.g., computers, high-speed video, **electromyography [EMG]**) has been invented and adapted to control the testing situation and to record and analyze the movements. Technology has allowed the use of real-world movements instead of movement invented in the laboratory just for research purposes. Motor behavior courses often include laboratory experiences so that students like you can copy the experiments you read about. This type of hands-on learning has demonstrated benefits to your learning.

Types of Studies

There are many ways to answer any particular question in research studies. Three experimental designs, or techniques, are used frequently in motor behavior research. The between-group design compares two or more groups that were exposed to different treatments (interventions) but tests them using the same task. The researcher could use this design to answer the question "Does practicing a simple movement increase its speed?" The research design compares two groups (randomly formed) on the same task (movement speed). The treatment in this example is practice: If two groups practice differing amounts, will their performance on the task be different? In the second design, the within-group design, all participants are exposed to two or more different treatments and are tested on the same task.

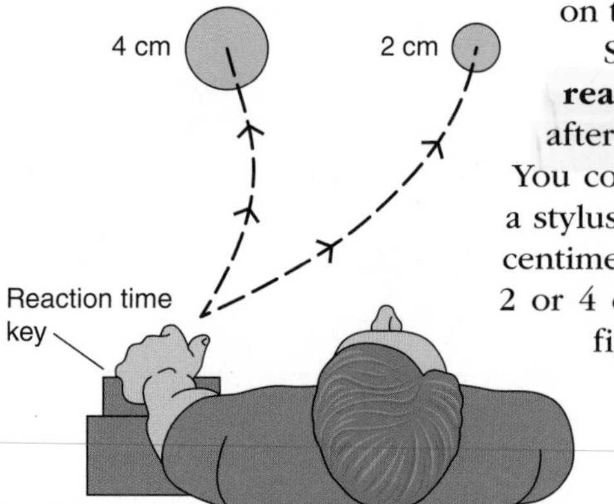

Suppose you want to study whether a participant's **reaction time**—how quickly the movement begins after a signal—varies because of the size of the target. You could ask the participant after you say "Go" to move a stylus (like a pencil) as rapidly as possible to a target 30 centimeters (11.8 inches) away. The target is a circle either 2 or 4 centimeters (0.78 or 1.57 inches) in diameter (see figure 8.2). Does the time between hearing the signal and beginning the movement (reaction time) change depending on target size? Using a between-group design, one group could move to the 2-centimeter target and the second to the 4-centimeter target. Using a single group (within-group design), half the

FIGURE 8.2 A reaction time study measures how fast a movement occurs after a signal and whether the size of the target object affects the movement.

Activity 8.1

Three Techniques

In this activity in your online study guide, you'll review three scenarios and identify the technique each describes.

participants could move to the 2-centimeter target first and then to the 4-centimeter target, and the other half could move to the 4-centimeter target first and then to the 2-centimeter target. In deciding which design to use, an investigator should consider whether changing target sizes interferes with performance. If so, then a between-group design is preferred; if not, the investigator will use a within-group design because he or she will not have to test as many participants.

The third type of experimental design used in motor behavior research is descriptive research. Here, the investigator measures or observes participants performing a task. Sometimes the investigator observes the same participants several times to trace changes. For example, the investigator may measure reaction time in the same children when they are 4, 6, and 8 years of age. In other cases, the performance of different groups is compared. For example, 4-, 6-, and 8-year-old children are tested for reaction time. Researchers often use this technique to describe age differences or differences between experts and novices. Descriptive research is different from research using the first two techniques because the participants receive no treatment.

Studying the Early Stages of Learning

Motor learning has used **novel learning tasks** to provide certainty that no participant has tried the task. Novel learning tasks were created for use in experiments to eliminate the advantage that some participants might have because they practiced the task before the experiment. Novel learning tasks tend to be simple—so simple that to make the task challenging, the participant is often asked to wear a blindfold (see figure 8.3).

FIGURE 8.3 This linear positioning task has been used to study early stages of learning because the task is novel. Learners must rely on what they remember about the movement (e.g., speed, distance, beginning and ending points) because vision is occluded. The task is both novel and simple so that the motor behaviorist can observe learning. The participant is not likely to have done the task before (i.e., it is novel) and will be able to master the task in a short time (it is simple).

You might question whether the results of these experiments can be applied to real-world, complex movement situations. You would have a good point! These simple tasks allow us to study improvement and have helped us understand a great deal about how movements are learned. Using simple tasks in research limits what the researcher can learn. One reason is that the outcome of the movement (product) rather than the nature of the movement (process) itself is studied (Christina 1989). Novel learning experiments are not helpful for researchers interested in physical activity or sport tasks in which performers have had thousands of trials (e.g., keyboard typing or baseball batting). Did you know that "typewriter" is the longest word you can type without changing the row of keys that you use?

Studying Expert Performers

To address the limitations of studying novel learning tasks, some scholars began to study experts. They asked the question, "What do expert performers do during practice and competition?" One way to answer the question was to compare expert performers to novice performers and evaluate how they differ on perceptual knowledge—particularly decision-making, skill, and game performance variables. In such studies, a criterion like national team membership or ranking in the sport establishes expertise.

The knowledge and skills of a sport are often unique to that sport (Thomas & Thomas 1994). The point is that knowing about and being skilled in tennis does not help one play soccer. Videotapes of badminton players have been used to determine how age and expertise influence how well players could predict where a birdie struck by an opponent would land. Participants looked at a video of a badminton player hitting a shot. The video had been altered (for example, the head, arm, or racket was erased, or certain motions were zeroed in on);

➤ Before the early 1980s, motor behavior research used simple, novel tasks to study early skill learning; this research helps us understand how beginners learn new motor skills.

sometimes information was erased that was needed to predict accurately where the birdie would land. As you might guess, expert players could make better, faster predictions, with less information, about where the birdie would land than the novice players could. Experts also looked at different body parts than novices did and therefore based their decisions on different sources of information than novices did. The advantage seemed to be a result of playing experience rather than age. You can see how this finding could help a coach, teacher, or player improve badminton play. The information also helps us understand about learning many skills, because researchers have found similar patterns in many sport and motor skills. This type of experiment would be nearly impossible to do in a motor learning class because the students might not represent the range of expertise necessary (novice to expert).

Measuring Movements

The tasks used in motor behavior research provide a number of ways to measure movements and their outcomes. For example, suppose that senior citizens perform a movement that involves reaching 30 centimeters (11.8 inches) and then grasping and lifting containers of different sizes and weights. We could simply count the number of times each senior citizen successfully lifted the containers (reaching and grasping are prerequisites to lifting in this task). The number of successful lifts is an outcome measurement. We might find that as objects were smaller or heavier, the number of successful lifts decreased. While that is an interesting finding, our long-term goal is probably to understand why or how changing the size or weight of the object influences success. So we probably need to examine the process of the movement. To examine process, we could use a high-speed video of the movement, taken with two cameras, and evaluate how this movement differs for 55-, 65-, and 75-year-old participants. From such videos, researchers can measure **kinematics** (location, velocity, acceleration—see chapter 10 on biomechanics). Other measures include muscle activity (EMG) of the arm and hand during the movement, the pinching force between the forefinger and thumb during the grasp, error in the movement, speed of the movement, reaction time before the movement begins, and accuracy of the movement. These process measures might help us understand how the reach, grasp, or lift changed for the various weights and sizes of objects. Using our hypotheses (educated guesses) we could begin to answer our original question. For example, we might hypothesize that the reaching and grasping phase would be the same for objects of the same size, even when weight varied, but that the lift would be different when the weight changed. Do you think reach, grasp, or lift would be different as we changed size but kept the weight the same? If we predict that reach or grasp changes, it is critical to measure the process.

Characteristics of Movement Tasks

In addition to deciding what to measure, motor behavior scientists must consider the characteristics of the movements they study (Gentile 1972). For example, some movements are more continuous (as when people perform a gymnastics routine or ride a bicycle), and some are more discrete, involving a short period with a distinct beginning and end (as when people strike a ball with a bat). Some movements are more open in character, whereas others are more closed. An example of an open movement is hitting a thrown ball—the environmental characteristics change from trial to trial because the batter must respond to the speed and location of the oncoming ball. A closed movement takes place in a more consistent environment, in which the performer tries to do the same thing each time (e.g., in archery or bowling).

Activity 8.2

What Kind of Movement?
In this activity in your online study guide, you'll review video clips of athletes in action, then identify the type of movement demonstrated.

Expertise Is More Than Good Stats

Suppose you wanted to research the talent on a women's softball team. You have available standard data that are kept on players—batting averages, RBIs, home runs, and fielding percentages. But you think these data don't always reflect the contributions that players make to the team. You have videotapes of five games that the women have played, taken with a wide-angle camera that shows the complete field. Identify the kind of data can you get from the videotapes to answer the following questions:

- How many times did each player hit the ball hard while at bat?
- How many times did each player make a special effort that doesn't show up in the scorebook (e.g., backed up a throw, saved a bad throw, moved up a runner)?

Researchers and teachers must identify the characteristics (open or closed, discrete or continuous) of the tasks they study for two reasons. First, the results of two studies may be different if the task characteristics are different. Second, the characteristics of the task used in the study must match the questions asked. For example, if you want to know why something happens in bowling, you would not want to use an open continuous task to answer the question. Teachers also must know how to draw these distinctions between task characteristics because skills of one type may require a different instructional strategy than skills of another type.

Measuring Learning and Transfer

The idea of learning—as determined by retention and transfer—is similar to the idea in the previous discussion about experts. That is, a motor skill can be examined at any point from the first attempt to well beyond mastery. The goal of most practice is learning, and learning is defined as retention and transfer. An example that has meaning for most students is related to the three most threatening words in college—"comprehensive final exam." At the end of the semester, how much do you remember when taking a comprehensive final exam? The purpose of the exam is to determine whether you learned what was taught in the course. For those who have really learned something, a comprehensive final exam is not a problem. As you will see in your motor learning class, retention is also used to measure learning in motor skills. Another way to measure learning is with transfer, in which you must do a slightly different version of the task. Experts are better at transferring information or skills than novices, and experts can perform well after periods without practice (called retention intervals). Many coaches are frustrated when the team "learns" a play during practice but has trouble executing during the game. This situation is an example of transfer, and it suggests that the team did not really learn the play. A great example of a task that is well learned based on both retention and transfer is riding a bike. You can ride successfully after a winter off, and you can negotiate new routes by transferring the skill.

We know from motor learning research what the reason is: Once you learned to ride a bike, you continued to practice until the skill was automatic and "overlearned." Comprehensive final exams are designed to get students to this same point. By now you should also understand that retention and transfer are the goals of most learning. So the variables that affect retention and transfer are important to motor learning. What is confusing is that everything we see is performance, but *some* of what we see is also learning. Performance variables can also be interesting to motor learning students and scholars, as you will see later in this chapter. Performance variables are things that influence performance but do not influence learning.

➤ To study motor skill acquisition, researchers must also study how well skills are retained and how they transfer to other, similar situations.

Overview of Knowledge in Motor Behavior

Now let's take a look at what motor behavior researchers have learned from years of study. Our goal is not to cover all of motor behavior, but to present six major principles that are well supported in the motor behavior literature. These principles are based upon research using the measurement and methods outlined in the previous section. These speak to the goals of motor behavior presented previously. In motor control, development and learning research and theory were formed around a concept of how the brain and central nervous system work to create movements. So before presenting the principles, we will briefly describe that concept.

To help scientists conceptualize the brain as the master controller in planning, organizing, selecting, and controlling movements, a model called information processing was adopted. This model represents the motor behavior system as a computer. General commands are sent from the brain (the central processing unit or CPU) through the spinal cord (wiring), which probably reduces the complexity of the information into relatively simple commands, to the muscles or muscle groups (printer, screen). (If your brain were really like a computer, when you accidentally touched a hot burner on a stove, a message would pop up asking "Are you sure?" before you could pull your hand away.) From this perspective, the goal of motor behavior is to explain response selection (how the skill to be used is selected) and response execution (how the skill that is selected is performed).

Motor Learning

➤ Motor learning is an internal state that is relatively permanent; practice is required in order for it to occur, and it is difficult to observe and measure.

The two principles we have selected to represent motor learning are that (1) correct practice improves performance and supports learning and (2) augmented feedback enhances practice and thereby learning. These principles were selected for several reasons. Numerous motor learning textbooks include, at a minimum, a chapter on each (Fischman 2007) of these variables. Further, these are likely the two most widely studied variables in motor learning, and scholars agree (Schmidt 1988) that practice and feedback are the two most important variables in motor learning. Two of our favorite historical quotations are "Practice the results of which are known makes perfect" and "Practice is a necessary but not sufficient condition for learning." Both of these quotes suggest that practice and feedback are independently important variables, and the relationship between them is also important.

Consider why these variables are important for you. Assuming that there is transfer from one learning situation to another, for example from learning a motor skill to learning the content in anatomy, understanding how to practice for learning could be helpful. As you move toward a career, you are going to have to help someone learn something.

The brain helps you plan, organize, select, and control movements when you're performing a skill.

Eyewire

While you may not be a teacher, understanding how to help someone learn will be to your advantage. For example, you may need to teach an employee how to do his or her job, want to assist your own child as he or she learns a new skill, have to lead an exercise class, or want to learn a new sport.

The skill acquisition process is an orderly progression. The learner begins by making many large errors while trying to understand the task. Early in learning, the cognitive demands are great; in fact, the task may be more cognitive than motor (Adams 1987; Fitts & Posner 1967; Gentile 1972). With practice the errors become more consistent; rather than making a different error on each trial, performers make the same errors repeatedly. At this point the demands are less cognitive and more motor, and the errors are smaller and less frequent—response execution is improving. When the learner can execute the skill with fewer and smaller errors and doesn't have to think about the skill while performing it, the skill is considered learned or automatic. Now the performer can think about the opponent or strategy when deciding which response to select and execute, instead of what each body part is doing (response execution).

Recall that motor learning is an effort to explain and predict conditions that will make skill acquisition easier or faster as well as make learning more permanent. Such conditions include individual differences in the learner like speed of movement and coordination. Task differences are also important conditions in skill acquisition, because tasks may be more open, as in batting, or closed, as in bowling. Environmental conditions that may affect learning include practice, **feedback (intrinsic** and **extrinsic),** and transfer. The idea of automatic or learned skill is especially important in the study of experts. Research suggests that expertise begins to emerge at 10,000 hours of practice (Ericsson 2003). One could guess you are an expert student, and that you will begin to become an expert at your job after about five years of full-time work!

At this point you might wonder how we know when something is learned. Clearly that is a challenge for motor learning research, theory, and application.

Although learning reflects the successful acquisition of a skill, performance reflects the degree to which someone can demonstrate that skill at any given time. Performance is the current observable behavior—in other words, what the learner is doing right now. Sometimes performance reflects learning, as when a player is able to demonstrate successfully his or her newly acquired skill, but at other times it does not. For instance, most of us have taken an exam and after turning it in were able to remember an answer that we knew but could not put on the paper during the exam. We would argue that we had learned the material but just could not produce it for the exam! In this case we are saying that performance does not represent learning.

One way to distinguish between performance and learning variables is to remember that performance variables have a temporary effect whereas learning variables have a relatively permanent effect. Knowing the difference between these effects is a critical part of motor learning. You may have trouble typing a term paper after you have been working and typing late into the night and are tired. Yet the next day, after some rest, you may type rapidly while making few errors. Fatigue depressed your performance, but you had learned the keyboard through practice and could demonstrate this learning later when you were rested.

Because performance does not always reflect learning, researchers prefer to measure learning with retention and transfer tests (Christina 1992; Magill & Hall 1990). In sport, retention refers to how much of a skill a performer can demonstrate after a period of no practice. Examples of retention include being able to swim at the beginning of the summer after learning the previous summer or typing after a vacation without typing. Transfer tests require that you use the information in a slightly different way from how the skill was originally practiced. Using a principle from physics to solve a biomechanical problem is an example. You use transfer in several ways—for example, you transfer skills that you learn in practice to a game, or you use your experience throwing a baseball when you throw a javelin for the first time.

➤ Because variables such as fatigue can influence motor performance, tests of retention and transfer provide the best measure of motor learning.

Practice

Practice is repeating, but of course there is more to understanding why practice is such an important factor beyond just repeating. More practice is usually associated with better performance and is a requirement for learning. However, practice must be organized differently depending on whether the objective is performance (short term) or learning (long term). Has a teacher, coach, or someone else ever said "Oh come on, it's easy" when you were struggling to learn a new skill? Hearing that remark can be maddening, because performing the skill is not easy. Practice that leads to learning is hard work. Good teachers and coaches understand this principle of motor learning. In your motor learning class you will study practice in two phases, comprising before-practice variables and during-practice variables, specifically:

Activity 8.3

Planning Practice
In this activity, you'll determine what the type and order of practice variables in two different scenarios should be.

Before Practice

- Goal setting (Locke & Latham 1985)
- Instructions
- Demonstrations or modeling (McCullagh 1993; Weiss & Klint 1987)
- Mental practice

During Practice

- Scheduling of practice
- Context of practice (Chamberlin & Lee 1993; Schmidt 1991)

Once you have successfully completed a motor learning course, you will understand the prepractice variables and how to schedule and organize practice. Some of these variables are presented in the example of practice organization. For example, constant practice involves doing the same thing over and over, and variable practice involves not doing the same thing on consecutive practice trials. The first step is to consider the learner and the skill, and the next step is to plan the type of practice.

Feedback—Knowledge of Results and Performance

> Knowledge of performance is given by an instructor who provides feedback about the nature or process of the movement. Knowledge of results is information about the outcome of the movements that a teacher, coach, or trainer may supply.

As you can see, feedback is an integral part of the practice regimen. Learning a skill correctly cannot occur without feedback. Feedback serves to guide the learner toward performing the task correctly and also reinforces correct performance. Feedback can be intrinsic or extrinsic (also called augmented). Intrinsic feedback is information about performance that you obtain for yourself as a result of the movement. Extrinsic feedback is information provided by an outside source such as a teacher or videotape.

Besides distinguishing between intrinsic and extrinsic feedback, motor behaviorists also categorize extrinsic feedback as knowledge of results (KR) and knowledge of performance (KP). Figure 8.4 presents examples of intrinsic and extrinsic KR and KP. Knowledge of results is information about the result (you missed the target); it helps advanced performers more than it does beginners because advanced performers understand how to make corrections based on KR. That is, they are more likely to understand how to change the movement to influence the outcome. A good typist who is told that he keyed in "Hybe" instead of "June" would know that the right hand was one key away from home. Because he is an advanced typist, he is able to use KR feedback. A beginner typist, however, might need a cue ("Check

KR and KP
Test your understanding of the two types of knowledge in this activity in the online study guide.

Example of Practice Organization

Your sister has asked you to teach your 7-year-old nephew to swim. He remembers choking after inhaling water when he was younger and has avoided swimming lessons since then. Fortunately, he looks up to you and is willing to give swimming a try. Unfortunately, you have not taught swimming before. So you face an important challenge and undoubtedly realize that to be successful you will need to do some planning.

At this point you know that practice and feedback are important to learning. We can consider planning practice in three parts: prepractice, practice, and postpractice. During practice you also need to think about feedback. We know that our participant is a novice. What kind of skill is swimming? "Continuous" and "relatively closed" are good descriptors. As you plan the first practice session, consider what the learner is thinking, what you—the teacher—are thinking, and what you should say. The following table demonstrates this process.

	The nephew is saying:	The teacher thinks:	The teacher says:
Prepractice (goal setting, simple instructions, modeling)	"I am afraid. This is going to be hard, I will probably fail."	"I better set a goal, demonstrate, and keep my instructions simple."	"It is OK to be afraid; you are smart to be afraid. The only way not to be afraid is to learn to swim. So today I want you to see how the water holds me up." (The teacher demonstrates a float.) "To float, you must put your face in the water. Once you get your face wet, then we will work until you keep your face in the water for 5 seconds."
Practice (constant practice, contextual practice, blocked practice, extrinsic feedback)		"So far, so good. He is not choking, and his face is in the water. Short practice trials with rest in between are working. I should show him what I want him to do. I can give him feedback when he comes up for air."	"That was great—you had your face in the water for 5 seconds. I am proud of you. Next time I want you to get your ears wet too, so put your face in far enough to get your ears wet. I know you can do it! Let me show you what I want." (The teacher demonstrates.)
Postpractice (verbal rehearsal, goal setting)	"That wasn't so bad. I can put my face in the water for 5 seconds, and I didn't choke. I could feel the water in my ears."	"Next I need to increase the difficulty, but I must also let him rest so that he doesn't get fatigued and make a mistake."	"You are making progress. Now we will practice walking in the water, then we'll try putting your face in again. Can you tell me what you've accomplished so far?"
	"I put my face and ears in the water."		"Good, you know it took a lot of practice before I learned to swim. If we keep working you will learn to swim just like I did."

(continued)

(continued)

A few days later the practice continues, as you can see in this table.

	The nephew is saying:	The teacher thinks:	The teacher says:
Prepractice (goal setting, demonstration, verbal rehearsal)	"So far this is easy. I am pretty good at this. I can put my face in the water, walk in the water, and float."	"He puts his face in the water, walks in the water, and can do each of the parts (float, kick, and arms). I think he is ready to begin to swim. So, I want to provide verbal cues that he can rehearse and use for the progressive part of practice."	"Today you are ready for something new. You can count to three—and there are only three things you need to remember and do to swim: (1) Put your face in the water and float, (2) kick your feet, and (3) move your arms. Let me show you while you count for me." (The teacher demonstrates.) "What do 1, 2, and 3 stand for? Now it's your turn."
Practice (blocked practice, extrinsic feedback after several trials)	"OK, 1 is face in to float, 2 is kick, 3 is arms."	"He did it! Now we need to repeat this several times with just a little rest between." "Should I change the practice? Let's see, this is a closed continuous skill, so constant practice is OK. Maybe we should practice in a different pool or at least a different place in this pool next time."	"Good job. You did all three. Now do it again." (Repeating several times.) "You are going about 10 feet before you stand up. That is good. You are working very hard. Let's see if you can go a bit farther. From where you are to where I am standing is about 12 feet. Try to get to me. I am sure you can do it!"
Postpractice (mental and verbal rehearsal, extrinsic feedback)	"I am totally awesome. 1, face in and float; 2, kick; 3, arms."	"He is swimming. We have made a lot of progress, but I am worried that the progress will slow down. I need to keep him motivated."	"You reached the goal—good work! Now say the three steps for me. Good. Can you think of the way it feels as you say the steps? Think about this and say the steps before next time."

DO it Activity 8.5

Learning and Performance

Differentiate among practice variables and identify appropriate practice settings in this activity in your online study guide.

your home position") to correct the error. This typist is using KP feedback. Sometimes we tell a beginning driver to "slow down" or "slow up"; why do those phrases mean the same thing? The goal of feedback is ultimately to help performers detect and correct their own errors. The study of feedback has focused on KP and KR with regard to frequency, precision, modality, and processing time. Feedback may also be reinforcing ("That was a good effort") or negative ("You didn't try very hard"). However, KR and KP are the forms of feedback discussed here.

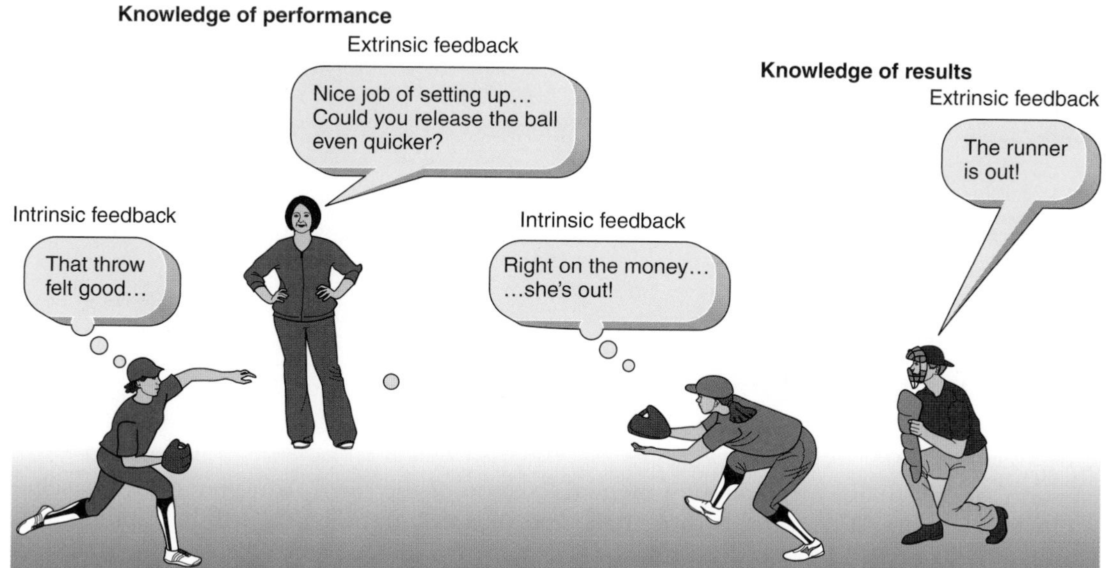

FIGURE 8.4 Four kinds of feedback for a shortstop.

Frequency of feedback refers to the percentage of the practice trials on which feedback is given (e.g., 50% or every other trial vs. 100% or all trials). On the surface, more would seem better, but the performer can become dependent when feedback is too frequent. If the goal is to detect and correct your own errors, then you do not learn the detection and correction processes when you receive feedback all the time. Knowledge of performance (or KP) should be given more often at the beginning of learning and then gradually reduced (Salmoni, Schmidt, & Walter 1984), a process often referred to as fading the feedback. Besides considering KP versus KR and the frequency of feedback, two other variables you will find out about in motor learning are precision and modality of feedback. Motor development considers the amount of time that children need to use the feedback, so you will likely learn more about this as well.

➤ Learning requires both practice and feedback. Extrinsic feedback (e.g., KR or KP) must be information that the learner could not obtain on his or her own, should be corrective, should be provided on about half of the trials, and must be followed by sufficient time to make corrections before the next attempt.

In baseball, the batter watches the pitcher for critical information that may give a clue about the pitch; the batter ignores the crowd's yelling and gathers only related advance information about the pitch and where to hit it. Figure 8.5 provides an example of how long the batter has to make decisions about swinging the bat after the pitcher releases the ball. The figure demonstrates that a batter has more decision time (150 milliseconds vs. 130 milliseconds) if he waits longer and speeds up the swing (140 milliseconds vs. 160 milliseconds). The example also demonstrates how critical advance information is to successful performance. Thus, it becomes obvious why batters want to know in advance whether a pitcher is likely to throw

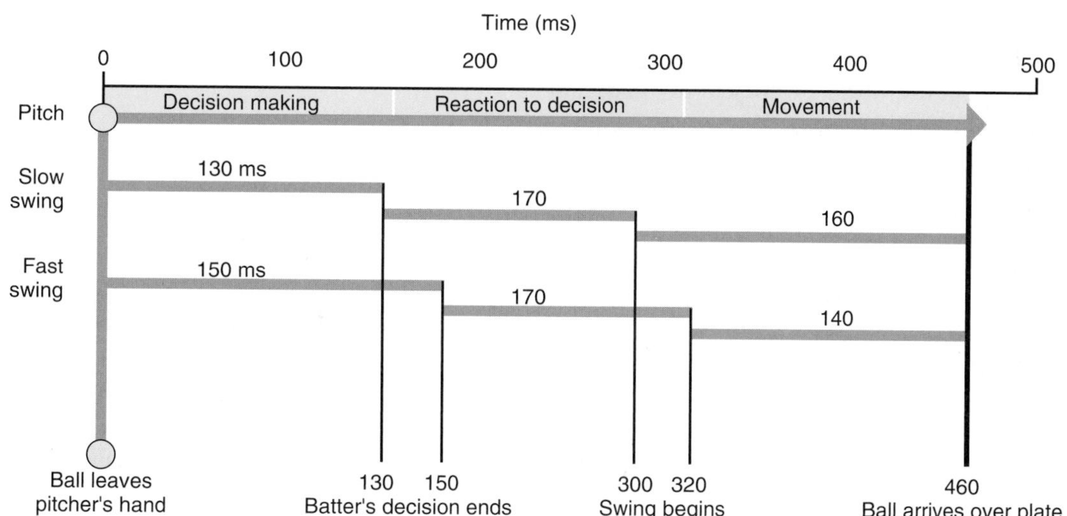

FIGURE 8.5 Time line showing the critical events in hitting a pitched baseball. The movement time is 160 milliseconds for the slow swing and 140 milliseconds for the fast swing.

Reprinted by permission from R.A. Schmidt and C.A. Wrisberg 2007.

Feedback in Golf

Here's a fun way to test yourself on how well you can identify types of feedback. Read the joke and try to categorize the caddy's feedback.

Golfer: "Do you think my game is improving?"
Caddy: "Yes, sir, you miss the ball much closer now."
Golfer: "I think I'm going to drown myself in the lake."
Caddy: "Do you think you can keep your head down that long?"
Golfer: "You've got to be the worst caddy in the world."
Caddy: "I don't think so, sir. That would be too much of a coincidence."

a curve or a fastball as well as the potential location of the pitch. You may hear a baseball commentator say, "The count is three balls and one strike; the batter will be looking for a fastball high and inside" (a certain pitch in a specific location). The batter is trying to reduce the decision to a go or no-go situation—he'll swing if the pitch is what he expects—rather than consider all possible options. The batter can do this because in this situation he can decide not to swing even if the pitch is a strike. If the count were three balls and two strikes, the batter would be less likely to look for a specific pitch.

Motor Control

➤ Motor programs are a proposed memory mechanism that allows movements to be controlled. As motor programs are developed, they become more automatic, allowing the performer to concentrate on the use of the movement in performance situations.

The first principle of motor control is that the brain uses the central nervous system to initiate and control muscles that make the movements. The second principle of motor control is that a goal of most movements is to rely on the decision-making centers in the brain as little as possible once the movement is initiated. There are two theories, and many research studies, that led us to select these two principles. First, motor programs (Schmidt 1975) are the theoretical explanation of how we successfully produce and control movements. We can compare a motor program to a computer program that does math problems. First, you select the program to use; this is response selection. The program can add, subtract, multiply, and divide. If you put in the numbers and indicate what math operations to perform, the program outputs the answer; this is response execution. A motor program is like that; you select the program (response selection) and indicate what it should do (operations). Then the program specifies how to do the skill and sends signals through the spinal cord to the muscles that perform the movement (response execution). Schmidt (1991) indicates that motor programs, at a minimum, must do the following five things:

- Specify the muscles involved in the action
- Select the order of muscle involvement
- Determine the forces of muscle contraction
- Specify the relative timing and sequences of contractions
- Determine the duration of contractions

If we had to remember every single movement we have made, our memory would be overloaded. Motor programs explain why we do not have a storage problem. Instead of storing in memory each movement that we have ever done, we store groups of movements with similar characteristics. These groups, called schema, are the foundation of motor program theory (Schmidt 1975).

Not all researchers agree with the idea of motor programs from schema theory, so they have proposed another theory called dynamical systems (Haken, Kelso, & Bunz 1985; Kelso 1995; Fischman 2007), which suggests a more direct and less cognitive link between information that the perceptual system picks up and the motor action that occurs. This direct link is called a coordinated structure. One characteristic of a coordinated structure is automated movement that relies on very little decision making or central control in the brain.

You may be wondering which theory—motor program or dynamical systems—more accurately represents the process of motor control. A decisive test for theories of motor control is how well they explain motor learning and the development of motor expertise (Abernethy, Thomas, & Thomas 1993). You will have a chance to examine and compare these theories in your motor learning class.

➤ Future research that contrasts predictions and key elements from a dynamical systems view of motor behavior or a motor program view is likely to be both exciting and controversial.

The study of motor control addresses five areas:

1. Degrees of freedom—coordination (Rosenbaum 1991)
2. Motor equivalency
3. Serial order of movements—coarticulation (Rosenbaum 1991)
4. Perceptual integration during movement (Rosenbaum 1991)
5. Skill acquisition (Rosenbaum 1991)

The brain initiates the planning of movements, and the nervous system then sends signals through the spinal cord to the muscles, which in turn make the movements. What is not known is how the brain represents the information to be sent to the muscles. The area of motor control is an effort to understand what the brain, nervous system, and muscles are doing to direct movements. Although skill acquisition—the notions of learning and improvement—is the focus of motor learning, it is critical in motor control as well. Skill acquisition accentuates the relationship between motor learning and motor control.

Once again, you may be wondering why understanding motor control is important. For some, the answer is obvious; it would be, for example, if you were interested in physical therapy, rehabilitative medicine, sports medicine, or athletic training. Anyone who wants to learn or teach a skill uses the principles of motor control by trying to use the simplest movement possible at the early stages of learning. Some of what you will learn about motor control is less intuitive and often misunderstood. For example, consider the baseball batting example from figure 8.5; here motor control tells us that the most rapid adjustments are made when we are not actually thinking about the flight of the pitch or the swing of the bat. That is, we turn the control of the skill over to a more automatic part of the system. The two theories of motor control explain this equally well. However, the idea of automatic responses is not well understood by many sport enthusiasts. When you hear coaches or parents say "Watch the ball all the way to the bat," you know they do not understand how to best use the sensory system to make rapid and accurate adjustments. We hear comments that allude to this shift from central (decision-making) control of movements to peripheral control when athletes say "It just felt right."

DO it Activity 8.6

Assess Your Expertise

In this activity in your online study guide, you'll reflect on your own physical activity experiences.

Understanding motor control is important because the range of performance of motor skills from normal to expert depends upon the shift from learning (central control, practice and feedback) to more automatic movements that are programmed ahead of the movement and controlled peripherally.

Developmental Motor Learning and Control

The two principles of motor development (developmental learning and control) are that (1) children are not miniature adults and (2) children are more alike than different (Thomas & Thomas 2008). After watching a baby and a child perform a motor task, you can see clear differences and similarities related to age. Infants demonstrate more random movements, have a smaller repertoire of movements, and tend to do the same movement repeatedly. Children have a larger repertoire of motor skills, do more voluntary skills, and exhibit greater skill than infants (Seefeldt & Haubenstricker 1982), although less than adults. Children are not miniature adults; their movements are not scaled-down versions of adult movements. One explanation for the relative inefficiency of children's movements may be their lack of practice. In addition, adults probably select and plan movements, a cognitive process, better than children do.

➤ If you are planning to study physical activity, you must understand the physiology, biomechanics, and motor behavior underlying the development of movement to address problems such as how children gain control of movement skills, how people achieve expertise, and how these skills deteriorate as people age.

Studying children is valuable because children typically change in an orderly fashion so that the order of change is the same across children while the rate varies. Hence the second principle, that children are alike, refers to gender, culture, and developmental age. Understanding how motor skills develop and the factors influencing development allows us to make predictions. Those predictions are important to pediatricians, teachers, and parents. Virtually all the questions in motor learning and motor control can be studied as part of motor development—by examining the same questions across age. In addition, some unique problems arise in studying infants, children, and persons who are elderly. Children differ from adults in several ways: physical growth, information processing, experience, and neurological development (Thomas, Gallagher, & Thomas 2001). Although some topics are unique to developmental motor learning and control, considerable overlap occurs with the previous two goals of motor behavior—motor learning and motor control (Ulrich 2007). In some programs, motor development is part of the motor learning and control course, and in other universities the class is separate. Motor development includes several topics that are not part of motor learning, for example physical growth and youth sport.

Developmental Changes in the Mechanics of Movement

The mechanics of movements are different at different ages partly because the body executing the skill is of a different size and has different proportions. For example, a baby's head is one-quarter of her height, whereas an adult's head is only one-eighth of her height. The legs also change dramatically during childhood. At birth the legs are typically less than 30% of body length; adult leg length is often over half of body length. Obviously, these physical characteristics will influence balance and locomotion.

Growth is also a factor in motor behavior; children are growing until around 13 to 14 years of age for females and around 18 to 20 years of age for males, whereas growth is not a factor in adult motor learning studies. Children grow in three physical dimensions: overall size, proportions (e.g., leg length, shoulder breadth, chest depth), and body composition (e.g., increases in muscle). Growth influences motor performance, in part, because children must contend with changes in their bodies, an issue that adults do not have to deal with. Further, the adult–child differences can be factors explaining child-to-child differences. For example, size and strength have a positive relationship: Larger children are stronger. As children grow, the increase in size produces increases in strength. But strength also increases because of neuromuscular efficiency. Because growth is not under our control, we must understand how growth influences performance to accommodate the challenges of physical growth.

Life Span Development

Besides addressing the effects of growth on motor performance, developmental motor learning and control also deals with the effects of aging on motor performance; thus, this area of study is sometimes referred to as life span development (Spirduso & MacRae 1990; Stelmach & Nahom 1992). Changes in strength and motor coordination are most rapid at the extremes of the age continuum. Growth is less important after adolescence, because the changes in physical parameters are less dramatic in adults and elderly people than in infants and young children.

Cognitive processes also change during childhood because of the developing nervous system and experience. Children use fewer cognitive processes and use them less effectively than adults do. For example, when asked to remember a movement, a 5-year-old child might "put on his thinking cap"—not a very effective strategy for remembering. A 7-year-old might do the movement repeatedly, whereas a 12-year-old could repeat the movement in a series composed of several movements. Adults and older children know that they must do something to learn and remember, whereas younger children may not recognize the importance of deliberate practice or may select an ineffective practice regimen. One of the reasons for the increasing interest in cognitive processing during motor skill acquisition in children is that the strategies used by adults can be given to children to improve performance. In other words, this is an area in which research enhances teaching.

Experience

Experience was presented as a factor in children's performance. Experience can help elderly people and can be a negative factor for children. In a situation in which experience is a benefit, those who are elderly have an advantage and children have a disadvantage because children lack experience and elderly people have more experience. Clearly, we can enhance experiences for children in two ways—quality and quantity. Research shows that children with experience can outperform adults with less experience, which means that practice and experience help. This finding has clear implications for pedagogy and curriculum development. In other words, curriculum and instruction need to provide experience to enhance sport performance in children. Further, we know that the quality of the experience can be improved too through use of what we know about information processing to help children get more out of the practice. So, if we provide children with adult learning strategies, they will benefit more from a given amount of experience. The key is that children need practice, and that to be effective, practice must help children retain information, skill, and decision making, which they would normally lose. To do this, we teach adult learning strategies to children.

DO it **Activity 8.7**

Younger vs. Older Players
Test your understanding of developmental motor learning and control in this activity in your online study guide.

Changing Neuromuscular Systems

The hardware (analogous to the computer itself and its disk drives and monitor), or neurology, of children is different from the hardware of adults. Infants have fewer synapses (connections between neurons in the nervous system) and less myelin (the covering of nerve cells with a fatty sheath that aids nerve impulse transmission). In addition, babies at birth have fewer neurons than adults do. The hardware is nearly complete as children begin school, but it may not be functioning as well as it does in older children and adults.

➤One of the most intriguing developmental questions in motor behavior is, How do the brain and nervous system adjust their control to increases in cognitive function, body size, and strength during childhood and to decreases in those variables as people age?

The deteriorating nervous system in persons who are elderly may negatively affect performance. As people age, they lose neurons in the brain and motor neurons in muscles. This change results in slowness and variability in movement control. Older people, however, benefit from experience and use it to compensate for the loss of speed, strength, and control. Understanding the changes in the central nervous system is important because we have little opportunity to accelerate or decelerate these changes.

Motor control issues are important in understanding adult motor behavior, and they are important in developmental learning and control. Critical issues are similar to those listed in the previous section on motor control (for a review, see Clark & Phillips 1991). One additional point is important: How does growth become integrated into the motor control system? As previously mentioned, children may change very rapidly, particularly at puberty (Malina 1984; Malina, Bouchard, & Bar-Or 2004). How does the motor control system account for changes in size, proportion, and mass? Consider a boy who played baseball from March until July at 12 years of age. By the next March he is 13 years of age, he has grown 4 inches (10.16 centimeters) in height, his arms have grown in length, and he weighs more. He has not practiced batting or throwing much since the last summer, yet he can still bat and throw successfully. How does the nervous system compensate for these changes to produce coordinated movements?

At the other end of the age continuum are persons who are elderly. How does a system of motor control that has functioned for years using one set of parameters account for loss in cognitive function, body mass, flexibility, and strength? No good explanation has been found for these important issues from the developmental aspects of motor learning and control, although some progress is being made (for reviews, see Thelen, Ulrich, & Jensen 1990; Thomas, Gallagher, & Thomas 2001).

Growth and Gender in the Development of Overhand Throwing

We have all seen examples of people of different ages and genders showing markedly different physical abilities. Can we document those differences? What are the reasons for them? Here we will look at what researchers have learned about overhand throwing. First, we need to examine physical growth and development. See figure 8.6 for a summary of growth in terms of stature and weight, and figure 8.7 for proportion. Then we will see how these factors affect overhand throwing skills.

Physical growth is rapid during infancy, constant during childhood, and rapid during the growth spurt associated with puberty (Malina 1984; Malina, Bouchard, & Bar-Or 2004). Children's limbs grow more rapidly than the torso or head does. The increase in proportion of total height attributed to leg length explains some of the improvement in running and jumping performance during childhood. The shoulders and hips become wider, with males' shoulders becoming much broader on average than females' shoulders, giving males an advantage in activities in which shoulder power is helpful, such as throwing. Growth is orderly, with length and then breadth and circumferences increasing. Children in upper elementary school grades often look awkward—their legs and arms appear to be long and skinny. Figure 8.7 shows the change in proportion from birth though adulthood. After most of the length of a limb has been attained, the limb gains thickness. The bone circumference and muscles (and fat!) increase to make the body look more proportional; children fill out. Men's shoulders and chests usually grow (bones and muscle) until about 30 years of age, whereas most women have stopped growing by their late teens.

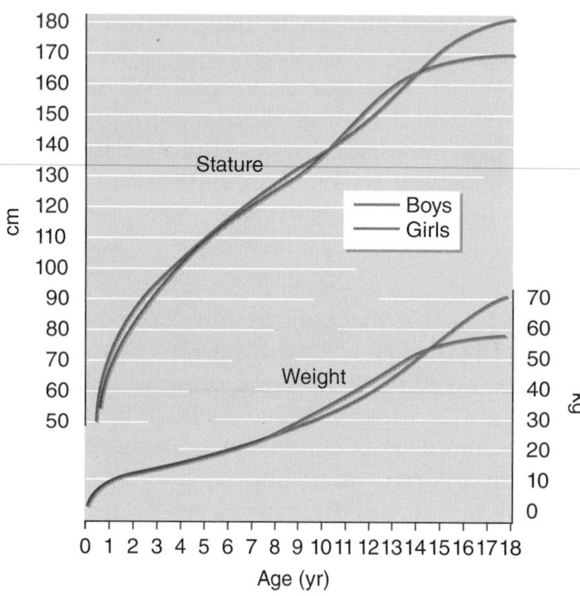

FIGURE 8.6 Average height and weight curves for American boys and girls.

Adapted from R.M. Malina 1984.

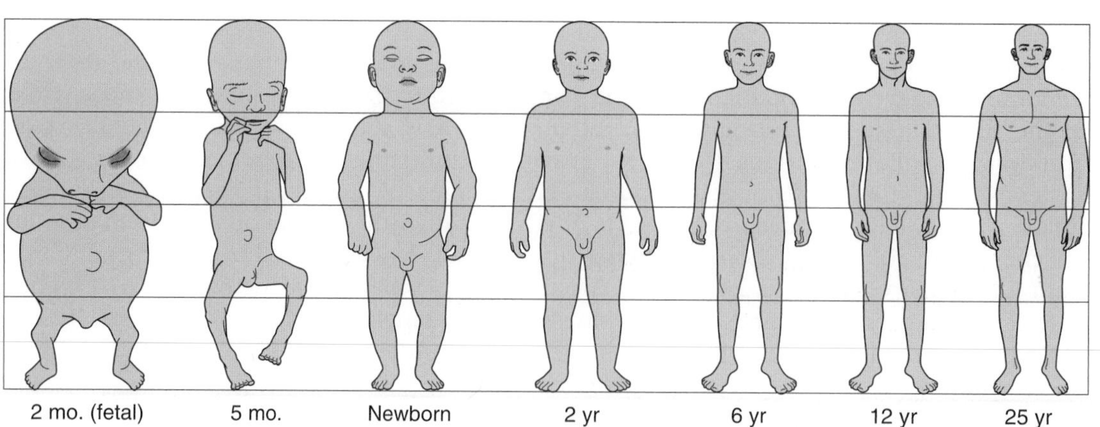

FIGURE 8.7 Changes in form and proportion of the human body during fetal and postnatal life.

Adapted from K.M. Newell 1984.

Development of Body Actions During Throwing

Level	Body actions
Preparatory arm backswing	
Level 1	The child's hand moves directly forward to release from where the child first grasps the ball.
Level 2	The forearm moves backward beside the head and then forward in ball release.
Level 3	The ball moves backward with a circular and upward backswing with elbow flexion.
Level 4	The ball moves backward with a downward swing below the waist and then up behind the head with elbow flexion.
Arm action	
Level 1	The upper arm moves backward and forward at an angle to the body. The forearm is linked to the upper arm movement and does not lag behind it.
Level 2	The upper arm is aligned with the shoulder and moves with it until near release, when the upper arm leads the shoulder. The forearm lags behind the upper arm (i.e., it appears to stay at the same point after the shoulder and upper arm have started forward).
Level 3	The upper arm lags behind the shoulder as the forward movement of the arm begins. The forearm lags even farther behind the upper arm and shoulder, reaching its final point of lag as the body is front facing.
Trunk action	
Level 1	Little or no trunk action occurs; the throw is accomplished by just the forward and backward action of the arm.
Level 2	The trunk rotation is in one block; the spine and hips move together in one rotating action.
Level 3	The hips lead in forward rotation followed by the spine (differentiated rotation).
Foot action	
Level 1	Feet do not move during throwing action.
Level 2	Step occurs but with the foot on the same side as the throwing hand.
Level 3	Step occurs but with the foot on the opposite side of the throwing hand.
Level 4	Same as level 3 except that the step is longer (about one-half of standing height).

Reprinted from M.A. Roberton, 1984, Changing motor patterns during childhood. In *Motor development during childhood and adolescence*, edited by J.R. Thomas (Minneapolis, MN: Burgess Publishing), 75. By permission of Jerry R. Thomas.

The changes in physical growth present three challenges: mechanical, adaptive, and absolute. The mechanics of movement change because of different body proportions; the individual must adapt to a rapidly changing body. These changes are especially problematic in seasonal sports. Consider the wrestler who experienced a rapid growth spurt from the end of one season to the beginning of the next. The center of gravity has changed, which may influence balance and the location of optimal points for exerting maximal force. Finally, absolute changes in size may influence performance. Females gain fat at puberty, which adversely influences performance in most physical activities; but males gain muscle, which has a positive influence on performance. During childhood, physical growth interacts with all the other factors that are developing (e.g., motor programs) and therefore must be considered for instruction and research.

A good way to examine structural and functional change and its influence on skill across childhood and adolescence is to consider overhand throwing. Figure 8.8 shows the changes in throwing for distance that occur across childhood and adolescence for girls and boys. The developmental nature of the overhand throw across childhood and early adolescence is detailed on page 233. This description is separated into various body actions during the throw (e.g., arm, trunk, foot).

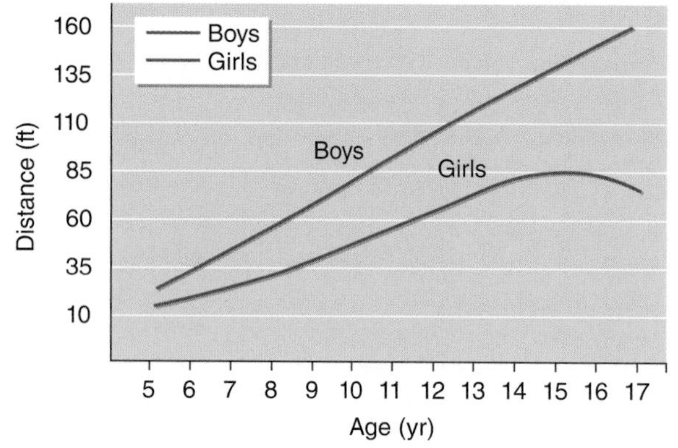

FIGURE 8.8 Age differences in throwing for distance.

Data from A. Espenschade 1960.

Given that prepubescent boys and girls are structurally similar, differences as large as those observed in throwing are unusual. Thomas and French (1985) noted that the gender differences in throwing at 5 years of age were three times as large as those for any of the 20 tasks that they reviewed. Although it seems likely that practice, encouragement, and opportunities account for a large part of these differences, throwing may be one of the few motor skills in which biological factors play a significant role in gender performance differences before puberty. Again we see that although we are unique, we (males and females) are more alike than different—both before and after puberty!

Student Learning Demonstrates Teaching Skill

How well did you do planning of swimming practice for your nephew? Compare the information in the sidebar on pages 225 and 226 to the sections on practice, feedback, and transfer. How can you judge whether practice sessions influence the learning of a skill? If your nephew could do what you presented in the first lesson the next time you met, you have evidence of retention. Remember that retention is one measure of learning. What characteristics of practice might disguise whether or not skills are being learned?

Wrap-Up

Although the area of motor behavior has produced important knowledge for human behavior since the late 1800s, it began to evolve as a significant subdiscipline in the field of physical activity in the 1960s.

Knowledge developed through motor behavior has become increasingly important in all aspects of society. Although we often think of learning and control of sport skills as the main issue, our society depends on human movement in many ways—babies learning to use a spoon, surgeons using a scalpel, pilots learning to control airplanes, children learning to control a pencil in writing, dentists using a drill. Understanding the development, learning, and control of these and other motor skills so that people can use them more effectively is the goal of motor behavior.

Knowledge of motor behavior is essential in several professions, including physical or occupational therapy, physical education, coaching, or working with children in community organizations such as the YMCA or YWCA or Boys and Girls Clubs. If you want to know more about this subdiscipline, you might want to read some of the journals that regularly publish research in motor behavior, such as *Journal of Motor Behavior, Motor Control,* or *Research Quarterly for Exercise and Sport.* Your school probably offers at least one undergraduate course in motor learning, motor control, and motor development.

KEY▸ Activity 8.8

Use the key points review in your online study guide as a study aid for this chapter.

www. Activity 8.9

These end-of-chapter questions and activities are also in your online study guide. Your instructor may ask you to complete them online and turn them in.

1. How does motor behavior differ from the psychology of sport?
2. Within motor behavior, explain the differences between motor learning and motor control.
3. Why is the change in motor learning and motor control across the life span of interest?
4. Think about the practice issues discussed in this chapter, such as feedback, retention, transfer, goal setting, and scheduling. Choose a sport that you are familiar with and discuss how the practice characteristics would influence your planning if you were a coach. Pick a specific age group or performance level, such as high school, college, or professional coaching.
5. Can you think of an example in which more difficult practice conditions result in better retention and transfer? Why does that happen? Can you plan practices to promote this? How?
6. Discuss when it might be best to provide either knowledge of performance or knowledge of results to a person learning a motor skill.

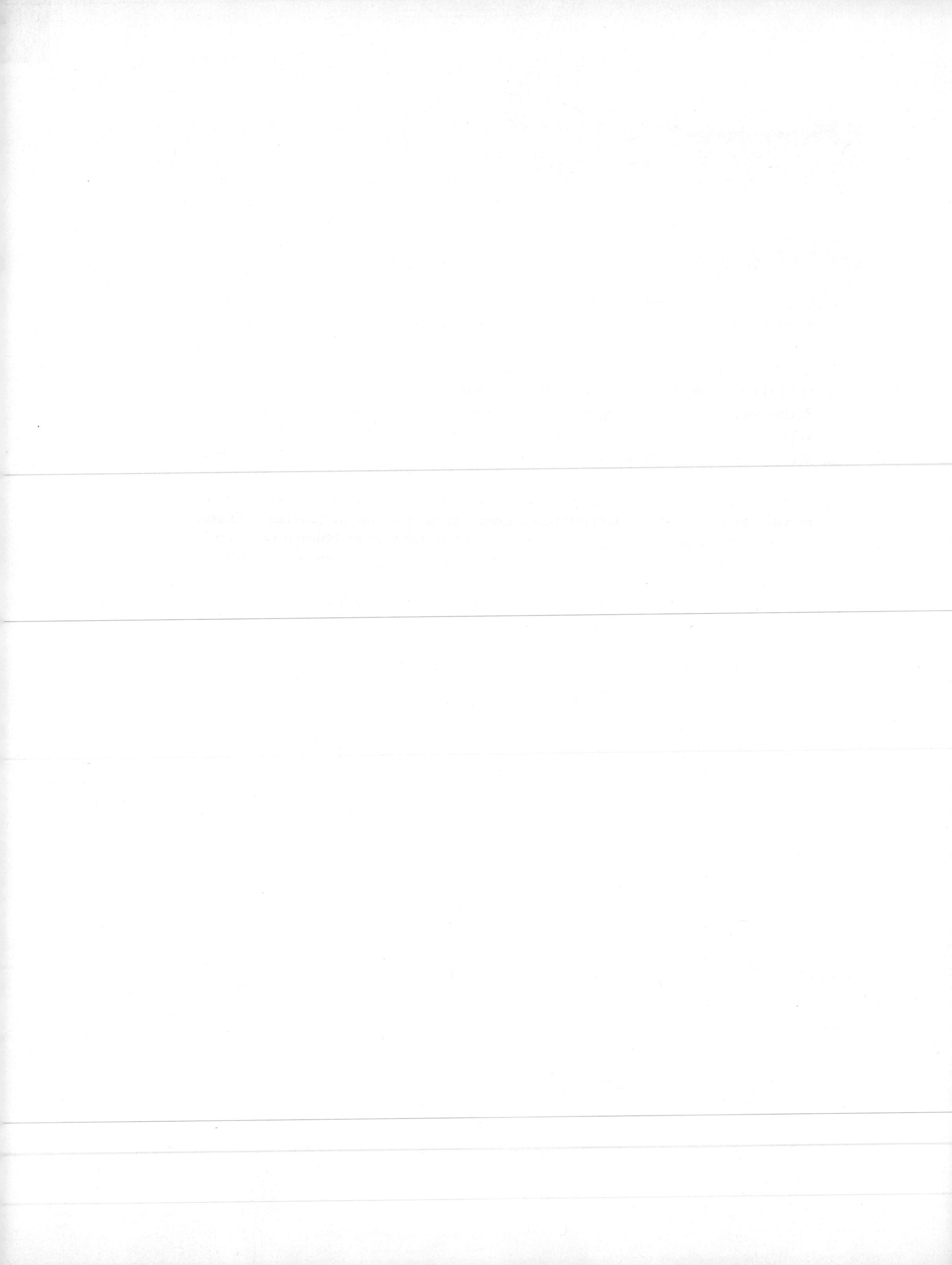

9

Sport and Exercise Psychology

Robin S. Vealey

Stockbyte

CHAPTER OBJECTIVES

In this chapter we will

- discuss what sport and exercise psychologists do;
- explain how sport and exercise psychology fits within kinesiology;
- describe how sport and exercise psychology evolved within the field of physical activity;
- help you understand how professionals in sport and exercise psychology engage in research and practice;
- discuss what research tells us about personality, motivation, energy management, and group processes in sport and exercise settings; and
- explain how intervention techniques are used to enhance participation in sport and exercise.

You are an elite swimmer poised on the starting block for the 50-meter freestyle race at the Olympic Games. In the other lanes are the top swimmers in the world who, like you, have trained for years to hone their bodies into finely tuned physical machines—all in preparation for this moment. Your muscles are coiled as tightly as springs waiting to explode into the water. What are you thinking? Should you even *be* thinking? Where is your attention focused? Do you have negative thoughts and fears about not performing well, or do you feel a confident excitement that you are ready to meet the challenge? Do you feel a relaxed inner calmness or a feverish emotional intensity? You have prepared yourself physically for this moment by swimming hours and hours for miles and miles. But have you prepared yourself mentally?

The psychologist William James wrote that "the greatest discovery . . . is that human beings, by changing the inner attitudes of their minds, can change the outer aspects of their lives." How true that is! Our participation in physical activity is greatly influenced by our "inner attitudes," or how we *think about* what we're doing. This mental aspect of physical activity is the focus of study in sport and exercise psychology. Specifically, sport and exercise psychology focuses on the mental processes of humans as they engage in physical activity.

If you're the Olympic swimmer in the opening scenario, systematic training with a sport psychologist has prepared you to think productively and focus optimally at this crucial moment. Your mental practice has prepared you to focus your attention on productive, energizing thoughts instead of worries and distractions. You have programmed your mind and body to respond optimally without distracting thoughts. You have planned and mastered the mental focus and physical readiness that you need for your best performance, and now your body and mind are one, waiting to explode from the platform for the performance of your life!

Most of us will not compete in the Olympic Games, but the mental aspects of physical activity will challenge us repeatedly as we go through our lives. Consider the following situations that you may encounter:

- You take up mountain biking and are amazed at how the activity enhances your self-esteem and confidence along with your physical health.
- You begin an exercise program to enhance your personal fitness, but you lack motivation and quit the program after two months.
- You really enjoy your weekly tennis league, but you become very nervous playing doubles because you don't want to let your partner down.
- As the coach of a local youth soccer team, you are attempting to decide the best ways to provide feedback and structure practices for children.
- Your father's physician recommends that your father begin a progressive aerobic exercise program, and your father asks you to help him establish a physical activity routine that he will enjoy.

➤ Physical activity always involves the mind as well as the body. The mental aspect of physical activity is the realm of sport and exercise psychologists.

All these examples illustrate the critical and fascinating link between mind and body inherent in physical activity.

What Is Sport and Exercise Psychology?

Sport and exercise psychology involves the study of human thought, emotion, and behavior in physical activity. Think of sport and exercise psychology as the study of the ABCs of physical activity. *A* represents the term *affect,* which means emotion. For example, the emotions of anxiety and anger are studied to see how they influence athletes' performance. Emotional moods are influenced by exercise, which has been shown to decrease depression and enhance feelings of well-being. *B* is for *behavior.* Understanding why people behave differently from one another when engaging in physical activity is fascinating. That is, why are some people so

Activity 9.1

Do You Know Your ABCs? Test your understanding in this activity in your online study guide.

committed to and persistent in their exercise routines? Why do some athletes train exhaustively to achieve excellence, whereas other talented athletes fail to reach their potential because they're lazy?

Finally, the *C* in the ABCs of physical activity represents *cognitions.* A cognition is simply a thought, so sport and exercise psychology studies how the thought processes of individuals influence and are influenced by their participation in physical activity. Attempting to understand why athletes choke under pressure requires an understanding of their thinking processes. Similarly, exercise psychologists are interested in how physical activity can help individuals develop self-esteem and confidence, or better ways to think about themselves.

Remember your ABCs to keep in mind that sport and exercise psychology is the study of affect (emotions), behavior (actions), and cognitions (thoughts) related to physical activity participation.

This chapter discusses the psychology of physical activity in relation to the specific areas of sport and exercise. Sport psychology and exercise psychology have developed into two distinct areas of study in the physical activity field. Exercise psychology is devoted to the study of the psychological aspects of fitness, exercise, health, and wellness. In contrast, sport psychology focuses more on the psychological aspects of competitive sport participation. Because the two areas are closely related and share a great deal of theoretical content, however, this chapter covers both.

➤ Sport psychology and exercise psychology, as subdisciplines of kinesiology, focus on the study of human thought, emotion, and behavior in physical activity.

Although sport and exercise psychology are areas of study in kinesiology, they have close ties to the discipline of psychology. Psychology is the science that deals with human thoughts, feelings, and behavior, so sport and exercise psychology studies thoughts, feelings, and behaviors of people involved in sport and exercise. For example, researchers attempting to understand why people begin or discontinue participation in fitness programs often apply motivational theories developed in the discipline of psychology.

In summary, sport and exercise psychology studies the ABCs of physical activity to learn how and why people feel (*A* for affect), act (*B* for behavior), and think (*C* for cognitions) in various ways as they participate in sport and exercise. These areas of study focus on the fascinating link between our minds and bodies.

An athlete's thoughts and feelings influence her behavior. Sport psychologists study the ABCs of physical activity to learn more about how those factors influence people as they participate in sport and exercise.

AP Photo/Mary Ann Chastain

What Does a Sport or Exercise Psychologist Do?

To understand what sport and exercise psychology is all about, it's helpful to learn about what kinesiology professionals in sport and exercise psychology do. As in any subdiscipline of physical activity, sport and exercise psychology professionals work in many career areas.

Exploring the ABCs of Physical Activity

Consider the following questions related to sport and exercise psychology:

1. Recalling that we've defined physical activity as intentional and purposeful movement (chapter 1), think of several forms of physical activity in which you participate. What do you think your motivation is for performing each of these activities?
2. Do you think that participation in physical activity has influenced your character? If so, what was the activity and how do you think it affected you? If not, why do you think that participation has not influenced your character?
3. Have you ever been involved in or watched any form of violence in sport? If so, what happened and why? What do you think caused the violence to occur?
4. Why do you think so few Americans participate in habitual exercise when it has been clearly shown to enhance health and longevity?

Kinesiology professors of sport and exercise psychology at universities fulfill the multiple roles of researcher, teacher, and service provider. A kinesiology professor specializing in exercise psychology may develop theory and conduct research on the psychological benefits of aerobic exercise, teach courses in exercise psychology, develop a health promotion program within the university, and conduct workshops for the community on motivation and stress management. A kinesiology professor specializing in sport psychology may conduct research on stress in sport and athlete burnout, teach courses in sport psychology, consult with athletes from the university and surrounding community, and offer regional workshops on coaching effectiveness.

Some sport and exercise psychology–trained kinesiologists focus more exclusively on practitioner or service provider roles. For example, a sport psychology specialist may work on performance enhancement, personal development, and lifestyle management issues of athletes within a university athletic department, a professional sport team, or a private sport training academy. Likewise, an exercise psychology specialist may focus on worksite health promotion for a large corporation, develop fitness and wellness programs for a community recreation department, or coordinate physical activity programs for all age groups through a YMCA. Sometimes sport and exercise psychology professionals become consultants for sports medicine or physical therapy clinics to help injured people through the psychological aspects of injury rehabilitation.

www. Web Search 9.1

Finding Your Path
Research SEP careers in this activity in your online study guide.

Differentiating between an exercise or sport psychologist trained as a physical activity specialist and one trained as a clinical or counseling psychologist is important. Clinical or counseling psychologists are licensed to provide psychotherapy and consultation for individuals with clinical conditions such as depression, phobias, or anorexia nervosa. The focus of kinesiology-trained practitioners of sport and exercise psychology within the physical activity field is on education or the teaching of skills to enhance the performance or personal fulfillment of individuals involved in sport or exercise. Although some clinical and counseling psychologists provide services for athletes and exercisers, the main focus of their practice is different from that of sport and exercise psychology as a subdiscipline of the physical activity field.

➤ Career opportunities in sport and exercise psychology include positions as university professors, performance enhancement specialists, fitness and health promotion specialists, and sports medicine consultants.

Goals of Sport and Exercise Psychology

The first goal of sport and exercise psychology is *to understand the social-psychological factors that influence people's behavior and performance in physical activity.* For example, why are some individuals motivated to exercise and others not? Why do some athletes choke in pressure situations in competition? Why do children drop out of youth sport programs? Is it better to exercise alone or with others?

The second goal is *to understand the psychological effects derived from participation in physical activity.* For example, does youth sport participation build character? Can weightlifting enhance self-esteem and decrease stress? Do ice hockey players learn to be violent and aggressive? Does running reduce depression?

The third goal of the subdiscipline is *to enhance the sport and exercise experience for those who participate in physical activity.* This third goal logically follows the first two as sport and exercise psychology professionals attempt to apply knowledge gained through research to implement sound practices. Examples might include using behavior management techniques to increase adherence in exercise programs, modifying children's sport for greater enjoyment and development of self-worth, and helping athletes engage in effective mental training to perform better.

How Sport and Exercise Psychology Fits Into Kinesiology

As shown in figure 9.1, sport and exercise psychology fits with the other scholarly areas of kinesiology to provide an integrative way of understanding physical activity. Motor behavior—the development, learning, and control of skilled movement—is certainly influenced by psychological (e.g., self-confidence, motivation) and social (e.g., teacher behavior, spectators) factors. We know that anxiety can negatively influence optimal biomechanical and physiological performance in sport. And we know that the ABCs of sport and exercise psychology must always be studied in relation to cultural ideas and practices (sociology). For example, why are certain sports viewed as more appropriate for males versus females, and how does this view affect the ABCs of athletes in these sports? Think of ice hockey and figure skating and how society views these two ice sports as more appropriate for male versus females.

It is also helpful to consider the ways in which knowledge from sport and exercise psychology is used by various kinesiology professionals. Physical education teachers must consider and use different forms of motivation to "hook" youngsters into a love of physical activity.

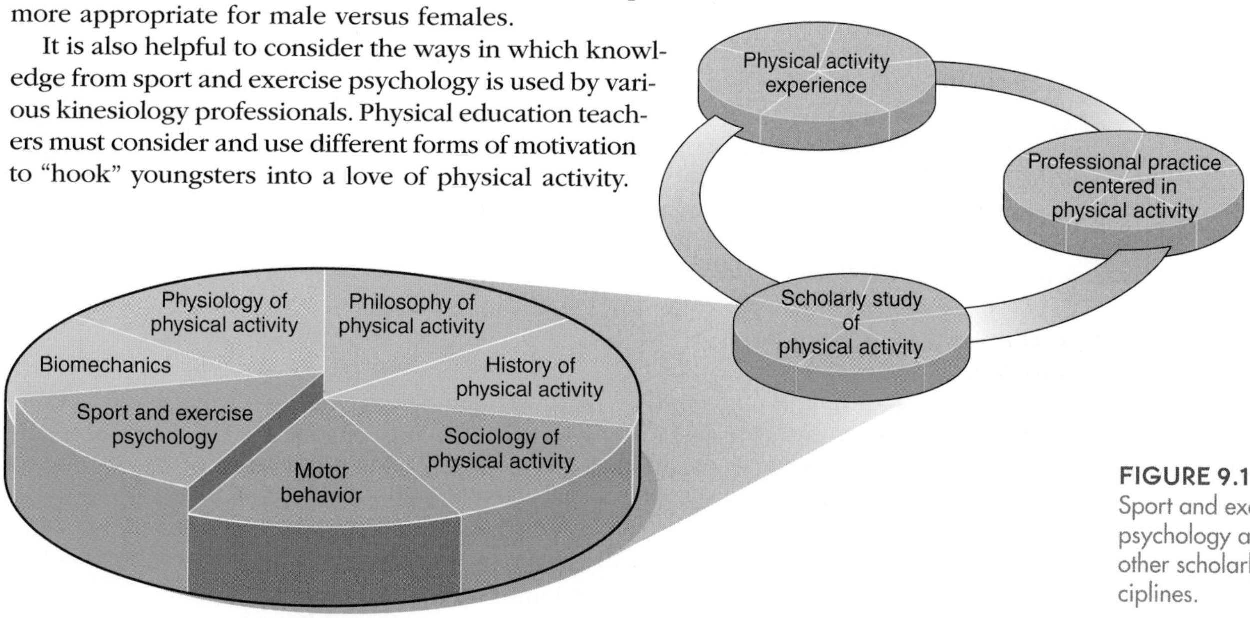

FIGURE 9.1 Sport and exercise psychology and the other scholarly subdisciplines.

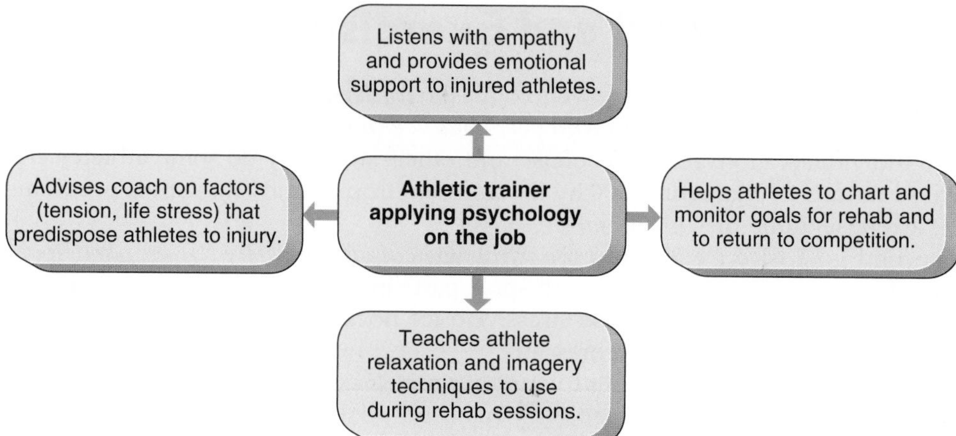

FIGURE 9.2 How an athletic trainer can put knowledge of sport and exercise psychology into practice.

Coaches have to be master psychologists to communicate effectively with athletes and nurture strong team chemistry. Athletic trainers understand that helping athletes recover from injuries requires not only physical, but also psychological, rehabilitation (see figure 9.2). Corporate and personal fitness coordinators must understand the complex personality and social factors that affect people's decisions to begin and continue physical activity programs. Recreational and elite sport managers have to practice effective leadership to successfully transform their programs. Seemingly, all professionals in kinesiology use psychological knowledge and experience in their jobs every day.

History of Sport and Exercise Psychology

Throughout history, the important connection between mental health and physical activity has been noted. However, sport and exercise psychology was not recognized as a subdiscipline of kinesiology until the 1960s.

Beginnings of Sport and Exercise Psychology

The relationship between exercise and psychological well-being was recognized as early as the 4th century B.C.E. (Buckworth & Dishman 2002). Around the turn of the 20th century, physical educators began to write about the psychological benefits of physical activity. Researchers began assessing the social influences of the presence of others on motor performance, a research area that later became known as social facilitation. The first person to examine these social influences was Norman Triplett, who in 1898 studied the effects of the presence of other people on bicycling performance. The effects of exercise on depression were reported in 1905 (Franz & Hamilton 1905).

Griffith Era in Sport Psychology

The true beginning of sport psychology in North America dawned with the work of Coleman Griffith, who, as a professor at the University of Illinois, engaged in the first systematic examination of the psychological aspects of sport between 1919 and 1938. Griffith established the Athletic Research Laboratory at Illinois, published numerous research articles, and wrote two classic books—*Psychology of Coaching* in 1926 and *Psychology and Athletics* in 1928. Griffith also interviewed sport celebrities of that time such as Red Grange and Knute Rockne about the mental aspects of their sports. Phillip Wrigley hired him in 1938 as a sport psychology consultant for the Chicago Cubs baseball team. During this work, Griffith developed psychological profiles for specific players, such as Dizzy Dean, and researched methods of building confidence and increasing motivation.

Unfortunately, Griffith may be thought of as a prophet without disciples because no one continued his significant work. Although sporadic publications about psychological aspects of sport and exercise appeared after Griffith's time, the subdiscipline largely lay dormant until a reawakening in the 1960s.

➤ Coleman Griffith began systematic research in sport psychology in the 1920s, but because his early work was not extended, the area was not recognized as an academic subdiscipline of physical activity until the 1960s.

1960s and 1970s

The 1960s saw an upsurge of interest in the social-psychological factors associated with physical activity. The research of this period focused primarily on personality traits related to sport participation and social facilitation or audience effects on motor performance. Bruce Ogilvie, a clinical psychologist at San Jose State University, began early work examining personality in athletes and began applied psychological interventions with athletes. Dorothy Harris at Pennsylvania State University is also considered a pioneer of this period because she began a systematic research focus on women in sport. The first meeting of the International Society of Sport Psychology (ISSP) was held in 1965; the first North American Society for the Psychology of Sport and Physical Activity (NASPSPA) conference was held in 1967; and the Canadian Society for Psychomotor Learning and Sport Psychology was founded in 1969. All these served as important forums for sharing knowledge in sport psychology.

The 1970s was the decade in which sport psychology was formally recognized as an established scientific subdiscipline within kinesiology. Systematic research programs were established at leading universities; graduate study became available; and in 1979, *Journal of Sport Psychology* began publication. Much of the research during this time was experimental research conducted in laboratory settings that involved testing theory from the parent discipline of psychology. Rainer Martens, a leading scholar during this time, pioneered the systematic study of competitive anxiety in sport.

Modern Sport and Exercise Psychology

Three trends marked the decade of the 1980s. First, exercise psychology separated from sport psychology into a distinct subdiscipline within kinesiology. Before this time, pockets of research investigating the psychological components of exercise existed, but it was during the 1980s that systematic research programs and graduate program offerings in exercise psychology were developed. William P. Morgan was a major force in promoting the field of exercise psychology and served as the first president of Division 47, Exercise and Sport Psychology, of the American Psychological Association. Daniel Landers also served as a leader in this area, particularly in pioneering psychophysiological research in the field. In 1988, *Journal of Sport Psychology* became *Journal of Sport and Exercise Psychology.*

The second trend of the 1980s was explosive growth in field research in sport and exercise psychology. Much of the research of the 1960s and 1970s occurred in laboratory settings, seemingly to gain academic credibility for the young field. But researchers realized the need to study sport and exercise in their actual settings to gain a better understanding of the psychological processes as they operated in competition and various forms of physical activity.

Logically following this move to the field, the third trend of the 1980s was increased interest in applied sport psychology or mental training with athletes. The establishment of two new applied journals, *The Sport Psychologist (TSP)* in 1987 and *Journal of Applied Sport Psychology (JASP)* in 1989, and a new organization, the Association for the Advancement of Applied Sport Psychology (AAASP) in 1986, was indicative of the expansion of the subdiscipline to include applied interests. Influential professionals during this period included John Silva, the founding president of AAASP and initial editor of *JASP;* Dan Gould, the founding editor of *TSP;* and Terry Orlick, whose mental training books and consulting work with Canadian Olympic athletes pioneered the era of applied sport psychology to come.

➤ The 1980s saw the emergence of exercise psychology, the growth of field research, and an explosion of applied mental training with athletes.

Activity 9.2

Development of SEP
Test your knowledge of significant events in the study of SEP in this activity in your online study guide.

During the 1990s, sport psychology and exercise psychology grew as separate subdisciplines through the accumulation of research and applied programs. In 1991, AAASP implemented criteria identifying minimum professional training standards for individuals to be certified to provide consulting services in sport and exercise psychology. Ethical standards and guidelines for consulting in sport and exercise settings were approved, and the United States Olympic Committee created a registry of certified professionals eligible to consult with Olympic teams and athletes.

Research Methods in Sport and Exercise Psychology

By now you should understand the nature and historical development of sport and exercise psychology as a scientific area of study. But how is science conducted in this subdiscipline? What methods do researchers use to ask important questions about the psychological aspects of physical activity participation? Researchers use six methods to systematically assess thoughts, feelings, and behaviors in sport and exercise psychology.

Questionnaires

Questionnaires are widely used in sport and exercise psychology. Questionnaires may be survey instruments that assess demographic variables such as age, sex, or socioeconomic status as well as general information such as the type, frequency, and duration of exercise engaged in during the past week. Most of the questionnaires used, however, are **psychological inventories,** which are standardized measures of specific forms of thoughts, feelings, or behaviors (Anastasi 1988). For example, psychological inventories are used to measure the amount of anxiety, motivation, and confidence an individual feels about exercising or competing in sport.

Consider for a moment the difficulty of accurately measuring the thoughts, feelings, and behaviors of people. Assessing individuals' levels of self-esteem is much different from measuring the amount of oxygen that they expend during a fitness test or how much weight they lose in the preseason. For this reason, psychological inventories must meet rigorous standards of uniformity of procedures with regard to their development, administration, and scoring so that researchers can gain a valid and reliable assessment of behavior. Therefore, psychological inventories used by researchers are not simply a list of questions thrown together; they are carefully constructed and tested assessment tools that have met specific standards set by experts in the subdiscipline.

Interviews

Interviews are used in sport and exercise psychology when the research question being pursued requires in-depth understanding of individuals' beliefs, experiences, or values. For example, interviews may be useful if one is attempting to understand why children drop out of youth sport programs because interviews allow children to explain things in their own words rather than respond to questionnaires. Like any other scientific method, however, interviews must be structured to be systematic, and researchers must be trained in the use of interviews in order for them to be effective and valid.

Have You Got Passion for Physical Activity?

Complete the following questionnaire by marking a response for each item as you think about your favorite physical activity.

Scoring your responses: totally agree = 7; agree = 6; sort of agree = 5; unsure = 4; sort of disagree = 3; don't really agree = 2; don't agree at all = 1.

	Totally agree	Agree	Sort of agree	Unsure	Sort of disagree	Don't really agree	Don't agree at all
1. I am completely taken with my activity.							
2. My activity reflects the qualities that I like about myself.							
3. My activity allows me to live memorable experiences.							
4. My activity is in harmony with other activities in my life.							
5. My activity is a passion that I manage to control.							
6. I cannot live without my activity.							
7. My mood depends on my being able to do my activity.							
8. I have difficulty imagining my life without my activity.							
9. I am emotionally dependent on my activity.							
10. I have an almost obsessive feeling for my activity.							

Add your scores for items 1 through 5 together to get your positive passion score. This represents how much you love to engage in your activity based on your own free will to choose to participate. You gain great joy from the activity, you don't feel you have to do it, and it doesn't control your life.

Add your scores for items 6 through 10 to get your obsessive passion score. This represents how much you feel you have to engage in your activity, as if your activity controls you and you *have* to do it (for others, for approval, for yourself).

Are you passionate about your physical activity? How much of that passion is positive and how much is obsessive? You want your positive passion to outweigh your obsessive passion, so you feel in control of your life and gain joy from participation, not pressure to perform. Consider ways you can lessen the obsessive passion you feel and increase your positive passion for your physical activity.

Adapted from R.J. Vallerand et al., 2003. "Les passions de l'ame: On obsessive and harmonious passion," *Journal of Personality and Social Psychology* 85, 756-767.

Observations

Observation of behavior is the third method used in sport and exercise psychology. For example, researchers often observe the behavior of coaches during practice or competition to assess the frequency of various types of feedback and communication that they provide to athletes. Research examining the motivation of children to engage in physical activity or vigorous play activities benefits from using behavioral observation. Typically, observation studies employ some type of behavior checklist or coding system to ensure that observation of the behavior is occurring within a particular set of parameters. Also, two observers are typically used to code behavior, and their results are checked against each other to ensure a consistent, reliable assessment.

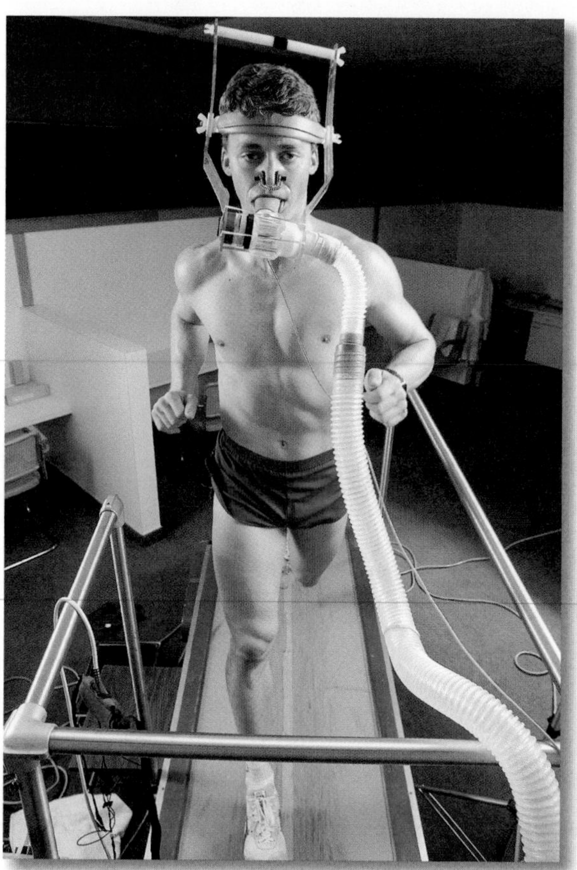

Physiological measures, or biofeedback, are used to assess the body's responses to stressors associated with physical activity.
Eyewire

Physiological Measures

Physiological measures of physical, mental, and emotional responses are sometimes referred to as biofeedback. For example, blood pressure and heart rate may be measured to assess the effects of psychological stressors on individuals. Exercise psychologists use measures such as these to study the effects of exercise on stress reactivity and existing anxiety levels. Sport psychologists might measure the amount of tension in muscles to assess how well athletes can learn to relax physically through mental training. Researchers can even measure brain waves to assess levels of attention or relaxation.

Biochemical Measures

Biochemical measures are used less frequently in sport and exercise psychology, but brief mention here will be helpful. This type of assessment involves drawing and analyzing blood or urine for chemicals from the body that represent responses to stressors or emotions. Examples are epinephrine and cortisol, which the adrenal gland releases in response to certain types of stressors.

Content Analysis

Researchers use content analysis to analyze written material from various sources such as government documents, newspapers or magazines, and even television programming. For example, a researcher could analyze several popular television shows to assess the levels of physical activity being modeled by the media to viewers. Written or dictated physical activity logs are useful to exercise researchers because they provide detailed accounting of all or selected types of physical activity performed within a given period.

Thus, researchers in sport and exercise psychology have various methods available to them. Often, researchers use two or more methods within a single research study. For example, a study examining exercise adherence in an adult fitness program could use

Interviews With Practicing Professionals

Photo courtesy of Michael DeCello

Michael DeCello

Varsity boys' basketball coach at Talawanda High School in Oxford, Ohio

Q: As a coach, how do you use sport psychology during practices and games to help your players improve?

A: You have to be a master psychologist to be a successful coach. The biggest obstacles in sport are the mental ones, so you had better be ready to coach the mental game. That means using repetition to ingrain habits that don't break down under pressure. It also means running practices at game speed, and simulating as much pressure as we can. Another thing we do is teach our guys the steps to follow in grooving their preshot routine at the free throw line. Kids today mimic what they see on television—I don't allow them to spin the ball like all the NBA players do. It's not a sound movement to lock in your shot. We make sure they gaze at the rim for at least two seconds, and visualize the ball dropping soft over the front. I read an article in a sport psychology class that explains that this "quiet eye" routine worked best for shooters compared to other routines. I said, "We're doing it!" We also had a sport psychology consultant work with our team to use imagery to mentally practice free throws and some of our plays. I thought it was great. An individual videotape of each player shooting free throws was made for them, and they would watch this tape (shot from behind them) and then lock in their images of perfect form, perfect rotation, and a perfect swish! We called the guy the "Shot Doctor"—even got him a T-shirt to wear to practice. We also do some team-building activities and pair up a senior with a younger player to create "performance partners," kind of a mentoring system. The mental game's tough, though, and I'm always looking for new ideas.

---

observation to assess participation rates, intensity of exercise, and instructor behavior; questionnaires to measure participants' self-confidence and perceived benefits in relation to exercising; and physiological measures of participants' resting heart rate and lung volume. But remember, all methods are designed to be used in a systematic manner to ensure that accurate and consistent measures are obtained. This systematic approach is the mark of science—and it is very different from the casual observations that we all make in everyday social interactions.

Activity 9.3

Which Method?

In this activity in your online study guide, you'll review two scenarios and determine which research method to use in each.

Remember that all scientific methods in sport and exercise psychology have limitations. Questionnaires typically provide a systematic measure of certain phenomena, but they lack the depth and richness of interviews. On the other hand, establishing consistency and a systematic approach is more difficult when one is using interviews as compared with questionnaires. Scientists in sport and exercise psychology look at the menu of research methods available to them and then select the method that best fits the recipe of their particular research study. This menu also includes other choices for researchers, such as whether to conduct the study in a laboratory or field setting and what type of participants to use in the research.

➤ Research methods in sport and exercise psychology include questionnaires, interviews, observation, physiological measures, biochemical measures, and content analysis.

Overview of Knowledge in Sport and Exercise Psychology

This section provides a brief overview of sport and exercise psychology topics to give you a glimpse of the knowledge that researchers and practitioners have produced. The topics in this section are divided into six main areas: personality, motivation, energy management, interpersonal and group processes, developmental concerns, and intervention techniques for physical activity enhancement.

Personality

One of the most popular issues in sport and exercise psychology concerns the relationship between personality and physical activity participation. We typically think of **personality** as the unique blend of the characteristics that make individuals different from and similar to each other. Our personalities determine our thoughts, feelings, and behaviors (ABCs) in response to our environment.

Personality and Sport

Despite popular opinion, no distinguishable "athletic personality" has been shown to exist. That is, no consistent research findings show that athletes possess a general personality type distinct from the personality of nonathletes. Also, no research has shown consistent personality differences between athletic subgroups (e.g., team athletes vs. individual sport athletes, contact sport athletes vs. noncontact sport athletes).

Research has identified several differences in personality characteristics between successful and unsuccessful athletes (e.g., Krane & Williams 2006). These differences, however, are not based on innate, deeply ingrained personality traits but rather result from more effective thinking and responding in relation to sport challenges as well as higher levels of motivation. Specifically, successful athletes, as compared to less successful athletes, are

- more self-confident;
- better able to use more effective cognitive strategies and coping mechanisms to retain optimal competitive focus in the face of obstacles and distractions;
- better able to self-regulate activation efficiently;
- more positively preoccupied with sport, in terms of thoughts, images, and feelings; and
- more highly determined and committed to excellence in their sport.

For example, the coping skills, confidence, and motivation of professional baseball pitchers were more important in predicting their success in baseball than their physical skills were (Smith et al. 1995). In addition, these mental skills were predictive of all players' survival in professional baseball two and three years later (Smith & Christensen 1995). Another example of the importance of personality in sport behavior and performance is perfectionism. Research has shown that positive perfectionism in athletes is related to success, whereas negative perfectionism is related to anxiety and burnout (Gould, Dieffenbach, & Moffett 2002; Gould et al. 1996). Positive perfectionists set high personal standards for achievement and are highly motivated to succeed. Negative perfectionists also set high standards of achievement, but they are overcritical in evaluating their performance and unable to accept mistakes.

Although most sport personality research has focused on the influence of personality on sport behavior, research has also examined the effects of sport participation on personality development and change. A belief commonly held in American society is that sport builds character, or that sport participation may develop socially valued personality attributes. Research shows, however, that competition serves to reduce prosocial behaviors such as helping and sharing, and that losing magnifies this effect. Sport participation has been shown to increase rivalrous, antisocial behavior and aggression. Sport participation has also been linked to lower levels of moral reasoning.

Nevertheless, the sport story has a positive side. Research in a variety of field settings has demonstrated that children's moral development and prosocial behaviors (cooperation, acceptance, sharing) can be enhanced in sport settings when adult leaders structure situations to foster these positive behaviors (Hellison 2003; Shields & Bredemeier 2007). Interventions with children were successful in building character when naturally occurring conflicts arose and were discussed with the children to enhance their reasoning and values about sport and life events. The moral of the story is this: Sport doesn't build character, people do!

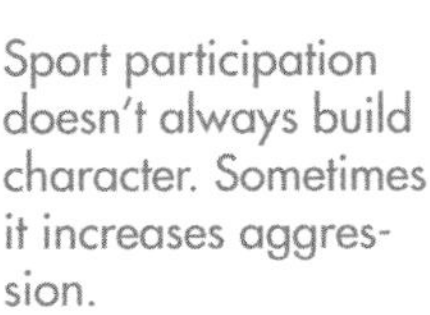

Sport participation doesn't always build character. Sometimes it increases aggression.

AP Photo/Michael Sohn

Personality and Exercise

As in sport, researchers have found no "exercise personality" or set of personality characteristics that predicts exercise adherence. Exercisers cannot be differentiated from nonexercisers based on an overall personality type. Two personality characteristics, however, are strong predictors of exercise behavior. Individuals who are more confident in their physical abilities tend to exercise more than those who are less physically confident. A second important predictor of exercise behavior, obviously, is self-motivation, with self-motivated individuals beginning and continuing exercise programs and less motivated individuals never getting started or dropping out.

A personality type termed "obligatory exercisers" describes individuals who participate in exercise at excessive and even harmful levels (Coen & Ogles 1993). For these individuals, exercise becomes the central focus of life, and their behavior becomes pathological in terms of their need to control themselves and their environment. Clinical evidence demonstrates a similar link between anorexia nervosa, a psychopathological eating disorder, and compulsive exercise. Specialists in exercise psychology attempt to help individuals plan and engage in exercise behavior that is healthy and noncontrolling to enhance total well-being.

➤ No set of underlying traits exists for an "athletic personality," but successful athletes possess self-perceptions that are more positive, have stronger levels of motivation, possess more adaptive perfectionism, and use more productive cognitive coping strategies as compared with less successful athletes.

Echoing the idea that sport builds character, exercise or fitness training has also popularly been associated with positive personality change and mental health (Landers & Arent 2007). The various psychosocial outcomes related to exercise participation are shown in the activity in your online study guide.

The personality characteristic that researchers have most frequently examined in this area is self-esteem. Self-esteem is our perception of personal worthiness and the emotions associated with that perception. Think of self-esteem as how much we like ourselves. Research has generally confirmed that fitness training improves self-esteem in children, adolescents, and adults. Research has also shown that exercise positively influences perceptions of physical capabilities, or self-confidence. Interestingly, the research indicates that these changes in self-esteem and self-confidence may result from perceived, as opposed to actual, changes in physical fitness. In addition, many aspects of intellectual performance have been related to physical activity, suggesting that cognitive functions respond positively to increased levels of physical activity.

DO it Activity 9.4

Ups and Downs

In this activity, you'll identify which personality characteristics increase or decrease because of physical activity.

Many people also associate exercise with changes in mood and anxiety. Most individuals say that they "feel better" or "feel good" after vigorous exercise, which emphasizes the important link between physical activity and psychological well-being. In addition, research documents that anxiety and tension decline following acute physical activity. This effect of exercise on anxiety begins within 5 minutes after acute exercise and continues for at least 2 hours. Reductions in anxiety are associated with activities involving continuous, rhythmic (aerobic) exercise rather than resisted, intermittent exercise. The greatest reductions in anxiety occur in exercise programs that continue for more than 15 weeks. Much research has been conducted to determine whether exercise or fitness reduces people's susceptibility to stress, and the generally accepted conclusion is that aerobically fit individuals demonstrate a reduced psychosocial stress response. A tentative explanation for this finding is that exercise either acts as a coping strategy that reduces the physiological response to stress or serves as an "inoculator" to foster a more effective response to psychosocial stress (Landers & Arent 2007).

➤ Sport has not been found to build socially valued attributes, or character, but exercise has been shown to produce several benefits including enhanced self-concept and psychological well-being and decreased anxiety and depression.

Prolonged physical activity is also associated with decreases in depression and a lessening of depressive symptoms in individuals who are clinically depressed at the outset of the exercise treatment. Explanations for these changes range from the distraction hypothesis, which maintains that exercise distracts attention from stress, to other hypotheses that focus on the physiological and biochemical changes in the body after exercise.

Motivation

In the previous section, you learned that motivation is an essential characteristic needed for success in sport and adherence to exercise programs. Therefore, sport and exercise psychologists are interested in understanding what motivates people to engage in physical activity. **Motivation** is a complex set of internal and external forces that influence individuals to behave in certain ways. The behaviors most typically associated with motivation are choice, effort, and persistence. That is, we assume that people are motivated when they make choices to join a fitness program, work intensely during the program, and continue to adhere to their training program when their lives become busy. Motivation directs and energizes our behavior in sport and exercise. Consider how you would answer the following two popular questions about motivation:

1. What is the best way to motivate people?
2. What makes some people motivated and others not?

Intrinsic and Extrinsic Motivation

The first question assumes that motivation is something that you give to others—like a cup of water. You might recall from chapter 4 that this type of motivation is termed **extrinsic motivation,** which means that people engage in a certain behavior to gain some external reward from that participation such as a trophy in sport or weight loss in exercise. Rewards and punishment are often used as incentives to motivate individuals to exercise or exert effort in sport. Gaining popularity in school or eliciting parental approval from sport achievement clearly motivates children to continue this behavior. Adolescent boys gain motivation to lift weights when their bodies become muscular and are admired by others.

Extrinsic motivation and rewards, however, serve to enhance motivation only in the short term and do not fuel long-term commitment to achievement in sport or fitness training. Thus, although extrinsic motivators are always present in society and offer powerful incentives, this

How's Your Motivation?

What is your favorite physical activity? What motivates you to participate in this activity? Is your motivation more internal or external? Explain your answer by including information about the choices you make in physical activity, the effort you expend, and the persistence with which you pursue it. If your motivation is external, describe the reward you're seeking. If it's internal, explain the pleasure and satisfaction you get from the activity.

type of motivation is short-lived. The enduring motivation necessary for achievement and success is not something that you can give to another person. Rather, enduring motivation comes from within.

Thus, the second question—why are some people motivated and others not?—is more interesting to sport and exercise psychologists. This motivation from within, called **intrinsic motivation,** involves engaging in behavior because you enjoy the process and gain pleasure and satisfaction from that participation. Intrinsically motivated athletes perform because they love the sport. People who exercise regularly do so because they enjoy physical activity. Intrinsic motivation serves as a long-term fueling process for commitment and achievement of important goals in sport and fitness training. Thus, sport and exercise psychologists seek to understand how to develop and enhance intrinsic motivation rather than focusing on the quick-fix use of rewards and gimmicks to motivate individuals extrinsically.

➤ Intrinsic motivation is self-fueling over the long term because it is based on controllable feelings of enjoyment and competence as compared with extrinsic motivation, which relies on external reinforcers from the social environment.

The answers to the two motivation questions, then, are the following:

1. The best way to motivate people is to help them develop or increase their intrinsic motivation.
2. People who are more motivated than others typically have higher intrinsic motivation to achieve in a certain activity.

But that brings up another question—how is intrinsic motivation developed?

Developing Intrinsic Motivation

Although numerous theories about motivation exist, it is generally accepted that people are motivated to feel competent and self-determining. From the time we are born, we all attempt to be competent in our environment—even toddlers are motivated to gain independence by learning to crawl and walk. As our lives continue, our need to be competent is channeled in various areas through socialization. Some people are motivated to achieve in sport or through fitness training, others in music, others in a career.

➤ The key to understanding motivation is realizing that all humans, regardless of their individual goals, are motivated to feel competent and self-determining.

Although we are all intrinsically motivated to be competent, competence means different things to different people. Research in sport and exercise psychology has shown that individuals have different goals for achievement in sport and exercise. Therefore, to understand motivation, we must understand how each person defines success or competence for him- or herself. Because people engage in exercise programs for diverse reasons, including social affiliation, personal mastery and fitness, or competitive

DO it **Activity 9.5**

Motivation: External or Internal?
Differentiate between the two types of motivation in this activity in your online study guide.

bodybuilding, exercise programs must assess the goals of participants to fuel individual motivation.

Researchers have found that young children participating in physical activity are motivated to have fun, affiliate with their friends and meet new friends, and develop skills. Although adults often think otherwise, these goals motivate children more than the goal of winning does! For this reason, physical activity programs should provide opportunities for children and youth to attain these goals. Practices and competitions should be structured to be challenging and enjoyable for all children so that they develop their skills, learn important cooperative and social skills, and experience the fun and enjoyment of physical activity. The formula is simple. If we like something, we do it more. If we do it more, we become better at it. When we become better at it, we like it more. Think of this as the positive cycle of motivation (see figure 9.3).

FIGURE 9.3 Positive cycle of motivation.

Using Rewards to Enhance Motivation

Rewards, a type of extrinsic motivation, can be used in a positive way to enhance people's feelings of competence, which then serve to increase their intrinsic motivation. Many potential rewards exist in the physical activity environment (e.g., trophies, money, status, weight loss, fitness, popularity). What is important is how individuals perceive rewards that they obtain. If people perceive rewards as controlling, then the rewards serve only to enhance extrinsic motivation and will actually weaken intrinsic motivation. In sport, extrinsic rewards that are used to control athletes such as college scholarships, trophies, and all-star selections often decrease intrinsic motivation. Incentives such as weight loss and prizes for attendance may be motivating to exercisers but could be problematic if used to coerce or manipulate individuals. Behavioral contracts and attendance lotteries are examples of incentives used for exercise adherence, but research indicates that these inducements alone have no lasting effect on exercise behavior. Rewards should be used not to control athletes or exercisers but rather to make them feel that they've earned praise and reward through their effort and competence. Moreover, teachers and coaches must praise and reward the right behaviors. When young children first attempt physical skills, their coaches or teachers should reinforce these mastery attempts. With any luck, children will learn to value individual improvement and mastery in physical activities, which serves to enhance their self-worth.

Energy Management in Sport and Exercise

As just discussed, motivation involves intensity of behavior and the urge to be competent and successful. This intensity and urge to be competent creates many different types of energy, or feeling states, in individuals as they participate in sport or exercise. Have you ever had your performance disrupted because you were tense or nervous? On the other hand, can you remember performing "in the zone" where time seemed to stand still and you experienced an almost spiritual feeling of focus, sense of control, and total immersion in your activity? These examples, gut-wrenching anxiety as well as the exhilaration of peak experience, represent two ways in which individuals experience and use energy in physical activity. **Arousal** is a state of bodily energy or physical and mental readiness. Thus, competitive energy or arousal results from the ways in which athletes' minds and bodies respond to competition, or the ways in which those involved in exercise respond to physical activity.

Different Feeling States Experienced by Athletes and Exercisers

In chapter 4 we discussed the subjective side of physical activity. From a psychological perspective we could say that we experience arousal as different types of feeling states, which may be low or high in intensity as well as pleasant or unpleasant. People also experience feeling states as thoughts (mental) as well as bodily feelings (physical). Figure 9.4 shows examples of mental and physical feeling states categorized by intensity and direction. Research has shown that participation in physical activity, particularly exercise, creates pleasant feeling states (e.g., joy, calmness) and diminishes negative feeling states (e.g., depression, anxiety).

	Unpleasant		Pleasant	
	Mental	Physical	Mental	Physical
High intensity	Worried Concerned Angry Nervous	Tense Jittery Tight	Focused Confident Excited	Explosive Energized Vigorous
Low intensity	Tired Lethargic Confused Sad	Fatigued Depleted Tired Exhausted	Calm Relaxed Sleepy	Loose Fluid Relaxed

FIGURE 9.4 Mental and physical feeling states categorized by intensity and direction.

All types of feeling states are potentially positive and negative in terms of their effects on performance and well-being. A feeling state can feel unpleasant to a person but still have a positive effect on his or her performance. Likewise, a feeling state can be pleasant (e.g., calmness) but have a negative influence on performance. The key is for individuals to understand their unique patterns of feeling states to know how to optimize their energy and focus for physical activity.

People usually assume that stress is a form of negative energy, but stress actually serves the important purpose of stimulating growth. **Stress** may be defined simply as a demand placed on a person. We stress athletes all the time through weight training and exhaustive physical repetitions of their sport skills, and a careful progression of physical and mental stress builds their stress tolerance so that they can withstand the stress of competitive performance. But stress often creates **anxiety,** which is an unpleasant, high-intensity feeling state that typically results from a demand or threat. The anxiety most widely studied in sport psychology is based on the threat of failure or evaluation, as when athletes compete in public. As discussed previously, anxiety is of interest in exercise psychology based on research demonstrating that physical activity decreases feelings of anxiety, thus contributing to mental and physical health. Research has also demonstrated that social physique anxiety, or apprehension about one's body and appearance, is predictive of exercise behavior (Crawford & Eklund 1994). For instance, the less apprehensive people feel about their appearance, the more likely they'll be to participate in activities that require them to reveal more of their bodies (e.g., gymnastics or swimming).

Overall, stress creates intense and unpleasant feeling states, but it often has a positive influence on performance. Stress becomes a problem only when individuals do not allow adequate recovery time from life or competitive demands. Chronic stress without adequate recovery burns out the energy reserves of athletes and others engaged in physical activity.

When individuals' natural resources of energy are at their optimal state, the experience of flow occurs (Csikszentmihalyi 1990b). Exercisers speak of "runner's high," that elusive state in which physical activity is effortless and immensely enjoyable. Athletes train to achieve the

Revisit Your Flow Experience

Think back to a time when you experienced flow. Describe the activity that you were doing and what the experience felt like. What made it different from other experiences? How long did it last, and why do you think it went away when it did?

ultimate high of "being in the zone" and enjoying the ultimate thrill of sport characterized by peak performance. Thus, helping physically active individuals identify the special recipe of feeling states that lead to their optimal energy zones and flow will certainly enhance their performance and overall experience of physical activity.

How Energy Influences Performance

Generally, people perform better at moderate levels of arousal. The optimal intensity of arousal, however, is unique for each person. Individual factors that may affect preferred arousal levels include personality characteristics, coping ability, and skill level. In addition, the optimal intensity of athletes' arousal differs according to situational demands of different activities. A weightlifter can be at a much higher arousal level and perform well as compared with a golfer, whose optimal level for performance is somewhat lower. Arousal influences performance through its effect on muscular tension and the ability of people to focus their attention properly. The muscles of athletes performing in sport often become tense and lose the smooth coordinated movement needed to perform optimally. Their attention tends to turn inward toward worry and thoughts of inadequacy, or their attention narrows to the point where they miss important competitive cues they need to read to perform optimally.

Yuri Hanin (2000) has pioneered the study of pleasant and unpleasant feeling states related to sport performance. Because of the personalized nature of optimal feeling states, athletes must understand which types of competitive energy work for them and then create and maintain their special recipes of feeling states that lead to their optimal energy zones. Optimal energy zones have been shown to be important predictors of individual athletes' performance. An example of an individually created optimal energy zone for a javelin thrower is shown in figure 9.5. The player selected the pleasant (P+) and unpleasant (U+) feeling states that energize him in the most positive ways, as well as the pleasant (P–) and unpleasant (U–) feeling states that negatively affect his performance. He then identified his preferred intensity levels for each feeling state on a scale from 0 to 10, and he plans and implements mental preparation strategies

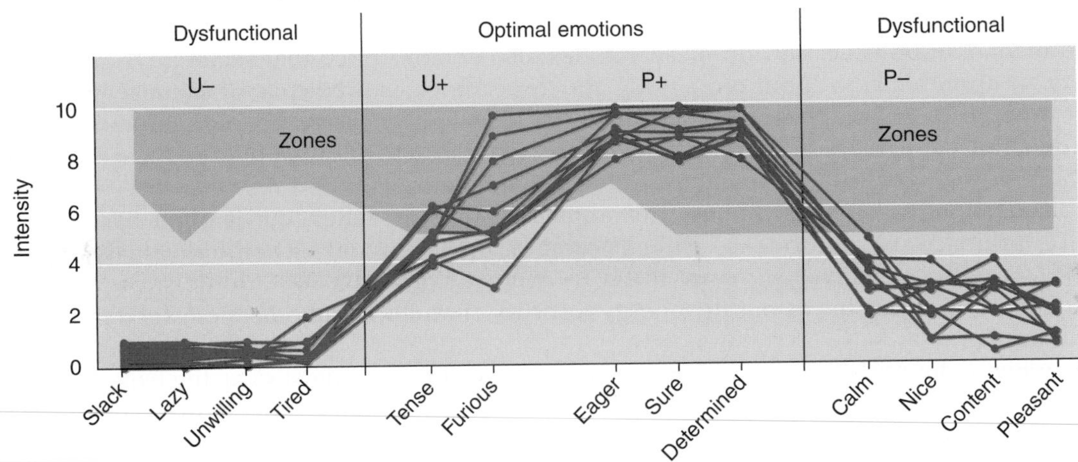

FIGURE 9.5 Optimal energy zone for an elite javelin thrower (Hanin 2000).

Reprinted, by permission, from Y. Hanin, 1997.

Activity 9.6

Your Optimal Energy Profile
In this activity, you'll identify which personality characteristics increase or decrease because of physical activity.

to create the feeling states that make up his optimal energy zone. He can also plot his feeling state levels for each competition and use this as feedback from which he can adjust mental preparation strategies.

Interpersonal and Group Processes

Our overview of knowledge has examined the three broad areas of personality, motivation, and energy as they relate to sport and exercise behavior. In this section, our focus is on interpersonal or group processes that influence individuals' behaviors in different ways, including the presence of others and group membership. In addition, we briefly explain the areas of aggression and gender socialization as behaviors or characteristics that result from interpersonal social processes.

Presence of Others

How is your behavior different when you are alone as opposed to when you are with other people? Why do some people prefer to exercise at home whereas some like to exercise with others in a public gym? Why do some athletes perform better in front of a big crowd? Does the home advantage really exist? Since the turn of the 20th century, researchers have been fascinated with the effects of an audience on human performance. Research has shown that the presence of other people increases our arousal, which then may hurt or help our performance. Generally, spectators have a negative effect on someone who is learning a skill and a positive effect on someone who is very skilled. Think about a beginning golfer who is learning the game and has to tee off in front of a large group on the first hole. The presence of spectators increases her arousal, and because her skills are not well learned, this arousal causes her to hit a bad shot. But consider a professional tennis player who makes the finals at Wimbledon for the first time. The huge crowd at center court is likely to inspire and elevate her performance.

Although this explanation sounds simple, the effects of others on our behavior are complex. For example, research has shown that it is not the mere presence of others that causes this effect but rather people's perceptions that others are evaluating them. We know from the previous section that every individual has a different optimal arousal zone, so the presence of others could influence performance differently based on individual responses to changes in arousal. One thing we do know from research in this area, however, is that people should avoid situations of excessive evaluation or analysis when learning sport skills; this pressure can hurt the learning process by adversely affecting beginners' arousal levels and quality of attention.

Researchers have also documented the home advantage, showing that teams playing at home sites win more often than do those playing at away sites. The reasons for this home advantage, however, are less clear and could even be attributed to expectancy, or the fact that athletes expect to play better at home because they believe this popular notion.

The presence of others applies a bit differently to exercise, because the people present are usually socially supportive workout partners or simply other individuals concurrently exercising on their own (e.g., while weight training at a public facility). Family social support is a strong predictor of exercise maintenance for women, and individuals who exercise with their spouses have higher rates of exercise adherence than those who exercise alone. Research has demonstrated that physical activity levels of children are related to the modeling of such activity, shared activities, social support, and encouragement within families. Several studies have demonstrated that active children perceive more parental encouragement than inactive children. The social isolation of many elderly people is problematic because social incentives to engage in physical activity decrease with age. Community-based programs can ensure the important social support for habitual physical activity participation for all ages.

➤ Family support and modeling are highly predictive of the physical activity participation of children, whereas elderly people tend to become less physically active because of social isolation.

Group Membership

The area of group dynamics focuses on how being a part of a group influences behavior, as well as how certain psychosocial factors influence collective group behavior. Are people more likely to adhere to exercise programs if they participate in groups rather than exercising individually? Why is it that some teams have better chemistry than others? And why do teams with better chemistry often perform better than more talented teams with poorer chemistry?

A group performs better and group members are more satisfied when the group is cohesive. **Cohesion** is the tendency for groups to stick together and remain united in pursuing goals (Eys et al. 2006). Thus, coaches and exercise leaders should strive to develop and nurture cohesion in their teams or groups in several different ways. Emphasizing uniqueness or a positive identity related to group membership facilitates cohesion. Nicknames such as the "Sweat Net" or the "Lunch Bunch" often reflect cohesion and group identification. Athletes sometimes shave their heads or wear team jackets to demonstrate their solidarity and commitment to the group. Cohesion also increases when individual members of teams understand and accept their roles within the group. Studies on adherence to exercise programs indicate that most people prefer to exercise with another person or in groups, an arrangement that enhances enjoyment through affiliation and social support and strengthens commitment to the program. Cohesion and friendship within the group constitute one of the strongest predictors of exercise adherence.

➤ Cohesion and group membership facilitate physical activity performance, but social loafing may occur unless individuals are monitored and their inputs are viewed as important to overall performance.

We also know that success breeds cohesion, so some early successes are crucial in the development of group dynamics. When groups experience success initially, cohesion develops accordingly. But we should realize that cohesion is a dynamic quality that is always changing within a group. For example, conflicts that require resolution may arise, but the group may still be cohesive. Also, the quest for cohesion can go too far, resulting in extreme cohesion that involves conformity and elitism. Although a group needs a strong, positive shared identity, the group should celebrate and respect diversity among its members.

Also of interest in group dynamics is how group membership influences individual performance. **Social loafing** refers to a decrease in individual performance within groups (Heuze & Brunel 2003). This decline occurs because individuals believe that their performance is not identifiable and that other group members will pick up the slack. Social loafing is not a conscious process—people do not decide to loaf. Rather, it is a psychological tendency when people perform in a group. But increasing the identifiability of individuals by monitoring their performances easily reduces social loafing. Common ways of accomplishing this in sport are video analysis of individual performance and individual statistics that break out a single athlete's performance from the total performance of the team. Social loafing can be monitored in exercise groups if performance totals (e.g., number of sit-ups, amount of weight lifted, distance run) are recorded so that progress can be checked. Research clearly demonstrates that identifying and monitoring individual efforts eliminates social loafing.

Group involvement and social support improve individual adherence to exercise programs.

Aggression

Sport competition in Western society often leads to aggression or violence. Why do fights break out in hockey matches? Why does crowd violence often erupt at soccer matches? **Aggression** is behavior directed toward inflicting harm or injury on another person. A main source of aggression in competitive situations is the inevitable presence of frustration.

Frustration often results when a person's goals are blocked; and in competitive sport, the main objective is to block the goal achievement of the opponent. Social learning theory views aggression as a learned behavior that develops because of modeling and reinforcement. Ice hockey players are glorified for fighting with opponents, and baseball players are encouraged and even expected to charge the mound and aggress against the pitcher after being hit by a pitched ball. Children learn aggressive behaviors at an early age. Research also links aggression to levels of moral reasoning, and athletes view aggression as more legitimate because of their lower levels of moral reasoning when compared with nonathletes (Shields & Bredemeier 2007).

Most people believe that competition or exercise reduces aggressive impulses in humans by providing a release or purging of aggression (called catharsis). But research does not support this claim. Instead, it shows that aggressive tendencies increase after competing, engaging in vigorous physical activity, or watching a competitive event. Thus, competitive and physical activity participation and spectatorship do not serve as a catharsis for aggressive responses.

Violence and aggression occur in sport because of frustration and social learning, such as watching sport role models engage in aggressive acts.

Gender Socialization

The social processes of **gender** formation and maintenance have been studied extensively, with important implications for physical activity behavior (Gill 2007). Gender is defined as social and psychological characteristics and behaviors associated with being male or female. A popular myth is that differences in the thoughts, feelings, behaviors, and physical performance capacities between males and females result from the biological sex differences of being born female or male.

This biological explanation for differences between males and females ignores the social complexity and variations in gender-related behavior and performance. For example, most people assume that males can naturally throw a baseball harder and farther than females because males are stronger. People also assume that males are better long-distance runners than females are because of greater muscular and cardiovascular endurance. But substantial overlap exists between male and female performance on all motor skills. Although the most highly trained male is stronger than the most highly trained female, some females are stronger than many males.

➤ Aggression, or behavior intended to harm another person, has been shown to be socially learned behavior and is related to lower levels of moral reasoning.

Although our society loves to assume that all females and males fall into stereotypical groups according to popular beliefs about limits of physical activity performance, males and females are more similar than they are different. Most gender differences apparent in physical activity behavior are based not on biology but on the differential socialization patterns of girls and boys, which typically benefit boys in terms of opportunity, support, and expectations for physical activity proficiency.

Besides perceived strength and motor proficiency differences, gender differences are also assumed and have been found in various psychological characteristics such as self-confidence, aggression, and competitiveness. As we saw in chapter 7, the documented gender differences develop over time and are influenced by rigid gender socialization and stereotypical expectations of conformity to sociocultural norms for distinct female and male behaviors. Much of the gender research has neglected to consider socialization and thus has reinforced existing and limiting gender stereotypes.

Developmental Concerns in Psychology of Sport and Exercise

Because human beings undergo constant physical and psychosocial development, examining the psychological implications of physical activity participation from a developmental perspective is important. As you saw in chapter 2, participation in physical activity tends to decline with age not only among adults but among children as well. A decrease of almost 50% occurs between ages 6 and 16. Ideally, a life span approach to the study of physical activity behavior allows us to understand how participation in physical activity relates to the psychological characteristics and responses of individuals at different stages in their lives. This knowledge can help physical activity professionals target different intervention strategies for different populations based on age and development.

Physical Activity in Children

➤ Successful physical activity programs for children emphasize fun, challenge, skill and fitness improvement, and social affiliation to match their participation motives.

Considerable research has examined the psychosocial aspects of children's participation in physical activity (Weiss 2004). In the United States, an estimated 25 million children under age 18 are involved in physical activity programs. These early sport experiences can have important lifelong effects on the psychological development of children. Parents' activity levels, beliefs about exercise, education, and encouragement are all strong determinants of exercise behavior in children. Most children cite multiple reasons for wanting to participate in physical activity, but most are intrinsic, such as having fun and learning skills. Most children withdraw from physical activity because of interest in other activities, yet they often cite some negative factors such as lack of fun and too much pressure. The sharpest age-related decrease in exercise participation occurs in late adolescence when young adults begin to make the transition to independent living. Physical fitness and aerobic activities have been linked to improving self-concept in children far more than participation in competitive sport has (Gruber 1986). A reason for this association may be that immediate feedback is derived from fitness training whereas a feeling of physical mastery is often a more long-term occurrence in sport participation. Physical activity programs that are beneficial for children emphasize fun, challenge, skill and fitness improvement, and social affiliation to match their participation motives.

Although most children who participate in physical activity do not experience excessive anxiety, stress can be a problem for certain children in specific situations. Children with low self-esteem, high trait anxiety, and frequent worries (about failure, social evaluation, and adult expectancies) are at risk for stress in their physical activity participation. Also, situations that maximize pressure and importance, such as championship events, may cause increased stress in children.

An important consideration for adult physical activity leaders is the modification of physical activity to make it developmentally appropriate. Modified competitive games have been shown to increase enjoyment and skill development of children. An example of a modified game is the coach-pitch variation in youth baseball. Children at a young age lack the skills to pitch a ball effectively, so a modified game in which the coach pitches to batters increases the number of hits, defensive activity, learning, and enjoyment. Modifying activities so that they are developmentally appropriate is an excellent way to ensure success. Positive experiences in physical activity as a child are crucial in developing the skills, knowledge, and habits for a healthy lifestyle in adulthood. Communities should provide safe places and facilities that are attractive to children to engage them in physical activity.

Early sport experiences help children's psychological development.
Photo courtesy of Robin Vealey

Physical Activity in Older Adults

Sport and exercise psychology specialists have begun to systematically study the psychological aspects of physical activity participation for older adults (Netz 2007). Research has documented the psychological benefits of physical activity for older adults. As with younger persons, physical activity has been shown to increase feelings of well-being and self-confidence and reduce anxiety and depression in older adults. In addition, physical activity has been shown to enhance the cognitive function of older adults, thought to occur through improved cerebral blood flow and energy metabolism and enhanced neurotransmitter activity. Physical activity programs should be specially adapted to meet the needs of older adults, who often seek social affiliation and need to retain a basic level of physical competence and a positive psychological outlook during this stage of their lives.

Intervention Techniques for Physical Activity

Often called mental training, intervention techniques are used to develop thinking skills (e.g., self-talk) and behavioral strategies (e.g., goal setting) that can enhance physical activity behavior. Intervention techniques may be used to increase exercise adherence, improve sport performance, develop important life skills for young people participating in physical activity, aid in rehabilitation from injury and disease, and enhance career transition and retirement from sport. Health promotion specialists most effectively administer exercise interventions, in both individualized and community-based interventions. Worksite-based exercise interventions, a promising approach to increasing levels of physical activity, have developed because of the rising cost of providing health insurance for employees. Professionals with graduate degrees in sport psychology typically conduct sport psychology interventions, yet a growing trend is the education of coaches in the use of intervention techniques. Coaches can then implement mental training with athletes as part of their overall physical training program.

➤ Intervention techniques in sport and exercise psychology can increase exercise adherence, enhance sport performance, develop life skills, aid in injury rehabilitation, and ease career transitions and retirement from sport.

DO it Activity 9.7

Intervention Techniques

Analyze training interventions in this activity in your online study guide.

Goal setting is a basic technique used to focus on specific attainable behaviors presented as reachable yet difficult. Goals are most effective if they are difficult and systematically monitored and evaluated. Simply setting goals at the beginning of a season or exercise program will not facilitate development unless these goals become the constant focus of attention. Other effective goal-setting practices include the use of short-term goals as progressive steps toward reaching a long-term goal and an emphasis on performance or controllable goals as opposed to outcome goals such as winning a race. The intervention checklist shown in "Helping People Get Started With Exercise" demonstrates how to set goals and monitor behavior to enhance commitment to physical activity.

Another popular intervention technique is self-talk, or personal statements that we all make to ourselves. The basic premise behind self-talk is that the things we say to ourselves drive our behaviors. The goal of effective self-talk is to engage in planned, intentional productive thinking that convinces our bodies that we are confident, motivated, and ready to perform. Athletes learn to identify key situations or environmental stressors that cause them to choke, and then plan and mentally practice a refocusing plan that they can use to focus attention appropriately in a given situation. Exercisers can use self-talk to strengthen their commitment to completing workouts by telling themselves that they feel strong, by telling themselves that they are gaining health benefits, or by seeing their bodies as finely tuned machines. Productive self-talk is critical in maintaining self-confidence, which is a major predictor of exercise adherence and sport performance.

Attentional control and focusing may be the most important mental skill for any type of physical activity. In sport, performance depends on the cues that athletes process from themselves and the social and physical environment. In exercise, learning how and when to focus on one's body (called association) as well as how to distract oneself (called dissociation), with music or visual imagery, for example, can enhance the exercise experience and help performance. Attentional control strategies such as centering (Nideffer & Sagal 2006) allow individuals to select relevant cues and design physiological (e.g., deep breathing) and psychological (e.g., feeling strong, quick, and confident) triggers to focus attention optimally.

Imagery, a mental technique that programs the mind to respond as programmed, uses all the senses to create or recreate an experience in the mind (Vealey & Greenleaf 2006). Jack Nicklaus, perhaps the greatest golfer of all time, says that hitting a golf shot is 50% mental picture. Research has demonstrated that imagery enhances motor performance, and although it cannot take the place of physical practice, it is better than no practice at all. Many elite athletes use imagery, and they often cite it as an important mental factor in their success. Novice athletes can use imagery to create positive mental blueprints of successful performance, and exercisers can use imagery to visualize their muscles firing and getting stronger when training for fitness. Exercisers can use imagery to dissociate, or to direct attention away, from the repetitive exercise activity to visualize motivational or pleasant things. Many runners like to use personal audio players when they run to help them create pleasant mental images.

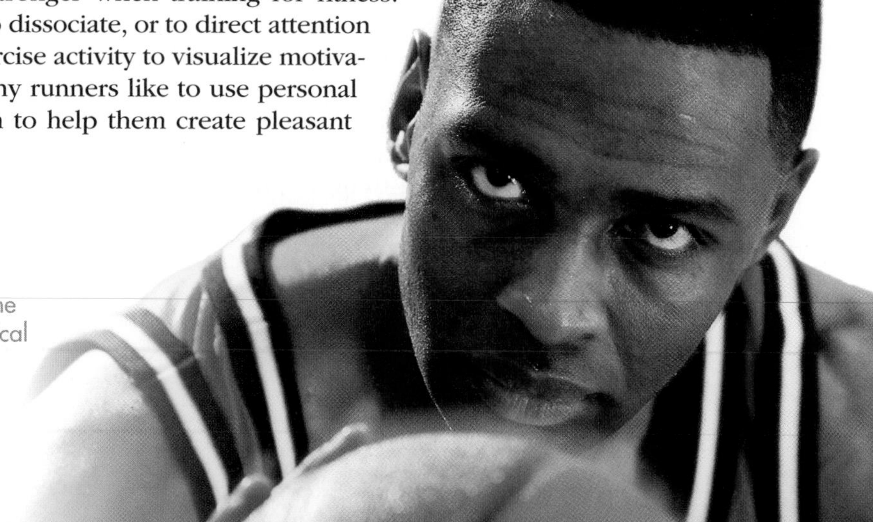

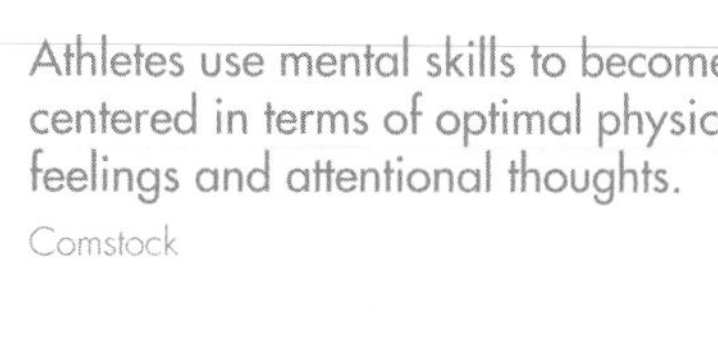

Athletes use mental skills to become centered in terms of optimal physical feelings and attentional thoughts.

Comstock

Helping People Get Started With Exercise

- Introduce people to light and enjoyable exercise (minimal risk of discomfort or injury).
- Help them establish personalized and realistic goals (avoid comparison with others).
- Use the shaping technique by rewarding small achievements along the way to reaching a goal (progressive, graduated exercise goals).
- Avoid situations in which the individual may feel vulnerable or have lapses in confidence (as when activity is viewed as inappropriate or sex typed).
- Use immediate, informational feedback; provide encouragement; and help people regulate their confidence by practicing productive self-talk.
- Physically guide them through movements in which they initially lack confidence.
- After they have gained skill and knowledge, encourage them to take gradual control of their exercise behavior through self-monitoring and self-regulation.

Adapted from R.A. Smith and S.J.H. Biddle, 1995, Psychological factors in the promotion of physical activity. In *European perspectives on exercise and sport psychology*, edited by S.J.H. Biddle (Champaign, IL: Human Kinetics), 85-108.

People use physical relaxation techniques to control their autonomic functions including muscular and hormonal changes that occur during physical activity. These techniques allow them to engage in physical activity with greater mastery and control over the responses of their bodies to environmental stimuli. Examples of physical relaxation techniques are breathing exercises, muscular tension–relaxation techniques, and various types of meditation. Momentary relaxation before exercising is effective in facilitating smooth, coordinated muscular effort. Individuals can learn how to regulate physiological arousal by reducing their heart and breathing rates to induce a more relaxed state. Obviously, people can use physical relaxation techniques in conjunction with goal setting, imagery, and self-talk to optimize both physical and cognitive readiness to engage in physical activity.

> The goal of physical relaxation training is to teach people how to control the responses of their autonomic functions when faced with stressors in physical activity environments.

DO it **Activity 9.8**

Experience Progressive Relaxation
Record the results of your experience doing this activity in the online study guide.

Concluding Your Overview of Knowledge

You've now read about several areas of study in sport and exercise psychology. Although your tour of knowledge didn't cover all the topics in these areas of study, you have had a glimpse of the knowledge that kinesiologists have developed over the years in sport and exercise psychology. You should now have a better idea about such things as how exercise influences our mental health, why intrinsic motivation is critical in fueling exercise and sport behavior, and how professionals in the subdiscipline can use intervention techniques to enhance sport performance and exercise adherence.

Burnout

To cap off the chapter, the next section focuses on a topic that is typically of great interest in sport and exercise psychology—burnout. The topic has become more important because of the pressure we feel in our society to perform, excel, and juggle many responsibilities at the same time.

"I don't know, I just feel burned out." How many times have you heard this statement from people? But what is burnout? Is it the same thing as stress? Is it real? Or is it a lack of mental toughness?

Burnout is real, and here's what it looks like. First, it involves *feelings of mental, emotional, and physical exhaustion*. Second, this exhaustion leads to *negative moods and feelings* (depression, despair) and a *negative change in responses to other people* (cynicism, acting aloof, lack of empathy). Third, people experiencing burnout feel a *lack of accomplishment,* which decreases their performance level and feelings of self-esteem. Finally, burnout causes people to become *disillusioned with their involvement* in an activity such as sport, their career, or an exercise program. Burnout is a complex condition that occurs when certain personality characteristics of people interact with life stressors.

DO it **Activity 9.9**

Burnout

Work through this activity in your online study guide to understand how and why burnout occurs.

To help prevent burnout, coaches and trainers should encourage challenge and variety in training. People in fitness programs should engage in different types of activities (cross-training) such as biking, hiking, rowing, and swimming. They could even consider activities like orienteering or Frisbee golf. Coaches should build variety into sport training to break the monotony of repetitive training. They should help individuals distinguish between overload and overtraining. As the social rewards for sport success and fit bodies have become more glamorous, a strong tendency is to think that more is better in training. People should focus on the quality of training and know when to push themselves and when to rest with adequate recovery time. Most people, especially children, should avoid intense specialization in one activity to the exclusion of all others. We enhance our mental health when we possess multifaceted identities or self-concepts with several areas of interest and expertise. Burnout is not a sign of weakness or a simple response to stress. Athletes and those involved in fitness training programs should watch for early signs of staleness and overtraining and should attempt to stay committed in a positive, passionate way, as opposed to feeling the pressure of *having* to engage in these activities.

Wrap-Up

Sport and exercise psychology, as a young science, has only begun to scratch the surface of understanding the thoughts, feelings, and behaviors related to participation in physical activity. But the knowledge base that has developed over the last four decades is impressive as researchers continue to study personality, motivation, energy and emotion, group processes, developmental concerns, and intervention techniques related to physical activity. Kinesiologists are committed to extending and applying this knowledge to enhance participation in physical activity for all people.

If you are especially interested in what you've learned so far about sport and exercise psychology, you might want to take an introductory sport or exercise psychology course and browse through some of the leading journals in the subdiscipline, such as *Journal of Sport and Exercise Psychology, Journal of Applied Sport Psychology,* and *The Sport Psychologist.* You might also want to talk to a professor about career opportunities in either sport or exercise psychology. We cannot all be Olympic athletes, but all of us can engage in meaningful physical activity to derive personal fulfillment and optimal health and well-being.

Activity 9.10

Use the key points review in your online study guide as a study aid for this chapter.

QA Activity 9.11

These end-of-chapter questions and activities are also in your online study guide. Your instructor may ask you to complete them online and turn them in.

1. What are the ABCs that kinesiologists in sport and exercise psychology study? Identify questions that kinesiologists in this area might study based on these ABCs.
2. What was significant about Coleman Griffith's early work in sport psychology? Why did the subdiscipline not emerge again until the 1960s?
3. Identify the six methods used in sport and exercise psychology and provide one example for how each method is used in the subdiscipline.
4. Does sport build character? Why or why not? Does exercise participation improve mental health? If so, how?
5. Explain why intrinsic motivation is a better source of motivation than extrinsic motivation. How can leaders in both exercise and sport contexts enhance intrinsic motivation in participants?
6. Explain cohesion in groups and discuss how cohesion can facilitate sport performance as well as exercise adherence. What are some ways in which cohesion can be developed and nurtured in groups?
7. Identify some special concerns regarding physical activity participation for children. Similarly, provide some examples of issues regarding physical activity for older adults.
8. Why is flow the goal of intervention strategies in physical activity? Describe how various intervention techniques can enhance flow and peak experiences in both sport and exercise settings.

Biomechanics of Physical Activity

Kathy Simpson

CHAPTER OBJECTIVES

In this chapter we will

- describe what biomechanics encompasses and its relevance to physical activity specialists;
- explain what biomechanists do and where they work;
- examine the goals of biomechanics;
- explore how human movement biomechanics emerged within the field of physical activity;
- observe how professionals in biomechanics and related fields engage in research and analyze movement; and
- discover what is known from biomechanics research about physical activity and bone strength.

Oscar Pistorius, the first person with two below-knee amputations to compete in the Olympics, is shown in the photo that opens the chapter. Imagine that you are serving in one of the following roles related to Oscar:

- Coach
- Physical therapist
- Sports medicine specialist
- Prosthetist
- Marketing director from the company that manufactures the prosthesis (artificial limb)
- Another role of interest to you

From observation of the runner, what specific elements of his technique might you select to determine whether his performance is effective and whether the prosthesis is functioning optimally? Do you find yourself assessing the positioning of the trunk? Arms? Legs? What criteria would you use to determine whether the runner appears to be running safely and effectively? If you watch a video clip (http://www.youtube.com/watch?v=eqg-2bHjjN4), you could look at other qualities, such as the smoothness of the movements or the absence of extra motions.

In this scenario, if you found yourself comparing the runner's body positioning to the "right" technique that elite athletes demonstrate, or if you instead looked for bad alignment of various segments (e.g., bow-legged alignment), then you can recognize that we've all been taught to think that there are optimal methods of moving our bodies. Thinking back on your physical activity experience, particularly if you were involved in organized sport or physical activity classes such as soccer or dance, you probably were taught that proper technique was related to your success at that activity.

Activity 10.1

Qualities of Movement
Evaluate the qualities of movement of another athlete in Activity 10.1 of your online study guide.

The idea of optimal movements is not limited to traditional physical activity or sport. Perhaps you have had a physically demanding job. Did your employer, who was concerned about worker productivity and costs because of worker injuries, teach you and other employees how to perform tasks? Or maybe you have engaged in volunteer work in a physical therapy setting. Using safe techniques for transferring patients from a wheelchair to a bed protects physical therapists' backs from injury.

Do *you* think that there is a "right" movement technique for a given skill? If so, isn't it possible that a person can exhibit unusual movements but still be effective, that is, accomplish the goal of the task with minimal risk of injury?

The following questions are investigated within the scope of biomechanics of physical activity.

- What principles underlie optimal movements?
- Why do we move similarly to one another yet demonstrate unique motions?
- How much variety of movement techniques exists among people that would be considered safe and effective?
- How do muscles, bones, and connective tissue produce various movements?

Biomechanics is the application of the mechanical laws of physics and engineering to the motion, structure, and functioning of all living systems, including plants and animals. In this chapter, however, we will restrict our interest to human movement biomechanics, as defined in the next section.

What Is Biomechanics of Physical Activity?

The subdiscipline of kinesiology in which we use mechanical laws that determine how we move during physical activity, and how forces act on our bodies to affect the structure and functioning of our bodies, is *biomechanics of physical activity* (see figure 10.1), also known as *human movement biomechanics*. The formal definition of **human movement biomechanics,** therefore, is the study of the structure and function of human beings using the principles and methods of mechanics of physics and engineering (see reviews in Atwater 1980; Hatze 1974; Winter 1985).

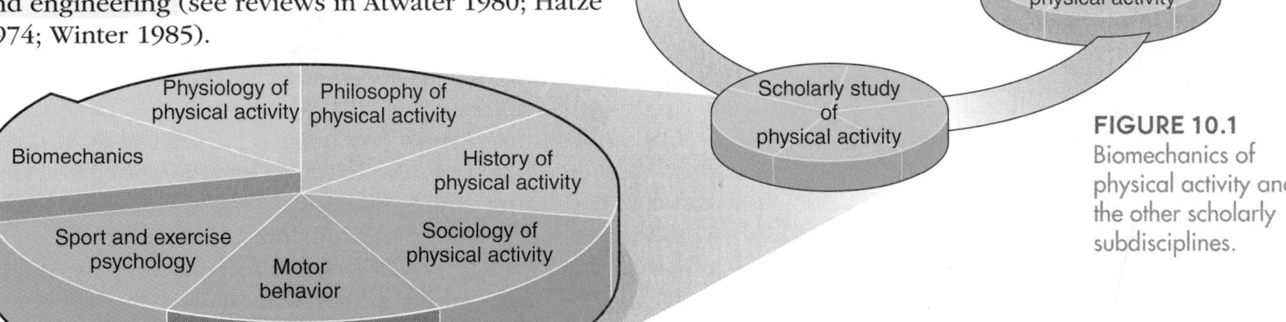

FIGURE 10.1 Biomechanics of physical activity and the other scholarly subdisciplines.

What do we mean by the "structure and function" of human beings? Structure is our anatomy; function is being able to accomplish some purpose. Physical function for a person post–hip replacement may mean being able to rise from a chair. Function for bone means being able to support our body weight during standing, protecting our internal tissues (e.g., our brain) from injury, and providing a rigid surface for our muscles to pull against.

Hence, the material in this chapter focuses on two main themes: (1) function—how we produce forces to generate movement during physical activity, as well as how forces (such as gravity) affect the functioning of body tissues, and (2) how forces affect the structure of our body.

Starting with the first main area, function, according to Newton's laws of physics (whether or not we move, speed up, or slow down our bodies), depends on how forces act on us. We cannot move our body a single inch unless a force acts on us. The runner shown at the beginning of this chapter would not be able to move forward without friction to propel him forward. Without gravity, when the runner pushed off the ground, he would never return back to Earth. Forces also must be applied to us to slow us down, whether we are running, riding in a car, or moving in a wheelchair. Therefore, we will later explore several biomechanical principles to show how these forces make us move in a particular manner.

With regard to the second theme of biomechanics, structure, while we are moving or even while we are just resting, forces are acting on our body. These forces therefore are affecting the structure, that is, the health, of our muscles, bones, and connective tissue. For example, in space, astronauts lose bone due to a lack of sufficient gravitational force. Forces can act to strengthen or conversely to injure our tissues. Hence, certain biomechanical principles have emerged to explain injury mechanisms. By applying certain engineering principles, we can determine the optimum level of magnitude, rate, and number of repetitions at which forces can be applied to human tissues in order to enhance the tissue's health without immediate or long-term damage.

Returning to the example at the beginning of the chapter, according to Newton's law of action-reaction (mechanical principle), each time the runner contacts the ground he applies force to the ground (the "action"). The ground, in turn, applies an equal force back onto the runner's foot (the "reaction"). During the initial ground contact, the ground applies a large

Biomechanists study anatomy and force to help engineers design aids for physical activity.

amount of force to the runner's foot. If the runner runs long distances every week with little rest, the cumulative effect of these forces could produce injuries to the lower limbs. And, because the residual limb of the amputated leg lacks some of the natural methods of absorbing force via the feet, lower leg muscles, and fat pads of the feet, the forces generated between the artificial and residual limb could create injury without proper equipment and training. Thus, biomechanists must understand how much force is absorbed by muscles or fat pads in the feet and how much force is applied to the limb during running. Such understanding has helped us design improved artificial feet that are shock absorbing. Understanding of the anatomy of the fat pads also has informed engineers how to design materials that mimic the fat pads to absorb force during other human movement collision situations for devices such as seat cushions, biking gloves, and shoe inserts.

Although biomechanics has been separated here into two themes, they often are interrelated. As an example, if you plan to work with improving the physical function of older adults, you'll learn that at older ages, our muscle *structure* changes, decreasing leg muscle power (muscle force × velocity of joint motion). Therefore, older adults who have been sedentary throughout life may have decreased *function* while moving about. If they start to fall, they may not have the ability to generate the quick leg power needed to catch themselves (Paterson, Jones, & Rice 2007).

You can see from these examples that all professionals in kinesiology who have training in biomechanics can use their knowledge of how the human body moves, and how it responds to forces acting on the body or within the body, to enhance performance, improve health, and reduce injury potential. These examples also illustrate how biomechanists and other kinesiologists need a good understanding of many areas of kinesiology. Knowledge of biomechanics usually is applied in conjunction with knowledge from a number of the spheres of scholarly study. For example, selecting an appropriate exercise machine requires understanding not only of biomechanics but also of anatomy and exercise physiology. A youth soccer coach needs to understand the interactions between the laws of physics and movement technique, the motor skills that children can learn at their age (motor development), and sport psychology to plan appropriate activities for the players.

What Does a Biomechanist Do?

Career opportunities in biomechanics include such positions as researcher, clinical biomechanist, performance enhancement specialist, ergonomist (industrial task analysis specialist), and university professor. Related positions are those of a certified orthotist or prosthetist.

Biomechanics researchers work in biomechanics laboratories, performing experiments on problems of interest to various industries or assisting with product development. For example, a biomechanist working at a footwear corporation would collaborate with the design engineers to understand the interaction between the structure of people's anatomy, the way they move, and the forces that act on them in order to design better footwear. A researcher in a biomechanics laboratory at a hospital may work with physicians and therapists to understand how best to help patients regain normal walking patterns after some type of treatment.

A performance enhancement biomechanist might work with elite athletes or professional teams or the coaching staff to improve performance. A biomechanics company or sports medicine clinic could have a facility in which biomechanists analyze the technique of athletes

DO it Activity 10.2

An Ergonomist in Action

Examine an ergonomist's approach and apply it to your own work space in this activity in your online study guide.

of any skill level and assess performance effectiveness or detect injury-related errors.

Biomechanists may also work as ergonomists. As you saw in chapter 2, ergonomists are experts in improving the efficiency and safety of employees. Some ergonomists work in research and development departments as part of a team of people who design equipment that humans use, such as gardening equipment, commercial airplanes, factory equipment, and sport equipment. Others may work for specialized ergonomics-focused corporations that perform job-site analyses. These analyses involve evaluating how and why employees are currently performing their work tasks. Next, the ergonomist generates data, then recommends appropriate modifications to the tasks or equipment, employee training, or incentives to encourage employees to modify their behaviors to improve their safety or efficiency. Also closely entwined with ergonomics is human factors design. **Human factors engineers** design technology based on how people process information and respond to the information when performing a task. For example, to design the brakes and steering wheel correctly, an automotive human factors professional would incorporate knowledge of how drivers naturally react when an unexpected object appears in the road. Meister (1999, p. 21) views human factors and ergonomics as one area and defines them as "everything relating the human to technology."

➤ Ergonomics is the use of a biomechanical approach to design the workplace (or equipment) by fitting the workplace (or equipment) to the worker (Chaffin & Andersson 1991). Human factors, an area closely intertwined with ergonomics, refers to designing the workplace based on how people process and respond to the information presented during the performance of a task.

Clinical biomechanists work in medical settings, such as research hospitals that include biomechanics laboratories, so that physical therapists and physicians can have biomechanical analyses performed for their patients or research participants. For example, the Shriners Hospital in Greenville, South Carolina, has a biomechanics laboratory. A team consisting of the surgeon, a physical therapist, and a biomechanist work together to assess whether a child with cerebral palsy requires surgery and, if so, to determine postoperatively if the surgery sufficiently improved the child's ability to walk. The biomechanist's responsibility is to perform and interpret the gait (walking) analysis and to communicate the findings to the surgeon and physical therapist (see the interview with Gene Jameson).

Although orthotists and prosthetists are not biomechanists, they use many biomechanical concepts and methods. Orthotists assess a client's anatomical structure to understand how the client's anatomy shapes his or her ability to complete tasks effectively. Anatomical deviations may be producing injury or pain; and in children they may be the result of abnormal bone growth, improper muscle functioning, neural defects, or some combination of these. The orthotist then fits the patient with an orthosis, a device that supports a limb, such as a molded plastic brace that holds a child's foot in the proper position. He or she then reassesses the effectiveness of the client's movements and watches for potential problems like injury, pain, or abnormal tissue changes such as

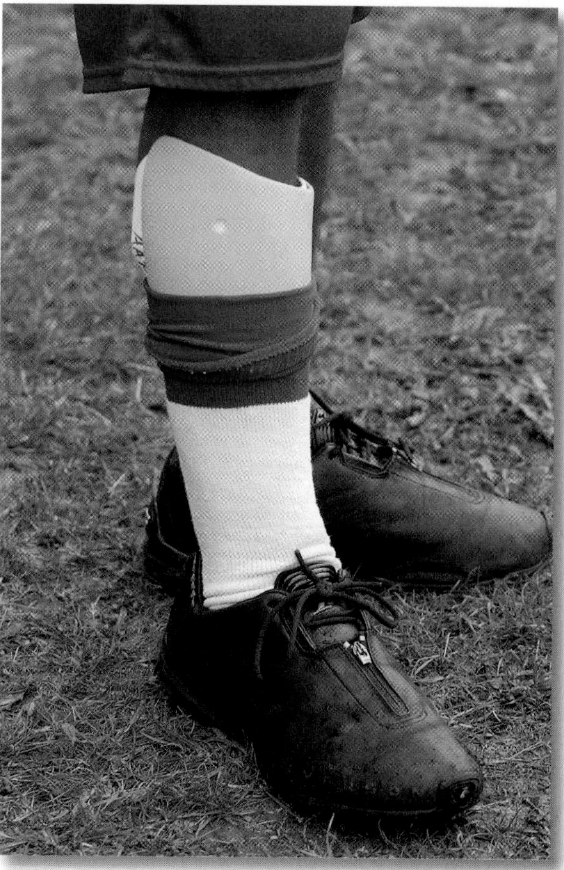

Orthotists help fit devices that support a limb.

Interviews With Practicing Professionals

Photo courtesy of Gene Jameson

Gene Jameson, MA
Biomechanist

Gene Jameson is a clinical and research biomechanist in the Motion Analysis Laboratory at Shriners Hospital for Children in Greenville, South Carolina. The primary focus of this laboratory is provide gait analyses of children with cerebral palsy to the physical therapists, physicians, and other members of the children's hospital team. These outcomes are used to assist the team in determining the best course of action to help improve the child's ability to walk.

Q: Please describe the research that is conducted in the Motion Analysis Laboratory.

A: We use three-dimensional motion analysis as well as electromyography and pedobarography (measurement of the pressure applied to the bottom of the foot during walking) combined with video analysis and clinical examination to assess the ambulatory capabilities of our patients. We then make recommendations for improvement that range from physical therapy interventions to orthotics assessments to suggestions for surgical options. One of the most rewarding aspects of my job is being able to compare preoperative data to postoperative data and see genuine improvements in a child's ability to walk.

Q: What are some of your day-to-day responsibilities?

A: I am responsible for daily clinical testing and interpretation of the results. In addition, I participate in research projects related to the clinical application of biomechanics to gait disorders. I believe that the research performed by the hospital team helps us understand what interventions are most effective, which also is rewarding. Outside of the laboratory, I currently participate in national and international conferences, presenting my work in the methodology and clinical application of biomechanics primarily as it relates to pediatric gait.

Q: Do you have prior work experience as a biomechanist?

A: Yes, in the field of sport biomechanics, I worked with baseball pitchers, boxers, triathletes, figure skaters, weightlifters, and archers at the American Sports Medicine Institute and the United States Olympic Training Center.

Q: In your opinion, what are some of the skills a successful biomechanist should have?

A: A biomechanist in my position must have a sound understanding of the fundamentals of physics and mechanics as they apply to human movement. With this knowledge, I am able to serve as a liaison between kinesiologists, engineers, and clinicians. The ability to make use of the knowledge and skills of biomechanics and other kinesiology areas comes into play daily in my job.

---

weakening muscles. Prosthetists are often certified orthotists, too. However, a prosthetist's main focus is to help a person obtain and use a replacement limb such as an artificial hand. By understanding biomechanics, both the orthotist and prosthetist can select and adjust the appropriate device and help the client function optimally during daily life and physical activity.

Many biomechanists work in universities, in departments or colleges ranging from kinesiology to engineering to medicine. The professor in a kinesiology-related department teaches biomechanics to a variety of physical activity and allied health professionals, and conducts research to help us better understand how the concepts related to generating and absorbing forces help us move and influence the health and structure of our tissues. Some biomechanists may also be clinicians and integrate the care of patients with research.

Biomechanics is a growing field. Many people who use the knowledge produced by human movement biomechanists are physical activity and allied health professionals. As shown in the next section, biomechanics can be applied to any movement situation, be it any daily activity (e.g., reading this book!), working, exercising, dancing, participating in sport, or engaging in physical or occupational therapy.

➤ Movements made during physical activity, the structure of our bodies, and the ability of our tissues to function effectively are shaped by forces and their effects. The human movement biomechanist uses principles from the mechanical branches of physics and engineering to discover how these principles can be used to improve the structure and functioning of our bodies.

Goals of Biomechanics

Physical activity and allied health specialists or other professionals apply human movement biomechanics to any situation involving human movement. They may ask the following practical questions of interest:

- How can I improve my student's sport performance?
- What can I recommend to my physical therapy patient to prevent another back injury from occurring during work?
- What exercise regimen will improve the bone mass of my clients?
- Which surgical technique during a hip replacement will enhance my patient's ability to walk up stairs?

If you work with "clients" in your profession, you will always be seeking answers to many practical questions of interest. However, your best choices of answers are based on our fundamental knowledge of how we produce movement and how the structure and functioning of our body can be changed. Therefore, *the first major goal of human movement biomechanics is to understand how people use and are affected by the fundamental principles of mechanical physics and engineering that explain how forces influence our structure and function.* Note that biomechanics is unique in relation to physics and engineering because humans are not just objects that act passively when a force acts on them. People's movements are affected by a variety of factors. First, people take in visual and other sensory input that provides information regarding the pressure applied to tissues, stretch applied to ligament, and their location in space. Performers make use of this input as well as their perceptions regarding features in the environment, such as the movements of the other performers or the body language of the disapproving coach or the hardness of the ground. For example, runners unconsciously hit the ground with less force when they perceive that the ground has become harder. Second, performers also bring their own movement experiences to the current situation, affected by what they have been taught. Third, people's movements also are influenced by the interactions among other myriad factors, for example cultural expectations, socioeconomic class, gender, and anatomy.

The following are some examples of fundamental knowledge currently being sought:

- What factors influence positively the composition and strength of damaged ligaments?
- How do the speed and location of impact between the head and ground affect the pattern of loads applied to the skull, neck muscles, and soft tissue?
- What are the neural, muscular, and mechanical factors that explain why humans display particular preparatory movements prior to generating maximal muscle power (e.g., we naturally choose to bend our knees just before jumping upward)?

Biomechanists analyze physics and engineering of movement and then apply that knowledge to help performers improve.
AP Photo/Steve McEnroe

www. **Web Search 10.1**

A Career in Biomechanics?
Search the Internet for career information and document it in this activity in your online study guide.

The second major goal of biomechanics is to apply to our theoretical understanding gained through attainment of the first major goal to determine how best to (a) improve performance effectiveness (improve function) or (b) increase the safety and health of those body tissues that are affected by forces or those tissues involved in physical activity (improve structure). This major goal of biomechanics can be applied to almost any situation that involves physical activity: the activities of daily living, repetitive work tasks, exercise, occupational or sport training, sport performance, music, dance, even lying in a bed. Thus, if we understand how living organisms can best exploit the mechanical laws that explain how motion is produced (or reduced), then we can intelligently select optimal movement techniques for a performer to use to complete a movement task.

From major goal 1, by understanding how forces influence the health and strength of tissues and how humans exploit mechanical laws to produce movement, we can use this theoretical knowledge also to improve the health and safety of tissues (major goal 2). Thus we can determine how to prevent baseball pitching injuries or develop more effective methods to repair injured tissues. Or, consider this injury prevention question: Does stretching before exercising prevent injury?

To demonstrate how we use the knowledge gained from the first major goal of biomechanics (generating fundamental knowledge) to accomplish the second major goal (application of that fundamental knowledge), let's again use the examples of fundamental knowledge questions just listed. By understanding how ligament healing is affected by forces and other factors (e.g., age) at the basic cellular level, we could develop more effective rehabilitation treatments. If we understood how the speed and location of impact of a head collision influenced the forces acting on the tissues of the head, we could better understand the mechanisms of head injuries that occur during football. If we understood the fundamental neural, muscular, and mechanical concepts that underlie maximal muscle power generation, we could teach people more effective jumping movements.

Clearly, we can see that the goals of biomechanics are applicable to many spheres of physical activity. Although biomechanics is a subdiscipline of kinesiology, it is closely related also to the subdisciplines within the fields of physics and engineering, called mechanical physics and mechanical engineering, respectively. What is unique about biomechanics is that it deals with applying the laws of physics, engineering, and biology to human beings. Branches of engineering closely allied to biomechanics (e.g., biological engineering) investigate problems focused on the design of objects, such as computer furniture, artificial body parts, prostheses, and exercise devices. Bioengineers also are interested in the materials that interact with the human body. Some examples are dental filling materials, hip replacement materials, and living-tissue materials (e.g., a burn patient's own skin, cultivated outside the body to replace destroyed skin). Separating the goals of biomechanics from the goals of this and other engineering areas is becoming increasingly difficult.

Goals of Biomechanics

Major goal 1: To understand how the basic laws of physics affect and shape the structure and function of the human body

Major goal 2: To apply our understanding of how the basic laws of physics affect and shape the structure and function of the human body in order to (a) improve the outcomes of our movements (i.e., performance effectiveness) or (b) increase or maintain the safety and health of our tissues

Tissue biomechanics, therefore, is an area of investigation that uses engineering techniques to investigate the properties of human and animal tissues, such as tendons, ligaments, muscles, and cartilage. In conjunction with engineers and orthopedic surgeons, biomechanists can make decisions about the surgical repair materials or implant materials that will best help bone adhere to an artificial joint implant.

To understand biomechanics and its relationship with these other areas, a brief look at the history of biomechanics will be helpful.

➤ Although biomechanics related to physical activity is closely tied to kinesiology, many of the guiding principles and concepts of biomechanics come from the subdisciplines of mechanical physics, mechanical and biological engineering, and biology. For medical applications, a great deal of overlap exists between biomechanics and biological engineering and other areas.

History of Biomechanics

Although relatively new to many people, biomechanics has been around for a long time—some say since humans began to make tools (Contini & Drillis 1966b)! The branch of biomechanics that is considered modern human movement biomechanics, however, is relatively young for a subdiscipline. Several historical events and individual scholarly activities led to the development of this field and still influence the methods that biomechanists use today to investigate human movement problems. Many people besides those included in this very condensed discussion have contributed to the progress of biomechanics.

Early Beginnings

One of the earliest known scientists who investigated biomechanical questions was Aristotle (384-322 B.C.E.), who made observations of animal motion and the walking patterns of humans (Fung 1968). The contributions of Leonardo DaVinci (1452-1519) show the influence of the scientific awakening that occurred during the European Renaissance. DaVinci is credited with developing the first systematic examination of mechanical principles of human and animal movement (Hart 1963). Not until the last quarter of the 19th century and the first quarter of the 20th century, however, when the industrial revolution brought many new inventions, were objective instruments such as cameras developed for measuring movements (Hart 1963). The development of biomechanics and kinesiology in physical education, sport, and dance applications also began in the very late 1800s and early 1900s. Two Swedish men (Posse 1890; Skarstrom 1909) interested in gymnastics were among the first in the United States to apply the term *kinesiology* to the analysis of muscles and movements in a physical education setting (Atwater 1980).

Researchers in the 1920s, including some from areas outside physical activity, had the tools and interest to investigate the mechanics of locomotion (e.g., Fenn 1929), efficiency of human movement in sport and industrial settings (Amar 1920), and energetics of running (Furusawa, Hill, & Parkinson 1927; Hill 1926). Mechanical analyses of basic movements were not emphasized as part of the formal training of dance and physical educators, however, until Ruth Glassow and her students at the University of Illinois in 1924 began classifying activities into categories such as locomotion, throwing, striking, and balance and applied fundamental principles of mechanics and physics to the skills in each category (Atwater 1980).

Several leaders in the 1930s supported the use of kinesiology to understand human movement in physical education and orthopedics. Glassow's work continued to expand at the University of Wisconsin (e.g., Glassow 1932). Thomas Cureton, a professor of applied physics and mechanics at the YMCA College in Springfield, Massachusetts, taught the mechanics of sport and physical activities at Springfield College, wrote articles to support the application of mechanical principles to movement (Cureton 1932), and performed research in swimming (e.g., Cureton 1930). Charles McCloy, a leading researcher in many areas of physical activity, identified specific mechanical principles that influence movement (see McCloy 1960 for review of his work). As a professor of orthopedic surgery, Arthur Steindler also profoundly influenced the study of kinesiology. According to Atwater (1980), one of Steindler's legacies was to coin the word *biomechanics* and give it a definition (Steindler 1942, November), although Nelson (1970) believes that Europeans may have used the word first. These early investigators produced

What Goals of Biomechanics Can Answer Your Question?

Imagine that you could hire a team of biomechanists. Each biomechanist has expertise in one of the goals of biomechanics presented earlier. Think of a question of interest to you, and then determine what each of your biomechanists would be able to contribute to answering your question. Here's an example to get you started: I'm interested in knowing if collegiate volleyball players should be required to wear ankle braces to prevent ankle injury. A biomechanist whose work is shaped around the general goals of biomechanics would provide information about underlying tissue mechanics that affect the strength of ligaments, such as the orientation of protein fibers that help provide strength to ligaments. Biomechanists whose work centers on the applied goals of biomechanics would be able to tell me how wearing preventive ankle braces would affect my performance by analyzing the design of the brace in relation to the structure and function of the ankle joint. Should volleyball players wear braces? In this case, the answer is likely yes, according to some investigators (e.g., Handoll et al. 2000; Pedowitz et al. 2008), but others who have reviewed all of the research believe that substantive proof is lacking (Thacker et al. 1999), and others believe that bracing mostly benefits players who have previously sprained an ankle (Handoll et al. 2000; Karlsson 2002).

a profound effect on the field of physical education and activity through their articles, books, and research because they were among the first to encourage physical education and activity professionals to identify and use biomechanical principles. Their influence was exponential because many of their students became the core of college professors who would teach future physical activity specialists and college professors (Atwater 1980).

Several societal events prompted the development of other forms of biomechanics. During World War I in the United Kingdom, there was high demand for ergonomic and human factors design analyses to assess the effectiveness of military industries (Meister 1999). Following World War I, researchers in France and Germany applied biomechanics to improve prosthetic devices for war veterans who had lost various limbs. This effort stimulated the beginning of the physical medicine areas, including physical therapy. Besides helping war victims of World War II, physical medicine research regarding pathological gait (walking) was used to help victims of the polio epidemics (Atwater 1980). Interestingly, the rebuilding of industry in Europe after World War II involved many ergonomic and human factors design analyses of work movements, particularly in Russia and Europe (Contini & Drillis 1966a).

In the 1950s, anthropometry (study of the physical dimensions of people) became even more important, as the new automotive, space, and aviation transport industries (Contini & Drillis 1966a) and the armed services (Thomas 1969) needed to know the measurements of people to design seats, cockpits, and instrument panels that fit these users. Thus, for white males, sizes, shapes, and weights of various body segments such as the trunk and limbs were determined.

➤ The era of contemporary biomechanics dates back to the 1960s, when it was known as kinesiology. In more recent years the term *biomechanics* was adopted to describe this area of inquiry. *Kinesiology* is the term now used to refer to the study of physical activity through experience, scholarly study, and professional practice.

The Era of Contemporary Biomechanics

The study of biomechanics of sport and physical activity continued to grow mainly out of college physical education programs. Biomechanical conferences were organized, professional organizations were created, and graduate-level university programs were established in the United States in the 1960s (Atwater 1980). Biomechanists and kinesiologists promoted the scientific study of human movement within the primary physical activity organization, the American Alliance for Health, Physical Education and Recreation (AAHPER). The Kinesiology Section (renamed the Kinesiology Council and now known as the Biomechanics Academy) finally received official recognition in 1965. Its purpose was to provide AAHPER with a theoretical framework for all foundation areas, such as biomechanics, exercise physiology,

Activity 10.3

Timeline

Match key events in the development of the field with the corresponding time periods in this activity in your online study guide.

and exercise psychology (Atwater 1980). The first international professional association (International Society of Electromyographical Kinesiology) was also formed. The 1970s saw a rapid expansion in the number and scope of national (e.g., American Society of Biomechanics) and international professional organizations (e.g., International Society of Biomechanics) and university programs in biomechanics. People in areas outside biomechanics also demonstrated interest in this area. Dance kinesiology evolved during this time as well.

Since then, university programs and professional organizations related to biomechanics have continued to expand. Many colleges and universities offer not only courses in biomechanics but also undergraduate- and graduate-level specializations in this area within a kinesiology-based, health allied, or engineering unit. Many biomechanics organizations exist across the world, including the International Society for Biomechanics.

Research Methods in Biomechanics

Biomechanics research has greatly benefited from technological advances. Even in the late 1990s it was still common for many biomechanists to spend hours manually tracking thousands of frames of video to analyze the motion of a performer. Now it is possible to use special cameras that can track the motion of the performer in just minutes. What other tools do biomechanists use to accomplish the goals of biomechanics? How, for example, do biomechanists perform gait analyses for patients, research the cushioning properties of newly designed shoes, analyze the performance of Olympic athletes to help them improve, and perform research to answer fundamental questions about how and why we move? Whenever an answer to a biomechanical question is sought, whether by an allied health professional, a coach, a biomechanist, or other party, a common analysis process is used to answer it.

A Model for Analysis

Researchers ask questions and then design experiments using systematic methods to answer those questions. Similarly, if you're a practitioner, you will have questions that you want to answer. Although the methods used by practitioners to solve physical activity problems are much different from the methods used in laboratory biomechanics research, they involve a systematic process. So, to give you a sense of how a teacher or physical therapist might proceed to answer a question of interest, we will go through the simplified steps of an analysis process (see figure 10.2). For our example, let us assume that we are working to determine whether a new prosthetic design will improve the gait of individuals who have lost part of a lower limb.

Step 1: Identify Your Question

As shown in the model, the first step is to identify the question of interest. Examples of questions are "Does this new shoe model control pronation?" and "Is this rehabilitation protocol effectively strengthening the client's lower back?" Let's assume that the question is, "Does prosthetic design XZ-2 or QR-8 allow the best but safest running performance for a long-distance runner who has had most of his leg amputated?"

Step 2: State Performance Goals

Next, identify the overall goals of the movement in conjunction with the participant's goals for performing the movement. A performer may wish to be able to jog 30 minutes for exercise,

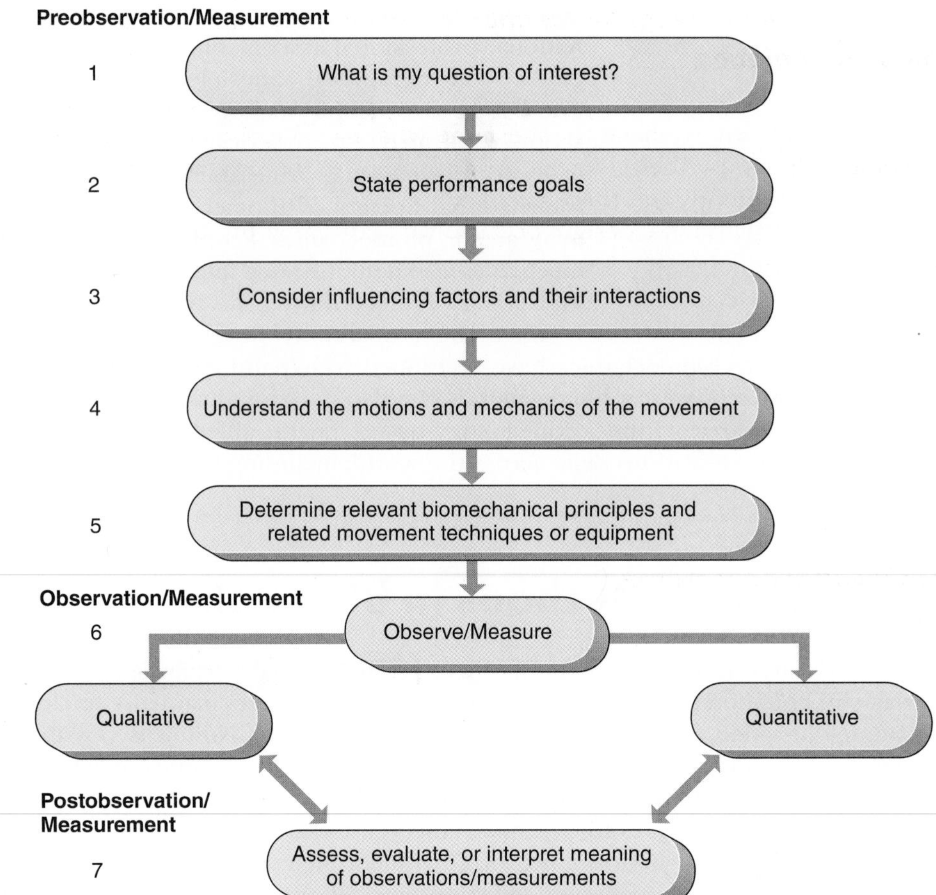

FIGURE 10.2 The movement analysis model used by practitioners or researchers to evaluate a question of interest. Each step of the analysis is labeled by number.

with the amputated limb performing as naturally and effortlessly as the intact limb. Our performer wants to run a 10,000-meter race as fast as possible without injury.

Step 3: Consider Influencing Factors

Next, for this step, we must understand the characteristics that are particular to our runner, that is, physical, emotional, and other characteristics. Thus, our performer may be motivated to run as fast as possible while trying out the new designs or may be hesitant if he perceives that he could be injured while wearing one of the new designs. The strength and length of the amputated limb are important physical characteristics for determining the proper fit of the prosthesis and how well the prosthesis functions. Working with our runner, we also need to understand other factors outside the performer. What type of shoe can he wear while wearing either design? What types of running surfaces are appropriate for these designs?

Step 4: Understand Motions and Mechanics

Execution of step 4 requires an understanding of the motions and mechanics involved in the movements used to perform the activity. Running requires performers to support their body weight on one foot and then propel themselves upward and forward through the air. When the leg is not on the ground, runners must pull it through in preparation for placing it back on the ground. Thus, our prosthetic design should allow performers to produce the force necessary to propel themselves forward, support their body weight without generating excessive pressure on the stump, and allow them to swing their prosthetic leg forward in a natural manner.

Step 5: Determine Relevant Biomechanical Principles and Movement Techniques

The next step involves identifying the mechanical principles most relevant to answering our question. In addition, we need to determine what movement techniques or equipment designs would help us use these mechanical principles to the best benefit. Newton's laws, principles related to energy, and fluid mechanics are examples of principles that we could choose from. For the prosthetic design, because we realize that the amputated limb is lacking muscles of the lower leg and foot, the prosthetic limb will have to provide some of the propulsive force needed. Thus, we might be interested in the principles related to storing and releasing energy, similar to what happens when a spring is compressed and then allowed to regain its shape. For the equipment, prostheses are built specifically for competitive running to take advantage of storing and regaining energy. Newton's third law of action-reaction is evident, too: For every force applied by one object onto another object (say, the performer's foot pushing backward and downward on the ground), there will be an equal and opposite reaction (the ground pushing back forward and upward on the performer's foot). Therefore, the ground can propel our runner forward and upward into the air.

Step 6: Observe or Measure

Finally, we've completed the preobservation or (pre–data collection) stage. Now, we must observe or measure the movements. There are two major methods to gather information. First, kinesiology specialists assess or evaluate the movement of a performer by use of observation. A movement specialist who observes the movement technique of a performer is doing a **qualitative analysis.** For example, if you were working with the runner who is trying out the two types of prostheses, you could watch to see whether his prosthetic limb swung excessively outward or listen to hear whether the prosthetic-limb side was hitting the ground too hard.

For the second method, the biomechanist who is performing research to determine what prosthetic design will best accomplish the client's goals needs to obtain numerical data using biomechanical instrumentation. Clinicians such as physical or occupational therapists or athletic trainers may also wish to obtain numerical data using biomechanical instrumentation to quantify their evaluation that the prosthesis selected is the most appropriate one for the client. The fitness instructor may

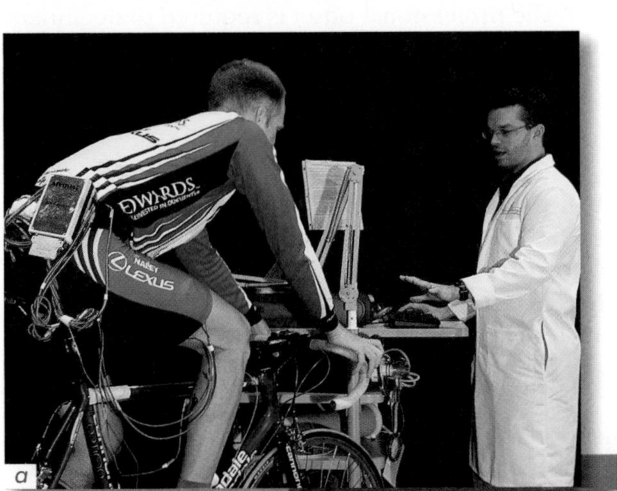

Orthopedic surgeons, rehabilitation specialists, coaches, and instructors use quantitative tools (left) and qualitative techniques (right) to analyze performers' movements.

© Kathy Simpson (left)

wish to measure the improvement in running speed of the performer over several months. If a tool is used to measure and evaluate quantities related to space, time, motion, force, or energy during a performance, the process is termed a **quantitative analysis.** (The instrumentation used for biomechanical research and other applications is discussed later in this chapter.)

Step 7: Assessment, Evaluation, and Interpretation

After obtaining the qualitative or quantitative data for your performer, you must interpret the meaning of the findings in relation to the question of interest, the underlying biomechanical principles, the influencing factors, and the performer's movement techniques. Assume that you were determining which of the two prosthetic designs (XZ-2 and QR-8) was optimal for your performer to enhance running velocity. For each design, the runner ran 50 meters (55 yards) as fast as possible. You could have measured how much the prosthetic knee flexed and how quickly it flexed and extended during various instants in time. Your findings: You observed that the knee flexion values and the maximum speed at which the runner could flex the knee of prosthetic design XZ-2 were much closer to the nonprosthetic limb values than were the values for design QR-8. The performer ran at the same speed wearing either design. Your assessment: The XZ-2 design allowed the performer to flex the knee more than QR-8 did during initial contact with the ground, allowing for more effective reduction of impact forces. Performance wasn't affected. Your recommendation: Try the XZ-2 design.

> DO it **Activity 10.4**
>
> **Movement Analysis Model**
> Order the steps in the problem-solving process in this activity in your online study guide.

In addition, you will need to share this information with a variety of people (e.g., your supervisor or accreditation board reviewers). The professional often is required to document the effectiveness of interventions, be it a teacher assessing a student's progress in a physical education class or a physical therapist submitting evidence to payment providers documenting the client's rehabilitative progress.

The last step in this process can also involve providing the performer qualitative or quantitative feedback. As a physical educator, fitness instructor, athletic trainer, occupational therapist, physiotherapist, physician, or other professional working to enhance performance of your client, you may provide feedback to the performer while the performer is learning or refining skills. Clinicians may choose to provide feedback to clients to enhance motivation as the client improves performance during rehabilitation.

Biomechanical Instrumentation and Other Tools

In step 6 of the analysis process, we mentioned that biomechanical instrumentation and other tools could be used during a quantitative analysis. We will now explore some of these various tools used to measure biomechanical quantities.

What simple tools besides a tape measure can you think of that can be used quantitatively to measure or observe time, movements in space, or force? The following are some of the instruments that may have come to mind: stopwatches or metronomes to measure time, barbells and free weights to measure how much force a person can exert, and a camera or camera-like device to capture the motion of a performer. From the motion images, a movement specialist could measure height, length, and distance that a limb traveled or the length of a dance step. A protractor could be used to measure how many degrees a limb rotated. For example, from a videotape of a fitness walker, we could evaluate with a protractor whether the walker was landing with the foot at a particular angle. Figure 10.3 shows various parts of instrumentation systems used in a biomechanics research laboratory.

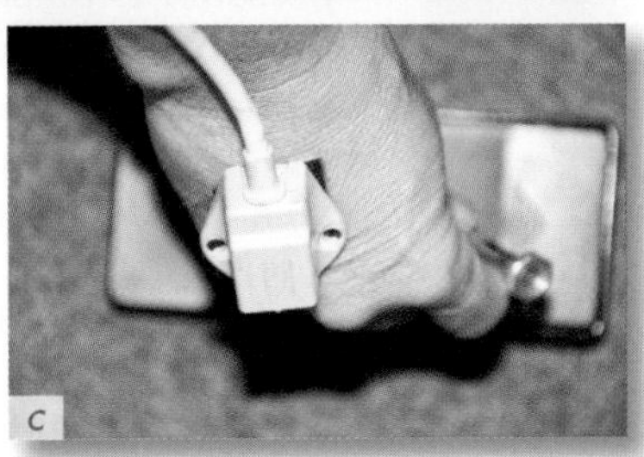

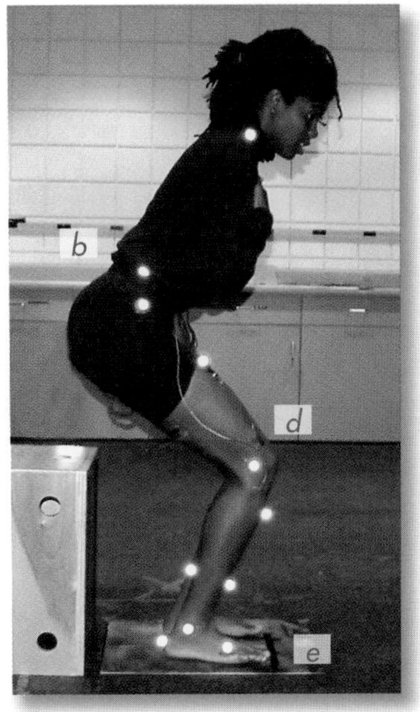

FIGURE 10.3 Biomechanical instrumentation. Components: for motion measurement, a digital, high-speed camera *(a)*, which tracks the location of reflective markers *(b)* and electromagnetic-sensitive marker *(c)*; for electromyography (EMG), the electrodes *(d)*; for ground reaction forces, the force platform *(e)*.

© Kathy Simpson

Motion Measurement Devices

The use of digital cameras and other motion-detection technology for recording motion in biomechanics is called motion measurement. Digital high-speed cameras can be used to trace the motion of reflective markers placed on points on the human body to track the motion of the various body segments. Other types of markers placed on the performer's body (e.g., light emitting diodes [LEDs]) emit signals that some type of sensor captures to track movement. Sensors attached to the performer that establish the position of a body part (e.g., the head) detect magnetic fields. The motions of any body segment (e.g., the jaw) or the entire body can be tracked and used to determine body positions, velocities, and accelerations. Feedback about performance techniques from motion analyses has been helpful to many elite and professional athletes in activities such as pole-vaulting, gymnastics, golf, and professional baseball. Perhaps you would like to have a private company that uses this technology analyze your golf stroke. Motion measurement equipment is also used to analyze movements observed in occupational and clinical settings (see www.univie.ac.at/cga/ for clinical gait analysis case studies).

> DO it **Activity 10.5**
>
> **Your Own Analysis**
>
> In this activity in your online study guide, you'll conduct an elementary analysis.

Force Measurement Devices

Measuring quantities related to force can tell us what is causing motion or injury. To measure how much force is being placed on a joint, ligament, or object requires attaching tiny force-measuring devices (force transducers) to tissues or artificial implants inside the body. The latter method is more commonly used in animal experiments because of the practical limitations of implanting devices inside of the body.

Transducers can also be used to measure how much force a patient can exert against a strength-measuring device. To measure ground reaction forces, force platforms have been used. A force platform is a metal plate with force transducers in the beams that hold up the plate. Usually a force platform is embedded solidly in the surface so that for gait analyses of patients, a person can walk along and step on and off it quite naturally. Ordinary or specialized force platforms can be used to diagnose postural and balance disorders. Knowledge of these ground reaction forces is helpful to biomechanists for other purposes, too, such as

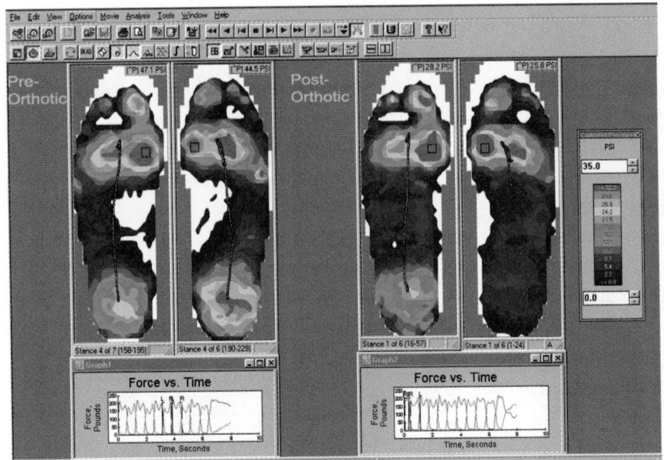

Pressure distributions of the bottom of the feet during walking at one instant in time.

Courtesy of Novel

the evaluation of new models of footwear or as a tool to help determine internal force loading on bones and cartilage.

Pressure is the amount of force applied to a given amount of surface area (see photo). Areas of high pressure against human tissue can indicate potential health problems. If too much pressure is placed on an area of the foot of a person with diabetes who cannot sense discomfort, for example, then the skin in that area can deteriorate and become infected. Finding the zones of high pressure from a pressure-sensitive insert placed inside the shoe allows a diabetic patient to have a shoe individually tailored to spread out the forces more evenly (Cavanagh, Ulbrecht, & Caputo 2000). Pressure devices can also be placed on wheelchair seats to test padding that will minimize pressure. Because zones of high pressure erode hip and knee joint replacements, researchers apply special sensors inside the replaced joint to determine pressure distributions.

Because of the difficulty of directly measuring muscle forces, motion and force platform data are used to calculate estimates of muscle forces via physics and engineering methods. Another method of estimating muscle forces involves measuring the electrical activity of the membranes of the muscle cells (fibers) when nerve cells stimulate them to contract. This method, called electromyography (EMG), requires placing electrodes on the bellies of muscles where the muscle fiber membranes are most likely to be stimulated by nerve cells. When the participant performs the movement of interest, for example walking, nerve cells send electrical signals to the muscle fibers that tell the fibers to generate force. The measured amount of electrical activity can be converted into the amount of force that the muscle produced during the movement.

Although obtaining accurate muscle force estimates using this method is difficult, EMG is being used for this purpose in complex mathematical models and for simulating movements. Electromyography also has been helpful in identifying the muscles that are active during a particular movement. In addition, this method helps identify what muscles to target for strength, power, or endurance training. Conversely, EMG studies of exercises can determine what muscles are being trained during an exercise. Electromyography also can be used clinically to determine whether a muscle is functioning correctly. People with movement disorders can learn how to substitute the use of other muscles for a nonfunctioning muscle. Knowledge about muscle activity has led to the development of stimulated walking, in which paralyzed patients wear a device that stimulates various muscles with electricity at the right time to produce the muscle forces necessary for the person to walk.

> DO it **Activity 10.6**
>
> **What and How to Measure**
>
> Continue the analysis you started in activity 10.5 of your online study guide.

All instrumentation described in the previous paragraphs measures some quantity, but mathematical techniques are required to calculate what cannot be directly measured. In addition, one method that is becoming readily usable is computer simulation. This method is based on mathematical modeling and uses the data of various instrumentation devices (EMG and motion measurement, for example), along with database information about the body such as dimensions and locations of muscle attachments. Simulations can help us understand basic mechanics at the cellular, molecular, or whole-body level. One example might be a simulation of knee joint motion, whereby the anatomy and forces involved with various structures, such as ligaments, tendons, cartilage, bones, and muscles, are modeled to enable understanding of typical knee joint motion. Various properties of the model can be changed. For example, moving an attachment of a particular muscle to a different location to observe whether the model predicts greater friction forces may help determine whether anatomical differences in muscle attachments underlie a mechanism of cartilage degeneration.

Being able to test this and other hypotheses that cannot be directly measured offers many possibilities. For example, King and Yeadon (2003) simulated an elite gymnast performing a back somersault in the air to learn how gymnasts control their movements when they make slight errors during the phases before the aerial phase. It certainly is less dangerous to have the computer-simulated gymnast make errors while performing complex aerial movements than to have real gymnasts do so.

Overview of Knowledge in Biomechanics of Physical Activity

To give you an idea of what biomechanists have learned through research, we will explore some fundamental concepts and research findings while incorporating several more steps of the movement analysis model (figure 10.2).

Biomechanical researchers follow a process similar to the movement analysis model when designing and carrying out an experiment. They ask a general question of interest such as "What is the most effective way to propel a wheelchair during an 800-meter sprint event?" A clinician might ask, "Is our 800-meter sprinter continuing to develop a shoulder injury, or is the sprinter improving due to our rehabilitation program?" Before the investigators can determine the biomechanical principles that may explain the answers, they must complete step 4 of the analysis process (understand the motions and mechanics of the movement). The mechanics of the movement are the forces being applied to the object that is being moved. Therefore, to determine the most effective method to propel a wheelchair, we need to know what outside or "external" forces are acting on the wheelchair and the performer, as both of them are being propelled. We also need to know these forces in order to understand how forces act to injure the shoulder region or in order to determine the tissues that need to be strengthened.

How Do External Forces Act on Performers?

Depicted in figure 10.4 are a road racer and a specially designed wheelchair. In the accompanying online study guide activity, see if you can identify all the external forces—that is, the forces that originate from outside the objects of interest—that act on the performer and chair.

➤ Some of the most common forces acting on a human performer include fluid resistance (air or water), gravity, friction, and ground reaction forces.

DO it **Activity 10.7**

External Forces

Work through this labeling activity in your online study guide to identify the forces.

We all know that gravity pulls us toward the earth. The force of gravity, known as weight, pulls the performer and wheelchair downward. The ground must therefore be holding up the wheelchair and performer ("pushing back"), suggesting that the ground can create force. Because the wheelchair and performer cause an action downward, the ground is reacting by pushing upward; hence, the ground is producing a **ground reaction force.** Also, notice that when the performer pushes on the wheel, the wheel pushes against the ground in the backward direction. The ground reacts by pushing

FIGURE 10.4 External forces act on a racer who uses a racing wheelchair.
Photodisc

forward, thus allowing the wheelchair and the performer to be propelled forward. This is another example of a ground reaction force.

A ground reaction force is generated any time you push against the ground (action) because the ground pushes back (reaction). Ground reaction forces (GRFs) can be created in any direction—sideways, vertically, forward, or backward. Just as in the example of the runner who uses a prosthetic leg and foot, whenever you're touching only the ground, without GRFs you couldn't go anywhere! You must push against the ground so that it can move you in the opposite direction. (Later in this chapter, we will explore how knowledge of GRFs has been used to understand how we can build up bone mass of adolescents and to investigate suggested causes of overuse injuries because of vertical impact forces.) Thus, when the racer pushes on the wheel, the wheel pushes against the ground, and the resulting GRF propels the chair and sprinter forward.

Are *friction* forces involved in the wheelchair-racing example? Definitely! When the wheel pushes backward against the ground, friction is the force needed to be sure that the racer doesn't slip when the ground pushes back. Whenever you push against the ground in a direction parallel to the ground—left, right, backward, or forward—the ground creates GRFs in these directions that are also friction forces.

Air (fluid) resistance also is acting on the sprinter and wheelchair, but this is often a negligible force. The faster a performer moves in the air or the water, however, the greater the fluid forces that act against the performer. This principle explains why a sprinter might wear special clothing to reduce air resistance but a long-distance recreational racer would not.

How Do Internal Forces Act on Performers?

Forces also can act inside our bodies and affect tissues. These are called internal forces. When a limb rotates, the inertia of the moving limb can stretch ligaments. When a person lands on the ground after jumping upward, the bones of the leg are compressed because of the vertical impact GRFs pushing upward on the body and the inertia of the body's momentum pushing the body downward. Internal forces, if excessive, could cause injury when they act on tissues, such as ligaments, tendons, and bones. Therefore, tissue biomechanists are interested in the internal force loading on body tissues in order to understand the etiology of injury and how to design appropriate prostheses and body replacement parts. To develop more effective ways of moving, biomechanists also explore how to increase force production by tissues such as muscles and tendons.

Shown in figure 10.5 are examples of some of the kinds of mechanical loading that internal forces produce, along with corresponding examples. Injuries can occur when the magnitude and rate of loading of the forces involved or the number of repetitions involved in various movements is too great. For *compressive loading* that occurs when forces act to push together or compress an object, we determined in our laboratory that when a dancer lands during a leap, the femur (upper leg bone) pushes against the tibia (shinbone), thereby producing compressive forces of up to 16.8 times body weight (Simpson & Kanter 1997)! To

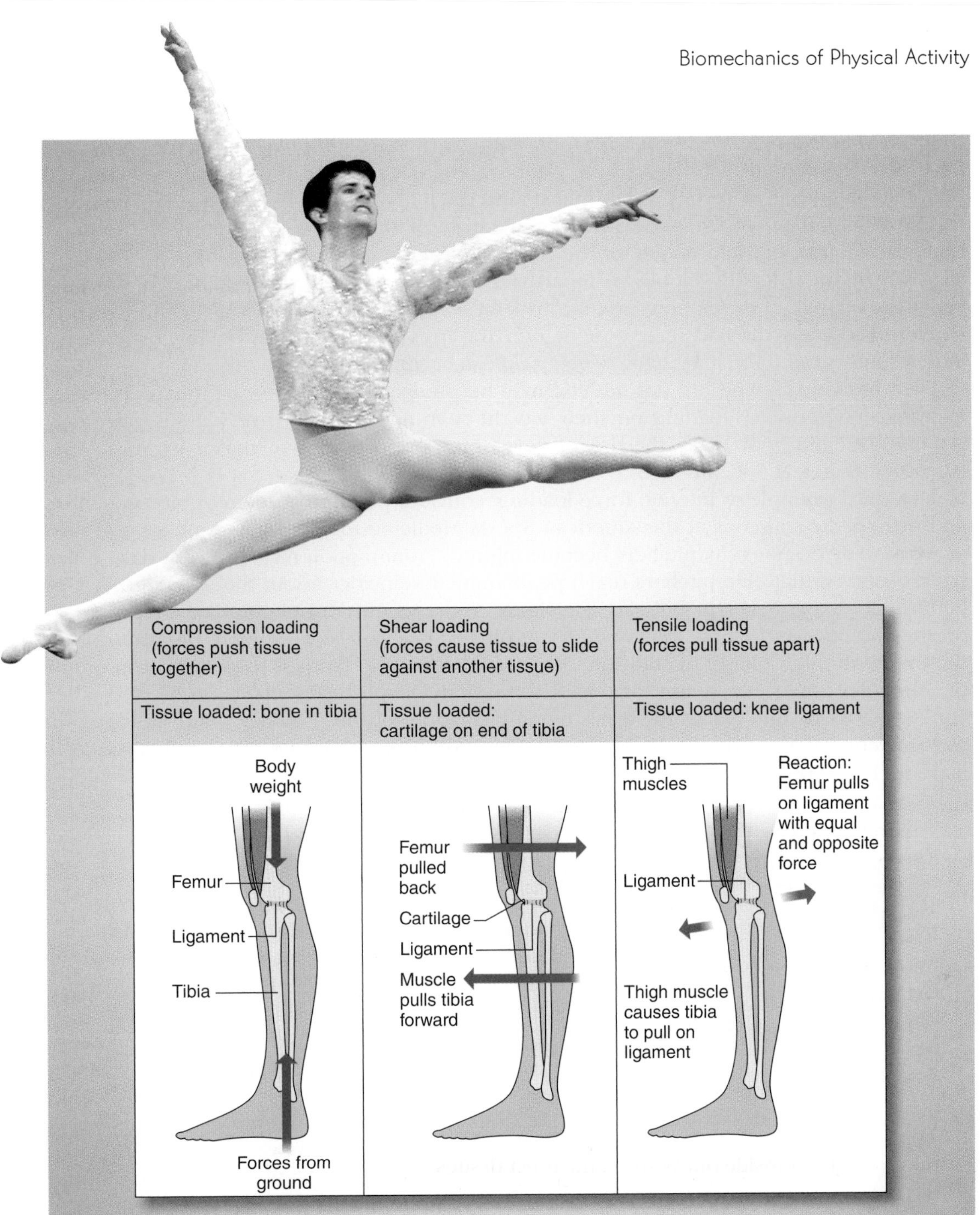

FIGURE 10.5 A few types of internal force loading applied to human tissues during landing from a dance leap. *Compression:* The shinbone (tibia) near the knee joint undergoes compression because of various forces that push this portion of the bone together. *Shear loading:* If the thighbone (femur) is pulled back and the tibia is pulled forward, the two bones are sliding past one another, causing shearing on the cartilage protecting the end of the tibia. *Tensile loading:* For the ligament that attaches to the tibia and femur, the front thigh muscles pull the tibia forward and other muscles pull the femur backward, causing the ligament to be stretched like a rubber band.

AP Photo/Steven Senne

generate these high compressive forces, the vertical GRFs push the tibia upward; the inertia and weight of the body push the femur downward; and the strong thigh muscles pull the two bones together. We also determined that the ends of the tibia and femur were sliding in the forward and backward direction against each other, which is called shear force loading. This was due partly to muscles pulling on the tibia, causing the tibia to slide forward. The maximum magnitude of this shear force was 5.1 times body weight (Simpson & Pettit 1997). To prevent the tibia from shearing too far forward, one of the ligaments attached to both the femur and tibia inside the knee resists this action. This knee ligament becomes stretched and undergoes tensile loading. With the magnitude of the forces being generated, it is easy to see how knees are injured so often in sport. To answer why shoulder injuries are occurring to our sprinter, we could look at the amount and types of loading occurring to the tendons and connective tissue of the shoulder region during propulsion of the wheelchair.

As another example of compressive force loading, runners and physical activity specialists have been concerned that, over a long period, the great number of foot strikes, combined with the magnitude of vertical GRFs that pound the body with each step, would contribute to the development of osteoarthritis (degeneration of the cartilage covering the ends of bones at a joint). But in a review of the research regarding the effect of exercise and joint structure on the risk of developing osteoarthritis, Lane (1995) observed that with low-impact activities, the risk of developing osteoarthritis in correctly aligned joints was not increased. However, athletes who had some type of joint abnormality or injury and highly competitive athletes appeared to be at risk for developing osteoarthritis.

Even individuals who are not athletic may be predisposed to osteoarthritis because of excessive compressive loading on their weight-bearing joints. For every 1 kilogram (2.2 pounds) of extra body mass, the risk of developing osteoarthritis increases by 9% to 13% (Cicuttini, Baker, & Spector 1996).

As a third example of internal force loading, consider pitching. A team of biomechanists and orthopedic surgeons at the American Sports Medicine Institute has done a good deal of work to determine why pitchers become injured. From motion measurement data, it has been observed that elite pitchers reach peak angular velocities about the shoulder joint of 7000 degrees per second (Dillman, Fleisig, & Andrews 1993)! Imagine the stress placed on the shoulder and elbow joints when the pitcher is trying to decelerate the arm during the follow-through phase. In addition, the inertia of the arm "wants" to pull the arm out of the shoulder joint. The amount of tensile force that acts on ligaments, tendons, and other connective tissue and resists the arm's being pulled out of the shoulder joint (Fleisig et al. 1996) is equivalent to the force required to hang off a bar with one arm while a friend is hanging from your legs.

FIGURE 10.6 When a runner wants to move forward and upward into the air, the runner pushes down and back (action). The ground creates a reaction by producing forces of equal magnitude that act in the opposite direction to the action. These forces produced by the ground are called ground reaction forces (GRFs).

Digital Vision

How Do Biomechanical Laws of Nature Shape Our Movements?

Now that we can identify what forces act on us and some that act inside us, we are ready to think about the next step in the analysis process, that is, selecting relevant biomechanical principles. Researchers select the relevant biomechanical principles to help identify their predictions of what will happen during their experiment. Whenever a physical activity practitioner or biomechanist is faced with improving performance, testing a theory about how we move, or evaluating a patient's movement, one of the most important questions that must be answered during the movement analysis process is, "What are the biomechanical principles that are most relevant to the performance of this movement?" Remember from step 5 of the analysis process described earlier that biomechanical principles refer to the mechanical principles and laws of nature that affect any living being. After identifying the most relevant principles for a given movement, a physical activity specialist, clinician, or biomechanist can determine how best to apply the principles to the individual performer in terms of movement techniques or equipment. These laws also apply to internal biomechanics, that is, what occurs inside the body.

To demonstrate how biomechanists use the mechanical laws of nature, let's examine two of Newton's laws of motion. The physical laws first described by Sir Isaac Newton affect the movement of any living being on earth. The law of inertia states that a body at rest or in motion will stay at rest or in constant motion until an external force acts on it. This law applies to any movement in which the performer must remain in a static position, as when a gymnast holds a pose on the balance beam. It also applies to many movements in which the performer is traveling through the air, for example executing any type of jump or leap or even running.

Because air resistance for many movements can be considered negligible, if the performer is moving through the air without touching anything else, no external forces are acting on the performer in the horizontal direction. Therefore, the performer will continue to travel horizontally through the air at the same speed until contact with the ground. How does the law of inertia apply to motion in the vertical direction? Because gravity is an external force acting on the performer, the vertical motion will not remain constant. The performer slows down until he or she reaches the peak of the flight path and then speeds up while falling back to Earth. So, even if a person just hops off the ground, if it weren't for gravity pulling down in the vertical direction, he or she would never come back down. The person would just keep moving forward and upward at the same speed and never come back to Earth.

➤An important skill for a physical activity specialist, biomechanist, or allied health rehabilitative specialist is to be able to choose the relevant mechanical principles that apply to a question of interest.

Since the performer's flight path is fixed because of the law of inertia and the force of gravity, performers who are trying to achieve maximal horizontal distance (e.g., long jumpers or leaping dancers) must concentrate on generating as much velocity as they can while they are still on the ground. Indeed, the velocity at takeoff is one of the most important factors in the success of movements when the goal is to achieve maximal horizontal distance (Hay 1993a).

As another example of a biomechanical principle, the law of action-reaction, described earlier, is important to any performer who wants to move. In fact, unless you have something to pull or push against, you won't have any forces pulling or pushing back on your body to make you move in any direction except for gravity (and maybe air resistance). Figure 10.6 shows how the law of action-reaction creates the GRFs described earlier. Later we will explore some ramifications of GRFs acting on us (see "Focus Topic: Making an Impact" on p. 289).

➤Examples of mechanical principles that apply to any movement include Newton's laws of motion (i.e., law of inertia, law of acceleration [not discussed], and law of action-reaction).

Several other principles help explain how to apply force to create motion or slow down motion. Mechanical principles can help us determine how to create rotation of a person or object. Whenever a force is applied so that it causes an object to rotate, the effect is called a **moment.** When muscles contract (see figure 10.7), they usually cause a limb or other body part to rotate; hence, muscles can create moments when they contract. The sprinter who uses a racing wheelchair must contract certain shoulder muscles to make the arms rotate about the shoulder joint in order to push the wheel forward. For lifting tasks, including lifting patients, children, or boxes in a factory, the weight of the lifted person or object "wants" to rotate the trunk forward and downward, that is, create moments. Back muscles and connective tissue must counteract these moments. Thus, for some lifting techniques, reducing these moments lessens the back muscle effort, which may help prevent back muscle and connective tissue strains.

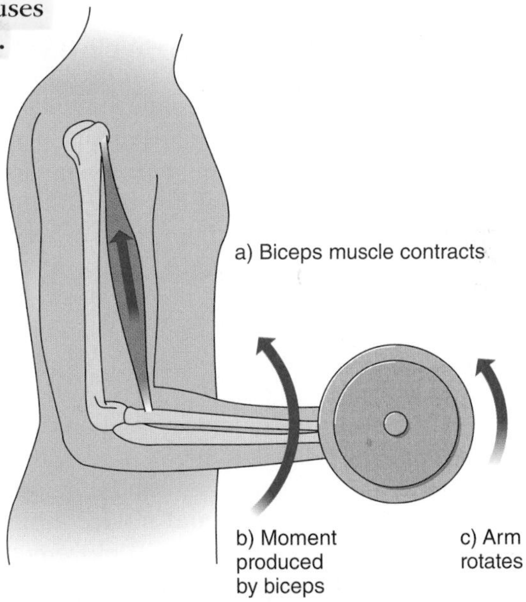

FIGURE 10.7 When muscles contract, they can make a limb rotate. Thus, they produce a moment on the limb.

Assessment and Evaluation of Performers: Biomechanical Profiles

How should a person move to accomplish a given task? Biomechanists continue to debate this question. What if a researcher wanted to determine whether a particular surgical technique helped a patient walk without limping? An overview of the assessment and evaluation of performers follows. Note that this issue corresponds to step 7, assessment and evaluation of performers.

Remember the issue raised at the beginning of this chapter: Is there a best way to perform a movement? Do elite athletes have better movement technique than novices do? If so,

➤ A force applied to a body that causes the body to rotate is called a torque or moment.

why? Portrayals of biomechanical and other characteristics of a group of performers—elite athletes, athletes with disabilities, or novices—are called profiles. By understanding how one group of people, elite athletes for example, move and produce force and otherwise exploit the laws of physics to produce optimal performance, we can direct attention to determining how to help novice performers move from rudimentary performance levels to more skilled levels. Sprinting performances by the most successful Paralympic competitors may be used to develop elite profiles for sprinters who compete in wheelchair race classified events, for example.

Use of profiles is also beneficial to the clinician, not just with reference to elite athletes but also for working with typical individuals. Profiles provide clinicians with typical data for comparison with individuals who have atypical or less than optimal movement effectiveness. If a clinician wants to know whether our sprinter has recovered sufficiently from a shoulder injury by looking at the client's graph of shoulder strength at a variety of arm positions, the clinician must be able to distinguish characteristics of a nontypical strength graph from those of a typical graph of a healthy shoulder.

Let's examine the two primary areas of profile applications: performance assessment in physical activities and performance assessment in clinical situations.

Profiles and Performance Assessment in Physical Activities

➤ *Profiles,* or sets of biomechanical and other performer-related characteristics of a given group of individuals, provide information for comparisons with other groups of individuals (e.g., individuals with dysfunctional movements vs. those with average functional movements, skilled vs. novice performers, injured vs. noninjured clients).

Everyone who wishes to win in an athletic competition or perform flawlessly during a juried dance performance wants to use the movement techniques that will ensure the best outcome. Typically, if you're trying to find the best technique, you might observe how the leading athletes perform and attempt to copy their movements.

Similarly, a biomechanist might investigate the biomechanics of elite athletes to determine the optimal techniques. This profile strategy is called the "elite athlete" model or highly skilled performer model because it is assumed that elite or highly skilled performers must be using ideal technique to achieve their success. Dancers, gymnasts, and divers, however, may have to adhere to highly prescribed movements. Even so, people try to emulate the most skilled of these performers. For most movements, people continue to try to improve on the prescribed movements or to invent new movements to be more successful.

As an example of use of the elite athlete model, a study by Hay (1993b) addressed the techniques of elite long jumpers and showed that when running toward the takeoff board they reach their maximal speed at the second-to-last step before takeoff. They slow down a little to get the body in the correct position for takeoff into the air. But jumpers don't want to slow down too much because the faster they can take off into the air, the greater the distance they can jump. For many competitive events that involve projecting the body (e.g., the long jump) or an object (e.g., a javelin), elite performers often generate more velocity to project the body or object farther than do less skilled performers (Berg & Greer 1995; Hay 1993b; Morriss & Bartlett 1996).

Although profile databases can be helpful in determining the biomechanics that underlie differences between elite and novice athletes, because movements are shaped by factors related to the performer (step 3, movement analysis model) it would be unreasonable to expect that the technique used for a given movement should be identical for people of all different sizes, strengths, and anatomical variations.

As an example, do you agree with the following (yes or no)? A very good long-distance runner would not want to perform wasted motions that would require energy but not help

Timely Technique

For an activity of interest to you, ask several people where they learned what makes up optimal technique. Were most people's answers based on the technique that highly skilled performers use? Why do you think people should or shouldn't use the same technique as top performers?

her or him move forward. Williams and Cavanagh (1987) assumed that highly skilled long-distance runners would adopt a running style that would eliminate unnecessary motions requiring additional energy. After filming runners using high-speed cameras and measuring oxygen consumption (energy used) while the runners ran on a treadmill at a set speed, Williams and Cavanagh were surprised to find that the best runners were not necessarily those who used the least amount of energy. But some evidence showed that the runners' techniques did affect their oxygen consumption. Runners who used less oxygen displayed less vertical oscillation, kept their arms from swinging in a large arc, and demonstrated other motion differences that may have represented subtle differences in muscle activity. The researchers concluded that each runner was naturally adopting a movement pattern based on individual anatomical and physiological characteristics. Therefore, although certain movement techniques should enable performers to accomplish their movement goals more effectively, the techniques must be selected with the individual in mind.

➤ Biomechanical profiles have helped researchers and coaches determine what movement techniques are related to performance effectiveness. Researchers also have found that the characteristics of each individual must be considered when attempting to enhance performance.

Profiles and Clinical Assessment

Looking at technique from a different perspective, clinicians are interested in assessing clients' biomechanics (e.g., movements, strength) to evaluate whether the client functions similarly to average people, whose values are the normative values. This method can be used to determine whether a dysfunction is present, whether rehabilitation or another intervention has been successful, or whether people are functioning at the level required. Using biomechanical values of the general population in a profile rather than values of highly skilled performers or elite athletes is a normative model.

Because walking is one of the most important and common movements that a client needs to be able to perform on a daily basis, we will use it as an example. For a given movement such as walking, normative values for movement characteristics such as speed, step length, cadence, and amount of knee and hip flexion are obtained via measuring the values of many people, averaging their scores, and determining the ranges of acceptable values. A medical rehabilitation team may want to know if a cerebral palsy patient should undergo surgery to improve walking ability. If the patient is found to be walking with shorter steps, at a slower cadence, and with reduced knee motion compared with the normative values of a population who are similar to the patient in age, then the rehabilitation specialists and the surgeon can investigate reasons and solutions such as surgery to improve the patient's walking ability.

Finding the right normative values for a given client can be difficult. To be meaningful, normative values must be based on a population that is very similar to the client. As Craik and Dutterer (1995) noted, many factors influence walking performance beyond the physical makeup of the individual's body: the purpose of walking, the environment, age, ethnicity, experience, and physical activity level. Older people may walk more slowly than younger adults do, but this difference can be due to many factors (Craik & Dutterer 1995) other than age, such as general health, medications, and strength. Last, as we mentioned earlier, people can show many subtle differences in their movement patterns yet be able to

Many factors influence walking performance beyond physical makeup, including the person's environment, age, ethnicity, experience, and physical activity level.
Digital Vision

walk at the same speed, cadence, and step length (Winter 1991). Thus, the biomechanically smart clinician looks for a pattern of unusual values (e.g., lack of hip flexion, short time on the ground while the affected leg is being supported, or shifting the weight toward the nonaffected side during walking) and seeks an explanation for them.

Athletic trainers who are trained to evaluate and rehabilitate injured athletes find that the normative values for athletes are different from the values for the general population. Dancers need more flexibility than the general population does, for example. Instead of using population norms, the trainer may be able to obtain preseason measures for strength and flexibility. Then, if an athlete becomes injured (e.g., a football player hurts his knee) the team doctor may allow the athlete to return to competition if the player's leg strength is within 90% of the athlete's preseason value.

Generating accurate and useful normative biomechanical values in clinical applications will remain an important biomechanical issue in the future. The same is true for improving profiles for occupational tasks to reduce the risk of job-related injuries. Ergonomists will continue to refine profiles for tasks such as lifting, keyboarding, and assembly line tasks.

Research in Review: Are Impact Forces Bad for You?

Using the maximum impact force value listed for a gymnast (see "Focus Topic: Making an Impact"), a 45-kilogram (100-pound) gymnast could hit the ground with a peak force of 635 kilograms (1400 pounds) during a back somersault. Placing this much stress on the body can be strenuous on the tissues of the performer over time, stimulating changes to the structure of weight-bearing tissues. Currently, controversy exists regarding the role of vertical ground reaction forces (VGRFs) in injury causation. This section focuses on the role of VGRFs as an integral element of various mechanisms surmised to produce or exacerbate particular lower limb injuries, diseases, and dysfunctions and, conversely, to help stimulate bone growth and increase strength of skeletal and connective tissue.

We do not have a good understanding of how VGRF loading directly influences bone growth or failure, but some evidence correlates VGRF and changes to bones. For triathletes who had trained for more than 10 hours per week for the previous three years, repetitive stresses were believed to have caused increased knee joint surface area (Eckstein et al. 2002). The investigators surmised that this change may have been beneficial, because increased surface area helped spread out the forces acting on the joint. A benefit of VGRF loading is to stimulate increased bone mass of children. It is surmised that the more bone a person builds during childhood, the older the person will be when osteoporosis (a disease in which the bones become brittle and break easily) develops. However, it is still unknown whether this benefit continues through later decades of life.

There is high interest in determining how to increase bone mass to prevent osteoporosis in later life. After reviewing the exercise-based intervention studies, Lanyon (1996) stated that "load-bearing is an important, if not the most important, functional influence on bone mass and architecture" (p. 1). Physical activity, especially up to early adulthood, therefore, can provide the mechanical loading stimulus necessary to cause the bone to respond, but only if performed under the right conditions. On the basis of a review of previous research (e.g., Frost 1999), we outline three conditions necessary to cause bone mass improvement and relate them to physical activity and our VGRF discussion. First, bone adaptation occurs only during dynamic loading, not during static loading (e.g., standing). Frost added that the stimulus for positive bone adaptation increases if the amount of the load and the rate at which bone deforms increase (because of the increased rate of forces applied to the bone). Thus, the highest value of VGRFs (F1 in figure 10.8) and the rate at which VGRFs are applied to the body for a particular movement are probably good indicators of how VGRFs may affect bone mass improvement of children (Bauer et al. 2001). The rate at which VGRFs are applied to the skeleton is the maximum slope shown in figure 10.8.

Second, only a short duration of mechanical loading is required to stimulate a beneficial, adaptive response. Further, as the duration of loading becomes longer, the adaptive response

Focus Topic: Making an Impact

Because most of us collide with the ground every day—walking, jogging, working (if we're on our feet), hiking, working out, playing a sport, and so on—you might find it useful to know something about the GRFs that act on you in the vertical direction. Let's look first at what the VGRFs look like during the phase of running and walking when the foot is on the ground (figure 10.8). The top curve, representing the GRF pattern of runners who often land on the heel first, demonstrates an impact peak that occurs soon after the runner touches the ground (F1). For most runners, who exhibit the middle curve, no initial impact peak occurs, often reflecting a more flat-footed landing. F1 represents the impact peak, that is, the highest amount of vertical force generated while the runner is colliding with the ground. The following are typical impact peak values for a few activities:

- Walking: 1.0 to 1.3 times body weight (BW) (Winter 1991)
- Running: 2.0 BW for jogging barefoot to 7.9 BW for sprinting with spikes (see review by Nigg 1983, p. 21)
- Gymnastics: 9 to 14 BW (Panzer 1987)

For walking, shown in the bottom curve of the figure, the first peak (F1) represents the walking impact peak. The steepest slope for each curve represents the highest rate at which impact forces are being applied to the performer. Understanding the rate at which the VGRFs are applied is desirable because the rate at which forces are applied may be correlated to bone failure (Radin et al. 1985, 1991), although a high rate of applying GRFs also is necessary to stimulate bone growth.

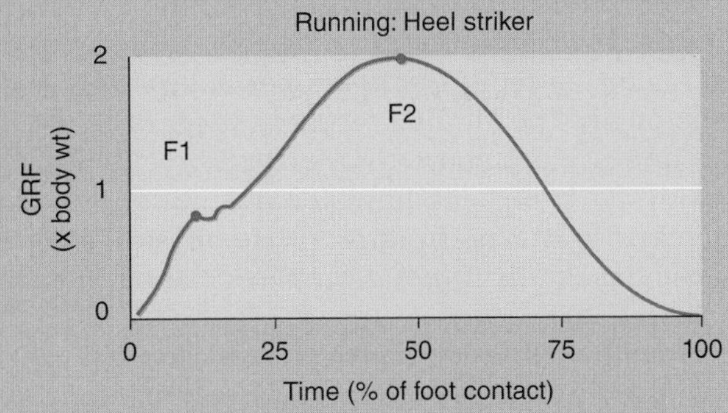

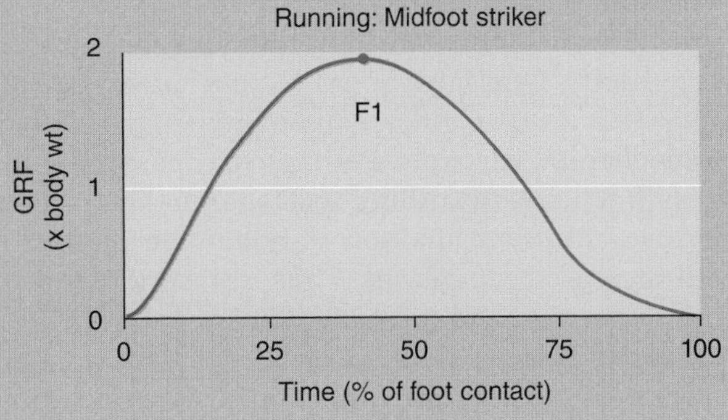

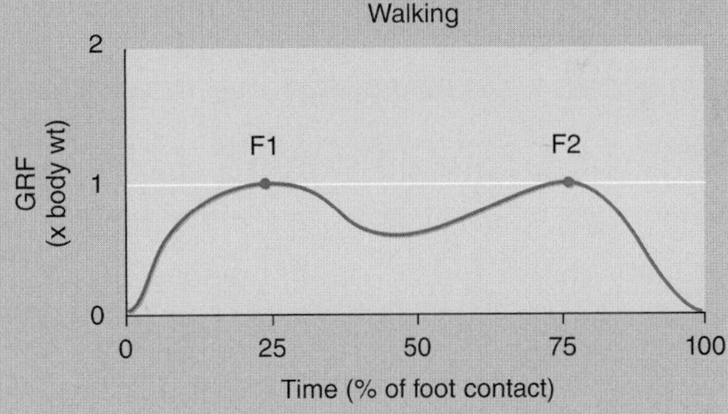

FIGURE 10.8 Vertical ground reaction force–time curves acting on one foot while it is on the ground. The top two curves are patterns produced by runners who typically strike the ground heel first (top) or more flat-footed (middle). The bottom graph represents walking. Biomechanists are interested in the maximum vertical impact force (F1), peak force (F2), and the maximum slope of these curves that occurs between foot strike and F1. The steepness of the slope indicates how rapidly impact forces are being applied to the body. Forces applied at a rapid rate to bone and joints during many repetitions, as occurs during a long run, have been associated with bone injuries and cartilage degeneration.

actually declines. Lanyon suggested (although this has not been substantiated) that no more than 50 repetitions of a high-impact activity were required per exercise session and that performing three sessions per week was probably sufficient.

Third, and perhaps most surprising, is that bone cells accommodate typical, daily loading so that routine loading is unlikely to stimulate bone adaptations. So someone who stands on his or her feet for 8 hours per day at work will not demonstrate increased bone mass or strength. This finding certainly suggests that participating in physical activities that load the bones much more than everyday activities do will cause bone to continue to adapt positively. At present, however, it is not known if preadolescent and adolescent girls who increase bone mass by appropriate physical activity will continue to maintain greater bone density as adults if they stop participating in such activity or athletics (Nichols, Sanborn, & Essery 2007).

➤ Weight-bearing activities that create impact forces of a magnitude greater than those produced by normal, daily activities and that are rapidly applied to the body can produce positive bone adaptations. It is likely that there is an optimum range of GRF magnitudes, rates of loading, and number of repetitions unique to each individual that will promote optimal bone health.

Other forces besides impact forces can stimulate increased bone mass during impact activities. Muscle force is an important stimulus for causing bone adaptations (Burr 1997; Frost 1999; Schönau et al. 1996). During moderate to intense physical activity, a muscle group can apply a tremendous amount of force to bone—actually, much more than the ground does (Simpson & Kanter 1997).

So, if impact forces are important due to their magnitudes and rapid rates of applying loads to the body, why have some biomechanists been so interested in trying to reduce impact forces that act on runners? Much interest in reducing impact forces has occurred in footwear biomechanics, certainly. Perhaps the reason for the attention is the mechanism underlying stress fractures. According to Donahue and colleagues (2000), microcracks occur in bone because of greater amounts of bone deformation (caused by greater forces) in combination with increased repetitions. Thus, someone who runs for long periods hits the ground with many repetitions. Additionally, damage accumulates with more repetitions. Bone strength then decreases. Bone remodeling will repair these microcracks if the rate of repair can outpace the rate of damage accumulation. If damage continues to accumulate, eventually a stress fracture will occur. This process is likely the reason for a high incidence of stress fractures in new military recruits who are not used to engaging in high amounts of marching, running, and hiking and who are given little time to allow bone remodeling to occur.

Hit the Ground Running

Select walking or running and calculate the maximum amount of vertical impact force that you might generate. To calculate, multiply your weight by the value selected on page 289. To get a sense of how much force that is, think about whether you could hold up that much weight with your arms.

➤ Footwear may help prevent certain types of injuries for those predisposed to injury. But for running, many other factors, such as the hardness of the surface, amount of training, and anatomical factors, are more likely to influence a person's predisposition to injury than footwear.

One method that we often think helps to reduce impact forces for high-repetition activities such as long-distance running is footwear. But is footwear effective? Making decisions about footwear can be frustrating because many factors influence a person's predisposition to injury. Most running injuries have been attributed to training errors (e.g., too much distance), anatomical factors, muscle imbalances, shoes, and surfaces (see review in Williams 1993). Indeed, to prevent injury, the training regimen and the surfaces that a performer runs on are apt to be much more important factors than the shoes. If a person runs on an extremely hard surface many hours each week, which shoes the runner wears won't matter; they won't prevent injury if the runner is predisposed to impact-related injuries. In summary, to avoid injury, people should buy a good pair of shoes designed for a particular activity and should train smart.

Identifying Causes of Injury

Find someone you know who has suffered from an overuse injury of the legs, for example a stress fracture in the foot or lower leg, or find someone who works with injured people. Ask the following: "What do you think are the major factors that caused, triggered, or contributed to the injury?" Are the answers consistent with the information that this section has presented?

Wrap-Up

Whether or not you consciously realize it, the physical laws of nature act on you at all times. Biomechanists as well as other physical activity specialists can apply the principles of these mechanical laws to enhance performance; reduce injury; evaluate the effectiveness of a movement; or select the proper sport equipment, tool, or occupational equipment. The theoretical knowledge regarding how mechanical principles influence our movements and the structure and function of the human body can be applied to occupational, sport, exercise, dance, daily task, and clinical rehabilitation situations. Furthermore, an understanding of the theoretical knowledge of biomechanics, as well as the use of biomechanical research methods, is helpful to professionals in other subdisciplines of physical activity ranging from motor control to pedagogy of physical education. If you're fired up to learn more about biomechanics in general, you might start by looking at the International Society of Biomechanics Web site (www.isbweb.org), which lists information such as job opportunities; graduate programs; links to sites focusing on a variety of biomechanical areas such as gait, prostheses, and sport; and abstracts of previous conferences.

As you come across someone touting a particular technique purported to improve performance or promoting a new piece of sport equipment, footwear, or exercise equipment, be aware of the evidence or rationale presented to convince you to try the technique or equipment. Is the evidence based on knowledge of mechanical principles and biomechanics research? (Skepticism is particularly warranted for techniques and products promoted on the World Wide Web, which contains many sites with official-sounding names. Beware if the sponsor is a corporation or if the site is just someone's personal Web site.) At the least, understanding biomechanics can help you make informed decisions.

Use the key points review in your online study guide as a study aid for this chapter.

QA Activity 10.9

These end-of-chapter questions and activities are also in your online study guide. Your instructor may ask you to complete them online and turn them in.

1. What is the definition of biomechanics?
2. How have various historical events shaped the progression of the subdiscipline of biomechanics?

3. Before use of the term *biomechanics* became prevalent, what was this subdiscipline called?
4. In what types of settings do biomechanists typically work?
5. What are the goals of biomechanics?
6. For each applied goal of biomechanics, provide an example of how this goal is relevant to an occupational setting of interest. First, list the specific description of the profession and the occupational setting, for example, profession—physical therapist; occupational setting—orthopedic outpatient clinic. Next, list the goals, and for each goal describe an example of how you would help clients achieve it in this career setting.
7. With a partner, go to a Web site of a professional biomechanics society and find the listing of jobs. (Go to International Society of Biomechanics [www.isbweb.org] first. Use the Related Links section to find other societies.) Categorize the jobs using a classification system of your choice. For example, one category of jobs might be research jobs in clinical settings; another might be ergonomic jobs in industry. With your partner, develop a short document or presentation about the job outlook in biomechanics: roughly how many jobs exist, what categories they fall in, what qualifications are necessary for different categories of jobs. Last, do any of these types of employment positions interest you?

11 Physiology of Physical Activity

Jennifer L. Caputo

Icon SMI

CHAPTER OBJECTIVES

In this chapter we will

- discover the key features of the subdiscipline of physiology of physical activity and the employment opportunities available to exercise science professionals;
- explain how the physiology of physical activity fits within the discipline of kinesiology;
- review the history and development of physiology of physical activity;
- identify the research methods used by kinesiologists working in exercise science; and
- examine how the body responds to physical activity and how these changes relate to physical performance and health.

Chelsea's parents have come to you to determine what precautions they should take to help increase the safety of their daughter's participation in her gymnastics club. They want to know if there are any restrictions on the type of training she should perform.

Mike is a sprinter who has come to you to see how he can optimize his performance. He wants to know if there is something he can do to alter his training sessions or an ergogenic aid he can take to help improve his running speed.

Jim suffered a heart attack and is now fearful of being too active. He wants to know what types of activities he can safely undertake. He also wants to know what benefits he can expect from being physically active.

These are common questions faced by individuals working in the subdiscipline of physiology of physical activity. Can you determine how the activity programs of these individuals will differ?

The questions posed in the chapter opening represent some of the areas of specialization in the kinesiology subdiscipline of physiology of physical activity. If you are curious about how your body and your health status are altered from either the performance of physical activity or a lack of physical activity, the subdiscipline of physiology of physical activity may be appropriate for you. With knowledge of physiology of physical activity you can help athletes achieve their peak performance, help people understand how to safely participate in physical activity, conduct research to help prevent and treat disease through physical activity and exercise, and even help those working in the extreme heat or cold or at different altitudes understand how to maximize their work capacity.

Understanding how your body responds to meet the immediate demands of being physically active, as well as how your body adapts to repeat bouts of exercise is a key feature of physiology of physical activity. When you chase down a shot on the tennis court or put on your running shoes to go for a run, your heart rate and breathing rate increase. Your blood pressure will rise and, over time, you will begin to sweat. These changes illustrate your body's way of meeting an increased demand for oxygen, energy, and temperature regulation. If you did these physical activities over several weeks, you might also begin to notice changes in your body weight and muscle mass. These physiological responses to physical activity and exercise are just a few of the topics that intrigue students of physiology of physical activity.

As shown in figure 11.1, physiology of physical activity, often called **exercise physiology** or **exercise science,** is a subdiscipline of the biophysical sphere, along with biomechanics of physical activity (discussed in chapter 10). Whereas biomechanists apply principles of physics and engineering to physical activity situations, exercise physiologists apply principles of biology and chemistry to understand how the body responds to physical activity.

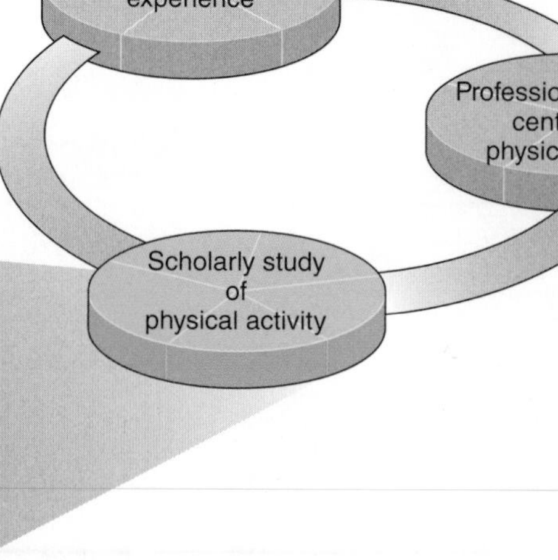

FIGURE 11.1 Physiology of physical activity and the other scholarly subdisciplines.

What Does a Physiologist of Physical Activity Do?

Many opportunities for employment exist within the subdiscipline of physiology of physical activity. You can choose the setting you most enjoy, such as a fitness or wellness center, hospital, corporation, academic institution, or research laboratory. You can work with different populations ranging from the young to the old and from the apparently healthy to those diagnosed with disease. The variety of jobs offers opportunity to those who have obtained various levels of education in the field.

Persons with academic knowledge of physiology of physical activity may also choose to work as exercise instructors in commercial and corporate fitness centers or as personal trainers and strength and conditioning coaches. Additionally, there are opportunities in clinical work in cardiac rehabilitation and pulmonary rehabilitation. The military and the National Aeronautics and Space Administration (NASA) also employ exercise physiologists to conduct research (e.g., on preventing heat illnesses during basic training or using physical conditioning in the space shuttle to prevent bone loss).

> www. **Web Search 11.1**
>
> **Careers in Exercise Physiology**
> Use the Internet to find career information and document it in this activity in your online study guide.

Often, the term exercise physiologist is associated with those who have obtained a master's degree, while teaching and conducting research in academic settings is typically reserved for those who have a doctoral degree in exercise science or exercise physiology. Faculty members teach courses in exercise physiology, anatomy and physiology, exercise testing and prescription, and sport nutrition; conduct laboratory research on problems such as the effects of conditioning programs on sport performance or ways of reducing the risk of chronic diseases; and write research grant proposals that they submit to federal agencies (e.g., the National Institutes of Health) or foundations (e.g., the Robert Wood Johnson Foundation) that support research on physical activity.

In addition to obtaining academic degrees in exercise science, there are opportunities to take certification exams. Certifications are offered through organizations such as the American College of Sports Medicine (ACSM), the National Strength and Conditioning Association (NSCA), the Aerobics and Fitness Association of America (AFAA), and the American Council on Exercise (ACE).

Goals of Physiology of Physical Activity and Kinesiology

The physiology of physical activity evolved from physiology, which is the study of how the body functions. Exercise science professionals study short-term changes in how the body functions during exercise and also the long-term changes due to training. For example, students discover that the heart beats faster as running speed increases and that resting and exercise heart rates decrease following endurance training. They also learn about the mechanisms that control these changes in the heart rate.

From chapter 1, we know that exercise is a form of physical activity engaged in to improve or regain performance, health, or bodily appearance. Exercise science professionals primarily focus on the exercise components of physical activity including training, health-related exercise, and therapeutic exercise. Early work in the discipline also helps us to understand the impact of physical activity on the anatomy and physiology of the body.

Enhance Sport Performance and Training

The application of physiological techniques to understand and improve human exercise performance has been a major goal of exercise physiology since its inception. Knowledge of the principles of physiology of physical activity serves as the foundation for the development of conditioning programs for various athletes. This area of exercise physiology is often called sport physiology and is the application of "the concepts of exercise physiology to training the athlete and enhancing the athlete's sport performance" (Wilmore & Costill 1994). Think about a track coach who needs to develop running programs for sprinters, middle-distance runners, and distance runners. The various speeds and distances covered by these athletes pose special challenges to the coach who must optimize the training program for each group of runners.

Some of the earliest studies conducted on the physiological responses to strenuous exercise and training came from the Harvard Fatigue Laboratory (Dill, Talbott, & Edwards 1930; Margaria, Edwards, & Dill 1933). Examples of the types of research currently being conducted by sport physiologists include studies of the impact of stretching and flexibility on performance, overtraining in swimmers, and the effects of dehydration on wrestlers. Sport physiologists also use information and techniques from other disciplines and subdisciplines to study ways of enhancing performance. One example is the relationship between carbohydrate intake and performance in endurance events. Sport physiologists may use information gleaned from nutrition research on the carbohydrate content of various foods to work with a sport nutritionist on manipulating the diets of athletes to enhance carbohydrate storage and improve performance. Sport physiologists are also concerned with the effects of the environment on sport performance. Data from sport physiology studies have been used to develop guidelines for the prevention of heat illness in sport and for the prevention of health problems brought on by prolonged immersion of scuba divers in water of great depth. Before the 1968 Summer Olympics in Mexico City (7218 feet [2200 meters] above sea level), leading exercise physiologists conducted numerous studies on the acute effects of and acclimatization to altitude. Sport governing bodies used the knowledge gained from these studies to better prepare their athletes for Olympic competition.

Strength and Conditioning Specialists

When working with athletes in the weight room, strength coaches need to understand how to alter conditioning programs to meet the needs of athletes from many sport teams. As exercise scientists, strength coaches learn how to produce varying results by altering the weight lifted, the number of repetitions and sets, and the length of the rest periods to produce appropriate responses in sprinters, basketball players, softball players, and football players.

Fitness

Understanding the physiological determinants of physical fitness and how training programs improve fitness is another important goal of exercise physiology. Two events in the 1950s—a study reporting that American children were less fit than European children and President Dwight D. Eisenhower's heart attack—stimulated interest in improving physical fitness in the United States (Berryman 1995). These two events led to the development of numerous training programs for improving fitness. Research conducted over the past four decades has resulted in recommendations on the optimal intensity, frequency, and duration of training programs. Individuals working in fitness centers need to understand how to adapt fitness programs to make them safe and appropriate for those who may be young or old, people who have been sedentary or are already training and want to increase their physical fitness level, and those with special conditions such as pregnancy or diabetes. There are pediatric exercise physiologists who specialize in studying and working with children and adolescents, and there are also exercise scientists who are interested in gerontology (the study of aging). The influence of heredity on physical fitness components and trainability has attracted considerable research interest among exercise physiologists. The work of Claude Bouchard and his colleagues with identical twins has stimulated much of this interest. The physiology of fitness is also linked with psychology and the kinesiology subdiscipline of exercise and sport psychology. Understanding why some people adhere to training programs and others stop exercising has led to various strategies for motivating people to become physically active.

Promote Health Through Therapeutic Benefits of Physical Activity

Exercise physiology also serves as a foundation for understanding why physical activity and exercise are beneficial in reducing the risk of disease and in treating some forms of disease. In 1996, the surgeon general of the United States emphasized in a report on physical activity and health that "significant health benefits could be obtained by including a moderate amount of physical activity on most, if not all, days of the week" (United States Department of Health and Human Services [USDHHS] 1996). There are many studies on the relationship of physical activity and physical fitness to coronary heart disease, and today we know much about the role of exercise in reducing the risk factors for coronary heart disease in men. We know less about the relationship of physical activity to heart disease in women. Research on the role of physical activity in preventing other diseases such as non-insulin-dependent diabetes, osteoporosis, and cancer is also under way. Sometimes exercise physiologists are interested in the big picture—how physical fitness can prevent disease in large populations. In such cases, exercise physiologists might collaborate with specialists in epidemiology, using two different types of studies—longitudinal (comparing the same group of people over several time periods) and cross-sectional (comparing different groups of people at the same time).

Fitness programs can be adapted for children by pediatric exercise physiologists.

Physiologists who study the role that physical activity plays in disease management and rehabilitation are known as clinical exercise physiologists. Students training in this area must become familiar with basic concepts in several medical specialties, such as cardiology and pulmonary medicine, and work under the direction of a clinician. For example, students interested in cardiac rehabilitation need to develop an understanding of the various forms of cardiovascular disease and the medical and pharmacological treatments of these diseases. Knowledge of the cardiovascular system and the responses to exercise, as well as knowledge of how to read an electrocardiogram (ECG, a tracing of the electrical activity of the heart), is also important. Cardiac rehabilitation specialists must understand how to modify exercise mode, intensity, and duration for those who have high blood pressure and for those who have had open heart surgery.

Understand Anatomical and Physiological Changes From Physical Activity

Early work in the exercise science discipline focused on the effects of exercise on the functions of organs and body systems. Exercise physiologists have examined the functional responses and structural adaptations of the cardiovascular, respiratory, muscular, and endocrine systems to different types of physical activity. Recently, research interests have focused on the

Interviews With Practicing Professionals

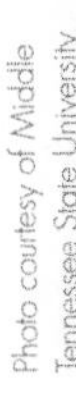

Adam Sayers, PhD, USSF "A" License

Division I Collegiate Soccer Coach

Adam Sayers is a soccer coach at Middle Tennessee State University. He received his PhD in human performance with a specialization in exercise science from Middle Tennessee State University in 2006 and obtained his MS in exercise and sport science from Eastern Kentucky University in 2002. He is a National Strength and Conditioning Association Certified Strength and Conditioning Specialist, and also holds the USSF "A" License, the highest coaching qualification offered by the United States Soccer Federation. He has been coaching soccer at the National Collegiate Athletic Association Division I level for four years and has been involved in coaching since completing a successful collegiate playing career in 2000.

Q: How has your knowledge of exercise science enhanced your coaching skills?

A: My ultimate goal is to be the most effective coach I can be. I believe that this encompasses a thorough understanding of all facets of the game, combined with the ability to apply this understanding in the appropriate situations in order for the team members to maximize their potential and achieve success. Through the knowledge and experience gained during my academic career, I now consider myself to be a vastly improved coach, which benefits the individual players and the team as a whole.

Given the intense physical nature of a collegiate soccer season, appropriate and accurate prescription and monitoring of physical activity during training and games are imperative in order to optimize physical performance and minimize injury rates. Direct application of my academic experience serves our team in several areas, including optimal warm-up and cool-down strategies, testing of all fitness components relative to the game, monitoring physical exertion during training, periodization of training, training of the specific energy systems compliant with the physical demands of the game, and recovery training.

The physical training aspect of the game of soccer is not the only key to success, but if not addressed appropriately, it can be responsible for the failure of a team. I draw from my exercise science knowledge on a daily basis in preparing my team for success.

---

responses of the reproductive, skeletal, and immune systems to acute and chronic exercise. There is a branch of exercise physiology closely linked to biochemistry (e.g., the chemistry of living things). Investigators looking to examine the fuel sources used by muscles during exercise, for example, have used research techniques developed in biochemistry and physical chemistry. The use of the muscle biopsy needle by Bergstrom and Hultman (Hultman 1967) to sample muscle tissue and examine muscle glycogen (stored carbohydrate) concentration during exercise resulted in one of the most important advances in exercise physiology. Biochemical techniques also have been used to develop our understanding of lactate production and use of energy stores during physical activity. Research in molecular biology, one of the newest subdisciplines of biology, has greatly enhanced our understanding of how cells function. Scientists use molecular biological techniques to determine how genes regulate protein synthesis. Exercise physiologists use molecular biological techniques to help them understand how muscle protein synthesis is turned on and off by increased and decreased muscular activity (Booth 1989). This molecular knowledge is important in understanding how muscles

Goals of Exercise Physiology

1. To understand how to enhance physical performance
2. To understand how to improve physical function in particular environments such as high temperatures and high altitude
3. To understand how physical activity and exercise improve health and fitness
4. To understand how exercise can be used in treating and preventing disease and alleviating symptoms of disease
5. To understand adaptations in human anatomy and physiology in response to physical activity

increase and decrease in size in response to increased and decreased activity, respectively, and in understanding how muscles are damaged and how they recover from injury.

History of Physiology of Physical Activity

Physiology of physical activity evolved from physiology in the 18th century after Antoine Lavoisier discovered that animals consume oxygen and produce carbon dioxide (USDHHS 1996). As a result, scientists began exploring how humans respond to physical activities and daily tasks (e.g., lifting loads).

Early Beginnings

Two of the early contributors to our understanding of physiology of physical activity were August Krogh of the University of Copenhagen and A.V. Hill of University College, London. Krogh developed one of the first cycle ergometers (exercise bikes), which he used to study the physiological responses to exercise. He received the Nobel Prize in Physiology in 1920 for his research on the regulation of microcirculation (Åstrand 1991). Hill received the Nobel Prize for his work on energy metabolism in 1921. Hill and Lupton presented many of the basic concepts of exercise physiology concerning oxygen consumption, lactate production, and oxygen debt (excess oxygen consumption following exercise) in a classic paper in 1923. Two laboratories that developed in physical education departments also had a major influence on research in exercise physiology. Peter V. Karpovich established an exercise physiology laboratory at Springfield College in 1927 (Kroll 1982). Karpovich became well known for his research on the effects of ergogenic aids on physical performance and as one of the founders of the American College of Sports Medicine. Thomas K. Cureton, Jr. established an exercise physiology laboratory at the University of Illinois in 1944. This laboratory became well known for research on physical fitness. Many of the leading investigators of the physiology of fitness received their training in Dr. Cureton's laboratory.

Significant Events Since 1950

One of the most important events that stimulated research on physiology of physical activity after 1950 occurred in England. In 1953, Jeremy Morris and colleagues published a study on coronary heart disease and physical activity in which they found that London bus drivers had a significantly higher disease risk than conductors on double-decker buses (Morris et al. 1953). This study was responsible for stimulating interest in epidemiological research on physical activity, physical fitness, and chronic disease (Paffenbarger 1994). Morris' research

Harvard Fatigue Laboratory

The Harvard Fatigue Laboratory, founded in 1927, was the brainchild of L.J. Henderson, a physical chemist. David Bruce Dill, pictured here, was the director of the lab. The lab included a room containing a treadmill borrowed from the Carnegie Nutrition Lab, a large gasometer, a room for basal metabolism studies, an animal room, and a climatic room (Dill 1967). Among the areas of research undertaken by members of this lab were environmental studies conducted at high altitude, in the desert, and in steel mills. Although the Harvard Fatigue Laboratory was in existence only until 1946, it had a profound effect on exercise physiology research in the United States and Europe during the second half of the 20th century. Many young investigators received their formative training in the laboratory as postdoctoral or doctoral students, including Ancel Keys, R.E. Johnson, Sid Robinson, and Steve Horvath. Among the many international scientists who spent time in the lab were Lucien Brouha, Belgium; Rodolfo Margaria, Italy; and E.H. Christensen, Erling Asmussen, and Marius Nielsen, Denmark (Dill 1967). All of these investigators went on to establish their own laboratories and were responsible for training many of the leading investigators in physiology of physical activity over the past 50 years.

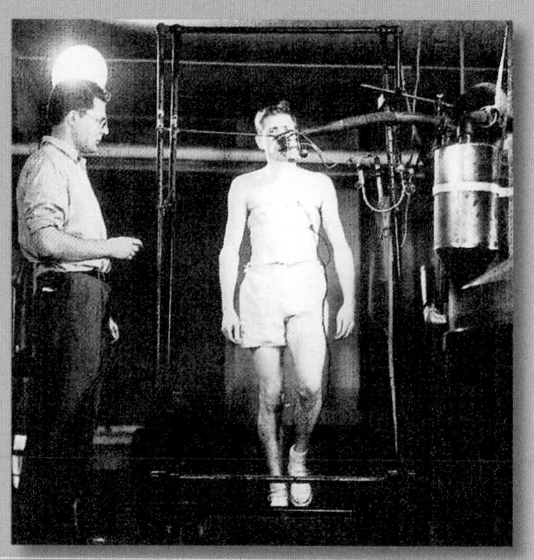

Reprinted, by permission, from J. Wilmore, D. Costill, and W.L. Kenney, 2008. *Physiology of sport and exercise*, 4th ed. (Champaign, IL: Human Kinetics), 6.

➤ Both the National Institutes of Health and the surgeon general concluded that moderate physical activity is beneficial in reducing the risk of chronic diseases (e.g., heart disease, diabetes, hypertension, and colon cancer).

also stimulated studies of the effects of fitness and endurance training on risk factors for coronary heart disease (e.g., serum cholesterol, blood pressure). Based on the results of many of these studies, U.S. government agencies released two official statements on the role of physical activity in the prevention of chronic diseases in the 1990s. The first followed a National Institutes of Health (NIH) Consensus Development Conference on Physical Activity and Cardiovascular Health held in December 1995. Among the consensus development panel's conclusions were that "physical inactivity is a major risk factor for cardiovascular disease" and that "moderate levels of physical activity confer significant health benefits" (NIH Consensus Development Panel on Physical Activity and Cardiovascular Health 1996). The other statement was the surgeon general's first report on physical activity and health by the U.S. Department of Health and Human Services in 1996 (see chapter 2). The report concluded that regular physical activity, in addition to reducing the risk of heart disease, reduces the risk of diabetes, hypertension (high blood pressure), and colon cancer and helps control body weight (USDHHS 1996).

Research Methods in Physiology of Physical Activity

Kinesiologists working in the subdiscipline of exercise science use many methods to help them measure and determine how the human body responds and adapts to physical activity. Work is conducted both within and outside of the laboratory setting.

Laboratory Work

Much of the equipment used to monitor and evaluate physiological responses is contained within exercise science laboratories. Working within a laboratory space is beneficial because there is greater opportunity to control factors that can affect responses to exercise such as temperature and humidity variations. It is important to be able to compare work conducted in different laboratories, so the use of standard protocols and techniques is encouraged.

There are many available techniques to assess each of the health-related components of fitness or those traits associated with lower risk of hypokinetic disease. For example, cardiovascular fitness is assessed either through direct measurement or through estimation of oxygen consumption. This requires measuring the volume of oxygen and carbon dioxide in expired (exhaled) air. Early investigators collected expired air in Douglas bags and analyzed the gas concentrations using chemical analyzers. They then measured how much gas was in the bag with a spirometer or volume meter. Such analysis is tedious and time-consuming. Now, investigators measure gas concentrations with electronic analyzers and gas volumes with flow meters. When interfaced with computers, these devices provide nearly instantaneous and continuous information on oxygen uptake as people rest or exercise.

By measuring oxygen uptake, researchers can obtain information on how the muscles use oxygen and how much energy is expended during physical activity. The most widely accepted method for determining the cardiovascular fitness of an individual is to increase the intensity of exercise progressively until the person can no longer maintain the required exercise intensity. This method is used to determine **maximal oxygen uptake** or **$\dot{V}O_2$max.** Physiologists also use submaximal exercise tests to estimate cardiovascular fitness. These tests do not require individuals to exercise at their maximal level; instead, prediction equations are used to estimate maximal oxygen uptake.

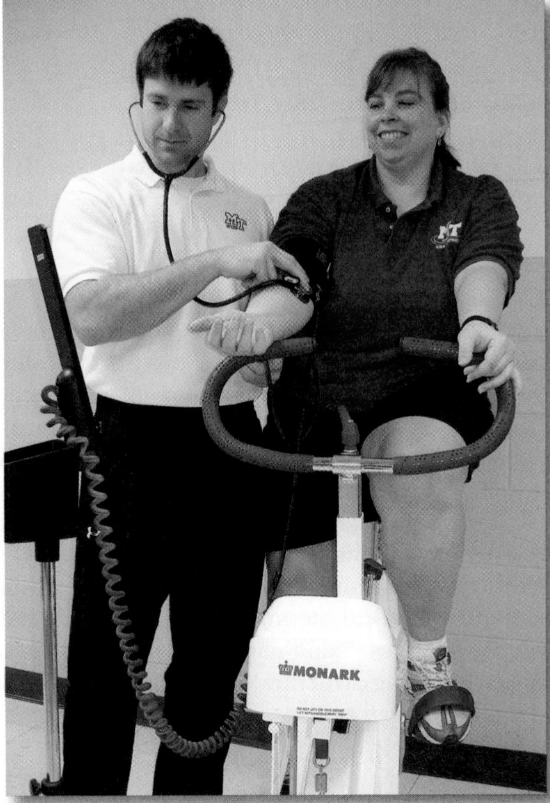

FIGURE 11.2 Blood pressure is easily monitored as a person rides a cycle ergometer.

Photo courtesy of Middle Tennessee State University

Ergometers are used to measure the work performed by muscles during exercise tests. Common ergometers used in exercise science include motorized treadmills and leg or arm cycles. Heart rate, blood pressure, breathing rate, oxygen consumption, and carbon dioxide production are often monitored during exercise. Blood samples may also be drawn.

Because body weight is supported during cycling, cycle ergometers are useful for comparing subjects of different body weights. Blood pressure measurements and blood samples during exercise are also easier to obtain during cycling because the arm is stationary (figure 11.2). The major disadvantages of the cycle ergometer are that oxygen uptake is generally lower and fatigue occurs earlier because the subject uses only the legs. Motorized treadmills allow subjects to be tested during walking and running at different speeds. Increasing the slope (grade) of the treadmill can increase exercise intensity at a constant speed. For studies of moderate to heavy exercise, treadmills are preferable to cycle ergometers because running is more strenuous than cycling for most people. But researchers have to account for body weight differences when using the treadmill. A heavier person walking or running at the same speed and grade as a lighter person will work harder because people must lift their body weight with every step. Measuring blood pressure and sampling blood are more difficult on the treadmill, especially during running, because the arm is moving. In addition to using treadmills and leg cycle ergometers, exercise physiologists can use arm ergometers to study upper body exercise. The swimming flume, developed in Sweden, has been used to study the physiological responses to swimming, and rowing ergometers are useful in studying oarsmen.

Body composition is also a common measure used by kinesiologists. This assessment is used to determine the percentage of body fat an individual has or the ratio of his or her lean tissue to body fat. The gold standard for determining body composition in humans is hydrostatic weighing, also known as **underwater weighing,** in which the individual is weighed while submerged under water. This technique makes use of Archimedes' principle, which states that the weight of the water displaced by an object is equal to the volume of the submerged object (density = mass ÷ volume). Once body density has been determined, equations are used to estimate the amount of body fat. Other techniques for estimating body density include measuring total body water using isotopes (e.g., deuterium), measuring the thickness of subcutaneous fat (lying beneath the skin) with skinfold calipers (see figure 11.4), determining tissue impedance with bioelectric impedance analyzers (BIA), and measuring body fat with dual energy X-ray absorptiometry (DEXA), as shown in figure 11.3.

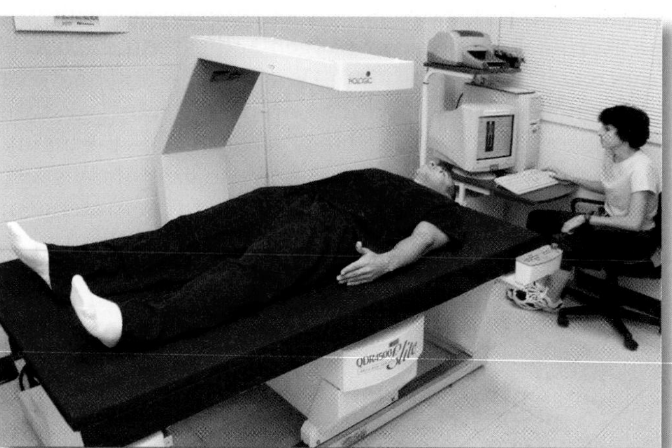

FIGURE 11.3 Body fat and bone mineral density can be assessed using dual energy X-ray absorptiometry in the laboratory.
Photo courtesy of Middle Tennessee State University

Biochemical Methods

Exercise physiologists also use biochemical methods to examine changes at the tissue and cellular levels during and following physical activity. These more invasive techniques include taking blood samples and muscle biopsies. Blood samples are obtained from either venipuncture of superficial arm veins or finger pricks. Monitoring changes in the concentration of blood constituents is useful in determining substrate (stored carbohydrate and fat) utilization, acid–base balance, dehydration, immune function, and endocrine responses. A commonly used biochemical technique in exercise physiology is the measurement of blood lactate from fingertip samples taken during exercise as an indicator of the use of anaerobic energy–producing systems. Physiologists use muscle biopsies obtained before and after exercise to examine changes in stored carbohydrate and fat, lactate production, and enzyme activity. Chemical methods used to analyze tissues help determine muscle fiber types in the biopsy samples. Examination of muscle tissue samples under an electron microscope is useful in determining structural changes following different types of training as well as structural damage after exercise.

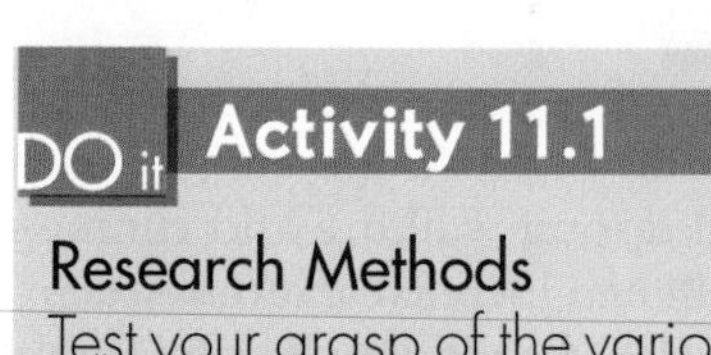

DO it Activity 11.1

Research Methods

Test your grasp of the various methods in this activity in your online study guide.

Animal Models

The effects of physical activity on some organs cannot be studied easily in humans. Examples include the brain, heart, and liver. In some instances, researchers use animals (e.g., rats) to examine both the functional and structural changes that occur in response to single and repeated bouts of exercise on animal treadmills and running wheels. One of the major advantages in using animals in physiological research is that both the subjects and the environment can be more carefully controlled than with humans. The use of animals with genetic abnormalities has also proven helpful for examining the effects of exercise on certain clinical disorders such as obesity and hypertension. Another advantage is that some experimental techniques not approved for human use can be used to study physiological responses. An example is the injection of radioactive isotopes of iron to study changes in tissue iron stores with training. One major criticism of animal research is that not all physiological and chemical mechanisms observed in animal species are identical to those observed in humans. There is some truth to this assertion. For example, growth hormone produced in humans is not identical to that of any other species. Changes in tissues because of training also may differ among species. Investigators must be careful to select the animal, usually a mammal, that most closely reflects the human responses to physical activity that they are interested in.

➤ Advances in technology continue to generate increasing opportunity to conduct research outside of the laboratory setting. However, it is the combination of thousands of research studies conducted over the past century, both inside and outside of the laboratory and using both humans and animals, that has generated our knowledge of physiology of physical activity.

Field Work

Exercise scientists often work with individuals and conduct research outside of the laboratory environment. Working in the field can present difficulties with respect to monitoring physiological responses, controlling exercise intensity, and controlling environmental conditions. When working in the field, it is helpful to use tests that require minimal equipment and tests that can be used to screen large numbers of subjects in a short time period.

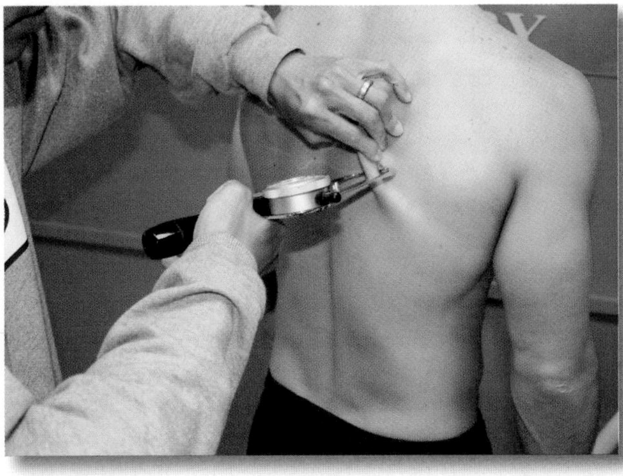

FIGURE 11.4 Body fat can be assessed through use of the skinfold measurement technique in the laboratory and in the field.

© Jennifer L. Caputo

Often, kinesiologists conduct research studies in school settings. It is not practical to bring a treadmill and a cycle ergometer to a school. However, alternative measures of cardiovascular fitness exist; for example, one can determine how quickly students can run or walk a mile marked off on the school field. The completion time is then used to estimate maximal oxygen consumption. Likewise, the PACER (Progressive Aerobic Cardiovascular Endurance Run), a multistage fitness test adapted from the 20-meter shuttle run, can be performed in school gymnasiums. In addition to allowing multiple students to be tested at a time, each of these tests requires minimal equipment and offers flexibility when space availability differs from one facility to another.

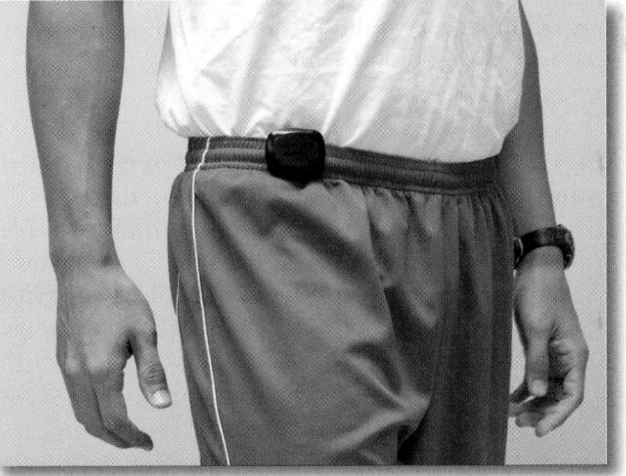

FIGURE 11.5 Pedometers can be used in the field to monitor step count.

© Jennifer L. Caputo

There are also alternative means to measure physiological response while working in the field. Whereas heart rate may be monitored with an ECG machine in the laboratory setting, low-cost battery-operated heart rate monitors allow values to be monitored outside of the laboratory. Some heart rate monitors allow information to be collected over several days before it is downloaded to a computer. These monitors are also useful in assessing and providing information on exercise intensity in the field.

As an alternative to self-report measures, which can be unreliable, motion sensors are now commonly used to measure physical activity. Pedometers and accelerometers, worn at the hip, ankle, or arm, are useful tools for measuring the quantity or intensity of physical activity (or both) and even estimating caloric expenditure outside of the laboratory. These instruments have become increasingly popular due their low cost, small size, and ease of use (figure 11.5). Whereas pedometers generally measure only step count, accelerometers are able to detect the intensity of movement using an acceleration-versus-time curve. Some pedometers and accelerometers have memory capacity that allows for monitoring of weekly activity and caloric expenditure goals.

Improving Sport Performance or Fitness

What one aspect of your sport performance or physical fitness would you like to improve? If you were to begin a training program to improve this aspect, which research method would you use to measure the physiological changes related to your performance or fitness level before training and after training?

Overview of Knowledge in Physiology of Physical Activity

Exercise physiologists have studied responses to single and repeated bouts of physical activity extensively over the past century. Much of this research has centered on three body systems: the muscular, cardiovascular, and respiratory systems. Physiologists have also examined the influence of environmental factors such as temperature, diet, and altitude on physiological responses and performance. In this overview, we will examine how physiological systems respond and adapt to physical activity; review factors that influence physiological responses; and then consider the relationship among fitness, activity, and health.

Skeletal Muscles

The body moves by contracting muscles. The muscles that you use when you are physically active, whether swimming or mowing the lawn, are skeletal muscles, which are under the control of the nervous system. Each muscle is composed of many muscle cells, which are called muscle fibers. Inside the muscle fiber, many myofibrils run the length of the fiber. The **myofibrils** contain the contractile elements that shorten to generate force and move your bones during physical activity. Human skeletal muscles contain three main types of muscle fibers.

Muscle Fiber Types

Did you ever wonder why some people can run extremely fast for short distances (e.g., the length of a football field) whereas other people are much better at running long distances (e.g., a marathon)? Muscle fiber types in the leg muscles of these individuals are likely very different. Muscle fiber types are classified according to the speed at which they contract as fast-twitch (FT) and slow-twitch (ST). Fast-twitch fibers are further subdivided according to the energy system they use (explained in greater detail in the next section). Fast-twitch fibers that use almost exclusively **anaerobic** (without benefit of oxygen) energy systems are called fast glycolytic (FG) fibers, whereas fast-twitch fibers that use both **aerobic** (with oxygen) and anaerobic (also called glycolytic) energy systems are called fast oxidative glycolytic (FOG) fibers. Slow-twitch fibers primarily use aerobic energy systems. During light- to moderate-intensity exercise such as walking to class or jogging, slow-twitch fibers are recruited first. Because these fibers fatigue slowly, light to moderate physical activity can be sustained for prolonged periods. As the intensity of activity increases, FOG fibers are recruited next, and the FG fibers are recruited at the highest intensities (e.g., in an all-out sprint). Although the FG fibers produce much greater force and power, they fatigue rapidly. Thus, we can sustain the highest intensities of physical activity for only a short time (less than 1 minute).

Energy Sources

For a muscle to contract, it must have an adequate energy supply. The primary source of energy for muscle contraction is a high-energy phosphate compound called adenosine triphosphate (ATP). When a muscle fiber contracts, ATP is split into adenosine diphosphate (ADP) and an inorganic phosphate (P). New ATP can be generated using several sources such as phosphocreatine, stored glucose or glycogen, and stored fat (triglycerides).

Three primary energy systems are used to generate the energy (ATP) that muscles need during physical activity:

- ATP-phosphocreatine system
- Glycolytic system
- Aerobic system

The ATP-phosphocreatine system is used during the first 10 to 15 seconds of activity; the glycolytic system is used for maximal exercise lasting 1 to 2 minutes; and the aerobic system is the primary source of energy in activities lasting more than 2 minutes.

ATP-Phosphocreatine System During rapid muscle contractions, ADP can combine with phosphocreatine (PC) to resynthesize a single ATP to generate energy for another muscle contraction. Because the quantities of ATP and PC stored in muscle fibers are relatively small, high-intensity exercise cannot be supported by the ATP-PC system for more than 10 to 15 seconds. We use this energy system when we sprint down a basketball court, run to first base, or do a vault in gymnastics. Some athletes take creatine supplements to increase the amount of muscle PC. Creatine supplements are most likely to improve performance in repetitive exercise bouts that last less than 30 seconds (Branch 2003).

Anaerobic Glycolytic System Because the ATP-PC system is depleted during the first 15 seconds of a run, to cover longer distances (e.g., around a 400-meter track) as fast as you can, your body will depend on an additional energy system. Muscle fibers also store glycogen (stored glucose). During anaerobic glycolysis, glucose is released from glycogen and is broken down to form ATP. This energy-producing pathway does not require oxygen and is therefore anaerobic. One of the end products of this energy system is lactic acid. Because muscles cannot tolerate excessive buildups of lactic acid, activities requiring anaerobic glycolysis can continue for only 1 to 2 minutes. Examples of activities that are likely to produce lactic acid are riding a bicycle uphill and running 800 meters as fast as possible.

➤ The anaerobic breakdown of glycogen produces lactic acid.

Aerobic System If, on the other hand, you were participating in prolonged physical activity, such as running a 10K race or hiking up a mountain, you would need energy from yet another energy system. The aerobic breakdown of glycogen and fats produces energy for these kinds of activities and takes place in the muscle **mitochondria.** The mitochondria are often referred to as the "power house" of the cell because large quantities of ATP are produced using the aerobic energy systems. The length of time muscle glycogen can continue to supply energy is determined by the amount of glycogen stored in the active muscles and the intensity of the activity. High-intensity activities (e.g., 77-93% of the maximal heart rate) deplete muscle glycogen stores more rapidly than moderate-intensity activities (e.g., 64-76% of the maximal heart

DO it **Activity 11.2**

Fueling Movement
Test your knowledge of these three sources of energy in your online study guide.

rate). Muscle glycogen stores are depleted in approximately 90 minutes during high-intensity exercise. Soccer is an example of a high-intensity sport in which muscle glycogen stores are very low at the end of a game. If the muscle glycogen in fast-twitch muscles is depleted, the athlete will not be able to sprint in the later part of the game. During light and moderate activity (e.g., walking), fat is a major contributor of energy for muscle contraction. The greatest source of energy stored in the body is stored as fat. Fats (triglycerides) are stored in adipose tissue located throughout the body, and small quantities of fat are located in the muscles. Because you have large amounts of stored body fat that can be mobilized and used for energy during light to moderate physical activity, you can walk or hike for many hours without becoming fatigued. The different energy systems used during sports and physical activities are summarized on this page.

Energy System Knowledge and Training Program Design An understanding of the way the body produces energy is fundamental for coaches and trainers when designing conditioning programs. Coaches must apply their knowledge of the anaerobic and aerobic energy systems when writing training programs because the adaptations and outcomes of a training program are specific to the muscles and energy system(s) that are targeted. This illustrates the training principle of specificity.

Sports such as football and short sprints in track rely predominantly on the ATP-PC pathway for energy; marathon running is overwhelmingly aerobic, and middle-distance running events such as the 800 meters or 1500 meters require a combination of anaerobic and aerobic energy sources.

Therefore, training to maximize ATP production by the ATP-PC system for an event such as the 100 meters would involve running short, high-intensity intervals less than 10 seconds in duration with short rest periods. This type of training increases the activity of enzymes (proteins that accelerate the rate of chemical reactions) involved in the ATP-PC and glycolytic systems. This leads to an increased capacity to resynthesize ATP during high-intensity exercise of short duration.

Different Energy Systems Used During Sports and Physical Activity

ATP-PC System

- Initial, high-intensity physical activity
- Dominates during the first 10 to 15 seconds
- Primary energy system in sprints up to 100 meters (e.g., basketball, soccer, football, field hockey, softball)

Anaerobic Glycolysis System

- Dominates once the ATP-PC system cannot support the activity
- Primary source of energy during first 1 to 2 minutes of high-intensity exercise
- Primary energy source in the following events: 400- and 800-meter runs, 100- and 200-meter swimming events, 500- and 1000-meter speed skating events, and alpine skiing events

Aerobic System

- Primary energy system used during prolonged, continuous light- and moderate-intensity activities
- Uses both carbohydrates and fats as energy sources
- Used in walking, jogging, distance running (e.g., 5000 meters and longer), cycling, swimming laps, triathlons, and cross-country skiing

Distance runners, on the other hand, benefit from long, slow, distance runs and high-intensity intervals of durations greater than 1 minute. This type of endurance training increases both the size and number of mitochondria in the muscle fibers. In addition, the activity of several enzymes important to the aerobic energy systems increases. This means that the muscle has a greater capacity to use the aerobic energy system and fats as a source of energy following endurance training. Another important adaptation is an increased blood supply to the muscle fibers as the number of capillaries increases. This adaptation increases both the oxygen and nutrient supply to the muscle fibers. The result is that aerobically trained muscles are better able to sustain physical activity for prolonged periods.

Resistance Training

➤ Muscles increase in strength due to progressive overload.

Muscular **strength** is defined as the maximal amount of force exerted by a muscle group. Muscular **power** is the product of the force times the speed of movement. The ability of a muscle to repeatedly exert force over a prolonged period is known as muscular **endurance.** Resistance training programs to improve muscular strength may use **isometric** (tension is produced without change in the muscle length), **isotonic** (muscle changes length without changing tension), or **isokinetic** (muscle changes length at a constant rate of velocity) exercises. An example of an isometric contraction is putting the palms of your hands against a wall and pushing as hard as you can. Lifting a free weight (e.g., barbell) is an example of an isotonic contraction. Resistance training equipment that allows you to set the speed of movement is isokinetic (see figure 11.6). Regardless of the type of exercise used, the exerciser must overload the muscle group in order for strength to increase. This is the training principle of progressive overload. There must be a gradual increase in the stress placed on the physiological system or tissue in order for improvements to be observed. Likewise, when you stop overloading the system, the training adaptation is lost. This is the training principle of reversibility.

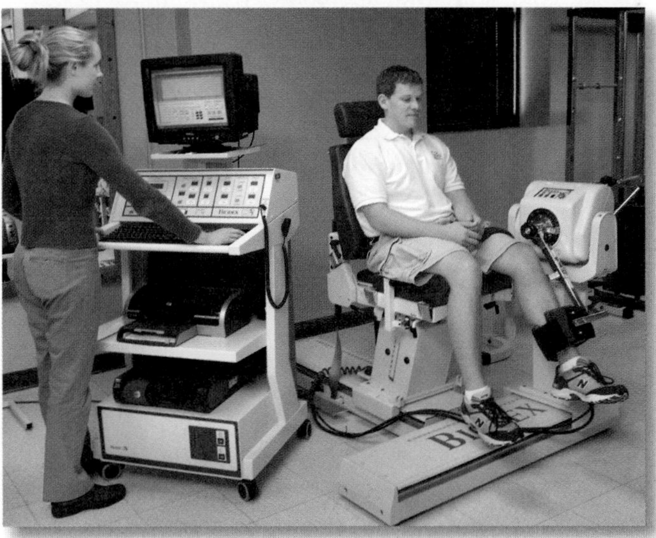

FIGURE 11.6 Isokinetic machines are often used in the rehabilitation of injuries.

Photo courtesy of Middle Tennessee State University

The principle of specificity also applies to resistance training. If you were interested in using resistance training to improve your performance in a given sport, the type of training that you select should match the sport techniques as closely as possible. Strength gains will be greatest over the range of movement and velocity of the resistance exercises used (Morrissey, Harman, & Johnson 1995). In some physical activities, the ability to develop force rapidly is critical to performance. For example, a volleyball player must rapidly extend the legs in jumping to block a spike at the net. If your sport requires you to move at high speed, higher-velocity isokinetic or isotonic training will be more beneficial than low velocities or isometric training.

Increases in muscular strength are believed to result from two factors: an increase in muscle size and neural adaptations. An increase in muscle size because of resistance training is due to the increase in size of individual muscle fibers **(hypertrophy)**. Although evidence from studies of resistance training in cats suggests that muscle fibers split and increase **(hyperplasia)** (Gonyea 1980), researchers have not observed increased numbers of fibers in other animal species (Gollnick et al. 1981; Timson et al. 1985). Neural adaptations that may occur include recruitment of additional muscle fibers (motor units), better synchronization of muscle fiber contraction, and reduction in neural inhibition.

Physical Activities and Resistance Training

Think about the physical activities that you participate in most often. Explain which activities require muscular strength and which activities require muscular endurance. Select one of these activities and describe the type of movement involved. Based on the speed and duration of the movement required, what type of resistance training would be most beneficial in improving your performance in this activity?

Cardiovascular System

Skeletal muscles draw from various energy systems to perform work. The aerobic system, which we use for sustained physical activities, is limited primarily by the availability of oxygen. The transport of oxygen to all tissues is a primary function of the cardiovascular system, which is composed of the heart, blood vessels, and the blood. To increase the supply of oxygen to skeletal muscles, the cardiovascular system responds immediately to physical activity in several ways. Both the volume and distribution of blood flow change during physical activity. In addition, chronic physical activity results in several important physiological adaptations in cardiac function that improve endurance.

Cardiac Output

Cardiac output, defined as the amount of blood pumped out of the heart each minute, is a function of both the heart rate (number of heartbeats per minute) and **stroke volume** (amount of blood pumped per beat). Resting cardiac output is remarkably constant in adult humans at approximately 5 liters (approximately 5 quarts) per minute. During physical activity, cardiac output increases as the muscles use more oxygen. Oxygen uptake (written as $\dot{V}O_2$), or the amount of oxygen used by muscle tissues, increases in direct proportion to the intensity of exercise until maximal oxygen uptake ($\dot{V}O_2max$) is reached. The amount of oxygen delivered to the tissues depends on how much oxygen is in the blood and how much blood the heart is pumping. As your muscles take up more oxygen, your cardiac output increases. At lower exercise intensities, this increase is due to increases in the amount of blood pumped per beat (stroke volume) and the number of beats per minute (heart rate). However, as your exercise intensity increases above 40% of $\dot{V}O_2max$, your stroke volume plateaus because you have reached the limit of how much blood your heart can pump per beat. Your heart rate continues to increase along with oxygen uptake, however, until you achieve your maximal heart rate (see figure 11.7). At this point, you have also reached your limit for oxygen uptake. So, as you can see, the capacity of the heart to pump blood, the maximal cardiac output, appears to be one of the primary factors limiting maximal oxygen uptake. Because $\dot{V}O_2max$ and maximal heart rate are reached simultaneously, the fitness instructor can use submaximal exercise tests to estimate $\dot{V}O_2max$. This method is useful in exercise testing and prescription because it does not require the client to exercise to fatigue.

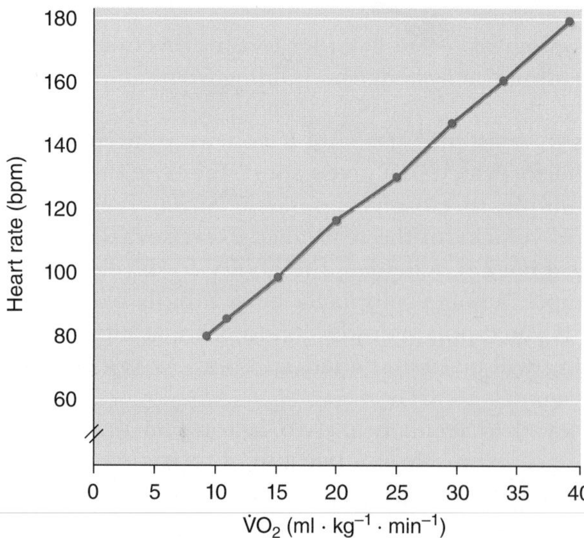

FIGURE 11.7 Relationship between heart rate and $\dot{V}O_2$ during a graded exercise test.

Blood Flow Distribution

When you are at rest, most of your blood flow (cardiac output) goes to your brain and internal organs (e.g., liver, kidneys), with only 15% going to your skeletal muscles. When you become physically active, your muscles need more blood to supply oxygen and nutrients and to remove waste products. At the onset of activity, your blood vessels constrict (vasoconstrict) in regions of your body that need less blood flow and dilate (vasodilate) in your skeletal muscles and heart. During heavy exercise, two-thirds of your blood flow goes to your skeletal muscles through a shift in blood flow away from your kidneys and digestive organs.

Cardiovascular Adaptations to Training

Your ability to exercise at moderate to heavy intensities for prolonged periods is referred to as your aerobic or cardiovascular endurance. You have learned that one of the best indicators of aerobic endurance is your $\dot{V}O_2$max, also known as maximal aerobic power. You can increase your maximal aerobic power with endurance training. Much of the improvement in your $\dot{V}O_2$max is due to an increase in stroke volume. The cardiac muscle of your heart actually increases in size and contracts more forcefully in response to endurance training. As a result of training, you will have increased stroke volume both at rest and during exercise, and your heart rate will decrease. In other words, you will be able to pump more blood with less work! Of course, you will want to know if your training regimen was successful. Exercise testing before and after an endurance training program is important for two reasons: (1) It establishes your baseline fitness level, and (2) it allows you to determine the effectiveness of the training program. A graded exercise test, in which intensity progressively increases, is one way to test the efficacy of your training program. At each stage of the test you should notice that your heart rate is lower than it was before you started training (see figure 11.8).

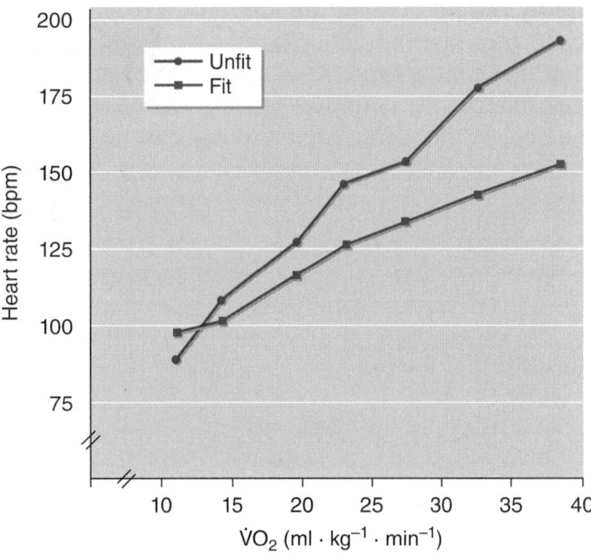

FIGURE 11.8 Comparison of the heart rate responses of fit and unfit subjects during a graded exercise test.

As we noted, your endurance training regimen will also result in an increase in the number of capillaries in your skeletal muscles. This change means that you can receive more blood and therefore more oxygen in your working muscle fibers during activity. Researchers have discovered that, following training, the difference in the amount of oxygen in the arteries (which move blood from the heart) and veins (which move blood back to the heart) is greater, suggesting that the muscles are extracting more oxygen from the blood. The increase in muscle capillaries may be responsible for the increase in oxygen extraction. Despite your best efforts at fitness training, you may find that others remain far ahead of you in gains, some seemingly without even trying. This point brings us to a discussion of the other factors that affect maximal aerobic power. One is your genetic makeup. Studies of the $\dot{V}O_2$max of identical and fraternal twins have shown greater variation between fraternal twin pairs than identical twin pairs (Bouchard et al. 1986). In other words, individuals with identical genes (e.g., identical twins) are more alike in maximal aerobic power, showing the strong genetic contribution to maximal aerobic power. Another factor that influences maximal aerobic power is age. $\dot{V}O_2$max begins to decrease after age 30 because of a decrease in maximal heart rate and therefore maximal cardiac output. The decline in $\dot{V}O_2$max with aging may also be due in part to a decrease in physical activity, as researchers have observed that athletes who maintain their training levels experience a slower decline in $\dot{V}O_2$max (Hagberg 1987).

Determining Your Resting Heart Rate

Determine your resting heart rate by counting your pulse for 15 seconds and multiplying by four. The average heart rate is 70 beats per minute. Is your resting heart rate less than 70 beats per minute? On the basis of the amount and types of physical activities that you participate in daily, can you explain why your resting heart rate is above or below 70 beats per minute?

Respiratory System

➤ Ventilation increases rapidly at the onset of physical activity and increases as a function of exercise intensity.

As you now know, the cardiovascular system transports oxygen to skeletal muscles. But where does that oxygen come from? The respiratory system regulates the exchange of gases (including oxygen) between the external environment (air) and the internal environment (inside the body). For this exchange to occur, air must move from the nasal cavity or mouth through the respiratory passages to the alveoli within the lungs. After the fresh air enters the alveoli, oxygen can diffuse into the blood in the pulmonary capillaries, and carbon dioxide can leave the blood and enter the lungs for exhalation into the environment. The process of moving air in and out of the lungs is known as **ventilation.** The amount of air exhaled per minute is known as the minute volume (V_e). It is the product of the amount of air exhaled per breath **(tidal volume)** and the number of breaths per minute (respiratory frequency). The minute volume of a person at rest is approximately 6 liters (6.3 quarts) per minute.

At the beginning of physical activity, you may notice that your breathing rate increases rapidly during the first minute until it reaches a plateau. Researchers believe that the stimulus to increase ventilation comes from sensory **receptors** in the moving limbs (e.g., muscle spindles, joint receptors) as well as the motor cortex (i.e., the part of the brain that stimulates muscles to contract). At low exercise intensities, the increase in minute volume (air exhaled) is due primarily to an increase in the amount of air exhaled with each breath; the number of breaths remains constant. At higher exercise intensities, however, both respiratory frequency and tidal volume increase—you take more frequent and bigger breaths.

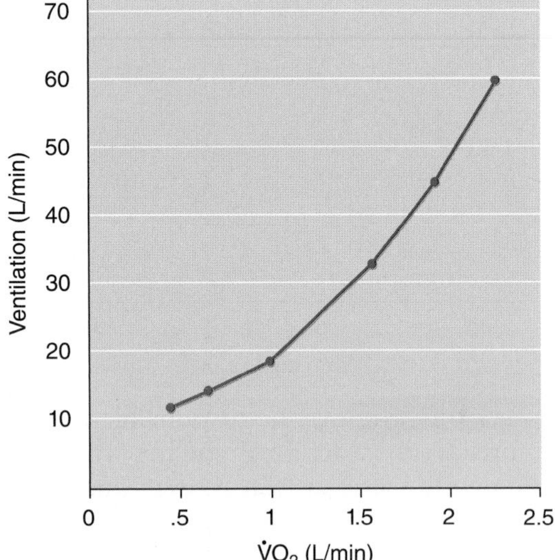

FIGURE 11.9 Relationship of ventilation to $\dot{V}O_2$ as exercise intensity increases progressively during a graded exercise test.

As your level of exercise intensity increases, your breathing also increases—steadily at first, and then much more rapidly at higher intensities (see figure 11.9). The point at which your breathing begins to increase rapidly is known as the **ventilatory threshold.** Ventilatory threshold occurs at exercise intensities between 50% and 75% of $\dot{V}O_2$max. After you have been training for a while, your maximal ventilation—the amount of air entering and leaving your lungs—increases when you exercise. Moreover, your ventilatory threshold occurs at a higher exercise intensity after training.

Temperature Effects

In addition to studying the three primary body systems affected by physical activity, exercise physiologists have also studied the influence of environmental factors on our experience of sport and exercise. One factor is temperature variation. Our bodies have complex and effective ways of dealing with temperature changes in the outside environment. Humans are able to regulate their internal (core) temperature so that it remains relatively constant over a wide range of environmental temperatures (Haymes & Wells 1986). When you begin to exercise, muscular contractions produce heat. Blood flowing through these regions of the body warms up and distributes heat to other regions of your body. As your body temperature rises, your skin blood vessels begin to vasodilate and you begin to sweat. In comfortable ambient temperatures, your body core temperature reaches a plateau in 20 to 30 minutes (see figure 11.10). The higher the intensity of your physical activity, the higher your core temperature will be when it reaches a plateau.

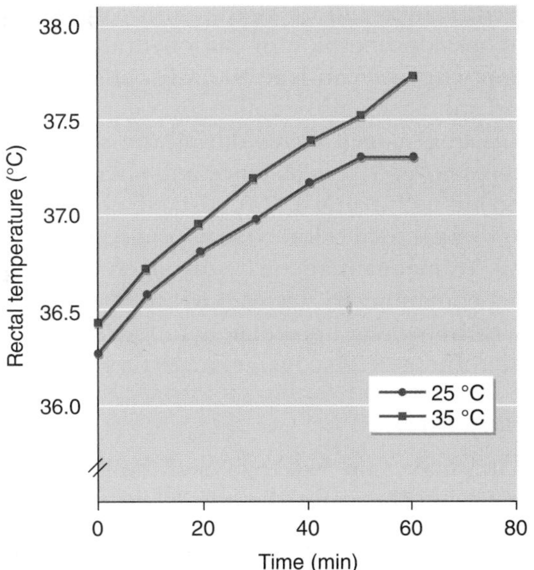

FIGURE 11.10 Increases in body core (rectal) temperature during exercise in comfortable (77 °F, 25 °C) and warm (95 °F; 35 °C) ambient temperatures.

Have you ever wondered why you sweat more profusely during physical activity in warm environments? When air temperature approaches or exceeds skin temperature, the body's mechanism of bringing warm blood to the surface to cool is not as effective. In this case, the evaporation of sweat becomes the major avenue of heat loss. In hot, humid environments, heat loss through evaporation is also limited. This results in a higher body temperature (see figure 11.10) and increased risk of heat illness (e.g., heat exhaustion, heatstroke). We tolerate physical activity in cool environments better because our bodies lose heat from blood close to the skin as well as from the evaporation of sweat. In cold water, however, heat loss from the skin increases dramatically because water is an excellent conductor of heat. Submersion in cold water stimulates shivering and an increase in **metabolic rate** (the rate at which the body uses energy). Swimming in cold water also elevates the metabolic rate, which may lead to an earlier onset of fatigue. Similarly, individuals whose clothing becomes wet during physical activity in cold environments lose heat more rapidly, begin to shiver, and are at risk of **hypothermia** (below-normal body temperatures).

➤ During physical activity, increased heat production by skeletal muscles stimulates vasodilation of skin blood vessels and sweating.

Nutritional Intake and Physical Activity

Nutritional intake is another factor studied by physiologists of physical activity. Although most of us have enough stored fat to sustain low-intensity activities for many days, we need shorter-term energy sources, such as carbohydrates, for moderately heavy endurance activities. Because the amount of carbohydrate stored in the body is less than 1 pound (0.45 kilograms), daily carbohydrate intake is crucial to athletic performance. We will explore the role of carbohydrates as well as that of two other nutrients, water and iron, in the following sections.

Carbohydrates

Carbohydrate is stored as glycogen primarily in skeletal muscles and the liver. During high-intensity physical activities, muscle glycogen stores are the primary source of energy. Normal muscle glycogen stores are depleted in approximately 90 minutes of continuous exercise at 75% $\dot{V}O_2$max. The amount of glycogen stored is directly related to the carbohydrate content of the diet. Coaches and athletes know that a diet high in carbohydrate (70% or more of the calories), consumed for two or three days, will increase muscle glycogen storage. Some endurance athletes use this technique, known as carbohydrate loading, before competition. Carbohydrate loading is most effective in activities lasting more than 2 hours (e.g., marathons). There is no apparent advantage to carbohydrate loading for activities lasting less than 1 hour.

Maintaining carbohydrate levels during prolonged exercise is important because low blood glucose is often associated with fatigue. Consumption of carbohydrate drinks containing 6% to 8% carbohydrate by weight during prolonged exercise has been shown to improve

➤ Body heat generated during exercise is lost through evaporation of sweat.

performance and delay the onset of fatigue (Coggan & Coyle 1991). Drinks containing a higher concentration of carbohydrate (greater than 10% by weight), however, delay gastric emptying and can lead to gastric distress (Davis et al. 1988). Intermittent physical activity and sports can also deplete glycogen stores. For example, soccer players often deplete their muscle glycogen stores during the second half of a game. Players with depleted glycogen stores covered less distance and spent more time walking and less time running during the second half (Saltin 1973). Also, some athletes with low carbohydrate intakes (i.e., less than 50% of the total calories) may progressively deplete their glycogen stores through daily training. To maintain adequate glycogen stores, physically active individuals should consume a diet containing 55% to 60% of the total calories as carbohydrate. Consumption of foods high in carbohydrate immediately following exercise is also beneficial because this increases the rate of muscle glycogen storage (Ivy, Katz, & Cutler 1989).

Fluid Intake

➤ Inadequate fluid intake during physical activity results in elevated body temperature, greater risk of developing heat illness, and possible decrements in performance.

Water makes up 55% to 60% of the human body. During physical activity, you lose some of this water through sweating. When you exercise on a warm day, even just walking to class, you will notice that you sweat even more. Because the volume of sweat lost can be substantial, fluid replacement is important. Sweat loss decreases body fluids within and between cells, as well as **plasma volume** (the fluid portion of blood). When plasma volume decreases, less blood returns to the heart, which reduces the amount of blood in each heartbeat (stroke volume). To compensate for the reduced stroke volume, heart rate increases. If fluid losses are not replaced, the sweat rate will fall, causing an increase in body temperature. This puts people at greater risk of developing heat illnesses, especially heat exhaustion.

Failure to adequately replace fluids can also lead to decrements in performance in some physical activities. For example, Armstrong, Costill, and Fink (1985) found that running velocities at distances ranging from 1500 to 10,000 meters (approximately 1 to 6 miles) fell 3% to 7% following a 2% decrease in body weight. Dehydration is less likely to affect short, high-intensity activities such as sprinting than more prolonged events. To ensure adequate fluid replacement during physical activity, you should drink fluids (e.g., water, carbohydrate-electrolyte drinks) at regular intervals (every 15 to 20 minutes). Fluid replacement is important during many types of physical activity including hiking, bicycling, aerobics classes, and working outdoors. If you play a sport such as soccer or field hockey, fluid replacement during play is difficult because the rules do not allow time-outs during competition except for injury. Drink additional fluid before a match and during halftime in these sports to reduce the fluid deficit. Following activity, you should continue to drink fluids even though you may not feel thirsty. Unfortunately, thirst is not an accurate indicator of the need for fluid. During activities lasting less than 1 hour, consuming carbohydrate-electrolyte drinks offers no advantage. In more prolonged activities, fluids containing a small amount of carbohydrate are more likely to be beneficial in enhancing fluid absorption and maintaining blood glucose.

Iron Intake

➤ Depletion of iron stores can lead to anemia, which reduces maximal oxygen uptake and endurance.

As you know, physical activity cannot take place without the transport of oxygen-carrying blood to the muscles. The blood is able to transport oxygen with the help of iron. Oxygen essentially rides on the iron atoms in the **hemoglobin** found in red blood cells. Each hemoglobin contains four iron atoms that bond with four oxygen molecules. A person with low iron cannot adequately synthesize hemoglobin. As hemoglobin concentrations fall, the amount of oxygen transported to the tissues decreases. Eventually, the person becomes anemic. **Anemia** is defined as a hemoglobin concentration below 12 grams per deciliter of blood in women and 13 grams per deciliter in men. Both maximal oxygen uptake and endurance are reduced in people who have anemia (Celsing et al. 1986).

Iron deficiency is one of the most common nutritional deficiencies in the United States, especially among adolescent girls and women. The most likely causes of iron depletion in physically active women are inadequate iron intake and excessive blood loss through the menses. Low iron intake in female athletes most commonly occurs in those who restrict their

DO it **Activity 11.3**

Create a Food Diary

Use this activity in your online study guide to document and evaluate your dietary intake.

caloric intake (e.g., gymnasts) and those who consume diets that are low in meat (e.g., runners) (Clarkson & Haymes 1995). Heme iron found in meat, fish, and poultry is more highly absorbed from the gastrointestinal tract than is the non–heme iron found in other food sources. People can typically avoid iron deficiency by consuming iron-rich foods on a daily basis.

Physical Activity, Fitness, and Health

➤ The recommended amount of physical activity for improvement in cardiovascular endurance is 20 to 60 minutes of activity at 70% to 94% of maximal heart rate, three to five days per week.

Strenuous exercise programs are not required to gain health benefits from physical activity. On the contrary, scientific evidence suggests that engaging in moderate-intensity physical activities on a regular basis confers health benefits (Pate et al. 1995). Moderate-intensity physical activities include brisk walking (3 to 4 miles per hour, or 4.8 to 6.4 kilometers per hour), cycling less than 10 miles per hour (16 kilometers per hour), swimming at moderate speed, badminton, table tennis, golf (pulling a cart), climbing stairs, and mowing the lawn. The 1996 report titled *Physical Activity and Health: A Report of the Surgeon General* recommended accumulating 30 minutes or more of moderate-intensity physical activity on most, preferably all, days of the week (USDHHS 1996). The activity need not be continuous for 30 minutes; it may include intermittent activities as well. The goal should be to accumulate activities that use approximately 150 to 200 calories over a 30-minute period. By emphasizing moderate instead of strenuous physical activities, experts hope that a higher percentage of the adult population, and women and older adults in particular, will regularly engage in physical activity.

Greater volumes and intensity of exercise will result in additional health and fitness benefits. To improve cardiovascular endurance, the American College of Sports Medicine (2006) recommends that adults take part in physical activity 20 to 60 minutes per day, three to five days per week, at 70% to 94% of maximal heart rate. Exercise intensities that are 64% to 76% of maximal heart rate are usually perceived as moderate exercise, and exercise intensities of 77% to 93% of maximal heart rate are considered hard or vigorous exercise. Activities that are aerobic and can be maintained for prolonged periods are best for improving cardiovascular endurance. These activities include running, hiking, walking, swimming, cross-country skiing, bicycling, aerobic dancing, stair climbing, and rowing. Sports such as soccer, field hockey, and tennis also help develop cardiovascular fitness because they are high-intensity, intermittent activities carried out over prolonged periods.

One of the most common reasons for stopping an exercise training program is injury (Pate et al. 1995). For this reason, exercise programs designed to improve fitness and health should begin with moderate-intensity activities and gradually increase the intensity and duration of the activities as fitness improves. For example, an exercise program for sedentary adults may begin with walking at a speed of 3 to 4 miles per hour (closer to the moderate intensity of 64% of maximal heart rate) and gradually increase the duration from 20 to 60 minutes. People can begin by walking, progress to alternating between walking and jogging, and advance to jogging.

Accumulating 30 minutes of moderate-intensity activity each day improves health.

Effects of Age on Fitness

Although research has shown that maximal oxygen uptake ($\dot{V}O_2max$) declines with age at the rate of approximately 10% per decade, the decrease is smaller (5% per decade) in individuals who remain physically active (Hagberg 1987). The decline in cardiovascular endurance in older individuals may be due in part to a reduction in the intensity and duration of physical activity. Master athletes who maintain their training intensity experience little change in maximal oxygen uptake as they age. $\dot{V}O_2max$, the primary indicator of cardiovascular endurance in adults, increases in absolute terms (liters per minute) with growth in children. Because body mass increases with growth, more oxygen is required to supply the active tissue. When $\dot{V}O_2max$ is expressed per kilogram of body weight, however, it remains relatively constant or decreases during growth (Bar-Or 1983). Results of several studies of children suggest that $\dot{V}O_2max$ per kilogram does not increase with training in prepubescent children until the peak of the growth spurt occurs (Zwiren 1989). This finding does not mean, however, that children's endurance does not improve before puberty. Improvements in children's endurance with training may be related to improvements in other factors such as running economy and anaerobic capacity (Rowland 1989).

Physical Activity, Fitness, and Coronary Heart Disease

➤ High levels of physical fitness lower the risk of developing and dying from cardiovascular disease. Regular participation in moderate-intensity physical activity also reduces the risk of cardiovascular disease and colon cancer.

In general, higher levels of fitness are associated with improved health status. This generally held assumption is based on research conducted by exercise physiologists. Blair and colleagues (1989) found that individuals who are the most fit have a lower relative risk of developing or dying from cardiovascular disease or cancer than those who are less fit. They based their determination of physical fitness on a cardiovascular endurance test. Participation in physical activities, as opposed to levels of fitness, also reduces the risk of cardiovascular disease and colon cancer. Paffenbarger (1994) estimated that engaging in moderate-intensity physical activities added approximately 1.5 years to life. Using the number of calories expended weekly in walking, stair climbing, and leisure-time sport as an index of physical activity, his study showed that Harvard alumni had a 46% lower risk of dying from cardiovascular disease if they expended 2500 calories or more per week in physical activities (Paffenbarger 1994). Participants in moderate-intensity activities (e.g., brisk walking, swimming, cycling) had a 37% lower risk of cardiovascular disease compared with those who did not participate in leisure-time sports.

Physical activity and fitness also play a role in reducing the risk of hypertension, or high blood pressure. In hypertensive individuals, an acute bout of moderate physical activity that lasts more

Health and fitness benefits can be achieved through physical activity across the age spectrum.
© Jennifer L. Caputo

Activity 11.4

Your Maximal and Exercise Heart Rates

Work through this activity in the online study guide to calculate your optimal HR range for exercise.

than 30 minutes reduces blood pressure for several hours (Hagberg 1989). Over time, endurance training lowers blood pressure in hypertensive individuals by approximately 10 mmHg (millimeters of mercury). Lower-intensity exercise programs (40-70% of $\dot{V}O_2max$) are as effective in reducing blood pressure as are those of higher intensity (Fagard 2001). Among people with normal blood pressure, those who are more physically active have lower blood pressures than those who are less active.

Physical Activity and Weight Control

➤ Energy expenditure during physical activity is directly proportional to the intensity and duration of the activity.

Participation in daily physical activity helps control body weight. Weight gain and weight loss are determined by the interplay between consumption of calories and expenditure of calories. An excess caloric consumption results in weight gain. Exercise physiologists have added to our understanding of caloric expenditure by investigating the process by which the body burns calories and uses energy. The metabolic rate—the rate at which the body uses energy—increases in direct proportion to the intensity of activity. Lower-intensity activities such as walking at 3 miles per hour (4.8 kilometers per hour) increase the resting metabolic rate threefold, a caloric expenditure of approximately 4 calories per minute. Higher-intensity activities such running at 7 miles per hour (11 kilometers per hour) increase the resting metabolic rate 10-fold, which is the equivalent of expending 12.5 calories per minute. Total energy expenditure also depends on the duration of physical activity. Moreover, energy expenditure does not return to resting levels immediately at the end of an exercise bout. Researchers have shown that the metabolic rate remains elevated during recovery from physical activity. Thus, the total energy expended from a single bout of exercise is somewhat greater than the energy cost of the activity.

Although some seek to maintain body weight through exercise, others are looking to lose body weight and body fat. The recommended exercise dose for weight loss is approximately 45 minutes of daily activity (American College of Sports Medicine 2001). When caloric intake exceeds caloric expenditure, the excess calories are converted to fat and stored in adipose tissue. Sedentary individuals are more likely to be overweight or obese, a condition that puts them at increased risk of developing coronary heart disease (Willett et al. 1995). There is also a higher risk of non-insulin-dependent diabetes (NIDDM) in those participating in low levels of physical activity (Paffenbarger 1994). Reducing body weight is effective in lowering serum triglycerides and reducing blood pressure in hypertensive individuals. Furthermore, participation in daily physical activities is beneficial not only in reducing the risk of NIDDM but also in helping individuals with NIDDM regulate their blood glucose levels.

Now that you have some basic knowledge about the physiological systems and some of their responses during physical activity, we will examine one of the most pressing problems facing physiologists of physical activity today: how to decrease obesity rates.

Recent large-scale population studies have shown that obesity has increased dramatically in the United States over the past 25 years. Obesity, defined as a body mass index (BMI) greater than 30 kg/m^2, is calculated using a person's weight and height: BMI = weight in kilograms divided by height in meters squared. The percentage of adults who were classified as obese increased twofold from 15% in 1975 to 30% in 2000. Adults are classified as overweight if the BMI is between 25 and 30 kg/m^2. More than 60 million Americans (approximately one out of every three people) are classified as being overweight. This problem of excess weight is also apparent for younger segments of the population. The proportion of children aged 2 to 5 years who are overweight increased from 5% to 13.9% between 1976 and 2004. During this same period, overweight increased from 6.5% to 18.8% among those 6 to 11 years of age and from 5% to 17.4% in adolescents 12 to 19 years of age. These trends are alarming, especially when we consider the health consequences of being overweight and obese. The risk of developing many chronic diseases is greater in individuals who are obese or overweight.

Obese men and women have a higher risk of developing heart disease, hypertension, and NIDDM. Recall that physical inactivity is also associated with these diseases. Impaired glucose tolerance, a preliminary stage in the development of NIDDM, was found in 25% of obese children and 21% of obese adolescents (Sinha et al. 2002). Overweight and obesity are also related to an elevated risk of some forms of cancer in adults, including endometrial (inner lining of the uterus) and breast cancer in women, and colon cancer in both women and men.

People can lose body weight by reducing caloric intake, increasing physical activity, or doing both. Exercise physiologists are particularly interested in studying the effects of increased physical activity or combined diet and exercise programs on reducing body weight. Researchers have found that creating a caloric deficit through a combined physical activity and diet program is as effective as dieting alone in reducing body weight. More important, however, individuals who combined physical activity and diet were more likely to maintain the weight loss after 18 months, while those who only dieted regained most or all of the weight lost (Pavlou, Krey, & Steffee 1989). Also, preschool children, ages 4 to 6 years, who participated in the highest amount of vigorous physical activity had less body fat than did children who were least active (Janz et al. 2002).

The challenge is how to get more people to participate regularly in physical activity. As you learned in chapter 2, although 22% of adults in the United States regularly participate in 30 minutes per day of moderate physical activity, a much larger proportion (54%) participate less frequently, and approximately one in every four adults participates in no leisure-time physical activity (Pate et al. 1995). Women are less likely to participate regularly in physical activity than men, and older adults are more likely to be sedentary than young adults. Several strategies have been proposed for increasing daily physical activity, including increasing the amount of physical activity in physical education classes; providing more opportunities for children and adolescents to participate in physical activity before, during, and after school; building walking and bicycle paths that are separated from roadways; and increasing opportunities for adult participation in physical activity at the worksite, in shopping malls, and in community facilities.

How Far Do You Need to Run to Lose One Pound?

An excess caloric expenditure of 3500 calories is needed to lose 1 pound (0.45 kilograms). If a person burns 100 calories per mile when running (jogging), how many miles would this individual need to run to lose 1 pound? How many days per week do you run? How many miles do you usually run each time? Assuming that your diet is constant, calculate how long it will take you to lose 1 pound from running.

Wrap-Up

Physiology of physical activity is a subdiscipline of kinesiology that centers on the acute and chronic changes to the physiology of the body in response to physical activity. Exercise scientists have studied these changes in an effort to help people be more physically active, healthy, and physically fit. Exercise science professionals work in many arenas including clinical, corporate, and commercial fitness programs and universities. There are also research opportunities in the military and other government organizations.

This chapter provided a summary of the research methods used in physiology of physical activity and an overview of the knowledge base of the subdiscipline. We learned about typical measures that are used to assess health and fitness and how many systems of the body are affected by physical activity. If you are interested in learning more about how the body responds and adapts to physical activity and exercise, you can begin taking courses in exercise physiology, anatomy, and physiology. You are also encouraged to visit the Internet sources listed in this chapter to learn more about the educational and certification opportunities available to you in exercise science.

KEY▸ Activity 11.5

Use the key points review in your online study guide as a study aid for this chapter.

QA Activity 11.6

These end-of-chapter questions and activities are also in your online study guide. Your instructor may ask you to complete them online and turn them in.

1. How does physiology of physical activity fit within the discipline of kinesiology?
2. What career path in exercise science is of most interest to you?
3. Describe the contributions that A.V. Hill and David Bruce Dill made to the physiology of physical activity.
4. Give an example of how maximal oxygen uptake can be measured in the laboratory and estimated in the field.
5. Why is knowledge of the energy-producing systems of the body important for coaches?
6. Explain how knowledge of exercise physiology can be used to help each of the following: a college athlete, a cardiac patient, and a person trying to lose weight.
7. Calculate your maximum heart rate and then measure your heart rate while you are performing each of three physical activities. Which of the activities meets the intensity guideline to help improve your cardiovascular endurance? Based on your goals, how many days per week and for how long should you perform the activity?
8. Get up and move at a high intensity for a few minutes. Make a list of all of the physiological changes that you notice. Can you explain why each of these changes occurred?

PART III

Practicing a Profession in Physical Activity

For some people, experiencing physical activity and studying about it are ends in themselves. Like psychology, English, or history majors, kinesiology majors may have chosen that major not because they plan to carve out a career in the field. As the discipline of kinesiology becomes more organized, its popularity as a liberal arts subject grows. This trend is understandable, given that the goal of a liberal education is to understand humanity in its totality and that physical activity plays a profound role in helping us express our humanity.

For most kinesiology majors, however, experiencing and studying physical activity are means toward ends: preparing for careers in the physical activity professions. This preparation requires mastery of a different type of knowledge than those described in parts I and II. Professional practice requires professional practice knowledge, the type that enables you to achieve specific goals by manipulating the physical activity experiences of those whom you serve. Remember that knowledge derived from professional practice often becomes incorporated into the discipline of kinesiology. When systematic knowledge—usually developed by careful observations of the effect of various manipulations of physical activity on clients (e.g., students, health club members, patients)—is included in kinesiology curriculums, it becomes part of the discipline. Because it is part of the curriculum, you will be required to master this knowledge of professional practice by demonstrating both your understanding of it and your ability to perform it.

You may have chosen your major with an eye toward working in a job centered on some particular aspect of physical activity. The array of careers open to kinesiology majors is vast: sport coaches; sport administrators; athletic trainers; physical education teachers; fitness trainers, consultants, and programmers; rehabilitation specialists in the fields of cardiology, physical therapy, occupational therapy; and many others. Each of these careers may require a different area of concentration in undergraduate course work, but all require understanding of the spheres of physical activity experience and the spheres of scholarly study of physical activity.

All those who work in the field of physical activity assume the responsibilities and obligations that belong to all professionals. Chapter 12 will stimulate your thinking about what it means to be a professional, the obligations and opportunities that belong to professionals, and the steps that you can take now to guarantee your success as a professional. Although experts in the discipline of kinesiology generally recognize seven spheres of professional practice, we will explore only five in this text: careers in health

and fitness, careers in therapeutic exercise, careers in physical education, careers in coaching and sport instruction, and careers in sport management. Chapters 13 through 17, written by experts in each of these spheres, focus on specific careers in detail. These chapters may be your first exposure to careers in physical activity. In other cases they may serve to help you consolidate your thinking and affirm your commitment to a long-anticipated career. To help you organize your thoughts about the types of work and worksites available to kinesiology majors, we have grouped the many different types of careers according to general objectives and goals, working environments, and qualifications for professionals.

This section of the book does not offer in-depth coverage of two of the seven spheres of professional practice—scholarly study and artistic expression. You might be surprised to learn that scholarly study can be an area of professional practice. Think back to the people who are responsible for compiling the spheres of scholarly study described in part II. Who discovered all this knowledge of physical activity? Most of these people were professionals who devoted their careers to studying, researching, and teaching about physical activity at an advanced level, usually at a college or university (as a college professor) but possibly at a research or scientific lab (as a research scientist). Your professors are examples of people who have embarked on such careers. These positions normally require a PhD in a specialized area in kinesiology. Because students usually make decisions about this type of career much later in their educational programs, we chose not to address them in this text. If, as you study about physical activity, you become interested in pursuing such work, don't hesitate to consult with your faculty advisor about steps that you might take to prepare yourself for this career.

The other sphere that this section does not cover includes physical activity careers in the area of artistic expression. Physical activity plays a central role in dance, drama, music, painting, sculpting, and other artistic forms; but training for careers in these areas normally is managed by art, music, dance, and drama departments rather than by kinesiology departments.

If you're like some of your peers, you may be somewhat uncertain about which physical activity career you will pursue—if you will pursue one at all. This uncertainty is normal, and it shouldn't cause you any anxiety at this point. In fact, one of the objectives of this section is to help you determine if you are suited to a career in kinesiology and, if you decide that you are, to help you chart a course for your professional future. Part III should move you closer to selecting a career. As you do this, keep in mind that this choice will be only the first of several career choices that you are likely to face over the course of your working years. Two, three, and even four changes in careers are likely for those currently entering the workforce. Nevertheless, you should make the most intelligent career choice possible, and this section will help you accomplish that important goal.

Shirl J. Hoffman

CHAPTER OBJECTIVES

In this chapter we will

- acquaint you with the characteristics of a profession,
- differentiate professional from nonprofessional work,
- explain the types of knowledge and skills essential for performing professional work,
- discuss what you need to do during your undergraduate years to gain entry into and succeed in your chosen professional field, and
- help you know if you are suited for a career in a physical activity profession.

Larry went home for Thanksgiving break during his senior year. He always enjoyed catching up with other high school buddies who were home on break, too. This year, the guys were talking about their plans for after college. "I don't know," Larry admitted, "right now life beyond graduation seems like the great black void." Thinking about the conversation later, Larry realized he'd been hesitant to commit to a particular career because he was afraid that, after preparing for it, he would decide he really didn't enjoy it.

Committing to a career at age 19 or 20 can be a scary proposition. (If this is your fear, it may help you to put it all in perspective if you remember that most people change careers or positions within careers several times during their working life span.) Students who identify a career path early, though, enjoy many advantages. They are better able to relate their course work to their anticipated career, they can gain experience in professional roles and settings, and they can begin to develop a network of associates in the profession. So, if you have already committed to a specific career in the field of physical activity, you are on track. You're likely to enjoy your study of kinesiology a bit more and derive more from it than students who have yet to make that commitment. If you are like Larry and still trying to decide on a career, this part of the text should help you narrow your career choices. In fact, even if you have already decided on a career, the chapters in this section should interest you as a budding kinesiologist who shares with his or her colleagues a special interest in physical activity.

In most people's minds the purpose of a college or university education is to prepare you for a job, but what kind of job do you want? You probably won't be satisfied with just any type of work. No doubt you are expecting your college degree to help you find a good-paying job that is respected by the community, offers good working conditions, and provides a reasonable amount of time off for leisure pursuits. You also probably want your job to count for something, to contribute to society in some substantial way. Your ideal job may be one in which other people benefit directly from your work, an aspect that always creates a good feeling. You may want a job that is so enjoyable that you willingly put in extra hours in the evenings or on weekends. Ideally, not just anybody should be able to do the job; it should require some thinking and skills that set you apart from those in the ordinary workforce. You may also want to be in a position to make most of the decisions concerning what you do and how you go about your work. (You surely remember previous jobs in which you didn't have much freedom to make decisions.) And last, but certainly not least, you want a job that involves physical activity.

Having the freedom to make decisions, often on your own without supervision, is one of the advantages of being a professional. But with freedom comes responsibility. The quality of your decisions will largely determine how successful you will be in your chosen profession. Will you be a good decision maker? Will the decisions you make as a professional be grounded in the knowledge and skills that you learned in your kinesiology program, or will you make decisions entirely by the "seat of your pants"? Will the decisions you make increase the health, well-being, and enjoyment of the students, clients, patients, or customers whom you serve, or will they result in wasted time and effort, and perhaps put in jeopardy the safety of those who depend on you? Learning how to make these decisions will require a lot of hard work on your part. In fact, this sums up what undergraduate education in kinesiology teaches you to do: to make correct decisions based on available knowledge.

There are many factors that affect this complex process of decision making. Here is a basic model of the process that, when informed by accurate knowledge and carried out systematically, can lead to good decisions:

1. Identify the problem.
2. Generate hypotheses regarding the cause of the problem.
3. Carefully evaluate each hypothesis using your knowledge base and past professional experience.

4. Plan how to implement what appears to be the best alternative solution.
5. Implement your plan.
6. Carefully monitor the results of your decision.

If this describes the position you hope to secure following graduation, then you might find a career in the physical activity professions appealing (see figure 12.1). This chapter is intended to help you decide whether a career in a physical activity profession is right for you. It also will explore the general nature of the curriculum used by colleges and universities to prepare you for your chosen field.

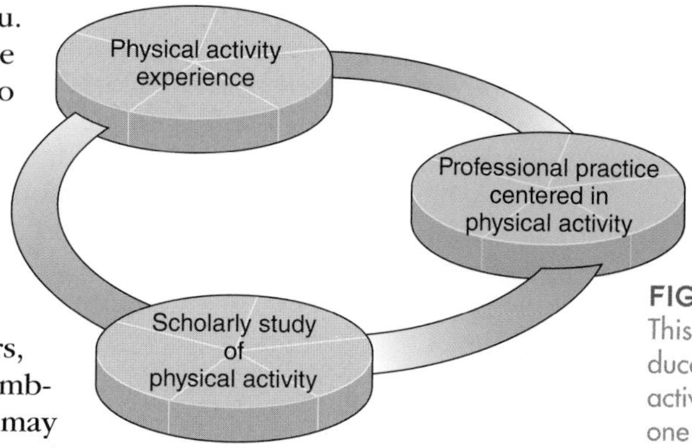

By now you probably know that not all workers are professionals. A profession is a particular line of work or occupation, and a professional is a particular type of worker. Teachers, doctors, lawyers, and physical therapists are professionals. The work of janitors, housepainters, and craftspersons such as plumbers, electricians, and carpenters, although they may describe themselves as professionals, lacks certain characteristics normally associated with the professions. Knowing why this is the case will help you better understand the distinctive roles and responsibilities of the type of work to which you aspire.

Physical activity experience
Professional practice centered in physical activity
Scholarly study of physical activity

FIGURE 12.1 This chapter introduces the physical activity professions, one of the three dimensions of kinesiology.

A degree in kinesiology is the foundation for a career in the *physical activity professions,* a term designating an assortment of specialized professions in which physical activity plays a central role, usually as a medium for therapy, education, health, recreation, or entertainment. Teaching physical education, coaching, being a fitness consultant or fitness director for a corporation, and serving as an athletic trainer or conditioning specialist for an athletic team are examples of positions in the physical activity professions. Some of these careers are similar with respect to general objectives, methods, educational requirements, working environments, and other factors. They can be grouped into relatively distinct **spheres of professional practice in physical activity.** Figure 12.2 depicts the major spheres of the physical activity professions. Keep in mind that these spheres represent general categories; each is made up of several different professional occupations.

The chapters that follow in this part of the book describe some of the most popular physical activity professions available to students graduating from kinesiology programs. The first step in deciding which of these exciting career paths you will follow is to familiarize yourself with what it means to be a professional. By joining the ranks of the professionals, you will share some of the attributes,

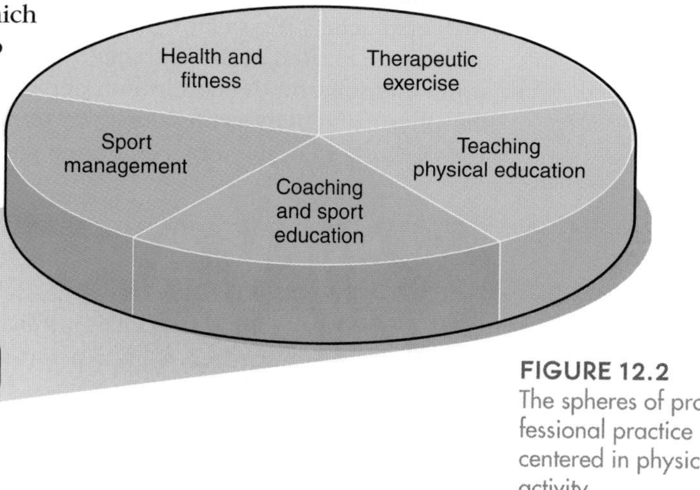

FIGURE 12.2 The spheres of professional practice centered in physical activity.

expectations, and obligations of those working in all professional fields, whether they are lawyers, counselors, clergy, physicians, or kinesiologists.

Saying that you're a professional is one thing, but becoming one is quite another. Do you know what sets professionals apart from nonprofessionals? Do you know the type of college preparation required of professionals? Do you know how someone becomes a respected member of a profession, remains in good standing, or merits dismissal from a profession? Do you know the benefits and obligations of being a professional? Are you able to recognize the traits that college professors and future employers look for in students to determine who are most likely to become respected, qualified professionals? These are important questions for anyone seeking admission to the professions; and, if you're like most of your peers, you probably haven't given them a great deal of thought.

Learning about the skills and knowledge required for specific occupations such as teaching or coaching, health and fitness promotion or consulting, community sport leadership, athletic training, cardiac rehabilitation, and other careers will come later. Right now it is time to pause and take stock of your own interests, talents, and commitments to find out if you have what it takes to enter the workforce as a professional in the physical activity field.

What Is a Profession?

➤ Those who are preparing for a position in the physical activity professions should know what a profession is, the type of work professionals do, how one gains entry to and acceptance in a profession, the obligations of professionals, and the most important factors to consider in preparing for a career in the physical activity professions.

Often we use the term *professional* in an informal way to describe the quality of a person's performance. The work of a cabinetmaker may be described as "very professional," or a friend who is an excellent golfer may be described as "a real pro." Sometimes we use the term to describe those who play or teach a sport as a full-time source of income (e.g., professional baseball players or tennis or golf professionals who teach at clubs) to distinguish them from those who play only part-time and receive no remuneration (amateurs). These usages, however, do not convey the strict, formal meaning of the term. Normally, we do not think of cabinetmakers or athletes as members of a profession in the same way that we think of lawyers, physicians, teachers, counselors, physical therapists, rehabilitation specialists, or clergy as members of a profession. Surely, members of a profession are known for doing their work well and for doing it full-time and for pay, but that is hardly the whole story.

A cluster of attributes has been used to describe professional types of occupations (Greenwood 1957; Jackson 1970). Although some characterizations of a profession may seem a bit idealistic, they provide a useful model for helping us understand the distinctions and responsibilities of this type of work. You will discover that most types of work aren't easily classified as being completely professional or completely nonprofessional. Although some workers such as physicians may be located at the extreme professional pole and day laborers may be located at the extreme nonprofessional pole, most types of work probably fall at some midpoint on the continuum depicted in figure 12.3. To underscore this point, some of the physical activity professions are termed *minor professions* or *semiprofessions* (Lawson 1984). Workers in the jobs positioned closest to the professional pole of the continuum

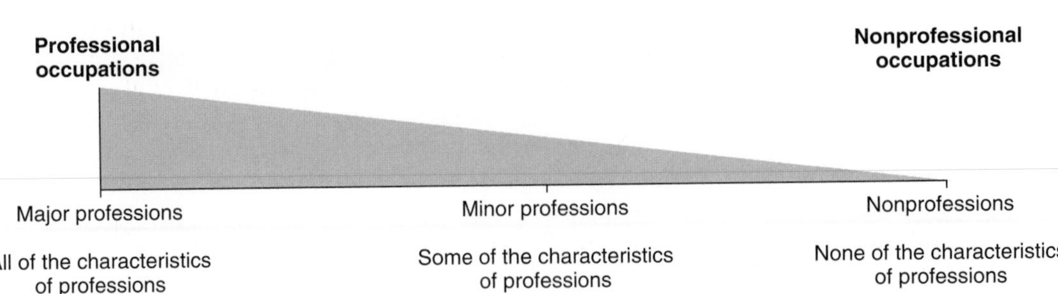

FIGURE 12.3 Occupations can be located along a continuum from professions to nonprofessions.

- master complex skills that are grounded in and guided by systematic theory and research;
- perform services for others, known as clients or patients;
- are granted a monopoly by the community to supply certain services to its members;
- are guided by formal and informal ethical codes intended to preserve the health and well-being of their clients; and
- meet the expectations and standards prescribed by their professional subcultures.

Let's look more carefully at each of these characteristics.

Professionals Have Mastered Complex Skills Grounded in Theory

The overriding hallmark of professional work is that it draws on a complex body of knowledge and theory that are developed through systematic research. Work that does not require mastery of complex knowledge (manual labor) is normally thought of as nonprofessional work. Because the workplace in general is becoming more technologically oriented, advanced specialized knowledge is becoming more important, even in occupations that once were not considered professional. For example, welding is generally considered to be a fabrication craft; but now, because it increasingly involves complex technology and sophisticated evaluation of the structure of materials, it is being touted as a profession (Albright & Smith 2006). Perhaps the overriding hallmark of professional work is that it draws on a complex body of knowledge and theory that are developed through systematic research. Work that does not require mastery of complex knowledge (manual labor) is normally thought of as nonprofessional work. This point deserves emphasis: Determining what is and what isn't a profession can be a difficult challenge.

Professionals are sometimes referred to as **practitioners** (derived from *practice*), a term signifying that they are not merely gifted thinkers but are also gifted *doers*. Professionals establish their reputations by being able to bring about predetermined outcomes efficiently and effectively, usually on behalf of others. Doing this requires skill; in fact, the expert skill of professionals is what sets them apart from other types of workers. We usually associate the concept of skill with expert motor performances—the speed of a world-class sprinter or the agility of a pro basketball player. But some professions also require deft motor skills—dentistry and surgery, for example. Motor skills are merely one category of human skill, however. Generally, the expertise of professionals is manifested in their cognitive, interpersonal, and perceptual skills. The **cognitive skills** of analysis, deduction, diagnosis, prescription, and high-level reasoning are particularly important. **Perceptual skills** that enable you to identify and recognize problems that have occurred or are likely to occur are also critical. And, because professionals always work closely with other people, **interpersonal skills** such as listening, communicating, and motivating are important too. In fact, development of these skills is essential if professionals are to make expert decisions regarding their professional practices.

For example, a workplace biomechanist must have a keen eye for movement (a perceptual skill); be able to analyze workers' movements to determine the reason for a high injury rate; be able to devise strategies for correcting the movements and reducing injury; and, as part of his or her bag of tricks (cognitive skills), must have the communication skills to explain the findings to workers and the supervisors who have hired him or her (interpersonal skills). Athletic trainers must possess perceptual skills that enable them to spot hazards and potential sources of injuries, cognitive skills of analysis and deduction through which they determine the correct emergency procedures to be used, and communication skills essential for relating their findings to physicians or for motivating athletes to adhere to their rehabilitation schedules. Becoming a professional, then, involves recognizing what factors must be taken into consideration in any situation; knowing the steps that must be taken to bring about the desired result; possessing the skill required to achieve the result; and communicating a plan of action to a client, student, or patient.

How do physical activity professionals acquire these important skills? In many cases, professional expertise is developed from knowledge gained from experience performing and watching others perform physical activity. This was discussed as physical activity experience in part I (see figure 12.4). For example, although not all soccer, basketball, or gymnastics coaches were elite performers of the sports they coach, most have spent time performing and watching the activity and are familiar with and comfortable in the contexts in which the sports take place. (We talk more about the contribution of physical activity experience to your professional development later in this chapter.)

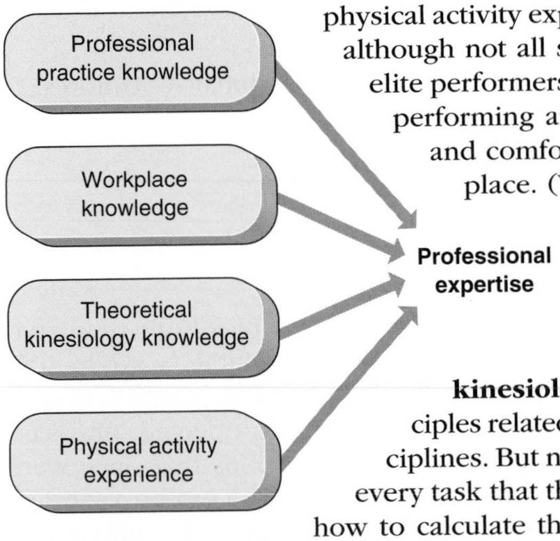

FIGURE 12.4 Many kinds of knowledge contribute to the development of professional expertise.

The skills of the physical activity professional also derive from knowledge of theory about physical activity. As we saw in part II, **kinesiology theory** is embedded in concepts and principles related to physical activity that are taught in the subdisciplines. But not all kinesiology theory is directly applicable to every task that the professional performs. For example, learning how to calculate the theoretically optimal angle of the projection of the shot put in biomechanics class may not translate into useful prescriptions for a track coach, just as learning the details of pulmonary function in exercise physiology class may not be directly applicable to the athletic trainer's work.

This isn't to suggest that kinesiology theory is totally irrelevant to practitioners. Because it focuses on broad principles about physical activity, kinesiology theory may have a potent, albeit indirect, effect on the way professionals think and the skills that they use to solve problems. Theory offers them a broad framework of knowledge that can be generalized. These frameworks of knowledge are important because they help professionals adapt knowledge to new situations, to integrate knowledge about the principles of physical activity, and to think in highly flexible ways. Contrast this, for example, with bricklaying and carpentry, respectable forms of work known as trades or crafts. These jobs certainly require use of cognitive and perceptual skills, but the skills are almost always applied to a stable set of problems and contexts. Because one brick wall is pretty much like another, even laypersons can buy books on home repair and, simply by reading, serve as their own bricklayers. The work of the physical activity professional entails solving a wide variety of complex problems with complex human beings. People cannot learn to do this by reading manuals.

For example, a physical education teacher may observe an error in a student's motor performance that he has never seen before. By systematically applying his knowledge of movement mechanics, motor learning, and pedagogy, as well as by using a diagnostic eye developed through practice, he will be able to devise practice experiences to correct the error. Likewise, an athletic trainer confronted in an emergency with a rare injury may rely on her in-depth knowledge of biomechanics, exercise physiology, and other kinesiological concepts to determine the best course of emergency action. Mastering theoretical knowledge of kinesiology helps professionals ask intelligent questions about their work and enables them to devise flexible strategies for solving problems, even those they may have never before confronted.

➤ Professionals develop a range of cognitive, perceptual, and motor skills, anchored in theoretical, workplace, and practice knowledge, that enable them to achieve predetermined outcomes efficiently and effectively.

But physical activity experience and theoretical knowledge by themselves are not sufficient to guide the skillful behavior of professionals. Professionals' reputations are determined largely by their wealth of **professional practice knowledge.** This is not knowledge that is acquired so much by reading, nor is it knowledge that can be tested on written exams. Professional practice knowledge is what some have called "knowledge in action" (Schon 1995, November/December). Professional practice knowledge is reflected in the physician's competency in diagnosing diseases or in the skill with which a lawyer assembles a convincing case for a court trial. Professional practice knowledge in kinesiology may be reflected in a physical education teacher's ability to help a class of 10-year-olds perform cartwheels, in a fitness leader's skill in prescribing effective exercises for a client who is recovering from a heart attack, or in an athletic trainer's skill in treating an on-the-field injury. Here it is in a nutshell: You demonstrate your command of professional practice knowledge not by answering questions on a true–false or multiple-choice test but by showing how well you can perform a series of professional skills. In many cases, highly competent professionals are unable to explain in any detailed fashion how they manage to bring about such effective results.

Consider, for example, a golf professional who is widely admired for her ability to correct swing errors in her clients. After carefully observing their swings, she is able to pinpoint

What Does Being a Professional Really Mean?

The term *professional* is used in many ways. Do you think the term is being used appropriately or inappropriately in each of the following examples?

- An award is bestowed on a physical education professor for having provided 25 years of *distinguished professional service* to the university.
- A director of a fitness center, aware that an employee is giving clients misinformation about nutrition and exercise, tells him that he is not behaving in a *professional manner.*
- A director of a community sport program tells her associate that he should join the North American Society for Sport Management because it is *the professional thing* to do.
- A physical therapist has a habit of coming to the clinic in dirty clothes and often fails to comb her hair or wear a clinical coat when treating patients. Her supervisor reprimands her for not conducting herself as a *professional.*
- The athletic director praises an athletic trainer for having the highest *professional standards.*
- A janitor at a local football stadium prides himself on the quality of his work, noting that he is regarded as a valuable member of the *janitorial profession.*

Notice the widely varying definitions used by those in health care management, sociology, and the insurance industry:

- "Professionalism . . . is the ability to align personal and organizational conduct with ethical and professional standards that include a responsibility to the patient and community, a service orientation, and a commitment to lifelong learning and improvement" (Garman, Evans, & Krause 2006, p. 219).
- Professionalism "exists when an organized occupation gains the power to determine who is qualified to perform a defined set of tasks, to prevent all others from performing that work, and to control the criteria by which to evaluate performance" (Friedson 2001, p. 12).
- "The attributes of professionalism include reliance on a high personal standard of competence in providing professional services; the means by which a person promotes or maintains the image of the profession; willingness to pursue development opportunities that improve skills; the pursuit of quality, competence, and ideals within the profession; (exuding) a sense of pride about the profession" (McGuigan 2007).

What similarities and differences do you see (either literal differences or implied) in these definitions of professionalism?

the causes of errors and quickly make corrections. If asked to explain how she does it, she may not be able to. We say that her knowledge of the process of identifying errors is "tacit," meaning that she is not aware of it. Skilled athletes, surgeons, counselors, or physical education teachers also operate on the basis of tacit knowledge. As they move through their workday they rely on tacit, hard-to-articulate knowledge that is only revealed in the competency with which they perform their professional duties. No doubt, physical activity experience and kinesiology theory contribute to their professional practice knowledge, but that knowledge is also the product of trial-and-error experience accumulated over many years of practice.

Finally, professionals also rely on **workplace knowledge,** usually to perform relatively mundane tasks associated with their jobs. This type of knowledge requires little formal education or training. For example, cardiac rehabilitation technologists must know how to maintain the electronic equipment they use, how to schedule patients for examinations, or the proper procedures for opening and closing the facility. Such tasks are not professional in their scope and do not require highly sophisticated knowledge, but they are important. A cardiac rehabilitation technician who does not maintain a clean testing facility or ensure that the equipment is kept in proper working condition may soon find herself unemployed, regardless of the quality of her professional skills.

Professionals rely on professional practice knowledge to carry out their daily tasks.

Professionals Perform Services for Clients or Patients

Most professional work takes the form of services that are performed for clients. **Service** connotes giving assistance or advantage to others, usually in a spirit of helpfulness and concern. Most professionals are conscious of their roles as service providers. Those in nursing, social work, family counseling, public health, recreation, and kinesiology increasingly refer to their professions as helping fields or **helping professions** (Lawson 1998b). The beneficiaries of a professional's services are known as clients. (In the allied medical fields, clients are patients; in teaching, they are students.) Professionals are in the "people business"; they like meeting and working with people, are comfortable in social settings, and are confident in their social skills. They derive great enjoyment from helping people meet their needs and accomplish important goals.

The notion of service comprises three important aspects. First, professionals are committed to helping others. One often hears professionals describe their career as a calling, suggesting that their primary motivation is to enhance the health, education, enjoyment, and general well-being of those they serve. In fact, concern for the well-being of clients often supersedes professionals' concern for their own comfort and satisfaction. You may have known physicians or lawyers who sacrificed lucrative practices to work for poor and disadvantaged people in inner cities or developing countries. The same spirit appears in the teacher who volunteers her time to give remedial instruction to a student on the weekend or the geriatric fitness leader who stays late following each session to give special attention to an elderly man too embarrassed to exercise with others present. Obviously, becoming an effective professional need not entail such extreme sacrifices, but possessing a spirit of wanting to help others is a prerequisite for professionals.

Second, professionals render expert service. Being a professional means more than simply being willing to help. Professionals know how to determine a client's needs and how to initiate and evaluate an appropriate course of action to meet those needs. Because of this, professionals are granted a certain amount of autonomy to carry out their work. Their opinions are respected, and they usually have a great deal of freedom to make decisions about how they will help clients. Some professionals have more autonomy than others do. Those in private practice tend to have more autonomy than do those working in institutions such as hospitals, school systems, universities, the government, or the military, where they may relinquish some degree of professional autonomy to satisfy the demands of the institutional bureaucracy. The personal trainer in private practice may enjoy much more autonomy in decision making than the fitness leader who works for a gym. The physical education teacher, whose actions may be limited by school board policies, the dictates of a school principal, and the curriculum guide, will enjoy less autonomy than the professional sport instructor who is self-employed.

The third aspect of service is the fact that clients are likely to become dependent to varying degrees on the professional. This circumstance is not always desirable, but it is inevitable. The dependence of clients can be undesirable when it is carried to extremes and results in professionals' adopting a paternalistic (parent-like) attitude toward those whom they serve rather than a collaborative attitude in which the professional and the client work together to achieve common goals. A dance student depends on his teacher to improve his technique just as a patient depends on her exercise therapist to design a program to help her recover from an injury. Unlike customers at a department store, who usually have a clear idea of what they want and can evaluate the products that they buy, clients usually do not have the knowledge essential for judging the quality of the professional's conduct. Consequently, they can only have faith that the professional will keep their best interests in mind when making decisions. One of the most egregious violations of professional conduct is exploiting the vulnerability of clients. For that reason, the coach who orders an injured player back into the game will quickly lose the respect of colleagues, as will a physical therapist who, not wanting to suffer financial loss, fails to refer a patient to a competitor even though the competitor may be better equipped to meet the patient's needs.

➤ The touchstone of professional work is the delivery of expert services to improve the quality of life for others, always with priority given to the client's welfare.

Taking Stock of Your Job History

Think about the various jobs that you have held and the different types of knowledge that you used in each. What job have you had that came closest to being professional work? What criteria did you use to determine this? Did you develop workplace knowledge on a job? For example, you might have learned how to use a time clock, operate specialized equipment, lock up a restaurant at closing, or how and where to store tools when not in use.

Have you held any position that required theoretical knowledge? (Review the discussion of theoretical knowledge before answering this.) Have you been employed in a situation that required professional practice knowledge? Maybe you taught swim lessons at a local pool and learned how to get young children to put their faces in the water. Or perhaps you volunteered at the local YMCA and observed how the program director organized activities, advertised new classes, or gave feedback to staff members. These types of knowledge will be essential as you develop in your profession.

Professionals Possess a Monopoly on the Delivery of Services

Society values professionals because they provide a needed service unavailable through any other occupational group. If you need a cavity filled, you go to a dentist. If you need spiritual guidance, you may go to a member of the clergy. If you need to rehabilitate an injured knee to play basketball again, you may go to a sport physical therapist.

The physical activity professions came into existence and continue to flourish because society recognized that people in such positions make a unique and valuable contribution to the health and well-being of society. Society needs specialists who can help people regain

➤ Professionals are granted monopolies because only they possess the knowledge and skills to meet particular needs of the community.

Kinesiology prepares professionals to meet a diverse range of client needs. Because certifications and licenses are needed to perform many of these services, the professionals possess a monopoly on them.

or maintain their health, become physically fit, increase their proficiency in motor and sport skills, recover from injuries, or direct and organize coaching sport programs. No discipline prepares people with the knowledge and skills needed to meet this diverse range of needs except kinesiology. In this sense, kinesiology and the physical activity professionals it prepares possess a *monopoly* on the delivery of physical activity services. Various certifications and licensing arrangements often protect the monopoly. Teachers, for example, usually are required to possess a certificate granted by the state or province in which they operate. The purpose of this certificate is to ensure that only professionals who have met the minimum standards of educational preparation will be granted a monopoly to practice the profession of teaching.

Professionals Collaborate With Colleagues to Ensure High Standards and Ethical Practices

Professionals recognize that society has granted them authority to regulate their own conduct and expects them to do it. When standards of conduct by all members of a profession remain high, the needs of society are satisfied and the monopoly of the profession is protected. Violation of accepted standards by a single professional threatens all members of the profession by bringing into question the legitimacy of the monopoly. Thus, professionals have a stake in maintaining or improving not only the quality of their own practices but also the quality of service delivered by all members of their profession. In this sense, although professionals may work at clinics, schools, hospitals, or other agencies, they view themselves primarily as representatives of the profession as a whole.

Sometimes the quality of professionals' practices suffers because they fail to keep abreast of developments in their fields. Because the knowledge base for the physical activity professions is expanding continually, exercise leaders, teachers, trainers, sport leaders, and other professionals can easily become outdated in their techniques. Professionals whose practices are not in accord with the latest information become a threat to all professionals in the field. A clinical exercise physiologist who fails to keep abreast of her field is in danger of recommending activities to clients that recent research has shown to be inferior or dangerous. Such a person may face accusations of acting unprofessionally.

As a way of ensuring that their knowledge and skills do not become outdated, competent practitioners join and participate in professional organizations; regularly read professional books and articles; and attend conferences, workshops, and other meetings sponsored by professional societies. Unlike salespersons, business executives, or others in the commercial marketplace, who tend to protect information from their competitors, members of a profession relate to each other as colleagues (literally *fellow workers*) and willingly collaborate and share information to serve the best interests of their clients. If you want to be a top-notch professional, you must commit to educating yourself continually. Plan to join professional organizations and read literature about your field regularly.

Operating on a less than adequate knowledge base is not the only way in which professionals breach standards of conduct. Professionals also can be sanctioned by their colleagues for unethical behavior in the workplace. Unethical conduct occurs when the behavior of a professional at the worksite comes into conflict with generally accepted standards of practices in the field. To ensure that members clearly understand ethical expectations, professional organizations often publish a **code of ethical principles and standards.** Review the code of ethics on this page published by the Association of Applied Sport Psychology. The code is intended to guide the practice of sport psychologists who consult with athletes, as well as the actions of scientists who conduct research in sport psychology. Likewise, the National Athletic Trainers Association has adopted a code of ethics for practitioners, which is available on its Web site.

The primary objective of such codes is to protect the rights of students, clients, and patients. Codes may instruct professionals to describe and advertise their competencies accurately, describe the nature of the relationship that should exist between clients and professionals and between subordinates and professionals, and clarify matters involving financial transactions with clients.

One way in which professions attempt to ensure that members understand the prevailing code of conduct and are competent to implement safe and effective practice is to license their members. A license is formal authority granted by law to practice a profession. Professions that require licenses can suspend a professional's license for any number of infractions, effectively terminating the professional's authority to practice.

➤ Professionals have a stake in maintaining high standards of conduct for all practitioners in their profession. To accomplish this, professional organizations make opportunities available for continued education by their members and publish guidelines specifying acceptable and unacceptable ethical conduct.

Although the medical and legal professions have made effective use of licensing provisions to control the quality of professional services, relatively few physical activity professions are licensed. An exception is physical therapy. Some states are now licensing athletic trainers, a trend that is moving across the nation. This doesn't mean that unethical or improper conduct by physical activity professionals goes unpunished. Usually, local entities administer the appropriate sanction at the local workplace, often in consultation with a professional organization. For example, the local school board may terminate a schoolteacher whose conduct violates ethical codes. A university may fire a conditioning coach who administers illegal anabolic steroids to athletes. Although the American College of Sports Medicine (ACSM) cannot suspend a cardiac rehabilitation therapist whose conduct puts a client at risk, the clinic director or other supervisor at the worksite may decide to terminate the worker after consulting ACSM guidelines.

Ethical Principles and Standards for Members of the Association of Applied Sport Psychology

Sport psychologists who work with athletes must carefully adhere to a strict code of personal and professional ethics. The Association of Applied Sport Psychology publishes a comprehensive ethics code for its members that covers a number of ethical issues, including those related to professional relationships, human differences, exploitation and harassment, misuse of members' influence, and fees and financial arrangements. The code has as its centerpiece five general principles, which we summarize as follows.

Competence: Members keep abreast of their field and recognize the limitations of their knowledge and skills. They are committed to protecting the welfare of all whom they serve.

Integrity: Members are honest and fair in reporting their qualifications, services, products, or fees and avoid improper dual relationships.

Respect for people's rights and dignity: Members respect the rights of those with whom they work to confidentiality, self determination, and autonomy.

Concern for others' welfare: Members work to resolve conflict that may arise with colleagues, and they do not exploit or mislead others.

Social responsibility: Members are aware of their responsibility to the community; they make their work public so as to benefit others and strive to prevent misuse of their work.

Adapted from "Ethics Code: AASP Ethical Principles and Standards" http://appliedsportpsych.org/resource-center/professionals.

Professionals Adhere to Standards of Their Professional Subculture

Although professionals work in a variety of different locales and contexts, they tend to hold high expectations concerning the way members present themselves and relate to others. For example, professionals emphasize respect and politeness not only to those whom they serve but to colleagues as well. They tend to be well organized in their work, are not clock watchers, and don't hesitate to work extra hours when required. They dress according to accepted occupational standards and are attentive to matters of personal grooming and hygiene. A sport program director who comes to a board meeting in dirty and disheveled clothes, a cardiac rehabilitation technician who forgets to wear her lab coat, and a physical education instructor who teaches classes in street clothes are not dressed professionally. The physical therapist who never combs or brushes his hair or the personal trainer who smells of body odor are guilty of presenting themselves in an unprofessional manner. These general expectations apply to professionals in any field.

In addition, physical activity professionals often face a more specific set of expectations. For example, fitness leaders or physical education teachers might be expected to model the physically active life, to refrain from smoking, and to maintain a level of fitness appropriate for their age.

> **DO it Activity 12.1**
>
> **What Is a Professional?**
> In the online study guide, you'll apply your knowledge about the characteristics of professionals to a career that interests you.

Different physical activity professions are likely to require unique types of knowledge, employ different types of skills, operate according to different standards and codes of conduct, require different languages and terminology, require different certification and continuing education procedures, and often involve different types of dress and on-the-job conduct. When an individual learns to adapt to these profession-specific expectations, he or she is said to have been socialized into the profession. Becoming **socialized** into a profession means learning how to conduct yourself according to the roles and responsibilities of the workplace. The ease, confidence, and efficiency with which professionals perform their daily activities and communicate with their clients and coworkers are good indications of how well they have been socialized into their professional roles.

Obviously, the sooner you become socialized into your profession, the more valuable you will be to an employer. For that reason, among others, growing numbers of students in professional programs are seeking opportunities for volunteering in clinics, YMCAs, schools, and community agencies. Although usually not counted as course work, such extracurricular experiences can give you a head start in learning how to assume professional roles. Similar growth and leadership opportunities are available in student major clubs, athletics, student government organizations, and other on-campus activities.

Adapting to the Professional Culture of the Physical Activity Professions

Adapting to the professional subculture of the physical activity professions may require some significant changes in your patterns of behavior. Is physical activity experience a daily part of your life? If not, you need to consider integrating it into your lifestyle. Do you smoke? Are you severely overweight? Obviously, your first consideration should be the physical damage that you are incurring by these unhealthy lifestyles, but you should also know that these and other unhealthy lifestyle habits do not blend well with the professional lifestyle of those who work in the field of kinesiology. Thus, they could be impediments to your gaining employment or promotion in a physical activity profession.

How Do Our Values Shape Our Professional Conduct?

Personal values affect how we lead our lives, including how we approach our work. This might be a good place to stop and reflect on what you learned in chapter 5, asking yourself about your values and how they might shape your professional philosophy and practices. Hal Lawson (1998a) has described two types of extreme value orientations that may characterize professionals in any field—**mechanical, market-driven professionalism** and **social trustee, civic professionalism.** Although these orientations are ideal types, which is to say that most professionals don't fit completely into one category or the other, they can serve as references in helping you determine how you want to operate as a professional.

➤ Your personal values will influence how you act out your professional role. Mechanical, market-driven professionals value the profession, profit, personal prestige, and status over the rights and needs of clients. Social trustee, civic professionals value clients and the social good more than themselves or their profession.

Mechanical, Market-Driven Professionalism

Mechanical, market-driven professionals tend to be fascinated with their professional methods and techniques, so much so that they sometimes lose sight of their clients' needs and desires. In many cases, their work becomes its own justification, valued by them and their colleagues regardless of whether it has any real benefits for society. The medical, nursing, and law professions sometimes are accused of this type of professionalism, but some of the physical activity professions can be guilty of it too.

Mechanical, market-driven professionals also are likely to serve their clients in a fragmented, compartmentalized fashion. That is, they view their professional contributions within the strict limits of their specialization and ignore the variety of relevant forces acting on the client, such as family disintegration, illness, or drug or alcohol dependency. They may dismiss these other problems as "somebody else's responsibility" and never go out of their way to see that proper referrals are made. Such professionals may treat clients humanely but with little feeling; they believe that an objective, arm's-length attitude is critical to professional success.

Because they are prestige conscious, mechanical, market-driven professionals devote much time to enhancing their status and competing with other professionals, even if they must sacrifice the quality of service that they give to their clients. They pride themselves on their expertise, view themselves as superior to those they serve, and often encourage clients to develop a dependency on them and their services.

Social Trustee, Civic Professionalism

Social trustee, civic professionals hold to a much different set of values. They believe that the worth of a profession is measured by its effectiveness in promoting the social welfare, enhancing social and economic development and democracy, and ensuring social responsibility and social justice. The actions of a social trustee, civic professional are guided not by a fascination with technology and technique or by a concern for status and prestige, but by a vision of the good and just society. They operate according to this rule of thumb: "Healthy people and a good society first, me and my profession second in service of this greater good" (Lawson 1998a, p. 7). They recognize

Social trustees put people, not status, first.

DO it Activity 12.2

Values

Listen to two professionals talk about their work lives and apply what you know about values in this activity in your online study guide.

that clients live in a multiplicity of worlds (e.g., work, school, family, church) and understand the ways in which worlds can interact to affect their clients' lives. Because of this they often work in teams with other specialists to achieve desirable goals. They make no pretense about being objective in their relationships with clients: Their professional practices reflect their personal values. They don't view clients as dependent on them, believing that the professional–client interaction can be a mutual growth experience in which each benefits from the other's knowledge and skill.

How Are Physical Activity Professionals Educated for the Workforce?

Now that you've examined some of the general characteristics of the professions, let's consider what formal education is required. By now you are aware that professional work requires advanced knowledge that you can obtain only through formal education at the undergraduate and graduate levels. Consequently, whether you will be qualified to enter the physical activity professions depends a great deal on your academic experiences in college. If you're like most college students, you may not always understand why certain courses are required or what the purposes are of various phases of the curriculum. You might be interested to learn that kinesiologists themselves are engaged in a debate concerning the structure of the kinesiology curriculum. Although the curriculum at your institution may not offer an exact replica of the curriculum described here, it probably is similar in terms of its general features. The curriculum model incorporates the following five different, yet related, types of academic experiences essential for preparing kinesiologists for their professional roles:

- Liberal arts and sciences
- Course work in physical activity
- Course work in theoretical and applied theoretical knowledge in kinesiology
- Course work in professional practice knowledge and skills for particular professions
- Apprenticeship or internship experience at the worksite

The best professional preparation programs immerse students in all academic experiences.

Liberal Arts and Sciences

All college graduates, regardless of their career aspirations, are assumed to be educated people. Used in this sense, the word "educated" describes someone who has struggled with the great works of literature, has learned about the historical development of human civilization, has developed critical skills essential for analyzing arguments logically, is able to determine the truthfulness of ideas and to identify the values embedded in them, has developed sufficient knowledge about the arts to appreciate the richness that they can add to one's life, and has developed insights into the problems and prospects of their own culture and the culture of others. Accomplishing this goal requires courses in science, philosophy, history, literature, sociology, mathematics, art, music, and the dramatic arts, among others. Usually, these courses are part of the "general education" or **liberal studies** requirements in college and are taken during the first two years of undergraduate study. They form the foundation for your professional education, which usually comes later.

Liberal education obtains its name because only by immersing yourself in the arts and sciences can you liberate, or free, yourself from dependence on the thinking of others. Liberally educated people have developed the capacity to educate themselves. They have developed a capacity for seeking the truth and for making intelligent choices when they confront life's problems and opportunities. Admittedly, these aims are grand. An educated person is aware that nobody is ever completely educated; everyone is always in the process of becoming educated. One of the hallmarks of educated persons is that they are aware of what they do not know and continually seek opportunities to expand their intellectual and cultural horizons by reading books and newspapers, attending concerts, patronizing the arts, and engaging in other lifelong educational activities.

Thus, the liberal education you receive in college is just the beginning. Whether this initial immersion in the liberal arts and sciences has accomplished its purpose will be reflected in your level of interest in continuing to educate yourself after graduation. What types of books will you read? What kind of music will you listen to? How will you speak and write? Will your analysis of the claims of politicians and other public figures be logical and based in fact? Your approach to these and hundreds of other life experiences should bear witness to the fact that you are a liberally educated person. Of course, a liberal education should have short-term effects too. The most important is to provide you with the broad-based knowledge and intellectual skills essential for undertaking more advanced study in your professional curriculum. A liberal education frees you from depending on others to do your thinking for you. It is the foundation on which all other educational experiences are constructed.

Course Work in Physical Activity

Kinesiology majors usually are required to enroll in physical activity courses, either as part of their liberal education requirements or in conjunction with their professional programs. As noted in part I, you entered college with a wealth of experiential knowledge about physical activity and no doubt found it to be a vehicle for learning much about yourself, others, and the world around you. But the physical activity experiences offered within the framework of a liberal or professional education should result in qualitatively richer experiences than the experiences you had before you came to college. Let's take a moment to consider how physical activity courses at the college level might affect you.

If you've already taken a physical activity course at your college, you may have discovered that the course was more intense and taught at a more advanced level than classes in high school. On the other hand, those who had an excellent physical education background in high school may have discovered that the college-level physical activity course wasn't much different from the courses they took in high school. In both cases the courses focused on sport, outdoor activities, or fitness. But even if the course of instruction

A liberal education expands your approach to physical activity by incorporating the totality of your personal experiences.
Photodisc/Getty Images

➤ College courses in physical activities are taken in conjunction with liberal studies and as an essential part of your professional preparation. As a result, you should develop the ability to examine your physical activity experiences from an informed, educated perspective and to integrate the knowledge learned from them into a coherent professional perspective.

seemed to be the same, what is different is you. Everything we learn occurs in a context of the totality of our personal experiences. As your liberal education continues to expand your horizons, your approach to physical activity experiences will change as well. You should be a more intelligent analyst of your performances and a more persuasive critic of the performances of others, and you should be able to play a more astute role in the planning and preparation for your own physical activity experiences. You should also be able to draw connections between your physical activity experiences and the world around you.

A course in martial arts should attune you to the nuances of pace, tempo, and rhythm; a course in fencing should foster an appreciation of beauty, force, and form in movement. Your course work in physical activity may sensitize you to ethical and moral questions that you might not have noticed before you were exposed to a liberal education.

Physical activity courses also are required as part of your kinesiology major. If you plan a career in teaching, coaching, or fitness leadership, the requirement may seem reasonable to you, but why does it exist for students in sport administration, physical therapy, or other professions in which physical activity performance is not part of the professional responsibilities? If you're struggling with this question, we suggest that you think back on the general model of physical activity described in chapter 1, recognizing that knowledge gained through physical activity provides one of the three legs of the stool on which the discipline of kinesiology rests. Whatever career you enter, performance of physical activity will be at center stage, and you should have acquired a wealth of firsthand physical activity experiences, not only to help you identify with physical activity problems confronted by your clients but also to broaden and enrich your comprehensive understanding of the phenomenon of physical activity.

Course Work in Theoretical Kinesiology

The chapters in part II gave you a general overview of the subdisciplines of kinesiology. Courses in these subdisciplines teach theoretical knowledge not only about sport and fitness but also about physical activity in general. Obviously, much of it is applicable to professional practice. For example, theoretical knowledge about how the body responds to exercise (drawn from the subdiscipline of physiology of physical activity) seems directly applicable to the work of a swimming coach or a cardiac rehabilitation specialist. Other theoretical knowledge, in sport history, for example, or biomechanics, may not seem applicable immediately; but over time, the knowledge may have a profound effect on how a professional thinks and thus acts.

Some professors would argue that students should not expect all theoretical kinesiology to be directly useful in their professional careers. This perspective views theoretical kinesiology as part of a liberal arts education, just as English, history, biology, literature, and other courses in liberal studies are. Disciplines such as literature, philosophy, biology, and physics are not valued because they prepare students for specific types of work but because we believe that they sharpen students' perspectives on themselves and the world around them and help them develop new and creative ways of thinking. One could argue that physical activity is also critical to human life, so much so that, like art, music, drama, and dance, it can teach us unique things about our culture and ourselves. Viewed in this light, a "discipline" of kinesiology (theoretical kinesiology) should not apply any more directly to professional practice than should other liberal studies such as poetry or medieval history. (You should know that kinesiology professors do not agree on this; some believe that kinesiology is strictly a professional field and that its only justification is that it prepares people for jobs in the physical activity field. A detailed discussion of this debate, however, is not within the scope of a textbook of this kind.)

Theoretical Kinesiology
Connect parent disciplines with ways of studying physical activity in this matching exercise in your online study guide.

Figure 12.5 shows how theoretical kinesiology as represented in the subdisciplines has its basis in the parent disciplines of the arts and sciences. Theoretical kinesiology is knowledge about physical activity that has been gleaned from traditional parent disciplines in the humanities; biological, behavioral, and sociological studies; and the arts. Generally, scholars in these parent disciplines have not been interested in studying the relationship between their fields and physical activity, and these neglected areas of the parent disciplines form the body of theoretical kinesiology. For example, whereas an introductory sociology course might offer scant coverage of the relationship between social class, gender, race, and sport, and an introductory physics course might ignore the mechanics of human motion, theoretical kinesiology (sociology of physical activity and biomechanics of physical activity) focuses precisely on these areas of the parent disciplines. By extracting this neglected knowledge about physical activity from parent disciplines, kinesiologists have created a stimulating body of knowledge for students planning to enter the physical activity professions. At present, isolated courses in exercise physiology, motor behavior, and so on represent this body of knowledge. The important challenge in front of us is integrating this knowledge in the way practitioners must do to make decisions in their work.

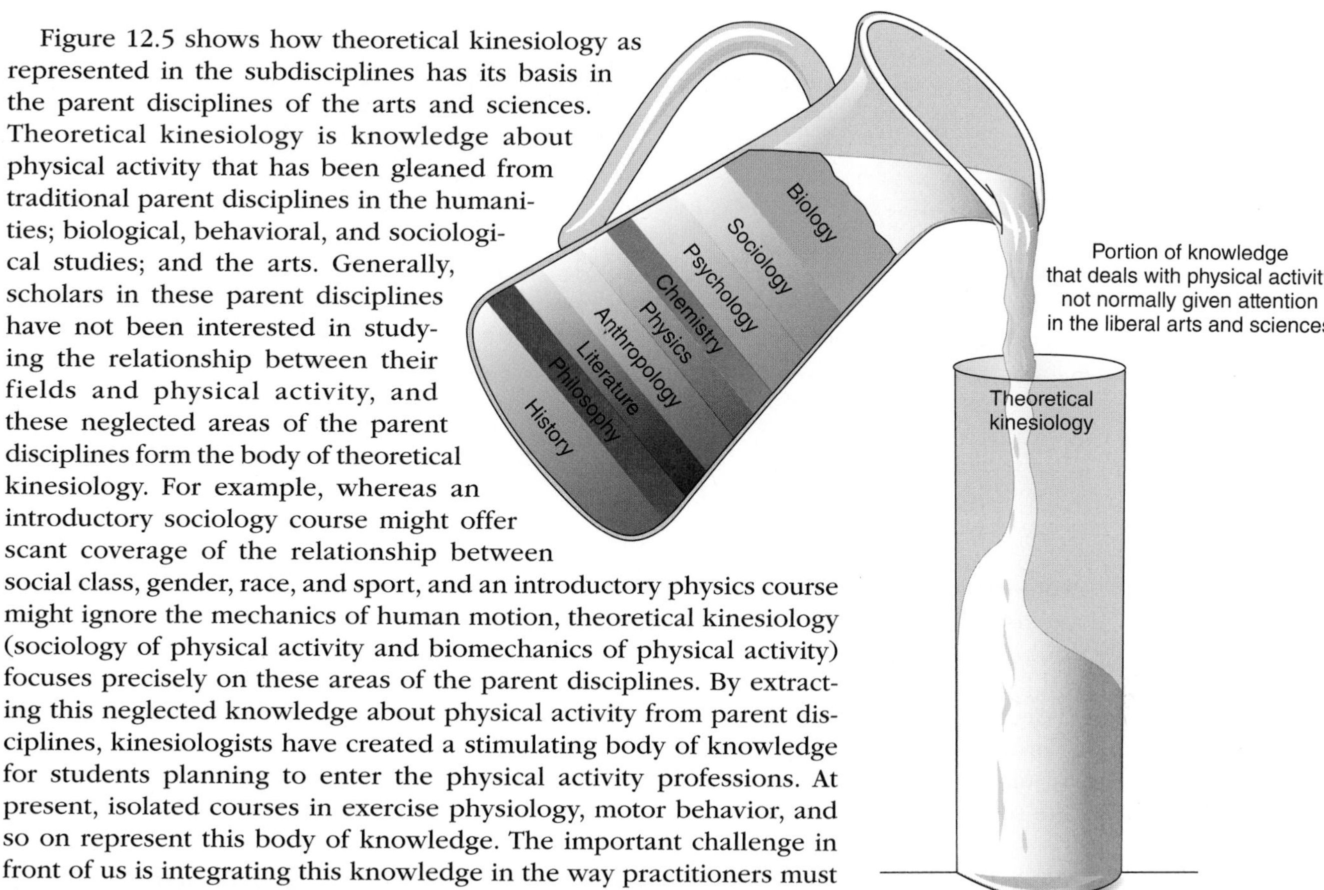

FIGURE 12.5 Much theoretical knowledge in kinesiology is derived from existing parent disciplines.

The evolution of theoretical knowledge of kinesiology has been an exciting development in the field of physical activity. But how much of this knowledge should be required of students embarking on careers in the physical activity professions? Should future athletic trainers, physical education teachers, or aquatic therapists be expected to master as much of the theoretical body of knowledge of kinesiology as those majoring in pure kinesiology? The sport management major may well ask why he must take courses in exercise physiology or biomechanics; and the student planning to be a conditioning coach may well ask why she should be expected to know about the history and philosophy of sport, especially when the concepts may not apply to the professional problems that she will face in the workplace.

The answer to such questions has been a matter of vigorous debate among scholars for over 20 years. On one side are those who argue that kinesiology theory is not especially helpful to professionals, especially when professors make no effort to apply it to professional problems (Locke 1990; Siedentop 1990). On the other side are those who contend that all undergraduate kinesiology majors, regardless of their career aspirations, should learn the body of knowledge of kinesiology (Newell 1990a). They believe that by mastering such knowledge, physical activity professionals set themselves apart from nonprofessionals, thereby bolstering their monopoly on the profession and protecting physical activity professionals against threats of deprofessionalization. For example, the personal trainer who understands the mechanics of energy systems or the physiology of muscle function will enjoy more professional respect than fitness instructors whose training consisted of a weekend workshop at the local gym.

Another argument for requiring students to master theoretical kinesiology is that this ensures that all professionals in the field are united around a common core of physical activity theory, without which kinesiology might fragment into a number of completely different fields. In addition, those who believe that all physical activity professionals should be required to study theoretical kinesiology argue that it might benefit future professionals in ways that we do not fully understand. Just as a liberal education alters students' ways of approaching

Becoming Familiar With Your Academic Program

One barometer of students' depth of engagement in their course of study is their familiarity with the requirements of the academic program in which they are enrolled. (If you aren't familiar with the requirements of your program, consult the bulletin or calendar for your institution that lists courses required in your major.) Do you know what core courses in the kinesiology department are required of all kinesiology or exercise science majors? Is the core requirement different for different career tracks? Why are these courses required? Do some courses have prerequisites? Why are these courses required? Are students in your department required to enroll in physical activity classes? Why are these courses required? Are you required to complete an internship (or student teaching) as part of your degree requirements? Why is this required? How do you apply for an internship placement? Does the internship have prerequisites? Must you maintain a minimum grade point average to enroll in higher-level courses? What is it?

problems and opportunities in their lives, so an immersion in the theoretical knowledge of the discipline may equip physical activity professionals with general concepts and ways of thinking about physical activity that will assist them in solving the myriad problems they will face in professional practice.

Because some disagreement still exists concerning the relationship of kinesiology theory to professional practice, course requirements in theoretical kinesiology tend to vary from institution to institution. Most undergraduate programs in kinesiology require students to take some courses in kinesiology theory regardless of their career aspirations.

Course Work in Professional Practice Knowledge and Professional Skills

Your preparation as a professional wouldn't be complete if you didn't acquire the professional practice knowledge essential for success in your career. The point made earlier bears repeating: Knowledge of kinesiology theory is not sufficient to prepare you for professional practice. Physical education teachers need to master the difficult skills of planning and organizing lessons, communicating with and motivating students, and evaluating students' performances. They must be able to operate efficiently and effectively in the social context of the schools. They learn this professional practice knowledge in courses that are substantially different from courses in kinesiology theory.

➤ Knowledge of the discipline itself will not equip you to perform the tasks required in your chosen profession. You develop the competencies required in a profession through course work that provides you with practice in performing these skills.

Whereas theories of kinesiology tend to focus on the performance of physical activity, mastering professional practices centers on clarifying clients' needs and desires, analyzing impediments to the realization of these needs and desires, and implementing actions appropriate to the context and other relevant factors. Professional practice knowledge is incorporated into **practice theories,** which are developed through applied research and professional experience. A practice theory may guide a professional in selecting the best way to perform cardiovascular tests on elderly persons at a nursing home with limited equipment, determining how to motivate a soccer player who is having academic difficulties and misses off-season training sessions, or selecting the types of balls and bats to use to speed the learning of children with autism.

Unlike kinesiological theory, practice theories and the skills of professional practice cannot be mastered simply through studying them; they are learned through action. Students acquire these theories and skills by hands-on practical experiences centered on solving real-life professional problems. You can learn *about* administering a stress test by reading a book, but you learn *how to* administer a stress test only by following up your reading with practical experience. By committing errors and receiving instructive feedback from your professors, you will gradually develop and sharpen your command of professional practice knowledge.

These initial attempts at putting practice theory into practice probably will be in relatively small classes on campus in which your performance will be evaluated more intensely than it might have been in kinesiology theory courses. Your classmates may serve as simulated clients, and your initial attempts to perform professional tasks will be in the safe environment of the classroom where your mistakes will not lead to serious consequences. If you perform up to departmental standards in these series of courses, you will be certified to enter the final stage of your professional education: the apprenticeship or internship.

Internships

An apprentice is a novice who learns how to perform a job by working closely with an experienced worker. Apprentices usually assist a veteran for little or no pay in return for the opportunity to learn the skills required to enter the full-time workforce. The term is used most commonly in connection with crafts or trades (e.g., plumbing, electrical contracting, carpentry) in which the apprenticeship is the only type of training offered for the job. In professional education programs, the apprenticeship phase is called an **internship,** and it is the culminating experience of four years of study. Unlike apprenticeships for the crafts or trades, internships for professional careers are preceded by a rigorous course of study.

➤ The internship is the culminating educational and evaluative experience for preprofessionals. As an educational experience, the internship provides you with an opportunity to apply the knowledge and skills that you learned in the undergraduate curriculum while working under the supervision of a trained professional at an off-campus site. The internship is also an evaluative experience in that it tests your level of preparedness to enter your chosen profession.

In the physical activity professions, the internship is known by a variety of names. It may be referred to as an internship in sport management and fitness instructor programs, as student teaching in physical education teacher education programs, and as clinical training in athletic training programs. Regardless of what it is called, the experience has the following two purposes:

- To teach you how to apply the knowledge and skills that you have learned in your professional program to a real-life situation
- To test your level of preparedness to enter professional practice

For most students, this is the first opportunity to assume the role of a member of their chosen profession. The internship usually is the most challenging experience of the undergraduate years, and the most enjoyable. For these reasons, most students feel that it is the most valuable experience of their undergraduate education.

The internship differs from traditional course work in three ways. First, it occurs at the worksite rather than at the college or university. This arrangement means that the subculture of the worksite rather than the subculture of your undergraduate department will dictate your conduct. You will be evaluated not only on the knowledge you display but also on how well you conform to the rules of the workplace. If you were engaged in an internship at a conditioning program for a Division I athletic department, for example, you would need to pay attention to the local rules that govern the conduct of athletes and conditioning coaches in the training room. Is loud music acceptable? What are the dress expectations? Does the director of conditioning expect you to arrive 15 minutes before your scheduled starting time? What local terms are used to describe interventions, locations, or procedures? What safety measures are you expected to follow? What are the established

Interns get on-the-job experience that prepares them for a career in their chosen profession.

Build the Pyramid
Test your understanding of the path to becoming a PA professional in this activity in the online study guide.

emergency procedures? Does the supervisor like interns to show initiative, or does she prefer that you ask for advice or permission before undertaking an action? You must consider these and other such issues to ensure success in the internship as well as in the profession.

A second way in which an internship differs from your regular course work is that an experienced professional who probably is not a member of the faculty in the department you are enrolled in will supervise your performance on a day-to-day basis. A university supervisor may visit the intern site occasionally, but the primary overseer and evaluator of the quality of your work will be the professional to whom you are assigned. Faculty have chosen these on-site supervisors because the faculty members view them as good models for future professionals. At the same time, you shouldn't be surprised if their opinions, philosophies, and methods of operation differ from those that you learned in your professional course work. Exposure to a variety of philosophies and methodologies will usually serve to expand your knowledge of the profession.

The third way in which the internship differs from typical course work is that you will be serving real, not simulated, clients. Your actions will have important consequences, and they can be serious. Serving real clients also means that you must learn the social graces expected of a professional. You will need to demonstrate poise, courtesy, alertness, and initiative. You will be expected to demonstrate a level of confidence that makes clients feel comfortable and encourages them to place their trust in you.

Internship requirements vary across institutions and professional specializations. If you are studying sport management, your internship may involve a semester of 20-hour weeks

The Challenge of the Internship

For most students, the internship is a challenging experience, the point in their college training in which they discover how much they've learned and how much they have yet to learn. Clay and Sara, students studying to be physical education teachers, offered the following accounts during their internships (student teaching).

Clay

I really had no idea. . . . The first couple of weeks was really hard for me to get adjusted to it. I feel more confident because I'm feeling a lot more respect from the kids. They're starting to treat me like a teacher. . . . I had just finished teaching a track unit and the last week we had a competition, and I didn't have to tell one person to dress out or show up to the class. Everybody did because they enjoyed having the competition and enjoyed doing the different events. A lot of them, whether or not they thought so, did improve. . . . I feel prepared to teach. Mainly, the students' response gives me the feeling like, hey, I did come into the right profession.

Sara

I finally started to be a little more relaxed and flexible. At the beginning, I was really regimented. We stuck to it [our plans] no matter what, whether it worked or not. That made me frustrated, it made the students frustrated, and I know that Ms. Green [the internship supervisor] was frustrated with me. But I finally kind of loosened up and at the end became a little more flexible. Ms. Green helped me develop my own style. It is not really set in concrete; I know it is going to change. I need to become more assertive and aggressive.

Student quotes reprinted from M.A. Solomon, T. Worthy, A. Lee, and J.A. Carter, 1991, "Teacher role identity of student teachers in physical education: An interactive analysis," *Journal of Teaching Physical Education* 10, 188-209.

working in an athletic administration office where you might be required (like most professionals) to perform a range of duties, including mundane tasks such as handling mailings in the ticket office or answering the phone in the information office. You also have measured amounts of responsibility in performing the types of tasks that will be required of you as a sport management professional. If you are undertaking a student teaching experience in physical education, you may be required to spend an entire semester of full days in which you assume major responsibility for teaching classes. If you are an athletic training intern, you will be responsible for logging a preestablished number of clinical hours specified by the accrediting regulations of the National Athletic Trainers Association (NATA), usually in a training room and during travel with an athletic team.

Are You Suited for a Career in the Physical Activity Professions?

Now that you know something about the type of work that professionals do and the educational experiences required to prepare you for such work, it's time to ask yourself how well suited you are for a career in the physical activity professions. The five-step approach deomonstrated in activity 12.5 in your online study guide will help you undertake this self-examination process.

Do My Attitudes, Values, and Goals Match Those of Professionals?

The first step is to consider how compatible you are with professional work generally, regardless of the particular field. Approach this task honestly and objectively. If you decide that you're not well suited to a professional life, you should consider alternative careers.

Taking stock of your strengths, weaknesses, and potentials in relation to the personal attributes and preferences likely to be required in most types of professional work is an important first step. Nobody is in a better position to judge our attitudes, attributes, and preferences than we are, but sometimes our perceptions of ourselves can become a bit clouded. One way to bring our own attitudes into sharper focus is to compare them with those of our friends and peers. The results of a recent survey, by the Higher Education Research Institute at UCLA of nearly 300,000 freshmen entering four-year institutions in the United States in 2006, offer a convenient backdrop against which you can assess your own attributes and preferences.

> **DO it Activity 12.5**
>
> **Selecting a Career**
>
> Work through the decision tree in this activity to clarify whether a career in the physical activity professions is right for you.

If you are a typical undergraduate student, chances are that you believe helping others in difficulty is an important life objective, and you probably performed volunteer work during the past year. This suggests that you and your cohort of students entering college may be attuned to their civic responsibilities of helping others and improving society. This is a good start for future professionals! At the same time, a relatively small percentage (27%) of your peers indicate that they plan to continue to do volunteer work or community service work in college. This raises a red flag. If you don't plan to continue volunteer work in college, you may not be well suited for a career in the professions. Employers and graduate school admissions officers usually interpret your continuing commitment to volunteer work throughout college, even if only for 2 or 3 hours per week, as an indication that you understand the serious service responsibilities that come with planning a career in the helping professions.

> ➤ If you decide that your attitudes, values, and life goals match those normally associated with a professional career, you are ready to proceed to the next question in the decision model (see activity 12.5). If you conclude that you are not a good match for a professional career, seek advice from a faculty member or the career counseling center at your institution.

Inevitably, this self-searching exercise will cause you to reflect on your major life goals. Based on the survey, chances are better than 50-50 that you decided to attend college to

DO it Activity 12.6

Assessing Your Interest
Take this self-assessment in your online study guide to determine how interested you really are in physical activity.

gain a general education and appreciation of ideas (64% of freshmen), learn more about things that interest you (77%), and receive training for a specific career (69%)—all of which are admirable life goals for anybody planning a career in the professions. (Let's hope you aren't one of the 4% who said that they attended college "because there was nothing better to do"!) But according to the Higher Education Research Institute survey, only 46% of your peers consider developing a meaningful philosophy of life an important goal of college, a surprising finding given that one's philosophy of life will inevitably shape not only important professional decisions but personal decisions as well (Higher Education Research Institute 2007, August 31).

The survey results also suggest that you, like your peers, have your sights set on obtaining a better job upon graduation (70%), being well off financially (73%), and becoming an authority in your field (58%). Generally, these are encouraging findings, although you should keep in mind that being driven obsessively by a need for financial gain can create conflicts of interest in professional practice. Professionals, especially social trustee, civic professionals, are much more likely to identify service to humanity as their highest priority.

The fact that an overwhelming majority of your peers aspire to high-paying jobs, whereas only 57% plan to continue in graduate or professional school and only 41% would like to have administrative responsibility for the work of others, indicates serious misunderstanding about life in the professional workplace. High-paying jobs usually come with some administrative and supervisory responsibilities, and advancement in the professions is usually linked closely to educational credentials. Generally, graduate degrees open doors of opportunity routinely closed to those who lack such credentials. Because educational standards for professional practice usually rise over time, increasing numbers of undergraduate students will discover that a master's degree has become the minimal educational qualification for entering many of the physical activity professions.

Professional work often can assume a frenetic pace that may push you beyond the normal 40-hour week. Many students (approximately 38%) feel overwhelmed during their undergraduate years, but professionals in the workforce often feel this way too ("Freshmen Drinking and Studying More" 2002). Thus, stamina and robust health are especially important personal attributes for physical activity professionals. Yet, as we have seen in earlier chapters, a crucial means of improving and maintaining health—involvement in physical activity—tends to decline among the typical college age group. Data collected on 25,000 freshmen from 109 four-year colleges showed that at the beginning of the freshman year, nearly 53% spent 6 or more hours per week exercising or playing sports, but that percentage had fallen to 34% by the end of the freshman year. On the other hand, the percentage of students who drank beer, frequently or occasionally, grew from 45.8% at the beginning of the year to 58.5% at the end of the year ("Freshmen Drinking and Studying More" 2002).

After giving the matter some thought, you may have concluded that you really

Improving and maintaining health needs to remain a priority throughout your college career.
Digital Vision

Sleep Deprivation and Academic Performance

Staying up all night, studying or partying, is part of the tradition of college life. Students estimate that they get (on average) 3 fewer hours of sleep at night during exam week and often resort to various prescription and nonprescription drugs or high-energy drinks to help them stay awake to study. You should know that sleep deprivation can exact a very steep price. A number of studies have shown sleep deprivation to be associated with poor academic performance. One study that tested the effects of sleep deprivation on college students' cognitive functioning showed that students who were sleep deprived performed significantly lower on a cognitive task than students who had normal amounts of sleep. Even though they performed worse in the tests, sleep-deprived students rated their effort and concentration higher than the nondeprived students, suggesting that students underestimate the negative effects of sleep on their academic performance (Pilcher & Walters 1997).

aren't cut out for a professional career. Perhaps you don't find working with people attractive, you don't particularly like helping to solve other people's problems, and you would rather work for a supervisor than take responsibility for acting on your own authority. Maybe the prospect of educating yourself continually to keep abreast of the professional literature or constantly attending to the ethical ramifications of your work behavior is not part of the vision you have for your career. If so, it seems unlikely that a professional career will interest you, regardless of the area of specialization. We suggest that you continue to collect more information by reading the other chapters in this section of the book, talking to faculty members, and consulting with the career counseling center on your campus.

If, on the other hand, you are convinced that you would like a career in the professions but you recognize some glaring deficiencies (fear of speaking before large groups or a lack of organizational skills, for example), consult with your faculty advisor about including academic experiences during your undergraduate years to bolster these weaknesses.

Am I Interested—Really Interested—in Physical Activity?

If what you have read about the professions excites you and matches your interests and attitudes, the next step is to consider how much you like physical activity. Those who enter the physical activity professions should be more than merely interested in physical activity; they should be fascinated by it. If you're not intrigued by performing, studying, and watching physical activity, if you aren't curious about the mysteries of body movement in all of its manifestations (biophysical, behavioral, and sociocultural), and if you're not prepared to apply your intellect and skills to exploring these mysteries, you may be better off selecting another profession.

Do My Attitudes, Interests, and Talents Lend Themselves to a Specific Physical Activity Profession?

The third step in this personal evaluation is to ask yourself if your attitudes, interests, and talents match the specific characteristics of the occupation that you are considering. For example, if you envision a career in some aspect of therapeutic exercise or physical therapy, you should ask yourself if you enjoy interacting with sick and injured people. Obviously, most people—even medical students—require some time before they feel completely comfortable in medical environments, but you need to understand, fairly early on, your degree of comfort with the environments in which allied health professionals work.

If you like being around vigorous, healthy, active people, a career in physical fitness counseling, personal training, coaching, or athletic training would probably appeal to you.

Attitudes, interests, and talents determine the kind of career you may be interested in. Do you want to work indoors or outdoors? In a structured or flexible environment? In a high-powered position, or one that's low-key?

AP Photo/The Star, Heather Charles

Do you have a special place in your heart for people with disabilities such as mental retardation, autism, blindness, or deafness? Would you find it challenging to help such people maximize their physical and mental potential? If so, a career in adapted physical education seems a reasonable goal. If you like being the center of attention, organizing and speaking to large groups, planning and monitoring and evaluating activities that others perform, and accepting responsibility for the actions of people younger than you, a career in teaching fitness, coaching, or physical education may be a good choice.

Does working at a desk dressed in business attire and adhering to a 9-to-5 workday appeal to you? If so, you might be well suited to a career in sport management. Are you more attracted to fluid and informal working environments in which your work schedule might be more flexible, even though work might intrude on your weekends and evenings? If your answer is yes, then you might be better suited to more entrepreneurial types of professional positions such as that of personal trainer, professional sport instructor, or university professor. Chapters 13 through 17 will give you more complete descriptions of various physical activity professions to help you assess which careers may best suit you.

If you become convinced that your needs, interests, and attitudes do not closely match the specific professional occupation you had in mind, don't hesitate to explore other career possibilities within the physical activity professions. If, on the other hand, you're not sure, consider taking advantage of opportunities to volunteer at clinics, commercial or governmental agencies, schools, or a variety of other physical activity settings. Volunteering will give you firsthand experiences with various working environments and will increase the probability that the career you choose will match your interests, talents, and preferences. If possible, coordinate these volunteer experiences with your faculty advisor.

Will My College or University Program Prepare Me Well?

The fourth step is to determine whether the program in which you are presently enrolled will prepare you well for your chosen career. If, for example, you are most interested in a career in teaching and coaching but your department offers only concentrations in exercise physiology, sport administration, and health promotion, the department's curriculum would seem a poor match for your career objectives. The same would be true if the program slants heavily toward the preparation of physical education teachers or professionals in the fitness industry and you are most interested in a career in sport management. If you have doubts about the fit between the course requirements of the program you are in and your career aspirations, don't hesitate to talk to a faculty member.

> www. **Web Search 12.1**
>
> **Are You Qualified?**
> Search for entry-level positions and assess your qualifications for a job in this activity in your online study guide.

Perhaps your career goals have changed since you enrolled in your undergraduate program. Although this would occur less often if students took the five-step approach recommended here, changing career goals during the undergraduate years is common. Remember that switching majors or concentrations within the department (e.g., from physical education to exercise therapy) can delay your graduation for a year or more. Changing also can be inconvenient. If you decide to switch majors to an entirely new field, you may push your graduation even farther into the future. Making such a drastic change also will require you to become socialized into another department, become accustomed to new professors, and form new alliances with classmates. Inconvenient though switching your major may be, it is better to endure delays and inconvenience than to continue training for a career in which you have little interest. If you are deep into your undergraduate program and find that your career interests have changed to another physical activity profession, the best action might be to complete your present program and follow it up with a master's degree in your area of interest. Faculty members in your department will be ready and willing to advise you, even if doing so means referring you to other departments or other institutions.

Asking questions about the pertinence of your program of study to your life goals is risky business. You may not like how you answer your own questions! But this vital exercise is one that students too often ignore. Remember that those who reach career decisions early in their undergraduate years are more likely to enjoy their undergraduate studies and appreciate their relevance, particularly in courses that they take in their major.

How Committed Am I to Preparing to Be the Best Professional Possible?

At the first practice of the season, coaches often rhetorically ask their teams, "How badly do you want it?" They refer, of course, to the team's willingness to invest the time, energy, and hard work essential for winning the league championship. Now the question for you is, How badly do *you* want it? How much are you willing to invest to achieve success in your chosen profession? This last step in the process may be the most important because it supersedes all other questions in the process. You may have had little difficulty answering the questions up to this point, but if you aren't committed to success, there's little point in continuing. Are you willing to commit to preparing yourself to be the best professional possible? Let's examine some of the ways in which such a commitment might manifest.

Excellence in Academic Work

Graduating with a superb academic record is regarded by most employers as a baseline indicator of your level of commitment to becoming an outstanding professional. Obviously, some students are more academically gifted than others are, but often students can overcome slight shortcomings in academic ability by special application of effort. Few things make a more indelible impression on professors than a student's willingness to work hard to achieve academic success.

➤ Perhaps the best predictor of your success in the physical activity professions is the level of commitment you make to preparing yourself to be the most knowledgeable and highly skilled practitioner possible.

What types of behaviors suggest a willingness to work hard? Attending class regularly and on time, visiting the library regularly, and reading unassigned journal articles and books about the topics that you are studying in class all are signs that you are taking your academic work seriously. Most professors view students who collar them after class with questions about the material in the day's lesson as more committed than those who avoid them at all costs.

Early Identification With the Professional Field

How early in the undergraduate program students identify with their chosen career is also a good indicator of commitment. In the chapters that follow, you will find many references to professional associations. Joining a professional association while still an undergraduate may be the most reliable indicator of commitment. Professional associations usually admit preprofessionals at reduced rates and offer reduced registration rates at conferences. Members receive the organization's publication and other information on a regular basis. Attending professional conferences is another indicator. Professional associations usually hold annual regional and national meetings that feature lectures, workshops, and exhibits of equipment

Thumbnail Sketch of Your Classmates

During their undergraduate years, your classmates say the chances are good that they will do the following:

Socialize with someone of another racial or ethnic group (65%)
Get a job to help pay for college (44%)
Communicate regularly with professors (33%)
Change career choice (13%)
Make at least a B average (61%)
Participate in study abroad (29%)

During the past year they have done the following:

Socialized with someone of a different racial or ethnic group (67%)
Drunk beer (42%)
Been bored in class (41%)
Attended religious service (78%)
Smoked cigarettes (frequently) (7%)
Felt overwhelmed (29%)

They somewhat agree or agree strongly about these issues:

Abortion should be legal (57%).
The death penalty should be abolished (35%).
Racial discrimination is no longer a problem in society (19%).
Colleges should prohibit racist or sexist speech on campus (60%).
Same-sex couples should have the right to legal marital status (61%).
Affirmative action in college admissions should be abolished (47%).
Undocumented immigrants should be denied access to public education (47%).

They feel the following objectives to be essential or very important:

Raising a family (76%)
Being well off financially (73%)
Having administrative responsibility for the work of others (41%)
Becoming an authority in their field (58%)
Influencing the political structure (23%)
Developing a meaningful philosophy of life (46%)

➤ Identifying early with the profession you plan to enter by joining appropriate professional organizations, attending conferences, establishing alliances with veteran professionals, and obtaining professional certifications will give you a head start on developing a successful career.

used in professional practice. Attending these can be expensive, but then again, taking a spring break trip to the Florida beaches is costly. Remember, when we make a commitment, we reveal our priorities.

Why is it important to identify with a profession while you are still an undergraduate? The answer is that students who view themselves as members of the profession tend to approach their studies with special excitement and vigor. Their orientation changes from that of a student who views courses as obstacles to obtaining a degree to that of a professional who tries to get out of each course the knowledge and skills that will help him or her develop into the best qualified professional possible. When you identify with your chosen profession, you will seek the advice of veteran professionals and observe them in practice.

You will begin to establish a communications network with practicing professionals and will establish the habit of reading the journals in your field. You will learn the language of the field and begin to feel comfortable around experienced professionals, although you are still in the preparation phase of your career.

Another way to identify early with a physical activity profession is to obtain certification in an area in which you plan to work. Certifications are available from such groups as the American College of Sports Medicine, the National Strength and Conditioning Association, the American Council on Exercise, and the Aerobics and Fitness Association of America.

➤ Becoming involved in activities within and outside your department is one indication of commitment to developing a successful career in the professions.

Becoming Engaged in College or University Life

Being engaged means being connected to what is happening around you. Next to academic performance, your level of engagement in your department, college or university, and community is one of the best indicators of your level of commitment to becoming a successful professional. Unfortunately, levels of student engagement on campuses appear to have declined in recent years, a trend that faculty and administrators find disturbing (Flacks & Thomas 1998, November 27). Who are the engaged students? Engaged students are those who take responsibility for their academic experiences and their professional futures. They are curious, and they constantly test their personal limits by seeking new ways to become involved in life experiences. They are active, energetic leaders who view the department as their department and seek ways to participate in its operational life. They are likely to be leaders in their major clubs or student government on campus or coordinators of charity events or other community activities outside the academic environment.

Besides becoming engaged in social and organizational extracurricular activities, you should become involved in academic-related activities. Does a professor need students to assist on a research project? Does your institution sponsor undergraduate research assistantships? Does your department sponsor community-based projects in which you can work? Will a professor sponsor you for work on an independent study of a topic of interest to you? Don't be afraid to register interest in such projects. Independent study not only reflects a high level of commitment on your part but also offers an excellent opportunity to develop the type of leadership skills and knowledge essential for success in the world beyond college.

Participating in Volunteer Services

We have seen that the driving force of professionals, especially those in the helping professions, is a spirit of wanting to improve the quality of life for those in the community. Although it is easy for preprofessionals to say that they want to serve others, nothing speaks more loudly than actions. If two students have approximately equal academic records, the advantage always will go to the student whose resume displays clear evidence of volunteer service to community agencies. These might include such programs as Meals on Wheels, Boys and Girls Clubs, local food banks, Boy

Volunteer experience in physical activity pursuits demonstrates your commitment to the profession.

Scouts and Girl Scouts, shelters for the homeless, church-based programs, walk-a-thons for various charities, American Red Cross activities, and other community-based programs. When your volunteer experience is in an activity that bears some resemblance to the profession for which you are preparing, it has the added benefit of socializing you into your future profession. Physical therapy and sports medicine clinics, YMCAs and YWCAs, youth sport organizations, schools, Special Olympics, and other agencies associated with the physical activity professions are some potential volunteer opportunities that you might want to explore.

Attending Graduate School

➤ Volunteering regularly for service to a community agency is an indication that you share in your profession's commitment to service. Seeking in-depth, advanced education in graduate school may also be a sign that you are investing your interest and personal resources in a career in the physical activity professions.

Evidence of commitment to a profession may also be reflected in plans to pursue advanced graduate work. In professions such as physical therapy, a master's degree is a minimal entry degree, and the athletic training profession may require a graduate-level degree in the near future. Permanent certification for teaching physical education in many states requires a master's degree or its equivalent. You may not be prepared to think about graduate school at this early point in your undergraduate preparation, or you may have decided to delay that decision until a few years after you have graduated. For many, this may be a sensible plan. Overall, however, making an early decision to continue with advanced graduate education in the physical activity professions is another indication of commitment.

Generally, master's programs in kinesiology offer advanced education in a number of specialized areas, although not all universities offer all specializations. A faculty member at your institution may be an excellent source concerning which institutions you might consider attending. In addition, you may find posters or brochures advertising graduate study at various

Whom Would You Hire?

You are the supervisor of physical education for a large school district and have just narrowed your search for a high school physical education teacher to two candidates. Here is a thumbnail sketch of each candidate's undergraduate record:

- Anthony earned a respectable 3.65 GPA in kinesiology (teacher education emphasis) while working 30 hours a week as a waitperson at a local restaurant during the school year and full-time during summers. He was awarded the Senior Academic Prize for the highest GPA in his class. Although his work schedule prevented him from becoming involved in many on-campus activities outside class, he was a member of the cheerleading squad at his university. As part of his application for the teaching position, he supplied letters of reference from his supervisor at the restaurant and the coach of the cheerleading squad, both of whom praised him as being a bright, enthusiastic person with "good people skills." He earned high marks from his supervising teacher during his student teaching experience. An additional letter from one of his professors predicted that with experience and strong mentoring, he would probably develop into a first-rate teacher.
- Kelli earned a slightly lower GPA (3.50) in kinesiology (teacher education emphasis) while working 10 hours each week in a local Boys and Girls Club, where she taught classes in exercise and sport skills. In the summer months, she worked as a counselor for a wilderness camp, specializing in canoeing, white-water rafting, and backpacking. She was secretary to the kinesiology majors club, organized a walk-a-thon to benefit homeless people, and spent 2 hours each week serving as a volunteer fitness leader for a program targeting disadvantaged children in an inner-city neighborhood. Kelli was a student member of the American Alliance for Health, Physical Education, Recreation and Dance and attended the national conference one year when it was held at a nearby city. Her letters of reference include one from a professor and one each from the directors of the Boys and Girls Club and the wilderness camp. All describe her as bright and energetic, able to motivate young people, and able to make good on-the-job decisions.

Whom would you hire and why?

institutions posted in your departmental office. Normally, master's degrees require from 30 to 36 credit hours of work depending on the specialization. In most cases, full-time master's students are able to complete the requirements of the degree in one and a half to two and a half years. Master's programs that require a thesis have slightly fewer hours of course work but include an in-depth research project that may take up to one year to complete. Some master's programs do not require a thesis but may include more course work than thesis programs do. Sometimes programs that do not require a thesis do require an internship. If you delay attending graduate school until after you have secured a position and plan to do your graduate work by attending classes in the evenings, you may need as long as four or five years to complete a master's degree.

The Decision-Making Process

By going through the decision-making steps in activity 12.5, you have embarked on an important process. If you answered yes to all five questions, you are well on your way to a successful career in a physical activity profession. The next five chapters will give you an in-depth look at some of the professions you have to choose from.

If, instead, you hesitated at some questions or were not sure of your answers, the next five chapters may help you make a decision. Use this as an opportunity to assess honestly and objectively whether you are suited to professional work, and especially to professional work in one of the physical activity professions.

Wrap-Up

Career decisions are among the most important choices that people make during their lifetime. Such decisions are not irrevocable; in fact, you likely will change your career several times over your working years. Nevertheless, you will gain nothing by delaying your commitment to a career, and you have much to lose if you delay. Students who make career commitments early are in a better position to benefit from their undergraduate education than are students who have only tentative plans about the type of work that they will do when they graduate. If you have decided to attain a degree in kinesiology by default (it didn't particularly excite you, but no other college major seemed any more appealing) and life beyond graduation looms like "the great black void," you should take steps to assess your compatibility with a career in the physical activity professions.

Doing this necessarily involves learning what it means to be a professional and the types of work and work environments associated with each of the physical activity professions. It also involves making a realistic assessment of your level of excitement and commitment to kinesiology and the physical activity professions. If this chapter achieved its goal, you will already have begun this self-examination process. Now it is time for you to learn more about the physical activity professions by studying the next five chapters. These chapters, written by experts in the professions being described, will provide specific information about what it is like to work in each of these areas. When you have finished, you should be in a good position to make this important decision about your future career.

Use the key points review in your online study guide as a study aid for this chapter.

QA **Activity 12.8**

These end-of-chapter questions and activities are also in your online study guide. Your instructor may ask you to complete them online and turn them in.

1. List three ways in which professional work differs from nonprofessional work.
2. Why is it important for professionals to attend professional conferences and read the professional literature?
3. List three differences that you might observe between a community sport program leader who adheres to a mechanical, market-driven professionalism and one who adheres to a social trustee, civic professionalism.
4. What is the value of the liberal arts and sciences major, and why are the liberal arts courses required of kinesiology majors important?
5. Why is the internship an important experience in preparing kinesiology students for professional practice?
6. What are five questions that all kinesiology students should ask themselves before deciding to major in kinesiology?
7. What evidence would you—as an employer—use to determine an applicant's general suitability for a professional position?
8. What would be the ideal work history of a kinesiology graduate seeking to enter the athletic training field? The health-fitness field? The sport management field?
9. If you decide to pursue a career as a physical education teacher during your second year of the kinesiology major after having entered as a sport management major, what is the best course of action to take?

13

Careers in Health and Fitness

Jeremy Howell and Sandra Minor Bulmer

CHAPTER OBJECTIVES

In this chapter we will

- acquaint you with the wide range of professional opportunities in the sphere of health and fitness;
- familiarize you with the purpose and types of work done by health and fitness professionals;
- explain how the sphere of health and fitness is evolving and progressing, especially because of marketplace trends;
- inform you about the educational requirements and experiences necessary for becoming an active, competent health and fitness professional; and
- help you identify whether one of these professions fits your skills, aptitudes, and professional desires.

An executive who is overweight and out of shape hires a personal trainer. The executive wants to begin training with the goal of competing in a first-ever 5K in six months' time and is relying on the personal trainer to guide her in losing weight, gaining strength, and improving her cardiovascular endurance safely.

A group exercise instructor meets with the class members of a new session today and sees that two of the participants are over age 55. One of them tells the instructor that he has hypertension.

A health and fitness director realizes that her facility should expand its programs next year to reach less physically fit populations more effectively. What additional staff will she need to hire? What should their qualifications be? Will her budget allow this expansion?

In an interview with a health and fitness counselor, a working mother mentions that she is having trouble fitting her exercise program into her busy lifestyle. She feels overwhelmed, irritable, and constantly exhausted. She also feels guilty about not taking time for herself to exercise. How will the health and fitness counselor begin to develop an exercise program that improves this client's quality of life?

Health and fitness professionals face these types of situations every day in their careers. In each case, they must make decisions that will improve the health and fitness of participants and clients safely, while keeping within the goals and objectives of the program or facility where they work.

Health and fitness, one of the spheres of professional practice (see figure 13.1), is made up of many dynamic professions. The sphere of health and fitness professions includes such positions as group exercise instructor; fitness instructor; health and fitness counselor; wellness and lifestyle coach; personal trainer; health and fitness director; and specialist positions such as certified health educator, clinical exercise physiologist, and registered dietitian.

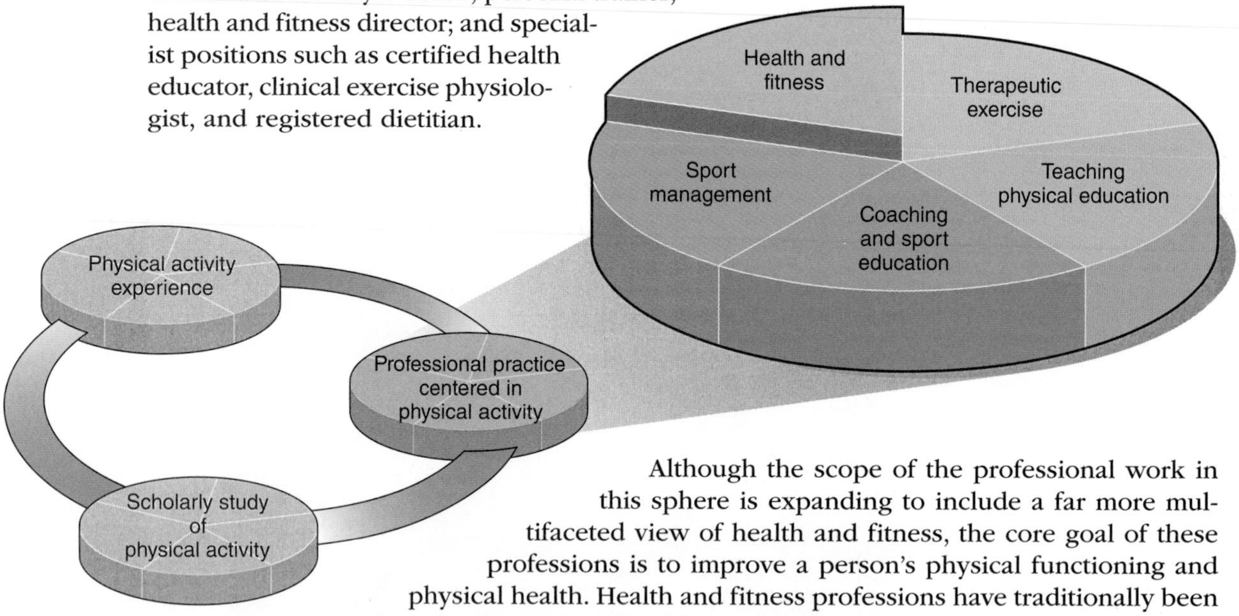

FIGURE 13.1 Health and fitness is one of the spheres of professional practice.

Although the scope of the professional work in this sphere is expanding to include a far more multifaceted view of health and fitness, the core goal of these professions is to improve a person's physical functioning and physical health. Health and fitness professions have traditionally been conducted in worksite, clinical, commercial, and community settings, but in recent years the lines between these settings have become blurred as each type of program or facility has expanded its offerings to reach more people and affect their lives in more ways.

This expanding scope is creating tremendous opportunities for kinesiology graduates and is making health and fitness positions some of the more exciting and dynamic careers people can enter. Thus, our purpose in this chapter is threefold. First, we want to describe the opportunities that exist for professionals in the health and fitness field. Exactly what do health and fitness professionals do and where do they do it? Second, we want to analyze why these opportunities are appearing at this time. What social forces are creating new possibilities for current kinesiology graduates? Third, we want to discuss how you can best take advantage of these opportunities as you complete your education and move into your chosen career. What are the educational requirements and experiences necessary for becoming an active, competent professional in this area?

Health and Fitness Professions

This is an exciting time for those involved in the health and fitness professions. Despite the many studies demonstrating that the American public is not nearly as active as it should be (see chapter 2), the marketplace for physical activity professionals has never been better. The health and fitness settings in which physical activities take place and the jobs that physical activity professionals perform are experiencing a remarkable expansion and transformation.

Not since the jogging boom of the late 1970s have we witnessed the dramatic growth that we are seeing today of a fitness and health field that has embraced and promoted the relationship between exercise, personal responsibility, and prevention in specific ways (Howell & Ingham 2001). Today, scientific research and popular culture have popularized the notion that regular exercise can increase lean body mass, heart and lung function, immune function, blood vessels, blood flow, muscular strength, flexibility, bone density, balance, and self-esteem. Perhaps most important is the research that links physical activity to lowering the risk of death because of coronary heart disease, the leading cause of mortality in the United States.

Indeed, numerous government reports now extol the undisputable virtues of exercise for good health. In 1996, the U.S. Department of Health and Human Services (USDHHS) published a document titled Physical Activity and Health: A Report of the Surgeon General. This was significant because for more than a century, the surgeon general has focused the nation's attention on the public health issues of the day. Previous reports on issues such as smoking, nutrition, violence, and HIV/AIDS heightened America's awareness and led to some innovative public health initiatives. In 2000, the USDHHS also published Healthy People 2010, a document that set out specific overall health objectives that the nation should achieve in the first decade of the new century. The objectives are based on a set of leading health indicators (LHIs), of which physical activity is one. Other LHIs are overweight and obesity, tobacco use, substance abuse, responsible sexual behavior, mental health, injury and violence, environmental quality, immunization, and access to health care. With regard to physical activity, Healthy People 2010 noted that in 1999 65% of adolescents in grades 9 to 12 engaged in the recommended 20 minutes of vigorous

As this boy grows older, will he continue to be physically active?

physical activity 3 or more days per week. In 1997 only 15% of adults aged 18 years or older engaged in the recommended 30 minutes of moderate physical activity 5 or more days per week. What might seem surprising is that 40% of adults engaged in no leisure-time physical activity at all. Consequently, Healthy People 2010 set objectives to increase the proportion of adolescents who engage in physical activity 3 or more days per week to 85% and to increase the proportion of adults who participate in daily exercise of 30 minutes to 30%.

Throughout the decade, the USDHHS, federal agencies, and other experts have been conducting a series of progress reports of Healthy People 2010 objectives to assess the status of the nation. Regarding physical activity, the results are discouraging, and little progress has been made. There has been little or no change in physical activity levels, moderate or vigorous physical activity for adults, or vigorous physical activity for adolescents (see figure 13.2).

Despite these discouraging statistics, physical activity objectives are now very much at the heart of U.S. government health intervention strategies. Consequently, in 2008 the USDHHS published The 2008 Physical Activity Guidelines for Americans. This document contains even more specific guidelines on the types and amounts of physical activity that provide substantial health benefits for people 6 years and older.

What all of these government documents show is that no longer do people perceive physical activity only as a recreational endeavor; it is firmly embedded in the national preventive health consciousness. However, as the Healthy People 2010 progress reports show, finding creative solutions to healthy living is not easy. Rather than just being asked to train, measure, and test people for improved athletic performance, health and fitness professionals have to develop individual and group exercise programs that connect to other physical, intellectual, emotional, social, and spiritual dimensions. Similarly, much better health and fitness strategies need to be in community, commercial, corporate, and clinical settings. This means there are enormous opportunities for kinesiology students at this time.

In this chapter, we first explain the traditional settings within which health and fitness professionals operate as well as the most common professions. Then we look closely at how the changing trends in society and culture are reshaping the health and fitness sphere of professional practice.

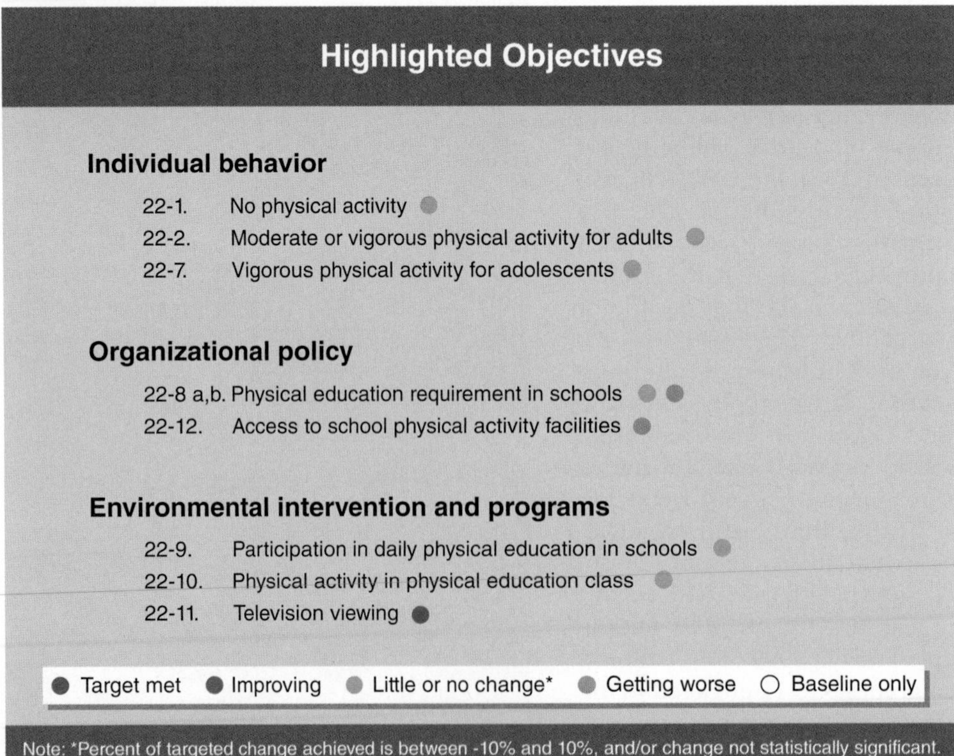

FIGURE 13.2 In 2000, the USDHHS published its Healthy People 2010 guidelines. This progress report, presented in June of 2008, shows discouraging results with regard to physical activity objectives.

Reprinted from Klein 2008.

Health and Fitness Settings

The health and fitness field is relatively young and is currently undergoing some exciting changes. These changes have occurred across all types of health and fitness settings and have resulted in expanded job descriptions and a greater number of positions for entry-level and experienced professionals. Traditionally, health and fitness professionals operate in four primary settings—worksite, commercial, clinical, and community. At one time, each of these settings had a distinct mission, target market, and set of objectives for programming. More recently, however, the various health and fitness settings have become more similar than dissimilar. Examples include commercial settings that offer worksite health promotion programs, clinical settings that operate for-profit fitness centers, and community settings that provide medically based fitness programs in partnership with area hospitals. In summary, the lines between the various settings are no longer distinct.

➤ Traditionally, health and fitness professions have operated in four settings: worksite, commercial, clinical, and community. Recently, the lines between these settings have blurred, resulting in expanded job descriptions and a greater number of positions.

DO it Activity 13.1

Job Skills

Match activities and skills with professions in this activity in your online study guide.

This section describes each type of health and fitness setting and provides an overview of the objectives, types of programs, and target markets for each setting. It also describes the changes that have occurred over time and the trends emerging in the future.

Worksite Settings

Worksite health and fitness programs (often referred to as worksite health promotion programs) have experienced tremendous growth during the past several decades. In the early 1980s, less than 5% of employers offered worksite health and fitness programs (O'Donnell & Harris 1994). According to Healthy People 2010, by 1994 61% of employees over 18 years of age participated in employer-sponsored health promotion activities. The Healthy People 2010 objective was to raise this to 75% (USDHHS 2000).

These early worksite programs focused primarily on physical fitness, nutrition, weight control, stress management, and smoking cessation and were available exclusively to employees of the sponsoring company. Pioneering corporations such as Johnson and Johnson, Coors Brewing Company, and Tenneco believed that health and fitness programs would help to reduce health care costs, increase productivity and morale, decrease absenteeism, and improve their corporate image. Since these early days, worksite programs have continued to evolve and many research studies have documented positive outcomes. Specifically, in a 2005 analysis of 56 research studies that had measured economic returns from worksite health promotion programs, it was found that "the evidence was very strong for average reductions in sick leave, health plan costs, and workers' compensation and disability costs of slightly more than 25%" (Chapman 2005).

The Health Enhancement Research Organization (HERO) has also provided strong evidence to support the economic value of worksite health promotion programs. Analysis of their database that includes 47,500 employees who have completed a health risk appraisal has revealed that there is a strong association between modifiable risk factors and health care expenditures (Goetzel et al. 1998). For example, employees with self-reported persistent depression had health care expenses 70% greater than did those who reported not being depressed. In addition, costs were 46% higher for those with uncontrolled stress, 35% higher for those with high blood glucose, 21% higher for those who were obese, and 10% higher for those with poor exercise habits. Additional analysis with this database in 2000 revealed that risk factors were associated with 25% of total health care expenditures (Anderson et al. 2000). Bungum and colleagues (2003) found that those who had healthier body weights, calculated by body mass index (BMI), reported fewer absent work days and decreased health care costs.

As worksite programs have continued to develop, additional services have been added to increase employee awareness and help employees make healthy lifestyle changes. Examples of

Some companies host health fairs to inform their employees about health care and health promotion options.

AP Photo/The Morning Sun, Kyle DeRodes

these services are health education classes, company health fairs, the distribution of educational materials, and the use of health risk appraisals to identify employee risk factors. More recently, worksite programs have expanded offerings even further to provide a work environment that further supports healthy lifestyles. In this expanded role, worksite programs have developed and implemented on-site work policies regarding smoking behavior, availability of nutritious food, and the use of **ergonomic workstations.** Work–life balance initiatives have also become popular as ways to further promote employee wellness in all dimensions of health. These initiatives include flexible work schedules, on-site child care, and leave-of-absence policies for caregiving and other life events. Other popular worksite programs include recreation activities, volunteer service opportunities, financial planning, and **employee assistance programs (EAPs).** Also, in an effort to reduce the medical costs of all individuals covered by the company health plan, some worksite programs now promote health and fitness services to employees' family members and company retirees.

In economically prosperous times, many companies built elaborate health and fitness facilities as a way to recruit and retain new employees. Examples are companies in the Silicon Valley near San Francisco, such as Oracle, EA Sports, and Applied Materials. Other companies have contracted with local health clubs to offer access to exercise facilities for free or at a reduced rate. In either case, the trend has been for companies to establish partnerships with outside organizations who have expertise in health and fitness to manage their worksite health promotion facilities and programs. For instance, Club One Inc., a San Francisco–based corporation, manages the fitness and health programs for 3Com Corporation, Advanced Micro Devices, AOL-Time Warner, Applied Materials, Chevron, Fujitsu, General Atlantic, and Netscape. They also own several commercial clubs. Another example is MediFit Corporate Services, based in New Jersey. This company has an extensive list of corporate clients in the Atlantic region of the country and provides a variety of contracted services including facility management, on-site programming, and Web-based programming for employees who may be working from home or in smaller satellite branch offices.

➤ Employers have found that health and fitness programs reduce health care costs, increase productivity and morale, decrease absenteeism, and improve their corporate image.

Perhaps the most innovative change in workplace health promotion has come about as the result of many large corporations becoming self-insured. Over the past decade or so, the relationship between soaring health care costs for both the employer and employee cannot be ignored. According to a 2009 large-scale employer survey by the Kaiser Family Foundation, the average annual health insurance premium for family coverage is increasing. In 2009 the figure was $13,375. Of that amount, $9,860 was employer paid and $3,515 was contributed by the employee. That total is a staggering 131% increase in premiums over the 1999 figure of $5791 (Kaiser Family Foundation 2009). Many large companies have now decided to operate their own health insurance plan as opposed to purchasing coverage from an insurance company. Usually the employer still pays a third party (such as an insurance company) to administer the plan they have designed, but they do so at a much lower cost. But while they save money on costs, the large company becomes responsible for all claims that are made. This raises the exposure of the company to greater risks. For self-insured companies, improvement in employees' health can save companies money; thus companies are interested in investing in employees' health.

As an example, the grocery store chain Safeway Inc. has created a new solution to providing their employees with health insurance. In a Wall Street Journal OpEd piece, CEO Steven Burd (2009) describes Safeway's Healthy Measures program as a completely voluntary insurance plan that covers 74% of the insured nonunion work force. As with most employers, Safeway's employees pay a portion of their own health care through premiums. However, in the Healthy Measures program, the amount of that premium is reduced based on the results of tests associated with their tobacco use, healthy weight, blood pressure, and cholesterol levels. To further encourage healthy behaviors, the company has financial incentive programs in place to promote cost-conscious health care decisions. Additional components of the program include a 24-hour information line and health advisor, wellness and lifestyle management programs, healthier options for food, large employee discounts for healthier food, and state-of-the-art fitness facilities at their corporate office. In 2006 Safeway reported that their total annual change in employer health care expense decreased by 13%. While the employer portion decreased by 5%, the overall employee portion decreased from 25 to 34% (Shachmut 2007).

Commercial Settings

Any description of the commercial setting must begin with an acknowledgment that many types of businesses fit into this category and not all of them offer quality health and fitness activities delivered by qualified personnel. Fortunately, one of the most significant trends in the commercial setting is an increase in the number of high-quality health and fitness services being offered to members. Commercial health and fitness facilities typically operate with the objective of generating a profit. The manner in which the facility generates this profit places it in one of two distinct categories: **sales-based facilities** and **retention-based facilities.** The following sections offer a comparison of these two business models.

Commercial facilities offer health and fitness programs similar to those offered in the workplace. Popular programs include fitness assessment and health risk screening, nutrition counseling, personal fitness training, massage therapy, group exercise classes, spa services, wellness education classes, and worksite facility and program management. In addition, specialized exercise programs for children, prenatal and postnatal women, older adults, and people needing physical therapy have increased in popularity. The nature of a for-profit business encourages innovation as a way to compete for customers in the marketplace. As a result, the commercial health and fitness business is dynamic and offers many opportunities for health and fitness professionals to become involved with a variety of programs.

➤ Because commercial health and fitness facilities need to make a profit, they compete for customers, and this requirement encourages innovative programming. Commercial health and fitness facilities are either sales based or retention based.

Sales-Based Facilities

Membership sales activities are the major focus in a sales-based health and fitness facility. These facilities rarely place limits on the number of memberships sold and often offer highly discounted rates as a way of attracting a large number of new customers. Another common characteristic of a sales-based facility is the selling of long-term membership contracts. Within the industry, sales-based facilities are often referred to as operating under a future service contract philosophy. To receive the best possible rate, new members must make a long-term commitment in advance rather than pay for their membership monthly. Because of this approach, sales-based businesses may not be highly motivated to meet members' long-term needs or focus on retention activities such as offering high-quality programs run by qualified personnel. Instead, they often allocate extensive resources to advertising campaigns that offer very low joining rates. Despite this, sales-based facilities might suit those who can regularly exercise at hours when the club is not full or do not need extensive assistance from expert personnel. Many of these facilities occupy a useful niche in the marketplace by providing low-cost access to fitness equipment and indoor facilities.

> **www. Web Search 13.1**
>
> **Commercial Facilities**
> Investigate fitness centers in your city in this activity in your online study guide.

Sales-based fitness facilities present two major barriers to improving the health and fitness of the general population: their marketing practices and sales strategies. These types of facilities often use billboards and newspaper advertisements featuring slender, muscular young models, suggesting that exercise is only for those who are already fit and healthy. The result of such advertising is that people who would benefit most from health and fitness programs—those who are older, have special health concerns, or need to lose weight—may feel discouraged from joining. Another challenge is that sales-based facilities often lure prospective members into facilities through the use of manipulative sales strategies that offer substantial membership discounts in exchange for long-term commitments. Making a long-term purchase and not having needs met can result in a negative first experience with fitness and can discourage people from getting involved with other facilities or programs in the future.

Retention-Based Facilities

➤ The credibility of the commercial fitness industry will improve when advertising begins to provide a more balanced image of health and fitness that includes all people regardless of age, race, ethnicity, body type, or current fitness level.

Selling memberships is an important aspect of any commercial facility, but retention-based facilities have a financial stake in making a sincere effort to meet the long-term needs of current members. These types of commercial facilities typically focus on delivering high-quality programs and services so that current members will continue their membership month in and month out and will be more likely to recommend the facility to others. Many in the industry refer to retention-based facilities as operating under a voluntary dues philosophy—that is, people pay dues voluntarily from month to month rather than pay one large membership fee up front.

Because retention-based facilities do not typically invest in extensive advertising, referrals from current members are essential to achieving the goal of profitability. Retention-based facilities place maximum limits on the number of memberships sold, usually based on a ratio of members to available space. If the membership base is at maximum, they engage in sales activities only to replace members who have left the club because of dissatisfaction, relocation, or job changes. The greatest emphasis in a retention-based business is always on high-quality facilities and programming that meets members' needs. Because of the additional cost of offering quality programs and hiring qualified personnel, monthly dues in retention-based facilities tend to be higher than those in sales-based facilities.

Unfortunately, retention-based facilities typically engage in less advertising than sales-based facilities and therefore limit the public's exposure to a credible and professional image of the fitness industry. Such facilities generally depend on word-of-mouth referrals to market their services, leaving the advertising market to the sales-based facilities. As a result, many people have an incomplete image of the commercial health and fitness industry.

Many of the most exciting and higher-paying job opportunities available in the health and fitness professions are in the commercial sector. Recently, retention-based facilities have expended more effort getting the word out to the public regarding the high-quality health and fitness programs that they offer. As a result, employment opportunities in this sector are likely to increase in upcoming years.

Fitness personnel in clinical settings often work with clients with specific medical conditions.

Clinical Settings

The **clinical setting,** which includes hospitals, outpatient medical facilities, and physical therapy clinics, has been one of the largest growth areas for health and fitness programs in the past decade. Today, clinical settings have clearly begun to recognize the value of providing health and fitness programs. The objective of these programs is to help patients manage medical conditions and keep the subscribers of their health insurance partners and programs healthy in order to avoid expensive medical procedures in the future. Increasingly, many of these facilities also offer memberships to the public similar to those offered by commercial clubs.

Examples of popular clinical programs are preventive screenings, physical therapy, cardiac rehabilitation, chronic disease management, water exercise therapy, childbirth and parenting education, weight management, and nutrition counseling. By definition, fitness professionals working within clinical settings will have more contact and work in partnership with both medically based professionals and clients with diagnosed medical conditions.

Current trends in the clinical setting are not unlike those in the worksite and commercial settings. Organizations in clinical settings are also building fitness facilities and offering a wide range of health and fitness programs delivered by qualified staff. In addition, these services and programs extend into other dimensions of health besides physical fitness. As an example, Eastern Connecticut Health Network provides health and wellness services reflecting the philosophy that the hospital exists not just to treat illness or injury but to help people stay well and engage in lifestyle behaviors that maintain health throughout life.

Much like worksite settings, some clinical settings have begun to develop partnerships with commercial facilities or agencies that provide expertise in the management of health and fitness facilities. This combination of commercial settings (with expertise in facility management) and clinical settings (with expertise in health promotion programming) looks promising for the future. One example is the partnership of HealthTrax and the Tully Health Center at Stamford Hospital in Stamford, Connecticut. This innovative wellness facility uses medically oriented personal improvement programs and offers health education and special programs for weight management, diabetes, stress reduction, cholesterol control, osteoporosis, and many other health risks.

DO it **Activity 13.2**

Health and Fitness Settings
Test your understanding of the four types of settings in this activity in your online study guide.

Community Setting

The **community setting** includes many types of health and fitness facilities. Local branches of nonprofit organizations such as the YMCA, the YWCA, the Jewish Community Center (JCC), and city parks and recreation departments make up one of the largest segments of the community setting. Another large segment of this setting is local health departments, state agencies, and voluntary health agencies such as the American Heart Association, the American Lung Association, and the American Diabetes Association. Community settings also include health and fitness facilities and programs that operate in churches, universities, and apartment complexes. Health and fitness programs offered in community settings may also be funded by government grants or other philanthropic donations and may be connected to specific research projects.

Typically, health and fitness programs in community settings seek to fill a specific community need. Most target a specific community group and attempt to reach as many individuals in that group as resources will allow. Many of these facilities receive outside funding and are classified as nonprofit organizations for tax purposes, allowing them to offer health and fitness programs at low cost and to provide financial assistance to individuals who could not otherwise afford to participate.

Commercial Fitness With a Medical Mission

In March 2005, Augie Nieto, the cofounder of Life Fitness, the world's largest commercial manufacturer of fitness equipment, received a diagnosis of ALS (amyotrophic lateral sclerosis, or Lou Gehrig's disease). ALS is a progressive disease that destroys the nerve cells controlling muscles, ultimately causing complete paralysis. Under the banner Augie's Quest, Augie has since dedicated his time and energy to fund-raising and research in order to find more effective treatments and, ultimately, a cure for ALS. The International Health, Racquet and Sportsclub Association (IHRSA), the trade association of high-quality commercial clubs, has assisted Augie by hosting an annual fund-raising gala at its annual conference and trade show. Many IHRSA member clubs have also participated via their own fund-raising programs. Since 2006, Augie's Quest has raised over $20 million. Through this funding and a partnership with the ALS Therapy Development Institute, scientists at ALS TDI are searching for the elusive treatment and cure. To learn more about Augie's Quest and the way in which a health and fitness community can come together in innovative ways, visit www.augiesquest.org.

The types of programs offered in community settings are as diverse as those offered in worksite, commercial, and clinical settings and might include park and recreation programs for seniors, community cardiac rehabilitation programs, swimming lessons for children, youth soccer leagues, strength training, aerobics, and yoga classes. Resources are typically more limited in the community setting, however, leading to greater reliance on volunteers to meet many staffing needs. In addition, some programs are offered on a for-profit basis as a way of generating revenue to deliver other community service programs at little or no cost to participants. Many YMCAs sell health club memberships to the public as a method of generating revenue to offset the cost of nonprofit programs such as youth summer camps or programs for individuals with special needs.

Community-based nonprofit organizations are not the only entities that offer community health and fitness projects. The mission to improve the health of the community has crept into worksite, commercial, and clinical settings in recent years. Although they do not typically operate as nonprofit entities, an increasing number of organizations in these settings have expanded their missions to include contributing to the health of the surrounding communities (see the sidebar on the previous page). The motives behind these activities are varied and can include a desire to help, improve public relations in the community, offer volunteer opportunities to employees, reduce tax exposure, or improve long-term business strength.

In summary, each of the four settings for health promotion programs has specific objectives, targets a specific population, and offers a particular menu of health and fitness programs. As a professional who may one day be seeking employment in one of these physical activity professions, you should look at each organization on its merits and not place too much emphasis on the label of worksite, commercial, clinical, or community setting. As we have shown, these settings overlap considerably, and all are rapidly expanding the scope of the health and fitness programs that they offer. According to IHRSA's *Profiles of Success* (2008), in 1982 there were just over 6000 health clubs in the United States. By 2007, that number had grown to over 29,000, and revenues were at $18.67 billion. Over 41.5 million people are members of facilities that IHRSA categorizes as commercial, not-for-profit (JCCs, YMCAs, hospital-based clubs, residential, municipal, university, and military), and miscellaneous (corporate fitness centers, resorts, spas, hotels, and country clubs) (see figure 13.3). Many exciting professional jobs are available in each setting. Your ideal job will be the one that best matches your goals as a professional.

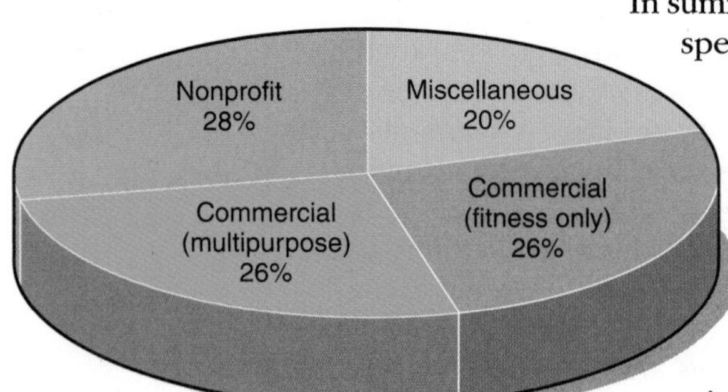

FIGURE 13.3 2005 health and fitness facility memberships (IHRSA, 2005).

Professional Roles in Health and Fitness

The types of jobs available in a health and fitness facility depend more on the specific programs being offered than on whether the facility is a worksite, commercial, clinical, or community setting. The number of organizations offering high-quality health and fitness programming in each type of setting continues to grow annually. This growth has resulted in fewer differences among the settings; many types of facilities now offer similar programs.

In this section we describe the types of jobs that are available across the four different health and fitness settings. More important, we explain how these jobs have evolved and continue to evolve in response to social forces and market demands. Note that strength and conditioning are a critical part of an overall fitness program, and many professional jobs in the health and fitness field require this expertise.

Group Exercise Instructor

Once called aerobics instructor, the position of group exercise instructor has changed greatly over the past decade. Group exercise instructors continue to lead aerobic exercise classes, but now they serve a broad population and provide instruction on a variety of activities. Group exercise instructors teach outdoor activities, aquatic fitness, and exercise classes for specific populations such as pre- and postnatal women, seniors, children, and clients who need medically supervised exercise. The aerobics studio is no longer an empty room with mirrors. Group exercise instructors use a variety of functional equipment such as steps, slides, spinning bikes, exercise bands, dumbbells, fitness and exercise balls, and barbells. In most places this position continues to be part-time and paid on an hourly basis.

The aging of the American population has brought an increasing number of older and higher-risk clients into the group exercise setting. We will discuss this in depth in the next section. For now, the important thing to realize is that what was once an aerobics studio filled exclusively with young adults performing complicated dance routines is now a studio filled with a diverse population. Group exercise instructors need to be knowledgeable and dynamic, have excellent leadership skills, and enjoy working with people in a group setting.

A recent trend in many health and fitness settings has been the emergence of yoga, tai chi, and Pilates. These activities are often combined to form a broader mind–body approach to physical activity. The Pilates programs available in most settings are vastly different from the form taught by originator Joseph Pilates. Although many instructors have formal training, other practitioners now have backgrounds in a variety of physical therapeutic, massage, and chiropractic techniques. Much Pilates training occurs in a one-on-one or very small group setting. Training usually involves use of highly specialized apparatus and dedicated studio space. But given the culture of group exercise programming in many health and fitness settings, larger-group Pilates-inspired activities that use mats, core boards, rotator disks, and resistance rings rather than large specialized equipment are now common. While Pilates is a relatively modern movement form, yoga has been around for centuries. Physically demanding, yoga is an inward practice meant to unite the body, mind, and spirit. According to Jason Crandell, editorial advisory board member of *Yoga Journal,* most American practitioners gravitate toward hatha yoga, the physical practice of postures (asanas), breathing exercises

Exploring New Careers

As you read this section, compare the job descriptions to the list that you created in activity 13.1. What positions can you add to your list? What job duties are different from those that you imagined? We hope that you will become as excited as we are about the numerous possibilities that exist for health and fitness professionals.

(pranayama), and relaxation techniques (http://www.sfbayclub.com/web/site/benefits/programs/yoga). However, as an ancient movement form, there are many practices throughout the world. While there are no national standardized educational requirements for teaching yoga, training with established and respected gurus, often through generational lineage and affiliation, is common. A review of the popular *Yoga Journal* will lead you to numerous organizations offering workshops for beginning, intermediate, and advanced practitioners.

Fitness Instructor

The fitness instructor position is predominantly an entry-level position for the fitness professional with a bachelor's degree in kinesiology or other discipline related to health and fitness. Fitness instructors work primarily with apparently healthy adults, although as the population ages, it is becoming more common for people with some functional limitations to seek assistance in physical activity programs. Job duties typically include conducting fitness and functional assessments and designing comprehensive individualized exercise programs that incorporate strength, flexibility, and aerobic fitness components and maximize safety and long-term results. This exciting job provides an opportunity to introduce many types of people to the benefits of exercise. For many sedentary adults who are joining their first health and fitness facility, fitness instructors are the initial contact person who can either excite or discourage them in their endeavor to become more healthy and fit.

Fitness programs for children—of any age—should focus on play, skill development, and energy expenditure for maintaining a healthy body weight.
Brand X Pictures

Most reputable health and fitness facilities now provide some type of health screening or fitness assessment for new participants. The fitness instructor is responsible for administering assessments of such things as blood pressure, body composition, aerobic capacity, muscular strength, muscular endurance, and flexibility. The results of these assessments are then used to design individualized programs that meet the client's specific health and fitness needs. Some fitness instructors stay in touch with participants as they progress with their programs. In other facilities, other professionals such as personal trainers or health and fitness counselors provide follow-up appointments. The fitness instructor position can be either part-time or full-time. Compensation is usually an hourly wage.

With the rise in childhood obesity rates, some fitness facilities have begun to offer programs targeted specifically at children and adolescents. These programs center on the unique needs of children, focusing on play while emphasizing a level of movement and activity that allows for conditioning, caloric expenditure, and skill development. Fitness professionals who also have training in elementary school physical education should find an increasing number of job opportunities across all types of health and fitness settings.

Health and Fitness Counselor

The health and fitness counselor position represents an exciting growth area in the health and fitness professions. In some facilities this position has replaced that of the fitness instructor. In others, it has led to the creation of a new category of employee called a wellness coach or lifestyle coach. The health and fitness counselor provides guidance to a diverse population on a broad range of health and fitness topics. Many of these jobs are full-time, salaried positions and include benefits.

Health and fitness counselors have job descriptions that are consistent with O'Donnell's (1989) definition of health promotion: "the art and science of helping people change their lifestyle to move toward a state of optimal health" (p. 5). As a result, job duties are broad and include such things as working with clients on behavior change, stress management, relaxation techniques, time management, smoking cessation, social participation, and weight management, besides prescribing exercise programs.

DO it Activity 13.3

What's My Profession?
Guess which profession is being described in this game in your online study guide.

Health screening and health and fitness assessment are important skills for health and fitness counselors, but counselors view these only as tools to assist them in working with clients. They do not apply these techniques universally to all clients during an initial appointment, as those in the fitness instructor position traditionally have. Instead, health and fitness counselors assess clients' readiness for change (Prochaska & DiClemente 1986), help them set short- and long-term goals, and assist them in moving at an appropriate pace toward a state of optimal health.

Sutter Health Partners, a division of Sutter Health Network, runs an extensive Live Well for Life program where wellness coaches work on lifestyle solutions for clients, patients, and employees who have specific risk factors such as diabetes, heart disease, high cholesterol, hypertension, and obesity. Sutter Health's wellness coaches have a myriad of undergraduate and graduate backgrounds in fields such as applied behavioral science, exercise and sport science, leisure and recreation, nutrition, gerontology, health education, business, sports medicine, health science, industrial and organizational psychology, and human development. Many have career training as nurse practitioners, registered dietitians, and certified personal trainers and group exercise instructors. With an emphasis on optimal health, the customized programs they design promote physical, emotional, intellectual, social, and spiritual health. As such, wellness coaches have an array of counseling skills to help individuals understand personal behavior and health risk factors, discuss obstacles, review lifelong habits, and provide personal guidance and unwavering encouragement to make lifelong changes.

Although educational credentials and certifications, along with constant continuing education, are essential for the health and fitness counselor, equally important is the ability to provide counseling on a broad range of health topics. Health and fitness counselors need to be organized, understand principles of behavior change, have excellent communication skills, and be effective in marketing and promoting programs and services. The health and fitness counselor position is increasingly becoming a salaried rather than hourly position.

Personal Trainer

Personal trainers do many of the activities of both the fitness instructor and the health and fitness counselor, but they typically provide these services on a fee-for-service basis. Some personal trainers work independently, traveling to individual clients' homes, conducting training sessions outdoors, or sub-contracting with a private gym. They can be paid hourly, weekly, or monthly. More typically, a facility employs personal trainers, and trainers market their services directly to members. Most facilities pay personal trainers on an hourly fee-for-service basis, at a rate significantly higher than the hourly rates for fitness instructors or health and fitness counselors. An increasing trend is the classification of personal trainers (e.g., from trainer to master trainer) based on skill, experience, sales volume, education, and certifications held. Usually, experienced trainers are able to charge higher rates for their services and therefore earn higher rates of pay from their facilities. Personal trainers are often substantially compensated in comparison with other health and fitness professionals.

The personal trainer position provides a unique opportunity for the health and fitness professional to develop an ongoing relationship with specific clients. This relationship allows the trainer to develop a clear picture of individual clients' needs and quickly implement new strategies for behavior change. Personal trainers work with clients on a number of health issues such as weight management, stress management, physical fitness, and sport conditioning. Current trends in this position include specialization in working with specific populations such as medically based clients, older adults, and children, as well as working toward specific client goals such as injury rehabilitation, running a 10K race, or completing a triathlon. Increasingly, personal trainers are starting to cater to these market segments by offering small-group specialty fitness programs as opposed to just individual services. In

Interviews With Practicing Professionals

Jennifer Beaton, MS
Vice President of Fitness, Western Athletic Clubs, Inc.

Q: You are currently vice president of fitness for Western Athletic Clubs, a San Francisco–based company that owns and operates sport resorts and urban health clubs. What are your responsibilities?

A: I oversee the operations, fiscal and capital budgets, P & L (profit and loss), and strategic development for personal training, Pilates, yoga, group exercise, and specialty fitness programs for all eight of Western Athletic Club's properties. I am responsible for achieving $10 million in yearly revenue generation and overseeing a $15 million fitness operations budget.

Q: What background do you have that has enabled you to move into senior management?

A: It really is a mix of academic, business, and practical training. I received a bachelor's degree in exercise science from Creighton University in 1996 and a master's degree in exercise and wellness from Arizona State University in 1998. I am a group exercise instructor and personal trainer, certified by ACE, NASM, and NSCA. I am also a fellow of the Grey Institute for Functional Transformation (GIFT). I have worked in cardiac rehabilitation at Creighton University Cardiac Center and pediatric endocrinology and diabetes at the University of Nebraska Medical Center. I have been a fitness and wellness director at the Native American Health Center in Oakland, the University of California at Irvine, Arizona State University, the Decathlon Club in Santa Clara, and the San Francisco Bay Club. Most recently, I was the director of Optimum Health for Western Athletic Clubs. This was an upstream health care business in that we had a team of doctors, therapists, nutritionists, and personal trainers all working together with a patient to combat conditions such as metabolic syndrome, cardiovascular disease, diabetes, and cancer.

Q: Given that you have worked in commercial, clinical, and community health and fitness settings, what advice do you have for undergraduate students in kinesiology?

A: First, take every opportunity given to you. To move up in management, you need to have a variety of experiences. Second, work hard—work really hard. Within any industry, you have to put in the time, the passion, and a lot of effort. Third, become really good at networking. A large part of moving up is through professional and social networks. Fourth, never stop learning. Actively seek out new information, people to debate with, and ways to practice what you learn. It will keep you excited about what you do. Fifth, don't be afraid to take risks and love what you do. If you find something you are really passionate about, your energy will be contagious. You need to wake up every morning and be inspired to go to work.

Q: What message do you have for prospective employees?

A: This industry is increasingly moving to a performance-based model. People are seeing smaller base salaries with huge upsides based on performance. This new model really forces people to perform rather than become complacent. The world is much more competitive now. There is more competition for clubs to find members, for trainers to find clients, and so on. Also, on the personal training side of the world, the expected caliber of trainers is increasing. People are demanding trainers who are well versed in all dimensions of wellness.

Personal trainers develop relationships with clients, which helps them develop new strategies to help the clients adopt or maintain healthy behaviors.
iStockPhoto/Ron Chapple Stock

this regard, specialty fitness programs bridge the gap between one-on-one personal training and group exercise. Programs are for small groups (2-6 people) for a defined time and usually focus on a specific goal (e.g., weight loss). They have gained in popularity because they allow people access to personalized training within a small group, thus providing a price point that is more appealing and accessible to a wider population (Jennifer Beaton, personal communication, October 17, 2009).

Personal trainers need a variety of skills in order to be successful. Besides having a strong knowledge base in fitness assessment and exercise prescription, personal trainers must be excellent communicators and counselors; have strong business skills in sales, marketing, administration, and time management; and be effective in working with a diverse client population. Personal trainers are the artists of the health and fitness professions. Besides evaluating individual client needs and designing individualized programs, personal trainers must continually present new and exciting ways to keep clients on track, interested, motivated, and excited about making changes and maintaining new health behaviors as part of their lifestyle.

Over the past decade, the public has become more knowledgeable about the types of credentials and certifications that are appropriate for the personal trainer position. Simply being fit is no longer an acceptable qualification. Personal trainers need to make the same effort that health and fitness counselors do in pursuing an interdisciplinary education.

Personal trainers are rarely hired directly from bachelor's degree programs unless they also have extensive experience working with clients on a variety of health and fitness issues. For this reason, an aspiring personal trainer will likely begin a career as a fitness instructor or health and fitness counselor and move into a personal trainer position after gaining considerable additional experience and a specialized certification. Personal trainers keep skills current by attending conferences, workshops, and other continuing education programs.

Specialist

Many health and fitness settings are now incorporating specialist positions into their programs, including the following (see also chapter 14):

- Clinical exercise physiologist
- Physical therapist
- Athletic trainer
- Sport coach
- Registered dietitian
- Population specialist (e.g., older adults, children, those with chronic disease)

Many specialist positions require graduate-level studies, certifications, or licensure along with a bachelor's degree in kinesiology or another health- and fitness-related discipline. One innovative program offered by the extended education division of California State University at Fullerton is a certificate in gerontology. This professional certification program provides 16 units of practical courses in topics related to aging and four units of field experience. Specialist positions provide services that go beyond those provided by health and fitness instructors, counselors, or personal trainers. These jobs can be exciting and challenging and allow for increased earning potential.

What are the differences between health and fitness professionals and specialists? Consider the provision of nutrition information to clients in a health and fitness facility. Health and fitness counselors or personal trainers are often called on to provide basic nutrition advice to healthy adult clients. Counselors or trainers would not be qualified, however, to prescribe a specific type of diet or supplement or make recommendations to individuals who have health risks or specific health conditions. Only registered dietitians have the appropriate qualifications to make such recommendations. Attaining that status requires additional course work beyond what the usual kinesiology department offers.

Injury rehabilitation is another example of the boundaries between various professionals and specialists in a health and fitness facility. Personal trainers are required to obtain a full injury history on each client and to consider this information when designing an exercise program. An appropriately designed exercise program can help people prevent injuries in the future. A personal trainer, however, should not design a specific program for specific injury rehabilitation. Physical therapists are the specialists with the appropriate knowledge and qualifications to provide injury rehabilitation services to clients. In an ideal scenario, a personal trainer would work in partnership with a physical therapist or registered dietitian. The trainer may supervise specific aspects of a program and communicate with the specialist regarding the client's progress.

As mentioned earlier, most specialist positions require graduate-level education in addition to a bachelor's degree in kinesiology or another health- and fitness-related discipline. If you are interested in pursuing a specialist position, check with your academic advisor to find out how you can best prepare for an appropriate graduate program.

Health and Fitness Director

➤ Health and fitness professionals must understand the ways in which social and cultural forces affect their industry.

The health and fitness director is a key position in most health and fitness settings. This individual is typically responsible for managing the health and fitness services and programs that take place in a facility. In larger facilities, the health and fitness director may be part of the general management team and supervise a team of managers who handle specific responsibilities within departments. Examples of these management positions include group exercise director, yoga director, Pilates director, personal training director, aquatics director, athletic and recreation director, and youth activities director. You should consider all these management positions as job opportunities for health and fitness professionals.

The job description for a health and fitness director position includes hiring, training, and providing support for a diverse staff of group exercise instructors, health and fitness counselors, personal trainers, specialists, and equipment maintenance personnel. In addition, health and fitness directors are responsible for business planning, managing budgets, planning facility renovations, selecting equipment, designing and marketing programs, and forecasting future trends in health and fitness programs. The health and fitness director must have a broad, interdisciplinary education that includes business planning, budgeting, marketing, and staff management besides a strong foundation in kinesiology.

Currently, educational requirements for the health and fitness director position emphasize academic degrees more than certifications. A degree in kinesiology is especially appropriate for students aspiring to be health and fitness directors. Many facilities require health and fitness directors to have a graduate degree in kinesiology or another health- and fitness-related discipline. Other requirements include experience in a position such as health and fitness counselor or personal trainer. Through experience in such positions, directors gain a full understanding of individualized program design and strategies for successful behavior change. In addition, these job experiences provide valuable insight into the management responsibilities of hiring, training, and supporting staff.

Health and fitness directors need to be effective leaders. Besides managing day-to-day operations, these individuals must be capable of articulating a vision and motivating other individuals to work toward specific goals. The health and fitness director is both a visionary who keeps current with the field of health and fitness and a mentor who supports staff members in moving toward individual goals and aspirations.

Marketplace Trends and Opportunities

Across every industry, the best leaders are those who do extensive reading in many fields and keep up to date on social and cultural trends and the latest research in their discipline. They continually ask questions about how the world around them affects their particular organization. Now that you have an understanding of the various positions available in the health and fitness professions, you should broaden your view by adding an understanding of the trends that affect these professions.

➤ Health and fitness programs increasingly address other dimensions of wellness besides physical fitness, drawing from a multidimensional model that includes physical, intellectual, emotional, social, and spiritual well-being.

As a future health and fitness professional, you should be aware of the ever-changing context in which these professions are immersed. This is a dynamic time in the health and fitness field; personal, social, cultural, and political issues have a powerful influence on the way we view health and fitness—and on the way we work in the profession. Although it is beyond the scope of this chapter to outline all these forces, we will deal with three major issues currently influencing these professions: increasing interest in a multidimensional model of wellness, health care reform, and emerging demographic trends.

DO it **Activity 13.4**

Dimensions of Wellness

In this activity in your online study guide, you'll test your understanding of the multidimensional model.

Multidimensional Model of Wellness

A trend across all health and fitness settings is the expansion of program offerings to address other dimensions of wellness in addition to physical fitness. The multidimensional model of wellness has five dimensions: (1) physical—the ability to conduct daily tasks with vigor; (2) emotional—the ability to control emotions and express them appropriately and comfortably; (3) intellectual—the ability to learn, grow from experience, and use intellectual capabilities; (4) spiritual—a guiding sense of meaning or value in life that may involve belief in some unifying or universal force; and (5) social—the ability to have satisfying interpersonal relationships and interactions with others (Dintiman

& Greenberg 1986) (see figure 13.4). Our previous description of clinical health and fitness settings included an example of a hospital group that offers a wide variety of wellness programs and services. Besides offering exercise programs, this hospital provides classes on topics such as meditation, parenting, and self-esteem. In our description of worksite settings, we discussed ways in which worksite programs have evolved over time to provide services that not only assist employees in making changes but also support them in maintaining healthy lifestyle behaviors. Besides offering employees state-of-the-art fitness facilities, many worksites now offer things such as family activities, volunteer service opportunities, and programs for facilitating personal growth and improving communication skills.

As facilities expand their program offerings, they will also expand their job descriptions. If you elect to focus exclusively on the physical dimension of wellness, you will continue to find many exciting job opportunities. But if you embrace the challenge of delivering services that address other dimensions of wellness, you are likely to find even more job opportunities.

As we have shown in our description of various job categories, many health and fitness positions have already expanded to involve a broader range of responsibilities than in the past. Health and fitness counselors and personal trainers continue to design exercise programs but also assist clients with issues such as stress management, personal growth, and involvement in social activities. In short, industry leaders in all settings are looking for bright people who can deliver a broad range of services and integrate concepts of exercise and fitness into a broader definition of health.

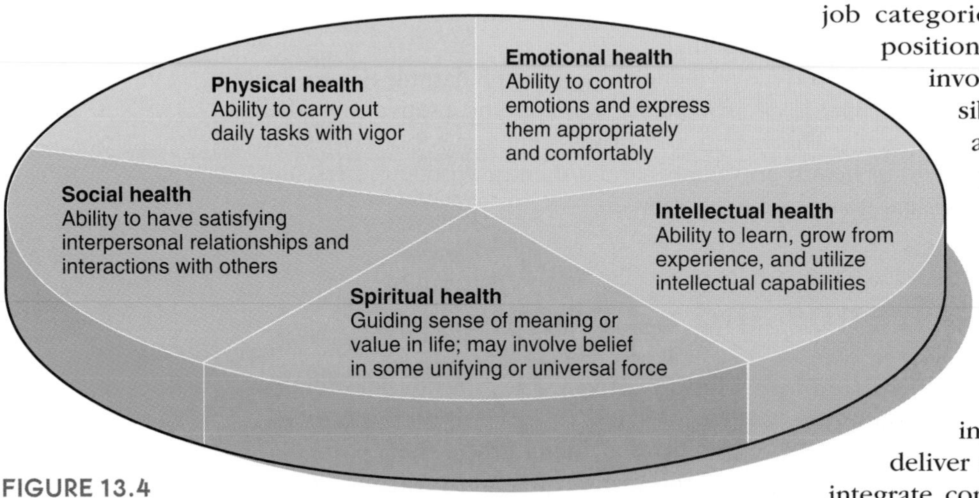

FIGURE 13.4 Dimensions of wellness.

A Multidimensional Approach to Fitness and Healing

The Santa Barbara Athletic Club was well ahead of its time when it created its Cancer Well-Fit program in 1994. Since that time, over 1,000 cancer survivors have enrolled in their program, and more than 66% of the participants believe that participating in the exercise-based program is part of a much broader multidimensional wellness and social network (Santa Barbara Athletic Club Cancer Well-Fit 2009). Similarly, from 2003 to 2008 Western Athletic Clubs (WAC) funded an exercise-based cancer treatment program called Integrating Medical Professionals and Certified Trainers (IMPACT) as part of its corporate philanthropic initiatives. The program was a unique partnership among WAC, the Golden Gate Center for Integrative Cancer Care at the University of California at San Francisco (UCSF), California Cancer Care in Redwood City and Marin, and the Bucholz Medical Group in Mountain View. Through IMPACT, WAC personal fitness trainers worked directly with participating oncologists and their patients, both at WAC facilities and the medical office. The program had both a 6-month treatment phase and a follow-up 6-month survivor phase (see Howell and Bulmer, 2007).

Health Care Reform

As mentioned earlier in the chapter, health care costs have skyrocketed over the past decade. Here is an example on how the payment system for insured patients works: Mike, a long-time employee at his company, feels a pain in his right knee. As an employee, Mike has the bulk of his monthly family health care insurance paid for by his company. He (the consumer) pays the difference by having money taken out of his paycheck before it is taxed every month. Because of his knee pain, Mike takes a sick day from work and is treated by a doctor at the local hospital. The provider (the doctor) then submits a claim to a third-party payor (a national insurance plan such as Aetna or a regional plan such as Blue Cross/Blue Shield). The payor processes the claim based on the patient's insurance coverage and also the payor's contract with the provider. Once the claim is approved, the payor sends its portion of the claim to the provider. Based on the amount covered by the payor, the provider then determines how much is owed by the patient. Subsequently, Mike receives a bill from the provider (doctor) that is the total cost of the treatment minus the amount covered by the payor (insurance plan). It is then Mike's (the consumer's) responsibility to pay the provider.

In describing this process, Finn, Pellathy, and Sinhal (2009) report that in 2007, estimates show that employers paid payors (insurance companies) $520 billion for health care insurance for their employees. Consumers, either as part of their workplace coverage or through a personal insurance plan, paid payors $250 billion. Consumers also paid $265 billion directly to providers (physicians, hospitals, and pharmacies). The U.S. government paid payors $820 billion. Having collected all of this insurance money, payors paid providers $1,415 billion for their contracted services. This leaves payors with approximately $175 billion for administrative costs and profits.

Obviously, there are enormous financial transactions in this system. Physicians, hospitals, and laboratories spend about $100 billion or more a year on managing submission and processing of claims. There is also about $45 to 65 billion in bad debt across the industry. According to the Centers for Medicare and Medicaid Services, Office of the Actuary, and the National Health Statistics Group, these expenses will continue to soar without some type of reform (see figure 13.5). Add to that the estimated 47 million uninsured Americans, and it is clear that reform will have to be significant and address health system redesign, coverage expansion, reimbursement reform, and insurance market reform (Finn et al. 2009).

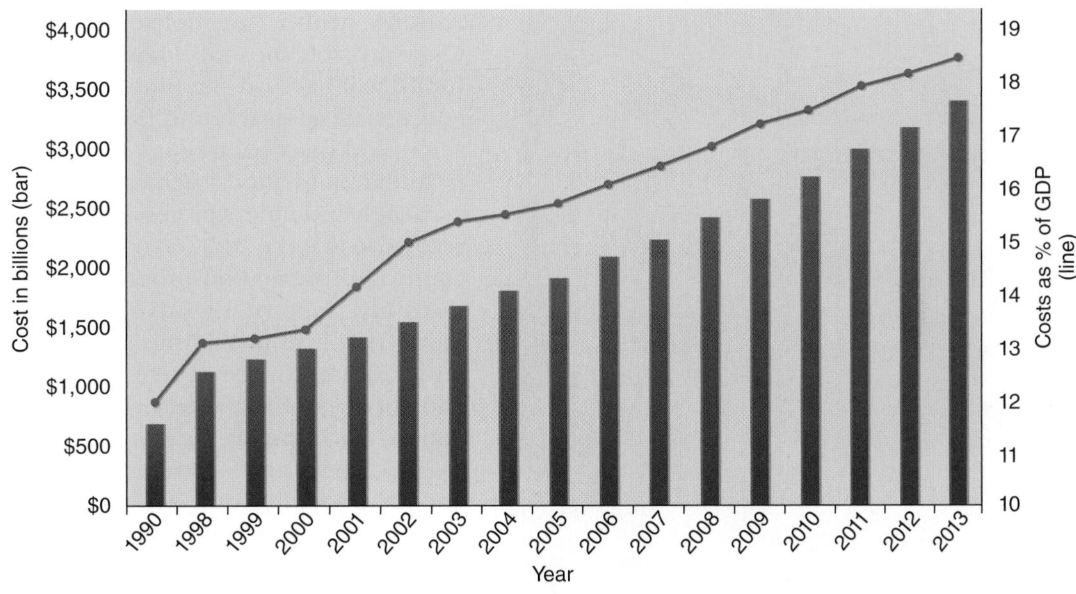

FIGURE 13.5 Projected increases in health care costs in the United States.

From Centers for Medicare and Medicaid Services, Office of Actuary.

It might be said that the best way to control health care costs is to stay healthy, and a good way to stay healthy is to adopt and maintain healthy lifestyle behaviors, including daily physical activity. But both the rising cost of medical care and the manner in which health care is paid for in the United States have slowed the progress toward offering more preventive services. Health care organizations are so busy developing cost-effective ways to treat people who are sick that they seem unable to focus on the development of better strategies for prevention. In addition, when employees leave their jobs, they tend to change health insurance providers. Health insurance providers will be motivated to invest money in preventive services only when they are confident that they will reap the benefits in future cost savings.

However, there is clearly a move in the right direction. Many health plans now offer an extensive menu of health and fitness programs free of charge to their members. In some cases, this effort has the goal of improving the overall health of potential subscribers in a geographic area. In other cases, it is a marketing strategy aimed at increasing the number of people who enroll in the health plan. Blue Shield of California was one of the first health plans to actively promote healthy lifestyles as part of their business model. Their online Healthy Lifestyle Rewards program helps people adopt healthy lifestyle choices related to diet, stress, and physical activity.

The previous discussion on health fitness counselors covers the role of wellness coaches in the Sutter Health Partners Program. Sutter Health, one of the nation's leading not-for-profit networks of community-based physicians, hospitals, and other health care providers, delivers high-quality care in more than 100 Northern California communities.

As another example, American Specialty Health (the largest network of its kind in the United States) provides wellness, fitness, and specialty managed care programs to health plans, insurance carriers, and employers. Their network provides lifestyle coaching, incentive programs, online wellness information, personal health risk assessments, exercise and walking programs, home fitness programs, worksite programs, Internet-based programs, and discounted access to alternative health care providers (e.g., chiropractors, acupuncturists, massage therapists, dietitians, and naturopathic doctors). This direct intervention by health networks and health plan providers certainly seems to be the wave of the future. For the first time, physical activity is increasingly seen as an integral component of the nation's health care delivery system. Quite possibly the model hospital of the future will have on-site fitness facilities and a staff of health and fitness counselors and personal trainers in addition to traditional medical personnel.

Aging baby boomers represent a fast-growing segment of health club membership in the United States.

Monkey Business/fotolia.com

Health and fitness professionals must recognize that the addition of insurance-sponsored prevention programs is not the only answer. Creative solutions mean not just providing individuals with access to preventive programs and services but also ensuring participation. Encouraging people to adhere to good diets and other health habits and to adopt physically active lifestyles is just the beginning. Developing and implementing effective outcome-driven programs will be the key.

Demographics

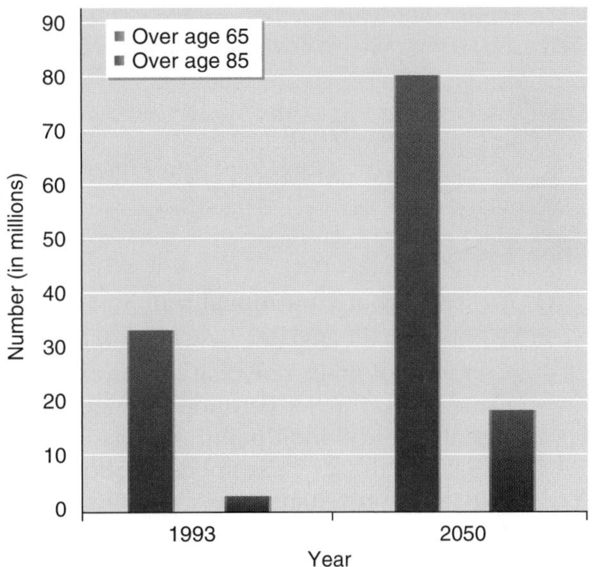

FIGURE 13.6 Aging of the American population.

Demographic trends are radically redefining the meaning of healthy lifestyles and healthy aging. According to the U.S. Census Bureau's data (2009), the median age of the total U.S. population increased from 23 years in 1900 to 30 years in 1950, then to 35.3 years in 2000 and to 36.8 in 2008. Driving this increase is the aging of the baby boom generation, the 77 million people born in the United States between 1946 and 1964. For the past quarter century, this generation has had a huge effect on the marketplace because of its sheer numbers.

This "age wave" (Dychtwald 1990) raises some interesting questions. By 2011, the first baby boomers will reach age 65. According to the U.S. Bureau of the Census (2009), in 2030, when all of the baby boomers have turned 65, about 20% of the population will be 65 or older. By 2050 this age group is expected to increase to 88.5 million, more than doubling the 2008 number of 38.7 million (figure 13.6). What does this aging of America mean for the health and fitness fields? What does it mean when people say that baby boomers will not just grow older chronologically but will age youthfully? Will middle age be reinvented? And what will happen when future customers, clients, and patients retire at 65 years of age and then have another 25 years of living to do? Consider the fact that the number of people over age 85 is expected to more than triple from 5.4 million in 2008 to 9 million in 2030 to 19 million in 2050. These demographic shifts will surely challenge our definition of physical activity as it relates to sport, competition, adventure, movement, pleasure, and health.

According to IHRSA (2008), people over the age of 55 represent the fastest-growing health club membership segment in the United States, having increased in health club membership by 411% since 1990.

In addition, many job opportunities are likely to arise because of the demographic patterns shaping our country. We will all witness a far more diverse nation in the coming years. By 2050, approximately 50% of the population will be non-Hispanic white; 16% will be black; 23% will be of Hispanic origin; 10% will be Asian and Pacific Islander; and about 1% will be American Indian, Eskimo, and Aleut. By 2015 and 2050, the Hispanic-origin population will double and quadruple its 1990 size, respectively (USBOC 2000).

Such trends are having an enormous influence on the way many American health and fitness businesses are developing their strategic plans for the next decade. Many settings are already taking advantage of this demographic shift in their hiring and programming practices. It is worth noting that new kinesiology graduates will likely be working in multi-generational work teams as the trend in new "encore careers" for individuals who are in later stages of their professional life continues.

Certification and Continuing Education

To be hired for a position in the health and fitness professions, you will need to fulfill specific requirements in your education and training. In addition, once you are hired, you will need to continue seeking information and developing new skills, because the health and fitness field is dynamic, and new information is continually becoming available.

Interviews With Practicing Professionals

Peter Toth
Owner, Apex Health and Fitness

Q. Describe your educational training, certifications, and job experiences.

A. I received my BS degree in exercise science human performance from Southern Connecticut State University, where I am currently completing my MS degree in the same field. I am a certified strength and conditioning specialist with the National Strength and Conditioning Association (NSCA) as well as a certified personal trainer and corrective exercise specialist, both with the National Academy of Sports Medicine (NASM). I started my career as an exercise physiologist at St. Vincent's Medical Center in Bridgeport, Connecticut, where I worked with postoperative cardiac patients. I also began to train clients at The Edge Fitness Clubs in Stratford, Connecticut. Once I established myself as a successful personal trainer, I developed my own personal training company and discontinued my work with cardiac patients. Today I specialize in weight reduction and lifestyle modification programs for the overweight and obese.

Q. Which aspects of your education and training have been most helpful for achieving success as a personal trainer?

A. Success as a personal trainer is defined by one thing: results. The only way to help your clients get those results is by understanding how the body works and how to manipulate it. While my university course work taught me the science behind how the human body moves and functions, it was my professional certifications that taught me how to create individualized fitness programs for people.

Q. What advice would you give to undergraduate students who are interested in pursuing careers in personal training?

A. Personal training is an extremely competitive field that takes an incredible amount of hard work, dedication, and time in order for success to happen. When you decide to become a personal trainer, you do not start off with 40 hours of sessions per week. You need to build a clientele and work your way toward a full schedule. Above all, become a personal trainer because you have a genuine passion for helping others.

Q. How do you keep yourself current in your profession?

A. The best way to stay current in any profession is to read. New equipment, supplements, exercises, and research are always being discovered.

Q. What industry trends have you observed in recent years, and have you taken any specific actions in your business to respond to these trends?

A. I have been incorporating kettle bell training into clients' routines for a while now, but due to the surge in popularity over the last year or so, I now also teach group kettle bell classes. To lift and control a kettle bell, the user really has to focus on proper form and technique. The movements are dynamic and explosive, and they require the simultaneous use of multiple muscle groups.

A. What other insights might be helpful to current undergraduate students who are pursuing a degree in kinesiology?

Q. Consider a minor in nutrition, business, marketing, or communications. Helping clients understand and practice proper nutrition is what elicits the greatest results. Enrolling in business and marketing courses will familiarize you with the tools for establishing a successful company or becoming successful in an already-established company. Finally, and I cannot emphasize this enough, know how to communicate and talk to people—a lot of people. It does not matter how knowledgeable you are in this field if you are unable to teach another person.

Certification

Investigate national certification programs in this activity in your online study guide.

Typically, corporate, commercial, and community organizations prefer to hire health and fitness instructors, health and fitness counselors, and personal trainers with a minimum of a bachelor's degree in kinesiology or a health- and fitness-related discipline. Indeed, many organizations require this degree. This is a positive move because club members, employees, and patients are demanding work with high-caliber educated professionals who have extensive knowledge of health and fitness. However, many talented professionals remain in the field who do not have kinesiology degrees but do have extensive work experience and a great deal of continuing education and practical training. In either case, most employers now also require individuals to have a certification from a nationally recognized organization.

The most common certifications are from the American College of Sports Medicine (ACSM), American Council on Exercise (ACE), National Academy of Sports Medicine (NASM), National Strength and Conditioning Association (NSCA), National Federation of Professional Trainers (NFPT), Cooper Institute (CI), and National Council on Strength and Fitness (NCSF). Each of these organizations has received third-party accreditation of its certification procedures and practices from the National Commission for Certifying Agencies (NCCA). A complete description of the accredited certifications and other programs offered by these organizations can found in Web search 13.2 in the online study guide.

A certification from Personal Training Academy Global (PTA Global, www.ptaglobal.com) has emerged that is gathering interest and traction from the commercial club industry. What makes this certification unique is that it is jointly designed, developed, and delivered by many of the best-known continuing education specialists in the United States, England, and Australia. PTA Global focuses on long-term career development as opposed to single certification examinations.

There are also several educational certificates that many health and fitness organizations recognize when hiring and promoting employees. The Gray Institute of Functional Training (GIFT) fellowship is a unique 10-month program in applied functional science at the Gray Institute in Adrian, Michigan (www.grayinstitute.com). In a strategic partnership with the American College of Sports Medicine (ACSM), Wellcoaches Corporation is attempting to set the gold standard for the newer profession of wellness and lifestyle coach (www.wellcoaches.com).

Needless to say, you must do your research when applying for health and fitness positions so that you will know what educational certifications and programs specific organizations require, recommend, and support. Another major benefit of becoming certified is that certification organizations require you to continue your education by taking recognized continuing education units (CEUs) in order to maintain that certification. It is professionally essential to keep up to date with new scientific material long after the completion of your undergraduate degree.

Joining a professional association is an important first step. Many associations offer discounted membership rates for students; provide journal subscriptions, educational materials, and training; and sponsor regional and national conferences. Important organizations in this profession include those listed previously as well as the American Alliance for Health, Physical Education, Recreation and Dance (AAHPERD), National Wellness Association (NWA), American Public Health Association (APHA), Society for Public Health Education (SOPHE), and Medical Wellness Association. In addition, IHRSA provides specific information on the commercial health and fitness field. New Internet-based companies also offer valuable networking opportunities for professionals in the field (e.g., www.ptonthenet.com). Finally, various governmental agencies within the Department of Health and Human Services (www.hhs.gov) provide a wealth of health data, materials, and resources for health and fitness. Information on how to contact these organizations to find out about membership, publications, workshops, and regional and national conferences is listed on page 376.

Checking Skills and Competencies Required in the Health and Fitness Profession

Take one final look at the list of job titles, job duties, and required skills and competencies that you created in activity 13.1. Now compare your list to that provided in the table of job descriptions. How did your list of knowledge and skills compare? Although this list is by no means complete, it illustrates that health and fitness professionals must have a broad range of skills and competencies.

Job title	Evolution	Job duties	Skills and competencies
Group exercise instructor	This position no longer exclusively involves teaching aerobic dance classes. The group exercise instructor now teaches a broad range of classes to a diverse population of individuals.	Leads group exercise classes for various population groups including • seniors, • children, • pre- and postnatal clients, and • medically based clients.	Musts: • Certification by a nationally recognized organization • Strong teaching skills Preferred: • Bachelor's degree in a health- and fitness-related field (e.g., kinesiology, physical education, or health education) Helpful: • Additional certifications in a specialized area
Health and fitness counselor	Having evolved from the more traditional fitness instructor position, the health and fitness counselor provides counseling on a broad range of health topics in addition to conducting fitness assessments and designing exercise programs.	Provides guidance to a diverse population of individuals in areas such as • behavior change, • stress management, • smoking cessation, • social participation, • weight management, and • exercise programming.	Musts: • Bachelor's or graduate degree in a health- and fitness-related field • Certification by a nationally recognized organization • Skills in counseling, behavior change, cultural diversity, and teaching Preferred: • Marketing and promotional skills
Personal trainer	This position has changed from exclusively providing individualized exercise programs to providing individualized services on a broad range of health topics.	Provides ongoing support and guidance to a diverse population of individual clients on topics such as • physical fitness, • weight management, • stress management, and • sport conditioning.	Musts: • Certification by a nationally accredited organization • Counseling and teaching skills • Business, marketing, sales, and promotion training • Background in exercise programming Preferred: • Bachelor's or graduate degree in a health- and fitness-related field

Job title	Evolution	Job duties	Skills and competencies
Specialist	Number and types of positions have expanded in recent years and moved from exclusively clinical settings to a variety of health and fitness settings.	Provides specialized health and fitness services to clients with special needs. Examples include • physical therapist, • certified health educator, • population specialist, and • sport coach.	Musts: • Graduate degree • Other specific experiences and skills required for each type of specialist position Preferred: • Additional certifications and licensure—may be required to practice in specific states
Health and fitness director	This position has evolved from a fitness instructor who had additional administrative responsibilities to a full-time manager who actively participates in all aspects of personnel and facility management.	Manages all aspects of a health and fitness department. Responsibilities include • departmental leadership, • staff management, • programming, and • all aspects of business administration.	Musts: • Bachelor's or graduate degree in a health- and fitness-related field • Additional skills in business administration, management, marketing, and promotion Preferred: • Previous experience in an entry-level health and fitness position

Advice for Health and Fitness Students

This is an exciting time for the health and fitness professions. An incredible transformation is currently under way as we move toward a broader understanding of health and fitness services. We are beginning to view physical activity as a key component in the health care delivery system. We are also experiencing a considerable demographic shift in population that challenges us to work with increasingly diverse groups of people.

Today's health and fitness professional needs to obtain a broad education across several scientific, behavioral, and humanity-based disciplines (see the job descriptions in the "Checking Skills and Competencies Required in the Health and Fitness Profession" table on p. 374). To develop the necessary skills and competencies, you should seek a combined degree in kinesiology and health with courses in teaching; behavior change psychology; communication; gerontology; marketing; and the sociology of race, gender, and ethnicity. Course work in theoretical kinesiology (part II of this text) will be extremely valuable, and other specialized health- and fitness-related courses will round out your professional practice knowledge. You must also gain practical experience counseling clients on a variety of health and fitness topics and be able to relate to a diverse population of clients from different age groups and ethnic backgrounds. Involve yourself in a health and fitness job setting through an internship, cooperative education experience, or work-study program to keep up to date on what is happening in the field.

When you graduate from college, you will find yourself competing for jobs with people who have extensive practical experiences and insightful life experiences. For this reason you should be sure to obtain practical skills as part of your education. In addition, be sure to read the relevant health and fitness journals and industry publications. Most of these publications are available through the associations listed in this chapter. Many are available to

Professional Health and Fitness Membership Associations

The following organizations provide membership opportunities for the purpose of advancing health and fitness professions.

American Alliance for Health, Physical Education, Recreation and Dance (AAHPERD)
1900 Association Dr.
Reston, VA 20191-1598
800-213-7193
www.aahperd.org

Member benefits: additional membership in two of AAHPERD's national associations; research journals; newsletters; district associations; advocacy; reduced fees for conventions and seminars; networking; career services; reduced rates for health, auto, liability, and life insurance

American College of Sports Medicine (ACSM)
P.O. Box 1440
Indianapolis, IN 46206-1440
317-637-9200
www.acsm.org

Member benefits: scientific journal, e-newsmagazine, reduced rates for ACSM meetings, access to group rate insurance, regional chapter membership, and educational resources

American Public Health Association
800 I St. NW
Washington, DC 20001-3710
202-777-APHA (2742)
www.apha.org

Member benefits: national annual meeting, publications and periodicals, networking, employment information, professional growth opportunities

IDEA Health and Fitness Association
10455 Pacific Center Ct.
San Diego, CA 92121-4339
1-800-999-4332
www.ideafit.com

Member benefits: conventions, health and fitness publications, career development resources, professional fitness education and continuing education, discounted liability insurance, member-only Web site

International Health and Racquet Sportsclub Association (IHRSA)
263 Summer St.
Boston, MA 02210
800-228-4772
www.ihrsa.org

Member benefits: provision of primarily membership benefits at the health club level; student membership benefit includes access to member-only Web site areas

Medical Fitness Association
P.O. Box 73103
Richmond, VA 23235-8026
804-897-5701
www.medicalfitness.org

Member benefits: e-newsletter, health and fitness publications, reduced rate to conferences, career listings on Web site, regional meetings, internship job board, member-only Web site, quarterly educational teleconferences

Medical Wellness Association
3211 Grand Cayman
Sugar Land, TX 77479
www.medicalwellnessassociation.com
Member benefits: member advocacy, communications, and informative updates; membership directory and reference guide; speakers network, education, and mentoring programs; discounts on educational and training materials; career services, support, and opportunities

National Strength and Conditioning Association (NSCA)
1885 Bob Johnson Dr.
Colorado Springs, CO 80906
719-632-6722
www.nsca-lift.org
Member benefits: industry and research publications, member-only Web site, career resources, specialized conferences, committee service, networking, scholarships and grants, insurance options

National Wellness Institute, Inc.
1300 College Ct.
P.O. Box 827
Stevens Point, WI 54481-0827
800-243-8694
www.nationalwellness.org
Member benefits: professional publications, access to assessment software, discounts on continuing education credits, access to member directory, networking opportunities

Society for Public Health Education (SOPHE)
10 G St. NE, Suite 605
Washington, DC 20002-4242
202-408-9804
www.sophe.org
Member benefits: journals and newsletter, reduced conference fees, access to online member directory, job bank service, opportunities for awards

students at substantially reduced rates, and some are even provided free of charge to health and fitness professionals. A number of online resources are available for health and fitness professionals who wish to share knowledge and ideas. Many have a membership subscription fee but offer continuing education workshops and provide workshops in the community (e.g., www.ptonthenet.com).

Given the competitive nature of the job market, you must maximize your educational experience while in college. Work with your advisor to create a degree plan that will prepare you for the current and future job market. Besides taking the courses required by your major, you will want to select elective courses that will prepare you for the type of job you want. Also, take advantage of as many internship and community service learning experiences as possible. In many cases, a degree in a health and fitness discipline alone will not be sufficient to get you the job you want. Besides having the credentials on paper, you must be able to demonstrate that you can apply the skills you have learned.

➤ You must maximize your educational experience while in college, being sure to select a broad spectrum of elective courses in your degree program.

Newly graduated students often show up at interviews, excited, with degrees in hand, but completely unprepared for the types of jobs available. In some cases, students interviewing for health and fitness counselor positions report that they do not have any teaching experience because they chose an area of emphasis that deals strictly with the biophysical sphere of physical activity. Because teaching courses fall within another area of emphasis (i.e., the behavioral sphere of physical activity), these students didn't believe that they needed them. Don't find yourself in this situation by limiting yourself to one area of emphasis at the expense of gaining skills that will help you be successful in the job you want.

Be sure to take special care in choosing elective courses in your degree program. If your department does not offer a course in teaching or counseling skills, then inquire what departments on campus might. Perhaps you can speak to your advisor and can have a faculty member with teaching or coaching expertise supervise you in an independent study. We can make a similar argument for classes based on business and marketing. Clearly, in our media-saturated culture, marketing skills are important for health and fitness professionals. After all, a big part of your job as a health and fitness professional will be to motivate people to take part in the programs that you offer and to promote the idea of making healthy lifestyle changes. Again, if your department does not offer a specific class in health and fitness marketing, another course on campus may provide general marketing information that you can apply to your discipline. Remember, what will matter most when you interview for a job is that you have knowledge and skills to do the job successfully. Do your homework early and make sure that you obtain the appropriate skills along the way.

Wrap-Up

In this chapter we have described the health and fitness field and summarized the types of settings and jobs that are currently available for health and fitness professionals. In addition, we have provided some insight into current and future trends in health and fitness and described ways in which these trends affect the field. Return to the situations described in the chapter opening. Consider how those in health and fitness careers are directly responsible and able to assist each of the individuals described there. The world of health and fitness is ever changing, which is partly what makes this profession so dynamic and exciting. If you are considering a career in the health and fitness professions, learn as much as you can about them. It is never too early to begin to tailor your education to your personal passion. Remember also to consider industry trends as you complete this phase of what should be an ongoing educational process. This discussion about the health and fitness professions should serve only as the point from which you begin your process of discovery. Enjoy the journey.

KEY▸ Activity 13.5

Use the key points review in your online study guide as a study aid for this chapter.

QA Activity 13.6

These end-of-chapter questions and activities are also in your online study guide. Your instructor may ask you to complete them online and turn them in.

1. Identify program objectives, target markets, and types of programs offered for each of the following health and fitness settings: (a) worksite, (b) commercial, (c) community, and (d) clinical.
2. Compare and contrast sales-based and retention-based commercial health and fitness business models.
3. Describe typical job duties for the following health and fitness positions: (a) group exercise instructor, (b) fitness instructor, (c) health and fitness counselor, (d) personal trainer, and (e) health and fitness director.
4. How will demographic trends in the U.S. population (e.g., aging baby boomers) affect the health and fitness professions?
5. List the knowledge, skills, and abilities that you need to obtain to be competitive in the health and fitness job market.
6. List specific physical activity-related companies, programs, equipment, resources, and services now available that were created in the past 10 years. Forecast what new companies, programs, equipment, resources, and services might be developed in the next 10 years.

14

Careers in Therapeutic Exercise

Chad Starkey

AP Photo/Tom Gannam

CHAPTER OBJECTIVES

In this chapter we will

- acquaint you with the wide range of professional opportunities in the sphere of therapeutic exercise;
- familiarize you with the purpose and types of work done by professionals in therapeutic exercise;
- inform you about the educational requirements and experiences necessary to become an active, competent professional in the area; and
- help you identify whether one of these professions fits your skills, aptitudes, and professional desires.

A high school football player fails to get up following a play. The athletic trainer arrives on the scene and finds the athlete's left tibia fractured.

Following a stroke, a 54-year-old chief financial officer is recovering from surgery. Along her road to recovery she will meet physical therapists, occupational therapists, and cardiovascular rehabilitation specialists. Each of these professionals will contribute to her return to a successful career.

Surgeons successfully repair an infant's clubbed feet, but the child is showing a delay in walking upright. The physician refers the child to a physical therapist for the rehabilitation necessary to achieve normal development.

A 60-year-old widow has a brain tumor. She is unable to perform the cognitive or motor functions (activities of daily living, ADLs) required for independent living, but she is hopeful of being discharged from an assisted living facility. An occupational therapist and therapeutic recreation specialist work with her to help her restore her mental abilities and gain the motor skills that she needs for independent living.

All these scenarios depict daily real-life situations that therapeutic exercise professionals face in work settings such as hospitals, laboratories, clinics, and sports medicine facilities. Although their patients and clients may differ, athletic trainers, clinical exercise physiologists, occupational therapists, physical therapists, therapeutic recreation specialists, and strength and conditioning specialists all apply exercise and movement experiences to improve a person's physical functioning.

Figure 14.1 illustrates the spheres of professional practice, including the sphere of therapeutic exercise. Physical activity professionals in this sphere help people restore lost function (rehabilitative therapeutic exercise) or acquire skills and functions considered normal or expected (habilitative therapeutic exercise).

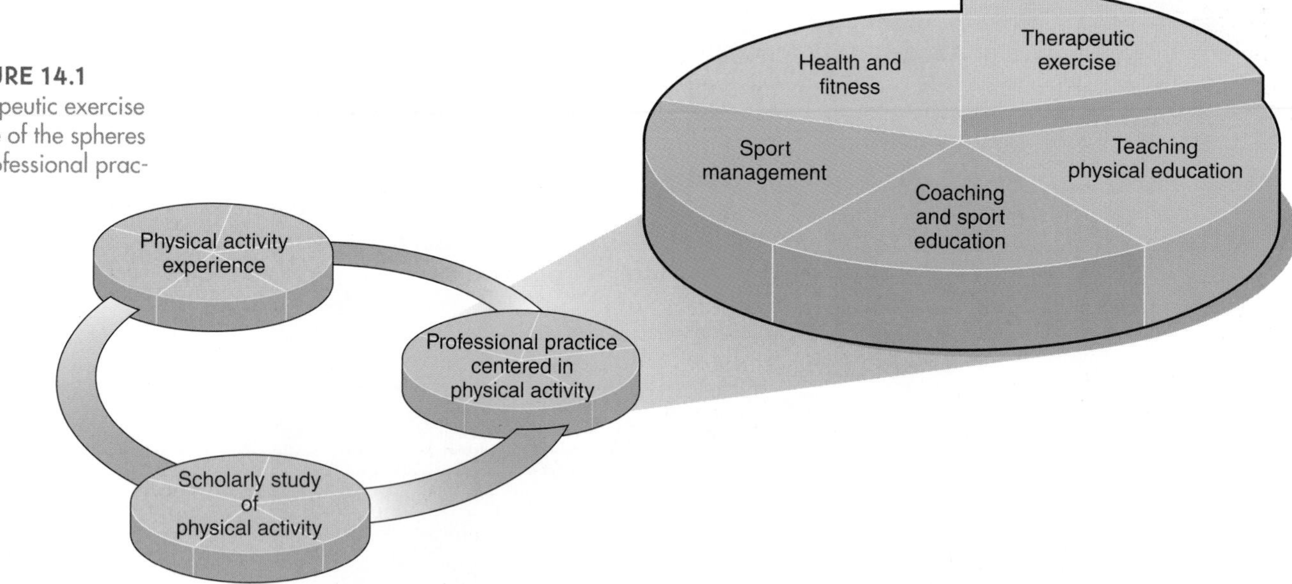

FIGURE 14.1 Therapeutic exercise is one of the spheres of professional practice.

The World of Therapeutic Exercise

Exercise therapists design and implement movement experiences for the purpose of restoring or improving motor function to a level that enables people to reach personal or career goals unencumbered by physical limitations. Reaching these goals requires the application of knowledge of human anatomy, human physiology, exercise physiology, and kinesiology in structured activity programs. To develop therapeutic goals, clinicians must call on their knowledge of the effects of exercise on the muscular, nervous, skeletal, and cardiovascular systems and relate those effects to the patient's needs. Depending on the patient, workplace, and conditions being treated, therapeutic goals may include restoring muscular function and strength, joint range of motion, proprioception, cardiovascular and pulmonary function, and metabolic function.

Therapeutic Exercise, Rehabilitation, and Habilitation Defined

What is therapeutic exercise? In technical terms, **therapeutic exercise** is the systematic and scientific application of exercise and movement experiences to *develop* or *restore* muscular strength, endurance, or flexibility; neuromuscular coordination; cardiovascular efficiency; and other health and performance factors. In practice, therapeutic exercise is programmed physical activity aimed at improving or restoring the quality of life. Thus, therapeutic exercise can be classified as being rehabilitative or habilitative. In this chapter we draw a distinction between these two aspects of therapeutic exercise to help you understand the different types of professional work associated with this area.

➤ Therapeutic exercise can be classified as rehabilitational (restoring lost function) or habilitational (helping to acquire normal function).

Broadly speaking, **rehabilitation** describes processes and treatments that restore skills or functions that were previously acquired but that have been lost because of injury, disease, or behavioral traits. If you have ever strained a muscle or broken a bone, then you know the value of rehabilitation in helping you regain lost functions. Although rehabilitation can take the form of ice, heat, electricity, ultrasound, or psychological counseling, we focus in this chapter on treatments that involve physical activity.

Rehabilitation specialists require a thorough knowledge of the pathological aspects of injury and disease, the limitations that they impose on human performance, and the types of treatments required to restore normal function. Because people are more than just muscles and bones, rehabilitation specialists must take into account the psychological effect of the injury and the subsequent rehabilitation program as well. The regimen used by an athletic trainer working to restore a running back's knee so that he can return to competition is an example of rehabilitative therapeutic exercise.

Generally, the concept of rehabilitation that practitioners use also includes the notion of **habilitation,** the processes and treatments leading to the acquisition of skills and functions that are normal and expected for an individual of a particular age and status (Dudgenon 1996). The standards or expectations that signal a need for habilitation may be vastly different for individuals of the same age. For example, we would expect different physical performance standards of a lawyer than we would of an athlete. The lawyer who is physically fit according to the definition in chapter 3 is probably not in need of habilitation—his state of physical fitness is normal for his age, and he does not need special physical abilities for his occupation. On the other hand, an athlete who is fit but lacks the cardiovascular endurance or strength to perform the tasks expected of her in field hockey is a candidate for habilitation involving intensive conditioning, training, and other skill-specific exercise.

Habilitative therapeutic exercise helps athletes develop skills and functions to improve their performance.
AP Photo/Tom Hevezi

In both rehabilitative and habilitative therapeutic exercise, the physical status of the individual regarding any permanent disabilities or impairments, such as blindness, amputation, or paralysis, must be considered. For example, a paraplegic's functional loss of use of the lower extremities would not be a cause for rehabilitation, but lack of upper body strength relative to that considered normal for a spinal cord injury would require care. Moreover, a physical therapist who is attempting to correct congenital postural problems is practicing **habilitational therapeutic exercise** because it involves bringing the client to a level of functioning not previously attained.

Obviously, heath and fitness programs targeted at unfit populations may involve habilitative therapeutic exercise because the populations served are usually performing at levels below that considered normal and may not have previously attained the level of health or fitness desired. Because we addressed professional opportunities in fitness and health promotion in chapter 13, we will not consider them here. We will, however, consider other types of professional services that include habilitative and rehabilitative therapeutic exercise.

Each of the professions described in this chapter is increasingly emphasizing the importance of preventing injury and disease. An athletic trainer, for example, helps to identify musculoskeletal problems such as lax joints, weak muscles, and family medical conditions that could predispose an athlete to injury or even death. Likewise, a physical therapist may assess a person's posture or evaluate a worker's biomechanics when lifting heavy objects to prevent job-related injuries. A clinical exercise physiologist who performs electrocardiograms and exercise stress tests is also practicing the art of preventive medicine by identifying people who are at risk of experiencing a heart attack. By developing appropriate exercise programs to remedy these problems, these specialists help reduce the likelihood of specific injuries and diseases.

➤ "An ounce of prevention is worth a pound of cure." Both rehabilitational and habilitational therapeutic exercise focus on developing the body's systems so that injury or disease is less likely.

Although preventive therapeutic exercise programs are usually habilitational, at times they may be considered rehabilitational. In habilitational therapeutic exercise, bringing a person's physical state up to expected standards is preventive because it reduces the risk of injury or illness. **Rehabilitational therapeutic exercise** may also be preventive; by restoring the patient's physical skills and functions, rehabilitation reduces the risk of reinjury and of suffering different types of injuries.

Considering a Career in Therapeutic Exercise?

All therapeutic exercise professions discussed in this chapter involve working closely with people. These professionals work with a number of clients or patients daily, routinely discussing protocols and treatment plans with colleagues or other physical activity professionals. Do you enjoy contact with people? Do you have good communication skills? Do you have good problem-solving skills? If you are considering a career in therapeutic exercise, in what areas might you need to seek out further instruction and education?

Rehabilitative Therapeutic Exercise

Physical dysfunction is characterized by the inability to use one or more limbs or the torso. If you have ever broken an ankle or sprained your wrist, you have experienced physical dysfunction. Physical rehabilitation specialists help individuals who are experiencing physical dysfunction stemming from traumatic injury, congenital defects, or disease to regain the use of the affected body part or compensate for its disability. Physical therapists, occupational therapists, and athletic trainers specialize in rehabilitating these conditions. They work closely with other health care professionals such as physicians, nurses, and dietitians.

Exercise Therapy for the Rehabilitation of Musculoskeletal Injuries

Strokes, spinal cord injuries, back problems, and other injuries from automobile, industrial, sport, and home accidents can all cause the nervous system or muscles to no longer function properly. Physical activity also carries with it the risk of trauma, as when repetitive motions lead to neuromuscular trauma (see chapters 2 and 10). Not only does injury to muscles, joints, and bones directly affect the traumatized tissues, but the accompanying immobilization and associated decreased physical activity can cause more widespread problems.

Following a significant injury that inhibits the ability of a limb to function, the resulting disuse may affect the limb and the activity of the entire body. Consequently, over time the efficiency of the heart and lungs may degrade. Perhaps you have experienced this effect after injuring a bone or joint; after regaining full motion you may have noticed that running up a flight of stairs or playing basketball left you more winded than such activities would have before your injury.

The goal of orthopedic rehabilitation programs is to restore symptom-free movement and restore the function of the cardiopulmonary system. Restoring the function of a limb consists of increasing joint range of motion, increasing muscular strength and endurance, and building cardiovascular endurance.

Although **neuromuscular injuries** directly affect the extremities, the resulting loss of function and inactivity decrease cardiovascular function. Most neuromuscular rehabilitation programs also emphasize restoring the function of the heart and lungs. Clinicians involved in this type of rehabilitation use passive and active exercise to restore the limb to function. Strength and muscular endurance training follows, which serves to protect the limb. Various forms of heat and cold, electrical stimulation, therapeutic ultrasound, and manual therapy techniques augment these exercises. Throughout this process, therapists attempt to maintain the patient's level of cardiovascular endurance. Most types of neuromuscular rehabilitation occur in outpatient physical therapy clinics, hospitals, and athletic training clinics.

Exercise Therapy for the Rehabilitation of Athletic Injuries Rehabilitation of athletic injuries deserves specific attention here because the field of physical activity usually offers the training for this area. You may have experienced or witnessed an athletic injury at one time or another—whether you broke a finger playing sports or anxiously waited during an injury time-out to find out how seriously your favorite football player was injured. Almost immediately following a sport injury, the athlete's rehabilitation begins. Restoring the body part to its prior level of function is only part of the rehabilitation process. Once the limb begins to regain strength, range of motion, and neuromuscular coordination, the athlete must be reintegrated into athletic activity. A unique aspect of athletic rehabilitation is the sport-specific functional progression. Here, traditional rehabilitation protocol merges with the skills and tasks needed to compete in a particular sport.

Sports medicine is an aspect of therapeutic exercise that is exclusively dedicated to the prevention, treatment, and rehabilitation of athletic injuries. But sports medicine is also distinct from other aspects of therapeutic exercise in that the rehabilitation protocol used is more aggressive than that used for the general population. The spectrum of sports medicine is broad and encompasses a range of activities that are beyond the scope of this text. Students who are interested in more information regarding sports medicine should explore sections on the athletic trainer, the physical therapist, and the strength and conditioning specialist that appear later in this chapter.

➤ Sports medicine is a generic term that refers to the practice of medicine, the art of rehabilitation, and the science of research as they relate to athletic participation (Prentice 2006).

No longer unique to the sports medicine setting, a new form of therapeutic exercise has emerged from this area: **prehabilitation.** Whereas rehabilitation occurs following an injury or after surgery, prehabilitation occurs

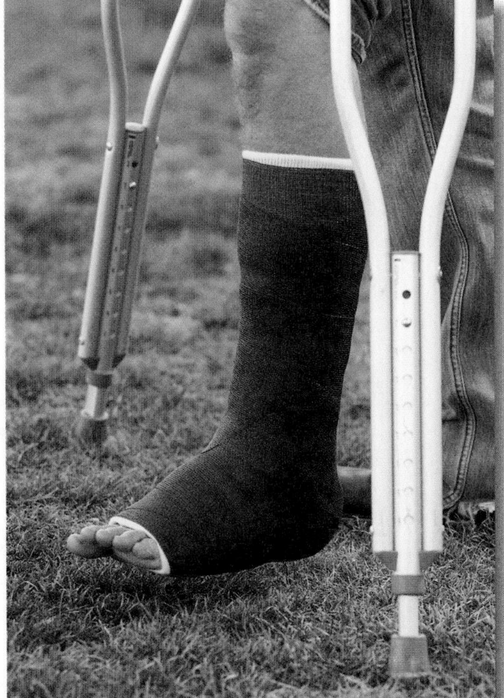

Rehabilitative therapeutic exercise programs restore the patient's functions to the preinjury level.

following the injury but before surgery. When a physician deems that surgery is necessary for an orthopedic injury, an anterior cruciate ligament (ACL) tear, for example, part of the rehabilitation process must focus on restoring lost function. Prehabilitation involves developing as much strength and range of motion as possible before surgery. Then, following surgery, the patient will have an increased functional level on which to build. This concept is sometimes also applied to people who will be undergoing cardiac surgery to strengthen the heart and improve circulation.

Exercise Therapy for the Rehabilitation of Postsurgical Trauma Although surgery is performed to restore a person's health, the process of surgery itself has detrimental effects on the body. The incision affects the involved muscles and nerves, and the bed rest associated with the recovery process leads to decreased muscle mass and function and decreased cardiovascular efficiency. Although this effect is most evident when surgery involves the limbs, the heart and lungs are also affected. Just as skeletal muscle wastes away when it is not used, cardiac muscle responds similarly. In most cases, people recovering from long-term disability must participate in neuromuscular rehabilitation to restore normal cardiopulmonary function.

Exercise Therapy in Cardiopulmonary Rehabilitation

Diseases of the cardiopulmonary system include coronary artery disease, arrhythmia, hypertension, heart attacks, and emphysema. Collectively, these diseases are the leading causes of death and long-term disability among adults in the United States. Undiagnosed heart disease is also the leading cause of death among young athletes (Maron 2007). Early identification of people at risk for these conditions, and subsequent intervention, will reduce the likelihood of their occurrence and increase the chance of survival following an episode.

Working closely with a physician, the exercise therapist may be responsible for planning, implementing, and supervising physical activity programs designed to help restore individuals to normal function. Therapeutic exercise specialists must be aware of certain conditions that can make rehabilitative exercise a potentially deadly activity. Asthma, for example, can be dangerous because the lungs cannot exchange oxygen and carbon dioxide at the rate needed to support the body's metabolism during exercise. Because the risk is greatest when exercise occurs in a hot, dry environment, therapeutic exercise specialists treating asthmatic clients will prescribe regimens that the client can more easily tolerate, such as aquatic exercise routines.

➤ Because of the nature of the diseases being treated, a physician supervises most cardiopulmonary rehabilitation programs.

Cardiovascular exercises to improve the efficiency of the heart and lungs include breathing exercises, walking, running, swimming, and resistance exercises such as weight training. These are prescribed at progressive levels of intensity and duration. Diagnostic tests such as stress testing, $\dot{V}O_2$max testing, electrocardiograms, and echocardiograms are used to monitor the patient's progress and adapt the level of activity accordingly. Stress tests and other diagnostic tests are usually conducted in hospitals or other medical environments. Cardiovascular exercise programs normally take place in clinics, health clubs, or wellness centers, with the patient also being prescribed independent programs for home use. A licensed physician must supervise all diagnostic tests and exercise prescribed to high-risk patients in **cardiopulmonary rehabilitation** programs.

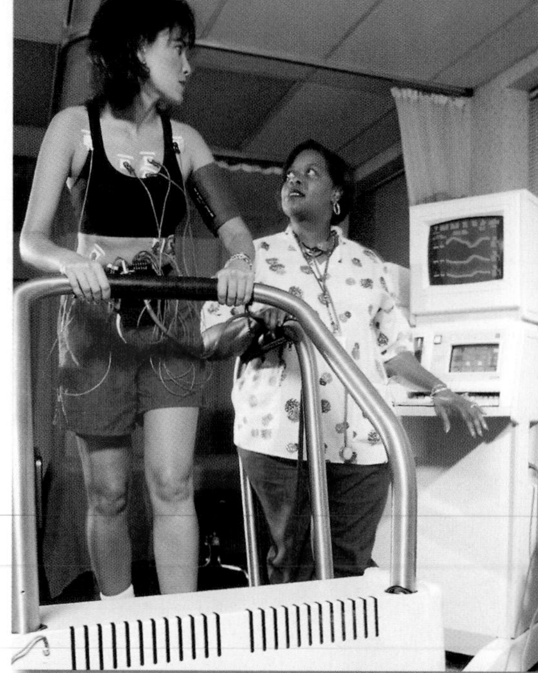

Cardiopulmonary care can begin before the onset of a patient's symptoms to identify those who are at risk of disease.
Brand X Pictures

Exercise Therapy for the Rehabilitation of Older Populations

Humans are living longer, and the percentage of the population aged 65 and over continues to grow. The natural decrease in strength and flexibility and the increase in age-related skeletal diseases such as arthritis hinder the quality of life for many people. As we saw in chapter 2, loss of ability to perform activities of daily living can have profound physical and emotional effects on people's lives. In addition, as a person enters the later years, the aerobic ability of the body decreases and the percentage of body fat increases, decreasing one's capacity to do physical work.

Although these conditions are inevitable, various forms of therapeutic exercise designed to increase strength, slow the loss of flexibility, and improve cardiovascular condition can delay their onset and minimize their effects, thereby helping seniors regain or maintain independent, active living. Programs that involve physical activity can decrease blood pressure, control cholesterol, and reduce the risks of experiencing a stroke or heart attack, and may lower the risk of certain types of cancer. Exercise in the older population, however, is often a double-edged sword. Although physical activity is required to maintain an appropriate level of fitness, the body is less capable of withstanding the forces of exercise. Clinicians working with clients who have arthritis, for example, whose joints cannot tolerate the physical stresses associated with extensive walking, jogging, or running, must modify the exercise program to minimize the amount of force exerted on the joints. In this case, water aerobics can maintain cardiovascular fitness without further injuring the joints.

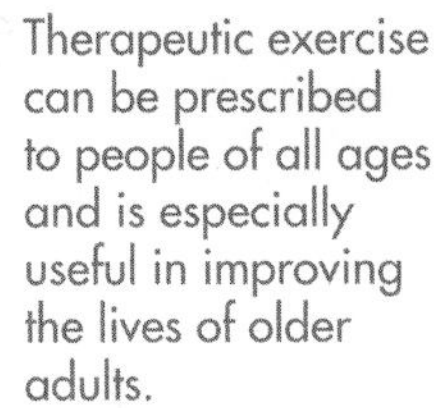

Therapeutic exercise can be prescribed to people of all ages and is especially useful in improving the lives of older adults.

Exercise Therapy for the Rehabilitation of Psychological Disorders

The physical benefits of exercise should now be apparent to you. But one area is often overlooked: the psychological and emotional benefits of physical activity. Exercise is a wonderful way to reduce stress and unwind from the toils of life. People who are involved in regular physical activity tend to sleep better and suffer fewer emotional disorders than those who are not. Twenty to 30 minutes of aerobic exercise has been found to produce changes that are on a par with standard forms of psychotherapy (Sims et al. 2006). In the healthy population, aerobic exercise can prevent the onset of some types of psychological maladies; for those with these types of conditions, exercise can often be considered a form of treatment.

Some psychotherapists prescribe exercise regimens as part of their patients' therapy programs. The older population referred to in the previous section can also gain psychological benefits from exercise. Elderly people who engage in physical activity have decreased signs of depression, increased self-image, and improved morale, all leading to better quality of life (Conn, Minor, and Burks 2003). As an aside, smoking and alcohol cessation programs often include exercise, although scientific data have yet to show that this is effective.

➤ Exercise is as important for the mind as it is for the body. Improving one's physical health through exercise also improves one's mental state.

Habilitative Therapeutic Exercise

Habilitative exercises help bring people in line with established physical standards such as those presented in chapter 3. The goal of this type of therapeutic exercise is to help people reach the expected level of physical fitness for their demographic classification. Throughout this text we have spoken about the relationship of physical activity to health and the kinesiologist's role of improving health through physical activity. Many of the professions described in chapter 13 also meet the definition of habilitative therapeutic exercise. The following sections describe some of the areas in which therapeutic exercise is used for habilitative purposes.

Exercise Therapy for Specialized Habilitation

Although not recognized as such, specialized habilitation is probably the single largest role of habilitative therapeutic exercise. Specialized habilitation involves bringing specific groups of people in line with standards that exceed rather than merely meet those of the general population. These may include running a 5-minute mile, having body fat within a certain range, or being able to lift a certain amount of weight. Preseason sport camps, military boot camps, police and firefighter academies, and even astronaut training programs are examples of settings in which specialized habilitative therapeutic exercise takes place. Exercise specialists in this area must have full understanding of the muscular and cardiovascular capabilities needed by people in these special groups. Often, the exercise specialist has a professional background in the specialty area. For example, a former professional baseball player with a degree in kinesiology may specialize in organizing habilitative exercise regimens for spring baseball training camps. Such experience can help keep the exercise program in line with physical expectations of the baseball players involved.

Exercise Therapy for the Habilitation of Obese Populations

Physical fitness, exercising, and losing weight are controversial American obsessions. Many people overemphasize weight loss, and others ignore all aspects of healthy eating or exercise habits. Ironically, both approaches are hazardous and potentially deadly.

Fad dieting, exercise addiction, and distorted body image have reached almost epidemic proportions among females of high school and college age. Medical conditions associated with such obsessions, especially anorexia and bulimia, are rising at alarming rates. At the other end of the spectrum is the overweight, sedentary portion of our population who, because of their lifestyles, are predisposed to cardiovascular disease and diabetes. The mass media continually bombards us with conflicting and confusing information on which diets do—and do not—work. These contradictions may leave most Americans unclear about what is meant by healthy eating.

Because of its role in bringing people in line with established standards, therapeutic exercise used to control weight and body mass is considered habilitational. Proper nutritional and exercise counseling are essential in promoting a healthy lifestyle and decreasing the incidence of disease. Specialists such as exercise physiologists, strength and conditioning specialists, and personal trainers work in cooperation with physicians, nutritionists, counselors, and other professionals to develop and implement programs that assist people in returning to normal, functional, healthy lifestyles.

Habilitational therapeutic exercise helps people reach minimal physical fitness levels based on their age, gender, and desired activity. This form of therapeutic exercise can begin shortly after birth and continue throughout the life span.

Exercise Therapy for the Habilitation of Children With Developmental Problems

In the not too distant past, children born with physical abnormalities faced living life with a functional handicap. Early identification of specific conditions followed by appropriate intervention has provided the opportunity of a fruitful life to countless people.

The goal of this type of habilitative exercise is to help the child adapt to, or compensate for, functional anatomical and physiological deficits. In certain cases the underlying condition may have been surgically corrected. Other cases may involve teaching the child how to use a prosthetic device or how to perform basic skills such as rolling over or walking, or eating in a modified manner. Exercise therapy may also be used to strengthen the muscles in children with cerebral palsy, thereby improving their walking performance.

Hab or Rehab?
Use this online study guide activity to distinguish between the two types of therapy.

Although working with children has a unique set of rewards, it likewise presents a unique set of challenges to therapists. Issues such as communication between the therapist and child, home care problems, and schooling are unique to this population. Regardless of the challenges facing the therapist, the ultimate goal of habilitative regimens for this population is to promote physical, social, and cognitive development at a rate near the norm for the child's age group.

Exercise Therapy for Habilitation for General Fitness

With all the benefits of exercise described in this text, the logical question is, Why doesn't everyone participate in exercise? The truth is that although more people are now involved in exercise than ever before, most of our population live sedentary lives. Although no one factor is singularly to blame for this lack of exercise, the longtime popularity of television and the recent growth in computer fixation both contribute to our "couch potato" society.

➤ During the industrial age, a larger proportion of the population was engaged daily in strenuous physical activity. The information age has created a relatively sedentary group of people. Even casual exercise can offer a more balanced lifestyle.

Research has shown that exercise can help prevent disease and enable people to reach their maximum life expectancy. Therapeutic exercise professionals play an important role in helping the general population achieve healthy standards of physical fitness by introducing them to exercise through hospital-based fitness centers, commercial gymnasiums and fitness centers, employer-based wellness centers, and community-based nonprofit agencies. Chapter 13 discussed many of these professions.

Therapeutic Exercise Settings

Therapeutic exercise professionals work in a variety of settings, including inpatient facilities (such as hospitals), outpatient clinics, athletic training clinics, and even clients' homes. Some are employed by others, and some work in their own private practices. Although the decision to pursue a given profession in the sphere of therapeutic exercise may dictate the setting in which you work, you may still have opportunities to work with professionals from a wide range of work settings. Most health care facilities employ the team approach in which representatives from several different professions collaborate in the planning and delivery of patient care.

Inpatient Facilities

Rehabilitation hospitals provide specialized care that strives to return individuals to their maximum level of function. People with severe conditions such as brain or spinal cord trauma, or those recuperating from severe disease, may require a long-term stay in a rehabilitation facility. **Custodial care facilities** such as nursing homes provide services to assist in activities of daily living as well as to meet people's specific medical needs. In both types of **inpatient facilities,** the level of patient disability usually dictates the degree of coordinated effort that physicians, rehabilitation specialists, and social services personnel must arrange.

Besides having the skills described in this chapter, therapeutic exercise professionals who work at inpatient facilities often must have knowledge of ambulation and patient transfer techniques, prevention of bedsores, and other skills unique to long-term care.

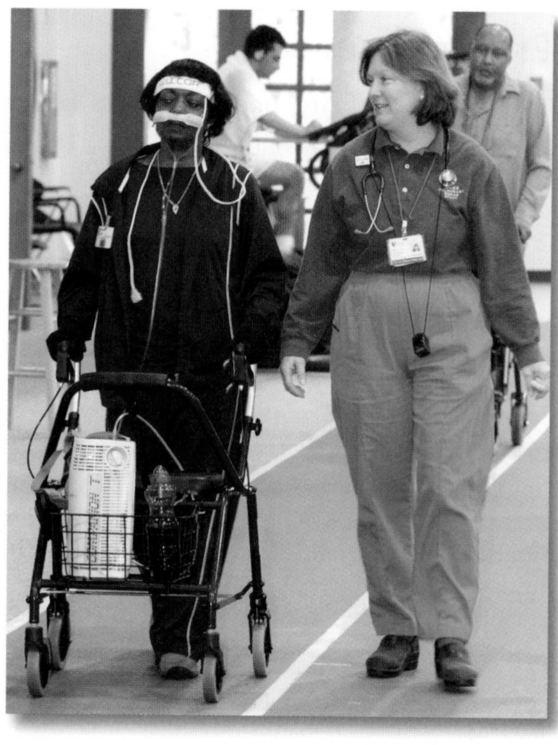

Therapeutic exercise professionals provide specialized care for people who have been severely injured to help them become as functional as possible.

Outpatient Clinic Settings

Outpatient clinic settings are characterized by short-term patient visits (patients do not stay overnight). These types of facilities comprise the most diverse settings for therapeutic exercise professionals and include physical therapy clinics, sports medicine clinics, and cardiac rehabilitation facilities. Outpatient facilities may have a wide range of specialized equipment to meet the needs of their clientele. The more diverse the range of people being treated, the broader the range of equipment. Most outpatient clinics consist of examination areas, specialized treatment and rehabilitation areas, hydrotherapy pools, and open-space exercise areas. Some outpatient clinics are specialized and accept only people with cardiovascular disease, hand injuries, or spinal conditions.

Social, economic, or other factors can often prohibit people from traveling regularly to an **outpatient facility** for treatment. As a result, many outpatient clinics send therapists to patients' homes to provide both rehabilitative and habilitative home care.

Sport Team Settings

Athletic training clinics represent a specific type of outpatient facility. Located exclusively in high schools, colleges, and professional team facilities, the first athletic training "rooms" were simply a table located in the corner of the locker room. Although the size and complexity may vary, athletic training facilities at major colleges and universities are state-of-the-art health care clinics.

Facilities typically include treatment and rehabilitation areas, examination rooms, hydrotherapy areas, and space dedicated to injury prevention such as taping, wrapping, and bracing. Most professional sports medicine facilities, and many of those in colleges and universities, have X-ray rooms and space where minor medical procedures can be performed.

Many sport team settings also offer weight rooms, cardiovascular fitness centers, swimming pools, and exercise physiology laboratories. These types of facilities expand the scope and breadth of therapeutic exercise programs that can be offered to habilitate and rehabilitate athletes.

Private Practice

The thought of being one's own boss is intriguing to many people. **Private practice** is an entrepreneurial venture in which the professional establishes his or her own place of work. Physical and occupational therapy are capital-intensive professions requiring expensive equipment and a lease on the physical facility, so a business loan, personal wealth, or investors are required for start-up costs. People must also consider other expenses such as utilities and support staff must before venturing into private practice.

Most professionals do not begin their careers in their own practice; they usually make the transition into private practice following employment in an established clinical practice. By working first in an established practice, beginning professionals can learn about the nuances of the business world and build a patient base and reputation. Before venturing into private practice, beginning therapeutic exercise professionals should first gain work experience, amass capital, and establish a strong patient base.

TE Settings

In the online study guide activity, you'll identify the characteristics of the different therapeutic exercise settings.

Up-Close Views of Professions in Therapeutic Exercise

Because of their background in the anatomical and physical principles of human movement, kinesiology students are well positioned to enter health care professions that use therapeutic exercise as a part of their habilitation or rehabilitational treatment regimens, including athletic training, cardiac rehabilitation, occupational therapy, physical therapy, and strength and conditioning. This section describes these professions, the settings in which they take place, and the educational requirements and credentials needed to practice in them.

You will notice that the professional roles discussed here often overlap. Indeed, many people pursue multiple credentials such as athletic training and physical therapy, or physical therapy and strength and conditioning. As competition for the health care dollar increases, and with the health care job market tightening, the need for multiskilled and multicredentialed individuals will increase in the future.

Many therapeutic exercise professions such as physical therapy and occupational therapy require postbaccalaureate education to enter the profession. Physical therapists currently must graduate from a master's degree program in physical therapy. Over 70% of physical therapy programs award the clinical doctorate in physical therapy (DPT) degree. Occupational therapy required an entry-level postbaccalaureate degree as of 2007. Currently, the entry-level education requirements for physical therapy assistants and occupational therapy assistants are unchanged and remain at the associate's degree level.

An undergraduate education in kinesiology will provide you with comprehensive knowledge of human movement, the effects of exercise, and normal anatomical and physiological function. Many graduate-level therapeutic exercise programs view an undergraduate degree in kinesiology as an ideal background for advanced studies. You should always check the current educational requirements for any profession you intend to pursue and use elective courses to fulfill any specific programmatic requirements.

The educational requirements of these professions vary depending on their focus, but most still have the common traits of a strong science base and an active clinical education component. Course work in theoretical kinesiology (part II of this text) will be valuable, and other specialized courses and clinical practicums will round out your professional knowledge. To combat the trend toward overspecialization, the Pew Health Professions Commission (1995) recommended that all health care professions, including those involved in therapeutic

Identifying Whom You Would Like to Work With

People exploring a career in therapeutic exercise often first identify the population with whom they wish to work. Take a moment to think about your favorite populations. Put a check mark by the following groups that seem appealing to you.

____ Infants	____ Athletes
____ Children	____ Nonathletes
____ Adults	____ Healthy
____ Senior citizens	____ Recovering
____ Injured	____ Requiring long-term care
____ Disabled	____ Other: ______________________

Professional Organizations

Profession	Professional organizations	Professional journals
Athletic training	**United States** National Athletic Trainers' Association, Inc. 2952 Stemmons Freeway Dallas, TX 75247-6196 214-637-6282 214-637-2206 fax www.nata.org **Canada** Canadian Athletic Therapists' Association Suite 402 – 1040 7th Ave. SW Calgary, AB T2P 3G9 403-509-CATA (2282) 403-509-2280 fax www.athletictherapy.org	*Journal of Athletic Training* *Athletic Therapy Today* *Journal of Sport Rehabilitation*
Cardiac rehabilitation specialist	American College of Sports Medicine 401 W. Michigan St. Indianapolis, IN 46202-3233 317-637-9200 317-634-7817 fax www.acsm.org American Association of Cardiovascular and Pulmonary Rehabilitation 401 N. Michigan Ave., Ste. 2200 Chicago, IL 60611 312-321-5146 312-527-6924 fax www.aacvpr.org	*Journal of Cardiopulmonary Rehabilitation* *Journal of Clinical Exercise Physiology*
Occupational therapy	**United States** American Occupational Therapy Association, Inc. 4720 Montgomery Lane Bethesda, MD 20824-1220 301-652-2682 301-652-7711 fax www.aota.org **Canada** Canadian Association of Occupational Therapists CTTC Building, Ste. 3400 1125 Colonel By Dr. Ottawa, ON K1S 5R1 800-434-CAOT 613-523-2552 fax www.caot.ca	**United States** *American Journal of Occupational Therapy* **Canada** *Canadian Journal of Occupational Therapy*
Physical therapy	American Physical Therapy Association, Inc. 1111 N. Fairfax St. Alexandria, VA 22314-1488 703-684-2782 703-684-7343 fax www.apta.org	*Physical Therapy* *Journal of Orthopedic and Sports Physical Therapy*

Profession	Professional organizations	Professional journals
Strength and conditioning	National Strength and Conditioning Association 1885 Bob Johnson Dr. Colorado Springs, CO 80906 719-632-6722 719-632-6367 fax www.nsca-lift.org	*Strength and Conditioning* *Journal of Strength and Conditioning Research*
Therapeutic recreation	American Therapeutic Recreation Association 207 3rd Ave. Hattiesburg, MS 39401 601-450-2872 601-582-3354 fax http://atra-online.com/cms/	*Annual in Therapeutic Recreation* *Therapeutic Recreation Journal*

exercise, should possess a common set of competencies by 2005. Your background in kinesiology will provide you with a well-rounded education including many of the skills that will be expected of tomorrow's health care professional.

Differing levels of preparation and regulation are found among therapeutic exercise professions. Health care professionals must pay close attention to state licensure requirements, especially when changing location. The professions discussed in this chapter and the professional organizations and journals that serve them are listed on pages 390-391. These additional resources will allow you to explore the professions in detail.

Athletic Trainer

High school, college, and professional athletes become injured at staggering rates, with as many as one-third of the 1 million high school football players suffering an injury each year (Fernandez, Yard, & Comstock 2007). Combining the excitement of athletics with the demands of health care, athletic training is the only profession that falls entirely within the realm of sports medicine. The title *athletic trainer* can be misleading. These professionals are not coaches, personal trainers, or other performance-improving personnel. Athletic training is a health care profession that addresses the prevention, evaluation, management, treatment, and rehabilitation of injuries and other conditions experienced by athletes and other physically active individuals. Athletic trainers (ATs) work under the direction of a licensed medical or osteopathic physician. Besides attending to the direct health care of the athlete, the AT coordinates referrals to appropriate specialists.

Employment Settings

Athletic trainers traditionally have been employed by high schools, colleges and universities, and professional sport teams. New roles are emerging for athletic trainers in hospitals, sports medicine clinics, industrial rehabilitation clinics, and other allied medical environments. The athletic trainer's roles and responsibilities vary depending on the work setting.

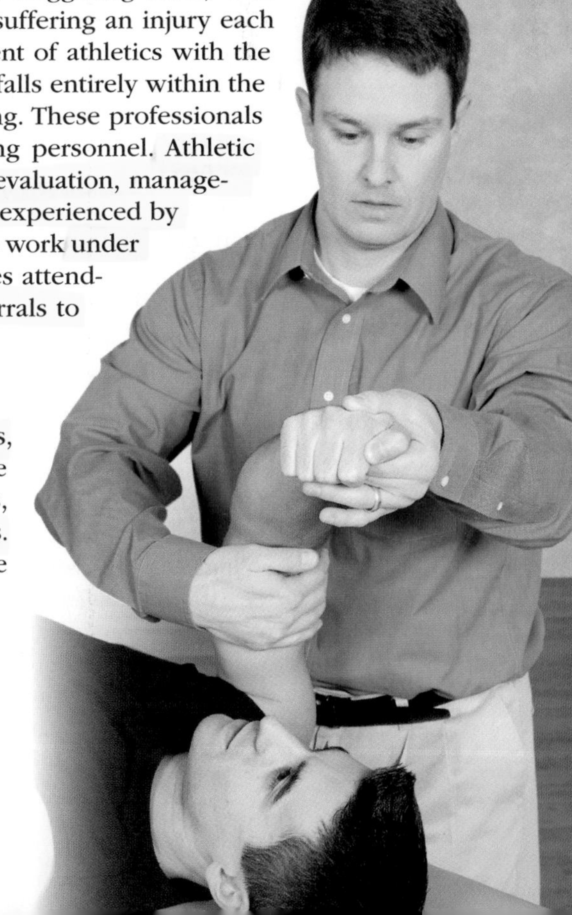

Athletic trainers use therapeutic exercise to return athletes and other patients to competition in the shortest time possible.

Sample Course Work for Athletic Training

- Acute care
- Administration
- Clinical diagnosis of injuries and illnesses
- Biomechanics*
- Chemistry
- Counseling
- Exercise physiology
- Gross anatomy
- Human anatomy
- Human physiology
- Injury prevention
- Nutrition
- Pathology
- Pharmacology
- Physics
- Statistics
- Therapeutic exercise
- Therapeutic modalities

*Still called kinesiology at some universities.

States are gradually beginning to mandate athletic training coverage of high-risk high school sports. The trend toward hiring full-time athletic trainers is growing, although many secondary schools that receive the services of athletic trainers do so by contracting services from local physical therapy and sports medicine clinics and hospitals. Colleges and professional teams hire their own full-time athletic trainers.

- **Athletic setting.** Athletic trainers are in a unique position to implement procedures to prevent or decrease the occurrence of sport injuries. They have the opportunity to work with the same athletes before an injury, examine the injury to form a clinical diagnosis, and then work with the patient during the rehabilitation process.
- **Clinical setting.** Community recreation leagues, road races, and personal workout regimens have exposed millions of Americans to the possibility of incurring athletic-related injuries. Combine this trend with the fact that most high school athletes do not receive the direct services of an athletic trainer, and you can understand the increase in the number of sports medicine and physical therapy clinics. The emergence of athletic trainers in these settings has spurred growth in other rehabilitation settings including corporate and industrial clinics, hospitals, and private practice. Athletic trainers' roles in these emerging settings may vary from those of athletic trainers in more traditional settings. These roles may also vary among states (Cormier et al. 1993). Athletic trainers are responsible for the prevention, evaluation, management, treatment, and rehabilitation of musculoskeletal injuries.

Education and Credentials

Subject matter required in an athletic training major accredited by the Commission on Accreditation of Athletic Training Education (CAATE) is listed later in the section. Academic course work is supplemented by clinical education that permits students to have affiliations with high school, college, and professional sport team athletic training rooms as well as the opportunity to learn in clinics and hospitals.

The Board of Certification, Inc. (BOC) conducts national certification testing of athletic trainers. An athletic trainer needs BOC certification to work as an athletic trainer in major colleges and universities, with professional sport teams, and with the U.S. Olympic Committee. Additionally, 40 states require BOC certification as a prerequisite for licensure, which is needed to practice athletic training.

In Canada, athletic trainers (referred to as athletic therapists) must complete an academic program that has been accredited by the Canadian Athletic Therapists Association and complete 1200 hours of clinical experience under the supervision of a certified athletic therapist (CAT(C)). Reciprocity for certification and licensure between the United States

Interviews With Practicing Professionals

Photo courtesy of Carrie Gladwell

Carrie Gladwell, ATC, LAT
Licensed Athletic Trainer

Carrie Gladwell is a licensed athletic trainer employed by Physio-Therapy Associates and contracted out to provide health care services to the student-athletes at Gahanna Lincoln High School in Columbus, Ohio.

Q: Carrie, what made you choose a career in athletic training?

A: I've been involved in athletics my whole life and I wanted a college degree in a health care profession, so athletic training was a logical choice. My high school athletic trainer was a major influence on my career path. I was always very impressed by his knowledge of the human body and his ability to accurately examine and rehabilitate orthopedic injuries. He definitely sparked my interest in this field.

Q: How did you train for your career?

A: When I was in college, I enrolled in Ohio University's entry-level athletic training program and was immediately taken by the opportunity to learn in an exciting and fun environment. I feel that I received a great education in the athletic training program I graduated from. In my professional career I have never felt that knowledge in a specific area was below what it should be or lacking in any way. Courses in orthopedic clinical diagnosis, therapeutic exercise (rehabilitation), and injury prevention helped me in the application of both habilitational and rehabilitational therapeutic exercise.

Q: Are you satisfied with your career choice?

A: I'm pretty outgoing, so I was naturally drawn to finding a job that continued to expose me to the energy of working with athletes. This setting is different from most professions; I work in a setting where everyone loves what they are doing and is full of energy and excitement. Matching the setting where I work to my own personality keeps my job satisfaction and enjoyment high. I love working with high school athletes: The athletes are always fun to be around and make me laugh on a daily basis, and even the coaches and administration in high school sports are doing something they love and look forward to every day.

There are some down sides, though. I work long hours, and frequently six days a week. This can be extremely tiring, especially working in a setting where a lot of the work is done behind the scenes and, often, alone. By the end of a school year I am ready for a break. But summer break is an added bonus that makes it all worthwhile!

Q: What are your plans for the future?

A: In the short term, I plan to continue my current career path, both working in the relative tranquility of a clinic and dealing with the kinetic demands of working in the high school and college athletic environment. Some day I would like to become more involved in the management and administration of athletic health care.

and Canada currently doesn't exist, but mutual recognition of credentials allows professionals who travel with a team from one country to the other to provide athletic training and therapy services during their stay. For example, the athletic trainer for the Cleveland Cavaliers basketball team can legally provide services to the team when the Cavaliers play in Toronto (and the athletic trainer for a Canadian team can do so when the team plays in the United States).

Employment Opportunities

The 2006-2007 Bureau of Labor Statistics projection is that jobs for athletic trainers will grow at a much faster rate than for other professions through 2014. School districts are replacing part-time athletic trainers with full-time positions and adding more athletic trainers to their existing sports medicine department to follow a collegiate athletic health care model. Employment opportunities in hospitals, physical therapy, and sports medicine clinics, where athletic trainers work as part of a rehabilitation team, may also increase. Additional job growth is also seen in the industrial setting, where athletic trainers care for line employees, and in physician's practices and hospitals, where they serve to extend the functions of a physician.

Clinical Exercise Physiologist

When supervised by clinical exercise physiologists (CEPs), exercise is an effective treatment for people with chronic cardiovascular, pulmonary, and metabolic diseases. Clinical exercise physiologists administer cardiopulmonary exercise tests in hospital exercise testing laboratories, administer fitness testing, and implement and deliver cardiovascular conditioning programs for individuals who are apparently disease free.

Exercise testing such as graded exercise tests and exercise prescription are the responsibility of the clinical exercise specialist (CES), who also provides patient education and counseling. Depending on the specialist's educational background and place of employment, job responsibilities range from taking vital signs (blood pressure, pulse, temperature, and so on) to performing echocardiograms and electrocardiograms before, during, and after exercise. Because the CES works closely with patients, excellent interpersonal communication skills are essential. The registered clinical exercise physiologist (RCEP) is responsible for developing and directing clinical exercise programs and for working in conjunction with a physician, who has overall control of the patient's rehabilitation program. The RCEP is also responsible for the administration of the rehabilitation center and the education of the cardiac rehabilitation staff, and is often engaged in research.

➤ Credentials similar to the American College of Sports Medicine's ES and RCEP certifications are available to people who want to work in the fitness setting.

A hospital might offer a cardiac rehabilitation exercise program for people recovering from heart surgery. The RCEP and physician, working as a team, collect the client's medical history and perform an evaluation of cardiopulmonary function. From these data the two will develop a structured exercise program. The exercise specialist (ES) will then work with the client in progressing through the exercise program. At regular intervals the program director, physician, and ES reevaluate the patient's progress.

Employment Settings

Registered clinical exercise physiologists and ESs are employed in hospitals, specialty clinics, health and fitness centers, and urgent care centers. Most CEPs work in rehabilitation centers where patients are recovering from injury, disease, or surgery. Many work in stand-

Sample Course Work for a Cardiac Rehabilitation Specialist or Clinical Exercise Physiologist

Advanced cardiac life support
Cardiopulmonary assessment
Cardiopulmonary disease
Cardiopulmonary physiology
Cardiovascular technology
Clinical biomechanics*
Echocardiography
Electrocardiography
Exercise physiology
Exercise prescription
Imaging devices
Nutrition
Pathophysiology
Pharmacology
Physics
Respiratory care

*Still called kinesiology at some universities.

alone cardiac rehabilitation centers where specialized exercise equipment and rooms are available in an outpatient facility. Some hold exercise classes at YMCAs, schools, or other external facilities. Generally, CEPs work closely with nurses, nurse practitioners, physicians, and physician assistants. Besides serving an important role in rehabilitation programs, ESs may work in prevention programs as discussed in chapter 13.

Education and Credentials

Typical course work required of people pursuing a career in clinical exercise physiology is listed in the sidebar on this page. Besides taking courses in these areas, students should pursue clinical rotations and fieldwork to supplement their learning experiences.

A bachelor's degree is required for certification as an ES, although a master's degree is preferred. The RCEP requires a master's degree and 1200 hours of experience working with a variety of people who have chronic diseases.

The American College of Sports Medicine also offers certifications for similar professions in fitness settings, such as health and fitness instructors.

Employment Opportunities

The U.S. Bureau of Labor Statistics does not collect specific data on exercise physiologists, but the aging of the U.S. population will continue to create a job demand for those involved in cardiovascular rehabilitation. The Bureau of Labor Statistics projects a 26% increase in the demand for cardiac rehabilitation specialists through 2016. Employment opportunities are increased when RCEP is combined with another degree or credential such as nursing.

Occupational Therapist

Occupational therapists (OTs) assist people with physical, emotional, or mental disabilities to restore or maintain as much independence as possible in daily living and work throughout their lives (note that the term *occupational* has roots in the word *activity*). Some of the physical care rendered by OTs closely mirrors some of the rehabilitative exercises used by physical therapists. Indeed, in many rehabilitation centers, a physical therapist and an OT work together on a single patient's case.

As with many health professions, the roots of occupational therapy can be traced to postwar veteran rehabilitation; pioneers in the profession taught craft skills to soldiers with disabilities (Ambrosi & Barker-Schwartz 1995). Today, OTs specialize in functional bracing and the modification of everyday items for the special needs of their patients. Functional bracing is the use of a supportive or assistive device that allows a joint to function despite anatomical or biomechanical limitations. Examples include knee braces and splints of various kinds, such as wrist and hand splints that allow people to eat with a fork. Occupational specialists also work with people to improve their concentration, motor skills, and problem-solving ability.

Occupational therapists may also specialize in task-specific, work-related rehabilitation, helping clients reacquire the motor and cognitive skill needed to return to work. This type of activity may range from teaching basic skills such as coordinated movement to specific skills such as hammering, typing, or driving a car. To assist people with disabilities, OTs may be called on to evaluate the layout of schools,

Occupational therapists help injured or ill individuals reach their maximum level of independence by emphasizing the acquisition and retention of functional skills.

AP Photo/Mike Derer

Sample Course Work for Occupational Therapy

Occupational Therapist

- Abnormal psychology
- Assistive technology
- Biology
- Biomechanics*
- Gross anatomy
- Human anatomy
- Human performance abilities
- Human physiology
- Neuroanatomy
- Occupational analysis
- Physical dysfunction
- Psychology
- Statistics

Occupational Therapy Assistant

- Biomechanics*
- Human anatomy
- Human development
- Human physiology
- Neurology
- Occupational therapy procedures
- Organization and administration
- Psychology
- Psychosocial dysfunction

*Still called kinesiology at some universities.

homes, and workplaces to suggest methods of eliminating functional barriers. Occupational therapists often employ clinical technicians called certified occupational therapy assistants (COTAs) to carry out rehabilitation plans.

Employment Settings

Occupational therapists and occupational therapy assistants (OTAs) work in hospitals, rehabilitation centers, nursing homes, and orthopedic clinics and provide outpatient occupational therapy service in secondary schools and colleges. In 2004 an estimated 92,000 OTs were in the work force (U.S. Bureau of Labor Statistics [USBLS] 2007). They instruct people with disabilities such as muscular dystrophy, cerebral palsy, and spinal cord injuries in the use of adaptive equipment like wheelchairs, walkers, and aids for eating and dressing. They may design special equipment for home or work use. They may work in cooperation with employers or supervisors to modify the work environment for the patient.

Schools often hire OTs to work with a particular age group or a group with a particular disability. Occupational therapists may evaluate a child's abilities, modify equipment, and help the child participate fully in school programs and activities. Others may work in mental health settings or in drug or alcohol rehabilitation settings. They may work in large rehabilitation centers in cooperation with physical therapists or provide home health care services.

Education and Credentials

Occupational therapists attain their education through professional four-year undergraduate programs or in entry-level master's degree programs. Beginning in 2007 a master's degree was the minimum educational requirement. Certification or licensure as an OTA requires successful completion of a two-year technical program (see the list on this page).

Occupational therapists and OTAs must be licensed in the states where they practice. The basic requirement for licensure is completion of an accredited program and passing the National Board for Certification in Occupational Therapy, Inc. examination. Those who pass the exam earn the occupational therapist registered title (OTR).

Professional reciprocity exists between the United States and Canada. Graduates from Canadian-accredited occupational therapy programs are eligible to sit for the U.S. examination (students in the province of Quebec must successfully complete three Tests of English as a Foreign Language [TOEFL] examinations). Graduates of accredited programs in the United States are eligible to practice in Canada.

Employment Opportunities

According to the U.S. Bureau of Labor Statistics, the employment picture for OTs looks promising through 2012. As the baby boom generation moves into middle age, an age category in which the incidence of heart attacks and strokes tends to increase, demand for OT services will grow. In addition, the steady growth in the population over age 75 will ensure that demand for therapeutic services will remain high. Hospitals will continue to employ a large number of OTs, including in their outpatient rehabilitation clinics.

The market for school OTs also looks promising as the population of schoolchildren grows and children with disabilities continue to receive specialized attention. The median annual income of OTs in 2006 was $54,660 (USBLS 2007).

Physical Therapist

Physical therapists (PTs) are educated to provide rehabilitative care to a diverse patient population with a wide range of injuries, illnesses, and diseases. Patients include accident victims; people with low back pain, arthritis, and head injuries; and those with congenital or acquired disease states such as cancer. A PT may treat individuals ranging from infants to seniors with conditions such as joint injury, burns, cardiovascular disease, and neurological deficits. Usually PTs are in daily contact with other health professionals such as nurses and physicians.

➤ Physical therapists administer the patient's program and perform the required functional evaluations. Physical therapy assistants assume much of the hands-on patient care, functioning under a PT's supervision.

Therapists work approximately 40 hours per week, including some evenings and weekends. Physical therapists combine diagnostic tests with therapeutic exercise; therapeutic modalities such as ice, heat, and electrical stimulation; and manual techniques such as joint mobilization to rehabilitate or habilitate their clients. The therapy used depends on the condition that they are treating. Therapy for orthopedic injuries relies heavily on resistance training and proprioception activities (activities that improve knowledge of position, weight, and resistance of objects in relation to the body), whereas therapy designed to treat people with disease states and neurological conditions tends to emphasize cardiovascular aspects and neuromuscular control.

Entry-level education prepares students to enter the workforce as generalists, but once a person begins to practice professionally, the tendency is to begin to specialize in one of the seven areas outlined on page 400.

Changes in the health care system, such as managed care and the limitation of reimbursement, have influenced the practice of physical therapy. More of the responsibility for patient treatment (the use of therapeutic modalities and therapeutic exercise) is moving to the physical therapy assistant (PTA), while the PT performs patient evaluations and assumes a broader administrative role.

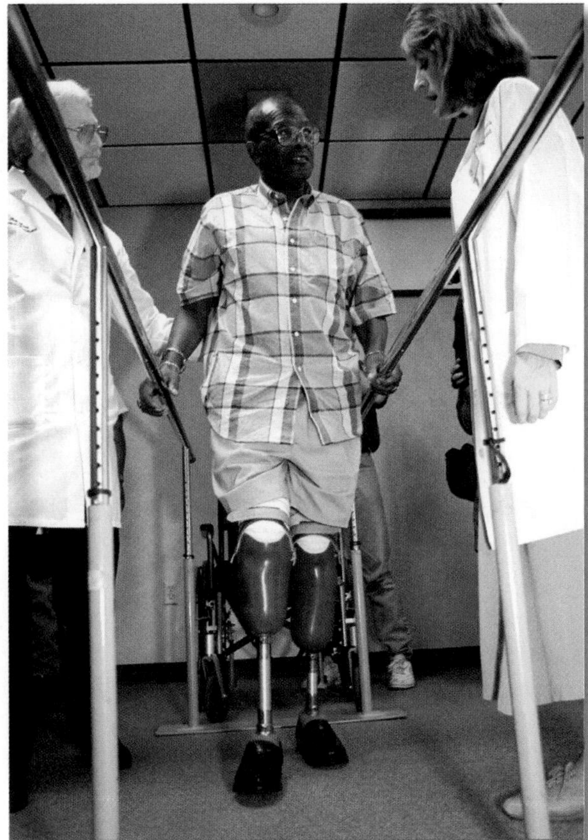

Physical therapists may practice rehabilitational or habilitational therapeutic exercise.

Employment Settings

The employment settings for PTs and PTAs are as diverse as their patient base. Because they tend to specialize in their practice, PTs may work in specialized private clinics or serve specialized (e.g., pediatric or neurological) roles in hospitals, although some professionals still fill the generalist role.

Currently, about two-thirds of PTs work in hospitals, skilled nursing facilities, health and wellness centers, or clinics. Physical therapists may also be in private practice or provide home care services. School systems, colleges, and universities employ PTs to provide on-site services to students. Some PTs hold more than one job—one in private practice and one in a health care facility.

Education and Credentials

The wide range of potential patients and the broad scope of conditions that PTs treat require a vigorous and demanding academic program. The academic program also includes clinical affiliations that expose the student to several work settings and patient types (see the list on p. 401). Entry-level physical therapy education is at the master's degree level, but approximately 80% of entry-level programs are clinical doctorates (DPT). The Commission on Accreditation in Physical Therapy Education (CAPTE) accredits physical therapy programs.

The American Physical Therapy Association (APTA) is the national professional organization for PTs. Although certification is not specifically required, the APTA offers certification for each specialty area. Licensed professionals may also seek advanced education in physical therapy at the master's and doctoral degree level. Physical therapists who specialize in sports medicine have a strong orthopedic interest and may choose to pursue specialty certification in orthopedic or sport physical therapy.

➤ Physical therapists are educated as generalists but tend to develop specialties while in the workforce.

Becoming a PTA involves successful completion of a two-year associate's degree program accredited by CAPTE, the same organization that accredits physical therapy programs. Many, but not all, states require licensure for the PTA to practice. Regardless of whether they are licensed, PTAs work under the supervision of a PT to carry out the prescribed protocol. The PT must perform regular reevaluations of patients who are under the care of a PTA.

Physical Therapy Specialty Certifications

Specialty	Description
Cardiopulmonary	Treatment of patients with acute or chronic diseases of the cardiovascular or respiratory system
Clinical electrophysiology	Measurement of normal and abnormal electrical activity within the human body
Geriatrics	Conditions related to aging or other problems associated with older members of the population
Neurology	Treatment of patients with injuries or diseases of the brain and nervous system
Orthopedics	Treatment of patients with injuries or diseases of the muscles, bones, and joints
Pediatrics	Treatment of children in health and disease during development from birth through adolescence
Sports physical therapy	Treatment of an athletic population, normally people who have incurred injuries as the result of competition

Sample Course Work for Physical Therapy

Sample Pre–Physical Therapy Course Work

Biology
Chemistry
Exercise physiology
Physics
Human anatomy
Human physiology
Psychology/Developmental psychology

Physical Therapist (Entry-Level Clinical Doctorate Degree)

Biomechanics*
Cardiopulmonary evaluation and treatment
Clinical medicine
Geriatrics
Gross anatomy
Management and reimbursement
Medical ethics
Musculoskeletal evaluation and treatment
Musculoskeletal imaging
Neuroanatomy
Neurological evaluation and treatment
Pathology and pathophysiology
Pediatric evaluation and treatment
Pediatric neurology
Pharmacology
Psychological and social aspects of disability
Statistics and research design
Therapeutic modalities

Physical Therapy Assistant (Two-Year Associate's Degree)

Biomechanics*
Clinical practice
Human growth and development
Physical disabilities
Physical therapy procedures
Therapeutic exercise
Therapeutic modalities

*Still called kinesiology at some universities.

Licensure of PTs and PTAs is contingent on the successful completion of the appropriate program (i.e., PT or PTA) and passing the state examination. Historically, the PT's patients were required to have a referral from a physician and physician oversight. Many states now permit direct physical therapy access, thus eliminating the need for a referral. Insurance reimbursement, however, often requires a referral or physician consultation.

The regulations for foreign-educated PTs vary from state to state. In general, these individuals must complete a course of study that is substantially equivalent to that of U.S. physical therapy education and be proficient in the English language. For more information, contact the Foreign Credentialing Commission on Physical Therapy (www.fccpt.org).

Employment Opportunities

The job outlook for PTs in the United States is encouraging. Employment is expected to grow much faster than the average for all occupations through 2014. As with OTs, the demand for PTs is likely to increase as the age of the general population increases.

With advances in medical treatments, a higher percentage of accident victims are likely to survive, increasing the demand for rehabilitative care. Employers, eager to cut down on work-related injuries, are likely to employ more PTs to serve as worksite evaluators and as safety instructors to limit the number of injuries (USBLS 2007). Pending federal legislation imposes limits on reimbursement to therapists and may affect job growth in the short term, but the long-term employment picture is quite healthy.

The median income of PTs was $60,180 in 2006. On average, PTAs earned $37,846. Physical therapists working in home health care settings earned the highest salaries (USBLS 2006).

Therapeutic Recreation Specialist

The physiological and psychological benefits of exercise and leisure activities are described throughout this text. Therapeutic recreation specialists, or recreation therapists, use these activities plus dance, drama, arts and crafts, animals, and community reintegration outings to help people with physical, cognitive, emotional, or behavioral disabilities restore function and reduce or eliminate the effects of disability and develop independence.

Recreation therapists work cooperatively with others on the treatment team to integrate functional activities into an overall rehabilitation program. Depending on the setting, the treatment team may consist of PTs, OTs, speech therapists, physicians, dietitians, and mental health counselors. The recreation therapist focuses on leisure activities to restore function. For example, recreation therapists can use leisure activities such as fishing, painting, or singing to restore limb dominance, to decrease anxiety, to introduce positive leisure values and alternatives, or simply to promote the individual's overall sense of well-being.

In many functional programs, goals are intended to achieve independent function and promote social integration of the patient (Wardlaw, McGuire, & Overby 2000). For example, a recreation therapist working in an outpatient program for people with severe and persistent mental illnesses may have a strong emphasis on

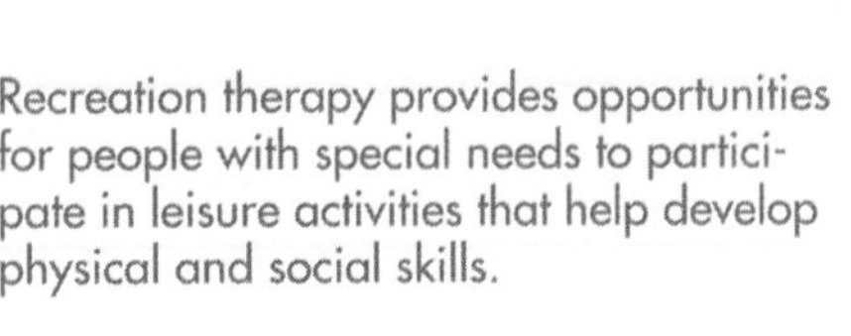

Recreation therapy provides opportunities for people with special needs to participate in leisure activities that help develop physical and social skills.

Sample Course Work for Therapeutic Recreation

Introduction to therapeutic recreation
Assessment and documentation in therapeutic recreation
Therapeutic recreation interventions and techniques
Leisure education
Program planning and design in therapeutic recreation
Professional trends and issues in therapeutic recreation
Human growth and development across the life span
Abnormal psychology
Anatomy and physiology
General recreation course work

working toward social skill development and leisure education. In contrast, a person working with older adults who have dementia may use leisure activities to help slow the decline in cognitive function.

Those considering a career in recreation therapy have an opportunity to work with a wide variety of people in many different settings. A strong belief in the role that recreation plays in creating a healthy lifestyle and an acceptance of people who have different ability levels serve as a personal foundation for entering this career. Recreation therapists learn diverse techniques for helping people meet their personal goals.

Employment Settings

Historically, recreation therapists have worked in inpatient and outpatient settings in psychiatry and physical rehabilitation. Increasingly, recreational therapists (as with other health care professionals) are working with older adults across their continuum of care in facilities such as subacute care settings, adult day care centers, and assisted living residences.

Additionally, many recreation therapists work in community settings, most often community recreation, but also in programs for people with continuing disabilities such as mental retardation or developmental disabilities, cerebral palsy, and so on. In these settings they often help people to continue their treatment programs postdischarge from inpatient programs and introduce people with disabilities to new or modified leisure options. Recreation therapists in community settings work both in segregated programs created specifically to meet the needs of people with disabilities and as inclusion specialists, helping to include people with and without disabilities in typical recreation programming.

Other facilities where recreation therapists are employed include substance abuse rehabilitation centers, residential facilities for people with mental retardation or developmental disabilities, camps for children or adults with disabilities, and public schools. Recreation therapists usually incorporate community settings into their patients' treatment plans. Recreation therapists are often the treatment team members within inpatient settings who determine readiness for discharge by observing the patient's functioning in a community setting during an outing toward the end of a hospital stay. Approximately one-third of recreational therapy jobs are based in hospitals, and another third are in nursing care facilities. Relatively few recreational therapists are self-employed (USBLS 2006).

As with all allied health professionals, recreation therapists assess patients to determine their needs and desires, create a treatment plan in conjunction with the patient, and implement and evaluate the plan. They must be familiar with limitations of medical conditions and be able to keep accurate documentation for medical records. Good interpersonal skills are necessary.

Interviews With Practicing Professionals

Todd Blind, CTRS

Certified Therapeutic Recreation Specialist, The Ohio State University Medical Center

Early in his career, Todd Blind served as volunteer with the Special Olympics, and that experience changed his life. Todd is a Certified Therapeutic Recreation Specialist (CTRS). He has progressed from an entry-level clinician serving stroke, spinal cord injury, traumatic brain injury, and general rehabilitation patients to being the stroke and general rehabilitation team leader at The Ohio State University Medical Center's Dodd Hall Inpatient Rehabilitation facility. Todd's position gives him the chance to work with a multidisciplinary team and offers him constant educational opportunities. The primary role of his position is to help his clients reintegrate into their communities, and provide adapted leisure skills training and leisure education.

Q: Todd, how did you come to your career in therapeutic recreation?

A: I explored several career paths, and I found I liked how this occupation embraces the value of leisure activities in improving a person's physical, cognitive, social, and emotional well-being. I learned that leisure needs to be present in everyone's life.

Many people claim they don't have time for leisure activities, but they don't recognize that some of their most cherished activities are recreation. In my job, I get to use what a person truly values in life as the foundation for developing a therapeutic tool for intervention. It's rewarding to see a person gain insight into their functional potential based on their therapeutic recreation clinical outcomes.

Q: What's the biggest challenge you see facing the profession you work in?

A. The profession is being underutilized. Administrators in most facilities fail to take the time to know the clinical value of a CTRS and what it brings to the health care team. Our role of looking at behavioral and functional change along with lifetime wellness encourages increased quality of life and minimizes a person's dependence on the medical system.

Q: What did you find most valuable in preparing for your career?

A: The courses and experiences that helped me learn to build rapport and trust with the patient, develop rehabilitation goals, and appreciate the many elements of diversity—social, cultural, educational, language, spiritual, financial, age—were instrumental.

Education and Credentials

To enter the therapeutic exercise profession, a bachelor's degree in therapeutic recreation or a degree in recreation with an emphasis in therapeutic recreation is required (see the list on p. 403). A limited number of institutions offer stand-alone therapeutic recreation programs; most commonly therapeutic recreation is an emphasis area in a general recreation program. Entry level for therapeutic recreation professionals requires a bachelor's degree in therapeutic recreation, a term (quarter or semester-long) internship under a certified recreation specialist, and successful completion of the national certification exam. Therapeutic recreational professionals—certified therapeutic recreation specialists (CTRSs)—are credentialed by the National Council for Therapeutic Recreation Certification (NCTRC).

Employment Outlook

According to the U.S. Bureau of Labor Statistics there were 25,000 working recreation therapists in 2006. The median salary for recreation therapists is $32,900 (USBLS 2006). Hospitals, which employ one-third of recreation therapists, are expected to reduce their staffs because of cost containment pressure. Some of these cuts will undoubtedly be recreation therapists. The most promising opportunities for recreation therapists are likely to be at nursing facilities, assisted living residences, and adult day care centers because of the rising population of older adults.

Strength and Conditioning Specialists

People need proper strength and conditioning to obtain maximum physical performance, reduce the frequency of injury, and decrease the possibility of cardiovascular disease. Strength coaches design weight training programs and cardiovascular conditioning programs based on the demands inherent to specific sports and the specific needs of individual athletes. Strength and conditioning specialists may develop individualized programs in conjunction with athletic trainers, PTs, or physicians for athletes who have been identified as having a specific deficit or who have completed their rehabilitation program.

Employment Settings

University athletic departments, professional sport teams, health clubs, and corporate fitness centers employ strength and conditioning specialists. Recognizing the contribution of training programs to athletic success, universities with large Division I-A athletic teams have invested heavily in strength and conditioning programs for athletes. These programs are commonly housed in facilities of more than 15,000 square feet (about 1,400 square meters) and may have a staff of four or more conditioning coaches, each of whom is responsible for one or more sports.

Given the large number of athletes participating in a variety of sports, these facilities tend to be in use year-round. Likewise, a growing number of high schools are recognizing the value of having strength and conditioning specialists on staff. Individuals who have other certifications, licenses, or other areas of specialization can often implement strength and conditioning principles into their current workplaces.

DO it Activity 14.3

What's My Profession?

Guess which profession is being described in this game in your online study guide.

Education and Credentials

Those certified as strength and conditioning specialists often have other professional credentials in areas such as athletic training, exercise physiology, physical therapy, and general medicine. Credentialing of strength and conditioning specialists is not associated with a degree program, but a candidate must have a bachelor's degree and current cardiopulmonary resuscitation (CPR) certification to take the examination sponsored by the National Strength and Conditioning Association (NSCA). A degree in kinesiology is preferred. To receive certification, candidates must successfully complete a two-part examination.

➤ A teaching degree is useful for gaining employment as a high school strength coach; college athletic departments often look for candidates with a degree in exercise physiology or a related field.

Passing both sections designates the person as a certified strength and conditioning specialist (CSCS). Credentialed professionals who become certified as strength and conditioning specialists can compete more strongly in the job market and increase their salaries. Most National Strength and Conditioning Association members hold a degree beyond the level of a bachelor's degree.

Individuals interested in becoming a strength coach for a collegiate or professional team should also possess a bachelor's degree in a related field such as exercise physiology or kinesiology. Although no uniform course of study is offered at the undergraduate level, an emphasis in fitness leadership with courses in performance assessment, program organization, and exercise leadership, in addition to the scientific core of courses normally offered in a kinesiology undergraduate program, provides excellent preparation. Obviously, an internship in a department of athletics conditioning facility is useful, as are appropriate volunteer experiences. Certification as a teacher would enhance the possibility of being hired as a high

Are You Clinically or Technically Oriented?

Do you prefer to devise solutions on your own, or would you rather follow a structured sequence of steps? Clinicians devise and develop therapeutic exercise plans and are therefore called on to solve problems and make decisions. Technicians, on the other hand, are experts at performing specific sets of skills.

Technical professions include PTAs and OTAs. Clinical professions include athletic training, physical therapy, occupational therapy, and higher levels of cardiac rehabilitation. Working with student-athletes as an athletic trainer is an option for those interested in more clinically oriented careers.

school strength coach. A graduate assistant position or internship position as a strength and conditioning coach also improves a person's marketability.

Employment Opportunities

No specific data on strength and conditioning specialists is available, but this trend follows the same pattern as that for fitness workers in general, whose job numbers are expected to increase faster than average for all occupations. More people are spending more time and money on fitness, and health clubs are providing more personal services to retain their members.

Inside Advice for Therapeutic Exercise Students

You should understand by now the importance of having keen knowledge of human anatomy, human physiology, and kinesiology, and of knowing the effects of exercise on healthy and unhealthy individuals when working in the sphere of therapeutic exercise. These areas, however, represent only a partial list of the skills and knowledge required to practice in this sphere.

Exercise is therapeutic only when it is performed in a safe and appropriate manner. Developing and delivering appropriate therapeutic exercise routines require the ability to solve problems, collect information regarding the patient's or client's condition, identify the therapeutic goals, and determine the appropriate course of action. To solve problems effectively, therapeutic exercise professionals must be able to access, manage, and interpret various forms of information and apply them to the current case. This task often requires the use of computers and an understanding of how to access Internet sites and perform literature searches.

Health care professionals also must be able to communicate effectively across demographic, sociocultural, and professional boundaries. Certain settings or roles require additional areas of knowledge. For instance, professionals who work in fee-for-service environments need to become familiar with record keeping and insurance billing systems. A business background would be useful for practitioners in private practice, and a person in any type of leadership position would benefit from course work dealing with management and administration.

Skills Analysis

If you are interested in entering a therapeutic exercise profession, you will want to work toward a degree in kinesiology as well as become certified in a particular area. What other skills do you need to learn? Do you have good problem-solving and communication skills? Will you enjoy the business sides to these professions—billing, record keeping, and so on? In what ways can you test the waters of these professions to see whether they are for you?

So how does one choose an area in which to work? Although reading about these professions can provide you with an overview of this group of professions, the only way to discover what a profession is really like is to see professionals at work. Many professionals allow students to shadow them throughout the course of a day. Perhaps a professor or other acquaintance can help you make this connection. You may also have an opportunity to do an internship or apprenticeship or find a part-time or summer job in one of these areas.

Earlier in this text we discussed ethical professional conduct; we must reiterate its importance here. Therapeutic exercise professionals must adhere to basic ethical and professional principles that are often reinforced through state laws and certification guidelines. Although the particulars vary from profession to profession and from state to state, some themes are common across the professions.

First, practitioners must protect patients from harm and maintain the confidentiality of medical records. Likewise, practitioners must adhere to the profession's scope of practice—the legal parameters that define the profession. Although your instructors and supervisors will identify these issues for you, it is ultimately your responsibility to seek out this information and assure that you remain in compliance. Failure to do so could result in criminal liability.

Wrap-Up

Professions arise to fulfill a societal void. The therapeutic exercise professions fill a void by either assisting people in obtaining their desired level of physical fitness (habilitation) or helping injured individuals regain lost function (rehabilitation). Part of the attractiveness of this sphere of professional practice is the wide range of the population that is served. Highly honed athletes, newborn infants, and the geriatric population may all rely on therapeutic exercise specialists.

Almost as diverse as the population being treated are the settings in which therapeutic exercise professionals work. From the structured environments of hospitals and laboratories to athletic training rooms with their hectic pace, these work settings appeal to a wide range of interests. This chapter covered some of the more prominent therapeutic exercise professions such as athletic training, cardiac rehabilitation, occupational therapy, and physical therapy. If you are interested in any of these professions, we encourage you to seek more information from your professors and to explore this area in greater depth.

Activity 14.4

Use the key points review in your online study guide as a study aid for this chapter.

Activity 14.5

These end-of-chapter questions and activities are also in your online study guide. Your instructor may ask you to complete them online and turn them in.

1. Describe how therapeutic exercise promotes healthy lifestyles.
2. Discuss the similarities and differences between habilitation and rehabilitation.
3. Identify the common overlaps among the therapeutic exercise professions described in this chapter.
4. What skills or attributes are unique to the professions presented in this chapter?
5. Identify courses that are common among the professions presented in this chapter.
6. People obtain multiple credentials or specialization to make themselves more marketable. Describe some possibilities for dual credentials and cross-training based on the descriptions of professions in this chapter. What benefit would dual credentials and cross-training provide? What professions not described in this chapter would also lend themselves to dual credentials?

Careers in Teaching Physical Education

15

Kim C. Graber and Thomas J. Templin

Eyewire

CHAPTER OBJECTIVES

In this chapter we will

- describe what a physical education teacher does,
- provide insights into exciting research,
- explain how to remain current in the field,
- describe settings in which physical educators teach,
- describe highly effective teachers, and
- provide information about career options.

The head of a large district's physical education program is leading the beginning-of-the-year in-service event for the district's 12 secondary schools, and he introduces the four new physical activity specialists who have joined the district's staff. It is an exciting time because they represent a new generation of teachers who were educated at their colleges and universities to address the obesity epidemic, different dimensions of wellness, and physical inactivity patterns that characterize an increasing number of adolescents in the nation. Each has the potential to make a significant contribution to the school in new ways.

Tara was hired because of her expertise in physical fitness. She will design an individualized fitness class that will be a requirement for all students prior to graduation. She will oversee all fitness facilities at her school and will supervise those physical educators who teach classes designed to improve students' health-related fitness levels. Robert is an exciting addition to the district's teacher pool because he has expertise in physical education curriculum development. He will integrate sport education and outdoor pursuits into the curriculum across the district and will host a series of workshops designed to train current teachers in the effective use of class time. Tessa will serve as the physical activity director at one of the district's middle schools, working with classroom teachers to encourage physical activity across the curriculum. She also will serve as a consultant to the district-wide nutritional staff, particularly as an advocate for providing students with healthy but appealing nutritional options in vending machines. Finally, Peter will serve as the community relations and before- and after-school physical activity director at the district's newest middle school. He will encourage a physical activity link between the school and community and will plan exciting physical activity options in which students can participate prior to and after school. If his program goes well, the district supervisor hopes to expand it to other schools.

As a participant in a physical education class, you probably were concerned primarily with improving your performance and fitness levels, playing well during competition, and enjoying the experience. You may have felt proud of your ability, hard work, and dedication when you had an especially successful performance. The truth of the matter is that if your performance improved, and if you enjoyed the experience, your teacher was largely responsible. That is, the teacher planned effective class sessions, executed appropriate instruction, kept participants actively engaged in suitable learning tasks, used proper technology, maintained a positive learning environment, and provided useful feedback.

➤ Good teachers work hard to create an environment in which all students can learn.

Effective teachers understand their subject and know how to convey it appropriately. Expert teachers in river rafting, for example, know which techniques should be used to propel a raft when the river is smooth as opposed to when it is at its most dangerous. They understand when to expect a change in conditions of the river and how to maneuver safely and effectively from one type of water to another. They also know which techniques are most appropriate for novices and which are best for experts. They convey instructions in a manner that is understandable and provides the best practice opportunities for learners to accomplish new skills.

As the scenario at the beginning of the chapter suggests, to teach others successfully requires a background in and knowledge about the curriculum and **pedagogy** of physical education. Without a background in pedagogy, physical activity instructors tend to "wing it," an approach that can sometimes work out fine in the short run but often leads to wasted time and effort—and stress for the teacher. In the opening scenario, the four teachers were hired because of their expertise as instructors in different dimensions of physical activity. Not only do they have good knowledge about curriculum development, but they are effective teachers with exceptional knowledge about pedagogy. They know what it takes to get students to learn.

- If you are a teacher, how will you get and keep the students' attention?
- How much time should you spend talking, and how much time should students spend practicing?
- Will your methods work?

- How will you structure class to ensure that students have adequate time during class to improve their fitness levels?
- How will you motivate students to engage in physical activity outside of the school setting and make appropriate nutritional selections?

A person with a background in pedagogy knows the answers to these questions. The knowledgeable teacher knows that the best way to get and keep students' attention is to establish rules and procedures for good behavior in the gymnasium and on the playing field. An effective teacher with a background in pedagogy knows that practicing a skill correctly and in game-like situations is more important than listening to a 20-minute lecture on basketball techniques. An effective teacher also understands the critical importance of physical activity to children's health. The teacher knows that today there are three times as many overweight children as there were in 1980 (U.S. Department of Health and Human Services 2001) and is concerned that this is the first generation of parents who may have a longer lifespan than their children, but believes that physical education can help reverse these negative trends.

Activity 15.1

Teacher- or Student-Focused? Use this online study guide activity to distinguish between the two types of instruction.

Webster's defines *pedagogy* as the art, science, or profession of teaching. For our purposes, we'll modify that definition a little and say that pedagogy focuses on *teaching behaviors* and *producing learning in students*. Although the distinction between teaching and producing learning might not be immediately clear, what it comes down to is focusing on the student instead of the teacher. Rather than approach each class with the question "What am I going to teach today?" effective teachers instead ask themselves, "What are my students going to learn today?" This approach can lead to improved interaction between the learner and the teacher and thus to a better education for students.

Ideally, then, those who pursue pedagogy as a career choice are concerned with producing some type of learning in students. Whether you want to become a physical education teacher, physical activity director, or high school coach, you need to learn the skills associated with effective pedagogy. Typically, prospective teachers gain pedagogical knowledge through study in an accredited agency or program, such as an education department at a university; other times they might acquire the required knowledge through working with a mentor, such as a cooperating teacher in a public school, or by attending a state teachers' conference. Ideally, though, all would-be teachers will supplement what they have learned about teaching with further information from the knowledge base that concerns itself with the study of the theories and methods of teaching physical education.

How Would You Rate Your Experiences in Physical Education Class?

Reflect on your own experience by answering the following questions:

1. Who stands out in your memory as the best physical education teacher you've ever had?
2. What qualities did this teacher have that made him or her effective?
3. Who stands out in your memory as the worst physical education teacher you've ever had?
4. What weaknesses did this teacher possess, or what qualities did he or she lack, that made him or her ineffective?
5. What do you think are the most important traits that a teacher should possess?

An Overview of the Teaching Profession

Research has long demonstrated that undergraduate students who are interested in teaching often believe that they already know all there is to know about teaching and have little more to learn (Lanier & Little 1986). This belief results from your experiences with and observations of professional practice while you participated in physical education or worked as a coach, camp counselor, or supervisor in a recreation center. Consider that you have spent approximately 13,000 hours as a student. Add to these the hours that you spent in other learning environments, such as summer camps, recreational sport programs through school or the park district, or private instruction, and you'll find that you've accumulated many strong beliefs about teaching (Lortie 1975; Schempp 1989).

➤Experiences and beliefs acquired during pretraining will influence all instructors, but the best instructors base their programs and teaching methods on sound pedagogical precepts.

Despite the logic of emulating someone whose qualities you admire, relying only on experience has some drawbacks. Because teachers influence different students in different ways, you can't assume that everyone shared your positive experience with a particular instructor. Nor can you assume that a teacher is effective because everyone liked him or her. For example, say you were a starter on your high school's basketball team. Because of your status, your physical education teacher may have allowed you to skip class on game days, singled you out to demonstrate new basketball skills to classmates, and given you more opportunities than others to play basketball during class. Such personal experiences might lead you to believe that the pedagogy was sound. You may believe that athletes *should* be released from class before a game, that better players *should* demonstrate their skills for the benefit of their classmates, and that higher-skilled students *should* receive more opportunities for activity than lower-skilled students receive. Despite your personal experience, however, all these beliefs may be mistaken, because they are based on unsound, unfair, or unproven teaching principles. But the scenario might have been different. Your teacher may have realized that not all students in class were as highly skilled as you were. In response, this teacher may have spent many hours planning to ensure that all students had an opportunity to improve their skill level and to experience success. This teacher's practices were grounded in effective pedagogy.

Although it is probably human nature to accept and imitate the behavior of people you respect and reject the behavior of those you do not respect when forming your own pedagogy, a better approach is to analyze the available research about teaching and learning and base your behavior and values on this analysis. To some degree you will always refer to personal experience when making decisions about instruction and designing curriculum, but this experience should never be your only criterion.

This chapter will introduce you to the study of pedagogy of physical education (see figure 15.1), sometimes referred to as **sport pedagogy.** As you read this chapter, think of your educational experiences thus far, asking yourself whether you received effective instruction and why or why not. Consider how

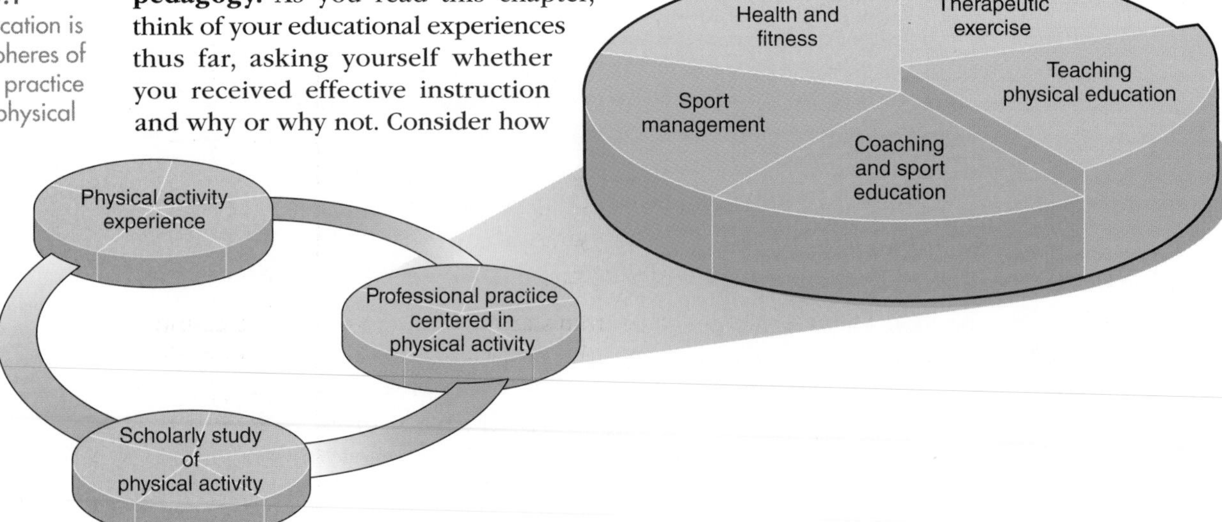

FIGURE 15.1 Physical education is one of the spheres of professional practice centered in physical activity.

you might incorporate some of the findings from research into your teaching, particularly if these findings contradict your experiences as a student. Because most of the research on pedagogy focuses on teaching in public schools, much of this chapter is about public school physical education. We will, however, try to make connections to other types of physical activity instruction, for example the instruction delivered by community recreation directors, coaches, or exercise instructors for older adults.

Similar to those in other specializations described in this book, experts in physical education teaching have many career options. Generally, most elect to obtain certification and become teachers and coaches in public schools. On the basis of state certification requirements, individuals elect a grade level in which they intend to teach and enroll for courses that specifically address how to instruct children of that age. Although many choose to work at the elementary level because they like interacting with young children and appreciate the enjoyment that youngsters experience while engaged in physical activity, others decide to work in middle or high schools because they also wish to coach. As a means of supplementing their salary in the summer months, many teachers also work as camp counselors, swim instructors, and recreation leaders.

Some people who enter departments of kinesiology for purposes of becoming physical activity instructors do not intend to teach in the public schools. Instead, they decide that they are better suited to teaching in private schools or community recreation centers. Some may elect to become specialists, such as wilderness experts or golf professionals. Other will choose to work at a college or university as an instructor or professor. Most, however, elect to become instructors of physical activity because they are interested in improving the health and wellness levels of the population, engaging people in activities that they can continue throughout their life span, and working with children in schools.

Certification and Continuing Education

As mentioned above, although the majority of students who are admitted into a department of kinesiology or physical education certification program intend to teach in the K through 12 schools, others may wish to become recreation specialists, fitness trainers, coaches, or outdoor education experts. Regardless of their career path, students must meet many program requirements before formal admission into a teacher education program and before certification. These requirements typically involve the following:

- Grades or grade point average: In most teacher education programs, students must maintain a minimum grade point average. In fact, most programs require a minimum grade point average just for admission into the program. Often, students must maintain a grade of B or higher in their professional education courses to continue in the program and graduate with an emphasis in teacher education.
- Praxis I: Before admission into certain courses (e.g., methods courses) or a particular phase in a program, students are required to complete a preprofessional skills test (PPST or Praxis I), entailing a series of tests developed to assess basic proficiency in reading, writing, and math. Students must achieve minimum scores in each area to progress in the program.
- Praxis II: Before admission into certain courses or student teaching, students must pass Praxis II. This exam evaluates specialty content knowledge and pedagogical knowledge (methods and curricular knowledge). The Educational Testing Service administers the test.

Although the requirements for certification vary from state to state and country to country, they usually involve a series of exams. In addition, the program in which you are enrolled is responsible for ensuring that you have developed adequate knowledge and have certain competencies. State officials periodically evaluate the curriculum in these programs. If a teacher education program fails to meet established state standards, the program is placed on probation and may lose its ability to offer teacher certification programs to students.

Web Search 15.1

Certification
Research requirements for a specific state of interest to you in this activity in your online study guide.

After a student meets all program and state testing requirements, he or she may apply for a teaching license. In many cases, reciprocity between states enables a graduate from a program in one state to teach in one or more other states throughout the country. Some states require teachers to complete continuing education courses on a periodic basis to maintain a teaching license. This requirement helps ensure that teachers remain current and knowledgeable about the most effective teaching practices. For example, students may be required to take 12 credits of continuing education courses every four years in order to remain certified. They can usually accomplish this by taking courses at a college or university, enrolling in a master's degree program, or attending approved professional conferences. Requirements, however, vary considerably from state to state.

Major Professional Associations and Journals for Sport Pedagogy

Professional Associations

- American Alliance for Health, Physical Education, Recreation and Dance (AAHPERD)
- National Association for Sport and Physical Education (NASPE)
- Curriculum and Instruction Academy (a division of NASPE)
- American Educational Research Association (AERA)
- Special Interest Group (SIG; a division of AERA that focuses on research on learning and instruction in physical education)
- National Association for Kinesiology and Physical Education in Higher Education (NAKPEHE)
- Association Internationale des Ecoles Supérieures d'Education Physique (AIESEP)

Journals

- *Action in Teacher Education*
- *American Educational Research Journal*
- *College Student Journal*
- *Educational Technology*
- *Elementary School Journal*
- *High School Journal*
- *International Journal of Physical Education*
- *Journal of Educational Research*
- *Journal of Physical Education, Recreation and Dance*
- *Journal of Research and Development in Education*
- *Journal of Teacher Education*
- *Journal of Teaching in Physical Education*
- *Physical Educator*
- *Quest*
- *Sport Education and Society*
- *Strategies*
- *Teaching and Teacher Education*

With regard to other physical activity positions like coaching, some schools mandate that coaches be certified teachers; others require only an undergraduate degree, and the remainder expect coaches to have specialized training or an endorsement from a group such as the American Sport Education Program. Similarly, fitness professionals are often expected to become certified as a personal trainer. Some students enter colleges and universities for their training whereas others obtain certification through online courses.

www. **Web Search 15.2**

Resources for Professionals
In this activity in your online study guide, you'll explore the NASPE Web site.

In the United States, the easiest way to remain current and informed is to join the National Association for Sport and Physical Education (NASPE). This association is one of five health-related associations that reside within the American Alliance for Health, Physical Education, Recreation and Dance (AAHPERD). By connecting to the AAHPERD Web site (www.aahperd.org), you can access additional information about the six associations and about NASPE specifically (www.aahperd.org/naspe/). Although dues occasionally fluctuate, the annual student fee is much lower than the fee for full-time professionals. For your annual dues, you will receive a monthly subscription to *Journal of Physical Education, Recreation and Dance* and will be able to attend the national convention at a reduced rate. In addition, you will be able to log on to a members-only Web site where you can access the most recent and important updates within the field.

➤ A large number of journals are geared specifically to teachers. They are designed to provide information about recent advances in the field, innovative teaching strategies, and results from research that teachers can easily interpret. To remain current, teachers should subscribe to and read as many journals as time permits. Good journals to begin with are *Journal of Physical Education, Recreation and Dance; Strategies;* and the Canadian *Physical Education and Health Journal.*

In Canada you will want to remain informed by joining the Canadian Association of Health, Physical Education, Recreation and Dance (CAHPERD). This association has representation in all provinces and is committed to advancing the field of physical education. The association offers *Physical Education and Health Journal,* which contains much useful information for teachers. AAHPERD and CAHPERD members often attend the same national conferences. For more information you can log on to the CAHPERD's Web site (www.cahperd.ca/eng/index.cfm).

Considering the cost of your undergraduate education, the annual fee for joining these associations is a modest means of remaining current. Becoming a member is a professional responsibility that informed professionals take seriously. Other professional associations and journals are listed on the previous page.

Examining Research to See What Effective Teachers Do

We can learn much about what teachers do by examining the wealth of research that has been conducted in the area. Research on teaching pertains to the teaching and learning process that occurs in educational settings such as public or private schools. Researchers in a typical study might try to determine whether students learn better through one method of instruction than another. They might look at how many times an instructor provides feedback to individual students during class and whether that feedback is specific or general. Or, they might investigate whether students who receive more practice opportunities during class learn more than those who receive fewer opportunities. Many of the findings about effective teaching are transferable from one setting, like a school, to another setting, like a recreation center.

Research on teaching is concerned with the scientific study of the processes of teaching and learning, and is deeply rooted in what happens in education generally. Over the past two decades, scholars have undertaken many different types of investigations and have examined many aspects of teaching.

Over the last several decades, researchers have studied the characteristics of effective teachers and their classrooms. Although the majority of research examining the effectiveness of physical activity instruction has occurred within the realm of physical education, many of the findings are easily transferable to other educational settings. As such, research on effective teaching in physical education can inform other types of physical activity instructors.

Teaching Expertise

➤ Teaching experience alone does not guarantee expertise.

We asked you earlier to consider how experience may or may not relate to effective teaching. Although experience can enhance teacher effectiveness, it doesn't guarantee effectiveness (Graber 2001). Unfortunately, some teachers become bored or burned out with teaching and lose interest in learning about ways in which they can improve their teaching or influence student leaning. Although we would like to believe that this group represents a small minority of teachers, boredom and burnout are problems in many schools, particularly at the high school level. These teachers no longer learn from experience, nor are they interested in reading professional journals, taking graduate classes, or attending professional conferences. On the bright side, thousands of effective teachers continue to learn from their experiences. Some of these teachers also have a high degree of expertise. Experts, and those who continue to learn about teaching by reading and by attending professional conferences, are probably less likely to experience burnout.

Berliner (1988) proposed that teachers progress through a series of stages as they gain expertise. Some scholars believe that expertise relates to being able to perform well in a particular domain. For example, Marcus Camby, Tiger Woods, and Michelle Kwan can be characterized as having high levels of performance skill in their respective sports of basketball, golf, and figure skating. Other scholars, however, argue that although high skill level is important, it is not adequate for ensuring that a physical education teacher has teaching expertise (Dodds 1994).

Appropriate Practice Experiences for Students

Your experiences in physical education, recreation, or camp settings were probably similar to those that many others had as children. If so, your teachers and counselors probably had you play dodgeball, kickball, "steal the bacon," and other such games. As you progressed into the middle and high school grades, you may have received a few days of instruction before playing softball, basketball, football, volleyball, soccer, and the like. If you are now interested in pursuing a career in the physical activity field, you probably had above-average skills in these games and sports—you might have enjoyed eliminating other students in dodgeball or scoring the winning goal in soccer. Such memories probably remain satisfying for you and likely influenced your interest in a career in physical education.

Unfortunately, probably not all your classmates shared your positive experience when engaged in physical activity. For those students who were less skilled—the ones you quickly eliminated from games such as dodgeball—physical activity time was, at best, supervised recess and, at worst, an opportunity for embarrassment. For most of those students, no instruction was available and thus they had little chance of improving their skills. Think about it: Did you acquire your athletic expertise by participating in physical education class?

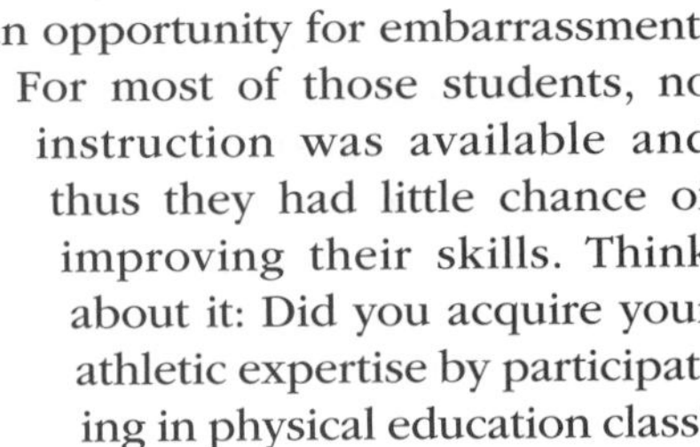

Kids with above-average skills often enjoy PE class.
AP Photo/Seth Perlman

Or did you acquire skill through extracurricular activities organized by coaches (or parents) who were determined to help you improve your playing ability?

Defending physical education class and nonstructured recreation programs as opportunities to improve skill level is sometimes difficult. For example, if a student spends one or two days practicing a volleyball bump pass before being expected to perform the pass correctly during a game, how much chance does the student have to succeed? Consider that serious volleyball players spend hundreds of hours developing their skills. For students to succeed at any skill, they must be exposed to *appropriate practice.* When designing activities, instructors need to consider whether drills simulate game situations. For example, the most appropriate method of practicing a volleyball set is to practice setting the ball to a target area from a position in which the volleyball is bump passed to the setter, not tossed. This drill closely resembles a game situation. (Remember the principle of quality and the principle of quantity discussed in chapter 3.)

Active Learning Time

Do you recall the many times as a student you had to wait in line for a turn? If you're like many students, you sometimes had to wait through half the class for an opportunity, say, to bounce once or twice on a trampoline. Or maybe you played volleyball with 11 others on your side of the net, waiting minutes at a time for a chance to touch the ball. If this sounds familiar, you're not alone in your experience. For various reasons, thousands of students each year receive inadequate opportunities to practice and learn new skills.

Research conducted in the classroom (Brophy & Good 1986) has clearly linked time on task to student achievement. Time on task, also called **engaged time,** is defined as the time students spend actually doing physical activity or sport. Engaged time in physical activity settings is traditionally low. In fact, research from several decades ago indicates that students in physical education classes spent only about 30% of class time engaged in physical activity (Anderson & Barrette 1978; Metzler 1979, 1989). Historically, students have spent 50% to 70% of the class waiting, in transition, and being managed by the teacher. Fortunately, today's teachers are being trained to engage students in activity for at least 50% of the time, which is what experts recommend (Siedentop & Tannehill 2000). There also is increased emphasis on making sure that students are *appropriately* engaged (performing correctly with frequent success). Such time is often called **academic learning** or **functional learning time.**

When students are idle in physical education class, little skill or health fitness is likely to develop.

Considering that physical education class is the only opportunity throughout the entire school day that some students have to be engaged in physical activity, physical education teachers must keep students active (and successful). Given the obesity epidemic and the high rate of physical inactivity that characterizes people throughout the world, physical education teachers must also be committed to keeping students active in appropriate learning activities for the majority of the class period.

Did You Receive Adequate Opportunities to Participate Actively in Physical Education Class?

Think back to your days in a structured physical activity setting and recall a class in which you engaged in physical activity for only a small percentage of the time. What might have accounted for the low percentage of active learning time? Lack of equipment? Poor class management? Listening to the instructor talk? Waiting in line?

Do you recall any classes in which you were engaged in active learning for a high percentage of class time? What did this instructor do differently to increase your active learning time?

Effective Class Management and Discipline

Probably the most common concern among new teachers is class management and discipline. *Class management* involves organizing students in such a way that learning is most likely to occur, whereas *discipline* involves teaching rules, enforcing them when they are broken, and rewarding exceptional behavior. Regardless of their experience level, teachers can never prepare for all the situations that they might encounter. How do you handle a student who refuses to take a time-out? How do you deal with a student who ridicules other students?

➤ As a rule, instructors can best assist students as they learn rules and routines by (1) having high expectations, (2) being firm but warm, (3) developing clear rules, and (4) describing how rules will be enforced.

The best that instructors can do is to learn to take proactive measures that tend to decrease potential difficulties. George Graham (2001) has suggested ways to create a positive learning environment. In general, he suggests using the first few classes of a new academic year or unit to establish and implement rules and procedures for good behavior. For example, in a physical education class, students need to learn the new teacher's signals, such as start and stop signals, and other signs to look or listen for during class that tell them to behave in a certain way or that a particular behavior choice is good or poor. If students learn such lessons early, the entire school year will run more smoothly for the class.

In a study of seven effective elementary specialists, instructors were observed teaching students the stopping and starting signals for class by using the signals a total of 346 different times during the first few days of instruction (Fink & Siedentop 1989). After the first few classes, most students were behaving appropriately; teachers promptly reprimanded those who did not follow the signals. By implementing this routine during the first few days of every unit, these teachers quickly constructed an environment in which learning could take place.

Accountability

Recalling your experiences as a student in physical education, do you remember being asked to perform a dozen push-ups, only to stop as soon as the teacher turned away to take attendance? Some students try to manipulate their teacher's expectations to suit their own purposes. In an eye-opening study of public school physical education, students were observed hiding from the teacher and acting as though they were actively engaged (Tousignant 1981). The investigator referred to these students as **competent bystanders**—well-behaved students who consistently avoided participation without attracting notice. Although this study occurred many years ago, if you walk into a high school physical education class, you will notice a new generation of competent bystanders.

➤ Students manipulate the learning environment when they engage in off-task behaviors or become competent bystanders. By ignoring off-task behavior, instructors encourage further manipulation.

DO it Activity 15.2

Is This Teacher Effective?
Observe a teacher in action and evaluate her effectiveness in this activity in your online study guide.

Students learn accountability through clearly stated and consistently enforced expectations. One instructor characteristic that promotes student accountability is called "with-it-ness." Teachers demonstrate their

with-it-ness by knowing what's happening in the learning environment and by displaying this awareness through oral or other communication with students. By holding students accountable for their actions, these teachers reduce the likelihood that students will try to manipulate the learning environment.

Specific Feedback

Besides providing clear instructions, teachers must also provide clear, specific, and immediate feedback (Rink 2006). Unfortunately, teachers often provide too many instructions, and students either become bored and stop listening or cannot remember all that the teacher said. In contrast, during practice, teachers seldom provide students an adequate amount of correct, prompt, and specific feedback.

The research and our own experiences indicate that teachers make the following mistakes when providing feedback.

- Feedback is often incorrect. For example, a teacher may not see the fundamental error committed by a student and focus instead on a secondary feature of performance.
- Teachers sometimes focus on an aspect of performance that does not require feedback while neglecting an area that does require feedback. For example, Stroot and Oslin (1993) found that teachers neglected to provide feedback on a component of the overhead throw because they were unable to diagnose deficits, choosing instead to provide feedback on aspects that students had already mastered.
- Some teachers do not time their feedback so that the learner receives prompt help when practice trials are defective in form or when they produce unsuccessful results. Practicing incorrect performance for an extended period has predictably negative results!
- Teachers provide less feedback during game play. Feedback provided during game play can be valuable; it gives all students information about ways to improve.

➤Teachers can increase the probability that instruction will be effective by providing appropriate learning activities; maximal active learning time; and correct, prompt, and specific feedback.

Alternative Curriculums

You and most of your classmates were likely exposed to a traditional curriculum in which instructors designed activities with little student input and employed a teacher-dominated approach. Although a traditional approach can be effective if used by a creative teacher, alternative curriculums offer strong promise and can be implemented in many different environments.

Physical education today is much more exciting than ever. Although the obesity epidemic is a serious health concern, it can be credited with motivating teachers to make improvements in their physical education programs. You may have heard the term "new physical education." Teachers who incorporate novel and exciting learning ideas in their classes, particularly those who emphasize fitness and nontraditional activities like cycling, rock climbing, and in-line skating, are said to be teachers of this new movement of physical education. They are sometimes profiled in the national media as having changed the face of physical education.

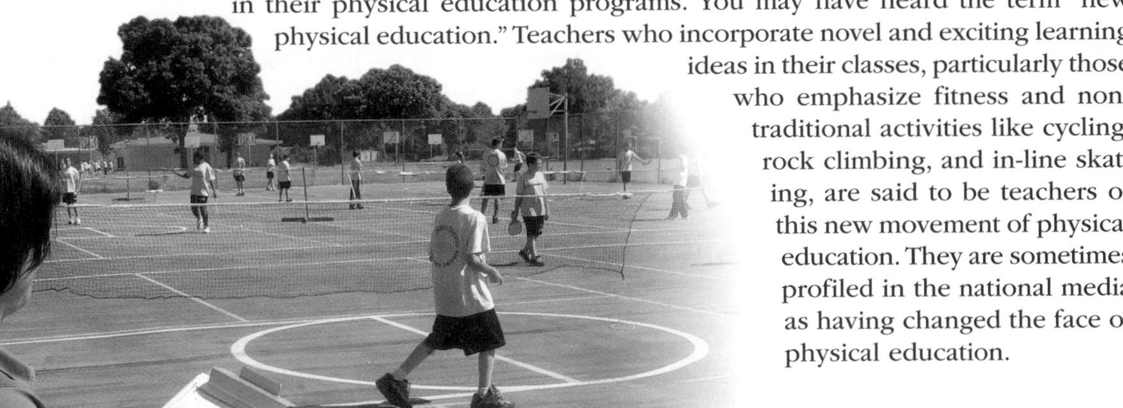

Effective teachers carefully monitor students throughout the entire lesson.

Alternative curriculums offer exciting possibilities for students who may not be interested in traditional activities.
AP Photo/Gene J. Puskar

- The *elective* curriculum allows learners to choose one activity from a wide selection of activities. They are encouraged to develop skills that will transfer into a lifetime interest in that activity. Many large-school programs offer the elective curriculum to students.
- The *fitness curriculum* emphasizes cardiorespiratory efficiency, muscular strength and endurance, flexibility, and body composition, particularly at the secondary level. Students participate in health-related fitness activities such as running, swimming, circuit training, cycling, in-line skating, and aerobics (Corbin et al. 2001). This type of curriculum is becoming increasingly popular because it is perceived as an opportunity to address such health concerns as the obesity epidemic during physical education. If implemented appropriately, it can provide students with an opportunity to become enjoyably engaged in moderate to vigorous physical activity. Physical educators must not forget, however, that it is equally important to teach skills to students so they can successfully engage in multiple activities throughout their life span.
- The *sport education* model appears to hold promise, particularly as a means of improving secondary physical education (Siedentop 2004). In this model, teachers treat students as athletes. Students are formed into teams and not only learn how to play but also assume some responsibility for roles as managers, coaches, trainers, officials, statisticians, and tournament administrators.
- The *wilderness and adventure education* curriculum introduces learners to such activities as canoeing, backpacking, camping, white-water rafting, skiing, first aid, ropes courses, climbing walls, and new and cooperative games. Evidence suggests that wilderness and adventure education can be highly appealing to students (Dyson 1995).
- The *social development model* is concerned with teaching students self-control and responsibility and draws heavily on the pioneering efforts of Hellison (1995). This model is particularly effective in environments in which students have not had opportunities to develop personal responsibility and positive social skills.
- The *teaching games for understanding* model has received increasing attention from both teachers and researchers. Whereas the traditional way of teaching games is first to ensure that students have acquired the motor skills necessary for game play, the teaching games for understanding model emphasizes teaching tactical awareness, game strategies, and game appreciation while placing less emphasis on initial motor skill development (Griffin, Mitchell, & Oslin 1997).

➤ Effective teachers are concerned with implementing curricular models that are interesting to students and produce the greatest opportunity for student learning.

Equity Issues and Student Needs

The learning environment in which physical education occurs is often an inequitable environment in which some students receive less (or more) attention from teachers because of race, gender, physical ability, physical appearance, or socioeconomic status. Despite measures such as Title IX and Public Law 94-142, students continue to be disadvantaged for reasons beyond their control.

In physical education settings, one of the most common forms of discrimination is a bias against students with less ability. Elimination games, for example, are a common activity. Students with the lowest levels of skill are usually the first eliminated, although they are the ones most in need of practice. Furthermore, classmates often ridicule these students. If you were not highly skilled, how would you feel when it was your turn to roll on a scooter during a relay race, particularly if your team was in the lead?

Another common means of discrimination is to single out obese and out-of-shape students in negative ways. Interestingly, the obese student who gets lapped while participating in the mile run may actually be working harder than the student who crosses the finish line first. For this reason, an increasing number of teachers are beginning to use technology such as heart rate monitors as standard practice to more fairly assess student performance.

➤ Teachers must ensure that low-skilled or obese students will not be ridiculed or embarrassed.

Similarly to low-skilled students, females are often treated in negative ways. In 1984, Patricia Griffin published the results of a classic study in which she described six patterns of female involvement during coeducational physical education classes. Whereas some females were highly skilled and participated vigorously, others had much greater difficulty when participating in game situations. Some barely participated in physical activity during class and instead assumed the role of "cheerleader," "femme fatale," "lost soul," or "system beater." If you are female, do you see yourself on this list? Although Griffin (1984) conducted the study over 20 years ago, her descriptions of female behaviors during physical activity are far from obsolete. It remains true that far too few instructors make efforts to ensure that girls have an equitable, enjoyable, and successful experience in physical activity.

Teachers must also consider other factors that might be affecting the students they teach. Students come to school with a host of problems that interfere with their ability to learn. Studies have shown, for example, that students' grades typically decline when their parents undergo conflict or divorce. In some areas such as inner cities, learners face a multitude of serious problems (gang involvement, poverty, lack of adequate nutrition) that get in the way of or even prevent learning (Tyson 1996). As a result, instructors often confront a need to address the emotional and material needs of students and have far less time for concern about physical activity (Ennis 1994).

➤ All learners carry personal problems into the learning environment at one time or another. Instructors need to be aware of this and be ready to adjust their teaching approach as necessary.

Expectations for Students

Teacher expectations can significantly influence the learning process. Expectations may concern a group of students or an individual, and they typically center on student performance or behavior. The curious thing about teacher expectations is that they can influence students to behave precisely in accordance with the expectations. This phenomenon is known as **self-fulfilling prophecy** or the **Pygmalion effect.** For example, if a teacher believes that a particular child cannot achieve and subsequently ignores the child, the lack of interaction with the teacher causes a void in the child's opportunity to learn. As a result, the learner may begin to believe that he or she is incapable, although this may be far from the truth. If the child performs poorly, the instructor believes that his or her initial impressions were correct. Even if the child performs well, the instructor may believe that luck was responsible. In contrast, if the instructor expects the child to be a high achiever, the teacher may provide additional support to assist the learner. The child may begin to believe that he or she is skilled and act accordingly. In essence, the Pygmalion effect can cut in either direction, with negative or positive results.

Social context is equally likely to create impressions about performance. Martinek and Karper (1986) discovered that physical education teachers held higher expectations for highly skilled students when those students were engaged in individual or competitive activities. They expected lower-skilled students to perform better in cooperative activities (and less well in individual activities). Furthermore, teachers provided more technical feedback to high-ability students during individual activities and communicated more empathy (and lower expectations) toward less skilled students during competition.

The effects of expectations can be particularly unfortunate for students who are perceived as low achievers (Martinek & Karper 1984). Feelings of helplessness often become a part of these students' personalities. When this occurs, students are referred to as having acquired "learned helplessness" and may be observed to exert little effort, become abusive, blame others, or quickly concede failure (Martinek & Griffith 1993, 1994; Walling & Martinek 1995).

➤ Teachers must continually assess their expectations and be cautious when communicating expectations to learners.

Teacher–Coach Role Conflict

➤ Both teaching and coaching are satisfying career choices, but people must be careful to fulfill the obligations of each role if they elect to engage in both simultaneously.

How well you are able to fulfill multiple roles along with teaching physical education may determine your ability to perform the teaching role competently. For example, a teacher may have a second job outside the school setting, may have family responsibilities, may volunteer in a variety of civic organizations, and so forth. Inside the school the teacher may assist in a variety of duties such as cafeteria monitor, recess supervisor, bus supervisor, or student organization sponsor.

Another primary role that most teachers of physical education assume is that of coach for one or more sport teams at the school. In fact, many people enter the teaching ranks, and more specifically physical education teaching, because of a strong desire to coach. Many physical education teachers competed on interscholastic or intercollegiate teams before becoming a teacher. Researchers have studied the dual roles of teaching and coaching over the last decade and have found that for some, if not many, a conflict in performance of both roles may arise (Locke & Massengale 1978). Specifically the research has focused on how one can be an effective teacher of physical education while also striving for success as a coach.

> DO it **Activity 15.3**
>
> **Conflicting Roles**
>
> Decide whether four different teacher-coaches show signs of role conflict in this activity in your online study guide.

Role conflict is defined as two or more incompatible roles that are difficult to perform simultaneously. For the teacher-coach, this definition suggests that an individual finds the concurrent roles and the expectations linked to the two roles difficult, if not impossible, to navigate. If one examines the many characteristics linked to teaching versus coaching, both compatibility and incompatibility between the two roles become clearer.

By recalling your experience of observing your former teachers and coaches and reflecting on each role, you should recognize that the successful performance of both roles may be difficult for some people. The research suggests that because of this difficulty, many individuals withdraw (called *role withdrawal* or *retreatism*) to one role over the other. Those who prioritize coaching over teaching will withdraw or retreat to coaching; those who prioritize teaching will concentrate on teaching. When this happens, one role suffers, and the clients (either students or athletes) will likely suffer the consequences of a less than satisfactory experience.

You will occupy many roles in your life, both personal and professional, but as you prepare for them and occupy them, think about your ability to perform them concurrently and the implications for your own success and that of others.

Teachers who also coach need to guard against role conflict.

Key Pedagogical Principles Based on Research

1. Begin to develop expertise by acquiring experience and new knowledge.
2. Provide appropriate practice.
3. Provide a high amount of academic learning time.
4. Always be concerned about class management and discipline.
5. Hold learners accountable.
6. Provide clear, specific feedback.
7. Develop knowledge about alternative curricular models.
8. Ensure an equitable learning environment that addresses the individual needs of all learners.
9. Consider how your expectations influence students.
10. Be mindful of teacher–coach role conflict.

What Instructors Need to Know About Effective Pedagogy

Although research has the potential to inform teachers how to teach most effectively, most teachers do not subscribe to the research journals. Instead, in many cases, teachers derive the knowledge that informs teaching practice primarily through their experiences as students in the public schools and other physical activity settings and through certification programs, informal discussions with other instructors, and in-service workshops and conferences focusing on the practitioner rather than the researcher.

➤ Although professional practice knowledge is a powerful source of information, such knowledge is not a substitute for thoughtful consideration of the available research literature on effective teaching.

Besides what they learn from experiences outside actual instructing time, teachers also acquire much of their pedagogical expertise in the form of **professional practice knowledge.** You already learned in chapter 12 that this type of knowledge underlies the practice of all professions; teaching physical education is no different. Over hours, weeks, and years of instructional experience, teachers learn (often the hard way!) which techniques work well with students and which do not. Instructors and coaches often refer to their professional practice knowledge when explaining their instructional decisions to others.

Nevertheless, as valuable as personal experience can be, it is a serious mistake for teachers or coaches to look to personal experience as their only guideline for how to teach or coach. Enough scholarly research has now been done that any teacher, no matter how experienced, can learn from research.

Although the available research sometimes seems inaccessible to instructors, because it can be difficult to understand or does not apply directly to their situations, scholarly study has significantly influenced the way in which instructors teach physical education. Instructors now have a vocabulary for discussion of such issues as management time, reducing off-task student behaviors, and promoting quick lesson transitions. This all indicates that teachers are gradually becoming aware of the research on effective teaching and making attempts to incorporate the results into their own practice.

Teaching Settings

We all have been exposed to excellent physical education settings. Typically, these settings have good facilities and outstanding teachers. They are a joy to experience and significantly contribute to our quality of life. Learning is fun, and we come away with an expanded set of physical skills. Teachers in this type of setting are highly motivated and thrive in a

The environment in which teachers work can facilitate their success. Gymnasiums that are cheerful and well equipped also send a message to students about the importance of the subject matter.
AP Photo/Ludington Daily News, Andy Klevorn

workplace that promotes their effectiveness. In essence, factors promoting a high quality of work life are present: respect from students, teacher participation in decision making, collegial support and stimulation, a high sense of efficacy among teachers, resources to enable effective performance, and a common sense of vision within the school (Stroot 1996). Regrettably, the quality of life for teachers in other settings may be just the opposite. Settings that are underfunded, poorly maintained, and underequipped and that have no vision often produce unmotivated teachers.

The quality of the instructional environment in a school often can determine the success and satisfaction of both teachers and students. Understanding the influence of the workplace or the context in which one is employed is extremely important to the teacher of physical education. Like other workplace settings, schools contain a host of contextual factors that one must understand and cope with to be successful. Social, psychological, political, economic, and other organizational factors may emerge within the individual work setting or class. For example, having access to a power base along with having the moral and financial support of colleagues and administrators within a school setting is critical for a teacher.

Larger influences outside the workplace may affect what transpires within the workplace as well. Issues related to family, population demographics, civil rights, health care, drugs and violence, educational reform and policy, and advances in information technology may influence what occurs in the school setting (Tyson 1996). How instructors cope with contextual factors influences their success as instructors and, ultimately, the success of learners.

Many of the research findings are interesting to all instructors of physical activity, regardless of the setting in which they work. Some findings are positive, whereas others offer unflattering portraits of the school environment.

Life and Work in the Physical Education Setting

Teacher effectiveness and satisfaction are often linked to the characteristics of the work setting. Understanding the specific nature of any workplace is important as one contemplates working within a given activity arena. An important early research study (Locke 1975) yielded three important results:

1. Diverse activity and isolation from other adults characterizes the work life of some physical education teachers.
2. Differences among students influence teacher effectiveness.
3. The curriculum of physical education is distinctly different from that of other subjects relative to space, activities, and relationships.

➤ In contrast to the exciting and profitable learning settings observed in most elementary schools, secondary physical education settings have been described as often being nothing more than a waste of time, in which embarrassment, humiliation, discomfort, and student inactivity appear to be the norm.

Other studies have illustrated the positive and negative aspects of school physical education—what life in the gym is like for teachers and students. They have shown that although elementary physical education is in a relatively healthy state, secondary physical education suffers from a variety of difficult social and political problems—some of which appear now to be reaching a point of crisis.

In reporting the results of an extensive study of high schools, O'Sullivan and colleagues characterized the state of secondary physical education in 1994 as follows. The information still represents the state of high school physical education in many schools throughout the United States.

- Teachers and students share similar views of the purpose of physical education—to expose students to lifetime activities. But differences emerge in relation to defining how fitness connects to physical education.
- Fitness activities have little relationship to the development of health- or skill-related fitness.
- Teachers spend too much time on class management activities, suggesting the absence of well-planned, meaningful, and engaging activities for students.
- Teacher autonomy is born out of benign neglect—physical education teachers are viewed as marginal figures in the overall school hierarchy.
- Teachers have limited goals for students.
- Teachers validate their sense of self-worth through activities outside physical education, such as coaching and officiating.
- Even in the absence of great levels of student achievement, gymnasiums can be well-managed and happy places if they include a teacher who expects appropriate student behavior.
- The curriculum of physical education is consistent among teachers; they most value traditional sports and activities.
- Instruction is casual; the premium on learning is mostly absent; and students with lower skill levels enjoy class less.
- Student evaluation is not based on the measurement of performance but on compliance with behavioral standards (e.g., following the rules and rituals established in the class).
- More than half the students like physical education less and perceive it as less important than other subjects.

Overall, these findings reflect the dysfunction of many secondary physical education sites and suggest that in these schools physical education may need to be restructured or alternative systems provided.

Other studies (Locke & Griffin 1986) vividly depicted the malaise that existed within American society relative to the role of secondary physical education in the mid-1980s. Cases of struggle appeared to be more commonplace than profiles of excellence. Griffin (1986) summarized the various obstacles to excellence at that same point in time:

- Lack of teacher or program evaluation
- Lack of formal incentive or reward for good teaching
- Lack of professional support for teacher and program development activities
- Inadequate facilities, equipment, and scheduling
- Failure to include teachers in decision making
- Higher value placed on compliance and smooth operations than on teaching competence
- Acceptance of mediocrity
- Isolation

In discussing these obstacles, Griffin stated, "In most cases, the schools did not intentionally limit what could be accomplished. . . . Instead, they practiced benign neglect. The tendency of everyone to ignore physical education presented what seems to have been the most formidable obstacle to excellence" (p. 58). In essence, physical education as a subject area at the secondary level seems to be perceived as superfluous to student learning, and the physical education teacher holds an occupational status that has been described as marginal.

Although much remains the same since these descriptions were written over two decades ago, there is some basis for optimism. Legislative acts such as Public Law 108-265, which

➤ Research on school physical education has revealed both positive and negative dimensions of the school environment.

requires every school in the United States that participates in the National School Lunch Act to implement a school-wide plan addressing physical activity and nutrition, give new hope for significantly improving the environment for physical education teachers. In fact, teachers have never been better positioned to assume a major role in leading the school initiative to improve the health and fitness levels of students (Graber & Woods 2007; Woods & Graber 2007). Physical education teachers who were once marginalized are now encouraged to assume roles such as that of school physical activity director (Castelli & Beighle 2007).

Workplace Conditions

Among the factors that can influence the teaching and learning process are political, organizational, and personal and social factors. For example, the economic conditions of a school often influence what facilities and equipment are at the disposal of teachers and students. As a result, some schools are able to offer a wide variety of curricular opportunities for students, whereas other schools are more limited. Figure 15.2 illustrates the factors most likely to influence workplace conditions.

Within the school context, teachers' success and satisfaction, or lack thereof, are linked to many variables. Class size, for example, has the potential to influence what material a teacher is able to cover, the amount of feedback he or she can provide to individual students, and the degree to which students have sufficient opportunities to practice. School policy may dictate how often students receive instruction or how they will be graded. Tyson (1996) suggests the following as conditions encountered by physical education teachers that negatively influence their work.

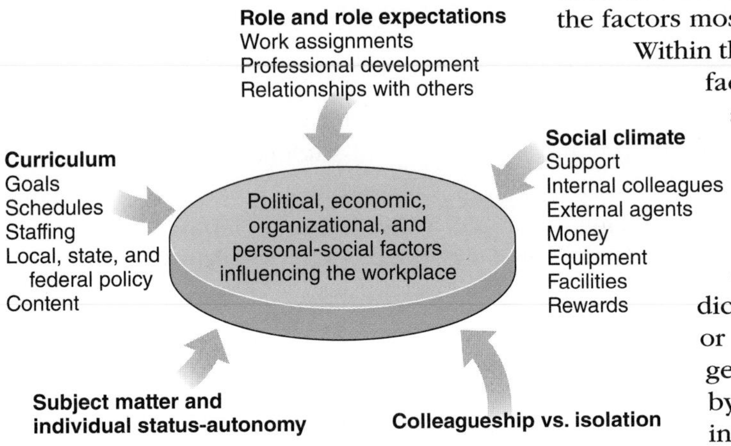

FIGURE 15.2 Workplace conditions influencing the teacher.

- **Curriculum content.** Teachers at both the elementary and secondary levels often teach too many physical activities and too many classes in a day to be effective. Teachers often feel that they have no authority to determine the curriculum in terms of time allocation and number of activities.
- **Class size.** Class size is generally quite large in physical education—well beyond the size of other classes.
- **Staffing.** Physical education teachers are often isolated and left alone to fulfill instructional demands.
- **Policy.** Local and state policies often require the minimum in physically educating students at the elementary and secondary levels. Grading policies dictate a teacher's instructional approach.
- **Students.** Physical education teachers who are preoccupied with the control of disruptive students feel less effective. Student gender, ability, and other characteristics greatly influence a teacher's effectiveness.
- **Status and rewards.** Physical education teachers are often perceived as marginal faculty relative to the overall goals of a school. Administrators, parents, and students perceive physical education as less important than other school subjects. Physical education teachers who receive strong support from administrators, colleagues, students, and parents feel more successful in their work. They often believe, however, that others appreciate them more for their work in coaching than for their work in teaching. Teachers do feel a sense of accomplishment based on student progress.
- **Facilities.** Teachers often express the need for better facilities and equipment to enhance their instruction.

➤ Workplace conditions have the potential to facilitate or constrain the instructor of physical activity.

Striving Toward Positive Work Settings

The degree to which you are able to thrive in an environment that is less than ideal often depends on your ability to employ various coping strategies. Compliance, compromise, and attempts to redefine the context in which you work often call for skilled decision making—sometimes at the risk of alienation from others. You may challenge the norms of the setting in trying to overcome problems related to a particular situation. Challenging the status quo is always risky, but persistence and a commitment to improve your teaching or the instructional setting may reap positive results. At the other extreme, you can give up, ignore the problem, or even relocate to another setting. In any case, you must learn to adapt to the conditions of the physical activity settings in which you work.

The complexity and shifting interactions of various conditions are what shape a stimulating, frustrating, challenging, and rewarding career. The bottom line is to be prepared to work and understand the challenges and rewards inherent in any workplace. The ideal setting is one in which teachers are rewarded for excellence. Such settings exhibit a social order that is caring, supportive, and politically positive—in which politics and economics do not hinder your ability to perform good work. The social order is healthy if people work collaboratively toward achievement of goals. Such a setting values and rewards teaching. Finally, a good work environment will facilitate both progress and the continuing struggle to improve. Positive work settings and good teachers usually go hand in hand for the benefit of students.

Professional Roles in the Teaching Profession

In contrast to some of the negative characterizations of school physical education presented earlier, there are positive examples of excellence in the schools. In fact, the recent wellness legislation, Public Law 108-265 (discussed earlier in the chapter), has great potential for improving the quality of elementary and secondary school physical education. As physical educators become acknowledged as school wellness experts, and as greater emphasis is placed on physical activity across the curriculum, it is likely that physical education teachers will become less marginalized and instead be perceived as school leaders in promoting physical activity and health through quality school programs (Graber & Woods 2007; Woods & Graber 2007). Of course, program improvement is not easy, and it will take dedicated individuals who have the ability to negotiate a variety of obstacles such as low budgets and sometimes less than ideal support from colleagues to begin the change process. The following are likely to be associated with teachers who are successful:

- Innovative instructional strategies
- Novel curriculums
- Integration of physical education with other subject matters
- Unique ways to promote learning progressions
- High-profile public relations programs
- Supportive colleagues and administrators
- Adequate funding
- Exemplary classroom management strategies
- Involvement in professional development activities
- After-school programs for students and adults
- Modeling of athletic skill and fitness
- Promotion of equitable learning settings

These teachers demonstrate a sincere interest in and enthusiasm for teaching, a genuine concern for students, and a continued desire to grow and develop as teachers. The settings in which they work are not unlike other physical activity settings (e.g., fitness clubs, private golf and tennis clubs, recreation centers) in which the welfare of participants or customer service is a high priority.

Typical Teaching Responsibilities

Before becoming a certified teacher, most people will have spent many hours in the schools observing and assisting experienced teachers during practicum and student teaching experiences. Although no experience can completely prepare new teachers for all that they will encounter once they enter the instructional setting as certified teachers, these experiences will have provided them with good exposure to students that will ease the transition into their new roles.

Teachers can expect to meet with students for approximately 180 days each school year. Although most schools operate on a nine-month calendar, some school districts convene throughout the year (year-round schooling). In the latter case, students attend school for the same number of days as those attending for nine months, but they have longer breaks more frequently throughout the year.

Teachers at the elementary level generally teach 8 to 10 classes per day for 30- to 45-minute periods. These teachers meet with students from one to five times per week and sometimes split their time between two schools. Those at the secondary level teach fewer classes (five to seven) for 45- to 60-minute periods. These teachers usually meet with students three to five times per week at the same school. Whereas physical education teachers at the elementary level may be the only specialists at their schools, those in secondary settings often have several colleagues in physical education.

Although national recommendations state that elementary school students should receive at least 150 minutes of weekly physical education and secondary students at least 225 minutes of physical education per week for the entire school year (NASPE 2006), the amount of school-based physical education instruction that students receive is far less than that in most states. For example, the percentage of students who receive physical education declined from 42% to 28% between 1991 and 2003. Fortunately, however, because of increased levels of concern about children's health, 44 states have introduced legislative acts since 2005 to improve or increase the amount of physical education that students receive (Cawley, Meyerhoefer, & Newhouse 2006).

With regard to content, teachers at the elementary level tend to teach short units of instruction that emphasize basic skill development and introduce children to concepts such as fast, slow, force, and direction. Teachers at the middle school level tend to teach longer units that introduce students to a variety of sports, games, and fitness and exploratory activities. Those at the high school level typically teach the longest units as a means of helping students acquire adequate skill to participate proficiently in a few activities throughout the life span. Teachers at all levels also may be required to teach classes specifically designed for children with special needs. As a result, most teacher education programs offer courses that provide future teachers with knowledge concerning the needs of special students and appropriate ways of structuring the curriculum.

Physical educators teach units in movement, sport instruction, and lifespan physical activity.
AP Photo/David Adame

Although teacher salaries are modest, the rewards of teaching can be great. Witnessing the satisfaction of a child who accomplishes a new skill or seeing a previously inactive student become an active participant can offer satisfaction that is difficult to find elsewhere.

Rewards for Outstanding Teaching

Often teachers labor for most of their careers without receiving adequate reward for their efforts, but local and national groups regularly recognize hardworking, competent teachers. Teachers of the year are those who have received state, regional, or national recognition for being among the best in their profession by displaying special characteristics linked to excellent teaching, outstanding curriculums, and the promotion of student learning. You may have wondered who these people are. Where do they teach? What and how do they teach? What led to their development as great teachers? We have all known great teachers throughout our lives. Can you identify the teachers whom you considered great and who had an influence on your life?

For every subject area, teachers across our county have received recognition on an annual basis for excellence in teaching. These teachers are selected through various award programs. For example, the Council of Chief Secondary School Officers (CCSSO) has sponsored a program since 1952. Through a nomination process (see www.ccsso.org), a nominee for and ultimate winner of this award should

- inspire students of all backgrounds and abilities to learn;
- have the respect and admiration of students, parents, and colleagues;
- play an active, useful role in the community as well as in the school; and
- be poised, be articulate, and possess the energy to withstand a taxing schedule.

USA Today recognizes an All-USA Teacher team. A number of organizations sponsor this program:

- National Association of Secondary School Principals
- National Middle School Association
- National Association of Elementary School Principals
- National Education Association
- American Association of Colleges for Teacher Education

This program honors 60 teachers, with 20 teachers honored on first, second, and third teams. The criteria for selection center on how a teacher addresses student needs and the influence of the teacher on students, the school, and community. Since 1997 the program has honored a few teachers of physical education. In 2002 Phil Lawler, a middle school teacher from Naperville, Illinois, was recognized as a member of the first team.

Excellence

In this activity in your online study guide, you'll identify behaviors that characterize excellent teachers.

More closely related to our discipline, the National Association for Sport and Physical Education, with funding from Sporttime International, sponsors a Teacher of the Year (TOY) program. The first NASPE Teacher of the Year was Regina McGill from Bettendorf, Iowa. The NASPE TOY selection process begins at the state level. State winners are then considered for district honors. From there, a special panel of judges considers district winners for the national awards at the elementary, middle, and high school levels. TOY nominees on all three levels are expected to meet the following criteria (www.aahperd.org/naspe/template.cfm?template=naspeAwards/criteria/toy.html):

- Conduct a quality physical education program that reflects the NASPE standards and guidelines for K through 12 physical education programs
- Use various teaching methodologies and plans for innovative learning experiences to meet the needs of all students

- Serve as a positive role model epitomizing personal health and fitness, enjoyment of activity, fair play, and sensitivity to the needs of students
- Participate in professional development opportunities
- Provide service to the profession through leadership, presentations, or writing

In a study of four Teachers of the Year from the state of Georgia, DeMarco (1999) discovered that they were usually given unique nicknames like Betty the Movement Educator Extraordinaire, Cara the Passionate Professional, Tommy the Gentle Giant, and Angela the Athletic Social Worker. In addition, they all displayed the qualities just listed. These teachers maximized student engagement in motor tasks and managed the physical education class with a high level of effectiveness (i.e., students were not disruptive). Also, DeMarco found that all four teachers had a strong belief in the value of their work and their ability to do that work, and they loved students and teaching. Furthermore, students, fellow teachers, administrators, and parents greatly admired these TOYs. Finally, all four teachers were active in professional organizations (in local teaching organizations and at the state and national levels of AAHPERD).

We can't all be teachers of the year, but we can learn from research what teaching behaviors are likely to have the highest payoff. Let's look at the most important factors that instructors should keep in mind when teaching physical education. Whether you are teaching baseball skills in the gymnasium or backpacking in a wilderness setting, each of these factors will be important in determining the success that your students will achieve. These factors are not mere suggestions; extensive research on teaching physical education has documented their value.

Interviews With Practicing Professionals

Individuals enter the teaching profession for a variety of reasons. Some elect to teach because they enjoy working with children or like working in a school setting. Others decide to teach because it provides them with an opportunity to coach, supervise intramurals, or also serve in an administrative capacity. The two teachers profiled here, both NASPE Teachers of the Year, share their stories about their career selection and daily work life.

Photo courtesy of Meg Greiner

Meg Greiner

NASPE 2005 Elementary Teacher of the Year

Q: Why did you become a teacher?

A: For me becoming a teacher was a calling, not a choice. I just knew that I was put on this earth to teach; it came very naturally for me. Physical education was a perfect fit. I loved sports and physical activity, even at a very young age, and I wanted to share my love of movement with the world, hoping to turn children on to physical education and activity and thus hoping to enhance their lives. If you don't have your health, you don't have anything!

Q: What have been the most important experiences in preparing you for your career?

A: The most important "Aha!" experience for me happened during my third year of teaching. That year I was particularly blessed to have an opportunity to take classes and workshops. It was during these workshops that I had an earth-shattering revelation that changed the way I thought about physical education and teaching forever. It revolved around three key individuals. Thank you, Dr. Robert Pangrazi, for introducing me to behavior management and the four-part lesson plan. Thank you, Dr. George Graham, for introducing me to "skill themes." And thank you, Don Hellison, for sharing your Levels of Responsibility model; this was the piece that brought everything together for me. Thanks to these three extraordinary individuals, I put everything together and came up with my own model for teaching physical education.

Q: What is the average workday like for you?

A: I teach 48 classes a week—that's 9 or 10 classes a day. I see my 1st through 4th grade students three times a week for 35 minutes, my kindergarten students twice a week for 30 minutes, and an adapted physical education class once a week. My special needs students also attend regular physical education classes three times a week.

A typical day looks like this. I usually arrive at work at 7:45 a.m. and prepare for TEAM Time. TEAM (Together Everyone Achieves More) Time is the way Independence Elementary School students, staff, and community start the day. When the bell rings for the beginning of school, approximately 400 students, staff, and community members meet in the gym to start their day with movement, dancing, team-building activities, signing, stretching, "Brain Gym," aerobics, and singing, followed by morning announcements and the Pledge of Allegiance. I choose the music, set my play list, and practice my dances.

As soon as I get there, my gym is basically "open" for indoor morning break. I put out various equipment including jump ropes, basketballs, juggling equipment, pogo sticks, flower sticks, stilts, and unicycles. During this time, until school starts, students can either play in the gym, practicing their skills, or go outside for recess.

At 8:30 a.m., school begins and everyone comes to the gym for TEAM Time. TEAM Time last 20 minutes, and then everyone goes back to class.

From 8:50 to 9:20, my gym is swept and I set up for physical education for the day.

From 9:20 to 9:50 I see kindergartners; from 9:50 to 10:25, 3rd grade; from 10:25 to 11:00, a 3rd-4th grade blend class; and from 11:00 to 11:35, 4th grade. I have lunch from 11:35 to 11:55 and teach kindergarten from 11:55 to 12:25. My afternoon consists of all 1st-2nd grade blends (four classes) taught back to back from 12:25 to 2:45, which is the end of our school day.

After school twice a week, I teach sport enrichment and circus art classes in our after-school program, called Prime Time, for 2 hours (1 hour to the 3rd and 4th graders and 1 hour to the 1st and 2nd graders).

Q: What do you like most about your job?

A: What I like most about my job are the students. I love interacting with them, getting to know them, and getting to spend five years watching them grow up. But the interaction doesn't stop there; many students continue visiting me after leaving our school and invite me to high school events they are participating in. I love watching them grow into amazing young adults and knowing that I might have had a small part in helping them along the way.

Q: What do you like least about your job?

A: The hectic schedule—I'm exhausted at the end of the day. Good teaching is hard work!

Q: What support do you receive to promote your success?

A: Since receiving the 2005 NASPE National Elementary Physical Education Teacher of the Year Award I have been given the opportunity to present across the United States, meeting many wonderful colleagues and sharing what I do. My school district and principal have been very supportive, allowing me the time off to present and to attend the national AAHPERD convention. I'm extremely appreciate of this gift and feel very blessed that they value my continued professional growth and that of other physical educators.

Q: What advice would you give to students who would like to pursue a career in physical education teaching?

A: Teaching physical education is a very challenging discipline physically and emotionally but truly worth every bit of sweat and toil when you see the effect you have on the life of a child. The greatest compliment that I can receive is when a parent says, "I wish you would have been my physical education teacher, I might have liked physical education [or movement and exercise]!"

Be passionate about your profession, physical education; get involved early in your career and keep learning throughout it. Participate in every learning experience that comes your way—you never know what doors this might open.

Be passionate about your students. They are why you are there. Encourage every one of them to always do their best, and be their cheerleader and number-one fan.

Photo courtesy of Gregg Agena

Gregg Agena
NASPE 2006 Middle School Teacher of the Year

Q: Why did you become a teacher?

A: In 1990, I had an opportunity to coach Little League baseball for Wahiawa's senior division (ages 13-15). We were the "underdog" in the league; and the year before my friend and I coached, our team had finished in last place. We lost our first regular-season game, but we won the rest of our games, clinching the championship before the season was over. My friend and I had the honor of coaching the all-star team, and we finished second in our district. This was the first time I had worked with kids, and this positive experience made me decide to be a teacher. I decided to teach specifically physical education since I had succeeded as a coach and because of my love for sports.

Q: What have been the most important experiences in preparing you for your career?

A: I have been fortunate throughout my teaching and coaching career to have been influenced by many outstanding individuals. I had the honor of coaching under the top high school baseball coach in the state of Hawaii, and I learned and embraced a Japanese philosophy that he believed in called *kaizen*.

According to this philosophy, you can always do it better, make it better, and improve it even if it's not broken, because if you don't you cannot compete with those who do. It is used by major corporations like Cannon and Toyota in the belief that the ideas of everyone, from the president to the assembly line worker, have merit.

I learned to never be complacent and to constantly pursue perfection.

Q: What is the average work day for you?

A: I teach six periods a day, and each class is 50 minutes long. I teach two of our physical education electives: physical fitness and team sports (three classes) and body conditioning (three classes). Each class starts off with taking attendance through a roll call number system, and students then dress out for physical activity (5 minutes). Next, students report to the basketball courts for warm–ups, which include a dynamic workout and static stretching led by our school's physical education leaders (10 minutes). Following warm-ups, students jog to their assigned area for their physical activity and participate in conditioning exercises, skill work, and assessment or modified games and activities or both (25 minutes). Finally, students are dismissed to the locker rooms to shower and change back into their school clothes (10 minutes).

Q: What do you like most about your job?

A: It is never boring. I have the opportunity every day to equip my students with the skills, knowledge, and desire to maintain a physically active lifestyle. I play a significant role in my students' mental, social, emotional, and physical development.

Q: What do you like least about your job?

A: I don't like when students come into my class with a negative attitude toward physical education because of their past experiences. I try to create a fun, positive, and nurturing environment so that my students will have a positive experience and a positive attitude toward physical activity.

Q: What support do you receive to promote your success?

A: My national recognition was the end result of many minds who contributed to our physical education program. I was influenced by fellow teachers, resource teachers, professors, administrators, coaches, counselors, cafeteria and custodial staff, and security. Many of the ideas I implemented came from non–physical education teachers, and I encourage others to seek advice also from people who are not in their content area or grade level. Many individuals whom I built a positive relationship with influenced not only my teaching, but also my personal life as a father and a husband.

Q: What advice would you give to students who would like to pursue a career in physical education?

A: When opportunity knocks, answer the door.

People who flourish in life excel because of their ability to understand and work well with others.

If you always do what you've always done, you'll get what you've always got.

You can't direct the wind, but you can adjust your sails.

See change as an opportunity to learn and grow.

Preparing for a Career

Although most students who major in physical education will ultimately teach in the public schools, a variety of options are available. Most teachers, particularly at the high school level, will also choose to coach. Some people will pursue graduate education and obtain a master's degree, which will enable them to be instructors at a college or university, or obtain a doctorate degree so that they can teach and conduct research in a higher education setting. Some may elect to work in a fitness facility or community recreation center. Others may simply decide to volunteer their time as a youth sport coach.

Physical Education Teacher in the Schools

Most people who become certified teachers elect to work in a public or private school at the elementary, middle, or high school level. Their university or college course work will have consisted of courses designed to teach them about the process of teaching (instructional strategies, teaching styles) and about the product of teaching (curriculum). Future teachers in most states will have to pass one or more state certification exams designed to ensure that they have basic skills in subjects such as reading and math, have acquired knowledge about the subject matter in which they elect to teach (physical education), and have the competence to teach effectively. As a means of helping university professors design an undergraduate curriculum that adequately prepares future teachers, NASPE has developed a set of national standards that most teacher education programs throughout the United States strive to achieve (see the next page). These standards relate to the knowledge and skills that you will be expected to apply as a teacher of physical education.

The 10 NASPE National Standards for Beginning Physical Education Teachers

- Standard 1: Content Knowledge. Understand physical education content and disciplinary concepts related to the development of a physically educated person.
- Standard 2: Growth and Development. Understand how individuals learn and develop, and provide opportunities that support physical, cognitive, social, and emotional development.
- Standard 3: Diverse Learners. Understand how individuals differ in their approaches to learning and create appropriate instruction adapted to these differences.
- Standard 4: Management and Motivation. Use and have an understanding of individual and group motivation and behavior to create a safe learning environment that encourages positive social interaction, active engagement in learning, and self-motivation.
- Standard 5: Communication. Use knowledge of effective verbal, nonverbal, and media communication techniques to enhance learning and engagement in physical education settings.
- Standard 6: Planning and Instruction. Understand the importance of planning developmentally appropriate instructional units to foster the development of a physically educated person.
- Standard 7: Student Assessment. Understand and use the varied types of assessment and their contribution to overall program continuity and the development of the physical, cognitive, social, and emotional domains.
- Standard 8: Reflection. Understand the importance of being a reflective practitioner and its contribution to overall professional development and actively seek opportunities to sustain professional growth.
- Standard 9: Technology. Use information technology to enhance learning and personal and professional productivity.
- Standard 10: Collaboration. Understand the necessity of fostering collaborative relationships with colleagues, parents/guardians, and community agencies to support the development of a physically educated person.

National Standards for Beginning Physical Education Teachers (2003) reprinted with permission from the National Association for Sport and Physical Education (NASPE), 1900 Association Drive, Reston, VA 20191-1599.

Teaching in Higher Education Settings

A career option that some individuals elect to pursue is to teach at a college or university. By obtaining a master's degree, you would be qualified to work as an instructor who might be responsible for teaching skills courses (e.g., basketball, aquatics, racket sports) or supervising student teachers. Although some people elect to remain employed as instructors throughout their life span, others believe that this career option is limited. Because these individuals do not have a doctorate, they cannot serve on important college or university committees and cannot teach or advise graduate students.

By obtaining a doctorate, which is the precursor to the professorial ranks, people gain additional career responsibilities that are often considered desirable. Most professors of pedagogy have worked at one point or another as teachers in the public schools. Many elect to work at the college or university level because they want to improve teacher education. An increasing number of professors also want to engage in research and mentor graduate students. They design studies intended to help us better understand the work of teaching and how to promote effective instruction that leads to student learning. In addition, professors present papers at professional conferences; collaborate with teachers in the public schools; remain informed about state certification requirements; and serve on departmental, college, and university committees.

➤ If you decide to become a professor of physical education, in which capacity you will likely prepare people to become teachers, you should acquire public school teaching experience and maintain ties to the public schools. This arrangement will provide you with greater credibility in the eyes of both your students and the teachers in the public schools who will supervise your students during their practicum teaching experiences.

Regardless of whether you elect to pursue graduate studies, you should have some public school teaching experience. If you are enrolled in a university teacher education program, would your instructor or professor have greater credibility if he or she had several years of experience as a teacher? In most cases, your answer will be yes. Public school experience can open the door to other interesting career possibilities. In many cases, it is a requirement if you wish to be hired at the college or university level.

Adapted Physical Education

Adapted physical education is a program of skill instruction and exercise that has been modified so that all students can participate. In the United States, federal Public Laws 94-142, 101-476, and 105-17 stipulate that students with disabilities will be provided with a program of physical education that includes motor skills; fundamental skills such as running, catching, and throwing; and skills in games, sports, and aquatics. The range of disabilities that fall under these laws is large, including Down syndrome, hearing disorders, spina bifida, muscular dystrophy, and a host of others. If a child has unusual restrictions, the instruction must be provided through an individualized program developed by the child's teacher, parents, and other interested parties. Where the disability permits, children must be mainstreamed into the regular classroom or gymnasium as a way of meeting Public Law 94-142, which means that the physical education teacher must be prepared to teach a curriculum to children with a diverse range of abilities.

In larger school districts, students with similar disabilities may be grouped for physical education instruction. Obviously, one cannot presume to teach students with serious disabilities without understanding something about disabilities and how they affect the child's performance. Approximately 14 states now have some sort of licensure or certification to teach children with disabilities, but some states are lax in this respect. Some colleges and universities offer an endorsement in physical education special education; others have full-blown certification programs. Your state probably requires at least one course in adapted physical education as part of your teacher certification program, but you may want to specialize in adapted physical education and focus exclusively on this exciting area.

Beyond the public schools, specialized schools for children with mental disabilities, those with severe orthopedic disorders, and those with limited or no capacity for seeing or hearing usually hire a staff of adapted physical education teachers. If working with children and adults who have been challenged by disease and injury is a career possibility that excites you, give some thought to continuing for a master's degree in adapted physical education, a background that will prepare you for almost any challenge that might present itself in this area.

Coaching

Many people who enter teaching do so because they desire a career in coaching. (The next chapter covers careers in coaching in depth.) That is, they are less interested in teaching and more interested in coaching.

Physical education teachers must be prepared to teach children with a diverse range of abilities.

Although a career in teaching physical education offers entry into the coaching realm, it can also result in a teacher who becomes less committed to teaching than to coaching (see previous discussion of teacher–coach role conflict). In such cases, students in the classes of these teacher-coaches often suffer.

Of course, many teacher-coaches do an outstanding job of balancing the workload of teaching and coaching. These people believe that physical education is critical because it offers an opportunity for students to acquire skills regardless of ability. They carefully plan lessons that promote skill development in all students. They also believe that athletics are important because they offer students an opportunity for an advanced level of competition. They carefully manage their time to ensure success in both teaching and coaching.

Those who elect to coach spend long hours at their schools each day. They often arrive at school before other teachers do to plan classes and practices. Although some receive an extra release period during the day to plan, many spend their evenings and weekends planning practices and scheduling competitive events. Immediately before or after school they spend several hours practicing with students, engaging them in drills and fitness activities that will enable them to compete successfully against other students. Some coaches coach only one sport during the year, whereas others elect to coach year-round and even during the summer months. Some become so engaged in the coaching role that they offer summer clinics or camps for students.

Although most physical education teachers elect to coach in the public schools, some prefer coaching at a more competitive level. These people choose to coach at colleges or universities. Most begin their careers as graduate teaching assistants enrolled in master's degree programs. As they gain experience and acquire name recognition, a few lucky individuals eventually obtain an assistant or head coaching position in higher education. At the most competitive level, Division I, these individuals only coach. Their career advancement is contingent on winning. As a result, they face enormous pressure to recruit and produce winning teams. For more information on coaching at a competitive level, see chapter 16.

Prospects and Opportunities for Careers in Teaching

If you intend to become a teacher of physical education, your prospects for locating employment are high. In many states, schools have a desperate need for physical education teachers. In Illinois, for example, students at all grade levels are required to take physical education daily. As a result, severe teacher shortages exist in many areas of the state. In Cook County, the largest county in the Chicago area, the need for teachers is so severe that undergraduate students can apply for a loan that will be forgiven if they agree to teach in Cook County for a certain length of time after they become certified. Rapidly growing states such as Nevada also have a tremendous need for quality physical education teachers. If you are willing to move to a part of the country that needs physical education teachers, you are likely to find employment quickly.

You can improve your employability by ensuring that you acquire as many skills as possible while enrolled in a teacher education program as an undergraduate. Although you are likely aware that physical education teachers who also are quality coaches are always in demand, you may be unaware that physical education teachers often locate employment because they also possess the following:

- Certification as an athletic trainer
- Certification to teach health education
- Ability to speak a second language fluently
- Endorsement to teach in additional subject areas
- Certification to teach driver's education

Selecting a Career

List the advantages and disadvantages of each of the following career options.

1. Physical education teaching in the public schools
2. Teaching in higher education settings
3. Adapted physical education
4. Coaching in the public schools
5. Coaching in higher education such as a Division I school

Schools that currently do not offer physical education will likely be adding it as a required curricular area in the future. The obesity epidemic has created greater awareness of the need for students to engage in physical activity on a daily basis. For some students, school is the only place where they have an opportunity to be active. As more schools add physical education or increase the time devoted to the subject, the need for teachers will increase. Never has there been a better time to enter the profession.

For individuals who wish to teach or coach outside of the K through 12 setting, the education they receive while enrolled in a teacher certification program will place them in good standing for whatever teaching-related position they may eventually assume. For example, have you ever observed community soccer coaches who allow students to stand in long lines waiting for a turn while one student at a time kicks the soccer ball? People who matriculated through a teacher certification program, regardless of whether or not they will ever teach in a K through 12 school, would not allow students they coach to wait in lines. Recreation supervisors know that positive and specific feedback is most useful, and outdoor recreation guides make a special effort to equitably include both males and females in activities.

Advice for Future Teachers

When effective teachers are asked to describe when they ultimately learned how to teach, their response is that they have never stopped learning to teach. That is, they believe that learning to teach is a lifelong process that continues well after one becomes certified and leaves the influence of a teacher education program. Our goal in writing this chapter was not only to acquaint you with the teaching profession but also to emphasize that learning to teach is an ongoing process. In providing advice, we encourage you to engage in professional discourse with colleagues, attend professional conferences, and become familiar with knowledge generated from research investigations. Just as you would expect your physician to remain current with the most recent advances in the field, parents and students will expect you to remain current.

Wrap-Up

The objective of this chapter was to introduce the knowledge base in the pedagogy of physical education. Exciting developments have occurred over the past two decades, and new knowledge will continue to emerge as researchers ask more sophisticated questions and develop new methods for collecting data. The information in this chapter is but a small sample of the types of knowledge currently available to those in pedagogy.

We hope that you will aspire to be like one of the Teachers of the Year—that is, someone who is committed to student learning, engages in effective teaching practices, works toward developing subject matter expertise, and is professionally involved and current. We also hope that you will receive the same satisfaction in your career as have the thousands of teachers who enter the workplace each day—with the belief that they can make a difference in the lives of children.

If you are considering embarking on a teaching career, you will need to know much more about pedagogy than we are able to discuss here. You will have greater success if you consistently and objectively reevaluate your current beliefs in light of new research-based knowledge about effective teaching. As a member of the next generation of teachers, you can make a difference.

Use the key points review in your online study guide as a study aid for this chapter.

QA Activity 15.6

These end-of-chapter questions and activities are also in your online study guide. Your instructor may ask you to complete them online and turn them in.

1. How do the experiences that you acquired as a student of physical activity connect to the examples of effective and ineffective teaching discussed in the chapter?
2. How can you develop expertise in one or more of the content areas of physical education?
3. Describe three ways in which a teacher you have observed could increase the amount of learning time available to students.
4. To reduce management time, what routines are most important for students to learn? Which routines would you emphasize? Describe how you would implement those routines.
5. Which curricular framework would you choose to implement as an instructor? Why?
6. How would you interact with a student who is experiencing learned helplessness?
7. How can you avoid the pitfalls of teacher–coach role conflict?
8. Explain how you would work to improve a high school physical education program.

Careers in Coaching and Sport Instruction

16

Shirl J. Hoffman and Joseph A. Luxbacher

CHAPTER OBJECTIVES

In this chapter we will

- acquaint you with the wide range of professional opportunities available in the sphere of coaching and sport instruction;
- familiarize you with the nature of the work and qualifications for coaching and professional sport instruction;
- inform you about the educational requirements and life experiences necessary to become a qualified and successful coach or sport instructor; and
- help you identify whether one of these professions matches your skills, aptitudes, and professional desires.

Consider the following scenario.

Dr. William Smith, a former college soccer player and high school physical education teacher, volunteered to coach his son's U-17 club soccer team during the regional tournament games prior to the boys' senior year in high school. This was an important time for high school players aspiring to play at the college level, as a number of college coaches would be attending the games to evaluate potential recruits for their programs. The U-17 team was an assemblage of all-star players from various high schools. All were hoping to play well in the tournament so as to impress the collegiate coaches on hand and possibly receive an offer of an athletic scholarship. Parents were well aware that participating on the club team would be quite expensive (the registration fee alone was more than $1000, and added to that would be all of the travel costs, hotels, meals, and so on associated with tournament play), but most considered it an investment in their child's future and the best possible avenue to obtain a college scholarship. The team played well and won all of its regional games to advance to the national final, an event that would attract more than 300 college soccer coaches.

On the evening prior to the championship match, the team's leading goal scorer was caught vandalizing hotel property, which presented Dr. Smith with a dilemma. The biggest game of the season was just ahead, and his star player had violated team rules as well as hotel regulations. Although winning a national championship would be the athletic achievement of a lifetime, Dr. Smith felt strongly that it was ultimately more important to demonstrate that no player is above the law or considered to be more important than the team. He decided to suspend his star player and not play him in the final game. Despite the player's absence, the team played brilliantly and won the championship. For coach Smith, this was undoubtedly the most exciting moment in a lifetime of sport, and he got to share in it with his son. Everything had worked out perfectly—he had stood by his principles in suspending the star player, and the team had responded by winning the championship. Things simply couldn't get any better—and they didn't!

Later that evening while relaxing in his hotel room, Dr. Smith heard a knock at the door. Upon opening the door he was confronted by the angry and irate parents of the suspended player. The father verbally accosted coach Smith and threatened to sue him for preventing their son from obtaining a college scholarship. He said that he and his wife had invested thousands of dollars in their son's soccer career, and that coach Smith's decision not to play him in the final had virtually guaranteed that all that money had been wasted. The father also stated that coach Smith would not have reacted in a similar manner if it had been his son who was involved in the mishap. The conversation abruptly ended as the father told Dr. Smith that he would be hearing from their lawyer.

In reality, this story is combination of several real-life situations rolled into one that serves to illustrate the fundamental nature of the coaching profession—intense emotional highs that are often counterbalanced by depressing emotional lows. Unlike what is seen in most other professions, there is often very little middle ground. While coaching can be a very exciting, satisfying, and enjoyable profession, it can also at times be cruel and punishing for coaches and players alike. Coaching by its inherent nature brings into close contact individuals with differing personalities who may also have different expectations of the sport experience, whose short- and long-term goals conflict, and who perceive the world of competitive athletics in different ways. As a consequence coaching can make for an interesting yet sometimes exasperating way to make a living.

Consider the following possible real-life scenarios that further serve to illustrate that point.

- The coach of a Little League baseball team, attempting to win the league championship, asks one of his less talented players to feign injury so he does not have to put the player into the game for the required number of innings. The player is embarrassed in front of his team and decides to quit playing the sport.
- A coach at a large swim club notices that a 10-year-old swimmer has bruises on her body that indicate physical abuse. The coach must conduct an investigation to find the appropriate social agency to which he can report the incident.
- A Division I baseball coach recently had 32 talented players try out for the baseball team. Even though all are capable Division I athletes, the coach will have to cut seven of those players to bring the sport into compliance with Title IX requirements. He must explain the situation to the players and their parents.

- The director of a Boys and Girls Club is currently seeking volunteer assistants to help him develop an after-school physical activity program for kids at his club. He is hoping to teach beginning swimming to 30 kids from the neighborhood, none of whom have been in a swimming pool before.
- An instructor at a gymnastics academy has a meeting with parents who think that their child is Olympic material. In the instructor's estimation, the child's performance is merely normal for her age and experience. The instructor's challenge is how she can relate this information to the parents and child without creating hard feelings.
- A college football coach is afraid he may be fired because his team has not performed well recently. He knows that his team will have a better chance of success if players become bigger and stronger. Although he is philosophically opposed to the use of performance-enhancing drugs, he also suspects that all the other teams in the conference are using illegal drugs and fears that his team will do poorly and he will lose his job as a result.

The preceding examples represent situations that regularly challenge those who coach and those who supply sport instruction in various nonschool settings. In some cases, such as those of the 10-year-old swimmer and the college baseball players who were cut from the team, the decision that a teacher or coach makes can have a lifelong effect on a young person. Although most decisions that coaches and sport instructors make have far less momentous repercussions, they almost always affect people, and for that reason it is important that coaches and sport instructors give them a great deal of thought and base these decisions on up-to-date knowledge.

In the previous chapter, we focused on careers in teaching physical education. Although teaching constitutes a major professional track in kinesiology, some students have set their sights on careers as coaches or as sport instructors in nonschool settings. Of course, physical education teachers in schools often take on coaching positions as additional duties, but a growing number of opportunities to coach in nonschool settings are available as well. In addition, as swimming, tennis, and golf facilities open, as sport camps and instructional sport academies continue to expand, and as public interest in community nonprofit sport initiatives grows, opportunities for employment in these nonschool-based careers will also grow. This chapter focuses on careers in these areas. They are part of the coaching and sport instruction sphere of the physical activity professions shown in figure 16.1.

All the professions in this sphere are concerned with developing either modest or high levels of motor skill among those being served. The participants served by coaches and instructors vary tremendously, ranging from young children to university students and athletes to professional sport figures. The settings in which coaches and professional sport instructors work vary greatly; they include public

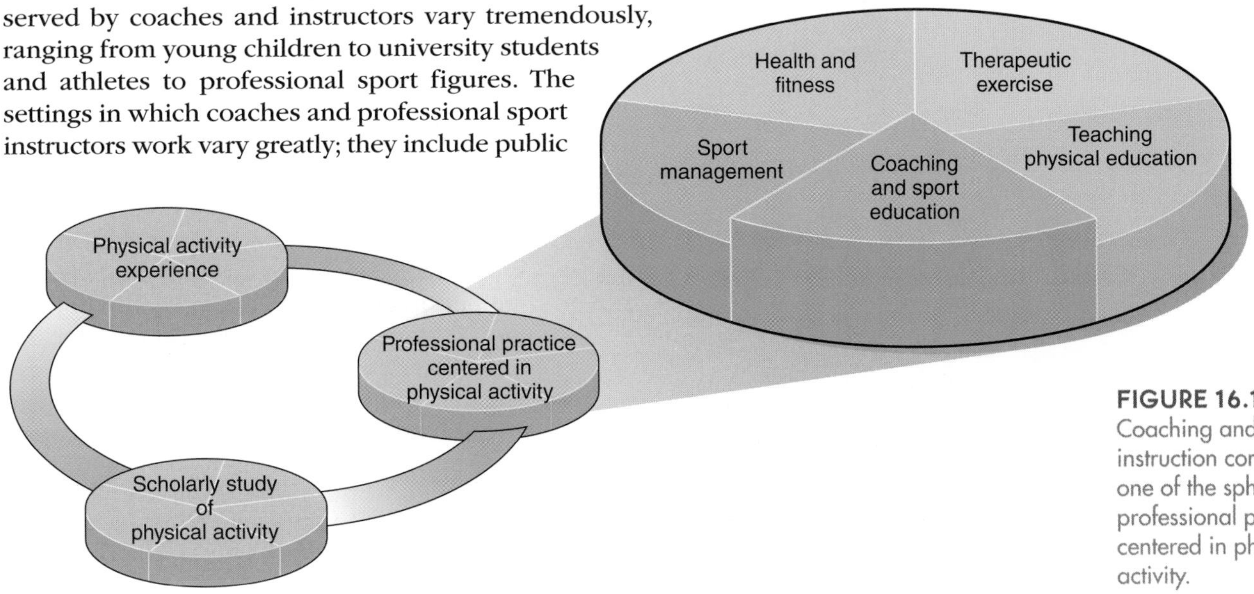

FIGURE 16.1 Coaching and sport instruction compose one of the spheres of professional practice centered in physical activity.

and alternative schools, community agencies and organizations, colleges and universities, sport clubs and academies, community agencies, and professional organizations operated as businesses. Although institutions and organizations employ most coaches and instructors, many are freelancers—self-employed instructors (in gymnastics, golf, baseball, tennis, or swimming, for example) who operate their own small companies.

An Overview of Coaching and Sport Instruction Professions

Coaching and sport instruction have a long and storied tradition, extending back as far as the ancient Greeks and Romans. We know, for example, that coaches hired by early Greek municipalities and supervised by public officials called gymnasiarchs taught young people how to compete in various forms of athletics. Although held in high esteem, coaches could be demanding taskmasters. One notable Greek vase depicts a trainer (or coach) flogging two pancratiasts (performing a vicious form of wrestling popular in the early Olympic Games) for gouging each other in the face while wrestling.

Now, many centuries later, coaches continue to be popular and well regarded, and many continue to have reputations as demanding taskmasters (although none flog their athletes as the early Greek trainer did!). Some ply their trade in the anonymous quarters of community gymnasiums and school athletic fields; others work in the glaring spotlight of the collegiate and professional ranks. Some earn very little; others earn millions of dollars for their efforts. Regardless of fame or fortune, coaching and serving as a sport instructor can be a fascinating and rewarding career. This chapter will help you determine whether it might be a career for you to consider.

You will note that this chapter includes both coaching and sport instruction. Indeed, coaching and sport instruction are similar in many ways. Both are performed by professionals who have mastered knowledge about physical activities and the skills and techniques required for efficiently and effectively transmitting that knowledge to other people. Both require expertise in designing practice experiences that stimulate learning and conditioning experiences to enhance performance. You will recall that chapter 3 addressed these two types of physical activity experiences.

But coaching and instruction also differ in significant ways. An instructor's efforts are usually—but not always—directed toward novices or those lacking a high level of proficiency, such as young children in a community youth sport development program, middle-aged women enrolled in a beginning swimming class at the local community pool, or people of all ages who want to improve their golf swings by consulting a teaching professional at the local driving range. Coaches at the high school, college, and professional levels, on the other hand, typically direct their planning and efforts toward a relatively elite population. Students with above-average physical proficiency for their age commonly compose the bulk of most school sport teams; college teams tend to be populated by an even more

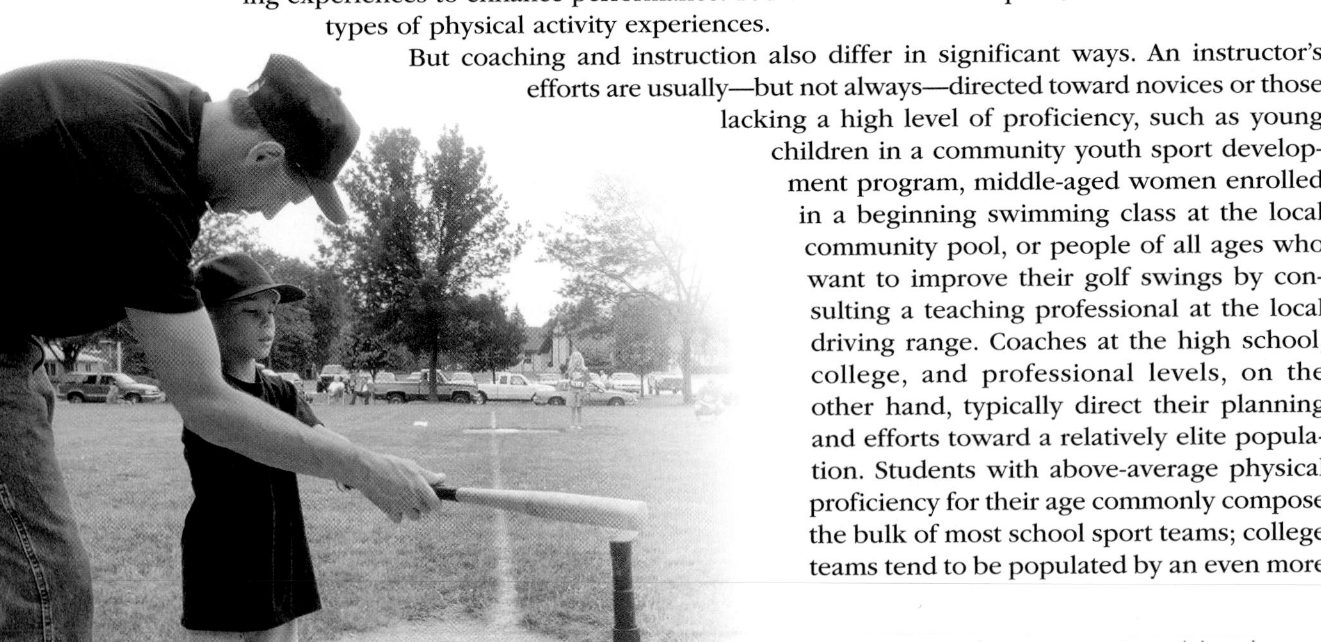

The younger and more inexperienced the athletes, the more the act of coaching becomes like teaching.

select group culled from the best of high school teams. Even youth sport programs, whether inadvertently or by design, tend to weed out those who are less gifted in their age groups, leaving a population of relatively skilled youngsters compared to others of similar age. And, of course, professional teams are an even more highly skilled group.

Still, teaching and coaching share many functions. Both involve instruction and dealing with people on a personal basis; in fact, the dictionary definition for coaching is close to that of teaching. Both the gymnastics teacher and the gymnastics coach try to impart knowledge and skill in certain routines, to instill in gymnasts a love of the activity, and to bring about improvements in their performances. Improvement comes through practice and conditioning and through explanations, instructions, and verbal feedback, as well as visual feedback in video replays and demonstrations.

Obviously, both teaching and coaching may take place in a variety of informal contexts, as when parents attempt to teach their children how to catch a ball or when an experienced bowler teaches a friend how to approach and release the ball. As discussed in chapter 12, such informal efforts by nonexperts are not examples of professional work, and for that reason we do not describe them in this chapter.

So how *do* we answer the question of what distinguishes instruction or teaching from coaching? A good way to address the question is to distinguish between the *acts* of teaching and coaching and between the *professions* of teaching and coaching. The acts of teaching and coaching both may be designed as attempts to alter the thinking, feelings, or behavior of a particular clientele by systematically exposing them to physical activity experiences (covered in chapter 3), along with appropriate verbal and visual experiences, to bring about predetermined outcomes.

Figure 16.2 shows the overlap in the acts of instruction and coaching as well as their uniqueness. The areas that do not overlap tend to be those directly related to the clientele served.

Instruction focuses primarily on novices, coaching on more advanced performers. Coaching is usually directed toward select or elite populations who to some extent have already acquired the skills, knowledge, and attitudes essential for performance. The overriding emphasis for coaches, particularly at higher levels of competition, must be on individual and team performance, as success is typically measured by wins and losses. Because of this, coaches, especially those in the college and professional ranks, tend to direct much of their effort toward improving skills already learned rather than teaching new or basic skills, and then meshing improved technical ability with sport-specific tactics and a high level of muscular fitness, aerobic fitness, or both.

The act of sport instruction

Activities common to acts of sport instruction and coaching

The act of coaching

Specific instructional activities for promoting learning in novices

Instructional activities designed to alter attitudes, knowledge, and behaviors with respect to physical activity

Specific instructional activities designed to improve performance in elite athletes

FIGURE 16.2 The acts of sport coaching and sport instruction. Note the considerable overlap of the two.

Thus, whereas coaches may spend a disproportionate amount of time motivating and conditioning athletes and refining and retaining acquired skills, sport instructors are likely to focus their efforts on helping their clientele acquire new skills and learn how to apply them in real-life settings.

But making too fine a distinction between the acts of teaching and coaching is probably not useful because the roles associated with a position can switch back and forth between those more characteristic of a coach and those more characteristic of an instructor. For example, in the coaching of young and inexperienced athletes, the activities of the coach become almost inseparable from those of the sport instructor. Youth sport coaches, for example, spend much of their time teaching children how to perform basic skills. The emphasis is not, and should not be, on who wins the games but rather on maximizing each player's level of expertise. The focus is on individual improvement as opposed to group performance, although that plays a part as well. High school varsity or college coaching, however, usually involves more of the unique aspects of coaching identified in figure 16.2. Sometimes professional sport instructors serve in a role that we might define as coach-instructor, as when they supply instruction on a one-to-one basis to professional golf and tennis players. Some of the more successful golfers regularly visit these instructors (also called teaching professionals or swing doctors) to correct flaws that have caused their performance to deteriorate. (When Tiger Woods' performance declined during the 2004 season, some suggested that the cause was his firing of his golf instructor in favor of two newer professionals, one to help him solve his problems off the tee and another to help him improve his short game.) Many professional tennis players have "coaches" in their entourage who perform the same function. Regardless of whether one calls them teaching professionals, golf or tennis instructors, or coaches, such people play fundamentally different roles and work in quite different settings than do coaches of scholastic athletics or college-level athletics or those who supply sport instruction to large groups in municipal recreation settings.

Although considerable overlap exists between the acts of coaching and instruction, *the same is not true between the professions of coaching and sport instruction*. By "profession," we refer to the entire range of duties of a coach or instructor, beyond the direct acts of coaching or instruction. Instructors and coaches have different professional responsibilities that they must carry out in substantially different occupational subcultures. Figure 16.3 shows the differences and similarities in the instruction and coaching professions. As you can see, similarities between the two professions are considerably fewer than similarities between the acts of instruction and coaching shown in figure 16.2. There are two major differences:

1. Instructors tend to spend relatively more of their time with on-task duties, by which we mean direct involvement in the dissemination of knowledge and molding of behavior of students. Coaches spend proportionately less of their time on this and more time on the off-task duties of recruiting, scouting, reviewing films, scheduling, budgeting, fund-raising, and so on.
2. The nature of the off-task duties of the two jobs differs. Whereas the instructor spends time maintaining records, repairing and maintaining equipment, advertising classes, and (if teaching in an institution) attending to institutional demands, coaches are more likely to become absorbed with the off-task duties described in figure 16.3.

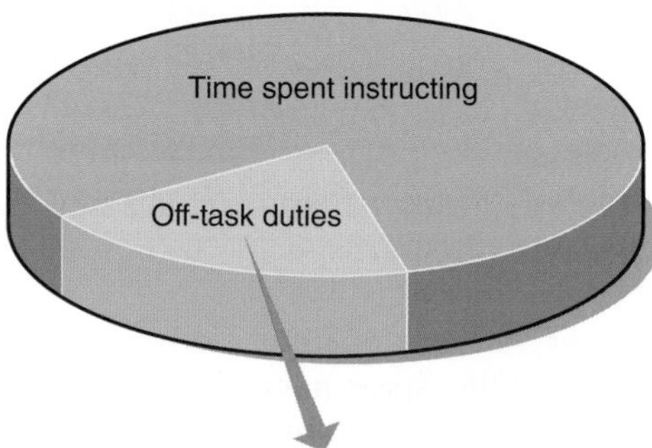

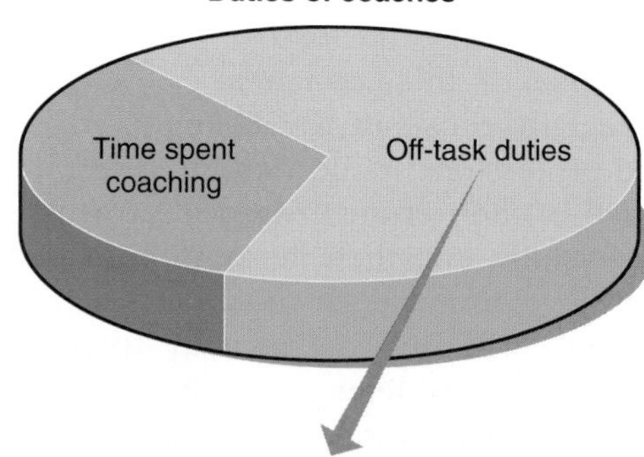

FIGURE 16.3 Differences and similarities in the coaching and sport instruction professions.

Based on the study of G.M. DeMarco 1999.

Activity 16.1

Coaching or Instructing? Test your understanding of the duties of coaches and instructors in this activity in your online study guide.

Clearly, then, the act of instruction is essentially the same as the act of coaching. In this narrow sense, all teachers are coaches and all coaches are teachers. But when we consider the professional responsibilities of the two roles, we see that they are really quite different. Even in cases in which people hold both roles in the same job, as when school or college physical education teachers also coach, the demands of the two roles are distinct. So, although we recognize similarities in many of the professional responsibilities of coaches and teachers, we also need to keep in mind that each operates in a distinctly different occupational subculture and each requires a unique set of professional knowledge and skills. Let's look more closely at the settings in which coaches and sport instructors work.

➤ The *acts* of sport coaching and sport instruction are more similar than distinct; the *professions* of sport coaching and sport instruction are more distinct than similar.

Coaching and Sport Instruction Settings

The setting in which one works is important because it usually defines the job expectations and duties of those who work there. Coaches and sport instructors work in many different settings, and their duties and responsibilities vary accordingly. Most settings fall into one of the following categories: community based, institutional, or commercial. Let's consider each of these in detail.

Community Settings

Community organizations offer a variety of settings for coaching and sport instruction. Because most of the organizations offering sport programs in these settings are nonprofit, volunteers fill many of the coaching positions, although some youth and adult sport leagues hire part-time coaches. Community organizations often hire professionals to be administrators of their coaching and instructional programs. Parks and recreation departments, for example, usually sponsor athletic leagues for both youths and adults. Other organizations such as the YMCA and YWCA, Young Men's Hebrew Association (YMHA) and Young Women's Hebrew Association (YWHA), American Youth Soccer Association, U.S. Volleyball Association, and U.S. Fencing Association provide similar opportunities to children and adults. Municipal recreation authorities often oversee community sites for coaching and sport instruction. Although recreation has developed into a specialized profession in its own right, the career aspirations of recreation graduates and kinesiology graduates overlap. If you are attracted to a career in municipal recreation, you should explore the possibility of obtaining an undergraduate minor in recreation or taking some courses in recreation leadership and management.

Coaching and administering youth sport programs can be rewarding.

Brand X Pictures

Community organizations are civic minded; they have as their first goal improvement of the social, physical, and moral development of the local community, often through physical activity, including sport, exercise, and fitness activities. (Think back to chapter 12 in which we talked about the two value systems that influence professional conduct. Coaches and instructors working in community programs are usually guided by a social trustee, civic professionalism orientation.) Usually these organizations rely on municipal funding, user fees, dues, and private and public contributions from agencies such as United Way, or on business or corporate sponsorships of athletic teams. Some are sponsored directly by tax dollars; most are funded through donations and corporate sponsors. The Boys and Girls Clubs of America, for example, is able to serve 4.8 million boys and girls at 4000 locations in 50 states with a trained staff of 49,000 only through the funding provided by major corporations in the country. The Amateur Athletic Union sponsors a number of sport programs and holds local as well as national competitions. The Joy of Sports Foundation in Washington, D.C., is one of the largest community programs to serve at-risk young people through sport. The foundation has both paid and volunteer staff.

Any or all of these organizations have resources for hiring instructors and an administrative staff to coordinate their programs. Instructors may work in pool areas teaching swimming classes, in gymnasiums or weight rooms facilitating adult group exercise programs, in testing rooms conducting stress tests and fitness evaluations, or in conference rooms engaging in health and nutrition consulting.

Typical worksites for coaches of community-based sport teams are on-site gymnasiums (for basketball or volleyball); skating rinks; soccer, football, softball, and baseball fields; and tracks. Generally, games are played on the organization's property or on that owned by municipalities or schools. Coaches are usually responsible for finding a gymnasium or field in which to practice. They may also be responsible for transporting equipment to and from games and practice, keeping records, and maintaining a flow of communication among team members.

Institutional Settings

Most of the sport instruction in institutional settings takes place in physical education class (see chapter 15). As an arm of education, physical education is part of the formal curriculum in public and private schools, colleges, and universities. But sport also is an important part of the extracurricular program in educational institutions. As a former high school student, you probably gained an appreciation of the importance that people in your school and community placed on interscholastic sports. As a college student, you are undoubtedly aware of how important sport programs have become in higher education. Many universities are identified more by their football and basketball programs than by their academic excellence.

Settings for Coaching

Community settings	Institutional settings	Commercial settings
YMCA or YWCA	Middle schools	National and Olympic facilities
Boys and Girls Clubs	Senior high schools	Private clubs, camps, sport academies
Municipal recreation facilities	Private secondary schools	Professional sport organizations
National youth sport programs	Community colleges	Freelance, entrepreneurial
Various nonprofit organizations	Colleges and universities	

Interviews With Practicing Professionals

Photo courtesy of Paul Besterman

Paul Besterman
Director of Recreation Department
Upper St. Clair Township, Pittsburgh, Pennsylvania

Q: I appreciate your willingness to share your experiences in the profession of recreation management, Paul. Would you provide our readers with a bit of information about yourself, your background, and some of your accomplishments?

A: I am married to my wife Amy and we have two children—Amelia, 8, and Sophia, 6. This without a doubt is my biggest accomplishment. I am the youngest of six children. I graduated from Slippery Rock University with a BS in parks and recreation business administration with an emphasis on community leisure services. My first job was at a YMCA two blocks from the house that I grew up in. I was a summer camp counselor. One of the staff members saw me "hanging out" at the Y, and he put his arm around me and said, "Why don't you take a job with us and try to make a positive difference in someone's life?" That moment changed my life forever.

I have been very fortunate. Parks and recreation has helped me procure employment with the Walt Disney World Corporation (intern); South Seas Plantation (a resort located on Captiva Island, Florida; intern); Shuman Juvenile Detention Center (youth care worker); YMCAs in the United States (youth and teen program director), Japan, and the Philippines (international program coordinator); and the municipalities of Cranberry Township (director of parks and recreation) and the Township of Upper St. Clair (director of recreation and leisure services).

My professional accomplishments are varied. I had the chance to build a $6,000,000 park in Cranberry Township that included three soccer fields, a baseball field, trails, a sledding area, pavilions, and an outdoor water park. Currently in Upper St. Clair we are building a new park that includes an environmental education center, trails, two football-soccer fields, two baseball-softball fields, and a community recreation center.

The most important accomplishment was taking teenagers from the United States and Japan to the Philippines to work within a community afflicted with leprosy. Many of the teens were unmotivated and felt that they deserved everything. They were going to help those in the Philippines. Little did they know that they were the ones who were going to get help. By the end of the work camp, these teens found meaning in their lives with a new appreciation of their families and what they had.

Q: You oversee a large and multifaceted recreation program in Upper St. Clair Township. From that perspective, what do you consider your primary responsibilities as director of recreation in a large suburban township?

A: Recreation is different things to different people. The job of our department is to try and get people to engage in a passion that re-creates the spirit. There is an old saying in the recreation field: "We work when you don't." I always thought that said a lot. We need to make sure that the facilities are safe and adequate, that the programs are varied and many, and that we provide the best customer service possible. Lifestyles are changing. We need to be prepared to address those quality of life issues.

Q: What is a typical workday for you?

A: There really isn't a typical day. For example, just this past Monday I came in at 7:45 a.m. and had a public meeting on the community recreation center until 1:00 the next day. Yes, that is true. Sometimes there are morning meetings, weekend programs or special events, or staff challenges that force you to do something you had no idea you were going to do as the day began. Always have a change of clothes.

Q: I've noticed that there are many different youth sport programs (soccer, football, basketball, etc.) in Upper St. Clair Township. How do you attract coaches for all of those teams? Is there any type of certification program or educational program in place to prepare volunteers to coach in the recreation program?

A: Upper St. Clair is very fortunate. There is just one athletic association. Many communities have baseball, soccer, football, and wrestling associations. Having one association improves communication among the groups and with residents and the Township. It is also up to the various groups in the athletic association to negotiate among themselves regarding the need for field space before they request a field permit. We currently have a waiting list to volunteer for some sports. Each sport has commissioners that oversee that particular sport. The commissioner has a volunteer in charge of recruiting and training other volunteers. This is done in groups, as well as individually with those who may need more assistance. Background checks are also done.

Q: What do you consider the most difficult aspect(s) of your job as recreation director?

A: Community recreation positions can be very challenging because you must serve the residents but also fulfill requests made by elected officials. Sometimes those two challenges don't meet—or may curtail or delay—your goal of accomplishing something you think is immediately critical.

Q: What do you consider the most satisfying or enjoyable part of the job?

A: People. You better like people. I really find enjoyment whenever a former youth camp participant comes back and wants to be a staff member. That little thing usually tells me we are doing something right.

Q: What were the most important experiences in preparing you for your career?

A: First and foremost was my relationship with the YMCA staff member previously mentioned. That person told me I could make a difference and I believed him. Second is education. You must have an education, and you should take it seriously.

Also, volunteer and intern experiences. I will never forget showing up my first day as a Disney intern and being told I needed to sweep the street in the park. This taught me to respect everyone and everyone's job; it taught me that I need to understand everything about an organization and the responsibilities of the people in the organization before I can really contribute or make change.

Q: What advice would you give an undergraduate kinesiology major who is interested in pursuing a career in municipal recreation management? What do you recommend in terms of education, graduate school, an education minor, and so on to competitively prepare for such a position? Is a graduate degree required?

A: I always recommend working in a field before you decide to get a degree in that field. Volunteer, take a summer job, work over your semester break. Go to a public meeting that involves a local recreation department. Today, the most basic requirement is an undergraduate degree. That is beginning to change. Employers for more and more positions are requesting graduate school experiences. If you work in municipal recreation, you will teach. You might teach a class, a volunteer, or your new employees. A minor in education will become even more desirable.

After you complete your education, be willing to do whatever job is required to allow you to get the experiences that will be in demand. I recently interviewed a prospective employee who had just graduated with a degree in elementary education. He wanted to start in management but had never done any work for a recreation department. I explained that we had positions in the summer working with children. He was not interested in starting at that level, and I have not heard from him since.

Q: Any final comments?

A: Be well rounded, listen, and be willing to make a difference in someone's life.

When sport instruction or coaching occurs in an institutional setting, it usually is only one of many activities sponsored by the institution. For that reason, the coach or instructor is usually evaluated with regard to the priorities and mission of the entire institution. In return for a regular salary (not guaranteed to freelance instructors or coaches), the professional must accept claims that the organization makes on his or her time. Colleges and universities tend to place fewer of these types of demands on coaches than secondary schools do. As you will see in the next section, "Professional Roles in Coaching and Sport Instruction," less structure does not necessarily mean less work. It simply means that college coaches tend to have greater flexibility in their schedules than public school coaches do.

➤ For middle and high school coaches, directing the team is merely one of many responsibilities they have at the school. For most, the primary responsibility is teaching classes.

High School and Middle School Settings

In public and private schools, coaches usually serve on the teaching faculty or in an administrative capacity as well, although some school districts hire part-time coaches as fewer full-time teachers opt to take on these responsibilities. Sometimes these part-time coaches, although they may have expertise in their chosen sport, have little training in kinesiology or education. Fortunately, school boards and superintendents increasingly are recognizing the need to hire only fully qualified coaches to work with student athletic teams. In most cases, coaching is an add-on responsibility to the primary responsibility as a teacher of physical education or other subject. Scholastic coaches, particularly in the sports of football and basketball, are often in the public eye. Because athletic teams are considered representatives of the school to the broader community, coaches must pay careful attention to the conduct of their athletes and, of course, their own conduct. Usually, practices and games take place on school property, although travel to other schools is necessary for away games. Coaches usually work during after-school hours, and in some sports the season can extend to five months or more. Coaches usually are among the most popular and admired teachers in the school. In many private schools, all students are required to be a member of at least one interscholastic team, and coaches must accept all willing and committed students as part of the team regardless of their athletic talent. Consequently, the range of talent among the sport team members in private schools may vary greatly. In these situations the coach must be able to deal effectively with both highly skilled and unskilled players.

College and University Settings

Work settings at the college level vary according to the type of institution and level of athletic competition. Coaches at community colleges usually coach one sport and teach physical education classes, although this can vary from National Junior College Athletic Association (NJCAA) Division I through Division III schools. Teaching and coaching take place in the same facility, which generally consists of an office complex, gymnasiums, weight rooms, swimming pools, and an athletic training room. Because the programs at many community colleges and technical colleges cater

Coaching at the college level can be a rich and rewarding career.

Florida Football and Basketball Coaches—Highest Paid?

Success translates into big bucks in Division I athletics. In June 2007, the University of Florida basketball coach, Billy Donovan, agreed to a contract that will reportedly earn him $3.5 million per year for each of the next six years, with an option for a seventh year. Florida football coach Urban Meyer agreed to a similar deal for 3.25 million per year. It is believed that these contracts make Donovan and Meyer the nation's highest-paid basketball-football coaching tandem in Division I athletics. The landmark contracts are rewards for National Collegiate Athletic Association (NCAA) championships in both sports, representing the first time in NCAA history that a university simultaneously held both the football and basketball national championships. The contracts will cost the University of Florida more than $40 million through 2013.

to part-time adult students, athletic programs sometimes do not receive strong emphasis. At these schools, athletic programs and facilities may be modest. Not surprisingly, the larger the school and the more emphasis the school places on sport, the larger and better equipped the facility will be. Athletic programs at larger community colleges usually participate as members of the NJCAA, which sponsors national championship competitions each year in a variety of sports. Currently over 500 institutions are members of the NJCAA, which offers sports at Divisions I through III to over 45,000 athletes.

The worksites at some small four-year colleges are similar to those of community colleges. Coaches at small colleges are likely to hold faculty posts in the physical education or kinesiology department in addition to their coaching responsibilities. A growing trend among small colleges, however, is to appoint coaches to full-time coaching positions. Where coaches also serve as instructors in the physical activity academic program, they must juggle the responsibilities assigned by the head of the kinesiology department with those assigned by the athletic director. Coaches at smaller colleges often become very involved in the social and academic life of the school. Usually, the pressure to win and the outside obligations at small colleges are not as burdensome as those that coaches at the highest levels experience, although self-imposed pressure to win is a potential cause of stress at all levels of competition. Sometimes coaches in small institutions gain tenure as faculty members.

Among larger American colleges and universities, coaches are commonly employed full-time to coach a single sport. At universities whose teams rank among the elite, a head coach often operates more as a CEO than as an instructor, overseeing a vast operation involving high-pressure recruiting and intense publicity. Coaches at large public universities are now often the highest-paid public employees in their states, earning several times more than the university president or the governor (see sidebar).

In the major-profile sports of football and basketball, the assistant coaches often handle the bulk of on-the-field coaching responsibilities. Varsity athletics at large universities tend to operate at a considerable distance from the academic and social life of the institution, although recent NCAA rules changes with respect to continuing eligibility and normal progression toward a degree have necessitated the creation of student-athlete academic support programs at most major universities. These "extended arms" of the academic community are complete with personal tutors and counselors. Because of the considerable investments made by universities in these athletic programs, the pressure on the coach to win is enormous. Failing to win enough games usually results in the firing of the head coach. When the head coach is fired, most of his or her assistants are fired as well.

➤ At community colleges and some smaller four-year colleges, physical activity personnel teach as well as coach. The roles of teacher and coach are usually separate at larger four-year colleges and universities.

Coaches in four-year colleges and universities work in an athletic facility consisting of office space, gymnasiums and weight and conditioning rooms, swimming pools, and an athletic training room. Schools offering football commonly have a separate stadium and office complex for the football staff. Coaches and athletic administrators (e.g., athletic directors and their assistants) are not the only ones who have offices in the athletic facility. Compliance officers, academic tutors and counselors, sport marketers and promotion personnel, media relations officers, facility directors, and event directors typically have offices there as well. Although the structure of university athletic departments varies, all athletic faculty report to an athletic director, who in turn reports directly to a vice president or to the university president.

Other Institutional Settings

Opportunities for coaching are also available in many other institutional settings, including residential and nonresidential schools for children with various disabilities (e.g., schools for the deaf or blind), residential juvenile institutions for troubled or disadvantaged youth, detention facilities, and military settings. In addition, increased opportunities are now available for coaches, instructors, and administrators in some large churches that have excellent facilities in which to conduct sport and exercise programs. In all these institutional settings, the general philosophy and special mission of the institution will guide coaches and coaching practices. Generally, sport in such settings is intended to accomplish an educational and or recreational mission rather than provide entertainment for spectators. Those choosing to coach in such settings should first make sure that they will feel comfortable working in the environment of the specialized institution, because coaching is usually highly integrated into the mission that the institution has been established to serve.

Commercial Sport Settings

Commercial sport settings are those in which sport instruction or coaching is offered as a product in a business venture, usually by self-employed professionals. These settings range from the huge arenas, stadiums, and practice facilities of professional teams to tennis, swimming, or squash facilities at swanky country clubs and health spas. An increasing number of human performance clubs are focusing attention on development of sport skills and improved athletic performance, in addition to exercise. Coaches and teaching professionals who work with elite athletes often have the most diverse worksites of all teaching and coaching professionals. Some instructors and coaches work at a fixed facility, whereas others may teach and coach at a variety of locations. Some itinerant golf or tennis teachers work out of their homes or the trunks of their cars. Other sport instructors own and operate their own martial arts studios, gymnastics schools, or swimming clubs. Coaches of professional sport teams in football, baseball, softball, hockey, and basketball work in fixed settings as employees of sport organizations. Professional coaches work at permanent facilities similar to those in colleges and universities; such facilities contain office space, practice and game areas, weight rooms, and athletic training rooms. In this type of organization, coaches usually report to a general manager and team owner.

> DO it Activity 16.2
>
> **Settings**
>
> Consider the challenges of working in different settings in this activity in your online study guide.

Elite athletes hire a select number of highly qualified coach-instructors. These professionals tend to be self-employed freelancers. Such coaches are often on the road, traveling to their clients' hometowns. For example, figure skating coaches coach high-caliber competitive figure skaters at the skaters' home rinks. Many tennis and golf coaches also travel to their clients, unless they work regularly with a nationally or internationally known athlete.

➤ In the elite sport setting, some teachers and coaches have a stable worksite, whereas others teach and coach at a variety of facilities. Still others are self-employed and may work out of their homes.

A few sport instructors who work with elite athletes have the luxury of a stable worksite. These sites are usually schools or academies with residences on or near the campus where budding teenage Olympic or professional athletes can stay for several months out of the year. For example, affluent parents who believe that their child has a promising career in basketball, baseball, ice hockey, golf, soccer, or tennis might consider enrolling their children in the IMG Academies (see next page). Aspiring young golfers may enroll full-time at the International Junior Golf Academy in Hilton Head, South Carolina, where they spend

Sport Instruction on a Large Scale: IMG Academies

IMG Academies is the largest, most successful multisport training and education institution in the world. More than 12,000 junior, collegiate, adult, and professional athletes from over 80 countries attend each year. Although IMG may be best known for its famous alumni, including Andre Agassi, Maria Sharapova, and Paula Creamer, you don't have to be a serious competitor to take advantage of the programs. IMG Academies is a gated, 300-acre campus in Bradenton, Florida, consisting of golf driving ranges and practice facilities, tennis courts, baseball fields, turf training fields, indoor basketball facilities, sport performance facilities, gyms, clubhouses, and dormitories. It has two cafeterias (one for adults and professionals and one for students) and operates an on-campus restaurant for students in the evening.

IMG Academies includes the following:

- Nick Bolettieri Tennis Academy
- David Leadbetter Golf Academy
- The Soccer Academy
- The Baseball Academy
- The Basketball Academy
- IMG Academies Performance Institute
- Crested Butte Mountain Sports
- The Officiating Academy
- Professional Fishing Academy
- IMG Academies Mental Conditioning
- Game On (communication skills development)
- Bollettieri Sports Medicine Center
- The Wellness Spa
- IMG Academies Golf and Country Club
- Evert Tennis Academy (Boca Raton, Florida)

Young athletes stay in campus dormitories and have access to a highly competent teaching and coaching faculty. For this, their parents pay over $30,000 per year. Tuition for the on-campus prep school and fees for private lessons and other amenities can cost an additional $35,000.

much of the day practicing and learning golf from highly skilled instructors and attending campus-based prep schools.

When we think of coaching and sport instruction settings, we usually think of the popular sports we see every day on television. Coaches today, however, work in a variety of settings. Rowing, kayaking, and canoeing instructors work on rivers. With competitive rock climbing becoming more popular, we find instructors at indoor climbing walls, in the great outdoors scaling mountains, and even in towns and cities climbing stone or brick buildings. As interest in eco-tourism and outdoor adventure experiences continues to expand, the need for instructors and coaches in outdoor recreation pursuits is expected to grow accordingly. Each summer, directors of camps, some with large budgets and impressive facilities and equipment, hire expert sport instructors to oversee instructional operations. Positions as sport instructors also are available at large resorts that sponsor extensive sport programs.

Professional Roles in Coaching and Sport Instruction

In the previous chapter we saw that physical education teaching is an exciting and rewarding career, although it often poses problems for the teacher. The challenge of teaching a large number of different skills to large numbers of students who possess a wide range of ability levels (some of whom would prefer not to be there), in less than optimal environments and often with inadequate time to accomplish instructional objectives, is somewhat daunting. By contrast, coaching and professional sport instruction are free of most of these disadvantages. Professional sport instructors usually work with relatively small numbers of highly motivated learners, athletes, or clients under optimal conditions in which attention can be focused exclusively on one skill or activity. Usually, coaches and professional instructors have adequate time available to accomplish their objectives. Of course, the extent to which coaching approaches this ideal depends on the setting in which one works. Let's examine more closely the duties and responsibilities of sport instructors and coaches.

Community Physical Activity Program Instructors

Community-based programs of physical activity classes are those offered by municipalities through departments of recreation and other nonprofit agencies. The nonprofit sector, in particular, has grown rapidly in size and significance over the past 20 years. Thousands of municipal recreation departments and over 1.5 million nonprofit agencies currently operate in the United States. Despite the great variety of organizations, of purposes, and of sources of funding, a common element prevails: These organizations have become critical to the quality of life in America. For the most part, nonprofit organizations are on the front lines of strengthening communities by providing needed services to individuals and families. Physical activity program leaders who work in nonprofit organizations can help meet this challenge. Some of the larger nonprofit organizations around the country that have a focus on physical activity instruction are the YMCA, the YWCA, and the Boys and Girls Clubs. These organizations may have national budgets of tens of millions of dollars per year, although smaller local community agencies have considerably smaller budgets. An estimated 50,000 professionals are needed annually to fill a variety of open positions in the nonprofit sector alone ("College Preview" 1997).

Community physical activity instructors also commonly teach older people in physical activity programs offered by community centers, recreation facilities, retirement centers, and other nonprofit agencies. Adult programs may include soccer, softball, tennis, and golf, as well as adult exercise, yoga, dance, and martial arts programs. The administrators of most of these agencies and facilities recognize the importance of physical activity to the continued vitality of adults through their later years. Working in nonprofit agencies such as these can be rewarding for professionals interested in serving an older clientele. As with employment in all nonprofit settings, financial compensation tends to be less than that available in commercial or other institutional settings, but knowing that your services are valued can be a form of compensation in its own right.

Traditional sport settings are not the only places in which an individual can lead physical activity programs. Many community organizations and nonprofit agencies need physical activity program leaders to provide quality instruction.
AP Photo/Elaine Thompson

Community Youth Sport Development Leaders

A particularly exciting area of community work focuses on underserved youth. Over the past two decades, many communities have experienced damaging social, economic, and political influences. These pressures have had a direct effect on the way in which human services are made available today. Public policy in the 1980s eliminated many of the governmental human services and resources that historically supported communities; subsequently, this change sparked growth in the number of community-based organizations. These organizations are stemming the tide of further deterioration of our nation's most underserved communities.

Working as a youth community sport instructor or coach can be challenging yet exciting.

Throughout the United States, communities are looking for a solution to the crises facing America's youth. The typical approach has been to focus on what is wrong with young people and then develop strategies and programs that will solve the problem. This traditional approach to helping youth, especially underserved youth, is well intended; but it focuses on kids' deficiencies and problems rather than on their capacities, skills, and assets. The new approach does not view young people as a burdensome problem; instead, it assumes that kids have knowledge, sufficient skills, and positive qualities that can be enhanced. Increasingly, community-based organizations are using this asset-based approach in their physical activity settings for youth.

Physical activity is a particularly effective way to focus on kids' capacities, skills, and assets. The unique energy, human interactions, and creativity that physical activity harnesses allow community-based physical activity program leaders to see kids' potentials and personal qualities that might otherwise go unnoticed. Furthermore, youth development philosophy holds that young people who are currently having difficulty in school or at home, or even those involved with the juvenile justice system, can, when given the proper opportunities, beat the odds and become contributing members of society.

Because of the assorted needs of community-based organizations, physical activity professionals at this level typically perform a range of tasks beyond instruction. Program instructors usually assume responsibility for directing the physical activity programs, which may require marketing the program, budgeting, and overseeing a staff. In addition, physical activity specialists in community organizations may be involved with a variety of managerial duties as well as tutoring, mentoring, literacy programs, and a host of other national community services. If you are a kinesiology major who wants to work with youth but don't want to be a public school teacher, becoming a community physical activity instructor may be an alternative. Salaries for some of these community youth development positions tend to be lower than those of public school teachers. But, as with all nonprofit ventures, the attraction is having the opportunity to make a difference in kids' lives and knowing that you are contributing to your community.

➤ Nonprofit organizations and park and recreation centers offer the majority of physical activity programs in community settings.

Interviews With Practicing Professionals

Photo courtesy of University of Pittsburgh

George Dieffenbach

Professional Sport Instructor, Division I Tennis Coach

George Dieffenbach has served both as a professional tennis instructor and as a Division I women's tennis coach and as such has firsthand knowledge of the similarities and differences between the two professions. His academic background includes an undergraduate degree in biological science and an MEd and EdD in science education. During his undergraduate days, George was a competitive Division I tennis player. Throughout his playing career George has held competitive rankings at local, regional, and national levels with the U.S. Tennis Association (USTA).

George is presently the women's tennis coach at the University of Pittsburgh and holds the Professional Level 1 Instructor Certification of the U.S. Professional Tennis Association (USPTA). George is also a competitive runner who has completed more than 20 marathons. In the 2005 U.S. Senior Olympics, he finished fifth in the nation in his age group for the 1500 meters.

Q: George, you have experience as both a professional tennis instructor and a Division I tennis coach. How do the jobs differ and how are they similar?

A: For one thing, the clientele is somewhat different. Most private instruction is set up on an hourly basis—you work with an individual whose experience can range from novice to highly competitive. During the lesson I typically focus on improving, or in some cases correcting, technique and biomechanics. I try to help students become more proficient at the sport, to make them more competitive within their group, and to get them to the next level, whether it be high school, college, or amateur competition. Competitive players generally return on a regular basis, so I serve as an ongoing consultant. At the opposite end of the spectrum, I also instruct novices who just want to get a workout, learn a new hobby, or simply become adequate at the sport. These clients don't usually come back on a consistent basis.

The job demands as a college tennis coach are different. A much smaller percentage of what I do deals with technique and biomechanics. I am not being paid by the hour—I am paid a salary to do a number of different things. I am trying to refine technique that players already have, to make them competitive at the college level. Much of what I do as a college coach deals with off-the-court issues—the academic side, compliance with NCAA rules, community service, and so on. As a college coach working in close association with a group of individuals over a long period of time, I must take a more global view of how I deal with athletes.

That said, the objectives when training an individual and a group of college athletes are generally the same—to get the best out of them.

Q: Which profession is more encompassing, requiring a greater investment of time and effort, and why?

A: Physically, the role of tennis instructor is more demanding. At times you may teach 8 to 10 hours a day, and you are hitting tennis balls the entire time. In all other aspects the responsibilities of a college coach are much more encompassing and time-consuming than those of a professional instructor.

Q: Describe an average workday in the life of a college coach.

A: Every day involves paperwork, mostly dealing with compliance and NCAA rules. Phone calls to prospective student-athletes, evaluating talent, planning the budget, planning and conducting our practice, coordinating workouts with strength and conditioning coaches. On days we compete I travel with the team and coach players within the match itself, which is very intense. As a Division I coach I am on the road a lot due to recruiting and our conference schedule. I would say that recruiting is probably the most important aspect of the job, in terms of sustaining a competitive program, and possibly the most difficult as well. The travel can be tiring, but I've found it to be interesting and enjoyable.

Q: What advice would you have for an undergraduate kinesiology major interested in pursuing a career as a professional tennis instructor? What do you recommend in terms of education, graduate school, a minor field of study, and so forth to competitively prepare for such a position? What about a career in college coaching? Are graduate degrees required for either or both?

A: For both professions I feel that competitive play at the college level is a must, so you realize what it takes and relate to the difficulty and adversity that players face. It is also important to attain a certification from the USPTA or USPTR (U.S. Professional Tennis Registry). As far as a graduate degree, I do not feel it is absolutely necessary, but I do believe it will enhance your chances of getting a job at the college level. I also believe it is important for you continue to compete after college—it will help you get a position and also help you relate to your players. If you are going to teach others, you must become a student of the game. You cannot always measure that knowledge with a degree—you must immerse yourself within the sport.

Q: Can you become a professional tennis instructor without first being a national-level competitor? What about a collegiate tennis coach?

A: Yes, I feel that you can become a teaching pro without being a national-level competitor. However, the higher the competitive level you've attained, the wider range of abilities you can comfortably teach. A national-level player would not come to a tennis pro for instruction if the instructor had not attained a high level of competition in his or her own career. A collegiate coach should have some experience competing at a high level.

Q: What have been the most important experiences preparing you for your career?

A: Playing college tennis is the single most important experience, although my education has equipped me to better understand the athletes that I coach. Becoming a certified tennis professional was also very important for my college position, as I had to really prepare for the certification and it helped in my understanding of the sport.

Q: What do you consider the most positive aspect(s) of your job as a professional tennis instructor? What is the most difficult or frustrating?

A: The financial rewards can be very good. On the flip side, you can put in long hours so you had better be very fit, as you will have to teach many hours at a high level of intensity. The physical demands are very challenging, which is probably the reason that the average age of the tennis pro is much lower than that of the college coach. Probably the most frustrating part of the job is the fact that you don't always see the results of your work with clients who do not stick with you on a regular basis.

Q: What do you consider the most enjoyable aspect(s) of your job as a Division I tennis coach?

A: Seeing my players compete, improve their games, and get a degree. You see athletes become better tennis players as well as better people. You don't necessarily have that experience as a professional instructor.

Q: What is the most difficult or frustrating part of the job?

A: Competing in Division I is very difficult, and there are many variables—some not within your control—that ultimately determine success or failure as measured by wins and losses. A college tennis coach has a very complex job, so you need good time management and organization skills to juggle many roles.

Q: How are job opportunities at the collegiate level? What is the best way to get your foot in the door?

A: The best way is to get a graduate assistant position at the college level and then move forward from there. The average tenure of a college tennis coach at any one school is rather short, so jobs open up all the time. The pay is not great, and you have a lot of responsibility. Financially, there is the potential to make more money as a teaching pro. However, if you want long-term security and a more stable lifestyle, a college tennis coach has it over the teaching pro.

Q: Any final comments?

A: If you are looking at the long term, I would say that college coaching is the route to follow. My recommendation would be to do both while you are young and ultimately phase yourself in to a full-time collegiate tennis coaching position. If you love tennis, you won't go wrong with either profession.

Professional Sport Instructors

We have all heard of the golf pro and the tennis pro, but who are they? Are they teachers, or are they coaches? *Teaching professional* and *professional sport instructor* are terms describing those who offer services (private or group lessons) in a specialized physical activity. The professional sport instructor is more similar to a teacher in his or her duties than to a coach. In fact, swim, golf, and tennis professionals who work either as freelance entrepreneurs or at clubs may also hold positions as members of the physical activity instruction faculty at colleges and universities. Professional teachers often instruct performers of all ability levels. They may offer lessons for beginners but also coach highly skilled individuals who are competing in leagues and tournaments. Within this profession are people who teach, people who coach, and some who do both. Outside the university, the teaching professional can be found at private, semiprivate, or public teaching facilities instructing beginning, intermediate, advanced, or elite players. Golf professionals, for example, may work at a municipal golf course giving lessons to the public and conducting golf camps or managing the pro shop or the clubhouse. And, of course, some serve as coaches or instructors to players on the Professional Golf Association and Ladies Professional Golf Association golf circuits.

Professional sport instructors typically are self-employed and are commonly hired as independent contractors. Employment security for certified coaches and instructors depends on the satisfaction of the clients and the owners or managers of the instructional facility. Thus, certified professional sport instructors, like professionals in other fields, have a high degree of control over their careers. Their continued employment depends on their ability to provide services that their clients find valuable and are willing to pay for. Unlike coaches at institutions, professional instructors usually are not covered by group insurance policies and retirement plans. Employment opportunities at the beginner or intermediate levels are more available than at the elite or professional levels. Although many people want to learn to swim, golf, rock climb, and sky dive, only a select few (in relative terms) want to work hard and pay an instructor to become proficient enough to perform at an elite competitive level. Employment opportunities for the professional sport instructor are more limited than are those for other teaching professions. Incomes for private instructors can range from only a few thousand dollars a year (the golf instructor whose office is the trunk of her car, for example) to over $100,000 a year for the instructor-entrepreneur (owner, operator, and instructor of a successful golf instruction school).

➤ The professional sport instructor is one who offers services (private or group lessons) solely in an activity of his or her expertise. More than any other teacher, the teaching professional is similar to a coach, although he or she is not saddled with all of the administrative duties commonly performed by the head coach of a team or group.

Certification and Education of Community Physical Activity Program Instructors and Professional Sport Instructors

Do you recall our discussion in chapter 12 about professionalism and the importance of certification procedures to ensure that only the most highly qualified people can serve in professional capacities? Community physical activity and sport programs require various types of credentials for the instructors who lead them.

Certification for Community Physical Activity Instructors

Currently no certification programs exist specifically for youth sport development work focused on underserved children. At least a bachelor's degree is required for most positions, preferably in a program that offers a minor specialization in youth sport development. Increasingly, colleges and universities around the country are beginning to offer bachelor's and master's degrees in kinesiology with a concentration in what is termed either community youth sport development or community leadership (Hellison et al. 2000). These programs are often multidisciplinary and teach students how to develop, implement, and evaluate community-based physical activity programs for young people. They also include courses on youth agency administration, principles of community organization, grant writing and fund-raising, social work, and community development. Elective courses in social work, counseling, and management may be helpful. Internships that involve working on a daily basis with children are an important part of these programs.

Many university kinesiology departments offer graduate specializations in adult fitness; some offer online certification programs in conjunction with the Senior Fitness Association and the International Council on Active Aging. Educational preparation of a broader nature is available through masters of education in adult and community education programs, which prepare personnel for broad-based roles in health and social service agencies as well as public institutions.

Certification for Professional Sport Instructors

Although certification in the form of college degrees is usually sufficient for coaches of the most popular sports (basketball, football, baseball) in the United States, certification also can be important in cases in which the coach or instructor works in a noninstitutional setting such as a sport club. For example, most coaches of age-group teams in large soccer clubs are certified (licensed) by the U.S. Soccer Federation or the National Soccer Coaches Association of America (NSCAA) or both. Certification occurs at both the state and national levels. Likewise, to teach swimming or to work as a lifeguard will in most cases require you to secure American Red Cross certification as a water safety instructor. This certificate will enable you to be employed by a private club, YMCA, or community center to teach swimming lessons and coach the swim or synchronized swim teams. If you are interested in teaching scuba diving, you will need to acquire the proper certifications from PADI (Professional Association of Diving Instructors), which start at the assistant instructor level and move to the open-water scuba instructor level.

PGA-certified instructors complete rigorous education and testing requirements.

The U.S. Fencing Association (USFA) has a rigorous certification program offered through Coaches College, in which would-be instructors move through five levels. At levels I and II they master skills in use of the three weapons (épée, saber, foil), and at levels

III through V they specialize in one of the weapons. To pass the first part of the certification program they must write a paper, propose a one-year training cycle, and serve as an intern at Coaches College. At the highest level, they must develop a fencer under their supervision. Only when the fencer places in the USFA ranking list or in the top 12 of NCAA fencers will the USFA grant certification.

Other ambitious certification programs are sponsored by the men's Professional Golf Association (PGA) and the Ladies Professional Golf Association (LPGA). One becomes a golf professional under either PGA or LPGA auspices only by passing an arduous education and testing program that may take up to eight years to complete. The PGA certification program requires study, training, experience, and testing in every aspect of golf—not only playing aptitude and teaching technique but also club repair, analyzing and correcting swing flaws, supervising tournaments, and handling the business side of golf including management of golf shop operations. The PGA recommends that those who are interested in obtaining this certification seek out opportunities to work at a golf course or facility so that they can acquire the necessary experience to undertake the certification program. Certification programs also exist for many other sports (see pp. 465-468). If you plan a career as an instructor or coach in one of these sports, you should begin working on the certification for that sport as soon as possible.

Coaching

➤ As with becoming a teacher, becoming a coach means building relationships that can change the life of another for the better.

We all have known people who have touched our lives as no one else could. Often the emotions associated with our memories of these people—the feelings of respect, awe, deference, and admiration—seem as vibrant and real as when we first experienced them. For many current and former athletes, a coach is in the forefront of their minds and hearts. Joe Paterno, head football coach at Penn State, had this to say about one of his high school coaches: "What may be his most important accomplishment is in the relationships he has had with his athletes. He [was] more than just a coach who want[ed] to win matches. Howard Ferguson provide[d] his student-athletes at St. Edward with the unique opportunity to become the best person they can possibly be—in sport and in life. He [was] a friend who [put] the time and effort into helping each one of them take advantage of every opportunity" (Ferguson 1990, foreword).

Many opportunities are available to those who want to coach. Individuals coach a variety of people in many different settings—young children to adults, recreational athletes to elite athletes. Some college and professional coaches are paid hundreds of thousands or even millions of dollars, while others do it simply for the joy it brings. Coaches can be found in sport programs in communities, secondary schools, colleges and universities, and professional sports. These programs have different coaching requirements, philosophies, and expectations, and the coaches in these programs often differ in their educational background and coaching experience.

Youth Sport Coaches

Municipal recreation departments usually offer a wide range of competitive youth and adult sports, for which the director or supervisor of recreation is responsible. The range of activities offered depends on the ingenuity of the director and the available budget and facilities. Community-based youth sport programs (for those from 3 to 5 years through 14 years old) range from age-group swimming programs conducted in public swimming pools to highly organized competitive programs in football, baseball, softball, soccer, basketball, tennis, and other popular sports. Generally, but not always, these programs tend to deemphasize winning and emphasize inclusiveness and fair play. Often, a "no cut and everyone plays" policy is in place. These types of programs often do not keep standings or play tiebreakers; and organizers attempt, at least in theory, to keep the playing ability of the teams equal. Volunteers recruited and trained by the director of the program usually staff community programs. The director also is responsible for hiring and training officials to oversee the contests. Games usually take place during evening hours, on weekends, or during the summer.

Some community coaches work on a voluntary basis for local nonprofit organizations such as the YMCA, park districts, community centers, or Dad's Clubs. Many are parents of the kids involved in the programs. These volunteers commonly dedicate a couple of evenings each week plus weekends to their coaching duties. The primary responsibilities of the youth sport coaches are teaching the basic skills and rules of the game, keeping game statistics, performing administrative team management, and providing an enjoyable and worthwhile experience for the young players. Coaches who have had formal education in kinesiology may serve as administrators of large community-based youth sport leagues or as instructors of coaching clinics that are offered to laypersons who coach in youth sport leagues.

Organizations that rely on volunteer coaches often face the difficult task of finding qualified, certified coaches. Because many states do not have specific standards or certification requirements for volunteer coaches, anyone interested in coaching is often deemed qualified. Generally, community organizations are short of volunteers, so anyone who wants to coach at this level can usually find a coaching position. Often a parent of one of the team members assumes the role of coach. If you are interested in pursuing a career in coaching, volunteering to coach a local youth sport team while still an undergraduate student will provide you with excellent ground-floor preparation. In addition, the experience will allow you to determine if coaching is the right professional occupation for you. Consider that the essentials of coaching at the youth level differ significantly from those required at the high school or college level of competition.

Community organizations cannot always meet the needs of the athletes, particularly those who excel at their chosen sport and wish to expand their opportunities, and private clubs fill this gap. Many of these private clubs have a philosophy similar to that of the community programs and play local, intraleague contests; but others are more competitive. In many private club sport programs, children join "travel teams" that, as the name implies, travel around a county or state to experience competitive interleague play. Some civic programs also offer competition on a regional or national scale. For example, age-group swimming programs compete at local, regional, and national levels, as do youth soccer programs. In theory the more gifted competitors are selected for the traveling teams, which offer the players opportunities to compete with others of the same ability. In actual practice, however, there are inherent problems with player selection in many of the travel programs, and the fact remains that the added competitive pressures imposed on young athletes are not always in their best interest. Little League baseball sponsors a national tournament that leads eventually to the Little League World Series in Williamsport, Pennsylvania, which is televised nationally.

The Little League World Series in Williamsport, Pennsylvania, is a local program with national interest.

AP Photo/Gene J. Puskar

Some of these programs hire professional coaches. Pop Warner Football and Amateur Athletic Union Basketball, which are private organizations, are two of several national programs for children that are more competitive than community-based programs. In general, coaches at this national competitive level spend more time performing the nonteaching duties associated with coaches such as scouting and recruiting new talent, viewing game films, and organizing team travel.

A growing problem confronting those committed to the educational benefits of youth sport programs is the tendency for organizations, media, and corporations to exploit them for commercial gain. Some may question, for example, whether the Little League World Series, televised around the world, is appropriate for young children. Others question the growing trend for professional sport organizations to sign young children who show athletic promise to professional contracts and commercial endorsements. In recent years, for example, sponsors have signed young children to lucrative multiyear contracts in the hope that they will mature into professional standouts. For example, Nike signed Freddy Adu, an exceptionally skilled soccer player, to a $1 million endorsement at age 13; Mountain Dew gave a lucrative contract to another 13-year-old, a snowboarder; and the parents of a 6-year-old skateboarder in Kirkland, Washington, have arranged sponsorships for their son with Jones Soda, Lego, and Termite Skateboards.

Coaching in Institutional Settings

Institutional coaches are those who work in educational institutions such as secondary schools (middle, junior high, or high schools), colleges, or universities. As mentioned earlier, institutional coaches may also serve in specialized residential institutions for young people who are disabled or socially disruptive. Some coaching opportunities are also available in the armed services. In each case the coach strives to evoke each athlete's potential and guide the athlete in overcoming personal challenges and achieving personal success within his or her limitations. Coaching duties include teaching physical skills; keeping individual and team statistics; scheduling practices, games, and tournaments; and managing and maintaining equipment. Working at an institution requires physical activity professionals to relinquish some autonomy in the carrying out of their duties. Generally, coaches working in institutions do not enjoy the freedom of the freelancer, although they do have the benefit of a steady and reliable income. Complying with institutional expectations and conducting oneself in accordance with established codes of conduct are essential. Ultimately, the institutional coach serves at the behest of the institution's administrator or its board.

> ➤ Community coaches volunteer in community-based sport programs that range from relatively noncompetitive to extremely competitive, and from programs that are educational and inclusive to those that emphasize performance and winning.

Coaching in Middle and Secondary Schools According to the National Federation of State High School Associations (NFSHSA), over 7 million high school students participated in athletic programs in 2006. An enormous number of people are required to coach these millions of students. For those who love sport and love being around young people and serving as a positive role model for them, coaching can be an exciting and rewarding career. At the same time, you should factor a certain dose of realism into your decision about whether to pursue coaching as a career. The most important point may be the fact that coaches work many hours for relatively low compensation. Even high school coaches of major sports such as football often earn only a small stipend on top of their regular teaching salaries. The committed coaches who love what they do would tell you that the money is merely icing on the cake; the rewards of getting close to the kids on their teams, of playing a substantial role in their development, and of having the respect of the community at large are more than adequate compensation. Yet you should weigh these considerations before you make your decision.

> ➤ Secondary-level school coaches typically spend their days teaching and receive an additional stipend for coaching.

Middle, junior high, and high school coaches usually are certified public school teachers who teach a full schedule and receive a stipend for taking on the extracurricular activity of coaching. Certified, part-time coaches who aren't members of the school faculty may be hired to coach specific teams. Although the qualifications for coaches in the public schools vary by state, a teaching or coaching certification and background check are typically required. The coach is one of the most visible representatives of the school. Unlike teachers who work in the confined spaces of the classroom, with only their students, parents, and principal assessing their performance, the coach performs in front of all the preceding as well as the

The Parent Problem

Parents who desperately want their children to succeed at sports can become a serious problem at the site of competition for youth sport coaches and organizations. Within the past few years we have witnessed a father's being sent to jail for beating and berating a coach because the coach took his 11-year-old son out of a game; a father (a dentist) who sharpened the nose guard of his son's football helmet so that the son could slash opponents; and a police officer who gave a Little League pitcher $2 to hit a 10-year-old player with a fastball during a game. In Massachusetts, a hockey player's father, who was supervising a practice session, was beaten to death by another father because of a disagreement concerning the rough play allowed during practice. Such incidents have led to the drafting and promulgation of sportsmanship codes for parents. In some states parents must sign the code before their child will be permitted to play. Perhaps this approach will help correct many of the problems; but the coaches, directors, and referees of such programs must remain vigilant, not hesitating to exclude parents who threaten to turn the fun of sport into the tragedy of violence. Organizations such as the National Alliance for Youth Sports (NAYS) and the Institute for the Study of Youth Sports (ISYS) at Michigan State have been instrumental in educating parents and coaches in this regard. The NAYS "Sport Parent Code of Conduct," and the "Bill of Rights for Young Athletes" written by Dr. Vern Seefeldt and Dr. Rainer Martens and distributed by the ISYS, have been widely adopted by youth sport programs across the country.

community at large. The daily newspapers cover his or her successes and failures. The coach's performance is often the center of conversation in local diners and coffee shops. Everybody is allowed to hold an opinion (and to voice it) about the effectiveness of the coach.

The number of contact hours between coaches and athletes varies depending on the governing body regulating public secondary school athletics. Typically, secondary school coaches spend approximately 8 to 10 hours a week in after-school practices. When competitive play begins, the number of practice hours may drop to 8 and game hours may increase to between 4 and 6. Some coaches have early morning practice schedules; others have late afternoon or early evening schedules.

Besides spending time in practices and games, coaches have many responsibilities behind the scenes (see figure 16.3). To varying degrees, all coaches are involved in organizing and scheduling team meetings and practices, ordering equipment, maintaining and overseeing inventory, checking athletes' eligibility, arranging transportation to and from events, talking to the media, preparing for home events, fund-raising, and organizing end-of-season banquets. The extent of coach involvement in these activities depends in part on the size of the coaching staff and the number of support personnel. For example, some schools have a full-time athletic director (AD) who is responsible for organizing home contests. Other schools may have a part-time AD, who must delegate that responsibility to another coach. Likewise, some sports (especially football) require many assistant coaches to whom head coaches can delegate specific duties.

The differences in coaching at the various educational levels tend to be a function of the athletes' developmental differences. Some middle school athletes are "diamonds in the rough" with the potential to be excellent athletes but are physically undeveloped (i.e., prepubescent and small) compared with others on the team. Others may have just gone through a growth spurt and need to relearn a given skill in their new-sized body.

Coaching requires good communication skills and an interest in young people.

Middle or junior high school coaches have the unique responsibility of introducing young athletes to the world of interscholastic sport, which requires athletes to balance the attention they give to their academics and their sport. The middle school coach's primary tasks are

- to help students interested in athletics develop a positive view of themselves as athletes and
- to teach them important, complex physical skills and the basic strategies needed to become competitive student-athletes.

These coaches also must introduce their athletes to a concept of competition in which winning takes on greater importance than it often does at the community level.

The high school coach is primarily involved with students who are interested in athletics and want to invest the time needed to become good athletes. The high school coach has the opportunity to build on the foundation set by the middle or junior high school coach by teaching more complex tasks and game strategies. Competition and the importance of winning take on new meaning at the high school level. The increased emphasis on winning necessitates that coaches choose their teams wisely. They spend more time reviewing game films, developing strategies specific to a given opponent, and scouting and recruiting potential team members.

Coaching in Colleges and Universities Winning is clearly extremely important to all intercollegiate coaches; the pressure to win at this level, especially at large universities, is much greater than that at the high school level. The enormous financial stakes of big-time college athletics can reduce it from an enjoyable, educational experience for everyone to a brutal exercise in survival. As anyone who reads the sport pages knows, the coach who doesn't win his or her share of games is soon fired.

> ➤ College coaches aim to maximize the athletic potential of already skilled athletes; but they also have many other responsibilities, including team and facility management, budgeting, monitoring academic progress, and recruiting.

The quest of college coaches is to maximize the athletic potential of already skilled athletes, build on the foundation established in high school, and fine-tune each athlete's ability to move that athlete toward becoming a member of the elite. Most college coaches have the advantage of working with a relatively small number of highly skilled athletes in well-equipped facilities. Given the popularity of basketball and football in American society, these two sports tend to receive most of the resources of the athletic department. Coaches of "minor" sports such as cross-country, volleyball, softball, soccer, tennis, and golf usually have modest budgets and resources at their disposal.

Like secondary-level coaches, college coaches have many behind-the-scenes responsibilities. To varying degrees all coaches are involved in team, facility and equipment management, and budgeting. At the college or university level, talent scouting, recruiting, fund-raising, public relations, and athletic eligibility take on greater significance. The degree of significance often depends on the type of college in which the coach works.

Employment opportunities at the college or university level are not as plentiful as they are at the secondary level, as the majority of teenage athletes do not compete at the collegiate level

Probability of Competing in Athletics Beyond High School

An NCAA study on the estimated probability of competing in athletics beyond high school provides sobering numbers to those who envision playing at the college and professional levels. In 2004, for example, 983,000 young men played high school football. Only about 5600 became NCAA student-athletes, and only 0.09% were eventually drafted in the National Football League (NFL). A similar trend is evidence in women's sports. In 2004 there were 456,900 women who played high school basketball. Of that group, 14,400 became NCAA student-athletes, with only 4100 making team rosters as freshmen, 3200 playing as seniors, and only 32 making it at the professional level. In short, the percentage of women going from high school to the professional level was 0.02%.

upon graduation from high school (see sidebar). Coaching at the college or university level is a goal of many high school coaches. Because there are no national or state qualifications or requirements for coaching at the college or university level, most colleges and universities searching for a coach look at applicants with an established track record in the sport. Typically, establishing a name requires college playing experience, coaching experience, a history of successful seasons, and being known by many other coaches and administrators. Many Division I coaches began their careers at the secondary level and moved through the lower divisions of college athletics. Others managed to secure an internship during their undergraduate years or sign on as team manager and, with a great deal of effort, work their way up to an assistant coaching position.

Coaching in smaller colleges offers many opportunities for coaches to get close to their players and to meet and interact with them in many different settings. Although pay generally is not high, the rich quality of life offers some compensation. Coaching in larger institutions where coaches' jobs, and their assistants' jobs, depend on the team's win–loss record usually requires an enormous commitment in time and energy, sacrifice of one's family and social life, and the fortitude to withstand the anxiety in years when the team fails to win many games. If you happen to be one of the few perennially successful coaches, however, coaching can be a lucrative and exciting career.

Coaching in Professional Sport

➤ Generally, the professional sport coach has many nonteaching and noncoaching duties, including administration, recruitment, and media appearances. Continued employment depends heavily on producing a winning team.

Earlier in this chapter we discussed the teaching professional. When one thinks of teaching pros, names such as Tim Gullikson (tennis), Jim Sutties (golf), and Jim Counsilman (swimming) may come to mind. Who comes to mind when you think of professional coaches? Such a question may evoke images of Joe Torre (baseball), Larry Brown (basketball), and Tony Dungy (football). Although the number of professional sport coaches is small compared with the number of secondary school and college coaches, they typically have national prominence.

As with other professional coaching positions, facilitating athletes' performances is central to the job of the professional sport coach. The position also includes elements of administration, athlete recruitment, and public relations events and media appearances. Generally, the professional sport coach has more nonteaching duties than does the teaching professional discussed earlier. For example, some sports, such as football and hockey, require coaches to spend a significant amount of time in coaching meetings and reviewing game films; continued employment depends heavily on the performance of their teams. The win–loss record in such sports is more important than the win–loss record in sports such as golf and tennis. Additionally, the professional sport coach deals solely with elite athletes, whereas the teaching pro is often involved with beginning and intermediate athletes.

Employment opportunities for the professional sport coach are few and far between (see sidebar). The number of professional sport teams is limited, and many individuals are qualified to coach at the professional level. As in the college and university setting, no national or

So You Want to Be a Professional Sport Coach. Good Luck!

Securing a head coaching position at the professional level may pose the most difficult job search imaginable. Consider the following: The five major professional sports in the United States have a total of 135 head coaching positions (National Football League, 32; Major League Baseball, 30; National Hockey League, 30, National Basketball Association, 30; Major League Soccer, 13). While it is true that the tenure of a professional sport coach may be brief and that openings do occur on a regular basis, it is also true that coaches already within the professional coaching community are often first in line to fill the most recent vacancy. This is evidenced by the fact that several coaches in each of the professional sports have coached two or three different teams within the league. They get fired or leave one job only to fill a similar position with another club.

state qualifications are needed for coaching at the professional level. Prospective professional sport coaches need an established name in the sport; most acquire one by being successful coaches at the college or university level.

Job security for the professional sport coach is tenuous at best. When you coach a team sport at the professional level, you have entered an industry where success relies in large part on public support. In professional sports, public support waxes and wanes relative to a team's win–loss record. Just as the organization's success depends on producing a winning team, so does retaining your job. Many professional sport coaches thrive on this aspect of the job. They bask in the challenge of creating or maintaining a winning tradition, and great rewards and personal satisfaction result from accomplishing that goal.

Certification and Continuing Education of Coaches

Coaching has been described as both an art and a science. Whatever the actual balance between the two, we can state with certainty that coaching is a highly complex process of applying knowledge about human performance, human development, motor learning, and exercise. Coaching also involves high-order cognitive skills such as analysis and diagnosis and basic counseling skills that enable the coach to communicate with the athlete. All successful coaches have interpersonal skills. They must effectively teach the techniques, tactics, and strategies of a given sport; use motor learning principles in teaching skills; and develop training regimens that ensure maximum learning. They must understand the fundamentals of exercise as it relates to training and conditioning. Managing any sport enterprise entails team planning, organization, appropriate staffing, leading others to team goals, monitoring progress toward team goals, and understanding the legal liability and risk factors associated with making decisions. In the last two decades, first aid and drug education have become significant concerns for coaches. As the sports medicine and athletic training fields have become more established, the coach's role has expanded to include injury prevention and emergency care facilitation. When injuries do occur, the coach should be knowledgeable enough to help trained medical personnel.

Certification of Youth Sport Coaches

Certification is becoming increasingly important as the problems afflicting youth sport have become more serious. Never before has the need for highly qualified coaches who operate on the basis of a sound coaching philosophy been more obvious. Coaches for most youth sport teams are parents and others who volunteer to serve. The kinesiology graduate is most likely to be the organizer and administrator of the program, not a coach. As such, you should recognize the value of certification for your coaches. A number of certification programs are available for youth sport coaches. The National Youth Sports Coaches Association (NYSCA) requires coaches to attend first-, second-, and third-year certification programs to qualify for and maintain membership in the organization. Sessions consist of lectures, videotaped presentations, and demonstrations by experienced coaches.

U.S. Soccer National Coaching Certification Program

Nearly 20 million boys and girls under the age of 19 are playing organized soccer in the United States, creating a huge demand for coaches at all levels. In an effort to educate coaches and provide a standard for coaching young soccer players, U.S. Soccer has instituted a national coaching certification program that offers both state and national coaching licenses for various levels of expertise and experience. Certification begins at the state level, which focuses on elementary principles of coaching and serves to prepare coaches for the 36-hour State "D" License course. National "A," "B," and "C" courses consist of seven days of instruction and two days of extensive oral, written, and practical examinations.

The American Sport Education Program (ASEP), active in coaching education for over 20 years, offers Rookie Coach, a training program that introduces the coach to coaching philosophy and focuses on instructional techniques and safety. Coaching for Young Athletes is a course for those who have passed Rookie Coach and includes about 5 hours of clinical instruction and provides further in-depth instruction. Sports Director, a program offered by ASEP, focuses on the skills and knowledge required to be a league administrator for youth sports.

A third certification program, offered by the United States Sports Academy, called the National Coaching Certification in Youth Sports, is available online, on-site, or through correspondence. A total of 20 clock hours of course work covers ethics and character in youth sports, administration, teaching and coaching techniques, conditioning and nutrition, and injury prevention and rehabilitation. All these programs require a modest tuition fee.

Various sport-specific certification agencies (e.g., U.S. Soccer; see sidebar on previous page) are also available for equipping youth sport coaches with the skills and knowledge required for successful coaching.

Education and Certification of Middle and High School Coaches

➤ The American Sport Education Program and the Coaching Association of Canada offer general coaching certification programs particularly for community and scholastic coaches. Certification requires knowledge of sport psychology, sport physiology, coaching philosophy, sport pedagogy, sport management, sport first aid, and drug education.

Your college or university probably offers a curriculum (usually as a minor rather than a major) to prepare you for coaching. Most regional school districts and local coaching societies sponsor continuing education programs and annual clinics for middle school, high school, and college coaches. In some states these clinics attract thousands of coaches and feature presentations by some of the most popular college and professional coaches. Some coaches, eager to learn more about the scientific side of coaching, enroll in master's degree programs that offer a concentration in coaching and sports medicine.

Coaching education experiences are also available through various national coaching associations. A good example is the certification program offered by the National Soccer Coaches Association, which sponsors a coaching academy that offers residential and nonresidential courses. The residential academy offers National, Advanced, and Premier Diploma course work. The National Diploma is a 50-hour course that includes fundamental coaching tactics. Graduates must pass oral, written, and practical tests. The Advanced Diploma is a 50-hour course for the experienced coach. A minimum of 10 years of coaching experience is required. The Premier Diploma is a 50-hour course that examines different systems of play, nutrition, sportsmanship, ethics, and personal and professional development. To enroll, a candidate must have passed the Advanced Diploma with a grade of distinguished pass. The courses are very informative but also quite costly (National Advanced Diploma costs $1000.00 for nonmembers residing at the academy).

ASEP's coaching curriculum provides certification programs to ensure coaches are well prepared for their responsibilities.

More and more school districts are relying on education and certification programs offered by the National Association for Sport and Physical Education (NASPE), ASEP, and the NFSHSA. These organizations have developed curricular guidelines to help institutions of higher education, organizations, and agencies design appropriate certification programs

Certification of Coaches in Canada

The Coaching Association of Canada sponsors an ambitious and comprehensive coaching certification program for coaches in all sports, conducted in cooperation with the many national sport federations in the country.

The program certifies coaches at five levels: Levels 1 through 3 are designed for community and school coaches who are working with developing athletes. Levels 4 and 5 are for coaches who are working with elite high-performance athletes and who are planning careers in coaching. Each level focuses on theory, technical aspects of coaching, and practical skills. The following is a snapshot of the certification and education program. For further information, search the Web for "Coaching Association of Canada."

	Hours of course work	Other requirements
Level 1		Must include a self-assessment of coaching performance
Theory	13	
Technical	8-14	
Practical	*	
Level 2		Must include a peer assessment of coaching performance
Theory	19	
Technical	12-21	
Practical	*	
Level 3		Must include an on-site evaluation of coaching effectiveness
Theory	28	
Technical	16-30	
Practical	*	

* Specific criteria are determined by national sport federation (in specific sport for which certification is desired).

Levels 4 and 5: Under the jurisdiction of national sport federations; include a detailed study of coaching encapsulated in 20 tasks or modules. For level 4 certification, the coach must complete at least 12 tasks, including tasks 9 and 12. Level 5 certification requires completion of all other tasks. The tasks are the following:

1. Energy systems
2. Strength training for elite athletes
3. Sport-specific performance factors
4. Nutrition for optimal performance
5. Environmental factors and performance
6. Recovery and regeneration
7. Psychological preparation for coaches
8. Psychological preparation for elite athletes
9. Practical coaching: skill training
10. Biomechanical analysis of advanced skills
11. Practical coaching: strategy and tactics
12. Planning and periodization
13. Analyzing performance factors
14. Practical coaching: training camps
15. Practical coaching: competitive tour
16. Athlete long-term development
17. Leadership and ethics
18. Individual studies
19. Canadian sport system
20. National team program

Adapted, by permission, from Coaching Association of Canada.

for coaches. Colleges and universities that offer a coaching minor may use the National Standards for Athletic Coaches, developed by NASPE, to create an appropriate course of study for their students. Additionally, guidelines are available from nationally based coaching education programs such as Coaching Principles, a course written by Rainer Martens and offered by ASEP (see www.asep.com/courses/ASEP_Previews/asep_300x_preview). Currently, ASEP is a popular certification program for interscholastic and club coaches in many states. Certification programs of ASEP highlight several areas of knowledge that encompass NASPE's national standards, including sport psychology, sport physiology, sport pedagogy, coaching philosophy, sport management and coaching philosophy, sport first aid, and drug education.

Education and Certification of College and Professional Coaches

Interestingly, while coaching at the high school level may require preparation in kinesiology, positions in coaching at the highest levels (college and university or professional teams) may require very little in the way of formal academic preparation. Mike Krzyzewski, the famous Duke basketball coach, for example, holds a BA from the U.S. Military Academy; University of Maryland coach Gary Williams earned a BS in business; and Bill Belichick, coach of the highly successful New England Patriot NFL team, holds a bachelor's degree in economics. None of these coaches availed themselves of a formal education in kinesiology. (Often the academic credentials of some of our most illustrious college coaches remain buried in university personnel files, unavailable for public inspection—something very curious for institutions that supposedly take academic qualifications very seriously.) Suffice it to say that if your goal is to be a coach at a very high level in football or basketball in the United States, getting your foot in the door to the coaching establishment at a fairly young age and showing you can produce winning teams are likely to benefit you much more than highly polished academic credentials.

Ethics and Coaching

Recently a coach at a major university was fired for hiring ghostwriters to write essays and term papers for many of his basketball players. In another celebrated case, a basketball coach was terminated for knowingly using an ineligible player, resulting in a scandal that eventually led to the dismissal of the institution's president and the suicide of the chairperson of the institution's board of trustees. Every year, middle school and high school coaches are fired for inappropriate conduct, ranging from unethical relationships with students to psychological and physical abuse of the athletes for whom they are responsible. Issues related to ethical conduct on the part of the coach have become an important consideration at all levels of coaching. Whether it takes the form of a coach's encouraging his or her players to cheat, pressuring other faculty to give passing grades to a star player, or performing antics during a game that incite players and spectators against officials, unethical behavior should have no part in the professional's life. The NFSHSA has promulgated a code of ethics for coaches that is intended to ensure that athletics are integrated with the total educational program and that coaches act with the athletes' welfare in mind at all times. The code of ethics is shown on page 469. If you plan to be a coach in a middle school or high school, memorize this code.

Activity 16.3

Professions in Coaching and Sport Instruction

Evaluate the qualifications and philosophies of three coaches in this activity in your online study guide.

In the final analysis, ethical conduct flows from a sound personal philosophy. Chapter 5 emphasized how important it is for professionals to have a solid base of values on which to make ethical decisions that affect the lives of those for whom they work. Without having developed this base and applied it to coaching, you stand a good chance of making decisions by the seat of your pants rather than from the larger perspective of the values that you hold sacred. Have you developed a personal philosophy that can guide you in making decisions as a coach?

National Federation of State High School Associations Coaches' Code of Ethics

1. The coach shall be aware that he or she has a tremendous influence, for either good or ill, on the education of the student-athlete and, thus, shall never place the value of winning above the value of instilling the highest ideals of character.
2. The coach shall uphold the honor and dignity of the profession. In all personal contact with student-athletes, officials, athletic directors, school administrators, the state high school athletic association, the media, and the public, the coach shall strive to set an example of the highest ethical and moral conduct.
3. The coach shall take an active role in the prevention of drug, alcohol, and tobacco abuse.
4. The coach shall avoid the use of alcohol and tobacco products when in contact with players.
5. The coach shall promote the entire interscholastic program of the school and direct his or her program in harmony with the total school program.
6. The coach shall master the contest rules and shall teach them to his or her team members. The coach shall not seek an advantage by circumvention of the spirit or letter of the rules.
7. The coach shall exert his or her influence to enhance sportsmanship by spectators, both directly and by working closely with cheerleaders, pep club sponsors, booster clubs, and administrators.
8. The coach shall respect and support contest officials. The coach shall not indulge in conduct which would incite players or spectators against the officials. Public criticism of officials or players is unethical.
9. Before and after contests, coaches for the competing teams should meet and exchange cordial greetings to set the correct tone for the event.
10. A coach shall not exert pressure on faculty members to give student-athletes special consideration.
11. A coach shall not scout opponents by any means other than those adopted by the league and/or state high school athletic association.

By permission of the National Federation of State High School Associations (NFHS). *The Coaches Code of Ethics,* www.osaa.org/osaainfo/coachescodeofethics.pdf.

Advice for Coaching and Sport Instruction Students

The first piece of advice that anyone who has been a coach or sport instructor will give to students who want to coach is for them to make sure that they are suited for a career in coaching. Take a moment to review the questions listed in the section "Are You Suited for a Career in Sport Instruction or Coaching?" If, after going through this exercise, you believe that you are suited for a career in this area, consider this second piece of advice: Carefully lay out a plan for developing a career in coaching. This process will entail at least four steps. The first step is to identify opportunities for taking coaching courses in your department or at neighboring colleges and universities. If your primary program of studies will accommodate

Are You Suited for a Career in Sport Instruction or Coaching?

If you are considering a career in coaching or sport instruction, ask yourself which setting most appeals to you. Will you enjoy working in a community, commercial, or institutional setting? Are you attracted to working with underserved youth in community youth development programs or with skilled or budding athletes in specialized sport programs? Do you enjoy spending many of your weekends "on the job," when most games are played? Do you have an entrepreneurial spirit that will support a venture as an independent sport instructor or teaching professional? Are you comfortable marketing yourself and your expertise? Ask yourself if you like dealing with people as much as you like playing your favorite sports. Are you an effective communicator? Do you think that you have what it takes to be a leader? Do you like to plan and organize activities? Do you like to engage in strategy? Are you an organized person? How committed are you to the sport you want to coach and to the young people who play it? Do you think that you have good judgment? Will you be happy with having both your successes and failures in the public eye? Will it bother you to have others evaluate your performance, second-guess you, and even criticize you? Do you have what it takes to look beyond the win–loss column and focus on the changes and experiences of the athletes who play for you? Are you willing to experience the emotional ups and downs that typically characterize a career in coaching? Answer these questions truthfully and you will get a better idea of whether you are emotionally and intellectually suited for a career in sport instruction, coaching, or both.

a coaching minor, arrange to do this minor. The second step is to seek out opportunities to gain coaching experience. This is a very important element in the overall process of preparing for a career in coaching. On-the-job learning, particularly at the collegiate and professional levels, is often viewed as more important than the completion of coaching courses or certification licenses.

Opportunities for coaching "internships" abound in most communities. Explore opportunities to become a volunteer coach in your local youth sport organization. Sport associations and recreation departments are always looking for volunteer coaches. You probably will start out as an assistant coach, so be patient; other opportunities will develop. If swimming, tennis, or martial arts are sports in which you have special interest, inquire about volunteer opportunities at local clubs or schools that offer classes. If golf is your passion, make inquiries at local country clubs or golf courses for opportunities to volunteer. Opportunities to gain experience may also be available in athletic programs in local school districts. Many coaches welcome offers by coaches-in-training to help with coaching duties. Be sure to check with private schools in your area; these schools often have openings for coaching positions that they fill with college students preparing for coaching careers.

➤ Successful teaching and coaching require skills that go beyond scientific knowledge. Most successful teachers and coaches have developed communication, leadership, and organizational skills that enable them to develop rapport with their participants and foster an environment that leads to a sense of community among participants.

The third step toward developing a career in coaching is to stay up-to-date on the latest trends and tactics of your chosen sport. Attend clinics, purchase coaching DVDs or tapes, and become a voracious reader of materials related to the sport you wish to coach. As a coach, people regard you as an expert in your sport. Read books, articles, and newsletters on coaching. Many of these will be available from national or local coaching associations connected to your sport of interest. Attend conferences, clinics, and other opportunities for coaching education in your sport. Teaching and coaching resources that you might find helpful are listed on page 471, along with organizations and associations that can help you develop as a coach. With some diligent Web site navigation (beginning with the sport that interests you), you will find many more. The fourth step is to begin, while you are still an undergraduate student, building a network of active coaches. You will find no quicker way of joining the fraternity or sorority of the coaching ranks than cultivating contacts who can serve as mentors. The importance of being connected in the coaching community should not be underestimated.

Important Resources for Coaches and Sport Instructors

Selected Coaching Resources

Amateur Athletic Union
http://aausports.org

National Association of Intercollegiate Athletics
http://naia.cstv.com

National Collegiate Athletic Association
www.ncaa.org

National Federation of State High School Associations
www.nfhs.org

National Junior College Athletic Association
www.njcaa.org

United States Olympic Committee
www.usoc.org

Women's Sport Foundation
www.womenssportsfoundation.org

National Soccer Coaches Association of America
www.nscaa.com

Institute for the Study of Youth Sports
http://ed-web3.educ.msu.edu/ysi/

Coaching Association of Canada
www.coach.ca

Important Resources for Coaches and Sport Instructors

Black Coaches and Administrators
http://bcasports.cstv.com

American Baseball Coaches Association
www.abca.org

Canadian Swimming Coaches and Teachers Association
http://csca.org

American Swimming Coaches Association
www.swimmingcoach.org

Sport Instruction Resources

Boys and Girls Clubs of America
www.bgca.org

National Association for the Education of Young Children
www.naeyc.org

National Association for Sport and Physical Education
www.aahperd.org/Naspe

YMCA of the USA
www.ymca.net

YWCA of the USA
www.ywca.org

Joy of Sports Foundation
www.joyofsports.org

Finally, take stock of your general personal abilities and work on developing those areas where you detect weaknesses. Coaching success depends a great deal on communication. A course in communication skills will help, but it cannot take the place of experience. Teachers and coaches can be highly knowledgeable about the principles of communication but not be effective communicators. Or they may know the principles of leadership but not be effective leaders. Being a successful teacher or coach requires that you have not only knowledge but also the skill to share that knowledge in such a way that people listen and act accordingly. This view is reflected in the fact that the greatest athletes on the field do not necessarily make the greatest coaches. While it is difficult to pinpoint why this is true, part of the reason may be that the sport came easily to them and as a consequence they never

Compare Organizations
Compare two sport organizations in this activity in your online study guide.

developed the communication skills needed to explain the reasons for their success to less talented individuals. Conversely, athletes endowed with less natural ability may have had to study the sport in greater detail and depth to achieve competitive levels of performance. They have experienced adversity and can understand the challenges confronting the average athlete, and can therefore relate and communicate more effectively with the group as a whole.

Successful coaching places a premium on organizational skills. Detailed planning of in-season practice sessions, off-season training and conditioning programs, and game preparation is an essential component of these jobs. Remember, for most middle and high school coaches, and for some college coaches, coaching is simply one part of their total job responsibilities. Fitting it all together requires a person to be organized. Do you possess good leadership skills? Coaching effectiveness hinges largely on your ability to inspire athletes to higher levels of performance. This undertaking requires leadership. Hundreds of books on leadership, many written by business executives and successful coaches, are on the market today. Begin a course of reading that will help you develop effective leadership skills. Examining your own style of communication, organization, and leadership will enable you to identify the weak spots and develop a plan for remediation.

Simply put, the more you do, the more you learn. Taking advantage of these opportunities will enhance your knowledge and skills, thus making you a more competent and effective teacher or coach when you enter the profession.

Wrap-Up

In this chapter you learned some basic information about the coaching and sport instruction professions. You also received an overview of possible careers within teaching and coaching and the work settings in which these careers take place. Both professions include instruction and offer opportunities to serve a wide range of people across many ages and ability levels. In fact, few professions offer such a wide array of populations from which to choose.

You have also met professionals (Paul Besterman and George Dieffenbach) in teaching and coaching and have been privy to an inside view of their impressions of their career choices. As you can see from the stories provided in the interviews in this chapter, a great range and number of opportunities are available to those who decide to enter a teaching or coaching career, not only to capitalize on personal interest but also to promote physical activity programs and serve society. As suggested by Paul Besterman, recreation director in Upper St. Clair Township, ask yourself the following question: Do you want to make a positive difference in someone's life? If the answer to that question is yes, then a career in sport instruction and coaching may just right for you.

Use the key points review in your online study guide as a study aid for this chapter.

QA **Activity 16.5**

These end-of-chapter questions and activities are also in your online study guide. Your instructor may ask you to complete them online and turn them in.

1. What distinguishes sport instruction from sport coaching?
2. Describe the differences and similarities between the act of instruction and the act of coaching.
3. Contrast the professions and subcultures in coaching and sport instruction.
4. Outline the variety of work settings for coaches and sport instructors. Provide a description of each setting, including the possible participants, students, or clientele.
5. List the three professions in coaching and sport instruction that are most appealing to you. Examine and discuss the educational requirements and qualifications for these careers.
6. Consider two professions from this chapter—one in sport instruction and one in coaching. Describe the duties and responsibilities for each, including their primary purposes.

17

Careers in Sport Management

Lori K. Miller and G. Clayton Stoldt

Clive Brunskill/Getty Images Sport

CHAPTER OBJECTIVES

In this chapter we will

- acquaint you with the wide range of professional opportunities in the sphere of sport management;
- familiarize you with the purposes and types of work done by professionals in sport management;
- inform you about the educational requirements and experiences necessary to become an active, competent professional in the field; and
- help you identify whether one of these professions fits your skills, aptitudes, and professional desires.

Consider the following scenarios representing decisions that sport managers confront on a routine basis.

- A sales representative for a major tennis event meets with a potential corporate sponsor. The topic of conversation is a proposal for the business to take advantage of a promotional opportunity that will benefit the sponsor's business and help draw fans to the event.
- Administrators in a college athletic program compile and analyze data regarding how their student-athletes are performing in the classroom. They will utilize these data in an academic progress report to the National Collegiate Athletic Association.
- The program director for a YMCA branch surveys current members regarding the appeal of adult soccer leagues. If interest is high enough, the program director will make plans to add adult soccer leagues to the array of services available to members.
- Members of the corporate communications department for a sporting goods company plan a charitable initiative that will benefit grassroots sport programs in a community where they have a manufacturing plant. The program is designed to demonstrate the company's commitment to the community and to cultivate future consumers.

In each of these scenarios, sport managers and employees of sport organizations make decisions that serve the consumer while enabling the sport organization to meet its financial goals.

Students with an interest in sport and physical activity may find a career in sport management appealing. Whether the job involves the use of marketing, program management, financial, or other administrative skills, managerial positions in sport usually prove to be both challenging and rewarding. Although competition for jobs in the sport management industry is often fierce, numerous career options are available, and the field is growing.

➤ The thriving sport management industry offers numerous viable options for students looking for careers in sport.

According to *Street & Smith's SportsBusiness Journal* (2007), sport is a $213 billion industry in the United States, making it twice the size of the auto industry. Included in that figure were $34 billion spent on sport equipment, apparel, and footwear; $32 billion spent by sport spectators; $14 billion spent on licensed products; and more than $3 billion spent on facility construction. Because of the many job opportunities in sport management, it makes up one of the spheres of professional practice centered in physical activity (see figure 17.1).

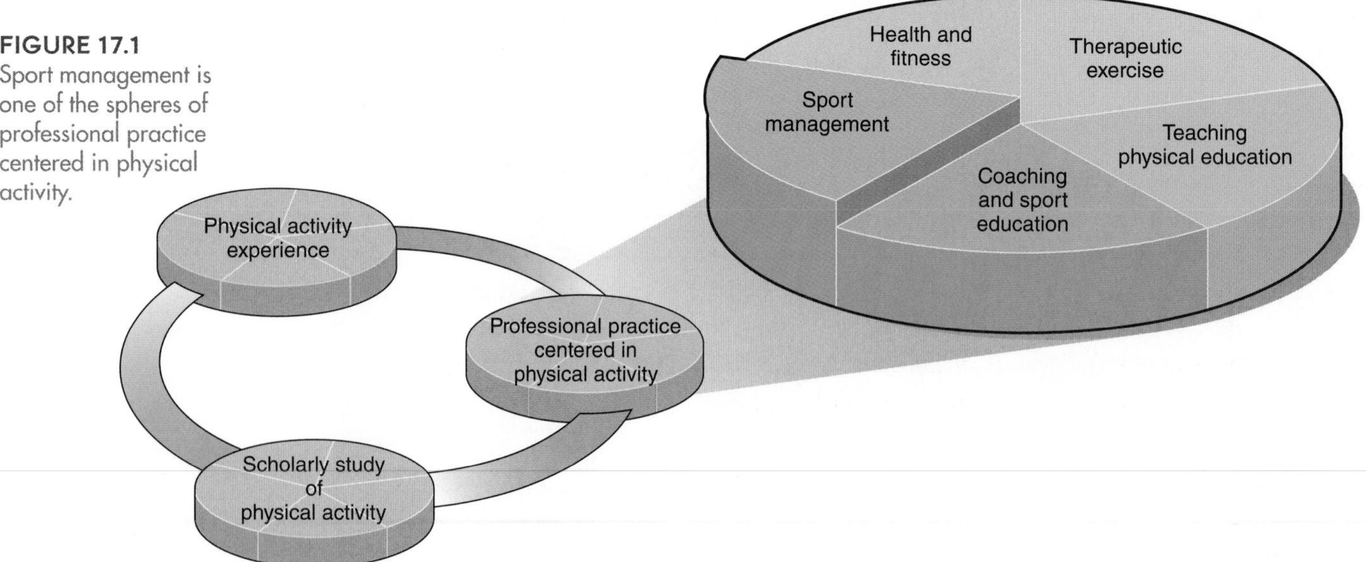

FIGURE 17.1 Sport management is one of the spheres of professional practice centered in physical activity.

Activity 17.1

Skills Self-Check

In this activity in your online study guide, you'll rate yourself on the skills needed to be successful in a career in sport management.

The purpose of this chapter is to examine professions within the sport management industry. The career opportunities within this industry are many, varied, and expanding. Job opportunities exist in professional sport, amateur sport, sport participation, sporting goods, and support services. Graduates with a well-rounded degree can choose from a variety of roles—from event management to human resource management to marketing management. We hope that this chapter helps you decide whether a career in sport management is right for you.

An Overview of the Sport Management Profession

The **sport industry** is composed of several distinct sectors, but it is clearly big business. One estimate showed sport as the 11th largest industry in the United States (Meek 1997). Sport management scholars have defined the term *sport industry* in a number of ways (Li, Hofacre, & Mahony 2001; Meek 1997; Pitts, Fielding, & Miller 1994). In this chapter we will use the following definition: "The market in which the products offered to its buyers are fitness, sport, recreation, and leisure related. These products include goods (e.g., baseball bats), services (e.g., sport marketing, health clubs), people (e.g., professional players), places (e.g., golf courses), and ideas" (Pitts, Fielding, & Miller 1994, p. 18). On the basis of this definition, the sport management sphere of professional practice centered in physical activity can intersect with career opportunities in three of the other spheres (health and fitness, coaching and sport instruction, and therapeutic exercise).

As defined in chapter 3, physical activity professionals are expert manipulators of physical activity experiences. In other words, their expertise lies in the manipulation of physical activity experiences to bring about improvement in performance and health. They may serve as personal trainers, teachers, athletic trainers, or coaches. Although many sport management professions relate indirectly to the manipulation of physical activity itself (e.g., fitness center administration), others focus instead on the manipulation of the elements supporting spectatorship (e.g., promotions, marketing, public relations) to maximize customer satisfaction. Generally, they provide support services, facilities, and other amenities to make physical activity and spectatorship possible. For example, a sport manager may be employed as the

- marketing coordinator for a local speedway (see the section titled "Sport Entertainment"),
- program coordinator for a YMCA (see the section titled "Sport Participation"),
- general manager for a golf course (see "Sport Participation"),
- athletic director in a high school (see "Sport Entertainment"),
- development officer for a college or university athletic department (see "Sport Entertainment"),
- event manager for a triathlon (see "Sport Services"), or
- manager of a sports medicine retail store (see "Sporting Goods").

Given the overlap between sport management and other spheres of professional practice, students interested in building careers as sport managers sometimes find themselves

Sport management offers a wide range of employment opportunities.

competing for jobs with students who have studied in related areas such as recreation, leisure, or sport tourism. Although similarities exist among these areas, so do distinctions.

As noted in other parts of this text, recreation or leisure involves physical activity that results in personal enjoyment, excitement, and fulfillment. Recreation commonly focuses on the consumer as an active participant (Gullion 1998). Students in recreation programs often prepare for careers in corporate fitness, nonprofit recreation, military recreation, parks and recreation, campus intramurals, and other forms of leisure services. Sport management students may have similar career interests, as the subsequent section in this chapter titled "Sport Participation" details. Graduates of sport management programs sometimes land jobs with parks and recreation departments, nonprofit fitness providers, and campus intramural programs.

Sport tourism involves people whose primary motivation in traveling outside their home communities is to participate in physical activity, watch sport events, or visit sport attractions, for example, a hall of fame (Gibson 2003). Students in sport tourism programs often prepare for careers as special event managers, travel coordinators, and sport commission administrators. Again, sport management students may pursue similar career paths and land jobs in sport tourism.

➤ The sport industry has grown significantly in both variety and complexity since the 1800s. Technological developments, greater discretionary monies and time, flexible work assignments, changing demographics and psychographics, evolving consumer needs and desires, and other societal influences have caused the sport industry to blossom.

So, what distinctions exist among the areas of study? Given the diversity in the way that sport management, recreation and leisure programs, and sport tourism programs are structured, it is difficult to define fixed differences. Still, we can say that recreation programs commonly offer greater depth in terms of specific forms of recreational programming and related considerations. Sport tourism programs address how to develop and manage sport-related travel opportunities and determine what the effect of such initiatives may be. Sport management programs, on the other hand, tend to focus on the sport-specific application of business principles such as management and marketing in varied for-profit and nonprofit settings (see the sidebar).

Although the academic study of sport management is relatively new, management of the sport industry has a long history. In the early 1800s, people planned how to promote boxing and billiards; and by the mid-1800s, those responsible for the promotion of baseball games scrutinized the placement of baseball fields outside urban areas. The 1880s and subsequent years brought about distribution issues that required the input of sport managers. During the 1890s, sporting goods manufacturers commonly hired professional advertising agencies (Fielding & Miller 1996). By 1920 the structure (e.g., management, marketing, legal issues and related regulations, financial structure, distribution) of organizations in the sport industry was well established.

Sport Management and Other Areas of Study

The following comparison of course descriptions may serve to illustrate distinctions among recreation, sport tourism, and sport management. The first set of courses is a part of undergraduate curriculum for the recreation, park, and leisure studies program at the University of Minnesota. The second set is from the concentration in tourism and hospitality management that is a part of the bachelor's program in recreation, parks, and tourism at the University of Florida. The third set is from the sport management program at Wichita State University.

Selected Courses in Recreation, Park, and Leisure Studies (University of Minnesota 2007)

- Orientation to leisure and recreation: Introduction to the history and development of the parks and recreation movement; sociological, economical, psychological, and political considerations of leisure and recreation in contemporary society; interrelationship between professional and service organizations; orientation to the professional field.
- Recreation programming: Various methods, skills, and materials needed for planning, developing, implementing, and evaluating professional recreation programs for diverse populations in various settings.
- Leisure and human development: Exploration of relevant issues concerning many roles of leisure in human development from influence on healthy fetal development to viability until death. Examination of diverse, multicultural perspectives on leisure, its centrality throughout history, and influence on how civilizations define themselves.

Selected Courses in Tourism and Hospitality Management (University of Florida 2007)

- Principles of travel and tourism: This class provides students with an overview of the travel and tourism industry. Course content covers historical, behavioral, societal, and business aspects of travel and tourism.
- Fundamentals of tourism planning: This course focuses on the planning of tourism services and facilities. The major topics include identification and planning of the use of the physical, social, and economic resources that are necessary to develop and support tourism.
- Hospitality management: Overview of the hospitality industry, including hotel management, food and beverage operations, business and leisure travel markets, convention services, hospitality trends, quest-based customer service strategies, and career opportunities.

Selected Courses in Sport Management (Wichita State University 2007)

- Organization and administration of sport: Discusses the fundamental aspects of management within any sport-related entity. Addresses management, marketing, facility management, human resources, legal issues, budgeting and finance, purchasing, and communication.
- Legal aspects of sport (I): Provides students with the knowledge, understanding, and application of how the following legal issues influence the sport industry. Specific content addressed includes the legal system, statutory law, risk management, tort law (negligence and intentional torts), contracts, and employment-related issues within the sport industry. A primary objective of this course is to enhance the decision-making and problem-solving ability of each individual student as it pertains to legal issues in sport and physical activity.
- Marketing sport and physical activity: Introduces tools and concepts used to market sport and physical activity. Emphasizes marketing strategies that are applicable to the sport administrator, teacher or coach, and exercise professional.

Sport Management Settings

➤ Sport managers are more involved in the activities and job responsibilities surrounding an event than they are in the sport or activity itself.

The sport industry can be divided into three segments, and all three segments share functional roles within the industry (see figure 17.2). For instance, professional sport franchises and nonprofit sport organizations both require people working in management, marketing, and so on. Likewise, both nonprofit organizations such as the YMCA and for-profit organizations such as ice-skating arenas need people with expertise in financial and legal services. A great deal of interdependence exists among the various sport industry segments, a point that will be illustrated later in the chapter. Furthermore, the tasks listed under the different functional areas (e.g., event management, financial management) are not exhaustive. Human resource management, for example, could also include motivating, compensating, retaining, evaluating, and so on.

The number of tasks performed in-house versus the number that are outsourced varies within the industry and within individual sport organizations. Similarly, industry segments differ in how each function is allocated among existing employees. Larger sport organiza-

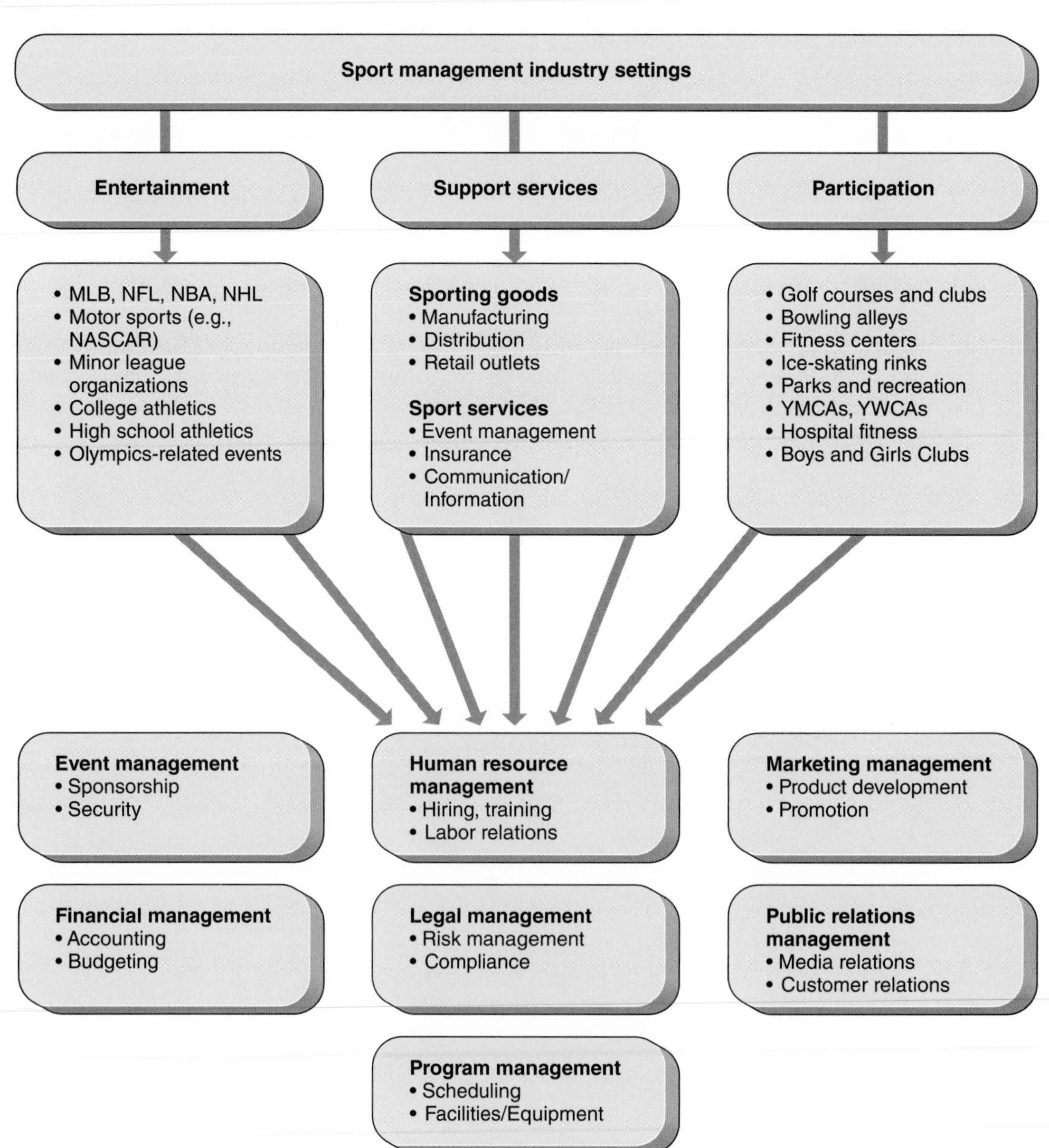

FIGURE 17.2 Sport management industry settings and some common functional roles in each segment.

Based on the work of A. Meek 1997.

Examining Your Interactions With Sport

To consider the prevalence and popularity of sport, you need only examine your interactions with sport and leisure during a typical day. Sport is an integral part of the lives of many people, and that is good news for sport managers because consumer demand for sport and leisure activities results in job opportunities for sport management students.

Think about your average week. How much time do you spend watching sport programming on television? How much time do you spend reading about sports online, blogging about them, or playing fantasy sports? How much time do you spend engaging in sport-related activities such as working out, golfing, or cycling? How much money do you spend on apparel featuring the logos of sport teams or the brand marks of sporting goods manufacturers? How much time do you spend in sport-related conversation?

tions often assign specialized responsibilities to individuals based on areas of expertise. On the other hand, people in smaller organizations are likely to be responsible for performing a variety of the job functions indicated.

Sport Entertainment

A proliferation of **sport entertainment** opportunities has created an abundance of jobs. At the professional level, a multitude of major and minor league teams as well as motor sports, golf, tennis, and mixed martial arts events entertain sport fans. Expansion teams, new professional leagues, and the professionalization of alternative sports (e.g., action sports) continue to fuel professional sport entertainment options and related career opportunities. Each entertainment production (i.e., the game or event) represents a tremendous undertaking, combining the skills, competencies, and knowledge of many people working in separate functional areas. As indicated in figure 17.2, each professional sport entertainment organization provides career opportunities in the following areas:

- Event management
- Financial management
- Human resource management
- Legal management
- Marketing management
- Public relations management
- Program management

The amateur sport entertainment sector mirrors professional sport in many ways. As with professional sport, the entertainment alternatives are vast and the career opportunities abundant in settings such as Olympics-related events and college athletics. Career opportunities in this industry segment often include the following:

- Development
- Public relations
- Marketing
- Operations
- Ticket sales
- Merchandising
- Compliance
- Student services

Sport Participation

Whereas managers in the entertainment sector enable consumers to enjoy the sport performances of others, those in the participation sector provide opportunities for customers to engage in sport activities themselves. The sport participation sector includes both for-profit and nonprofit organizations, and the following sections address both.

For-Profit Participation Segment

For-profit sport participation organizations include fitness and health clubs, bowling alleys, roller-skating rinks, miniature golf courses, golf courses, ice-skating rinks, sport parks, and more. As indicated in earlier chapters, the health and fitness industry has many facets. An individual pursuing a career as a personal trainer, for example, would have different professional responsibilities and educational training than a person employed in a sport management capacity. The International Health, Racquet and Sportsclub Association (2007a) reported that more than 29,000 health clubs were in business during 2005. The IHRSA's 1996 estimate was 13,400, so the number more than doubled over a 10-year period.

Of particular interest to potential sport managers is the breadth of interest in sport participation (see chapter 3). The sport industry is continuing to diversify, with consumers continuing to embrace traditional activities such as golf, reengaging in games like dodgeball that they may have played as children, and participating in new activities such as bicycle polo and cardio tennis (Consilience Group 2007).

The for-profit sport participation segment has also recognized growth within existing product mixes, supplementing core products such as resistance training with newer offerings such as spin classes. In addition, more and more businesses are catering to specific target markets such as unfit women and time-pressed professionals. Employment opportunities vary, typically according to the size of the organization; but job opportunities relate to a variety of functional areas, including

- management,
- marketing and promotion,
- program planning,
- human resources, and
- risk management.

For-profit sport organizations are in an enviable position as their target markets and product lines expand.

Nonprofit Participation Segment

As discussed in chapter 16, the nonprofit sport participation segment is replete with jobs for qualified sport management professionals. Nonprofit organizations providing participation

Considering Sport Entertainment Management

Think about the last time you attended or watched on television a professional or amateur sport event. Identify as many of the peripheral activities or elements (e.g., concessions) associated with the main event (i.e., the game) as you can. Are you interested in managing events or venues? If so, begin developing organizational, time management, leadership, and communication skills. All of those are necessary for a successful career in sport entertainment management.

www. **Web Search 17.1**

Careers in Sport Participation Management
Research for- and nonprofit organizations in this activity in your online study guide.

opportunities in sport include YMCAs, YWCAs, Boys and Girls Clubs, and hospital-affiliated fitness centers. As in other industry segments, individual jobs in these organizations may be specialized in one of the functional areas identified in figure 17.3 or may be more general. For example, a YMCA located in a major metropolitan area may offer job opportunities in specialized functional areas, whereas a Boys and Girls Club may employ one person to be responsible for all recreational-related areas.

Like many other areas in sport, the nonprofit sport participation sector offers significant opportunities. The IHRSA (2007b) reports that nearly 20% of all U.S. health club consumers are members in YMCAs or YWCAs. The YMCA had 1700 branches and just under 1000 member associations in the United States in 2006, and it generated $5.62 billion in revenue that year (YMCA 2007). The "Y" also has a presence in Canada, with 57 branches in operation (YMCA Canada 2007).

Other nonprofit organizations are also common in the field. Combined, facilities based in hospitals, universities, municipalities or towns, and churches or faiths hold a 15% market share in the United States (IHRSA 2007b).

Support Services

The support services segment includes an even wider array of sport organizations than do the previous two segments, sport entertainment and sport participation. Included in the sport support services segment are both the sporting goods industry and a variety of service providers. The following sections explore each of these vast categories.

Sporting Goods

The sporting goods industry represents a significant portion of the overall sport industry. As previously noted, *Street & Smith's SportsBusiness Journal* (2007) estimated that $34 billion was spent on sporting goods in the United States in 2006. Couple that with another $14 billion spent on goods licensed to display the names or logos of sport properties, and you'll see that the industry sector is sizable. Sport managers play a vital role in ascertaining what sport products people need and want. Employment opportunities in the sporting goods industry include the following:

- Research analysts
- Retail store managers
- Manufacturing sales representatives
- Accounting managers
- Licensing administrators

A nonprofit provider of sport participation opportunities, the YMCA offers numerous job opportunities for qualified sport managers.

Sport Services

As a result of the enormous growth in the sport industry, sport-related services have become more specialized. Today, sport managers can specialize in sport insurance, sport event management, or sport marketing. This segment of the industry is known as **sport services.** The following are examples of organizations that offer sport services:

- Sports & Fitness Insurance Corporation. Provides insurance for health clubs, fitness centers, martial arts studios, personal trainers, and dance schools.
- *Team Marketing Report.* Provides information services on sport consumers, ways to increase ticket and sponsorship revenues, individuals making sponsorship decisions, and more.
- Collegiate Licensing Company. Manages the licensing programs of universities and colleges.
- Exclusive Sports Marketing. Promotes and manages events such as triathlons and other recreational sporting events.
- Anthony Travel. Service provider specializing in working with university athletic departments, professional sport organizations, individual athletes, and fans.

The sport service segment is likely to realize continued growth as organizations outsource job functions to reduce personnel costs while at the same time receiving specialized services for seasonal sport product offerings.

Professional Roles in Sport Management

People perform many similar professional roles regardless of the sport industry segment in which they are employed. Rather than list separate professions in the sphere of sport management, as in previous chapters, in this chapter we describe the sport industry in terms of functional roles and related tasks that may be assigned to one or more employees within various sport organizations.

Event Management

➤ Sport event management is a complex function because it requires skills specific both to securing, organizing, and executing the event and to creating partnerships with other organizations that may have a stake in the event.

In all physical activity professions, the staging of an event is a tremendous undertaking. Many organizations hire people with specific expertise in the area of **event management.** These employees are responsible for the many facets associated with hosting an event including risk management, security, venue setup, concessions, scheduling, game presentation, customer service, and many more. While a single person may cover all of those responsibilities for a small-scale event, large staffs are commonly employed for a major event. For instance, the host organization for the 2007 Super Bowl had a 24-person board of directors, a set of eight committees overseeing

The smooth running of a sport event, including ticket sales, concessions, and crowd control, largely depends on sport managers who specialize in event management.
AP Photo/David J. Phillip

diverse aspects of the event, a host committee staff of 12 people, and 15 interns (Super Bowl XLI Host Committee 2006). Ultimately, the event attracted 74,512 game attendees and 436,850 nongame event attendees (Sport Management Research Institute 2007).

Clearly, event management requires sport managers to consider a multitude of factors: traffic control, crowd control, resource availability, management and coordination of paid personnel and volunteers, and more. Opportunities in this area are widespread and vary from the Super Bowl to an evening of Little League baseball at a local park.

Sport commissions—local entities that attempt to attract sporting events and franchises to their cities to boost the local economy—have seen an enormous increase in the United States in recent years. The National Association of Sports Commissions (NASC) has grown from 15 members in 1992 to more than 400 members today (SportAccord 2007). Similarly, 100 Canadian municipalities have joined the Canadian Sport Tourism Alliance in an effort to cultivate travel to their towns and cities for sporting events (CSTA 2007). Clearly, employment opportunities in this sector of the field are abundant.

Once an event is scheduled, sport commission personnel collaborate and cooperate with other personnel (e.g., team owners, employees, and volunteers) in the sport industry to ensure that the event is a success. A sport commission that secures a bid to bring a bowling championship to its city, for example, would work closely with the practitioners in the local bowling community to ensure that adequate facilities are available, customer service is exemplary, scheduling is organized, volunteers are gathered and committed, and so on.

Considering Career Options in the Sporting Goods Industry

Do you have a special interest in sport equipment and apparel? Perhaps you like to explore the newest merchandise on the market or imagine designing your own. Maybe you like mining data and researching trends to predict the newest fad in athletics. You might be great at getting people excited about new products or services. All of those skills and interests could contribute to a career in sporting goods.

Financial Management

Financial management is critical to the operation of most sport-related organizations. Financial managers may administer budgets, oversee income allocation, pursue development opportunities, and handle investments. Managing financial situations for sport organizations requires knowledge and understanding of financial tools (e.g., financial statements, budgeting procedures) as well as a solid understanding of the sport product. Without having an understanding of the sport product, a person is not equipped to capitalize on potential revenue opportunities.

➤High demand for sport-related products creates a positive environment for the financing of a variety of sport facilities including health clubs, swimming pools, arenas, and stadiums.

Two areas of sport financial management that provide a significant number of opportunities are accounting and facility finance. All sport organizations, and almost all employees in the sport industry, are involved in some manner of accounting and budgeting. Sport marketers, for example, who are responsible for promotions, advertising, and sponsorships, must ensure that expenses don't exceed revenues. Similarly, sport event managers must balance expenses associated with facility rental, ushers, concession purchases, parking attendants, and so on, with anticipated revenues generated from sources like tickets and sponsorships.

Financing facilities has become an important issue in the entire sport industry. Nearly $6 billion will be spent on facility construction in 2009, according to *Street & Smith's SportsBusiness Journal* (Muret 2007). Sport organizations and their public constituents have used a variety of mechanisms to fund their projects, including owner contributions, commercial loans, corporate partnerships, government loans, taxes, and various types of bonds (Fried, Shapiro, & DeSchriver

2003; Howard & Crompton 2005). The growth in facility construction has provided interested individuals with a number of finance-related jobs in the sport industry.

Human Resource Management

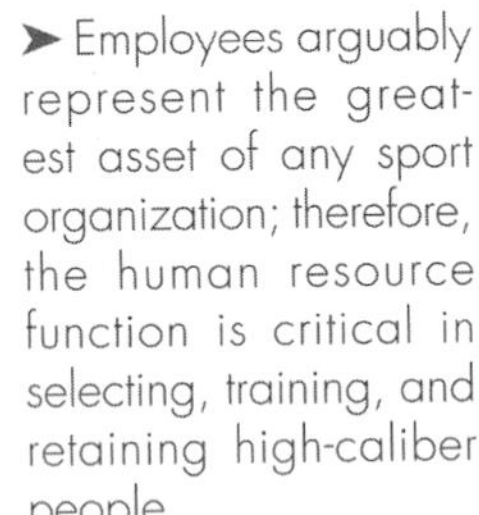

➤ Employees arguably represent the greatest asset of any sport organization; therefore, the human resource function is critical in selecting, training, and retaining high-caliber people.

Human resource management, also known as personnel management, may be one of the most important job functions in the sport industry (Chelladurai 2006; Miller 1997). Employees of sport organizations are arguably their organization's greatest asset. Happy and content employees are more loyal to the organization, resulting in reduced costs associated with turnover and in employees who are more experienced. Employees are responsible for the entire sport product delivery process including customer service, operations, and other crucial functions. A gap in any job functional area dilutes the quality of the sport product. Low-quality products eventually fall below the financial break-even point, causing sport organizations to go out of business.

Human resource managers work to combine employees' talents and desires with the needs of the sport organization to create a pleasant and legally defensible work environment. They have traditionally dealt with the hiring, training, maintenance, and dismissal of employees. Recent trends, however, reveal a far broader scope associated with human resource management. Today, human resource managers may organize and manage such benefits as day care for employees' kids, employee stock ownership plans, professional development opportunities, fitness facilities, and counseling services, all designed to attract and keep high-quality employees.

Human resource managers in sport organizations with unions have added responsibilities. Unions negotiate agreements with management regarding compensation, working conditions, and other issues. As evident from previous professional sport work stoppages, union and management relations are critical to the profitability of sport organizations. Sport managers employed outside the professional sport domain also encounter union issues when dealing with concession and merchandising vendors, security personnel, and officials and umpires.

Human resource professionals help ensure workplace conditions and policies contribute to a happy, effective work force.

© iStockphoto/Miodrag Gajic

Risk Management

➤ Risk management involves taking steps to reduce losses of all kinds. Risk management requires the support of all employees, from top management down to the front line.

As society has grown increasingly litigious, sport administrators have become more concerned about **risk management.** Many issues facing sport managers today are complex and warrant advice from attorneys and other legal experts, but all sport managers should have knowledge necessary to minimize risks within their organizations.

Risk management evolved from a predominant concern about financial losses to a more encompassing concern that includes losses of all kinds (Sharp, Moorman, & Claussen 2007). Today, the sport industry views risk management as including, but not limited to, losses resulting from inadequate or improper security, food service distribution, warnings, supervision, instruction, and facility and equipment design. Many sport organizations have designated risk management personnel to protect against losses of all kinds. Risk management groups assist sport entities in identifying potential hazards, suggesting and implementing ways to manage potential losses, evaluating existing risk management plans, and continually improving existing risk management efforts.

Risk management is especially important in the amateur and professional sport entertainment industry segments. Besides being attentive to the concerns already listed, the entertainment industry segment must comply with complex rules and regulations established by governing organizations. Fearing losses to their athletic organizations and programs,

Activity 17.2

What's in Your Closet?

Gauge your personal interest in sports apparel and equipment in this activity in your online study guide.

some colleges have responded by using outside legal assistance in their compliance efforts. Many others have designated particular administrators as compliance officers.

The mass media has chronicled the consequences of compliance-related failures, and such situations are unfortunately common. As of fall 2007, 40 National Collegiate Athletic Association (NCAA) institutions were on probation, some for rules violations in several sport programs (NCAA 2007a). Major violations occur frequently. As of fall 2007, according to the NCAA, 271 member institutions had been sanctioned at least once for major infractions (NCAA 2007b). A university compliance officer strives to keep the athletic department in compliance with the rules and regulations detailed in the lengthy NCAA compliance manual; in university policies and procedures; and in local, state, and federal legislation and ordinances.

Considering Your Experiences as an Employee

Think of an organization where you work or have worked in the past. Did that organization create a friendly work environment conducive to productivity and efficiency? What did the organization do well? What could it have done to improve the work environment? Does tackling such issues interest you? Can you see yourself being an instrument of change in an organization, recruiting others, establishing workplace policies, or ensuring company compliance with employment guidelines? If so, human resource management might be the profession for you.

Marketing Management

Marketing management represents a burgeoning career opportunity for qualified students. Some even suggest that effective marketing is the key to success for any sport product, whether it is a new running shoe, a sport clinic, or a sporting event. A great sport product, offered at an affordable price and available in a favorable location, generates little revenue unless people are aware of it. Sport marketers may spend time working in a variety of marketing management areas that include, but are not limited to, the following:

> Students of sport management who are educated in the area of sport marketing often find that they have a great deal of appeal to employers in all sport industry segments. A background in sport marketing, coupled with related job experience, provides people with lateral as well as vertical career opportunities.

- Research and development can range from sophisticated analyses of consumer opinions, economic impact studies, and so on to focus groups or brainstorming sessions on how to package and present products more effectively.
- Sport promotion comprises the activities that sport marketers undertake to heighten interest in products with the purpose of increasing consumption (e.g., sales, viewership). **Promotions** are tactics that marketers use to communicate with and attract buyers. Promotions come in a wide range of forms, including product-sampling opportunities, premium giveaways, price reductions, player appearances, camps, and clinics (Irwin, Sutton, & McCarthy 2002; Mullin, Hardy, & Sutton 2007).
- Sport sponsorship involves an exchange between a company and a sport organization. For example, a sponsoring company may provide either cash or in-kind resources (e.g., products, services) in exchange for title sponsorship of an event such as the AT&T USA Track and Field Championships. Sponsorships represent powerful promotional platforms that build brand awareness for sponsoring companies.

- Advertising is "any paid, nonpersonal (not directed to individuals), clearly sponsored message conveyed through the media" (Mullin, Hardy, & Sutton 2007, p. 237). Many sport-related organizations purchase **advertising** to promote their events or services and sell it as well, offering signage, for example, as a way to generate revenues.
- Merchandising efforts extend far beyond apparel. Customers who visit sport specialty stores may choose from an array of licensed products, including apparel, toys, automobile accessories, and even pasta in the shape of team mascots. Sport trading cards continue to be big business, with some of the rare or high-demand cards commanding thousands of dollars, and the market for sport video games is thriving. Some sport organizations conduct their own **merchandising** programs, while others sell licenses to other businesses to place their names and logos on various products.
- Distribution connects the sport product or service with the ultimate consumer. For sporting goods manufacturers, **distribution** issues include building relationships with distributors and retailers, securing optimal shelf space, and maintaining low-cost product transportation systems. For sport managers in the entertainment or participation sector, distribution concerns may include facility design, accessibility, flexibility, and attractiveness because the sport product (e.g., game) is produced and consumed in the same location (e.g., stadium). Broadcasting arrangements may also be considered a distribution-related issue because they focus on getting the sport product to consumers (Mullin, Hardy, & Sutton 2007). The National Football League reaps approximately $3.7 billion in television rights fees each year (Isidore 2005, April 22), illustrating the potential magnitude of such agreements in the sport entertainment industry segment.

Recent Research: Super Bowl XL

The December 2006 issue of *Sport Marketing Quarterly* focused entirely on Super Bowl XL with articles addressing a range of topics, from the marketing tactics employed by the Detroit Super Bowl bid committee (Schneider & Bradish 2006) to related corporate social responsibility ventures (Babiak & Wolfe 2006) to associated marketing activities in nearby cities such as Windsor, Ontario (Taks, Girginov, & Boucher 2006), and more. The issue opened with an article describing how business leaders, government officials, and others worked together to develop the winning bid to host the game in spite of weather- and reputation-related obstacles (Cuneen & Fink 2006).

Public Relations Management

Public relations management in sport seeks to achieve positive relationships between the sport organization and its most important constituents (Stoldt, Dittmore, & Branvold 2006; Stoldt, Pratt, & Jackson 2003). The function involves diverse activities ranging from media relations management to community relations to investor relations.

A number of people within sport organizations work in media relations. Almost all professional and college sport organizations assign at least one person to media relations. These public relations practitioners, sometimes known as sport information directors, produce media guides and programs, manage organizational Web sites, track statistics, arrange interviews, supply story ideas, and generally service media requests. These individuals provide an integral link between the organization and its stakeholders.

Although the media may be more integrally involved in sport than in any other industry in the United States, **public relations management** involves much more than just media relations. Sport entertainment organizations often employ people specifically to work in fan relations, and sporting goods businesses generally realize the advantages of having customer relations employees who provide a variety of services ranging from handling complaints to giving special instructions regarding the assembly of products.

Finally, some sport managers work in community relations. These public relations professionals may also seek opportunities for organizational representatives to make public appearances or speeches. Additionally, they may coordinate donations of cash or sport-related products to worthy organizations.

Media management is increasingly important at all levels of sport.
Photo courtesy of Wichita State Media Relations

Program Management

Program management is common in all sport industry segments. Health clubs, parks and recreation departments, athletic programs, and sport services all are involved with the provision and delivery of sport programs. A parks and recreation department, for example, may offer a youth sport program, an aquatics program, and a fitness program. Each program, in turn, is staffed with its own employees who deliver desired activities (e.g., lessons, clinics, tournaments).

Program management activities vary across functional areas and sport industry segments. Two of the more common task areas are scheduling and maintenance of facilities and equipment.

Scheduling

Scheduling is one of the primary concerns for program managers. Many a Little League coordinator has dealt with scheduling complications stemming from too many teams with access to too few facilities. Sport entertainment administrators must also address scheduling and related considerations. Scheduling games, events, and even practices can be complicated when other events in the community affect public demand or when one facility must accommodate multiple demands. Scheduling travel can also be a complex procedure. Some athletic programs have administrators whose primary function is to coordinate travel.

Facility and Equipment Maintenance

Maintenance of facilities and equipment represents a crucial task for program managers. Those working in sport participation organizations must maintain and update facilities to meet the needs and demands of consumers. Included is the responsibility to keep facilities and equipment clean, safe, and operative. In addition, sport managers must monitor the external environment (e.g., competitors, new product developments, and changing consumer demographics and psychographics) for new products and other developments that may enhance consumer offerings and satisfaction.

Considering Career Options in Sport Program Management

Think of a sport organization with which you are familiar. What programs does the organization offer? Do you have the skills and aptitudes necessary to run such programs? If you are interested in program management, what skills would you need to develop? After you've identified those skills, think about how you can start to develop them. What opportunities are available to you now? Which will require some research and planning to get involved in? Make a plan for your skill development.

Prospects and Opportunities for Careers in Sport Management

➤The 21st century provides many exciting employment opportunities for qualified individuals interested in sport management careers. The proliferation of amateur and professional sport teams, society's growing interest in health and fitness, progressive technology, and eclectic consumers with varying needs and demands explain, in part, the promising growth.

The sport management industry offers a plethora of job opportunities in a variety of industry segments and functional areas. Objectively determining the most popular industry segments is difficult, because they all appear to be hiring qualified students of sport management. A review of various Web sites and trade publications provides insight regarding available job opportunities. For example, TeamWork Online advertises administrative jobs available in most of the major professional sport leagues as well as in large sport entertainment companies.

Weighing Career Options in Sport Management

As you read the preceding descriptions of careers in sport management, did you put yourself in the picture? What positions felt the most comfortable to you? What tasks excited you? Perhaps you found yourself thinking, "I'd hate to do that!" Now that you have some idea of where your interests lie, what kinds of educational and practical experiences do you think will help you move toward a specific career?

Data from the U.S. Department of Labor also illustrate the vast array of job opportunities available in sport management. Table 17.1 offers information regarding the number of jobs and median salaries for selected positions and sectors of the industry.

Table 17.1 Mean Salaries for Selected Jobs in Sport

Industry sector	Job type	Annual mean salary
Spectator sport	Chief executive	$164,880
	Financial manager	$98,340
	Sales representative	$44,900
	Public relations specialist	$41,580
Fitness and recreational sport centers	Chief executive	$125,060
	Sales manager	$119,100
	Supervisor, personal services	$36,190
	Recreation worker	$21,390

From U.S. Department of Labor 2006.

Education and Qualifications

The North American Society for Sport Management (NASSM) has a Web site listing more than 250 colleges and universities with sport management programs at the bachelor's, master's, or doctoral level or more than one of these levels (NASSM 2007). More than 200 of those programs are located in the United States, but the site also lists programs in Canada, Europe, Australia, New Zealand, India, and Africa.

Some of the programs listed on the NASSM Web site are designated as "approved," meaning that they have met standards established by the National Association for Sport and Physical Education (NASPE) and NASSM (NASPE-NASSM, 2000). In 2008, a new chapter in sport management program review was opened with the activation of the Commission on Sport

Management Accreditation (COSMA). The new organization, working in conjunction with NASPE and NASSM, developed a comprehensive program of accreditation.

One of the many factors the COSMA planned to consider when reviewing programs are curricula that "reflect the mission of the institution and its academic unit, and are consistent with current, acceptable practices and principles of professionals in the academic and sport marketplace communities" (COSMA, n.d.). Programs seeking accreditation at the undergraduate level are expected to feature curricula covering seven broad areas (COSMA, n.d.). They are:

1. The social, psychological and international foundations of sport;
2. The management of sport, which includes not only sport management principles but sport leadership, sport operations management and event and venue management, and sport governance;
3. Ethics in sport management;
4. Sport marketing;
5. Principles of sport finance, accounting and economics;
6. Legal aspects of sport;
7. An integrative experience such as the development of a strategic management plan or policy manual, an internship, or some other capstone experience (e.g., thesis comprehensive exam).

Internships and Field Experience

Although all the content areas listed are critical to the attainment of a quality sport management education, the internship and other field experiences (e.g., practicum) deserve special emphasis. Sport organizations look for people with practical, real-world sport experience. Because the sport management industry is extremely competitive, the references, knowledge, and work experience that a student can gain from internship or practicum site supervisors are invaluable. Consequently, the internship or field experience is essential to the student's ability to secure a job.

➤ Sport management students should capitalize on available internship opportunities with credible sport management organizations that provide learning opportunities.

Students may obtain internships working in any one of the sport industry segments discussed earlier. Responsibilities associated with internships tend to be situation specific. For example, an internship at a Division I institution may be highly specialized. In other words, the student intern may work specifically in one area such as compliance, marketing, facilities, operations, or tickets. On the other hand, an internship at a Division III institution may combine roles. In this situation, the student intern may work in a variety of areas including marketing, facilities, operations, and tickets.

Sport promotions, like the pink bats for breast cancer awareness, are an increasingly important part of sport management.
AP Photo/Eric Miller

Figure 17.2 either directly or indirectly identifies all the competency areas recognized by the COSMA. As you can see, each functional area is somewhat dependent on, and linked to, the others. Moreover, many benefits can be gained from an understanding of each area. For example, if you worked in a youth sport program, you would be better able to provide successful program offerings if you had an understanding of budgetary limitations, managerial issues, consumer behaviors, risk management principles, and societal implications. Similarly, if you worked with a professional hockey franchise, you would have a better understanding of the intricacies involved and communication required of individuals working in different functional areas of a large event. You would learn, for example, that executing promotions requires the close cooperation of the people working in the area of facilities. Sound effects (e.g., music, clapping hands), lighting, fixtures, and so on must all be coordinated so that the fans are entertained, the sponsor's promotion is flawlessly executed, and promotion participants are not injured.

➤ Multifaceted individuals can make decisions that are in the best interest of the sport organization as a whole rather than decisions that benefit only particular business functions of the organization.

When you understand how various issues and functional areas influence your product, be it a sport event, clinic, or apparel item, you are better able to communicate with colleagues, employees, and customers. Your conceptual understanding makes you an invaluable asset to your sport organization. In addition, by being knowledgeable in a variety of competency areas, you enhance your marketability in the profession.

Sport management students must study different curriculums than do students studying for careers as teachers, coaches, personal trainers, athletic trainers, and so on. As you have seen, sport management students are often not required to take courses in motor learning, exercise physiology, or anatomy and physiology. Course work in theoretical kinesiology, however, can significantly enhance a person's knowledge and marketability if he or she wishes to work in the sport participation sector. For example, students working in the health club industry can improve their marketability by combining their sport management degree with course work in appropriate areas of theoretical kinesiology. This well-rounded education will enable a sport manager to make better hiring decisions and provide better supervision.

Preparing for a Career in Sport Management

As you know from reading about other physical activity professions in previous chapters, simply meeting educational requirements will not ensure your success as a sport management professional. The interview with Koni Daws, assistant athletic director and senior woman administrator, addresses this and other subjects (see the sidebar).

> **DO it Activity 17.3**
>
> **What's My Profession?**
> Test your understanding of the sport management professions by playing this game in the online study guide.

Besides completing the required courses and obtaining your degree, you will need practical experience in your area of interest. Such experience is invaluable in helping you improve your communication skills and enhance your problem-solving skills. Also, do not underestimate the value of a mentor, someone in the field who can give you practical insights into your new career. Finally, be sure to keep up with the profession by contacting sport management organizations and reading professional journals; resources to get you started are listed on pages 497-498. The following sections offer details on strategies that you can pursue to prepare yourself for a career in sport management.

Practical Experience

You should always be at work building your resume. Practical experience is a necessity in the competitive job market of the sport industry. Almost all programs have internship requirements, and we have already described the benefits of internships. The best internships,

Interviews With Practicing Professionals

Koni Daws

Assistant athletic director, Southern Methodist University

Koni Daws is the assistant athletic director and senior woman administrator at Southern Methodist University (SMU) in Dallas. She has a bachelor's and a master's degree in accounting and a master's degree in sport administration.

Q: What are your primary job responsibilities?

A: As an assistant athletic director, my primary responsibilities entail being the sport administrator for four sports, NCAA and conference championships coordinator, admissions liaison, academic monitoring, and NCAA cabinet member. I also monitor summer school and posteligibility financial aid requests and am a member of university committees and subcommittees. As the senior woman administrator, I monitor the implementation of the university's gender and racial equity plan, assist with the EADA (Equity in Athletics Disclosure Act) reporting, and act as the liaison with the conference and NCAA offices.

Q: What is a typical workday like for you?

A: My workday always begins with reviewing and answering e-mails and phone messages. I monitor academic reports and discuss academic issues with appropriate coaches and learning center counselors. Of course, there is always some issue with a sport or two that needs to be resolved. So I work with all areas of the athletic department to get the necessary answers and report back to the coach. Quite often I serve as the mediator or advocate, depending on the situation. Then I research and complete any special projects that the athletic director has assigned to me. As an administrator, I must wear many different uniforms every day. Therefore, I have to know at least a little bit about finances, public relations, marketing, sales, facility management, and counseling so I can serve my coaches and other staff effectively.

Q: What are the most and least enjoyable aspects of your job?

A: By far the most enjoyable part of my job is working with the student-athletes. I love to see the look of pride and satisfaction on their faces when they accomplish something you knew all along they could do and knew that all you had to do was show them you believed in them. There is nothing more satisfying to me then watching a student succeed. I also enjoy attending the competitions and cheering for the students. They appreciate it so much when they see you there supporting and believing in them. The least enjoyable part of my job is when I have to be the bearer of bad news to a student-athlete and his or her parents. Unfortunately, due to budget or scholarship constraints (or both), not all student-athletes can get financial assistance. Also it is not enjoyable trying to explain to parents why their son or daughter did not get to play much during a particular competition. This is a very personal and emotional issue for them, so they do not want to believe that it was because the skills of another student-athlete were needed more during that particular competition and not a personal decision. Also, working through the many different levels of bureaucracy to get something accomplished is not enjoyable at the time. But once I succeed, it is very satisfying.

Q: What have you found to be the most surprising part of working in college athletics?

A: One of the things that surprised me the most when I entered the world of college athletics from the corporate world was the amount of time it takes to get things done and the number of people that have to be consulted and have to approve before you can implement an idea. Another thing that tends to surprise me (or amaze me—however you want to look at it) is the lack of communication between the areas of the athletic department. Even though everyone usually works in close proximity to each other, they do not take time to really communicate with each other.

Q: What have been the most important experiences in preparing you for a career in college athletics?

A: When I was getting a master's in accounting, I taught accounting and I discovered that I loved working with students. At that point I decided to combine two areas that I was passionate about: working with young adults and sports. My accounting and business experience helped me to understand finance, which has helped me a great deal in college athletics. My teaching experience prepared me to be able to explain difficult subjects and taught me the patience to find new ways of explaining a situation if it is not understood the first time. This has helped me a great deal in working with student-athletes, coaches, and staff members.

Q: What are the most important qualities you look for when hiring a new employee or intern?

A: When I interview a candidate, I look to see if there is a passion for the position and if there is a belief that he or she can make a positive difference in the lives of the student-athletes and in the athletic program. After being passionate about the job, I believe the next most important qualities are being dedicated, hardworking, ethical, and having a great set of morals. Every staff member in an athletic department must be a good role model and set a good example for the student-athletes.

Q: What advice do you have for students who would like to pursue careers in sport management?

A: Working in sport management requires a passion to help the student-athletes. The student-athletes are the reason you have a job, so every day has to be about educating those student-athletes and making sure they are having the best possible experience they can have. The sport management career requires selfless dedication because it often entails long hours and very little recognition. But when one of the student-athletes or one of the coaches accomplishes a goal and you see the person's happiness, it is all worth it. Make sure that you pursue this career for the right reason—to help make the student-athletes great citizens.

---

however, often go to students with the most job experience. Try to find part-time work in the sport industry and volunteer for special events that may arise (e.g., fun runs, triathlons, Special Olympics). Such experiences will allow you to apply what you learned in the classroom to the job setting while simultaneously gaining valuable resume material. The interview with Brian Hargrove, director of events and marketing for the Greater Wichita Area Sports Commission, addresses this point (see the sidebar).

Mentors

Mentors are people working in a student's chosen profession who seek to provide students with guidance, learning opportunities, and contacts in the field. Students with mentors gain insights that they cannot obtain in the classroom, and they have an advocate working on their behalf when they seek future jobs. Some mentor relationships occur spontaneously when someone with education and experience takes an interest in a student. You can also arrange for a mentor directly by asking someone to be your mentor. A variety of people can serve as mentors, including job supervisors, university professors, and family friends. You can even have more than one mentor. Having multiple mentors increases your exposure to knowledge and opportunities.

Interviews With Practicing Professionals

Photo courtesy of Brian Hargrove

Brian Hargrove

Director of events and marketing, Greater Wichita Area Sports Commission

Brian Hargrove is the director of events and marketing for the Greater Wichita Area Sports Commission. He has a bachelor's degree in sport administration.

Q: What are your primary job responsibilities?

A: My main responsibilities include organizing, directing, facilitating, and managing multiple events during the year. I also assist the president/CEO in pursuing and obtaining funding through corporate sponsorships, membership programs, room commissions, in-kind donations, and event proceeds. Our goal is increasing sporting events and activities in the Wichita area by enhancing existing sport programs, attracting competitive sporting events to the area, and contributing to the development and improvement of athletic facilities.

Q: What have been the most important experiences in preparing you for your job?

A: There have been a few important experiences that have prepared me for my job. First, going to school and learning all that I could about the sport industry. The hands-on experience that I got while going to school was very important, as well as the mentoring program. And finally, my internship at USA Basketball really prepared me for my current job. Taking in everything from people that are successful is very important.

Q: What is a typical workday like for you?

A: Our hours are 8:30 until 5:00. However, when you work in sports, there is no such thing as a set schedule. When you have events, you could work 18 hours a day including on weekends. In just a normal day, though, my time is spent researching events that I feel would be good in the city, going through budgets, and seeing if I feel something would be successful in this city.

Q: What are the most and least enjoyable aspects of your job?

A: I love everything about my job. I love working in sports. I love the people I work with. Needless to say, I am very happy with what I am currently doing. Probably the most enjoyable part of the job is seeing something start from scratch and then seeing the final result. When you are in on something from the beginning stages and it all comes together and in the end is very successful, it is very satisfying. As far as least enjoyable aspects, there are none. I enjoy everything about what I'm doing.

Q: What are the most important qualities students need to address in order to effectively prepare themselves for careers in sport management?

A: The hands-on experience is the most important. Whether that is through a mentoring program or an internship, get as much experience as you can. One important aspect of all the things already listed is to put your full effort toward it. If you have an internship, do the best job that you can. You will then make an impression on your employer that will certainly help you down the road.

Q: What advice do you have for students who would like to pursue careers in sport management?

A: Volunteer, volunteer, volunteer! Get yourself out there and get some hands-on experience. This is very important. We use a lot of volunteers at the Wichita Sports Commission, and it really does give the students a chance to see behind the scenes at an event and everything that goes into it. It is also a great resume builder and a great way to meet people who can help you down the road.

Q: What changes do you anticipate seeing in the field over the next 10 years?

A:. Increased competition. With the entertainment market constantly growing and more choices for people out there, where will people spend their discretionary income? This is why you must make sure that your choice is the best choice, or you won't have a job for long.

Communication Skills

Sport managers working in virtually every functional area discussed in this chapter need strong communication skills. Most undergraduate curriculums require students to take one or more classes to facilitate development of these skills. Because most jobs in the sport industry are service oriented, however, we encourage sport management students to take more communications-related courses than the minimum number required. The ability to communicate effectively with others, write well, and use new communication technologies is essential.

Problem-Solving Skills

Sport managers making hiring decisions want qualified employees with an educational and employment background that will allow them to be successful on the job. A characteristic that often distinguishes successful practitioners is their ability to analyze a problem effectively and generate creative solutions. We encourage you to enroll in courses that will help you develop such skills and to seek work experiences that require such skills. For example, by volunteering to work in a fitness center's membership campaign, you will gain experience that is more valuable than what you would gain by simply checking in members at the front desk.

Assessing Your Sport Management Aptitude

If you want to pursue a career in sport management, do you have the skills and aptitudes mentioned in the text? Do you communicate clearly and confidently with people? Are you adept at problem solving? What types of experiences would be useful to pursue during your undergraduate years? Would you like to have a mentor? Have you considered seeking volunteer opportunities or part-time work within sport organizations to gain practical experience?

Generalization

Sport-related jobs are in high demand. As previously mentioned, by working in a variety of functional areas, you increase your marketability. Students with broad backgrounds have more options than those who specialize at the undergraduate level. Graduate school is often a more appropriate time to seek specialization in one of the many professional roles.

Sport Management Resources

Publications

Journal of Sport Management
Human Kinetics
P.O. Box 5076
Champaign, IL 61825-5076
http://www.humankinetics.com/JSM/journalAbout.cfm
ISSN: 0888-4773

Sport Marketing Quarterly
Fitness Information Technology, Inc.
International Center for Performance Excellence
School of Physical Education, West Virginia University
275G Coliseum, WVU-PE,
P.O. Box 6116
Morgantown, WV 26506-6116
Phone: 304-293-6888
Fax: 304-293-6658
E-mail: fit@fitinfotech.com
www.fitinfotech.com
ISSN: 1061-6934

Fitness Management Magazine
Leisure Publications
4130 Lien Rd.
Madison, WI 53704
Phone: 608-249-0186
Fax: 609-249-1153
E-mail: sales@fitnessmgmt.com
www.fitnessmanagement.com
ISSN: 0882-0481

Club Industry's Fitness Business Pro
9800 Metcalf Ave.
Overland Park, KS 66212
Phone: 913-341-1300
http://fitnessbusinesspro.com
ISSN: 0747-8283

Athletic Business
Athletic Business Publications Inc.
4130 Lien Rd.
Madison, WI 53704
Phone: 800-722-8764
Fax: 608-249-1153
www.athleticbusiness.com
ISSN: 0747-315X

International Journal of Sport Management
American Press Publishers
60 State St., #700
Boston, MA 02109
Phone: 617-247-0022
E-mail: americanpress@flash.net
http://americanpresspublishers.com
ISSN: 1546-234X

International Journal of Sport Communication
Human Kinetics
P.O. Box 5076
Champaign, IL 61825-5076
E-mail: webmaster@hkusa.com
www.humankinetics.com/IJSC/journalAbout.cfm
ISSN: 1936-3915

Athletic Management
Momentum Media Sports Publishing
31 Dutch Mill Rd.
Ithaca, NY 14850
Phone: 607-257-6970
Fax: 607-257-7328
E-mail: info@momentummedia.com
www.momentummedia.com
ISSN: 1041-5432

Street & Smith's SportsBusiness Journal
120 West Morehead St., Ste. 310
Charlotte, NC 28202
Phone: 800-829-9839
Fax: 704-973-1401
E-mail: help@sportsbusinessjournal.com
www.sportsbusinessjournal.com
ISSN: 1098-5972

Seton Hall Journal of Sports and Entertainment Law
Seton Hall School of Law,
One Newark Center
Newark, NJ 07102
Phone: 973-642-8239
Fax: 973-642-8540
http://law.shu.edu/journals/sportslaw/
ISSN: 1059-4310

(continued)

(continued)

Marquette Sports Law Review
Marquette University Law School
Sensenbrenner Hall
1103 West Wisconsin Ave.,
P.O. Box 1881
Milwaukee, WI 53201-1881
Phone: 414-288-7090
Fax: 414-288-6403
www.mu.edu/law
ISSN: 1533-6484

Journal of Legal Aspects of Sport
Sport and Recreation Law Association
Lori Miller, Executive Director
c/o Mary Myers, Wichita State University
Campus Box 127, 1845 Fairmount
Wichita, KS 67260-0127
Phone: 316-978-5445
Fax: 316-978-5451
E-mail: mary.myers@wichita.edu
ISSN: 1072-0316

Organizations

North American Society for Sport Management (NASSM)
West Gym 014
Slippery Rock, PA 16057
Phone: 724-738-4812
Fax: 724-738-4858
E-mail: nassm@sru.edu
www.nassm.com

Sport Marketing Association
President, Richard L. Irwin
Department of Human Movement Sciences and Education
FH 212, University of Memphis
Memphis, TN 38152
Phone: 901-678-3476
Fax: 901-678-5014
E-mail: rirwin@memphis.edu
www.sportmarketingassociation.com

Job- and Sport Management–Related Internet Sites

www.jobsinsports.com
www.onlinesports.com
www.sports-forum.com
www.teamworkonline.com

Wrap-Up

The sport management professions offer a broad range of exciting career opportunities for qualified individuals. In areas from sporting goods retailing to commercial bowling, sport managers in managerial and administrative positions are looking for individuals with a sport management education. Education, work experience, professionalism, volunteerism, networking, a good work ethic, and a positive attitude are characteristics that sport management employers seek. Individuals moving into sport management careers often find themselves holding jobs that are interesting because of the variety they offer, exciting because of the activities they involve, and challenging because of the ever-changing nature of the field. Not surprisingly, many sport administrators report high levels of job satisfaction. If you are looking for a fulfilling job and have a keen interest in sport, the sport management professions may be for you.

Use the key points review in your online study guide as a study aid for this chapter.

QA Activity 17.5

These end-of-chapter questions and activities are also in your online study guide. Your instructor may ask you to complete them online and turn them in.

1. Discuss the breadth of the sport industry and the related career opportunities it provides with specific reference to the industry segments discussed throughout this chapter.
2. Explain how the following would help an individual succeed in the sport industry: (a) good communication skills, (b) problem-solving skills, (c) practical experience, and (d) a mentor.
3. What type of course work and related academic content does a quality sport management program provide?
4. Identify three functional areas within the sport industry and elaborate on the types of jobs that might be specific to those professional roles.
5. Identify two existing sport organizations for each segment of the sport industry.
6. Elaborate on the type of sport management–related job you would be most attracted to and why.

GLOSSARY

abilities—Genetically endowed perceptual, cognitive, motor, metabolic, and personality traits that are susceptible to little or no modification by practice or training.

academic learning time—Time in which students are actively learning.

activities of daily living (ADLs)—Self-sufficient physical activities involving personal grooming, dressing, eating, walking, and using the toilet.

activity experience—Training in, observation of, practice of, or participation in physical activity.

adenosine triphosphate (ATP)—High-energy phosphate compound used by cells to do work.

advertising—"Information placed in the media by an identified sponsor that pays for the time or space" (Cutlip, Center, & Broom 1994, p. 10).

aerobic—In the presence of oxygen.

aesthetic experience—A subjective experience in which sensations appeal to our senses of beauty, grace, and artistic appreciation.

aggression—Behavior intended to inflict harm or injury on another person.

alveoli—Small hollow air sacs in the lungs where gas exchange between the air and blood occurs.

anaerobic—In the absence of oxygen.

anemia—A hemoglobin concentration below 12 grams per deciliter in women and 13 grams per deciliter in men.

anxiety—An unpleasant, high-intensity feeling that typically results from a demand or threat.

arousal—A state of bodily energy or physical and mental readiness.

ascetic experiences—Physical activity experiences that involve either discomfort, pain, or suffering.

athletic training clinic—A health care facility, usually associated with high school, collegiate, or professional athletic teams, that specializes in the health care needs of athletes.

attitudes—Relatively stable mind-sets toward physical activity.

behavior modification—The use of extrinsic reinforcers such as reward and punishment to shape human behavior.

biochemistry—The chemistry of living organisms.

biomechanics, human movement—Study of the structure and function of human beings using the principles and methods of mechanics of physics and engineering (see reviews in Atwater 1980; Hatze 1974; Winter 1985).

body culture—Societal conceptions of the human body including body image, body ideals, and body practices.

breadth of capacity—Achieving a low to moderate level of competence in a wide range of physical activities.

burnout—An extreme state of mental, emotional, and physical exhaustion that occurs because of chronic stress.

cardiac output—The amount of blood pumped out of one side of the heart per minute.

cardiopulmonary rehabilitation—Exercise programs designed to improve the function of the heart, lungs, and vascular system.

carpal tunnel syndrome—A type of cumulative trauma disorder to the hand and wrist, usually suffered by carpenters, typists, packers, assembly line workers, and others who repeat the same movements for several hours each day.

clinical settings—Health and fitness facilities that operate within hospital or medical facilities.

closed skill—A motor skill in which performers must coordinate their movements with a predictable, usually stationary, environment.

code of ethical principles and standards—Usually published and disseminated by professional associations to their members to ensure that the welfare of clients always receives top priority from practitioners.

cognitive skills—Human acts that require complex modes of thought, including rational analysis and problem solving, to achieve a predetermined goal. Most professionals rely on highly developed cognitive skills.

cohesion—The tendency for groups to stick together and remain united in pursuing goals.

community settings—Health and fitness facilities that operate within community centers, churches, or other nonprofit organizations such as YMCAs or Jewish Community Centers.

competent bystanders—Well-behaved students who avoid participation without attracting notice.

competition—A principle or framework for organizing physical activity in which participants compare their performances to each other's or to a standard for the purpose of increasing enjoyment.

conditioning—The temporary end state of training reflected in the performer's possessing an adequate level of strength, endurance, and flexibility to carry out desired tasks.

convection—The exchange of heat between the body and a moving fluid (water or air).

critical and poetic reasoning—A method of reflection that questions the validity of traditional philosophical thinking, combined with tentative, suggestive, and often speculative analyses.

cumulative trauma disorders—Injuries to muscle, tendons, nerves, and ligaments brought about through repetitive motion of a body part.

custodial care facility—Long-term care facility that caters to the medical, rehabilitational, and specialized needs of a patient, including assistance with the activities of daily living.

deductive reasoning—A method of reflection that starts with one or more broad, general principles and moves toward specific conclusions.

degrees of freedom—The number of dimensions in which a joint can move; for example, the elbow has one degree of freedom (flexion–extension), whereas the hip has three (flexion–extension, lateral, rotational); often used to explain movement complexity by summing all the degrees of freedom of the joints involved in a movement.

depth of capacity—Achieving a high level of competence in a narrow range of physical activities.

descriptive and speculative reasoning—A method of reflection that describes the essential qualities of one example of an object or event; also a technique that allows the philosopher to go beyond actual experience or observation.

developmental skills—Skills performed in nonsport settings where rules and competition are irrelevant.

discipline—Organized body of knowledge considered worthy of study, usually studied in a college or university curriculum.

disinterested spectating—A form of watching sport contests in which the observer is nonpartisan in his or her feelings about the outcome.

dispositions—Short-term, highly variable psychological states that may affect our enjoyment of physical activity.

distribution—The process of connecting the sport product or service with the ultimate consumer.

dualism—When applied to the nature of persons, a doctrine that emphasizes the radical distinctiveness and independence of mind and body.

electromyography (EMG)—A system for monitoring and recording the electrical activity in muscles.

emblems—Body movements such as hand signals that can be translated easily into explicit messages.

embodiment—A description of our human condition; usually a holistic conception emphasizing the impact of physicality on everything that people are and do.

emotional expression—The tendency to reveal something about our internal emotional states through physical activity.

emotions—Subjective reactions to changes in internal or external states.

employee assistance program (EAP)—Programs supported by business and industries to help employees with personal problems that may affect their work performance.

endurance—The ability to sustain physical activity for prolonged periods.

engaged time—Time that students spend engaged in activity.

ergonomics—The use of a biomechanical approach to design the workplace by fitting the workplace to the worker.

ergonomic workstation—A safe and effective work environment in which individual worker characteristics have been considered in the design and placement of equipment. Computer keyboards, chair adjustments, desk heights, and lighting are examples of equipment considerations in the creation of an ergonomic workstation.

ergonomists—Engineers who seek to improve the safety and efficiency of work through analysis of workers' movements and conditions in the workplace.

ethical values—Conceptions about right and wrong actions of individuals.

event management—The planning, organizing, execution, and evaluation of sport-related events (e.g., Super Bowl, Little League baseball tournaments).

exercise—Physical activity intended to improve one's health or alter the appearance of one's body.

exercise physiology—Another term for physiology of physical activity; also called exercise science.

experiential knowledge—Self-knowledge and knowledge about physical activity derived from performing or watching physical activity.

expressive movements—Movements employed in a physical activity as a way of expressing something of one's own emotion or personality; differentiated from instrumental movements.

extrinsic approach to physical activity—Valuing physical activity because of the benefits that come from participating.

extrinsic motivation—Motivation derived from pursuing and obtaining rewards outside the activity itself, such as money and status.

face-to-face contact activities—Contests such as football, wrestling, basketball, and soccer in which individuals interact with opponents' attempts to achieve the goal by physically manipulating their movements.

face-to-face noncontact activities—Contests such as volleyball, tennis, and baseball in which individuals interact with opponents to maximize their own chances of winning but do not physically manipulate their opponents.

feedback (intrinsic and extrinsic)—Information about the movement provided to the learner during and after a movement; this information may come from external sources (e.g., instructor, videotape) or internal sources (muscles, joints, through the nervous system).

financial management—Tasks involving financially related activities such as the prudent investment and use of assets, monetary development opportunities, and short-term and long-term budgeting.

focus group—A group interview for people's shared understandings. It involves an organized discussion with selected individuals on their views and experiences. Focus groups are used for sociological research and marketing purposes.

free time—Personal time that has not been encumbered with obligations; also called discretionary time.

functional learning time—Time in which students are appropriately engaged.

game spectator knowledge—Knowledge about the game one is watching, including the players, strategies, and competitive tactics.

gender—Social position based on a set of norms or expectations about how we should behave that are linked to societal understandings of sexuality and procreation.

gerontology—The study of aging.

gestures—Movements used to communicate our intentions to others; they may be illustrators, emblems, or regulators.

goal setting—Establishment of objectives for motor performance (either short term or long term).

ground reaction force (GRF)—An external force from the ground acting on a body. Based on the law of action–reaction, a GRF is created in the opposite direction and of equal magnitude as a response to a force created by the body acting against the ground. Hence, when a person stomps against the ground, the ground will create a GRF acting on the person. A GRF can typically act on a body in three directions—mediolateral (sideways), anterior-posterior (forward–backward), and vertical (upward).

habilitation—The processes and treatments leading to the acquisition of skills and functions that are normal and expected for an individual of a particular age and status.

habilitational therapeutic exercise—Processes and treatments leading to the acquisition of skills and functions that are considered normal and expected for an individual of a certain age, status, and occupation.

health maintenance organization (HMO)—Alternative systems of health care in which a certain population makes payment in advance on a fixed-contract fee basis.

health-related exercise—Exercise undertaken to develop or maintain a sound working body.

health-related fitness—Developed through physical activity experience, it refers to capacities and traits associated with low risk of hypokinetic diseases.

helping professions—Professions primarily committed to providing services.

hemoglobin—An iron-containing protein found inside red blood cells that carries most of the oxygen in the blood.

high-density lipoprotein (HDL) cholesterol—Cholesterol carried in the blood bound to high-density lipoproteins.

histology—The study of tissue.

holism—When applied to the nature of persons, a position that underscores the interdependence and interrelatedness of thought and physicality.

home maintenance activities—Self-sufficient activities intended to improve or repair conditions of living in one's apartment or house.

homophobia—Fear or dislike of gay, lesbian, bisexual, or transgendered individuals.

human agency—A theory suggesting that people are actively involved in developing or "constructing" their own sports.

human factors engineers—*See* ergonomists.

human resource management—Tasks involving, for example, the recruitment, orientation, retention, and evaluation of employees as well as benefit program planning and design and implementation of interval grievance procedures.

hyperplasia—An increase in muscle mass because of the splitting of muscle fibers.

hypertension—Elevated blood pressure, usually defined as a systolic pressure greater than 160 mmHg or a diastolic pressure greater than 90 mmHg.

hypertrophy—An increase in muscle mass because of the enlargement of the muscle fiber.

hypohydration—Failure to replace loss of body fluids adequately.

hypokinetic diseases—Diseases such as heart disease, obesity, and high blood pressure that are directly associated with low levels of daily physical activity.

hypothermia—A condition in which the body temperature falls below 95 °F (35 °C), causing heart rate and metabolism to slow; can be life threatening.

illustrators—Gestures used to demonstrate or complement what is being said.

imagery—Using all the senses to create or re-create an experience in the mind.

impersonal competition—Physical activities such as mountain climbing, long-distance swimming, and so forth, in which an individual attempts to better an established record in an activity that does not involve opponents.

inductive reasoning—A method of reflection that starts with specific cases or examples and moves toward broad, general conclusions.

inpatient facility—Rehabilitation hospital, nursing home, or other institution where patients spend extended periods of time for purposes of receiving medical or other health-related treatment.

instrumental activities of daily living (IADLs)—Less personal self-sufficient activities such as shopping, telephoning, cooking, or doing laundry.

instrumental movements—Movements employed in a physical activity to accomplish the goal of the action; differentiated from expressive movements.

insulin—Hormone that is essential for the metabolism of carbohydrates.

internalization—The process by which an activity gradually comes to be valued for its intrinsic qualities.

internship—Culminating educational experience for kinesiology majors planning professional careers. It involves extended work at one or two professional sites where students work under the supervision of a veteran professional.

interpersonal skills—Skills that enable the physical activity professional to listen to and communicate with clients, students, or patients.

intrinsic approach to physical activity—Valuing physical activity because of the subjective experiences embedded within the activity itself.

intrinsic motivation—Motivation derived from the rewards inherent or within the actual activity, such as enjoyment and feelings of accomplishment.

intuition—A process by which we come to know something without conscious reasoning.

intuitive knowledge—Knowledge gained through physical activity that doesn't depend on rational or conscious processes.

isokinetic—An exercise in which the muscle changes length at a constant rate of velocity.

isometric—An exercise in which tension is produced by the muscle without any change in length.

isotonic—An exercise in which the muscle changes length while maintaining a constant tension.

kinematics—A description of a movement, based on mechanical physics, in which movement characteristics like position, velocity, and acceleration are recorded.

kinesiology—The discipline or body of knowledge that studies physical activity through performance, scholarly analysis, and professional practice.

kinesiology theory—Theoretical knowledge about physical activity as embodied in the subdisciplines of kinesiology.

leadership—A behavioral process of influencing individuals and groups toward set goals.

learning—A permanent alteration in the functioning of the nervous system that enables performers to achieve a predetermined goal consistently.

leisure—A state of being in which humans find deep satisfaction and contentment, often accompanied by feelings of wonder, celebration, excitement, and creativity.

leisure activities—Physical activities that nourish or maintain the disposition of leisure.

leisure studies or recreation—An area of study or department in a college or university that focuses on preparing individuals for careers in the leisure industry.

liberal studies—Also known as the liberal arts and sciences, it is regarded as the core of a higher education and is responsible for inducing students to love the truth and to develop independence in thought and action.

lived body—The human body as it is immediately experienced in all varieties of life projects; the body as felt, as residing in the background. This stands in contrast to the body as objectified, an item in the foreground.

low-density lipoprotein (LDL) cholesterol—Cholesterol carried in the blood bound to low-density lipoproteins.

marketing management—Tasks involving the production, pricing, promotion, and distribution of a sport-related product or service in a way that inspires transactions between the sport organization and its consumers.

maximal oxygen uptake or **$\dot{V}O_2$max**—The highest rate of oxygen uptake during heavy dynamic exercise.

mechanical, market-driven professionals—Professionals for whom technique, methodology, profit, and prestige assume priority over clients' wants and needs.

memory drum theory—Developed by Franklin Henry (1960), it proposed that rapid and well-learned movements are not consciously controlled but are run off automatically (as an older computer uses a memory drum to store and retrieve data).

mentors—Professional people who can provide less-experienced professionals or preprofessionals with learning opportunities, responsibility, knowledge, and contact people who can contribute to the attainment of a successful career or continued improvement within an existing career.

merchandising—The sale of goods and services related to the sport organization (e.g., various team apparel or other products with team logos such as coffee mugs and note pads). Merchandising provides sport managers opportunities to enhance brand image, brand loyalty, and customer satisfaction, and to generate revenue.

metabolic rate—The rate at which the body uses energy.

mitochondria—The part of the cell where aerobic metabolism occurs and ATP is produced.

mixed methods—A research approach that involves collecting numerical data (e.g., via instruments) and gathering textual data (e.g., via interviews) so that both quantitative and qualitative methods are used. This approach provides a more complete picture of the phenomenon being studied.

modernization theory—A theory emphasizing that the rise of modern sport occurred during the industrial revolution as American society shifted away from being agricultural and locally oriented and developed city-based industries rooted in science and technology.

moment—The twisting, spinning, or rotational effect that occurs when a force is applied and rotation occurs.

motivation—A complex set of internal and external forces that influence individuals to behave in certain ways.

motor expertise—Proficiency held by a person with a high level of skill, usually determined by some criterion (e.g., national ranking, Division I football player).

motor performance fitness—A capacity developed through physical activity experience that enables people to perform daily activities with vigor. This type of fitness often incorporates an element of skill.

motor program—A cognitive mechanism that controls movement; analogous to a phonograph record that contains all the commands to the muscles.

motor skill taxonomy—A classification system that categorizes motor skills according to their common critical components.

motor skills—Human acts that require efficient, coordinated movements to achieve a predetermined goal. Some professionals such as dentists and surgeons rely on motor skills as well as cognitive skills.

movement—Any change in the position of body parts relative to each other.

myofibril—The part of the muscle fiber that contains the contractile elements.

mystical knowledge—Knowledge about another dimension of reality apprehended through participation in sport and exercise.

neuromuscular injury—Injuries to nerves and/or muscles that cause pain and limit or prevent the performance of physical activities.

novel learning tasks—A movement task with which the subject does not have prior experience; usually a simple movement like linear positioning or tracking.

observational learning (modeling)—Watching either oneself (on videotape) or another person (live or on videotape) perform a skill so that one can improve his or her performance.

occupational biomechanics—*See* ergonomics.

open skill—A motor skill in which performers must coordinate their movements with an unpredictable, usually moving, environment.

outpatient facility—A short-term care facility where patients do not stay overnight.

peak experiences—Special types of mystical experiences that runners and others engaged in strenuous sports and exercises frequently experience.

pedagogy—The art, science, and profession of teaching.

perceived freedom—Feeling free to participate in an activity without a nagging sense that you have to or that you should be doing something else.

perceptions—Meaningful constructs or messages based on the interpretations of sensations from past subjective experiences.

perceptual skills—Skills that enable you to identify and recognize problems that have occurred or are likely to occur in professional settings.

performance capacity—Qualities of physical activity such as flexibility, muscular endurance, cardiovascular endurance, and strength that are developed through training rather than learning.

personality—The unique blend of psychological characteristics and behavioral tendencies that makes individuals different from and similar to each other.

phenomenology—A method of reflection that examines the content of consciousness and gives credence to differences encountered in normal subjective life.

physical activity—Movement that is voluntary, intentional, and directed toward achieving an identifiable goal.

physical fitness—A capacity developed through exercise enabling one to perform the essential activities of daily living, engage in an active leisure lifestyle, and have sufficient energy remaining to meet the demands of unexpected emergencies.

plasma volume—The volume of extracellular fluid found in the blood.

power—(a) Ability to do what you want without being stopped by others; (b) the rate at which work is done (work ÷ time or force × velocity).

practice—(a) A type of physical activity experience that involves cognitive processing and leads to improvement in skill (learning); (b) repeating a task, often with an instructor's guidance or feedback, to promote learning.

practice theory—Knowledge concerning the client, the method, and the outcomes that guides practitioners in performing their duties as professionals.

practitioners—Those who use knowledge to bring about predetermined objectives. Professionals are practitioners.

prehabilitation—Occurs following an injury but before surgery; involves developing as much strength and range of motion as possible before surgery.

preprofessional—Professionals-in-training whose orientation to undergraduate studies is to become highly competent practitioners.

principle of quality—Experiences that engage us in the most critical aspects of an activity are most likely to improve our capacity to perform that activity.

principle of quantity—Increasing the frequency of experiences that engage us in the critical components of a physical activity will lead to increases in our capacity to perform that activity.

private practice—An entrepreneurial venture in which a professional establishes his or her own workplace with its own client pool.

professional practice knowledge—Knowledge derived from integration of performance, scholarly study, and practical experience about appropriate ways to deliver professional services.

program management—Those activities such as scheduling and equipment maintenance that are necessary for various sport organizations to address to ensure operational effectiveness.

promotions—Those activities such as personal selling, sales promotions, advertising, and publicity that are designed to attract consumers to sport-related products and/or services.

psychoanalytic self-knowledge—Knowledge about one's deep-seated desires, motivation, and behavior gained through participation in sport and exercise.

psychological inventory—Standardized or objective measure of a specific sample of behavior, typically in the form of a questionnaire.

public relations management—Tasks in areas such as media relations and community relations that are designed to achieve positive relationships between the sport organization and its most important constituents.

Pygmalion effect—Effect of teacher expectations on student performance.

qualitative analysis—Evaluating aspects of a movement without use of a measuring instrument (e.g., through visual observation).

quantitative analysis—Evaluating movement based on numerical measures obtained using an instrument (e.g., a stopwatch).

rational knowledge—Knowledge about facts, concepts, and theories gained through reason, logic, and analysis.

reaction time (simple and choice)—The speed of response to a light or sound; simple reaction time would involve pressing a button after seeing a signal; choice reaction time would involve choosing from among more than one button to press, depending on which signal one sees.

receptor—A specialized nerve ending found at the end of sensory neurons that detects changes in the environment.

reflection—A generic term used to describe the distinctive nature of philosophic research; often contrasted with the methods of science such as data gathering or empirical observation.

regulators—Hand and body movements used to guide the flow of conversations such as in greetings or when parting company.

rehabilitation—Physical treatment, exercise, and educational or counseling sessions that lead to a person's attaining maximum function and personally acceptable level of independence.

rehabilitational therapeutic exercise—Processes and treatments designed to restore skills or functions that were previously acquired but have been lost to disease, injury, or behavioral traits.

retention-based facilities—Commercial (for-profit) health and fitness facilities that make a large percentage of profits from ongoing monthly membership dues. These facilities engage in activities directed at helping members meet their goals and have a positive experience in the facility so that they will continue to pay for their membership on a month-by-month basis.

risk management—The prevention of loss associated with, for example, inadequate financial planning, employee management, facility and equipment maintenance, customer service, or short- or long-term planning efforts.

rituals—Physical activity employed to express symbolically some experience, truth, or value held deeply by a particular group.

sales-based facilities—Commercial (for-profit) health and fitness facilities that make a large percentage of profits from membership initiation fees and prepaid dues. In following this business model, sales-based facilities do not have significant incentive to satisfy current members. This emphasis on sales can result in overcrowding and lower-quality programs and services.

schema theory—Schmidt's (1975) explanation for how a motor program acquires a general set of rules from practice of similar movements.

self-efficacy—How confident one feels in one's ability to perform a physical activity.

self-esteem—Perception of personal worthiness.

self-fulfilling prophecy—Influence of teacher expectations on student performance.

self-reflection—A process whereby one experiences the subjective experiences of an activity performed in the past.

sensation-seeking activities—Physical activities that involve high speed, danger, or disorientation of the body in space.

sensations—Raw, uninterpreted information collected through sensory organs.

service—Human acts intended to improve the quality of life for others.

side-by-side competitive activities—Contests such as golf, swim racing, and so forth in which individuals do not directly interact in striving to accomplish the goal.

skill—The quality of physical activity experience that underlies the performance of motor skills.

skilled movement—Performances that require accuracy of direction, force, and rhythm or timing to accomplish predetermined goals.

social facilitation—The effects of the presence of an audience on human performance.

social loafing—A decrease in individual performance within groups.

social trustee, civic professionals—Professionals who adhere to the creed "Healthy people and a good society first. Myself and my profession second in service of this greater good" (Lawson, 1998a, July).

socialization—The process by which someone becomes accustomed to the norms and expectations of a subculture. Preprofessionals are judged to have been socialized into the profession when they understand the roles and responsibilities associated with the specific subculture of that profession.

Socratic self-knowledge—Knowledge about our capacities and limitations that enables us to perform physical activity safely within the range of our abilities.

specificity of practice—The finding that only practice conditions similar to game performance will benefit future game performance.

speculative reasoning—Uses inductive, deductive, descriptive, or imaginative reasoning in making claims that may or may not be true, but that are extremely difficult to demonstrate or defend.

spheres of physical activity experience—Various dimensions of everyday life in which physical activity plays an important and distinctive role.

spheres of professional practice—Categories of physical activity professions that are similar with respect to general objectives, methods, educational requirements, working environments, and other factors.

spheres of professional practice in physical activity—Groupings of different career paths in the physical activity professions according to similarities in the types of preparation required, clients served, and contexts in which the work is performed.

sport—Physical activity in which movement is performed to achieve a specific goal in a manner specified by established rules.

sport entertainment—Sport industry sector comprised of organizations that seek to attract consumers to watch athletes compete (e.g., NASCAR, NFL, NCAA institutions).

sport industry—The sport industry is "the market in which the products offered to its buyers are fitness, sport, recreation, and leisure related. These products include goods (e.g., baseball bats), services (e.g., sport marketing, health clubs), people (e.g., professional players), places (e.g., golf courses), and ideas" (Pitts, Fielding, & Miller 1994, p. 18).

sport pedagogy—The study of the teaching of physical education.

sport services—Sport industry sector composed of businesses providing needed services to other sport organizations (e.g., event management, representation services).

sport spectacles—Staged competitions designed and promoted for audiences and intended to evoke an entire range of human emotions by virtue of their grandeur, scale, and drama.

sports medicine—A field of medicine and therapeutic exercise that specializes in the treatment, prevention, and rehabilitation of athletes and others who are involved in sports and other forms of strenuous exercise. Sports medicine also involves the investigation of training methods and practices.

strength—The maximal amount of force exerted by a muscle group.

stress—A process in which individuals perceive an imbalance between their response capabilities and the demands of the situation.

stroke volume—The amount of blood pumped out of one side of the heart per beat.

subdisciplines—A way of dividing the scholarly study of physical activity to facilitate teaching and research. The divisions represent extensions of established disciplines such as psychology, physiology, and history.

subjective experience—Individual reactions, feelings, and thoughts about events.

task analysis—The systematic examination of a particular physical activity for purposes of disclosing its critical components.

technical definitions—Specialized meanings of terms used to convey information to others within a technical field. The technical definition of physical activity, for example, differs from how people define and use the term in everyday language.

theoretical knowledge—Knowledge of concepts and principles and the research strategies used to discover them. Theoretical knowledge in kinesiology is knowledge about physical activity, embedded in the subdisciplines, acquired by formal studying of the subdisciplines.

therapeutic exercise—The systematic and scientific application of exercise and movement to develop or restore muscular strength, endurance, and flexibility; neuromuscular coordination; cardiovascular efficiency; and other health performance factors.

thermogenesis—The generation of body heat by increasing the metabolic rate.

tidal volume—The amount of air inhaled or exhaled per breath.

training—Physical activity carried out for the express purpose of conditioning for performance in an athletic or other type of event.

underwater weighing—The procedure in which a person's body weight is measured while completely submerged to determine the person's body volume.

ventilation—The process in which gases are exchanged between the atmosphere and the alveoli of the lungs.

ventilatory threshold—The point during a graded exercise test at which ventilation begins increasing at a faster rate than $\dot{V}O_2$ does.

vertigo—The sensation that comes from disorientation of the body in space, often experienced in conjunction with dangerous activities.

vicarious participation—Feeling as though one is engaged in a sport contest that one is watching.

with-it-ness—Teacher awareness of all events transpiring in a learning environment.

workplace knowledge—Practical, mundane knowledge not grounded in theory that people use to perform everyday tasks in the workplace. Knowing where items are stored or how to clean, repair, or calibrate equipment are examples of workplace knowledge.

REFERENCES

Abbate, G. (2002). Gap between rich, poor widening. *Globeandmail.com*. July 7. www.globeandmail.com/serlet/ArticleNews/printarticle/gam/20020719/UGAPPCC.

Abernethy, B., & Sparrow, W.A. (1992). The rise and fall of dominant paradigms in motor behavior research. In J.J. Summers (Ed.), *Approaches to the study of motor control and learning*. Amsterdam: Elsevier.

Abernethy, B., Thomas, K.T., & Thomas, J.R. (1993). Strategies for improving understanding of motor expertise (or mistakes we have made and things we have learned!!). In J.L. Starkes & F. Allard (Eds.), *Cognitive issues in motor learning* (pp. 317-356). Amsterdam: Elsevier.

Acosta, R.V., & Carpenter, L.J. (1994). The status of women in intercollegiate athletics. In S. Birrell & C.L. Cole (Eds.), *Women, sport, and culture* (pp. 111–118). Champaign, IL: Human Kinetics.

Acosta, R.V., & Carpenter, L.J. (2008). Women in intercollegiate sport. A longitudinal, national study: Thirty one year update, 1977-2008. www.acostacarpenter.org.

Adams, J.A. (1987). Historical review and appraisal of research on the learning, retention, and transfer of human motor skills. *Psychological Bulletin, 101,* 41–74.

Adelman, M.L. (1986). *A sporting time: New York City and the rise of modern athletics, 1820–70.* Urbana, IL: University of Illinois Press.

Ainsworth, B.E., Haskell, W.L., Whitt, M.C., Irwin, M.L., Swartz, A.M., Strath, S.J., O'Brien, W.L., Bassett, D.R., Schmidtz, K.H., Emplaincourt, P.O., Jacobs, D.R., & Leon, A.S. (2000). Compendium of physical activities: An update of activity codes and MET intensities. *Medicine and Science in Sports and Exercise, 32,* S498–S516.

Albright, C.E., & Smith, K. (2006, March). Welding—trade or profession? *Techniques, 81*(3), 38.

Amar, J. (1920). *The human motor.* New York: Dutton.

Ambrosi, E., & Barker-Schwartz, K. (1995). The profession's image, 1917–1925, Part I: Occupational therapy as represented in the media. *American Journal of Occupational Therapy, 49*(7), 715–719.

American College of Sports Medicine. (2001). Position stand: Appropriate intervention strategies for weight loss and prevention of weight regain for adults. *Medicine and Science in Sports and Exercise, 33,* 2145–2156.

American College of Sports Medicine. (2006). *ACSM's guidelines for exercise testing and prescription.* Baltimore, MD: Lippincott, Williams, & Wilkins.

American Physical Therapy Association. (2007). *Physical Therapy Factsheet.* www.naahp.org.

American Recreation Foundation. (2007). *2006 Outdoor Recreation Participation Study.* www.funoutdoors.com.

Anastasi, A. (1988). *Psychological testing.* New York: Macmillan.

Anderson, D., Whitmer, R., Goetzel, R., Ozminkowski, R., Wasserman, J., & Serxner, S. (2000). Relationship between modifiable health risks and health care expenditures: A group level analysis of the HERO research database. *American Journal of Health Promotion 15*(1), 45–52.

Anderson, W.G., & Barrette, G.T. (1978). *What's going on in gym—descriptive studies of physical education classes* [Monograph No. 1]. Newtown, CT: Motor Skills.

Andrews, D. (Ed.). (1996). Deconstructing Michael Jordan: Reconstructing postindustrial America [Special issue]. *Sociology of Sport Journal, 13*(4).

Argyle, M. (1988). *Bodily communication* (2nd ed.). London: Methuen.

Armstrong, L.E., Costill, D.L., & Fink, W.J. (1985). Influence of diuretic-induced dehydration on competitive running performance. *Medicine and Science in Sports and Exercise, 17,* 456–461.

Arnot, R.B., & Gaines, C.L. (1984). *Sportselection.* New York: Viking Press.

Åstrand, P.-O. (1991). Influence of Scandinavian scientists in exercise physiology. *Scandinavian Journal of Medicine and Science in Sports, 1,* 3–9.

Atwater, A.E. (1980). Kinesiology/biomechanics: Perspectives and trends. *Research Quarterly for Exercise and Sport, 51,* 193–218.

Babiak, K., & Wolfe, R. (2006). More than just a game? Corporate social responsibility and Super Bowl XL. *Sport Marketing Quarterly, 15,* 214–222.

Baker, W.J. (1988). *Sports in the Western world.* Urbana, IL: University of Illinois Press.

Bar-Or, O. (1983). *Pediatric Sports Medicine for the Practitioner.* New York: Springer-Verlag.

Bauer, J.J., Fuchs, R.K., Smith, G., & Snow, C.M. (2001). Quantifying force magnitude and loading rate from drop landings that induce osteogenesis. *Journal of Applied Biomechanics, 17,* 142–152.

Bayley, N. (1935). The development of motor abilities during the first three years. *Monographs of the Society for the Research in Child Development* (Whole No. 1, pp. 1–26).

Berg, W.P., & Greer, N.L. (1995). A kinematic profile of the approach run of novice long jumpers. *Journal of Applied Biomechanics, 11,* 142–162.

Berkowitz, L. (1969). *Roots of aggression: A reexamination of the frustration-aggression hypothesis.* New York: Atherton Press.

Berliner, D.C. (1988). *The development of expertise in pedagogy.* Washington, DC: American Association of Colleges for Teacher Education.

Bernstein, N. (1967). *The coordination and regulation of movements.* London: Pergamon.

Berryman, J.W. (1973). Sport history as social history? *Quest, 20,* 65–73.

Berryman, J.W. (1989). The tradition of the "six things non-natural": Exercise and medicine from Hippocrates through ante-bellum America. *Exercise and Sport Sciences Reviews, 17,* 515–559.

Berryman, J.W. (1995). *Out of many, one: A history of the American College of Sports Medicine.* Champaign, IL: Human Kinetics.

Betts, J.R. (1952). Organized sport in industrial America. *Dissertation Abstracts, 12*(1), 41. (University Microfilms No. 3322)

Bhatt, R. (July 21, 2006). *San Francisco Business Times.*

Birrell, S., & McDonald, M.G. (2000). Reading sport, articulating power lines. In S. Birrell & M.G. McDonald (Eds.), *Reading sport: Critical essays on power and representation* (pp. 3–13).

Black, S.J., & Weiss, M.R. (1992). The relationship among perceived coaching behaviors, perceptions of ability and motivation in competitive age-group swimmers. *Journal of Sport and Exercise Psychology, 14*(3), 309–325.

Blair, S.N., Kohl, H.W., Paffenbarger, R.S., Clark, D.G., Cooper, K.H., & Gibbons, L.W. (1989). Physical fitness and all-cause mortality: A prospective study of healthy men and women. *Journal of the American Medical Association, 262*(17), 2395–2401.

Blair, S.N., Mulder, R.T., & Kohl, H.W. (1987). Reaction to "Secular Trends in Adult Physical Activity: Exercise Boom or Bust?" *Research Quarterly for Exercise and Sport, 58*(2), 106–110.

Blix, M. (1892–1895). Die lange und spannung des muskels. *Skandinavische Archiv Physiologie, 3,* 295–318; *4,* 399–409; *5,* 150–206.

Boden, B.P., Tacchetti, R.L., Cantu, R.C., Knowles, S.B., & Mueller, F.O. (2007). Catastrophic head injuries in high school and college football players. *American Journal of Sports Medicine, 35,* 1075-1081.

Boeck, S., & Staimer, M. (1996, December 6). NFL drug suspensions. *USA Today,* p. 1C.

Booth, F.W. (1989). Application of molecular biology in exercise physiology. In K.B. Pandolf (Ed.), *Exercise and Sport Sciences Reviews* (Vol. 17, pp. 1–27). Baltimore: Williams & Wilkins.

Bouchard, C., Lesage, R., Lortie, G., Simoneau, J.A., Hamel, P., Boulay, M.R., Perusse, L., Theriault, G., & Leblanc, C. (1986). Aerobic performance in brothers, dizygotic and monzygotic twins. *Medicine and Science in Sports and Exercise, 18,* 639–646.

Bouchard, C., Malina, R.M., Perusse, L. (1997). *Genetics of fitness and physical performance.* Champaign, IL: Human Kinetics.

Boyle, R.H. (1963). *Sport—Mirror of American life.* Boston: Little, Brown.

Bradley, B. (1977). *Life on the run.* New York: Bantam Books.

Branch, J.D. (2003). Effect of creatine supplementation on body composition and performance: A meta-analysis. *International Journal of Sport Nutrition and Exercise Metabolism, 13,* 198–226.

Brill, P.A., Burkhaulter, H.E., Kohl, H.W., Blair, S.N., & Goodyear, N.N. (1989). The impact of previous athleticism on exercise habits, physical fitness, and coronary heart disease risk factors in middle-aged men. *Research Quarterly for Exercise and Sport, 60,* 209–215.

Brophy, J.E., & Good, T.L. (1986). Teacher behavior and student achievement. In M.C. Wittrock (Ed.), *Handbook of research on teaching* (3rd ed., pp. 328–375). New York: Macmillan.

Brown, W.M. (1980). Ethics, drugs, and sport. *Journal of the Philosophy of Sport, 7,* 15–23.

Buchman, A.S. et al. (2007a). Physical activity and leg strength predict decline in mobility performance in older persons. *Journal of the American Geriatric Society,* 55(8), August.

Buchman, A.S. et al. (2007b). Physical activity and motor decline in older persons. *Muscle and Nerve, 35*(3), 354–362.

Buckworth, J., & Dishman, R.K. (2002). *Exercise psychology.* Champaign, IL: Human Kinetics.

Bungum, T.J., Satterwhite, M., Jackson, A.W., & Morrow, J.R., Jr. (2003). The relationship of body mass index, medical costs, and job absenteeism. *American Journal of Health Behavior, 27,* 456–462.

Burd, S. (2009). How Safeway is cutting healthcare costs. *Wall Street Journal.* http://online.wsj.com/article/SB124476804026308603.html.

Burr, D.B. (1997). Muscle strength, bone mass and age-related bone loss. *Journal of Bone and Mineral Research, 12*(10), 1547–1551.

Canadian Sports Tourism Alliance. (2007). CSTA members. www.canadiansporttourism.com/eng_memlist.cfm.

Carey, A.R., & Mullins, M.E. (1997, June 16). Toning up. *USA Today,* p. 1C.

Carlston, D.E. (1983). An environmental explanation for race differences in basketball performance. *Journal of Sport and Social Issues, 7*(2), 30–51.

Carr, J.H., & Shepherd, R.B. (1987). *A motor relearning programme for stroke* (p. 103, figure 4). Rockville, MD: Aspen.

Cary, K. (2005, March 27). Passion lost for the game. *Charlotte Observer,* p. 6C.

Castelli, D.M., & Beighle, A. (2007). Rejuvenating the school environment to include physical activity. *Journal of Physical Education, Recreation, and Dance, 78*(5), 25–28.

Cauley, J.A., Donfield, S.M., LaPorte, R.E., & Warhaftig, N.E. (1991). Physical activity by socioeconomic status in two population based cohorts. *Medicine and Science in Sports and Exercise, 23,* 343–352.

Cavanagh, P.R., Ulbrecht, J.S., & Caputo, G.M. (2000). New developments in the biomechanics of the diabetic foot. *Diabetes/Metabolism Research and Reviews, 16*(Suppl. 1), S6–S10.

Cawley, J., Meyerhoefer, C., & Newhouse, D. (2006). Not your father's PE: Obesity, exercise, and the role of schools. *Education Next, 4*(Fall), 61–66.

Cawley, J., Meyerhoefer, C., & Newhouse, D. (2007). The impact of state physical education requirements on youth physical activity and overweight. *Wiley Interscience,* February 27. www3.interscience.wiley.com/cgi-bin/abstract/114129506/ABSTRACT?CRETRY=1&SRETRY=0.

Celsing, F., Blomstrand, E., Werner, B., Pihlstedt, P., & Ekblom, B. (1986). Effects of iron deficiency on endurance and muscle enzyme activity in man. *Medicine and Science in Sports and Exercise, 18,* 156–161.

Centers for Disease Control. (2002). Nonfatal sport- and recreation-related injuries treated in emergency departments—United States, July 2000-June 2001. *CDC Morbidity and Mortality Weekly Report, 51,* 736–740.

Centers for Disease Control and Prevention (CDC) & President's Council on Physical Fitness and Sports. "Healthy people, midcourse review." (2005). www.healthypeople.gov/data/midcourse/pdf/fa22.pdf.

Centers for Disease Control and Prevention (CDC), National Center for Health Statistics. (2007). Exercise/Physical Activity. Hyattsville, MD: Division of Data Services. www.cdc.gov/nchs/fastats/exercise.htm.

Chaffin, D.B., & Andersson, G.B. (1991). *Occupational biomechanics* (2nd ed.). New York: Wiley.

Chamberlin, C., & Lee, T. (1993). Arranging practice conditions and designing instruction. In R.N. Singer, M. Murphey, & L.K. Tennant (Eds.), *Handbook of research on sport psychology* (pp. 213–241). New York: Macmillan.

Chapman, L.S. (2005). Meta-evaluation of worksite health promotion economic return studies: 2005 update. *American Journal of Health Promotion,* 19(6):1-11.

Chelladurai, P. (2006). *Human resource management in sport and recreation* (2nd ed.). Champaign, IL: Human Kinetics.

Christina, R.W. (1989). Whatever happened to applied research in motor learning? In J. Skinner et al. (Eds.), *Future directions for exercise science and sport research* (pp. 411–422). Champaign, IL: Human Kinetics.

Christina, R.W. (1992). The 1991 C.H. McCloy Research Lecture: Unraveling the mystery of the response complexity effect in skilled movements. *Research Quarterly for Exercise and Sport, 63,* 218–230.

Cicuttini, F.M., Baker, J.R., & Spector, T.D. (1996). The association of obesity with osteoarthritis of the hand and knee in women—a twin study. *Journal of Rheumatology, 23*(7), 1221–1226.

Clark, J.E., & Phillips, S.J. (1991). The development of intralimb coordination in the first six months of walking. In J. Fagard & P.H. Wolff (Eds.), *The development of timing control and temporal organization in coordinated action* (pp. 245–257). New York: Elsevier Science.

Clark, J.E., & Whitall, J. (1989). What is motor development? The lessons of history. *Quest, 41,* 183–202.

Clarkson, P.M., & Haymes, E.M. (1995). Exercise and mineral status of athletes, calcium, magnesium, phosphorus, and iron. *Medicine and Science in Sports and Exercise, 27,* 831–843.

Clough, P., Shepherd, J., & Maughan, R. (1989). Motives for participating in recreational running. *Journal of Leisure Research, 21*(4), 297–309.

Coakley, J.J. (1992). Burnout among adolescent athletes: A personal failure or social problem. *Sociology of Sport Journal, 9,* 271–285.

Coakley, J.J. (1998). *Sport in society: Issues and controversies* (6th ed.). New York: McGraw-Hill.

Coakley, J. (2007). *Sports in society: Issues and controversies* (9th ed.). Boston: McGraw-Hill.

Coen, S.P., & Ogles, B.M. (1993). Psychological characteristics of the obligatory runner: A critical examination of the anorexia analogue hypothesis. *Journal of Sport and Exercise Psychology, 15,* 338–354.

Coggan, A., & Coyle, E. (1991). Carbohydrate ingestion during prolonged exercise: Effects on metabolism and performance. In J.O. Holloszy (Ed.), *Exercise and Sport Science Reviews* (Vol. 19, pp. 1–40). Baltimore: Williams & Wilkins.

College Preview. (1997). American humanics: Training for feel-good careers, 10–13. Vol. XII, No. 4.

Commission on Sport Management Accreditation. (2006). *Accreditation manual* (Draft). http://iweb.aahperd.org/naspe/cosma/template.cfm?template=manuals.html.

Commission on Sport Management Accreditation. (n.d). *Philosophy of accreditation.* http://iweb.aahperd.org/naspe/cosma/template.cfm?template=philosophy.html.

Conn, V.S., Minor, M.A., & Burks, K.J. (2003). Sedentary older women's limited experience with exercise. *Journal of Community Health Nursing, 20*(4), 197–208.

Consilience Group. (2007). *Physical activity trends: Business and policy implications.* Bethesda, MD: Author.

Contini, R., & Drillis, R. (1966a). Applied biomechanics. In H.N. Abramson, H. Liebowitz, J.M. Crowley, & S. Juhasz (Eds.), *Applied mechanics surveys.* Washington, DC: Spartan Books.

Contini, R., & Drillis, R. (1966b). Kinematic and kinetic techniques in biomechanics. In F. Alt (Ed.), *Advances in bioengineering and instrumentation.* New York: Plenum Press.

Corbin, C.B., Lindsey, R., Welk, G.J., & Corbin, W.R. (2001). *Fundamental concepts of fitness and wellness.* New York: McGraw-Hill.

Cormier, J., York, A., Domholdt, E., & Keggeris, S. (1993). Athletic trainer utilization in sports medicine clinics. *Journal of Orthopedic and Sports Physical Therapy, 17*(1), 36–43.

Craik, R.L., & Dutterer, L. (1995). Spatial and temporal characteristics of foot fall patterns. In R.L. Craik & C.A. Oatis (Eds.), *Gait analysis: Theory and application* (1st ed., pp. 148–158). St. Louis: Mosby.

Crawford, S., & Eklund, R.C. (1994). Social physique anxiety, reasons for exercise, and attitudes toward exercise settings. *Journal of Sport and Exercise Psychology, 16,* 70–82.

Csikszentmihalyi, M. (1990a, January). *What good are sports?* Paper presented at the Commonwealth and International Conference of Physical Education, Sport, Health, Dance, Recreation, and Leisure, Auckland, New Zealand.

Csikszentmihalyi, M. (1990b). *Flow: The psychology of optimal experience.* New York: Harper & Row.

CSTA members. (2004). Ottawa, Ontario: Canadian Sport Tourism Alliance. www.canadiansporttourism.com/cng_memlist.cfm.

Cuneen, J., & Fink, J.S. (2006). Marketing Motown: Detroit sold cold, and Super Bowl XL was a winter blast. *Sport Marketing Quarterly, 15,* 200–205.

Cureton, T.K. (1930). Mechanics and physiology of swimming (the crawl flutter kick). *Research Quarterly, 1,* 87–121.

Cureton, T.K. (1932). Physics applied to physical education. *Journal of Health and Physical Education, 1,* 23–25.

Czaja, S.J. (1997). Using technologies to aid the performance of home tasks. In A.D. Fisk & W.A. Rogers (Eds.), *Human factors and the older adult* (pp. 311–334). New York: Academic Press.

Daily, The. (2002, October 18). National Longitudinal Survey of Children and Youth: Childhood obesity: 1994 to 1999. (StatisticsCanada, Cat. No. 11-001-XIE).

Davis, J.M., Burgess, W.A., Sientz, C.A., Bartoli, W.P., & Pate, R.R. (1988). Effects of ingesting 6% and 12% glucose/electrolyte beverages during prolonged intermittent cycling in the heat. *European Journal of Applied Physiology, 57,* 563–569.

De Grazia, S. (1962). *Of time, work and leisure.* Garden City, NY: Doubleday.

DeMarco, G.M. (1999). Physical education teachers of the year: Who they are, what they think, say, and do. *Teaching Elementary Physical Education, 10*(2), 11–13.

Dennis, W. (1938). Infant development under conditions of restricted practice and a minimum of social stimulation: A preliminary report. *Journal of Genetic Psychology, 53,* 149–158.

Dennis, W., & Dennis, M. (1940). The effect of cradling practices on the age of walking in Hopi children. *Journal of Genetic Psychology, 56,* 77–86.

Dietz, W.H. (1990). Children and television. In M. Green & R.J. Hagerty (Eds.), *Ambulatory pediatrics IV* (pp. 39–41). Philadelphia: W.B. Saunders.

Dill, D.B. (1967). The Harvard Fatigue Laboratory: Its development, contributions, and demise. In C.B. Chapman (Ed.), *Physiology of muscular exercise* (pp. 161–170). New York: American Heart Association.

Dill, D.B., Talbott, J.H., & Edwards, H.T. (1930). Studies in muscular activity. VI. Responses of several individuals to a fixed task. *Journal of Physiology, 69,* 267–305.

Dillman, C.J., Fleisig, G.S., & Andrews, J.R. (1993). Biomechanics of pitching with emphasis upon shoulder kinematics. *Journal of Orthopaedic & Sports Physical Therapy, 18*(2), 402–408.

Dintiman, G.B., & Greenberg, J.S. (1986). *Health through discovery.* New York: Random House.

Dishman, R.K., & Sallis, J.F. (1994). Determinants and interventions for physical activity and exercise. In C. Bouchard, R.J. Shephard, & T. Stephens (Eds.), *Physical activity, fitness, and health: International proceedings and consensus statement* (pp. 214–238). Champaign, IL: Human Kinetics.

Dishman, R.K., Sallis, J.F., & Orenstein, D. (1985). The determinants of physical activity and exercise. *Public Health Reports, 100,* 158–171.

Dodds, P. (1994). Cognitive and behavioral components of expertise in teaching physical education. *Quest, 46,* 153–163.

Dodson, J. (1996). *Final rounds: A father, a son, the golf journey of a lifetime.* New York: Bantam Books.

Donahue, S.W., Sharkey, N.A., Modanlou, K.A., Sequeira, L.N., & Martin, R.B. (2000). Bone strain and microcracks at stress fracture sites in human metatarsals. *Bone, 27*(6), 827–833.

Dong, N.L., Block, G., & Mandel. (2004, February 12). Activities contributing to total energy expenditure in the United States: Results from the NHAPS study. *International Journal of Behavioral Nutrition and Physical Activity, 1*(4). www.ijbnpa.org/content/1/1/4.

Donnelly, P. (1977). Vertigo in America: A social comment. *Quest, 27,* 106–113.

Dotingg, R. (2008). Pay for college chiefs rising fast. *Christian Science Monitor,* January 3.

Douglas, D.D., & Jamieson, K.M. (2006). A farewell to remember: Interrogating the Nancy Lopez farewell tour. *Sociology of Sport Journal, 23*(2), 117–141.

Dudgenon, B.J. (1996). Pediatric rehabilitation. In J. Case-Smith, A.S. Allen, & P.N. Pratts (Eds.), *Occupational therapy for children* (3rd ed., pp. 777–795). Baltimore: Mosby-Yearbook.

Duncan, M.C., Messner, M.A., & Willms, N. (2005). Gender in televised sports: News and highlight shows 1989-2004. www.la84foundation.org/9arr/ResearchReports/tv2004.pdf.

Duncan, M.C., & Robinson, T.T. (2004). Obesity and body ideals in the media: The health and fitness practices of young African-American women. *Quest, 56,* 77–104.

Dunkle, R.E., Kart, C.S., & Lockery, S.A. (1994). Self-care. In B.R. Bonder & M.B. Wagner (Eds.), *Functional performance in older adults* (pp. 122–135). Philadelphia: Davis.

Dunning, E., Murphy, P., & Williams, J. (1988). *The roots of football hooliganism: An historical and sociological study.* London: Routledge.

Dychtwald, K. (1990). *Age wave.* New York: Bantam Books.

Dyson, B.P. (1995). Students' voices in two alternative elementary physical education programs. *Journal of Teaching in Physical Education, 14,* 394–407.

Ebihara, O., Ideda, M., & Myiashita, M. (1983). Birth order and children's socialization into sport. *International Review of Sport Sociology, 18,* 69–89.

Eckstein, F., Faber, S., Mühlbauer, Hohe, J., Englmeier, K.-H., Reiser, M., & Putz, R. (2002). Functional adaptation of human joints to mechanical stimuli. *Osteoarthritis and Cartilage, 10*(1), 44–50.

Eitzen, D.S. (2006). *Fair and foul: Beyond the myths and paradoxes of sport* (3rd ed.). New York: Rowman & Littlefield.

Eitzen, D.S., & Sage, G.H. (1993). *Sociology of North American Sport* (5th ed.). Madison, WI: Brown & Benchmark.

Eitzen, D.S., & Sage, G.H. (1997). *Sociology of North American Sport* (6th ed). Madison, WI: Brown & Benchmark.

Eitzen, D.S., & Sage, G.H. (2003). *Sociology of North American Sport* (7th ed.). Boston: McGraw-Hill.

Ennis, C. (1994). Urban secondary teachers' value orientations: Social goals for teaching. *Teaching and Teacher Education, 10,* 109–120.

Ericsson, K.A. (2003). Development of elite performance and deliberate practice. In J.L. Starkes & K.A. Ericsson (Eds.), *Expert performance in sports* (pp. 49–84). Champaign, IL: Human Kinetics.

Eys, M.A., Burke, S.M., Carron, A.V., & Dennis, P.W. (2006). The sport team as an effective group. In J.M. Williams (Ed.), *Applied sport psychology: Personal growth to peak performance* (5th ed., pp. 157–173). Boston: McGraw-Hill.

Fagard, D. (2001). Exercise characteristics and the blood pressure response to dynamic physical training. *Medicine and Science in Sports and Exercise, 33,* S484–S492.

Fenn, W.O. (1929). Mechanical energy expenditure in sprint running as measured by moving pictures. *American Journal of Physiology, 90,* 343–344.

Ferguson, H. (1990). *The edge: The guide to fulfilling dreams, maximizing success, and enjoying a lifetime of achievement.* Cleveland: Getting the Edge.

Fernandez, W.G., Yard, E.E., & Comstock, R.D. (2007). Epidemiology of lower extremity injuries among U.S. high school athletes. *Academic Emergency Medicine, 14*(7), 641–645.

Fielding & Miller. (1996). Historical eras in sport marketing. In B.G. Pitts and D.K. Stotlar (Eds.), *Fundamentals of sport marketing.* Morgantown, WV: Fitness Information Technology.

Fine, G.A. (1988). Good children and dirty play. *Play and Culture, 1,* 43–56.

Fink, J., & Siedentop, D. (1989). The development of routines, rules, and expectations at the start of the school year. *Journal of Teaching in Physical Education, 8,* 198–212.

Finn, P., Pellathy, T., & Sinhal, S. (2009). U.S. Healthcare payments: Remedies for an ailing system. *The McKinsey Report.* www.mckinsey.com/clientservice/Financial_Services/Knowledge_Highlights/Recent_Reports/~/media/Reports/Financial_Services/US_healthcare_payments_Remedies_for_an_ailing_system1.ashx.

Fischman, M.G. (2007). Motor learning and control foundations of kinesiology: Defining the academic core. *Quest, 59,* 67–76.

Fitness programs are paying off, companies discover. (2004, July). *Industrial Safety and Hygiene News, 38*(7), 8.

Fitts, P.M., & Posner, M.I. (1967). *Human performance.* Pacific Grove, CA: Brooks/Cole.

Flacks, R., & Thomas, S.L. (1998, November 27). Among affluent students a culture of disengagement. *Chronicle of Higher Education,* p. A48.

Fleishman, E.A., & Hempel, W.E. (1955). The relationship between abilities and improvement with practice in a visual reaction time discrimination task. *Journal of Experimental Psychology, 49,* 301–312.

Fleisig, G.S., Barrentine, S.W., Escamilla, R.F., & Andrews, J.R. (1996). Biomechanics of overhand throwing with implications for injuries. *Sports Medicine, 21*(6), 421–437.

Fox, L.D., Rejeski, W.J., & Gauvin, L. (2000). Effects of leadership style and group dynamics on enjoyment of physical activity. *American Journal of Health Promotion, 14,* 277–283.

Franz, S.I., & Hamilton, G.V. (1905). The effects of exercise upon the retardation in conditions of depression. *American Journal of Insanity, 62,* 239–256.

Fried, G., Shapiro, S.J., & DeSchriver, T.D. (2003). *Sport finance.* Champaign, IL: Human Kinetics.

Friedson, E. (2001). *Professionalism: The third logic* (p. 12). Chicago: University of Chicago Press.

Freshmen drinking and studying more, exercising and volunteering less, a survey finds. (2002, December 13). *Chronicle of Higher Education (Notebook), 49*(16), p. A40.

Frost, H.M. (1999). An approach to estimating bone and joint loads and muscle strength in living subjects and skeletal remains. *American Journal of Human Biology, 11,* 437–455.

Fung, Y.C. (1968). Biomechanics—its scope, history, and some problems of continuum mechanics in physiology. *Applied Mechanics Reviews, 21,* 1–20.

Furusawa, K., Hill, A.V., & Parkinson, J.L. (1927). The dynamics of sprint running. *Proceedings of the Royal Society of London, 102B,* 29–42.

Galton, F. (1876). The history of twins as a criterion of the relative power of nature. *Anthropological Institute Journal, 5,* 391–406.

Game plans. (1994, March 19). *Economist,* p. 108.

Garcia, A.W., Broda, M.A.N., Frenn, M., Coviak, C., Pender, N.J., & Ronis, D.L. (1995). Gender and developmental differences in exercise beliefs among youth and prediction of their exercise behavior. *Journal of School Health, 65,* 213–219.

Garman, A.N., Evans, R., & Krause, M.K. (2006). Professionalism. *Journal of Healthcare Management, 51*(4), 219.

Gentile, A.M. (1972). A working model of skill acquisition with application to teaching. *Quest,* Monograph XVII, 2–23.

Gerber, E.W. (1971). *Innovators and institutions in physical education.* Philadelphia: Lea & Febiger.

Gesell, A. (1928). *Infancy and human growth.* New York: Macmillan.

Giamatti, A.B. (1989). *Take time for paradise: Americans and their games.* New York: Summit Books.

Gibbs, A. (1997). Focus groups. *Social Research Update.* www/spc/sirreu/ac/il/sru/SRU19.thml.

Gibson, H. (2003). Sport tourism. In J.B. Parks & J. Quarterman (Eds.), *Contemporary sport management* (2nd ed., pp. 337–360). Champaign, IL: Human Kinetics.

Gill, D.L. (2007). Gender and cultural diversity. In G. Tenenbaum & R.C. Eklund (Eds.), *Handbook of sport psychology* (3rd ed., pp. 823–844). Hoboken, NJ: Wiley.

Gill, F.B. (1989). *Ornithology.* New York: W.H. Freeman.

Glassow, R.B. (1932). *Fundamentals of physical education.* Philadelphia: Lea & Febiger.

Goetzel, R., Anderson, D., Whitmer, R., et al. (1998). The association between 10 modifiable risk factors and health care expenditures. *Journal of Occupational and Environmental Medicine, 40*(10) 1–12.

Goldstein, J.H., & Arms, R. (1971, March). Effects of observing athletic contests on hostility. *Sociometry, 34,* 83–90.

Gollnick, P.D., Timson, B.F., Moore, R.L., & Riedy, M. (1981). Muscular enlargement and number of fibers in skeletal muscles of rats. *Journal of Applied Physiology, 50,* 936–943.

Gonyea, W.J. (1980). Role of exercise in inducing increases in skeletal muscle fiber number. *Journal of Applied Physiology, 48,* 421–426.

Gorn, E., & Goldstein, W. (1993). *A brief history of American sports.* New York: Hill & Wang.

Gorn, E.J. (1986). *The manly art: Bare-knuckle prize fighting in America.* Ithaca, NY: Cornell University Press.

Gould, D., Dieffenbach, K., & Moffett, A. (2002). Psychological characteristics and their development in Olympic champions. *Journal of Applied Sport Psychology, 14,* 172–204.

Gould, D., Udry, E., Tuffey, S., & Loehr, J. (1996). Burnout in competitive junior tennis players: I. A quantitative psychological assessment. *Sport Psychologist, 10,* 322–340.

Graber, K.C. (2001). Research on teaching in physical education. In V. Richardson (Ed.), *Handbook of research on teaching* (4th ed., pp. 491–519). Washington, DC: American Educational Research Association.

Graber, K.C., & Woods, A.M. (2007). Stepping up to the plate: Physical educators as advocates for wellness policies—Part 2. *Journal of Physical Education, Recreation and Dance 78*(6), 19, 28.

Graham, G. (2001). *Teaching children physical education: Becoming a master teacher* (2nd ed.). Champaign, IL: Human Kinetics.

Greendorfer, S.L. (1979). Childhood sport socialization influences of male and female track athletes. *Arena review, 3,* 39–53.

Greenwood, E. (1957). Attributes of a profession. *Social Work, 2,* 45–55.

Griffin, L., Mitchell, S., & Oslin, J. (1997). *Teaching sports concepts and skills: A tactical games approach.* Champaign, IL: Human Kinetics.

Griffin, P.S. (1984). Girls' participation patterns in a middle school team sports unit. *Journal of Teaching in Physical Education, 4,* 30–38.

Griffin, P.S. (1986). Analysis and discussion: What have we learned? *Journal of Physical Education, Recreation and Dance, 57*(4), 57–59.

Griffith, C.R. (1926). *Psychology of coaching.* New York: Scribner.

Griffith, C.R. (1928). *Psychology and athletics.* New York: Scribner.

Gruber, J.J. (1986). Physical activity and self-esteem development in children: A meta-analysis. *American Academy of Physical Education Papers, 19,* 30–48.

Gruneau, R. (1983). *Class, sports, and social development.* Amherst, MA: University of Massachusetts Press.

Gullion, L. (1998). Recreational sport. In L.P. Masteralexis, C.A. Barr, & M.A. Hums (Eds.), *Principles and practice of sport management* (pp. 452–475). Gaithersburg, MD: Aspen.

Guttmann, A. (1978). *From ritual to record: The nature of modern sports.* New York: Columbia University Press.

Guttmann, A. (1984). *The games must go on: Avery Brundage and the Olympic movement.* New York: Columbia University Press.

Guttmann, A. (1991). *Women's sports: A history.* New York: Columbia University Press.

Hagberg, J.M. (1987). Effect of training on the decline of $\dot{V}O_2max$ with aging. *Federation Proceedings, 46,* 1830–1833.

Haken, H., Kelso, J.A.S., & Bunz, H. (1985). A theoretical model of phase transitions in human hand movements. *Biological Cybernetics, 51,* 347–356.

Handoll, H.H., Rowe, B.H., Quinn, K.M., & de Bie, R. (2000). Interventions for preventing ankle ligament injuries. *Cochrane Database of Systematic Reviews (online update 3),* CD000018.

Hanin, Y. (1997). Emotions and athletic performance: Individual zones of optimal functioning model. *European Yearbook of Sport Psychology* 1: 29–72.

Hanin, Y.L. (Ed.). (2000). *Emotions in sport.* Champaign, IL: Human Kinetics.

Hardy, S. (1982). *How Boston played: Sport, recreation, and community 1865–1915.* Boston: Northeastern University Press.

Hargreaves, J. (2001). *Heroines of sport: The politics of difference and identity.* London: Routledge.

Hart, I. (1963). *The mechanical investigations of Leonardo da Vinci* (2nd ed.). Berkeley, CA: University of California Press.

Hartwell, E.M. (1899). *On physical training.* Report of the commissioner of education for 1897–1898, Vol. 1. Washington, DC: U.S. Government Printing Office.

Hatze, H. (1974). The meaning of the term "biomechanics." *Journal of Biomechanics, 7,* 189–190.

Hay, J.G. (1993a). *The biomechanics of sport techniques* (4th ed.). Englewood Cliffs, NJ: Prentice Hall.

Hay, J.G. (1993b). Citius, altius, longius (faster, higher, longer)—the biomechanics of jumping for distance. *Journal of Biomechanics, 26*(Suppl. 1), 7–21.

Haymes, E.M., & Wells, C.L. (1986). *Environment and human performance.* Champaign, IL: Human Kinetics.

Helander, M. (2006). *A guide to human factors and ergonomics* (p. 220). Boca Raton, FL: Taylor & Francis (quote is personal communication from Sanders, 1980).

Hellison, D. (1995). *Teaching responsibility through physical activity.* Champaign, IL: Human Kinetics.

Hellison, D. (2003). *Teaching responsibility through physical activity* (2nd ed.). Champaign, IL: Human Kinetics.

Hellison, D., Cutforth, N., Kallusky, J., Martinek, T., Parker, M., & Stiehl, J. (2000). *Serving underserved youth through physical activity: Toward a model of university community collaboration.* Champaign, IL: Human Kinetics.

Heltne, P.G. (1989). Epilogue: Understanding chimpanzees and bonobos, understanding ourselves. In P. Heltne & L. Marquardt (Eds.), *Understanding chimpanzees* (pp. 380–384). Cambridge, MA: Harvard University Press.

Henry, F.M. (1964). Physical education: An academic discipline. *Journal of Health, Physical Education, and Recreation, 35*(7), 32–33, 69.

Henry, F.M., & Rogers, D.E. (1960). Increased response latency for complicated movements and a "memory drum" theory of neuromotor reaction. *Research Quarterly, 31,* 448–458.

Heuze, J.P., & Brunel, P. (2003). Social loafing in a competitive context. *International Journal of Sport and Exercise Psychology, 1,* 246–263.

Heyman, S. (1994). The hero archetype and high-risk sports participants. In M. Stein & J. Hollwitz (Eds.), *Psyche and sports* (p. 198). Wilmette, IL: Chiron.

Higher Education Research Institute. (2007, August 31). Attitudes and characteristics of freshmen at 4-year colleges, fall, 2006. *Chronicle of Higher Education, 54*(1), 18.

Hill, A.V. (1926). Scientific study of athletes. *Scientific American, 134,* 224–225.

Hill, A.V., & Lupton, H. (1923). Muscular exercise, lactic acid, and the supply of oxygen. *Quarterly Journal of Medicine, 16,* 135–171.

Howard, D.R., & Crompton, J.L. (2005). *Financing sport* (2nd ed.). Morgantown, WV: Fitness Information Technology.

Howell, J., & Bulmer, S. (2007). The IMPACT program: Evaluation of a cancer and exercise community intervention program (writers; presenters). United States: IHRSA (producer).

Howell, J., & Ingham, A. (2001). From social problem to personal issue: The language of lifestyle. *Cultural Studies,* 15, 326–351.

Hultman, E. (1967). Physiological role of muscle glycogen in man, with special reference to exercise. *Circulation Research, 21*(Suppl. 1), 99–114.

Human Resources and Social Development Canada (HRSDC). (2005). Advancing the inclusion of people with disabilities: Executive Summary. www.hrsdc.gc.ca/en/hip/odi/documents/advancingInclusion05/summary.shtml.

Hyland, D.A. (1990). *Philosophy of sport.* New York: Paragon House.

Income gap is widest since '40s, agency says. (1996, June 20). *Greensboro News and Record,* p. A6.

Institute of Medicine. (2007). *Adequacy of evidence for physical activity guidelines development: Workshop summary.* Washington, DC: National Academies.

International Health, Racquet and Sportsclub Association. (1992). *The economic benefits of regular exercise.* Boston: Author.

International Health, Racquet and Sportsclub Association. (2005). *Profiles of Success.* Boston: Author.

International Health, Racquet and Sportsclub Association. (2007a). The scope of the U.S. health club industry. http://cms.ihrsa.org/index.cfm?fuseaction=Page.viewPage&pageId=18853&nodeID=15.

International Health, Racquet and Sportsclub Association. (2007b). Health club membership by club type. http://cms.ihrsa.org/index.cfm?fuseaction=page.viewPage.cfm&pageId=18890.

International Health, Racquet and Sportsclub Association. (2008). *Profiles of success.* Boston: Author.

Irwin, R.L., Sutton, W.A., & McCarthy, L.M. (2002). *Sport promotion and sales management.* Champaign, IL: Human Kinetics.

Isidore, C. (2005, April 22). NFL's rights might. *CNNMoney.com.* http://money.cnn.com/2005/04/22/commentary/column_sportsbiz/sportsbiz/.

Ivy, J.L., Katz, A.L., & Cutler, C.L. (1989). Muscle glycogen resynthesis after exercise: Effect of time on carbohydrate ingestion. *Journal of Applied Physiology, 64,* 1480–1485.

Jackson, D.Z. (1989, January 22). Calling the plays in black and white. *Boston Globe,* pp. A30, A33.

Jackson, D.Z. (1996, March 27). Chasing spirits down the court at NCAA tourney. *Charlotte Observer,* p. 17A.

Jackson, J.A. (1970). *Professions and professionalization.* London: Cambridge University Press.

Janz, K.F., Levy, S.M., Burns, T.L., Willing, M.C., & Warren, J.J. (2002). Fatness, physical activity, and television viewing in children during the adiposity rebound period: The Iowa Bone Development study. *Preventive Medicine, 35,* 563–571.

Johnson, L.D., Delva, J., & O'Malley, P.M. (2007, October). Sports participation and physical education in American secondary schools: Current levels and racial/ethnic and socioeconomic disparities. *American Journal of Preventive Medicine, 33*(4), Suppl. 1, S195–S208.

Kaiser Commission on Medicare and the Uninsured. October 2006. Washington, DC: The Kaiser Commission.

Kaiser Family Foundation. (2003). *Employer health benefits 2003: Annual survey.* Menlo Park, CA: Author.

Kaiser Family Foundation. (2009). *Employer health benefits: 2009 summary of findings.* http://ehbs.kff.org/pdf/2009/7937.pdf.

Kalakanis, L., Goldfield, G.S., Paluch, R.A., & Epstein, L.H. (2001). Parental activity as a determinant of activity level and patterns of activity in obese children. *Research Quarterly for Exercise and Sport 72*(3), 202–209.

Kane, R.L., Ouslander, J.G., & Abrass, I.B. (1994). *Essentials of clinical geriatrics* (3rd ed.). Boston: Allyn & Bacon.

Karlsson, J. (2002). Ankle braces prevent ligament injuries. *Lakartidningen, 99*(36), 3486–3489.

Karpovich, P.V., Morehouse, L.E., Scott, M.G., & Weiss, R.A. (Eds.). (1960). The contributions of physical activity to human well-being. *Research Quarterly, 31*(2), part II [special issue].

Katz, S., Ford, A.B., Moskowitz, R.W., Jackson, B.A., & Jaffee, M.W. (1963). Studies of illness in the aged: The index of ADL: A standardized measure of biological and psychological function. *Journal of the American Medical Association, 185,* 914–919.

Kelso, J.A.S. (1995). *Dynamic patterns: The self-organization of brain and behavior.* Cambridge, MA: MIT Press.

Kenyon, G.S. (1968). A conceptual model for characterizing physical activity. *Research Quarterly, 39,* 96–104.

King, A.C., Blair, S.N., Bild, D., Dishman, R.K., Dubbert, P.M., Marcus, B.H., Oldridge, N.M., Paffenbarger, R.S., Powell, K.E., & Yeager, K.Y. (1992). Determinants of physical activity and interventions in adults. *Medicine and Science in Sports and Exercise, 24*(6), S221–236.

King, C.R. (2004). This is not an Indian: Situating claims about Indianness in sporting worlds. *Journal of Sport & Social Issues, 28,* 3–10.

King, M.A., & Yeadon, M.R. (2003). Coping with perturbations to a layout somersault in tumbling. *Journal of Biomechanics, 36*(7), 921–927.

Kinkema, K.M., & Harris, J.C. (1998). MediaSport studies: Key research and emerging issues. In L.A. Wenner (Ed.), *MediaSport* (pp. 27–54). London: Routledge.

Kleinman, S. (1968). Toward a non-theory of sport. *Quest, 10,* 29–34.

Kochman, T. (1981). *Black and white styles in conflict.* Chicago: University of Chicago Press.

Kordtlandt, A. (1989). The use of stone tools by wild-living chimpanzees. In P. Heltne and L. Marquardt (Eds.), *Understanding chimpanzees* (pp. 146-147). Cambridge, MA: Harvard University Press.

Krane, V., & Williams, J.M. (2006). Psychological characteristics of peak performance. In J.M. Williams (Ed.), *Applied sport psychology: Personal growth to peak performance* (5th ed., pp. 207–227). Boston: McGraw-Hill.

Kranz, L. (2002). *Jobs rated almanac* (6th ed.). Fort Lee, NJ: Barricade Books.

Krawthwohl, D.R., Bloom, B.S., & Masia, B.B. (1964). *Taxonomy of educational objectives: Handbook II: Affective domain.* New York: David McKay.

Kretchmar, R.S. (1975). From test to contest: An analysis of two kinds of counterpoint in sport. *Journal of the Philosophy of Sport, 2,* 23–30.

Kretchmar, R.S. (1985). "Distancing": An essay on abstract thinking in sport performances. In D.L. Vanderwerken & S.K. Wertz (Eds.), *Sport inside out: Readings in literature and philosophy* (pp. 87–103). Forth Worth, TX: Texas Christian University Press.

Kretchmar, R.S. (1994). *Practical philosophy of sport*. Champaign, IL: Human Kinetics.

Kretchmar, R.S. (1996). Philosophic research in physical activity. In J.R. Thomas & J.K. Nelson (Eds.), *Research methods in physical activity* (pp. 277–290). Champaign, IL: Human Kinetics.

Kretchmar, R.S. (2004). Walking Barry Bonds: The ethics of the intentional walk. In E. Bronson (Ed.), *Baseball and philosophy: Thinking outside the batter's box* (pp. 261–272). Chicago, IL: Open Court.

Kretchmar, R.S. (2005). *Practical philosophy of sport and physical activity* (2nd ed.). Champaign, IL: Human Kinetics.

Kroemer, K., Kroemer, H., & Kroemer-Elbert, K. (1994). *Ergonomics: How to design for ease and efficiency*. Englewood Cliffs, NJ: Prentice Hall.

Kroll, W.P. (1982). *Graduate study and research in physical education*. Champaign, IL: Human Kinetics.

Kruzenga, L. (2004, March 5). Report confirms health disparities for Aboriginals, poor. *The First Perspective*. www.firstperspective.ca/story_2004_03_05_report.html.

Kufahl, P. (2003, August). Set the stage for Pilates. *Club Industry,* 14–19.

LaBarre, W. (1963). *The human animal*. Chicago: University of Chicago Press.

Landers, D.M., & Arent, S.A. (2007). Physical activity and mental health. In G. Tenenbaum & R.C. Eklund (Eds.), *Handbook of sport psychology* (3rd ed., pp. 469–491). Hoboken, NJ: Wiley.

Lane, N.E. (1995). Exercise—a cause of osteoarthritis. *Journal of Rheumatology, 22*(Suppl. 43), 3–6.

Lanier, J.E., & Little, J.W. (1986). Research on teacher education. In M.C. Wittrock (Ed.), *Handbook of research on teaching* (3rd ed., pp. 527–569). New York: Macmillan.

Lanyon, L.E. (1996). Using functional loading to influence bone mass and architecture: Objectives, mechanisms, and relationship with estrogen of the mechanically adaptive process in bone. *Bone, 18,* 37S–43S.

Lapchick, R., with Brenden, J. (2006). The 2005 racial and gender report card: College sports. www.bus.ucf.edu/sport/.

Lapchick, R. (2007). *The 2006 racial and gender report card: Major League Baseball*. Institute for Diversity and Ethics in Sport, University of Central Florida. www.tidesport.org.

Lasch, C. (1979). *The culture of narcissism: American life in an age of diminishing expectations*. New York: Warner Books.

Lawson, H.A. (1984). *Invitation to physical education*. Champaign, IL: Human Kinetics.

Lawson, H.A. (1998a, July). Globalization and the social responsibilities of citizen-professionals. Unpublished paper. Address to AIESEP International Conference, Adelphi University, p. 7.

Lawson, H.A. (1998b). Here today, gone tomorrow: A framework for analyzing the invention, development, transformation and disappearance of helping fields. *Quest, 50,* 225–237.

Leonard, W.M. (1998). *A sociological perspective of sport*. Needham Heights, MA: Allyn & Bacon.

Levine, P. (1992). *Ellis Island to Ebbets Field: Sport and the American Jewish experience*. New York: Oxford University Press.

Lewko, J.H., & Greendorfer, S.L. (1988). Sex differences and parental influences in sport socialization of children and adolescents. In F.L. Smoll, R.A. Magill, & M.J. Ash (Eds.), *Children in sport* (3rd ed., pp. 287–300). Champaign, IL: Human Kinetics.

Linnan, L., Bowling, M., Childress, J., Lindsay, G., Blakey, C., Pronk, S., Wieker, S., & Royall, P. (2008). Results of the 2004 National Worksite Health Promotion Survey. *American Journal of Public Health, 98*(8): 1503-1509.

Li, M., Hofacre, S., & Mahony, D. (2001). *Economics of sport*. Morgantown, WV: Fitness Information Technology.

Locke, E.A., & Latham, G.P. (1985). The application of goal setting to sports. *Sport Psychology Today, 7,* 205–222.

Locke, L.F. (1975). *The ecology of the gymnasium: What the tourists never see*. Amherst, MA: University of Massachusetts. (ERIC Document Reproduction No. ED 104-823).

Locke, L.F. (1990). Commentary: Conjuring kinesiology and other political parlor tricks. *Quest, 42,* 323–329.

Locke, L.F. (1996). Dr. Lewin's little liver patties: A parable about encouraging healthy lifestyles. *Quest, 48*(3), 422–431.

Locke, L.F., & Griffin, P. (1986). Profiles of struggle. *Journal of Physical Education, Recreation and Dance, 57*(4), 32–63.

Locke, L.F., & Massengale, J.D. (1978). Role conflict in teacher/coaches. *Research Quarterly for Exercise and Sport, 49,* 162–174.

Longman, J. (2006, July 9). France's aging magician conjures a final trick. *New York Times,* p. 1.

Lorber, J. (1994). *Paradoxes of gender.* New Haven, CT: Yale University Press.

Lorenz, K. (1966). *On aggression.* New York: Harcourt, Brace & World.

Lortie, D. (1975). *Schoolteacher: A sociological study.* Chicago: University of Chicago Press.

Lortie, G., Simoneau, J.A., Hamel, P., Boulan, M.R., Landry, F., & Bouchard, C. (1984). Responses of maximal aerobic power and capacity to aerobic training. *International Journal of Sports Medicine, 5,* 232–236.

Loy, J.W., McPherson, B.D., & Kenyon, G. (1978). *Sport and social systems.* Reading, MA: Addison-Wesley.

Lucas, J.A., & Smith, R.A. (1978). *Saga of American sport.* Philadelphia: Lea & Febiger.

Madrigal, R. (1995). Cognitive and affective determinants of fan satisfaction with sporting event attendance. *Journal of Leisure Research, 27*(3), 205–227.

Magill, R.A., & Hall, K.G. (1990). A review of the contextual interference effect in motor skill acquisition. *Human Movement Science, 9,* 241–289.

Majors, R. (1990). Cool pose: Black masculinity and sports. In M.A. Messner & D.F. Sabo (Eds.), *Sport, men, and the gender order: Critical feminist perspectives* (pp. 109–126). Champaign, IL: Human Kinetics.

Malina, R.M. (1984). Physical growth and maturation. In J.R. Thomas (Ed.), *Motor development during childhood and adolescence* (pp. 2–26). Edina, MN: Burgess International.

Malina, R.M., Bouchard, C., & Bar-Or, O. (2004). *Growth, maturation, and physical activity* (2nd ed.). Champaign, IL: Human Kinetics.

Mandelbaum, M. (2004). *The meaning of sports* (p. 177). New York: Public Affairs.

Manton, K.G., & Gu, X. (2001). Changes in the prevalence of chronic disability in the United States black and nonblack population above age 65 from 1982 to 1999. *Proceedings of the National Academy of Sciences of the United States of America 98*(11), 6354–6359.

Margaria, R., Edwards, H.T., & Dill, D.B. (1933). The possible mechanisms of contracting and paying the oxygen debt and the role of lactic acid in muscular contraction. *American Journal of Physiology, 106,* 689–715.

Maron, B.J. (2007). Hypertrophic cardiomyopathy and other causes of sudden cardiac death in young competitive athletes, with considerations for preparticipation screening and criteria for disqualification. *Cardiology Clinics, 25*(3), 399–414.

Martin, T.W., & Berry, K.J. (1974). Competitive sport in post-industrial society: The case of the motocross racer. *Journal of Popular Culture, 8,* 107–120.

Martinek, T., & Griffith, J.B. (1993). Working with the learned helpless child. *Journal of Physical Education, Recreation and Dance, 64*(6), 17–20.

Martinek, T., & Griffith, J.B. (1994). Learned helplessness in physical education: A developmental study of causal attributions and task persistence. *Journal of Teaching in Physical Education, 13,* 108–122.

Martinek, T., & Karper, W. (1984). Multivariate relationships of specific impression cues with teacher expectations and dyadic interactions in elementary education classes. *Research Quarterly for Exercise and Sport, 55,* 32–40.

Martinek, T., & Karper, W. (1986). Motor ability and instructional contexts: Effects on teacher expectation and dyadic interactions in elementary physical education classes. *Journal of Classroom Interaction, 21,* 16–25.

Massengale, J.D., & Swanson, R.A. (Eds.). (1997). *The history of exercise and sport science.* Champaign, IL: Human Kinetics.

McAuley, E., & Courneya, K.S. (1994). The subjective exercise experiences scale: Development and preliminary validation. *Journal of Sport and Exercise Psychology, 16,* 163–177.

McAuley, E., & Jacobson, L.B. (1991). Self-efficacy and exercise participation in sedentary adult females. *American Journal of Health Promotion, 5,* 185–191.

McAuley, E., Wraith, S., & Duncan, T.E. (1991). Self-efficacy perceptions of success and intrinsic motivation for exercise. *Journal of Applied Social Psychology, 16,* 139–155.

McCloy, C.H. (1960). The mechanical analysis of skills. In W.R. Johnson (Ed.), *Science and medicine in exercise and sports.* New York: Harper & Brothers.

McCullagh, P. (1993). Modeling: Learning, developmental, and social psychological considerations. In R.N. Singer, M. Murphey, & L.K. Tennant (Eds.), *Handbook of research on sport psychology* (pp. 106–126). New York: Macmillan.

McDonald, M.G. (1996). Michael Jordan's family values: Marketing, meaning, and post-Reagan America. *Sociology of Sport Journal, 13,* 344–365.

McGraw, M.B. (1935). *Growth: A study of Johnny and Jimmy.* New York: Appleton-Century-Crofts.

McGraw, M.B. (1939). Later development of children specially trained during infancy: Johnny and Jimmy at school age. *Child Development, 10,* 1–19.

McGuigan, P.J. (2007, February). The ongoing quest for professionalism: Consumer decision processes for professional services. *Casualty and Property Insurance Underwriters Journal, 60*(2), 1–8.

McIntyre, N. (1992). Involvement in risk recreation: A comparison of objective measures of engagement. *Journal of Leisure Research, 24,* 64–71.

Meek, A. (1997). An estimate of the size and supported economic activity of the sports industry in the United States. *Sport Marketing Quarterly, 6*(4), 15–22.

Meier, K. (1980). An affair of flutes: An appreciation of play. *Journal of the Philosophy of Sport, 7,* 24-45.

Meister, D. (1999). *The history of human factors and ergonomics.* Mahwah, NJ: Laurence Erlbaum.

Metheny, E. (1965). *Connotations of movement in sport and dance.* Dubuque, IA: Brown.

Metheny, E. (1968). *Movement and meaning.* New York: McGraw-Hill.

Metzler, M. (1979). *The measurement of academic learning time in physical education.* Unpublished doctoral dissertation, Ohio State University, Columbus.

Metzler, M. (1989). A review of research on time in sport pedagogy. *Journal of Teaching in Physical Education, 8,* 87–103.

Miller, L.K. (1997). *Sport business management.* Gaithersburg, MD: Aspen.

Morgan, W. (1982). Play, utopia, and dystopia: Prologue to a ludic theory of the state. *Journal of the Philosophy of Sport, 9,* 30-42.

Morgan, W.J. (1994). *Leftist theories of sport: A critique and reconstruction.* Urbana, IL: University of Illinois Press.

Morris, D. (1994). *Bodytalk: The meaning of gestures.* New York: Crown Trade.

Morris, J.N., Heady, J.A., Raffle, P.A.B., Roberts, C.G., & Parks, J.W. (1953). Coronary heart disease and physical activity of work. *Lancet, 2,* 1111–1120.

Morriss, C., & Bartlett, R. (1996). Biomechanical factors critical for performance in the men's javelin throw. *Sports Medicine, 21,* 438-446.

Morrissey, M.C., Harman, E.A., & Johnson, M.J. (1995). Resistance training modes: Specificity and effectiveness. *Medicine and Science in Sports and Exercise, 27,* 648–660.

Mullin, B.J., Hardy, S., & Sutton, W.A. (2007). *Sport marketing* (3rd ed.). Champaign, IL: Human Kinetics.

Muret, D. (2007, February 26). Building toward a record year. *Street & Smith's SportsBusiness Journal, 9*(42), 1.

Murphy, W. (1995). *Healing the generations: A history of physical therapy and the American Physical Therapy Association.* Lyme, CT: Greenwich.

Nagel, T. (1987). *What does it all mean? A very short introduction to philosophy.* Oxford: Oxford University Press.

NASPE–NASSM. (2000). *Sport management program standards and review protocol.* Reston, VA: Association of the American Alliance for Health, Physical Education, Recreation and Dance.

National Association for Sport and Physical Education. (1995). *Moving into the future: National physical education standards.* New York: Mosby.

National Association for Sport and Physical Education. (2003). *National standards for beginning physical education teachers* (2nd ed.). Reston, VA: NASPE Publications.

National Association for Sport and Physical Education. (2004). *Moving into the future: National physical education standards* (2nd ed.). New York: McGraw-Hill.

National Association for Sport and Physical Education and the American Heart Association. (2006). *2006 shape of the nation report: Status of physical education in the USA.* Reston, VA: National Association for Sport and Physical Education.

National Center for Chronic Disease Prevention and Health Promotion (NCCDPHP). (2003). *Health topics: Physical activity.* (Strategies: School Programs.) United States Department of Health and Human Services. www.cdc.gov/nccdphp/dash/physical activity/promoting_health.

National Center for Health Statistics (NCHS). (2002). Centers for Disease Control and Prevention. www.cdc.gov.nchs/products/pubs/pubd/hestats/3and4/overweight.htm; also: www.cdc.gov/nchs/products/pubs/hestats/3and4/sedentary.htm.

National Center for Health Statistics (NCHS). (2007). *Health, United States, 2007, with chartbook on trends in the health of Americans.* Hyattsville, MD: National Center for Health Statistics, Division of Data Services.

National Collegiate Athletic Association. (2007a, October 26). NCAA member institutions on probation. https://web1.ncaa.org/pdf/convert?pdfurl=http://goomer.ncaa.org:2020/wdbctx/LSDBi/lsdbi.lsdbi_mi_rpts.currentprobationrpt.

National Collegiate Athletic Association. (2007b, October 26). Institutions with major infractions. https://web1.ncaa.org/pdf/convert?pdfurl=http://goomer.ncaa.org:2020/

wdbctx/LSDBi/lsdbi.lsdbi_mi_rpts.mostinfractionsrpt.

National Federation of State High School Associations. (2007). *Athletics participation survey totals*. www.nfhs.org.

National Federation of State High School Associations. (2007). High school sports participation increases again; girls exceeds three million for first time. www.nfhs.org/web/2007/09/high_school_sports_participation.aspx.

National Sporting Goods Association (NSGA). (2004). *Sports participation in 2002: Series I and II, April, 2003*. www.nsga.org/public/pages/index.cfm?pageid=732.

National Sporting Goods Association (NSGA). (2007). Sports participation 2006. www.nsga.org/i4a/pages/index.cfm?pageid=1.

National Sporting Goods Association (NSGA). (2007). Sports participation reports, "2006 vs. 2001 participation ranked by percent change." www.nsga.org.

Neal, M.A. (2006). *New black man*. New York: Routledge.

Nelson, R.C. (1970). Biomechanics of sport: An overview. In J.M. Cooper (Ed.), *Selected topics on biomechanics: Proceedings of the C.I.C. Symposium on Biomechanics* (pp. 31–37). Chicago: Athletic Institute.

Netz, Y. (2007). Physical activity and three dimensions of psychological functioning in advanced age: Cognition, affect, and self-perception. In G. Tenenbaum & R.C. Eklund (Eds.), *Handbook of sport psychology* (3rd ed: pp. 492–508). New York: Wiley.

Neufeldt, V. (Ed.). (1988). *Webster's new world dictionary* (3rd college edition). New York: Simon & Schuster.

Neulinger, J., & Raps, C. (1972). Leisure attitude of an intellectual elite. *Journal of Leisure Research, 4,* 196–207.

Newell, K.M. (1990a). Physical activity, knowledge types, and degree programs. *Quest, 42,* 243–268.

Newell, K.M. (1990b). Kinesiology: The label for the study of physical activity in higher education. *Quest, 42,* 269–278.

Newell, K.M. (2007, February). Kinesiology: Challenges of multiple agendas. The Academy Papers. *Quest, 59*(1), 5-24.

Nichols, D.L., Sanborn, C.F., & Essery, E.V. (2007). Bone density and young athletic women—an update. *Sports Medicine, 37*(11), 1001–1014.

Nideffer, R., & Sagal, M. (2001). Concentration and attention control training. In J.M. Williams (Ed.), *Applied sport psychology: Personal growth to peak performance* (4th ed., pp. 312–332). Mountain View, CA: Mayfield.

Nideffer, R.M., & Sagal, M. (2006). Concentration and attentional control. In J.M. Williams (Ed.), *Applied sport psychology: Personal growth to peak performance* (5th ed., pp. 349–381). Boston: McGraw-Hill.

Nigg, B.M. (1983). External force measurements with sports shoes and playing surfaces. In B.M. Nigg & B.J. Kerr (Eds.), *Biomechanical aspects of sport shoes and playing surfaces*. Calgary, AB: University Printing.

Nightmare in the backfield. (2006, November 17). *The Week,* 52-53. Reprinted from M. Lewis, 2006, *The blind side*. New York: Norton.

NIH Consensus Development Panel on Physical Activity and Cardiovascular Health. (1996). Physical activity and cardiovascular health. *Journal of the American Medical Association, 276,* 241–246.

North American Society for Sport Management (NASSM). (2007). Sport management programs. www.nassm.com/InfoAbout/SportMgmtPrograms.

Novak, M. (1976). *The joy of sports*. New York: Basic Books.

O'Donnell, M. (1989). Definition of health promotion. Part III: Expanding the definition. *American Journal of Health Promotion, 3*(3), 5.

O'Donnell, M.P., & Harris, J.S. (1994). *Health promotion in the workplace*. Albany, NY: Delmar.

Ogilvie, B. (1973, November). The stimulus addicts. *Physician and Sports Medicine,* 61–65.

Osterhoudt, R.G. (1991). *The philosophy of sport: An overview*. Champaign, IL: Stipes.

O'Sullivan, M., & Dyson, B. (1994). Rules, routines, and expectations of 11 high school physical education teachers. In M. O'Sullivan (Ed.), High school physical education teachers: Their world of work [Monograph]. *Journal of Teaching in Physical Education, 13,* 361–374.

Paffenbarger, R.S. (1994). 40 years of progress: Physical activity, health and fitness. In *40th anniversary lectures* (pp. 93–109). Indianapolis: American College of Sports Medicine.

Panzer, V.P. (1987). *Dynamic assessment of lower extremity loading characteristics during landing*. Unpublished doctoral dissertation, University of Oregon, Eugene.

Park, R.J. (1980). The *Research Quarterly* and its antecedents. *Research Quarterly for Exercise and Sport, 51*(1), 1–22.

Park, R.J. (1981). The emergence of the academic discipline of physical education in the United States. In G.A. Brooks (Ed.), *Perspectives on the academic discipline of physical education* (pp. 20–45). Champaign, IL: Human Kinetics.

Park, R.J. (1987a). Physiologists, physicians, and physical educators: Nineteenth century biology and exercise, hygienic and educative. *Journal of Sport History, 14*(1), 28–60.

Park, R.J. (1987b). Sport, gender and society in a transatlantic Victorian perspective. In J.A. Mangan & R.J. Park (Eds.), *From "fair sex" to feminism: Sport and the socialization of women in the industrial and post-industrial eras* (pp. 58–93). London: Frank Cass.

Park, R.J. (1989). The second 100 years: Or, can physical education become the renaissance field of the 21st century? *Quest, 41*(1), 2–27.

Parker, S.T. (1984). Playing for keeps: An evolutionary perspective on human games. In P.K. Smith (Ed.), *Play in animals and humans* (pp. 271–293). New York: Blackwell.

Pate, R.R. (1988). The evolving definition of fitness. *Quest, 40,* 174–179.

Pate, R.R., Pratt, M., Blair, S.N., Haskell, W.L., Macera, C.A., Bouchard, C., Buchner, D., Ettinger, W., Heath, G.W., King, A.C., Kriska, A., Leon, A.S., Marcus, B.H., Morris, J., Paffenbarger, R.S., Patrick, K., Pollock, M.L., Rippe, J.M., & Wilmore, J.H. (1995). Physical activity and public health: A recommendation from the Centers for Disease Control and the American College of Sports Medicine. *Journal of the American Medical Association, 273,* 402–407.

Paterson, D.H., Jones, G.R., & Rice, C.L. (2007). Ageing and physical activity: Evidence to develop exercise recommendations for older adults. *Canadian Journal of Public Health, 98,* Suppl 2, S69–S108.

Patient page: Fitness. (2005, December 21). *Journal of the American Medical Association 294*(23), 3048. http://jama.ama-assn.org/cgi/content/full/294/23/3048#JPG1221F1#JPG1221F1.

Pavlou, K.N., Krey, S., & Steffee, W.P. (1989). Exercise as an adjunct to weight loss and maintenance in moderately obese subjects. *American Journal of Clinical Nutrition, 49,* 1115–1123.

Pedowitz, D.I., Reddy, S., Parekh, S.G., Huffman, G.R., & Sennett, B.J. (2008). Prophylactic bracing decreases ankle injuries in collegiate female volleyball players. *American Journal of Sports Medicine, 36*(2), 324–7.

Peiss, K. (1986). *Cheap amusements: Working women and leisure in turn-of-the-century New York.* Philadelphia: Temple University Press.

Perry, C. (1983). Blood doping and athletic competition. *International Journal of Applied Philosophy, 1*(3), 39–45.

Petrie, A. (1967). *Individuality in pain and suffering.* Chicago: University of Chicago Press.

Pew Health Professions Commission. (1995). *Critical challenges: Revitalizing the health professions for the twenty-first century. The third report of the Pew Health Professions Commission.* San Francisco: Author.

Phenix, P.H. (1964). *Realms of meaning* (p. 171). New York: McGraw-Hill.

Pieper, J. (1952). *Leisure: The basis of culture.* New York: Pantheon Books.

Pilcher, J.J., & Walters, A.S. (1997). How sleep deprivation affects psychological variables. *Journal of American College Health, 46*(3), 121–126.

Pitts, B.G., Fielding, L.W., & Miller, L.K. (1994). Industry segmentation theory and the sport industry: Developing a sport industry segment model. *Sport Marketing, 3*(4), 15–28.

Pope, S.W. (1997). Introduction: American sport history—toward a new paradigm. In S.W. Pope (Ed.), *The new American sport history: Recent approaches and perspectives* (pp. 1–30). Urbana, IL: University of Illinois Press.

Posse, N. (1890). *The special kinesiology of educational gymnastics.* Boston: Lothrop, Lee and Shepard.

Poulton, E.C. (1957). On prediction in skilled movements. *Psychological Bulletin, 54,* 467–479.

Prentice, W.E. (2006). *Arnheim's principles of athletic training: A competency-based approach* (12th ed.). New York: McGraw-Hill.

Prochaska, J.O., & DiClemente, C.C. (1986). Toward a comprehensive model of change. In W. Miller & N. Heather (Eds.), *Treating addictive behaviors.* New York: Plenum Press.

Rader, B.G. (1990). *American sports: From the age of folk games to the age of televised sports.* Englewood Cliffs, NJ: Prentice Hall.

Radin, E.L., Martin, R.B., Burr, D.B., Caterson, B., Boyd, R.D., & Goodwin, C. (1985). Mechanical factors influencing cartilage damage. In J.G. Peyron (Ed.), *Osteoarthritis: Current clinical and fundamental problems* (pp. 90–99). Paris, France: CIBA-Geigy.

Radin, E.L., Yang, K.H., Riegger, C., Kish, V.L., & O'Connor, J.J. (1991). Relationship between lower limb dynamics and knee joint pain. *Journal of Orthopedic Research, 9,* 398–405.

Raedeke, T.D. (1997). Is athlete burnout more than just stress? A sport commitment perspective. *Journal of Sport & Exercise Psychology, 19,* 396–417.

Ramlow, J., Kriska, A., & LaPorte, R. (1987). Physical activity in the population: The epidemiologic spectrum. *Research Quarterly for Exercise and Sport, 58*(2), 111–113.

Ravizza, K. (1984). Qualities of the peak experience in sport. In J.M. Silva & R.S. Weinberg (Eds.), *Psychological foundations of sport* (pp. 452–462). Champaign, IL: Human Kinetics.

Regalado, S.O. (1992). Sport and community in California's Japanese American "Yamato Colony," 1930–1945. *Journal of Sport History, 19*(2), 130–143.

Reid, H.L. (2002). *The philosophical athlete* (p. 52). Durham, NC: Carolina Academic Press.

Rice, E.A., Hutchinson, J.L., & Lee, M. (1969). *A brief history of physical education.* New York: Ronald.

Rink, J.E. (2006). *Teaching physical education for learning* (5th ed.). Boston: McGraw-Hill.

Roberton, M.A. (1984). Changing motor patterns during childhood. In J.R. Thomas (Ed.), *Motor development during childhood and adolescence* (p. 75). Edina, MN: Burgess International.

Roberts, D.F., Foehr, U.G., Rideout, V.J., & Brodie, M. (1999). *Kids and media at the new millennium: A comprehensive national analysis of children's media use.* Menlo Park, CA: Kaiser Family Foundation.

Rosenbaum, D.A. (1991). *Human motor control.* San Diego: Academic Press.

Ross, H. (1996, March 10). Waiting for the call . . . . *Greensboro News and Record*, p. C1.

Rowland, T.W. (1989). Oxygen uptake and endurance fitness in children: A developmental perspective. *Pediatric Exercise Science, 1,* 313–328.

Rudolph, F. (1962). *The American college and university: A history.* New York: Vintage.

Sachs, M.L. (1981). Running addiction. In M.H. Sacks & M.L. Sachs (Eds.), *Psychology of running* (pp. 116–125). Champaign, IL: Human Kinetics.

Sage, G.H. (1997). Sport sociology. In J.D. Massengale & R.A. Swanson (Eds.), *History of exercise and sport science* (pp. 109–141). Champaign, IL: Human Kinetics.

Sallis, J.F., Haskell, W.L., Fortnam, S.P., Vranizan, M.S., Taylor, C.B., & Solomon, D.S. (1986). Predictors of adoption and maintenance of physical activity in a community sample. *Preventive Medicine, 15,* 331–341.

Sallis, J.F., & Hovell, M.G. (1990). Determinants of exercise behavior. In K.B. Pandolf (Ed.), *Exercise and sport science reviews* (Vol. 18, pp. 307–330). Baltimore: Williams & Wilkins.

Salmoni, A.W., Schmidt, R.A., & Walter, C.B. (1984). Knowledge of results and motor learning: A review and critical reappraisal. *Psychological Bulletin, 95,* 355–386.

Saltin, B. (1973). Metabolic fundamentals in exercise. *Medicine and Science in Sports, 5,* 137–146.

Santa Barbara Athletic Club Cancer Well-Fit program. (2009). www.cancerwellfit.com.

Sartre, J.-P. (1956). *Being and nothingness* (p. 59). Translated by Hazel E. Barnes. New York: Philosophical Library. Quoted in H.L. Reid, 2002, *The philosophical athlete* (p. 31). Durham, NC: Carolina Academic Press.

Scanlan, T.K., Stein, G.L., & Ravizza, K. (1988). An in-depth study of former elite figure skaters: II. Sources of enjoyment. *Journal of Sport Psychology,* 65–83.

Schempp, P.G. (1989). Apprenticeship-of-observation and the development of physical education teachers. In T.J. Templin & P.G. Schempp (Eds.), *Socialization into physical education: Learning to teach* (pp. 13–38). Indianapolis: Benchmark.

Schmidt, R.A. (1975). A schema theory of discrete motor skill learning. *Psychological Review, 82,* 225–260.

Schmidt, R.A. (1988). *Motor control and learning: A behavioral emphasis* (2nd ed.). Champaign, IL: Human Kinetics.

Schmidt, R.A. (1991). *Motor learning & performance: From principles to practice.* Champaign, IL: Human Kinetics.

Schneider, R., & Bradish, C.L. (2006). Location, location, location: The marketing of place and Super Bowl XL. *Sport Marketing Quarterly, 15,* 206–213.

Schoeni, R.F., Martin, L.G., Andreski, P.M., & Freedman, V.A. (2005). Persistent and growing socioeconomic disparities in disability among the elderly: 1982-2002. *American Journal of Public Health, 95*(11), 2065–2070.

Schon, D.A. (1995, November/December). Knowing in action: The new scholarship requires a new epistemology. *Change,* 27–34.

Schönau, E., Werhahn, E., Schiedermaier, U., Mokow, E., Schiessl, H., Scheidhauer, K., & Michalk, D. (1996). Influence of muscle strength on bone strength during childhood and adolescence. *Hormone Research, 45*(Suppl. 1): 63–66.

Seefeldt, V., & Haubenstricker, J. (1982). Patterns, phases, or stages: An analytical model for the study of developmental movement. In J.A.S. Kelso & J.E. Clark (Eds.), *The development of movement control and co-ordination* (pp. 309–318). Chichester, UK: Wiley.

Shachmut, K. (2007). Safeway: Business call to action for healthcare reform. Paper presented at Sutter Health Governance Symposium, San Francisco, October 19.

Shafer, R.J. (Ed.). (1980). *A guide to historical method.* Homewood, IL: Dorsey.

Sharp, L.A., Moorman, A.M., & Claussen, C.L. (2007). *Sport law: A managerial approach.* Scottsdale, AZ: Holcomb Hathaway.

Sheehan, G. (1978). *Running and being: The total experience.* New York: Warner Books.

Sherrington, C.S. (1906). *The integrative action of the nervous system.* New Haven, CT: Yale University Press.

Sherrington, C.S. (1940). *Man on his nature* (p. 107). Cambridge University Press.

Shi, J., & Ewing, M. (1993). Definition of fun for youth soccer players. *Journal of Sport and Exercise Psychology* (NASPSPA Abstracts), *15,* S74.

Shields, D.L., & Bredemeier, B.J. (2007). Advances in sport morality research. In G. Tenenbaum & R.C. Eklund (Eds.), *Handbook of sport psychology* (3rd ed., pp. 662–684). Hoboken, NJ: Wiley.

Shropshire, K. (1996). *In black and white: Race and sports in America.* New York: New York University Press.

Shulman, J.L. (2001). *The Game of Life.* Princeton, NJ: Princeton University Press.

Siedentop, D. (1990). Commentary: The world according to Newell. *Quest, 42,* 315–322.

Siedentop, D. (1996). Valuing the physically active life: Contemporary and future directions. *Quest, 48,* 266–274.

Siedentop, D. (2004). *Sport education* (2nd ed). Champaign, IL: Human Kinetics.

Siedentop, D., & Tannehill, D. (2000). *Developing Teaching Skills in Physical Education* (4th ed.). Mountain View, CA: Mayfield.

Silva, J.M. (1990). An analysis of the training stress syndrome in competitive athletics. *Journal of Applied Sport Psychology, 2,* 5–20.

Simon, R.L. (2004). *Fair play: The ethics of sport* (2nd ed.). Boulder, CO: Westview Press.

Simonsick, E.M. (2005). Just get out the door! Importance of walking outside the home for maintaining mobility: Findings from the women's health and aging study. *Journal of the American Geriatric Society, 53*(2), 198–203.

Simpson, K.J., & Kanter, L. (1997). Bone-on-bone forces at the ankle and knee joints during dance landings-I: Factors influencing axial force parameters. *Medicine and Science in Sports and Exercise, 29*, 916–927.

Simpson, K.J., & Pettit, M. (1997). Forces at the ankle and knee joints during dance landings-II: Factors influencing shear force parameters. *Medicine and Science in Sports and Exercise, 29*, 928–937.

Sims, J., Hill, K., Davidson, S., Gunn, J., & Huang, N. (2006). Exploring the feasibility of a community-based strength training program for older people with depressive symptoms and its impact on depressive symptoms. *BMC Geriatrics, 30*(6), 18.

Sinha, R., Fisch, G., Teague, B., Tamborlane, W.V., Banyas, B., Allen, K., Savoye, M., Rieger, V., Taksali, S., Barbetta, G., Sherwin, R.S., & Caprio, S. (2002). Prevalence of impaired glucose tolerance among children and adolescents with marked obesity. *New England Journal of Medicine, 345,* 802–810.

Skarstrom, W. (1909). *Gymnastic Kinesiology.* Springfield, MA: Bassette.

Slusher, H.S. (1967). *Man, sport and existence: A critical analysis.* Philadelphia: Lea & Febiger.

Smith, J.K. (1979). *Athletic training: A developing profession.* Unpublished master's thesis, Brigham Young University, Provo, Utah.

Smith, R.A. (1988). *Sports and freedom: The rise of big-time college athletics.* New York: Oxford University Press.

Smith, R.E., & Christensen, D.S. (1995). Psychological skills as predictors of performance and survival in professional baseball. *Journal of Sport and Exercise Psychology, 17,* 399–415.

Smith, R.E., Schutz, R.W., Smoll, F.L., & Ptacek, J.T. (1995). Development and validation of a multidimensional measure of sport-specific psychological skills: The athletic coping skills inventory-28. *Journal of Sport and Exercise Psychology, 17,* 379–398.

Smoll, F.L., & Schutz, R.W. (1980). Children's attitude toward physical activity: A longitudinal analysis. *Journal of Sport Psychology, 2,* 137–147.

Snyder, E., & Spreitzer, E. (1976). Correlates of sport participation among adolescent girls. *Research Quarterly, 47,* 804–809.

Solomon, M.A., Worthy, T., Lee, A., & Carter, J.A. (1991). Teacher role identity of student teachers in physical education: An interactive analysis. *Journal of Teaching Physical Education, 10,* 188–209.

Spirduso, W.W., & MacRae, P.G. (1990). Motor performance and aging. In J.E. Birren & K.W. Schaie (Eds.), *The handbook of psychology of aging* (3rd ed., pp. 184–197). San Diego: Academic Press.

SportAccord. (2007). Don Schumacher, executive director, National Association of Sports Commissions. www.sportcentric.com/vsite/vcontent/page/custom/0,8510,5035-179611-196829-39857-264474-custom-item,00.html.

Sport Management Research Institute. (2007). 2007 SBXLI market analysis and economic impact investigation. http://media.miamiherald.com/smedia/2007/05/23/22/2007_SBXLI_Economic_Impact_Investigation_Executive_Summary.source.prod_affiliate.56.pdf.

Staley, S.C. (1937). The history of sport: A new course in the professional training curriculum. *Journal of Health, Physical Education and Recreation, 8,* 522–525, 570–572.

State-Specific Prevalence of Obesity Among Adults—United States, 2005. (Sept. 15, 2006). *Morbidity and Mortality Weekly Report, 55*(36), 985–988.

Staurowsky, E.J. (1999, November). American Indian sport imagery and the miseducation of Americans. *Quest, 51,* 382–392.

Staurowsky, E.J. (2004). Privilege at play: On the legal and social fictions that sustain American Indian sport imagery. *Journal of Sport & Social Issues, 28,* 11–29.

Stein, P.J., & Hoffman, S. (1978). Sports and male role strain. *Journal of Social Issues, 34,* 136–150.

Steindler, A. (1942, November). What has biokinetics to offer to the physical educator? *Journal of Health and Physical Education,* 507–509, 555–556.

Stelmach, G.E., & Nahom, A. (1992). Cognitive-motor abilities of the elderly driver. *Human Factors, 34*(1), 53–65.

Stephens, T. (1987). Secular trends in adult physical activity: Exercise boom or bust? *Research Quarterly for Exercise and Sport, 58*(2), 94–105.

Stephens, T., & Caspersen, C.J. (1994). The demography of physical activity. In C. Bouchard, R.J. Shephard, & T. Stephens (Eds.), *Physical activity, fitness, and health: International proceedings and consensus statement* (pp. 204–213). Champaign, IL: Human Kinetics.

Stoldt, C., Pratt, C., & Jackson, J. (2003). Public relations in the sport industry. In J.B. Parks & J. Quarterman (Eds.), *Contemporary sport management* (2nd ed., pp. 211–230). Champaign, IL: Human Kinetics.

Stoldt, G.C., Dittmore, S.W., & Branvold, S.E. (2006). *Sport public relations.* Champaign, IL: Human Kinetics.

Strasburger, V.C. (1992). Children, adolescents, and television. *Pediatrics in Review, 13,* 144–151.

Street & Smith's SportsBusiness Journal. (2007). The sport industry. www.sportsbusinessjournal.com/index.cfm?fuseaction=page.feature&featureId=1492.

Stroot, S. (1996). Organizational socialization: Factors impacting beginning teachers. In S. Silverman & C. Ennis (Eds.), *Studying learning in physical education: Applying research to enhance instruction* (pp. 339–366). Champaign, IL: Human Kinetics.

Stroot, S.A., & Oslin, J.L. (1993). Use of instructional statements by preservice teachers for overhand throwing performance of children. *Journal of Teaching in Physical Education, 13,* 24–45.

Struna, N.L. (1996a). Historical research in physical activity. In J.R. Thomas & J.K. Nelson (Eds.), *Research methods in physical activity* (pp. 251–275). Champaign, IL: Human Kinetics.

Struna, N.L. (1996b). *People of prowess: Sport, leisure, and labor in early Anglo-America.* Urbana, IL: University of Illinois Press.

Struna, N.L. (1997). Sport history. In J.D. Massengale & R.A. Swanson (Eds.), *The history of exercise and sport science* (pp. 143–179). Champaign, IL: Human Kinetics.

Suits, B. (1978). *The grasshopper: Games, life and utopia.* Toronto: University of Toronto Press.

Super Bowl XLI Host Committee. (2006). *Super Bowl XLI host committee media guide.* Miami Gardens, FL: Author.

Swanson, R.A., & Spears, B. (1995). *History of sport and physical education in the United States.* Madison, WI: Brown & Benchmark.

Taks, M., Girginov, V., & Boucher, R. (2006). The outcomes of coattail marketing: The case of Windsor, Ontario, and Super Bowl XL. *Sport Marketing Quarterly, 15,* 232–242.

Thacker, S.B., Stroup, D.F., Branche, C.M., Gilchrist, J., Goodman, R.A., & Weitman, E.A. (1999). The prevention of ankle sprains in sports: A systematic review of the literature. *American Journal of Sports Medicine, 27*(6), 753–760.

Thelen, E., Ulrich, D., & Jensen, J.L. (1990). The developmental origins of locomotion. In M.H. Woollacott and A. Shumway-Cook (Eds.), *Development of posture and gait: Across the lifespan.* Columbia, SC: University of South Carolina Press.

Thomas, C.E. (1983). *Sport in a philosophic context.* Philadelphia: Lea & Febiger.

Thomas, J.R. (1997). History of motor behavior. In J.D. Massengale & R.A. Swanson (Eds.), *History of exercise and sport sciences.* Champaign, IL: Human Kinetics.

Thomas, J.R. (2006). Motor behavior: From telegraph keys and twins to linear slides and stepping. *Quest, 58,* 112–127.

Thomas, J.R., & French, K.E. (1985). Gender differences across age in motor performance: A meta-analysis. *Psychological Bulletin, 98,* 260–282.

Thomas, J.R., & Thomas, K.T. (1989). What is motor development: Where does it belong? *Quest, 41,* 203–212.

Thomas, J.T. (1969). Keynote address—the Army looks at biomechanics. In D. Bootzin &

H.C. Muffley (Eds.), *Biomechanics.* New York: Plenum Press.

Thomas, K.T., Gallagher, J.D., & Thomas, J.R. (2001). Motor development and skill acquisition during childhood and adolescence. In R.N. Singer, H.A. Hausenblas, & C.M. Janelle (Eds.), *Handbook of sport psychology* (2nd ed., pp. 20–52). New York: Wiley & Sons.

Thomas, K.T., & Thomas, J.R. (1994). Developing expertise in sport: The relation of knowledge to performance. *International Journal of Sport Psychology, 25,* 295–312.

Thomas, K.T., & Thomas, J.R. (2008). *Elementary School Journal, 108,* 181–195.

Thompson, P.B. (1982). Privacy and the urinalysis testing of athletes. *Journal of the Philosophy of Sport, 9,* 60–65.

Timson, B.F., Bowlin, B.K., Dudenhoeffer, G.A., & George, J.B. (1985). Fiber number, area, and composition of mouse soleus muscle following enlargement. *Journal of Applied Physiology, 58,* 619–624.

Todd, T. (1987). Anabolic steroids: The gremlins of sport. *Journal of Sport History, 14,* 87–107.

Torres, C. (2000). What counts as part of the game? A look at skills. *Journal of the Philosophy of Sport, 27,* 81–92.

Torres, C. (2002). *Play as expression: An analysis based on the philosophy of Maurice Merleau-Ponty* (Doctoral dissertation, The Pennsylvania State University, 2002).

Torres, C., & McLaughlin, D. (2003). Indigestion? An apology for ties. *Journal of the Philosophy of Sport, 30,* 144–158.

Tousignant, M. (1981). *A qualitative analysis of task structures in required physical education.* Unpublished doctoral dissertation, Ohio State University, Columbus.

Tyson, L. (1996). Context of schools. In S. Silverman & C. Ennis (Eds.), *Studying learning in physical education: Applying research to enhance instruction* (pp. 55–80). Champaign, IL: Human Kinetics.

Ulrich, B. (2007). Motor development: Core curricular concepts. *Quest, 59,* 77–91.

Ulrich, B., & Reeve, T.G. (2005). Studies in motor behavior: 75 years of research in motor development, learning, and control. *Research Quarterly for Exercise and Sport, 75,* S62–S70.

University of Florida. (2007). 2006-07 undergraduate catalog. www.registrar.ufl.edu/catalog0607/programs/courses/tourism.html.

University of Minnesota. (2007). Twin cities courses. http://onestop2.umn.edu/courses/courses.jsp?designator=REC&submit=Show+the+courses&institution=UMNTC.

U.S. Bureau of Labor Statistics (USBLS). (1961). *Occupational outlook handbook* (Bulletin No. 1300). Washington, DC: U.S. Government Printing Office.

U.S. Bureau of Labor Statistics (USBLS). (1982). *Occupational outlook handbook* (Bulletin No. 2200). Washington, DC: U.S. Government Printing Office.

U.S. Bureau of Labor Statistics (USBLS). (2008). *Occupational outlook handbook, 2008-09.* www.bls.gov/oco/home.htm.

U.S. Bureau of the Census (USBOC). (1961). *Statistical abstract of the United States, 1961.* Washington, DC: U.S. Government Printing Office.

U.S. Bureau of the Census (USBOC). (1966). *Statistical abstract of the United States, 1966.* Washington, DC: U.S. Government Printing Office.

U.S. Bureau of the Census (USBOC). (1996). *Statistical abstract of the United States, 1996.* Washington, DC: U.S. Government Printing Office.

U.S. Bureau of the Census (USBOC). (2002a). Selected spectator sports: 1985 to 2000. *Statistical abstract of the United States, 2002* (p. 754). Washington, DC: Government Printing Office.

U.S. Bureau of the Census (USBOC). (2002b). Participation in various leisure activities: 1997. *Statistical abstract of the United States, 2002* (p. 751). Washington, DC: Government Printing Office.

U.S. Bureau of the Census (USBOC). (2002c). National health expenditures—summary, 1960 to 2000, and projections, 2001 to 2011. *Statistical abstract of the United States, 2002* (p. 91). Washington, DC: Government Printing Office.

U.S. Bureau of the Census (USBOC). (2002d). Participation in selected sports activities: 2000. *Statistical abstract of the United States, 2002* (p. 756). Washington, DC: Government Printing Office.

U.S. Bureau of the Census (USBOC). (2007a). Participation in selected sport activities: 2004. *Statistical abstract of the United States* (p. 256). Washington, DC: Government Printing Office.

U.S. Bureau of the Census (USBOC). (2007b). Participation in selected sport activities: 2005. (Table 1222). www.census.gov/compendia/statab/tables/08s1222.pdf.

U.S. Bureau of the Census (USBOC). (2007c). *Statistical abstract of the United States, 2007.* Washington, DC: U.S. Government Printing Office.

U.S. Bureau of the Census. (2009). Census Bureau Estimates Nearly Half of Children Under Age 5 are Minorities: *Estimates find nation's population growing older, more diverse.* www.census.gov/Press-Release/www/releases/archives/population/013733.html.

U.S. Department of Health and Human Services. (1990). *Healthy People 2000.* (DHHS Publication No. [PHS] 91-50213). Hyattsville, MD: Public Health Service.

U.S. Department of Health and Human Services. Healthy People 2010: Understanding and *Improving Health.* 2nd ed. Washington, DC: U.S. Government Printing Office, November 2000.

U.S. Department of Health and Human Services (USDHHS). (1996). *Physical activity and health: A report of the surgeon general.* Atlanta: U.S. Department of Health and Human Services, Centers for Disease Control and Prevention, National Center for Chronic Disease Prevention and Health Promotion.

U.S. Department of Health and Human Services (USDHHS). (1997, July). *Musculoskeletal disorders (MSAs) and workplace factors: A critical review of epidemiologic evidence for work-related musculoskeletal disorders of the neck, upper extremity, and low back.* B.P. Bernard (ed). Cincinnati: National Institute for Occupational Safety and Health. www.cdc.gov/niosh/ergosci.

U.S. Department of Health and Human Services (USDHHS). (2000, November). *Healthy People 2010: Understanding and improving health* (2nd ed.) Washington, DC: U.S. Government Printing Office.

U.S. Department of Health and Human Services (USDHHS). (2001). *The surgeon general's call to action to prevent and decrease overweight and obesity.* Washington, DC: U.S. Government Printing Office.

U.S. Department of Health and Human Services (USDHHS). (2002). *Healthy people 2010* (Vol. 2). McLean, VA: International Medical.

U.S. Department of Health and Human Services (USDHHS). (2003). *Health, United States, 2003: Trends in the health of Americans.* Hyattsville, MD: National Center for Health Statistics, Division of Data Services.

U.S. Department of Health and Human Services (USDHHS). (2005). Summary health statistics for U.S. adults: National Health Interview Survey, 2005. www.cdc.gov/nchs/data/series/sr_10/sr10_232.pdf.

User fees, insurance rate increases and access to recreation facilities. (2003, November). *Physical Training.* http://ejmas.com/pt/ptart_orfa_1103.html.

Van Dalen, D.B., & Bennett, B.L. (1971). *A world history of physical education: Cultural, philosophical, comparative.* Englewood Cliffs, NJ: Prentice Hall.

van der Smissen, B. (1990). *Legal liability and risk management for public and private entities.* Cincinnati: Anderson.

Vealey, R.S., & Greenleaf, C.A. (2006). Seeing is believing: Understanding and using imagery in sport. In J.M. Williams (Ed.), *Applied sport psychology: Personal growth to peak performance* (5th ed., pp. 306–348). Boston: McGraw-Hill.

Vertinsky, P. (1990). *The eternally wounded woman: Women, doctors and exercise in the late nineteenth century.* Manchester, UK: Manchester University Press.

Walling, M., & Martinek, T. (1995). Learned helplessness: A case study of a middle school student. *Journal of Teaching in Physical Education, 14,* 454–466.

Wankel, L.M. (1985). Personal and situational factors affecting exercise involvement: The importance of enjoyment. *Research Quarterly, 56*(3), 275–282.

Wankel, L.M. (1988). Exercise adherence and leisure activity: Patterns of involvement and interventions to facilitate regular activity. In R.K. Dishman (Ed.), *Exercise adherence: Its impact on public health* (pp. 369–396). Champaign, IL: Human Kinetics.

Wankel, L.M., & Krissel, P.S.J. (1985). Methodological considerations in youth sport motivation research: A comparison of open-ended and paired comparison approaches. *Journal of Sport Psychology, 7,* 65–74.

Wardlaw, F.B., McGuire, F.A., & Overby, Z. (2000). Therapeutic recreation: Optimal health treatment for orthopaedic disability. *Orthopedic Nursing, 19,* 56–60.

Weiss, M.R. (Ed.). (2004). *Developmental sport and exercise psychology: A lifespan perspective.* Morgantown, WV: Fitness Information Technology.

Weiss, M.R., & Klint, K.A. (1987). "Show and tell" in the gymnasium: An investigation of developmental differences in modeling and verbal rehearsal of motor skills. *Research Quarterly for Exercise and Sport, 58,* 234–241.

Weiss, P. (1969). *Sport: A philosophic inquiry.* Carbondale, IL: Southern Illinois University Press.

Wertz, S. (1991). *Talking a good game: Inquiries into the principles of sport* (p. 190). Dallas: Southern Methodist University Press.

Whorton, J.C. (1982). *Crusaders for fitness: The history of American health reformers.* Princeton, NJ: Princeton University Press.

Wichita State University. (2007). 2007-08 Wichita State University undergraduate catalog. www.collegesource.org/displayinfo/catalink.asp?pid={5E9F17E9-28A8-45B9-AED9-459110B4A572}&oig={78AC3888-D5B8-4B8D-8F51-BCB5769AFEB3}&vt=5.

Wiggins, D.K. (1980). The play of slave children in the plantation communities of the old south, 1820–1860. *Journal of Sport History, 7*(2), 21–39.

Willett, W.C., Manson, J.E., Stampler, M.J., Colditz, G.A., Rosner, B., Speizer, F.E., & Hennekena, C.H. (1995). Weight, weight change, and coronary disease in women. *Journal of the American Medical Association, 273,* 461–465.

Williams, J.F. (1964). *The principles of physical education* (8th ed.). Philadelphia: Lea & Febiger. (Original work published 1927.)

Williams, K.R. (1993). Biomechanics of distance running. In M.D. Grabiner (Ed.), *Current issues in biomechanics* (pp. 3–32). Champaign, IL: Human Kinetics.

Williams, K.R., & Cavanagh, P.R. (1987). Relationship between distance running mechanics, running economy, and performance. *Journal of Applied Physiology, 63*(3), 1236–1245.

Willis, J.D., & Campbell, L.F. (1992). *Exercise psychology.* Champaign, IL: Human Kinetics.

Willis, P. (1982). Women in sport in ideology. In J. Hargreaves (Ed.), *Sport, culture and ideology* (pp. 117–135). London: Routledge & Kegan Paul.

Wilmore, J.H., & Costill, D.L. (1994). *Physiology of sport and exercise.* Champaign, IL: Human Kinetics.

Wilson, B., & Sparks, R. (1996). "It's gotta be the shoes": Youth, race, and sneaker commercials. *Sociology of Sport Journal, 13,* 398–427.

Wilson, K.S., & Spink, K.S. (2006). Exploring older adults' social influences for physical activity. *Activities, Adaptation and Aging, 30*(3), pp. 47–60.

Wilson, W. (1977). Social discontent and the growth of wilderness sport in America: 1965–1974. *Quest, 27,* 54–60.

Winter, D.A. (1985). *Biomechanics and motor control of human gait.* Waterloo, ON: University of Waterloo Press.

Winter, D.A. (1991). *The biomechanics and motor control of human gait: Normal, elderly, and pathological* (2nd ed.). Waterloo, ON: University of Waterloo Press.

Women's Sports Foundation Web page. (2008). Issues and actions. www.womenssportsfoundation.org/cgi-bin/iowa/issues/article.html?record=1017.

Woods, A.M., & Graber, K.C. (Eds.). (2007). Stepping up to the plate: Physical educators as advocates for wellness policies—Part I [Special Feature]. *Journal of Physical Education, Recreation & Dance, 78*(5), 17–28.

Xiang, P., McBride, R., & Bruene, R. (2003). Relation of parent's beliefs to children's motivation in an elementary physical education running program. *Journal of teaching in physical education, 22*(4), 410–425.

YMCA. (2007). About the YMCA. www.ymca.net/about_the_ymca/.

YMCA Canada. (2007). Frequently asked questions. www.ymca.ca/eng_faq.htm.

Zaichowsky, L.B. (1975). Attitudinal differences in two types of physical education programs. *Research Quarterly, 46,* 364–370.

Zakarian, J.M., Hovell, M.F., Hofstettere, C.R., Sallis, J.F., & Keating, K.J. (1994). Correlates of vigorous exercise in a predominately low SES and minority high school population. *Preventive Medicine, 23,* 314–321.

Zillman, D., Bryant, J., & Sapolsky, B.S. (1979). The enjoyment of watching sport contests. In J.H. Goldstein (Ed.), *Sports, games and play: Social and psychological viewpoints* (pp. 297–336). Hillsdale, NJ: Lawrence Erlbaum.

Zillman, D., & Cantor, J.R. (1976). A disposition theory of humor and mirth. In T. Chapman & H. Foot (Eds.), *Humor and laughter: Theory, research, and applications.* London: Wiley.

Zweigenhaft, R.L. (1977). The empirical study of signature size. *Social Behavior and Personality, 5*(1), 177–185.

Zwiren, L.D. (1989). Anaerobic and aerobic capacities of children. *Pediatric Exercise Science, 1,* 31–44.

INDEX

Note: The italicized *f* and *t* following page numbers refer to figures and tables, respectively.

D

R

ABOUT THE EDITOR

Shirl J. Hoffman, EdD, is professor emeritus of exercise and sport science at the University of North Carolina at Greensboro, where he served as head of the department for 10 years. He has taught at all levels of education, beginning his career as an elementary physical education teacher before moving on to positions as head basketball coach at Westchester Community College in New York and then as assistant professor in a liberal arts institution and later at a comprehensive state university (University of Nebraska at Omaha). He taught for 13 years at University of Pittsburgh, where he also was director of graduate studies, followed by a 22-year career at University of North Carolina at Greensboro. He has an extraordinarily broad background in the field that spans motor learning and performance, sociology of sport, and sport philosophy. His work has appeared in *Journal of Motor Behavior, Perceptual and Motor Skills, Research Quarterly for Exercise and Sports, Journal of Physical Education, Recreation and Dance, Quest,* and *Journal of Sport Philosophy.* He has spoken to such wide-ranging groups as The Canadian Psychomotor and Sport Psychology Symposium, the Research Consortium of AAHPERD, The North American Society for Sport Psychology and Physical Activity, International Congress on Physical Education, Sport Sociology Academy (of AAHPERD), National Conference on New Religions and Revitalization Movements, Joint Conference for the North American Society for Sport Sociology and the International Association for the Philosophy of Sport, The Popular Culture Association, the Second International Conference on Sport and Religion, and recently gave a keynote address at the Inaugural International Conference on Sport and Spirituality at York St. John University in England.

Hoffman has been a frequent contributor to the national dialogue on problems in kinesiology and higher education. He is a former editor of *Quest* and former associate editor of the *Chronicle for Physical Education in Higher Education*. He was named Distinguished Scholar by the National Association for Kinesiology and Physical Education in Higher Education and gave the Dudley Sargent Lecture to that group as well. He is a fellow emeritus of the American Academy for Kinesiology and Physical Education.

He has a special interest in the relationship between sport and religion and has spoken and published widely on the topic. He is editor of *Sport and Religion* (published in 1992) and has been featured in a number of televised documentaries on sport and religion that have aired on CBS ("Sport and Ethics"), ESPN ("Time to Pray, Time to Play"), and Channel 4 in Britain ("Praying to Win"), and nationally aired broadcasts on NPR ("A Whole New Ballgame"), BBC, and the Canadian Broadcasting Company ("Inside Track"). He has recently completed a book on sport and evangelism to be published by Baylor University Press.

Hoffman and his wife, Claude Mourot, reside in Greensboro, North Carolina, where he enjoys golf, swimming, traveling, and hiking in the nearby mountains.

ABOUT THE CONTRIBUTORS

Jennifer L. Caputo, PhD, is an associate professor and serves as co-coordinator of the undergraduate and graduate exercise science programs at Middle Tennessee State University. Her research interests focus on pediatric health and fitness and she teaches classes in exercise physiology, health and fitness assessment, and exercise prescription. Dr. Caputo is a certified strength and conditioning specialist through the National Strength and Conditioning Association and is a licensed medical bone densitometer operator. Dr. Caputo received her doctoral degree in exercise physiology from the University of North Carolina at Greensboro. Her master's degree is in sport and exercise psychology.

Margaret Carlisle Duncan, PhD, is a professor in the department of human movement sciences at the University of Wisconsin at Milwaukee. Dr. Duncan's ongoing research interest relates to portrayals of female athletes and women's sports in the media. She has also studied media depictions of women's bodies and body practices. Another line of her ongoing research focuses on perceptions of physical disability and people with disabilities. Currently Dr. Duncan is studying the intersections of race, class, and gender and how they shape one's experience of one's body. A new interest is in the social construction of obesity and the discourses of overweight and obesity. Dr. Duncan is a former president of the North American Society for the Sociology of Sport (NASSS), a former president of The Association for the Study of Play (TASP), and a former editor of the journal *Play & Culture*. She also serves on several editorial boards. She is a fellow of the American Academy of Kinesiology and Physical Education (AAKPE), was named a fellow of Wisconsin Teaching and a fellow for the interdisciplinary Center for 20th Century Studies (at UWM) on two different occasions. She has published several scholarly articles in sociology, sport sociology, media, and communication studies journals. She is also a coauthor (with Michael A. Messner) of four sport media studies commissioned by LA84, a nonprofit foundation dedicated to serving youth through sport and to increasing knowledge of sport and its impact on people's lives.

Kim C. Graber, EdD, is an associate professor in the department of kinesiology at the University of Illinois at Urbana-Champaign, where she coordinates the undergraduate and graduate programs in pedagogy. She received her bachelor's degree from the University of Iowa, her master's from Columbia University, and her doctorate from the University of Massachusetts at Amherst. Her research focuses on teacher education, children's wellness, and the scholarship of teaching and learning. She has served as president of the National Association for Sport and Physical Education, chair of the Curriculum and Instruction Academy, and is a fellow in the Research Consortium. Dr. Graber is currently serving on the review boards of the *Journal of Teaching in Physical Education* and *Quest*. She has published numerous articles in peer-refereed journals and books and has presented her work at dozens of national and international conferences. She was recently named a University of Illinois Distinguished Teacher Scholar, a title that will remain with her throughout her career at the university.

Janet C. Harris, PhD, is a professor and the director of the school of exercise and nutritional sciences at San Diego State University. Prior to this, she was a professor and the director of the school of kinesiology and nutritional science at California State University at Los Angeles (1998-2006). Dr. Harris was located at the University of North Carolina at Greensboro for a major portion of her career, where she was a professor of exercise and sport science (1980-1998). Her research has focused on symbolic and expressive aspects of sports in American society, with work on topics such as athletic heroes of children and teenagers, boys in youth baseball, marketing slogans of collegiate football programs, and sports and the mass media. Her research is published in journals such as *Sociology of Sport Journal, Research Quarterly for Exercise and Sport, Quest,* and *Exercise and Sport Sciences Reviews.* She was coeditor of the first edition of *Introduction to Kinesiology* (2000) and also coeditor of *Play, Games and Sports in Cultural Contexts* (1983). She has an ongoing interest in the discipline of kinesiology and related physical activity professions and has published several papers on these topics. Dr. Harris has served as president of both the North American Society for the Sociology of Sport and the AAHPERD Research Consortium. She has been the editor of *Quest* and the *International Review for the Sociology of Sport* and the section editor of sociology and cultural anthropology for the *Research Quarterly for Exercise and Sport.* Her recreational interests include tennis, workouts at the gym, eating out, attending musical performances, and gardening. Dr. Harris earned her BS and MS degrees at the University of California at Los Angeles and her PhD at the University of California at Berkeley.

Jeremy Howell, PhD, is associate professor in the exercise and sport science department and the sport management graduate program at the University of San Francisco. He writes and lectures on exercise, culture, and commerce and is on the editorial board of the Journal of Sport & Social Issues and the International Journal of Sport Management and Marketing. Dr. Howell has extensive industry experience and consults with numerous local and national for-profit and not-for-profit organizations. In 2008 Dr. Howell received the University of San Francisco's St. Ignatius Award, given to the faculty person who best personifies the meaning of service to his or her community, profession, and university.

Katherine M. Jamieson, PhD, completed her doctorate in kinesiology at Michigan State University in 1999. She is an associate professor in exercise and sport science and director of women's and gender studies at the University of North Carolina at Greensboro. Her teaching and research interests center on critical cultural analyses of gender as it links with race, sexuality, and social class both in and beyond U.S. geographies and sociopolitical conditions.

Scott Kretchmar, PhD, is a professor of exercise and sport science at Penn State University. He is a founding member of the International Association for the Philosophy of Sport (IAPS) and has served as its president. He has been editor of the *Journal of the Philosophy of Sport,* is a fellow in the American Academy of Kinesiology and Physical Education, and has authored a popular text in the philosophy of sport. He has been named Alliance Scholar by AAHPERD, Distinguished Scholar by NAKPEHE, and on two different occasions, the Fraleigh Distinguished Scholar by IAPS. He is currently the editor of the *Journal of Intercollegiate Sport* and serves as Penn State's faculty representative to the NCAA.

Joseph Luxbacher, PhD, has more than thirty years' experience in the fields of health, fitness, and competitive athletics. He holds a PhD in health, physical and recreation education and is employed in the department of athletics at the University of Pittsburgh. Dr. Luxbacher has authored more than a dozen books and numerous articles in the areas of sport (soccer), peak athletic performance, fitness, and weight control. He is presently the head men's soccer coach at the University of Pittsburgh and has twice been named Big East Athletic Conference Soccer Coach of the Year. He was honored as a Letterman of Distinction by the University of Pittsburgh in 2003 and was inducted into the Western Pennsylvania Sports Hall of Fame in 2005. Dr. Luxbacher, his wife, Gail, and their children, Eliza and Travis, live outside of Pittsburgh.

Lori K. Miller, EdD, JD, is a professor in the sport management department at Wichita State University in Wichita, Kansas. She has earned five academic degrees, including an undergraduate degree in business, a master's degree in education, an MBA, an EdD, and a JD. Dr. Miller enjoys learning and sharing learned knowledge with her undergraduate and graduate students in a manner that allows for content appreciation and understanding, as well as professional application and benefit. The law is Dr. Miller's primary area of academic passion, and her research addresses various legal issues and their respective impact on sport industry operations. Dr. Miller began her full-time career in higher education in 1989, and she has served in a variety of administrative and faculty roles during the past 20 years (e.g., faculty member, department chair, associate dean). Dr. Miller is an active member in numerous professional associations, and she currently serves as the executive director for the Sport and Recreation Law Association.

Sandra Minor Bulmer, PhD, is a professor in the department of public health at Southern Connecticut State University, where she teaches and conducts community-based research. She was previously the director of fitness operations for Western Athletic Clubs in San Francisco. She has her doctorate in health education from Texas Woman's University and her MS in exercise physiology from the University of Oregon. She is also a certified health education specialist (CHES), ACSM health/fitness specialist, and ACSM health/fitness director. She is active in several professional associations and serves on the board of trustees for the Society for Public Health Education (SOPHE).

Kathy Simpson, PhD, is an associate professor in the department of kinesiology at the University of Georgia and director of the biomechanics laboratory. She received a doctorate in biomechanics from the University of Oregon. Her research focuses on determining how people adapt their movements to varying demands (for example, impact forces or foot prostheses) and how these adaptations affect the functions and structure of the lower extremity. She has applied this research to areas such as the improvement of the sport performance of athletes with lower-extremity amputations and the performance of daily activities by individuals with hip or knee joint replacements or who use a lower-extremity prosthesis. Currently, she is serving on the editorial board of the *International Journal of Applied Sports Sciences*. She has served as the biomechanics section editor for *Research Quarterly for Exercise and Sport,* cochair of the biomechanics Interest Group in ACSM, the Executive Board of the American Society of Biomechanics (ASB), and chair of the Biomechanics Academy of AAHPERD.

Chad Starkey, PhD, ATC, is an associate professor and coordinator of the graduate athletic training program at Ohio University in Athens, Ohio. A graduate of West Virginia University, Dr. Starkey received his master's and doctoral degrees from Ohio University. Dr. Starkey has served on the board of directors for the NATA board of certification and was the chair of the NATA Education Council. He has authored several textbooks focusing on sports medicine, orthopedic evaluation, and therapeutic modalities.

G. Clayton (Clay) Stoldt, EdD, is an associate professor and serves as chair of the department of sport management at Wichita State University. He teaches classes in sport public relations and sport marketing. Dr. Stoldt is the coauthor of *Sport Public Relations: Managing Organizational Communication,* and his research activities have focused on sport public relations issues such as crisis communications, the roles of sport public relations professionals, and the application of advanced public relations practices in the field. As department chair, he administers a bachelor's program with more than 100 students and a master's program with more than 80 students. Dr. Stoldt received his doctorate of education from the University of Oklahoma in 1998. He has a master's degree in sport management and a bachelor's degree in journalism/mass communication. Prior to coming to Wichita State, he worked in the athletic department at Oklahoma City University, where he served as sports information director, radio play-by-play broadcaster, and development officer. He also served as an adjunct instructor at both Oklahoma City and the University of Oklahoma, teaching courses in sport management and mass communication. During his tenure as sports information director, Dr. Stoldt earned several national awards for sports information publications. His background also includes experience as a radio sportscaster and sales executive.

Richard A. Swanson, PhD, is a professor emeritus of exercise and sport science at the University of North Carolina at Greensboro, where he was dean of the school of health and human performance for 12 years. He previously served in faculty and administrative roles at Wayne State University and San Francisco State University. His publications include two books and several book chapters and research articles on the history of physical education, exercise science, and sport in the United States, as well as articles and book chapters on administration in higher education. Dr. Swanson is a charter member of the North American Society for Sport History (NASSH). He is currently pursuing research on the historical relationship between sport and religion in the United States. He has served as archivist for the National Association for Kinesiology and Physical Education in Higher Education (NAKPEHE) for more than 25 years and has served extensively as a member of the governing board since 1973. He was the Dudley Allen Sargent Lecturer for that organization in 1997 and was awarded the Distinguished Scholar Award in 2008. Dr. Swanson is also the recipient of the Distinguished Service Award (1992) from the College and University Administrator Council of the American Alliance of Health, Physical Education, Recreation and Dance (AAHPERD). He received his BS and MEd degrees from Wayne State University and his PhD from The Ohio State University. He resides with his wife, Lita, in Greensboro, North Carolina, and enjoys spending time with his two adult children and four grandchildren, playing golf, reading, traveling, and volunteering in the community.

Thomas J. Templin, PhD, is a professor and former head of the department of health and kinesiology at Purdue University. He also holds a courtesy appointment as a professor in the department of curriculum and instruction in the college of education. He received his education at Indiana University (BS, 1972; MS, 1975) and at the University of Michigan (PhD, 1978). He has been a member of the Purdue faculty since 1977 and served as an administrator from 1988 to 2006. Dr. Templin has received various honors through his career. Recently, he received the 2008 Curriculum and Instruction Honor Award from the C&I Academy within the National Association for Sport and Physical Education. In 2006, he was inducted as a fellow into the American Academy of Kinesiology and Physical Education. He was the Doris Drees Distinguished Scholar Lecturer at the University of Dayton and was the 100th Lansdowne Scholar for the faculty of education at the University of Victoria in the spring of 2003. He received the Outstanding Teacher Educator Award from the Indian Association of Colleges for Teacher Education in 1986.

Dr. Templin has focused his research on the lives and careers of physical education teachers. He is well known as an author and editor/reviewer of numerous journal publications and books in physical education, including *A Reflective Approach to Teaching Physical Education* with Don Hellison and *Socialization into Physical Education: Learning to Teach* with Paul Schempp. Along with Russell Carson and Howard Weiss, Dr. Templin completed research funded by the Spencer Foundation to study emotion and affective events in teaching. This research is ongoing. He has served in leadership roles and has presented numerous national and international papers for various professional organizations including the American Alliance of Health, Physical Education, Recreation and Dance (AAHPERD) and the American Educational Research Association (AERA). Dr. Templin is a former President of the National Association for Sport and Physical Education that represents 17,000 physical education and sport professionals. An avid golfer, he serves as a consultant with the Professional Golfers' Association of America through its university program, Play Golf America University, and as chair of the association's scholarship committee.

Jerry R. Thomas, EdD, has taught children's motor development and research methods for more than 35 years. He is a professor of kinesiology and dean of the college of education at the University of North Texas. Dr. Thomas also has been a professor at Florida State, Louisiana State, Arizona State, and Iowa State Universities. He has published more than 200 papers, book chapters, and textbooks. Dr. Thomas is former president of the American Academy of Kinesiology and Physical Education (AAKPE), the North American Society for the Psychology of Sport and Physical Activity (NASPSPA), and Research Consortium. He is the current president of the American Kinesiology Association. In addition, his scholarly work in physical activity has earned him the titles of C.H. McCloy Lecturer for children's control, learning, and performance of motor skills, Alliance Scholar for AAHPERD, and Distinguished Scholar for NASPSPA. His textbook, *Research Methods in Physical Activity,* is in its fifth edition and has been translated into 6 other languages.

Katherine T. Thomas, PhD, is an associate professor of health and human performance at Iowa State University, where she teaches a variety of teacher education and motor development courses. Dr. Thomas also has taught at Arizona State University, Southeastern Louisiana University, and Southern University at Baton Rouge. Her research and numerous publications focus on skill acquisition in sport and exercise and the relation of physical activity to health. She has external grant funding in excess of $1,000,000 to study physical activity and is the physical activity consultant for the USDA's Team Nutrition. Dr. Thomas is a member of the American Alliance for Health, Physical Education, Recreation and Dance (AAHPERD) and the North American Society for the Psychology of Sport and Physical Activity (NASPSPA). She received her doctorate in physical education from Louisiana State University.

Cesar R. Torres, PhD, is an associate professor of physical education and sport at The College at Brockport, State University of New York. He received his early professional training in Argentina and obtained his PhD from The Pennsylvania State University. A philosopher and historian of sport, Dr. Torres has written on the relationship between skills and the structure of games, and the ethics of sport contests as well as the development of Olympism and sport in Latin America. He has published numerous articles in academic journals, edited collections, and newspapers, both in English and Spanish. Prior to moving to the United States, Dr. Torres spent ten years teaching K-12 physical education. In 2007, the International Society for the History of Physical Education and Sport awarded him the Reinhard Sprenger Junior Award.

Robin S. Vealey, PhD, is a professor in the department of kinesiology and health at Miami University in Ohio. Dr. Vealey's focus area is sport psychology, and she is particularly interested in self-confidence, mental skills training, and coaching effectiveness. She has authored two books, *Coaching for the Inner Edge* and *Competitive Anxiety in Sport.* She has served as a sport psychology consultant for the U.S. Nordic Ski Team, U.S. Field Hockey, elite golfers, and athletes and teams at Miami University. Dr. Vealey is a Fellow, Certified Consultant, and past president of the Association for the Advancement of Applied Sport Psychology and former editor of *The Sport Psychologist.* She is also a Fellow of the American Academy of Kinesiology and a member of the United States Olympic Committee Sport Psychology Registry. A former collegiate basketball player and coach, she now enjoys the mental challenge of golf.

The editor acknowledges these contributors to previous editions:

- Janet C. Harris (chapters 5, with R. Scott Kretchmar; 6, with Richard A. Swanson; and 7, with Margaret Carlisle Duncan)
- Emily M. Haymes (chapter 11)
- James Kallusky and Lavon Williams (chapter 16)
- P. Greg Comfort (chapter 17, with Lori K. Miller and G. Clayton Stoldt)